Passive, Active, and Digital Filters

The Circuits and Filters Handbook

Third Edition

Edited by

Wai-Kai Chen

Fundamentals of Circuits and Filters

Feedback, Nonlinear, and Distributed Circuits

Analog and VLSI Circuits

Computer Aided Design and Design Automation

Passive, Active, and Digital Filters

The Circuits and Filters Handbook
Third Edition

Passive, Active, and Digital Filters

Edited by

Wai-Kai Chen
University of Illinois
Chicago, U. S. A.

CRC Press is an imprint of the
Taylor & Francis Group, an **informa** business

CRC Press
Taylor & Francis Group
6000 Broken Sound Parkway NW, Suite 300
Boca Raton, FL 33487-2742

© 2009 by Taylor & Francis Group, LLC
CRC Press is an imprint of Taylor & Francis Group, an Informa business

No claim to original U.S. Government works
Printed in the United States of America on acid-free paper
10 9 8 7 6 5 4 3 2 1

International Standard Book Number-13: 978-1-4200-5885-7 (Hardcover)

Library of Congress Cataloging-in-Publication Data

Passive, active, and digital filters / edited by Wai-Kai Chen.
　　　p. cm.
　　Includes bibliographical references and index.
　　ISBN-13: 978-1-4200-5885-7
　　ISBN-10: 1-4200-5885-1
　　1. Electric filters, Digital. 2. Electric filters, Passive. 3. Electric filters, Active. I. Chen, Wai-Kai, 1936- II. Title.

TK7872.F5P375 2009
621.3815'324--dc22
　　　　　　　　　　　　　　　　　　　　　　　　　　　　　　　　　　　2008048117

Visit the Taylor & Francis Web site at
http://www.taylorandfrancis.com

and the CRC Press Web site at
http://www.crcpress.com

Contents

SECTION II   Active Filters

SECTION III   Digital Filters

Preface

As circuit complexity continues to increase, the microelectronic industry must possess the ability to quickly adapt to the market changes and new technology through automation and simulations. The purpose of this book is to provide in a single volume a comprehensive reference work covering the broad spectrum of filter design from passive, active, to digital. The book is written and developed for the practicing electrical engineers and computer scientists in industry, government, and academia. The goal is to provide the most up-to-date information in the field.

Over the years, the fundamentals of the field have evolved to include a wide range of topics and a broad range of practice. To encompass such a wide range of knowledge, this book focuses on the key concepts, models, and equations that enable the design engineer to analyze, design, and predict the behavior of large-scale systems that employ various types of filters. While design formulas and tables are listed, emphasis is placed on the key concepts and theories underlying the processes.

This book stresses fundamental theory behind professional applications and uses several examples to reinforce this point. Extensive development of theory and details of proofs have been omitted. The reader is assumed to have a certain degree of sophistication and experience. However, brief reviews of theories, principles, and mathematics of some subject areas are given. These reviews have been done concisely with perception.

The compilation of this book would not have been possible without the dedication and efforts of Professor Rashid Ansari and Dr. A. Enis Cetin, and most of all the contributing authors. I wish to thank them all.

Wai-Kai Chen

Editor-in-Chief

Wai-Kai Chen is a professor and head emeritus of the Department of Electrical Engineering and Computer Science at the University of Illinois at Chicago. He received his BS and MS in electrical engineering at Ohio University, where he was later recognized as a distinguished professor. He earned his PhD in electrical engineering at the University of Illinois at Urbana–Champaign.

Professor Chen has extensive experience in education and industry and is very active professionally in the fields of circuits and systems. He has served as a visiting professor at Purdue University, the University of Hawaii at Manoa, and Chuo University in Tokyo, Japan. He was the editor-in-chief of the *IEEE Transactions on Circuits and Systems, Series I and II*, the president of the IEEE Circuits and Systems Society, and is the founding editor and the editor-in-chief of the *Journal of Circuits, Systems and Computers*.

He received the Lester R. Ford Award from the Mathematical Association of America; the Alexander von Humboldt Award from Germany; the JSPS Fellowship Award from the Japan Society for the Promotion of Science; the National Taipei University of Science and Technology Distinguished Alumnus Award; the Ohio University Alumni Medal of Merit for Distinguished Achievement in Engineering Education; the Senior University Scholar Award and the 2000 Faculty Research Award from the University of Illinois at Chicago; and the Distinguished Alumnus Award from the University of Illinois at Urbana–Champaign. He is the recipient of the Golden Jubilee Medal, the Education Award, and the Meritorious Service Award from the IEEE Circuits and Systems Society, and the Third Millennium Medal from the IEEE. He has also received more than a dozen honorary professorship awards from major institutions in Taiwan and China.

A fellow of the Institute of Electrical and Electronics Engineers (IEEE) and the American Association for the Advancement of Science (AAAS), Professor Chen is widely known in the profession for the following works: *Applied Graph Theory* (North-Holland), *Theory and Design of Broadband Matching Networks* (Pergamon Press), *Active Network and Feedback Amplifier Theory* (McGraw-Hill), *Linear Networks and Systems* (Brooks/Cole), *Passive and Active Filters: Theory and Implements* (John Wiley), *Theory of Nets: Flows in Networks* (Wiley-Interscience), *The Electrical Engineering Handbook* (Academic Press), and *The VLSI Handbook* (CRC Press).

Contributors

Phillip E. Allen
School of Electrical Engineering
Georgia Institute of Technology
Atlanta, Georgia

Yucel Altunbasak
School of Electrical and
    Computer Engineering
Georgia Institute of Technology
Atlanta, Georgia

Rashid Ansari
Department of Electrical and
    Computer Engineering
University of Illinois at Chicago
Chicago, Illinois

Andreas Antoniou
Department of Electrical and
    Computer Engineering
University of Victoria
Victoria, British Columbia,
    Canada

Gonzalo R. Arce
Electrical and Computer
    Engineering Department
University of Delaware
Newark, Delaware

Tuncer C. Aysal
School of Electrical and
    Computer Engineering
Cornell University
Ithaca, New York

Kenneth E. Barner
Department of Electrical and
    Computer Engineering
University of Delaware
Newark, Delaware

Benjamin J. Blalock
Department of Electrical
    Engineering and Computer
    Science
The University of Tennessee
Knoxville, Tennessee

Bruce W. Bomar
Department of Electrical and
    Computer Engineering
The University of Tennessee
    Space Institute
Tullahoma, Tennessee

Phakphoom Boonyanant
National Electronics and
    Computer Technology Center
Pathumthani, Thailand

A. Enis Cetin
Department of Electrical and
    Electronics Engineering
Bilkent University
Ankara, Turkey

Chalie Charoenlarpnopparut
Sirindhorn International
    Institute of Technology
Thammasat University
Pathumthani, Thailand

Wai-Kai Chen
Department of Electrical and
    Computer Engineering
University of Illinois at
    Chicago
Chicago, Illinois

A. G. Constantinides
Department of Electrical
    and Electronic
    Engineering
Imperial College of Science,
    Technology and Medicine
London, England

Artice M. Davis
Department of Electrical
    Engineering
San Jose State University
San Jose, California

M. H. Er
School of Electrical and
    Electronic Engineering
Nanyang Technological
    University
Singapore

Joseph B. Evans
Department of Electrical
    Engineering and Computer
    Science
University of Kansas
Lawrence, Kansas

Igor M. Filanovsky
Department of Electrical
   Engineering and Computer
   Technology
University of Alberta
Edmonton, Alberta, Canada

Norbert J. Fliege
Department of Electrical
   Engineering and Computer
   Engineering
University of Mannheim
Mannheim, Germany

Bahadir K. Gunturk
Department of Electrical and
   Computer Engineering
Louisiana State University
Baton Rouge, Louisiana

Nick G. Kingsbury
Department of Engineering
Trinity College
University of Cambridge
Cambridge, United Kingdom

Yong Ching Lim
School of Electrical and
   Electronic Engineering
Nanyang Technological
   University
Singapore

Wasfy B. Mikhael
Department of Electrical
   Engineering and Computer
   Science
University of Central Florida
Orlando, Florida

Stephen W. Milam
RF Micro-Devices
Greensboro, North Carolina

Timothy R. Newman
Department of Electrical
   Engineering and Computer
   Science
University of Kansas
Lawrence, Kansas

Nasir M. Rajpoot
Department of Computer
   Science
University of Warwick
Coventry, United Kingdom

Jaime Ramirez-Angulo
Klipsch School of Electrical
   and Computer
   Engineering
New Mexico State University
Las Cruces, New Mexico

Jose Gerardo Rosiles
Electrical and Computer
   Engineering Department
The University of Texas at
   El Paso
El Paso, Texas

Edgar Sánchez-Sinencio
Department of Electrical
   and Computer Engineering
Texas A&M University
College Station, Texas

Tapio Saramäki
Institute of Signal Processing
Tampere University
   of Technology
Tampere, Finland

Rolf Schaumann
Department of Electrical
   Engineering
Portland State University
Portland, Oregon

Jose Silva-Martinez
Department of Electrical and
   Computer Engineering
Texas A&M University
College Station, Texas

Mark J. T. Smith
School of Electrical and
   Computer Engineering
Purdue University
West Lafayette, Indiana

F. William Stephenson
Department of Electrical and
   Computer Engineering
Virginia Polytechnic Institute
   and State University
Blacksburg, Virginia

Sawasd Tantaratana
Sirindhorn International
   Institute of Technology
Thammasat University
Pathumthani, Thailand

David B. H. Tay
Department of Electronic
   Engineering
Latrobe University
Bundoora, Victoria, Australia

Roland G. Wilson
Department of Computer
   Science
University of Warwick
Coventry, United Kingdom

Xiaojian Xu
School of Electronic and
   Information Engineering
Beihang University
Beijing, China

Zhen Yao
Department of Computer
   Science
University of Warwick
Coventry, United Kingdom

I

Passive Filters

Wai-Kai Chen
University of Illinois at Chicago

1

General Characteristics of Filters

Andreas Antoniou
University of Victoria

1.1 Introduction

An electrical filter is a system that can be used to modify, reshape, or manipulate the frequency spectrum of an electrical signal according to some prescribed requirements. For example, a filter may be used to amplify or attenuate a range of frequency components, reject or isolate one specific frequency component, and so on. The applications of electrical filters are numerous, for example,

- To eliminate signal contamination such as noise in communication systems
- To separate relevant from irrelevant frequency components
- To detect signals in radios and TV's
- To demodulate signals
- To bandlimit signals before sampling
- To convert sampled signals into continuous-time signals
- To improve the quality of audio equipment, e.g., loudspeakers
- In time-division to frequency-division multiplex systems

- In speech synthesis
- In the equalization of transmission lines and cables
- In the design of artificial cochleas

Typically, an electrical filter receives an input signal or *excitation* and produces an output signal or *response*. The frequency spectrum of the output signal is related to that of the input by some rule of correspondence. Depending on the type of input, output, and internal operating signals, three general types of filters can be identified, namely, continuous-time, sampled-data, and discrete-time filters.

A continuous-time signal is one that is defined at each and every instant of time. It can be represented by a function $x(t)$ whose domain is a range of numbers (t_1, t_2), where $-\infty \le t_1$ and $t_2 \le \infty$. A sampled-data or impulse-modulated signal is one that is defined in terms of an infinite summation of continuous-time impulses (see Ref. [1, Chapter 6]). It can be represented by a function

$$\hat{x}(t) = \sum_{n=-\infty}^{\infty} x(nT)\delta(t - nT)$$

where $\delta(t)$ is the impulse function. The value of the signal at any instant in the range $nT < t < (n+1)T$ is zero. The frequency spectrum of a continuous-time or sampled-data signal is given by the Fourier transform.*

A discrete-time signal is one that is defined at discrete instants of time. It can be represented by a function $x(nT)$, where T is a constant and n is an integer in the range (n_1, n_2) such that $-\infty \le n_1$ and $n_2 \le \infty$. The value of the signal at any instant in the range $nT < t < (n+1)T$ can be zero, constant, or undefined depending on the application. The frequency spectrum in this case is obtained by evaluating the z transform on the unit circle $|z| = 1$ of the z plane.

Depending on the format of the input, output, and internal operating signals, filters can be classified either as analog or digital filters. In analog filters the operating signals are varying voltages and currents, whereas in digital filters they are encoded in some binary format. Continuous-time and sampled-data filters are always analog filters. However, discrete-time filters can be analog or digital.

Analog filters can be classified on the basis of their constituent components as

- Passive *RLC* filters
- Crystal filters
- Mechanical filters
- Microwave filters
- Active *RC* filters
- Switched-capacitor filters

Passive *RLC* filters comprise resistors, inductors, and capacitors. Crystal filters are made of piezoelectric resonators that can be modeled by resonant circuits. Mechanical filters are made of mechanical resonators. Microwave filters consist of microwave resonators and cavities that can be represented by resonant circuits. Active *RC* filters comprise resistors, capacitors, and amplifiers; in these filters, the performance of resonant circuits is simulated through the use of feedback or by supplying energy to a passive circuit. Switched-capacitor filters comprise resistors, capacitors, amplifiers, and switches. These are discrete-time filters that operate like active filters but through the use of switches the capacitance values can be kept very small. As a result, switched-capacitor filters are amenable to VLSI implementation.

This section provides an introduction to the characteristics of analog filters. Their basic characterization in terms of a differential equation is reviewed in Section 1.2 and by applying the Laplace transform, an algebraic equation is deduced that leads to the *s*-domain representation of a filter. The representation of analog filters in terms of the transfer function is then developed. Using the transfer function, one can

---

* See Chapter 4 of *Fundamentals of Circuits and Filters*.

obtain the time-domain response of a filter to an arbitrary excitation, as shown in Section 1.3. Some important time-domain responses, i.e., the impulse and step responses, are examined. Certain filter parameters related to the step response, namely, the overshoot, delay time, and rise time, are then considered. The response of a filter to a sinusoidal excitation is examined in Section 1.4 and is then used to deduce the basic frequency-domain representations of a filter, namely, its frequency response and loss characteristic. Some idealized filter characteristics are then identified and the differences between idealized and practical filters are delineated in Section 1.5. Practical filters tend to introduce signal degradation through amplitude and/or delay distortion. The causes of these types of distortion are examined in Section 1.6. In Section 1.7, certain special classes of filters, e.g., minimum-phase and allpass filters, are identified and their applications mentioned. This chapter concludes with a review of the design process and the tasks that need to be undertaken to translate a set of filter specifications into a working prototype.

1.2 Characterization

A linear causal analog filter with input $x(t)$ and output $y(t)$ can be characterized by a differential equation of the form

$$b_n \frac{\mathrm{d}^n y(t)}{\mathrm{d}t^n} + b_{n-1} \frac{\mathrm{d}^{n-1} y(t)}{\mathrm{d}t^{n-1}} + \cdots + b_0 y(t) = a_n \frac{\mathrm{d}^n x(t)}{\mathrm{d}t^n} + a_{n-1} \frac{\mathrm{d}^{n-1} x(t)}{\mathrm{d}t^{n-1}} + \cdots + a_0 x(t)$$

The coefficients $a_0, a_1, \ldots, a_n$ and $b_0, b_1, \ldots, b_n$ are functions of the element values and are real if the parameters of the filter (e.g., resistances, inductances, etc.) are real. If they are independent of time, the filter is time invariant. The input $x(t)$ and output $y(t)$ can be either voltages or currents. The order of the differential equation is said to be the *order* of the filter.

An analog filter must of necessity incorporate reactive elements that can store energy. Consequently, the filter can produce an output even in the absence of an input. The output on such an occasion is caused by the initial conditions of the filter, namely,

$$\left. \frac{\mathrm{d}^{n-1} y(t)}{\mathrm{d}t^{n-1}} \right|_{t=0}, \left. \frac{\mathrm{d}^{n-2} y(t)}{\mathrm{d}t^{n-2}} \right|_{t=0}, \ldots, y(0)$$

The response in such a case is said to be the *zero-input response*. The response obtained if the initial conditions are zero is sometimes called the *zero-state response*.

1.2.1 Laplace Transform

The most important mathematical tool in the analysis and design of analog filters is the Laplace transform. It owes its widespread application to the fact that it transforms differential into algebraic equations that are a lot easier to manipulate. The Laplace transform of $x(t)$ is defined as*

$$X(s) = \int_{-\infty}^{\infty} x(t) e^{-st} \mathrm{d}t$$

where s is a complex variable of the form $s = \sigma + j\omega$. Signal $x(t)$ can be recovered from $X(s)$ by applying the inverse Laplace transform, which is given by

---

* See Chapter 3 by J. R. Deller, Jr. in *Fundamentals of Circuits and Filters*.

$$x(t) = \frac{1}{2\pi j} \int\limits_{C-j\infty}^{C+j\infty} X(s)e^{st}\,ds$$

where C is a positive constant. A shorthand notation of the Laplace transform and its inverse are

$$X(s) = \mathscr{L}x(t) \quad \text{and} \quad x(t) = \mathscr{L}^{-1}X(s)$$

Alternatively,

$$X(s) \leftrightarrow x(t)$$

A common practice in the choice of symbols for the Laplace transform and its inverse is to use upper case for the s domain and lower case for the time domain.

On applying the Laplace transform to the nth derivative of some function of time $y(t)$, we find that

$$\mathscr{L}\left[\frac{d^n y(t)}{dt^n}\right] = s^n Y(s) - s^{n-1}y(0) - s^{n-2}\frac{dy(t)}{dt}\bigg|_{t=0} - \cdots - \frac{d^{n-1}y(t)}{dt^{n-1}}\bigg|_{t=0}$$

Now, on applying the Laplace transform to an nth-order differential equation with constant coefficients, we obtain

$$\left(b_n s^n + b_{n-1}s^{n-1} + \cdots + b_0\right)Y(s) + \Psi_y(s) = \left(a_n s^n + a_{n-1}s^{n-1} + \cdots + a_0\right)X(s) + \Psi_x(s)$$

where
  $X(s)$ and $Y(s)$ are the Laplace transforms of the input and output, respectively
  $\Psi_x(s)$ and $\Psi_y(s)$ are functions that combine all the initial-condition terms that depend on $x(t)$ and $y(t)$, respectively

1.2.2 Transfer Function

An important s-domain characterization of an analog filter is its *transfer function*, as for any other linear system. This is defined as the ratio of the Laplace transform of the response to the Laplace transform of the excitation.

An arbitrary linear, time-invariant, continuous-time filter, which may or may not be causal, can be represented by the convolution integral

$$y(t) = \int\limits_{-\infty}^{\infty} h(t-\tau)x(\tau)\,d\tau = \int\limits_{-\infty}^{\infty} h(\tau)x(t-\tau)\,d\tau$$

where $h(t)$ is the impulse response of the filter. The Laplace transform yields

$$Y(s) = \int\limits_{-\infty}^{\infty} \left[\int\limits_{-\infty}^{\infty} h(t-\tau)x(\tau)\,d\tau\right]e^{-st}\,dt$$

$$= \int\limits_{-\infty}^{\infty}\int\limits_{-\infty}^{\infty} h(t-\tau)e^{-st}x(\tau)\,d\tau\,dt$$

$$= \int\limits_{-\infty}^{\infty}\int\limits_{-\infty}^{\infty} h(t-\tau)e^{-st} \cdot e^{s\tau} \cdot e^{-s\tau}x(\tau)\,d\tau\,dt$$

Changing the order of integration, we obtain

$$Y(s) = \int_{-\infty}^{\infty} \int_{-\infty}^{\infty} h(t-\tau)e^{-s(t-\tau)} \cdot x(\tau)e^{-s\tau} dt \, d\tau$$

$$= \int_{-\infty}^{\infty} \int_{-\infty}^{\infty} h(t-\tau)e^{-s(t-\tau)} dt \cdot x(\tau)e^{-s\tau} d\tau$$

Now, if we let $t = t' + \tau$, then $dt/dt' = 1$ and $t - \tau = t'$; hence,

$$Y(s) = \int_{-\infty}^{\infty} \int_{-\infty}^{\infty} h(t')e^{-st'} dt' \cdot x(\tau)e^{-s\tau} d\tau$$

$$= \int_{-\infty}^{\infty} h(t')e^{-st'} dt' \cdot \int_{-\infty}^{\infty} x(\tau)e^{-s\tau} d\tau$$

$$= H(s)X(s)$$

Therefore, the transfer function is given by

$$H(s) = \frac{Y(s)}{X(s)} = \mathscr{L}h(t) \tag{1.1}$$

In effect, the transfer function is equal to the Laplace transform of the impulse response.

Some authors define the transfer function as the Laplace transform of the impulse response. Then through the use of the convolution integral, they show that the transfer function is equal to the ratio of the Laplace transform of the response to the Laplace transform of the excitation. The two definitions are, of course, equivalent.

Typically, in analog filters the input and output are voltages, e.g., $x(t) + v_i(t)$ and $y(t) + v_o(t)$. In such a case the transfer function is given by

$$\frac{V_o(s)}{V_i(s)} = H_V(s)$$

or simply by

$$\frac{V_o}{V_i} = H_V(s)$$

However, on occasion the input and output are currents, in which case

$$\frac{I_o(s)}{I_i(s)} \equiv \frac{I_o}{I_i} = H_I(s)$$

The transfer function can be obtained through network analysis using one of several classical methods,[*] e.g., by using

---

[*] See Chapters 18 through 27 of this volume.

- Kirchhoff's voltage and current laws
- Matrix methods
- Flow graphs
- Mason's gain formula
- State-space methods

A transfer function is said to be *realizable* if it characterizes a stable and causal network. Such a transfer function must satisfy the following constraints:

1. It must be a rational function of s with real coefficients.
2. Its poles must lie in the left-half s plane.
3. The degree of the numerator polynomial must be equal to or less than that of the denominator polynomial.

A transfer function may represent a network comprising elements with real parameters only if its coefficients are real. The poles must be in the left-half s plane to ensure that the network is stable and the numerator degree must not exceed the denominator degree to assure the existence of a causal network.

1.3 Time-Domain Response

From Equation 1.1,

$$Y(s) = H(s)X(s)$$

Therefore, the time-domain response of a filter to some arbitrary excitation can be deduced by obtaining the inverse Laplace transform of $Y(s)$, i.e.,

$$y(t) = \mathcal{L}^{-1}\{H(s)X(s)\}$$

1.3.1 General Inversion Formula

If

1. the singularities of $Y(s)$ in the finite plane are poles,* and
2. $Y(s) \to 0$ uniformly with respect to the angle of s as $|s| \to \infty$ with $\sigma \leq C$, where C is a positive constant, then [2]

$$y(t) = \begin{cases} 0 & \text{for } t < 0 \\ \frac{1}{2\pi j} \int_{C-j\infty}^{C+j\infty} Y(s)e^{st}ds = \frac{1}{2\pi j} \int_{\Gamma} Y(s)e^{st}ds & \text{for } t \geq 0 \end{cases} \tag{1.2}$$

where Γ is a contour in the counterclockwise sense make up of the part of the circle $s = Re^{j\theta}$ to the left of line $s = C$ and the segment of the line $s = C$ that overlaps the circle, as depicted in Figure 1.1; C and R are sufficiently large to ensure that Γ encloses all the finite poles of $Y(s)$.

From the residue theorem [3] and Equation 1.2, we have

$$y(t) = \begin{cases} 0 & \text{for } t < 0 \\ \frac{1}{2\pi j} \int_{\Gamma} Y(s)e^{st}ds = \sum_{i=1}^{K} \operatorname*{res}_{s=p_i} Y_0(s) & \text{for } t \geq 0 \end{cases}$$

---

* Such a function is said to be meromorphic [2,3].

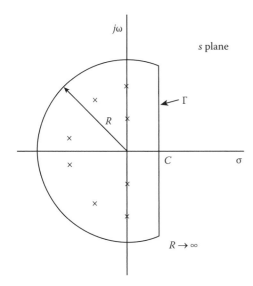

where $Y_0(s) = Y(s)e^{st}$ and K is the number of poles in $Y(s)$. If $Y_0(s)$ has a pole p_i of order m_i, the residue can be obtained by using the general formula [3]

$$\operatorname*{res}_{z=p_i} Y_0(s) = \frac{1}{(m_i - 1)!} \lim_{s \to p_i} \frac{d^{m_i-1}}{ds^{m_i-1}} [(s - p_i)^{m_i} Y_0(s)]$$

Note that complex poles yield complex residues. Hence, like the poles of $Y_0(s)$, its residues occur in complex–conjugate pairs. For this reason, $y(t)$ is found to be a real function of t, as can be easily verified.

Condition 1 listed previously may not be satisfied sometimes, for example, if

$$\lim_{s \to \infty} Y(s) = A_0$$

FIGURE 1.1 Contour Γ for the evaluation of the inverse Laplace transform.

where A_0 is a constant. In such a case, we can express $Y(s)$ as

$$Y(s) = A_0 + Y'(s)$$

where $Y'(s)$ satisfies conditions 1 and 2. Thus,

$$y(t) = A_0 \delta(t) + \mathscr{L}^{-1} Y'(s)$$

The inverse Laplace transform of $Y'(s)$ can now be obtained by using the inversion formula.

1.3.2 Inverse by Using Partial Fractions

The simplest way to obtain the time-domain response of a filter is to express $H(s)X(s)$ as a partial-fraction expansion and then invert the resulting fractions individually. If $Y(s)$ has simple poles, we can write

$$Y(s) = A_0 + \sum_{i=1}^{K} \frac{A_i}{s - p_i}$$

where A_0 is a constant and

$$A_i = \lim_{s \to p_i} [(s - p_i) Y(s)]$$

is the residue of pole $s = p_i$. On applying the general inversion formula to each partial fraction, we obtain

$$y(t) = A_0 \delta(t) + u(t) \sum_{i=1}^{K} A_i e^{p_i t}$$

where $\delta(t)$ and $u(t)$ are the impulse function and unit step, respectively.

1.3.3 Impulse and Step Responses

The response of a filter to an impulse $\delta(t)$ designated as

$$y(t) = \mathcal{R}\delta(t) \equiv h(t)$$

where $\mathcal{R}$ is an operator, is of considerable importance. Its absolute integrability guarantees the stability of the filter* and its Laplace transform, namely, $H(s)$, is the transfer function as has been shown in the section on the transfer function.

For an Nth-order, causal, linear, and time-invariant filter

$$H(s) = \frac{a_0 + a_1 s + a_2 s^2 + \cdots + a_M s^M}{b_0 + b_1 s + b_2 s^2 + \cdots + b_N s^N}$$

where $M \leq N$.

The step (or unit-step) response is the output of a filter to the signal

$$u(t) = \begin{cases} 1 & \text{for } t \geq 0 \\ 0 & \text{for } t < 0 \end{cases}$$

The Laplace transform of $u(t)$ is $1/s$. Hence, the step response of an arbitrary filter is obtained as

$$y(t) = \mathcal{R}u(t) \equiv y_u(t) = \mathcal{L}^{-1}\left[\frac{H(s)}{s}\right]$$

1.3.4 Overshoot, Delay Time, and Rise Time

Three time-domain parameters of a filter are usually associated with the step response [4], namely, the overshoot, delay time, and rise time. The *overshoot* γ is the difference between the peak value and the asymptotic value of the step response in percent as $t \to \infty$. The *delay time* τ_d is the time required for the step response to reach 50% of the asymptotic value. The *rise time* τ_r is the time required for the step response to increase from 10% to 90% of the asymptotic value. These three parameters are illustrated in Figure 1.2, where $K = a_0/b_0$ is a scaling constant that normalizes the asymptotic value of the step response as $t \to \infty$ to unity.

The delay and rise times defined in terms of the step response entail quite a bit of computation. Alternative definitions of these parameters that are easier to use have been proposed by Elmore [4]. These are based on the impulse response and give accurate results if the overshoot is small. The delay time is defined as

$$\tau_D = \int_0^\infty t h(t)\,dt$$

and the rise time assumes the form

$$\tau_R = \left[2\pi \int_0^\infty (t - \tau_D)^2 h(t)\,dt\right]^{1/2} = \sqrt{2\pi}\left[\int_0^\infty t^2 h(t)\,dt - \tau_D^2\right]^{1/2}$$

---

* See Section 22.1 of *Fundamentals of Circuits and Filters*.

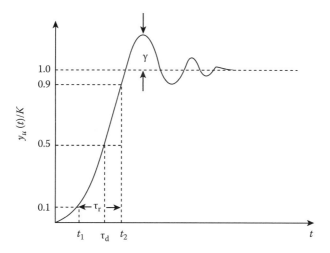

FIGURE 1.2 Overshoot, delay time, and rise time.

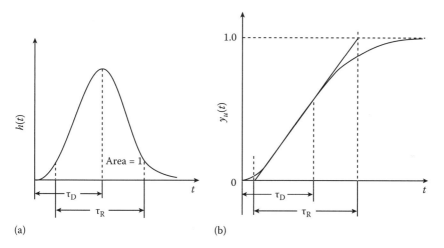

FIGURE 1.3 Physical interpretation of Elmore's definitions of delay and rise times: (a) impulse response $h(t)$ and (b) unit-step response $y_u(t)$.

The physical interpretation of these parameters is illustrated in Figure 1.3a and b. If the overshoot is small, say less than 1%, then

$$\tau_D \approx \tau_d \quad \text{and} \quad \tau_R \approx \tau_r$$

The simplification brought about by Elmore's definitions can be easily demonstrated. Consider a filter whose step response approaches unity as $t \to \infty$. Such a filter has a transfer function of the form

$$H(s) = \frac{1 + a_1 s + a_2 s^2 + \cdots + a_M s^M}{1 + b_1 s + b_2 s^2 + \cdots + b_N s^N} \tag{1.3}$$

that is, $a_0 = b_0 = 1$. From the definition of the Laplace transform,

$$H(s) = \int_0^\infty h(t)e^{-st}dt$$

$$= \int_0^\infty h(t)\left(1 - st + \frac{s^2t^2}{2!} - \cdots\right)dt$$

$$= \int_0^\infty h(t)dt - s\int_0^\infty th(t)dt + \frac{s^2}{2!}\int_0^\infty t^2 h(t)dt - \cdots$$

$$= \int_0^\infty h(t)dt - s\tau_D + \frac{s^2}{2!}\left(\frac{\tau_R^2}{2\pi} + \tau_D^2\right) - \cdots \qquad (1.4)$$

Alternatively, from Equation 1.3, direct division gives

$$H(s) = 1 - (b_1 - a_1)s + \left(b_1^2 - a_1 b_1 + a_2 - b_2\right)s^2 + \cdots \qquad (1.5)$$

Now by comparing Equations 1.4 and 1.5, we deduce

$$\int_0^\infty h(t)dt = 1, \quad \tau_D = b_1 - a_1$$

and

$$\tau_R = \left\{2\pi\left[b_1^2 - a_1^2 + 2(a_2 - b_2)\right]\right\}^{1/2}$$

The previous definitions are based on the assumption that the unit-step response approaches unity as $t \to \infty$. If this is not the case, i.e., coefficients a_0 and b_0 are not equal to unity, then we can write

$$H(s) = KH'(s)$$

where $K = a_0/b_0$ and

$$H'(s) = \frac{1 + a_1's + a_2's^2 + \cdots + a_M' s^M}{1 + b_1's + b_2's^2 + \cdots + b_N' s^N}$$

Using the coefficients of $H'(s)$ in the formulas for τ_D and τ_R yields approximate values for the delay time and rise time, since these parameters are independent of the absolute value of the step response.

1.4 Frequency-Domain Analysis

The frequency response of an analog filter is deduced by finding its steady-state sinusoidal response, as we shall now demonstrate.

1.4.1 Sinusoidal Response

Consider an Nth-order analog filter characterized by a transfer function $H(s)$. The sinusoidal response of such a filter is

$$y(t) = \mathcal{L}^{-1}[H(s)X(s)]$$

where

$$X(s) = \mathcal{L}[u(t)\sin \omega t] = \frac{\omega}{(s+j\omega)(s-j\omega)} \tag{1.6}$$

The product $H(s)X(s)$ satisfies conditions 1 and 2 imposed on the general inversion formula of Equation 1.2. Hence, for $t \geq 0$, we have

$$y(t) = \frac{1}{2\pi j} \int_{\Gamma} Y(s)e^{st}ds = \sum \mathrm{res}[H(s)X(s)e^{st}] \tag{1.7}$$

where Γ is a contour enclosing the poles of $H(s)$ and $X(s)$ as in Figure 1.1.

Assuming simple poles for the transfer function, Equations 1.6 and 1.7 give

$$y(t) = \sum_{i=1}^{N} X(p_i)e^{p_i t} \mathop{\mathrm{res}}_{s=p_i} H(s) + \frac{1}{2j}\left[H(j\omega)e^{j\omega t} - H(-j\omega)e^{-j\omega t}\right] \tag{1.8}$$

If the filter is assumed to be stable, then the poles are in the left-half s plane, i.e., $p_i = \sigma_i + j\omega_i$ with $\sigma_i < 0$.[*] As a consequence

$$\lim_{t\to\infty} e^{p_i t} = \lim_{t\to\infty} \left(e^{\sigma_i t} \cdot e^{j\omega_i t}\right) = 0$$

and since the residues of $H(s)$ are finite, the steady-state sinusoidal response is obtained from Equation 1.8 as

$$\tilde{y}(t) = \lim_{t\to\infty} y(t) = \frac{1}{2j}\left[H(j\omega)e^{j\omega t} - H(-j\omega)e^{-j\omega t}\right] \tag{1.9}$$

Equation 1.9 was deduced on the assumption that the poles of the transfer function are simple. However, it also applies for transfer functions with higher-order poles.

Now from the definition of the Laplace transform

$$H(s) = \int_{-\infty}^{\infty} h(t)e^{-st}dt$$

and hence

$$H(-j\omega) = \int_{-\infty}^{\infty} h(t)e^{j\omega t}dt = \left[\int_{-\infty}^{\infty} h(t)e^{-j\omega t}dt\right]^{*} = H^{*}(j\omega) \tag{1.10}$$

---

[*] See Chapter 22.1 of *Fundamentals of Circuits and Filters*.

If we write

$$H(j\omega) = M(\omega)e^{j\theta(\omega)} \tag{1.11}$$

where

$$M(\omega) = |H(j\omega)| \quad \text{and} \quad \theta(\omega) = \arg H(j\omega) \tag{1.12}$$

the steady-state sinusoidal response of the filter is obtained from Equations 1.9 through 1.12 as

$$
\begin{aligned}
\tilde{y}(t) &= \frac{1}{2j}\left[M(\omega)e^{j\theta(\omega)}e^{j\omega t} - M(\omega)e^{-j\theta(\omega)}e^{-j\omega t} \right] \\
&= M(\omega)\frac{1}{2j}\left[e^{j[\omega t + \theta(\omega)]} - e^{-j[\omega t + \theta(\omega)]} \right] \\
&= M(\omega)\sin[\omega t + \theta(\omega)]
\end{aligned}
$$

The preceding analysis has shown that the steady-state response of an analog filter to a sinusoid of unit amplitude is a sinusoid of amplitude $M(\omega)$, shifted by an angle $\theta(\omega)$. In effect, for a given frequency ω, the filter introduces a *gain* $M(\omega)$ and a *phase shift* $\theta(\omega)$.

As functions of frequency, $M(\omega)$ and $\theta(\omega)$ are known as the amplitude (or magnitude) response and phase response of the filter, respectively. The transfer function evaluated on the imaginary axis, namely, $H(j\omega)$ is the frequency response and, as was shown, its magnitude and angle are the amplitude response and phase response, respectively.

Two other quantities of a filter, which are of significant interest, are its *phase* and *group delays* These are defined as

$$\tau_p(\omega) = -\frac{\theta(\omega)}{\omega} \quad \text{and} \quad \tau_g(\omega) = -\frac{d\theta(\omega)}{d\omega}$$

respectively. For filters, the group delay is the more important of the two. As a function of frequency, $\tau_g(\omega)$ is usually referred to as the delay characteristic.

1.4.2 Graphical Construction

Consider a filter characterized by a transfer function of the form

$$H(s) = H_0\frac{N(s)}{D(s)} = H_0\frac{\prod_{i=1}^{M}(s - z_i)}{\prod_{i=1}^{N}(s - p_i)^{m_i}} \tag{1.13}$$

where H_0 is a constant. The frequency response of the filter is obtained as

$$H(j\omega) = M(\omega)e^{j\theta(\omega)} = \frac{H_0\prod_{i=1}^{M}(j\omega - z_i)}{\prod_{i=1}^{N}(j\omega - p_i)^{m_i}}$$

By letting

$$jw - z_i = M_{z_i} e^{j\psi_{z_i}} \tag{1.14}$$

$$jw - p_i = M_{p_i} e^{j\psi_{p_i}} \tag{1.15}$$

we obtain

$$M(\omega) = \frac{|H_0| \prod_{i=1}^{M} M_{z_i}}{\prod_{i=1}^{N} M_{p_i}^{m_i}} \tag{1.16}$$

and

$$\theta(\omega) = \arg H_0 + \sum_{i=1}^{M} \psi_{z_i} - \sum_{i=1}^{N} m_i \psi_{p_i} \tag{1.17}$$

where $\arg H_0 = \pi$ if H_0 is negative.

The gain and phase shift $M(\omega)$ and $\theta(\omega)$ for some frequency $\omega = \omega_i$ can be determined graphically by using the following procedure:

1. Mark the zeros and poles of the filter in the s plane.
2. Draw the phasor $s = j\omega_i$, where ω_i is the frequency of interest.
3. Draw a phasor of the type in Equation 1.14 for each simple zero of $H(s)$.
4. Draw m_i phasor of the type in Equation 1.15 for each pole of order m_i.
5. Measure the magnitudes and angles of the phasors in steps 3 and 4 and use them in Equations 1.16 and 1.17 to calculate the gain $M(\omega_i)$ and phase shift $\theta(\omega_i)$, respectively.

The amplitude and phase responses of a filter can be determined by repeating the preceding procedure for frequencies $\omega = \omega_1, \omega_2, \ldots,$ in the range 0 to ∞. The procedure is illustrated in Figure 1.4.

It should be mentioned that the modern approach for the analysis of filters is through the use of the many circuit analysis programs such as SPICE.* Nevertheless, the above graphical method is of interest and merits consideration for two reasons. First, it illustrates some of the fundamental properties of filters. Second, it provides a certain degree of intuition about the expected amplitude or phase response of a filter. For example, if a filter has pole close to the $j\omega$ axis, then as ω approaches the neighborhood of the pole, the magnitude of the phasor from the pole to the $j\omega$ axis decreases rapidly to a very small value and then increases as ω increases above this value. As a result, the amplitude response will exhibit a large peak in the frequency range close to the pole. On the other hand, a zero close to or on the $j\omega$ axis will lead to a notch in the amplitude response when ω is in the neighborhood of the zero.

Other situations are of interest, for example, if the poles of a filter are located in a band of the s plane below the horizontal line $s = \omega_c$ and its zeros are located above this line, then the filter will pass low-frequency and attenuate high-frequency components since $M_{zi} < M_{pi}$ if $\omega > \omega_c$ for all i. Such a filter is said to be a *low-pass* filter. If the zeros are located below the line $s = \omega_c$ and the poles above it, then the filter will pass high-frequency and attenuate low-frequency components, i.e., the filter will be a *high-pass* one.

---

* See Chapter 8 of *Computer Aided Design and Design Automation*, contribution of J.G. Rollins.

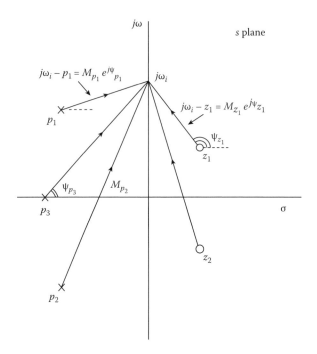

FIGURE 1.4 Graphical method for the evaluation of the frequency response.

1.4.3 Loss Function

Quite often, it is desirable to represent a filter in terms of its loss function. Consider a filter represented by the voltage transfer function

$$\frac{V_o(s)}{V_i(s)} = H(s) = \frac{N(s)}{D(s)}$$

where

$V_i(s)$ and $V_o(s)$ are the Laplace transforms of the input and output voltages, respectively

$N(s)$ and $D(s)$ are polynomials in s

The loss (or attenuation) of the filter in decibels is defined as

$$A(\omega) = 20 \log \left| \frac{V_i(j\omega)}{V_o(j\omega)} \right| = 20 \log \frac{1}{|H(j\omega)|} = 10 \log L\left(\omega^2\right) \qquad (1.18)$$

where

$$L\left(\omega^2\right) = \frac{1}{H(j\omega)H(-j\omega)}$$

$A(\omega)$ as a function of ω is the *loss characteristic*.

With $\omega = s/j$ in Equation 1.18, the function

$$L(-s^2) = \frac{D(s)D(-s)}{N(s)N(-s)}$$

can be formed. This is called the *loss function* of the filter and, as is evident, its zeros are the poles of $H(s)$ and their negatives, whereas its poles are the zeros of $H(s)$ and their negatives.

1.5 Ideal and Practical Filters

An ideal low-pass filter is one that will pass only low-frequency components. Its loss characteristic is given by

$$A(\omega) = \begin{cases} 0 & \text{for } 0 \leq \omega < \omega_c \\ \infty & \text{for } \omega_c < \omega < \infty \end{cases}$$

The frequency ranges 0 to ω_c and ω_c to ∞ are the *passband* and *stopband*, respectively. The boundary between the passband and stopband, namely, ω_c, is the *cutoff frequency*. An ideal high-pass filter will pass all components with frequencies above the cutoff frequency and reject all components with frequencies below the cutoff frequency, i.e.,

$$A(\omega) = \begin{cases} \infty & \text{for } 0 \leq \omega < \omega_c \\ 0 & \text{for } \omega_c < \omega < \infty \end{cases}$$

Idealized loss characteristics can similarly be identified for bandpass and bandstop filters as

$$A(\omega) = \begin{cases} \infty & \text{for } 0 \leq \omega < \omega_{c1} \\ 0 & \text{for } \omega_{c1} < \omega < \omega_{c2} \\ \infty & \text{for } \omega_{c2} \leq \omega < \infty \end{cases}$$

and

$$A(\omega) = \begin{cases} 0 & \text{for } 0 \leq \omega < \omega_{c1} \\ \infty & \text{for } \omega_{c1} < \omega < \omega_{c2} \\ 0 & \text{for } \omega_{c2} \leq \omega < \infty \end{cases}$$

respectively.

Practical filters differ from ideal ones in that the passband loss is not zero, the stopband loss is not infinite, and the transition between passband and stopband is gradual. Practical loss characteristics for low-pass, high-pass, bandpass, and bandstop filters assume the forms

$$A_{LP}(\omega) \begin{cases} \leq A_p & \text{for } 0 \leq \omega < \omega_p \\ \geq A_a & \text{for } \omega_a \leq \omega \leq \infty \end{cases}$$

$$A_{HP}(\omega) \begin{cases} \geq A_a & \text{for } 0 \leq \omega \leq \omega_a \\ \leq A_p & \text{for } \omega_p \leq \omega < \infty \end{cases}$$

$$A_{BP}(\omega) \begin{cases} \geq A_a & \text{for } 0 \leq \omega \leq \omega_{a1} \\ \leq A_p & \text{for } \omega_{p1} \leq \omega \leq \omega_{p2} \\ \geq A_a & \text{for } \omega_{a2} \leq \omega \leq \infty \end{cases}$$

and

$$
A_{\text{BS}}(\omega) = \begin{cases} \leq A_{\text{p}} & \text{for } 0 \leq \omega \leq \omega_{\text{p1}} \\ \geq A_{\text{a}} & \text{for } \omega_{\text{a1}} \leq \omega \leq \omega_{\text{a2}} \\ \leq A_{\text{p}} & \text{for } \omega_{\text{p2}} \leq \omega \leq \infty \end{cases}
$$

respectively, where ω_{p}, ω_{p1}, and ω_{p2} are passband edges, ω_{a}, ω_{a1}, and ω_{a2} are stopband edges, A_{p} is the maximum passband loss, and A_{a} is the minimum stopband loss. In practice, A_{p} is determined from the allowable amplitude distortion (see Section 1.6) and A_{a} is dictated by the allowable adjacent channel interference and the desirable signal-to-noise ratio.

It should be mentioned that in practical filters the cutoff frequency ω_{c} is not a very precise term. It is often used to identify some hypothetical boundary between passband and stopband such as the 3 dB frequency in Butterworth filters, the passband edge in Chebyshev filters, the stopband edge in inverse-Chebyshev filters, or the geometric mean of the passband and stopband edges in elliptic filters.

If a filter is required to have a piecewise constant loss characteristic (or amplitude response) and the shape of the phase response is not critical, the filter can be fully specified by its band edges, the minimum passband and maximum stopband losses A_{p} and A_{a}, respectively.

1.6 Amplitude and Delay Distortion

In practice, a filter can distort the information content of the signal. Consider a filter characterized by a transfer function $H(s)$ and assume that its input and output signal are $v_{\text{i}}(t)$ and $v_{\text{o}}(t)$. The frequency response of the filter is given by

$$
H(j\omega) = M(\omega)e^{j\theta(\omega)}
$$

where $M(\omega)$ and $\theta(\omega)$ are the amplitude and phase responses, respectively.

The frequency spectrum of $v_{\text{i}}(t)$ is its Fourier transform, namely, $V_{\text{i}}(j\omega)$. Assume that the information content of $v_{\text{i}}(t)$ is concentrated in frequency band B given by

$$
B = \{\omega\colon \omega_L \leq \omega \leq \omega_H\}
$$

and that its frequency spectrum is zero elsewhere.

Let us assume that the amplitude response is constant with respect to band B, i.e.,

$$
M(\omega) = G_0 \quad \text{for } \omega \in B \tag{1.19}
$$

and that the phase response is linear, i.e.,

$$
\theta(\omega) = -\tau_{\text{g}}\omega + \theta_{\text{o}} \quad \text{for } \omega \in B \tag{1.20}
$$

where τ_{g} is a constant. This implies that the group delay is constant with respect to band B, i.e.,

$$
\tau(\omega) = -\frac{d\theta(\omega)}{d\omega} = \tau_{\text{g}} \quad \text{for } \omega \in B
$$

The frequency spectrum of the output signal $v_o(t)$ can be obtained from Equations 1.19 and 1.20 as

$$V_o(j\omega) = H(j\omega)V_i(j\omega) = M(\omega)e^{j\theta(\omega)}V_i(j\omega)$$
$$= \left[G_0 e^{-j\omega\tau_g + j\theta_0}\right]V_i(j\omega) = G_0 e^{j\theta_0}\left[e^{-j\omega\tau_g}V_i(j\omega)\right]$$

and from the time-shifting theorem of the Fourier transform

$$v_o(t) = G_0 e^{j\theta_0} v_i\left(t - \tau_g\right)$$

We conclude that the amplitude response of the filter is flat and its phase response is a linear function of ω (i.e., the delay characteristic is flat) in band B, then the output signal is a delayed replica of the input signal except that a gain G_o and a constant phase shift θ_0 are introduced.

If the amplitude response of the filter is not flat in band B, then *amplitude distortion* will be introduced since different frequency components of the signal will be amplified by different amounts.

If the delay characteristic is not flat in band B, then *delay* (or *phase*) *distortion* will be introduced since different frequency components will be delayed by different amounts.

Amplitude distortion can be quite objectionable in practice and, consequently, in each frequency band that carries information, the amplitude response is required to be constant to within a prescribed tolerance. The amount of amplitude distortion allowed determines the maximum passband loss A_p.

If the ultimate receiver of the signal is the human ear, e.g., when a speech or music signal is to be processed, delay distortion is quite tolerable. However, in other applications it can be as objectionable as amplitude distortion and the delay characteristic is required to be fairly flat. Applications of this type include data transmission, where the signal is to be interpreted by digital hardware, and image processing, where the signal is used to reconstruct an image that is to be interpreted by the human eye. The allowable delay distortion dictates the degree of flatness in the delay characteristic.

1.7 Minimum-Phase, Nonminimum-Phase, and Allpass Filters

Filters satisfying prescribed loss specifications for applications where delay distortion is unimportant can be readily designed with transfer functions whose zeros are on the $j\omega$ axis or in the left-half s plane. Such transfer functions are said to be minimum-phase since the phase response at a given frequency ω is increased if any one of the zeros is moved into the right-half s plane, as will now be demonstrated.

1.7.1 Minimum-Phase Filters

Consider a filter where the zeros z_i for $i = 1, 2, \ldots, M$ are replaced by their mirror images and let the new zeros be located at $z = \bar{z}_i$, where

$$\text{Re } \bar{z}_i = -\text{Re } z_i \quad \text{and} \quad \text{Im } \bar{z}_i = \text{Im } z_i$$

as depicted in Figure 1.5. From the geometry of the new zero-pole plot, the magnitude and angle of each phasor $j\omega - \bar{z}_i$ are given by

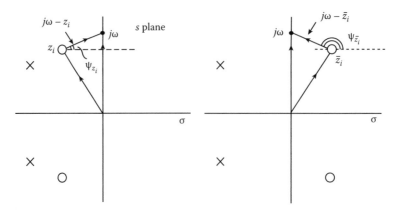

FIGURE 1.5   Zero-pole plots of minimum-phase and corresponding nonminimum-phase filter.

$$M_{\bar{z}_i} = M_{z_i} \quad \text{and} \quad \psi_{\bar{z}_i} = \pi - \psi_{z_i}$$

respectively. The amplitude response of the modified filter is obtained from Equation 1.16 as

$$\overline{M}(\omega) = \frac{|H_0| \prod_{i=1}^{M} M_{\bar{z}_i}}{\prod_{i=1}^{N} M_{p_i}^{m_i}} = \frac{|H_0| \prod_{i=1}^{M} M_{z_i}}{\prod_{i=1}^{N} M_{p_i}^{m_i}} = M(\omega)$$

Therefore, replacing the zeros of the transfer function by their mirror images leaves the amplitude response unchanged.

The phase response of the original filter is given by Equation 1.17 as

$$\theta(\omega) = \arg H_0 + \sum_{i=1}^{M} \psi_{z_i} - \sum_{i=1}^{N} m_i \psi_{p_i} \qquad (1.21)$$

and since $\psi_{\bar{z}_i} = \pi - \psi_{z_i}$, the phase response of the modifier filter is given by

$$\bar{\theta}(\omega) = \arg H_0 + \sum_{i=1}^{M} \psi_{\bar{z}_i} - \sum_{i=1}^{N} m_i \psi_{p_i}$$

$$= \arg H_0 + \sum_{i=1}^{M} (\pi - \psi_{z_i}) - \sum_{i=1}^{N} m_i \psi_{p_i} \qquad (1.22)$$

that is, the phase response of the modified filter is different from that of the original filter. Furthermore, from Equations 1.21 and 1.22

$$\bar{\theta}(\omega) - \theta(\omega) = \sum_{i=1}^{M} (\pi - 2\psi_{z_i})$$

and since $-\pi/2 \leq -\psi_{z_i} \leq \pi/2$, we have

$$\bar{\theta}(\omega) - \theta(\omega) \geq 0$$

or

$$\bar{\theta}(\omega) \geq \theta(\omega)$$

As a consequence, the phase response of the modified filter is equal to or greater than that of the original filter for all ω.

A frequently encountered requirement in the design of filters is that the delay characteristic be flat to within a certain tolerance within the passband(s) in order to achieve tolerable delay distortion, as was demonstrated in Section 1.6. In these and other filters in which the specifications include constraints on the phase response or delay characteristic, a nonminimum-phase transfer function is almost always required.

1.7.2 Allpass Filters

An allpass filter is one that has a constant amplitude response. Consider a transfer function of the type given by Equation 1.13. From Equation 1.10, $H(-j\omega)$ is the complex conjugate of $H(j\omega)$, and hence a constant amplitude response can be achieved if

$$M^2(\omega) = H(s)H(-s)\big|_{s=j\omega} = H_0^2 \frac{N(s)}{D(s)} \times \frac{N(-s)}{D(-s)}\bigg|_{s=j\omega} = H_0^2$$

Hence, an allpass filter can be obtained if

$$N(-s) = D(s)$$

that is, the zeros of such a filter must be the mirror images of the poles and vice versa. A typical zero-pole plot for an allpass filter is illustrated in Figure 1.6. A second-order allpass transfer function is given by

$$H_{AP}(s) = \frac{s^2 - bs + c}{s^2 + bs + c}$$

where $b > 0$ for stability. As described previously, we can write

$$M^2(\omega) = H_{AP}(s)H_{AP}(-s)\big|_{s=j\omega}$$
$$= \frac{s^2 - bs + c}{s^2 + bs + c} \times \frac{s^2 + bs + c}{s^2 - bs + c}\bigg|_{s=j\omega} = 1$$

Allpass filters can be used to modify the phase responses of filters without changing their amplitude responses. Hence, they are used along with minimum-phase filters to obtain nonminimum-phase filters that satisfy amplitude and phase response specifications simultaneously.

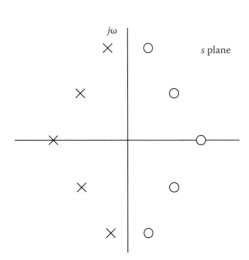

FIGURE 1.6 Typical zero-pole plot of an allpass filter.

1.7.3 Decomposition of Nonminimum-Phase Transfer Functions

Some methods for the design of filters satisfying amplitude and phase response specifications, usually methods based on optimization, yield a nonminimum-phase transfer function. Such a transfer function can be easily decomposed into a product of a minimum-phase and an allpass transfer function, i.e.,

$$H_N(s) = H_M(s)H_{AP}(s)$$

Consequently, a nonminimum-phase filter can be implemented as a cascade arrangement of a minimum-phase and an allpass filter.

The preceding decomposition can be obtained by using the following procedure:

1. For each zero in the right-half s plane, augment the transfer function by a zero and a pole at the mirror image position of the zero.
2. Assign the left-half s-plane zeros and the original poles to the minimum-phase transfer function $H_M(s)$.
3. Assign the right-half s-plane zeros and the left-hand s-plane poles generated in step 1 to the allpass transfer function $H_{AP}(s)$.

This procedure is illustrated in Figure 1.7. For example, if

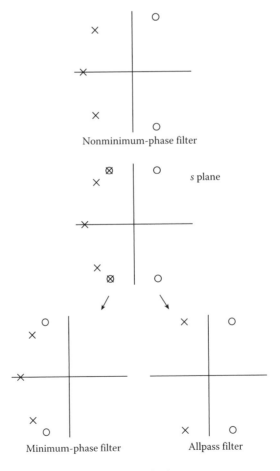

FIGURE 1.7 Decomposition of nonminimum-phase transfer function.

$$H_N(s) = \frac{(s^2 + 4s + 5)(s^2 - 3s + 7)(s - 5)}{(s^2 + 2s + 6)(s^2 + 4s + 9)(s + 2)}$$

then, we can write

$$H_N(s) = \frac{(s^2 + 4s + 5)(s^2 - 3s + 7)(s - 5)}{(s^2 + 2s + 6)(s^2 + 4s + 9)(s + 2)} \times \frac{(s^2 + 3s + 7)(s + 5)}{(s^2 + 3s + 7)(s + 5)}$$

Hence,

$$H_N(s) = \frac{(s^2 + 4s + 5)(s^2 + 3s + 7)(s + 5)}{(s^2 + 2s + 6)(s^2 + 4s + 9)(s + 2)} \times \frac{(s^2 - 3s + 7)(s - 5)}{(s^2 + 3s + 7)(s + 5)}$$

or

$$H_N(s) = H_M(s)H_{AP}(s)$$

where

$$H_M(s) = \frac{(s^2 + 4s + 5)(s^2 + 3s + 7)(s + 5)}{(s^2 + 2s + 6)(s^2 + 4s + 9)(s + 2)}$$

$$H_{AP}(s) = \frac{(s^2 - 3s + 7)(s - 5)}{(s^2 + 3s + 7)(s + 5)}$$

1.8 Introduction to the Design Process

The design of filters starts with a set of specifications and ends with the implementation of a prototype. It comprises four general steps, as follows:

1. Approximation
2. Realization
3. Study of imperfections
4. Implementation

1.8.1 The Approximation Step

The approximation step is the process of generating a transfer function that satisfies the desired specifications, which may concern the amplitude, phase, and possibly the time-domain response of the filter.

The available methods for the solution of the approximation problem can be classified as closed-form or iterative. In closed-form methods, the problem is solved through a small number of design steps using a set of closed-form formulas or transformations. In iterative methods, an initial solution is assumed and, through the application of optimization methods, a series of progressively improved solutions are obtained until some design criterion is satisfied. Closed-form solutions are very precise and entail a minimal amount of computation. However, the available solutions are useful in applications where the loss characteristic is required to be piecewise constant to within some prescribed tolerances. Iterative methods, on the other hand, entail a considerable amount of computation but can be used to design filters with arbitrary amplitude and phase response characteristics (see Ref. [1, Chapter 14]) for the application of these methods for the design of digital filters). Some classical closed-form solutions are the so-called Butterworth, Chebyshev, and elliptic* approximations to be described in Chapter 2 by A.M. Davis.

In general, the designer is interested in simple and reliable approximation methods that yield precise designs with the minimum amount of computation.

1.8.2 The Realization Step

The *synthesis* of a filter is the process of converting some characterization of the filter into a network. The process of converting the transfer function into a network is said to be the realization step and the network obtained is sometimes called the realization.

The realization of a transfer function can be accomplished by expressing it in some form that allows the identification of an interconnection of elemental filter subnetworks and/or elements. Many realization methods have been proposed in the past that lead to structures of varying complexity and properties. In general, the designer is interested in realizations that are economical in terms of the number of elements, do not require expensive components, and are not seriously affected by variations in the element values such as may be caused by variations in temperature and humidity, and drift due to element aging.

1.8.3 Study of Imperfections

During the approximation step, the coefficients of the transfer function are determined to a high degree of precision and the realization is obtained on the assumption that elements are ideal, i.e., capacitors are lossless, inductors are free of winding capacitances, amplifiers have infinite bandwidths, and so on. In practice, however, the filter is implemented with nonideal elements that have finite tolerances and are often nonlinear. Consequently, once a realization is obtained, sometimes referred to as a *paper design*, the designer must embark on the study of the effects of element imperfections. Several types of analysis are usually called for ranging from tolerance analysis, study of parasitics, time-domain analysis, sensitivity analysis, noise analysis, etc. Tight tolerances result in high-precision filters but the cost per unit would be high. Hence the designer is obliged to determine the highest tolerance that can be tolerated without violating the specifications of the filter throughout its working life. Sensitivity analysis is a related study that will ascertain the degree of dependence of a filter parameter, e.g., the dependence of the amplitude response on a specific element. If the loss characteristic of a filter is not very sensitive to certain capacitance, then the designer would be able to use a less precise and cheaper capacitor, which would, of course, decrease the cost of the unit.

1.8.4 Implementation

Once the filter is thoroughly analyzed and found to meet the desired specifications under ideal conditions, a prototype is constructed and tested. Decisions to be made involve the type of components and packaging, and the methods are to be used for the manufacture, testing, and tuning of the filter. Problems may often surface at the implementation stage that may call for one or more modifications in the paper design. Then the realization and possibly the approximation may have to be redone.

---

* To be precise, the elliptic approximation is not a closed-form method, since the transfer function coefficients are given in terms of certain infinite series. However, these series converge very rapidly and can be treated as closed-form formulas for most practical purposes (see Ref. [1, Chapter 5]).

1.9 Introduction to Realization

Realization tends to depend heavily on the type of filter required. The realization of passive *RLC* filters differs quite significantly from that of active filters which, in turn, is entirely different from the realization of microwave filters.

1.9.1 Passive Filters

Passive *RLC* filters have been the mainstay of communications since the 1920s and, furthermore, they continue to be of considerable importance today for frequencies in the 100–500 kHz range.

The realization of passive *RLC* filters has received considerable attention through the years and it is, as a consequence, highly developed and sophisticated. It can be accomplished by using available filter-design packages such as FILSYN [5] and FILTOR [6]. In addition, several filter-design handbooks and published design tables are available [7–10].

The realization of passive *RLC* filters starts with a resistively terminated *LC* two-port network such as that in Figure 1.8. Then through one of several approaches, the transfer function is used to generate expressions for the *z* or *y* parameters of the *LC* two-port. The realization of the *LC* two-port is achieved by realizing the *z* or *y* parameters. The realization of passive filters is considered in Section I.

1.9.2 Active Filters

Since the reactance of an inductor is ωL, increased inductance values are required to achieve reasonable reactance values at low frequencies. For example, an inductance of 1 mH which will present a reactance of 6.28 kΩ at 1 MHz will present only 0.628 Ω at 100 Hz. Thus, as the frequency range of interest is reduced, the inductance values must be increased if a specified impedance level is to be maintained. This can be done by increasing the number of turns on the inductor coil and to some extent by using ferromagnetic cores of high permeability. Increasing the number of turns increases the resistance, the size, and the cost of the inductor. The resistance is increased because the length of the wire is increased [$R = (\rho \times length)/Area$], and hence the *Q* factor is reduced. The cost goes up because the cost of materials as well as the cost of labor go up, since an inductor must be individually wound. For these reasons, inductors are generally incompatible with miniaturization or microcircuit implementation.

The preceding physical problem has led to the invention and development of a class of inductorless filters known collectively as active filters. Sensitivity considerations, which will be examined in Chapter 4 by I. Filanovsky, have led to two basic approaches to the design of active filters. In one approach, the active filter is obtained by simulating the inductances in a passive *RLC* filter or by realizing a signal flow graph of the passive *RLC* filter. In another approach, the active filter is obtained by cascading a number of low-order filter sections of some type, as depicted in Figure 1.9a where Z_{o0} is the output impedance of the signal source.

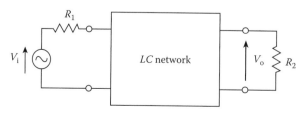

FIGURE 1.8 Passive *RLC* filter.

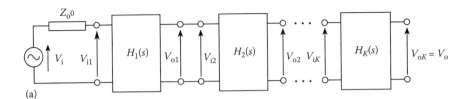

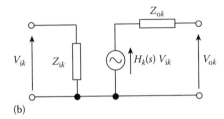

FIGURE 1.9 (a) Cascade realization and (b) Thévenin equivalent circuit of filter section.

Each filter section is made up of an interconnection of resistors, capacitors, and active elements, and by Thévenin's theorem, it can be represented by its input impedance, open-circuit voltage transfer function, and output impedance as shown in Figure 1.9b. The voltage transfer function of the configuration is given by

$$H(s) = \frac{V_o}{V_i}$$

and since the input voltage of section k is equal to the output voltage of section $k-1$, i.e., $V_{ik} = V_{o(k-1)}$ for $k = 2, 3, \ldots, K$, and $V_o = V_{oK}$ we can write

$$H(s) = \frac{V_o}{V_i} = \frac{V_{i1}}{V_i} \times \frac{V_{o1}}{V_{i1}} \times \frac{V_{o2}}{V_{i2}} \times \cdots \times \frac{V_{oK}}{V_{iK}} \tag{1.23}$$

where

$$\frac{V_{i1}}{V_i} = \frac{Z_{i1}}{Z_{o0} + Z_{i1}} \tag{1.24}$$

and

$$\frac{V_{ok}}{V_{ik}} = \frac{Z_{i(k+1)}}{Z_{ok} + Z_{i(k+1)}} H_k(s) \tag{1.25}$$

is the transfer function of the kth section. From Equations 1.23 through 1.25, we obtain

$$H(s) = \frac{V_o}{V_i} = \frac{Z_{i1}}{Z_{o0} + Z_{i1}} \prod_{k=1}^{K} \frac{Z_{i(k+1)}}{Z_{ok} + Z_{i(k+1)}} H_k(s)$$

Now if

$$|Z_{ik}| \gg |Z_{o(k-1)}|$$

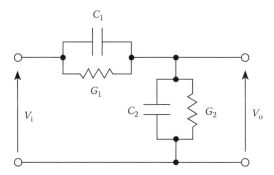

FIGURE 1.10 First-order *RC* network.

for $k = 1, 2, \ldots, K$, then the loading effect produced by section $k+1$ on section k can be neglected and hence

$$H(s) = \frac{V_o}{V_i} = \prod_{k-1}^{K} H_k(s)$$

Evidently, a highly desirable property in active filter sections is that the magnitude of the input impedance be large and/or that of the output impedance be small since in such a case the transfer function of the cascade structure is equal to the product of the transfer functions of the individual sections.

An arbitrary Nth-order transfer function obtained by using the Butterworth, Bessel, Chebyshev, inverse-Chebyshev, or elliptic approximation can be expressed as

$$H(s) = H_0(s) \prod_{k=1}^{K} \frac{a_{2k}s^2 + a_{1k}s + a_{0k}}{s^2 + b_{1k}s + b_{0k}}$$

where

$$H_0(s) = \begin{cases} \frac{a_{10}s + a_{00}}{b_{10}s + b_{00}} & \text{for odd } N \\ 1 & \text{for even } N \end{cases}$$

The first-order transfer function $H_0(s)$ for the case of an odd-order can be readily realized using the *RC* network of Figure 1.10.

1.9.3 Biquads

From the above analysis, we note that all we need to be able to realize an arbitrary transfer function is a circuit that realizes the biquadratic transfer function

$$H_{BQ}(s) = \frac{a_2 s^2 + a_1 s + a_0}{s^2 + b_1 s + b_0} = \frac{a_2(s + z_1)(s + z_2)}{(s + p_1)(s + p_2)} \tag{1.26}$$

where zeros and poles occur in complex conjugate pairs, i.e., $z_2 = z_1{}^*$ and $p_2 = p_1{}^*$. Such a circuit is commonly referred to as a *biquad*.

After some manipulation, the transfer function in Equation 1.26 can be expressed as

$$H_{BQ}(s) = K \frac{s^2 + (2\text{Re } z_1)s + (\text{Re } z_1)^2 + (\text{Im } z_1)^2}{s^2 + (2\text{Re } p_1)s + (\text{Re } p_1)^2 + (\text{Im } p_1)^2}$$

$$= K \frac{s^2 + (\omega_z/Q_z)s + \omega_z^2}{s^2 + (\omega_p/Q_p)s + \omega_p^2}$$

where

$K = a_2$

ω_z and ω_p are the zero and pole frequencies, respectively

Q_z and Q_p are the zero and pole *quality factors* (or Q factors for short), respectively

The formulas for the various parameters are as follows:

$$\omega_z = \sqrt{(\text{Re } z_1)^2 + (\text{Im } z_1)^2}$$

$$\omega_p = \sqrt{(\text{Re } p_1)^2 + (\text{Im } p_1)^2}$$

$$Q_z = \frac{\omega_z}{2 \text{ Re } z_1}$$

$$Q_p = \frac{\omega_p}{2 \text{ Re } p_1}$$

The zero and pole frequencies are approximately equal to the frequencies of minimum gain and maximum gain, respectively. The zero and pole Q factors have to do with the selectivity of the filter. A high zero Q factor results in a deep notch in the amplitude response, whereas a high pole Q factor results in a very peaky amplitude response.

The dc gain and the gain as $\omega \to \infty$ in decibels are given by

$$M_0 = 20 \log |H_{BQ}(0)| = 20 \log \left(K \frac{\omega_z^2}{\omega_p^2} \right)$$

and

$$M_\infty = 20 \log |H_{BQ}(j\infty)| = 20 \log K$$

respectively.

1.9.4 Types of Basic Filter Sections

Depending on the values of the transfer function coefficients, five basic types of filter sections can be identified, namely, low-pass, high-pass, bandpass, notch (sometimes referred to as bandreject), and allpass. These sections can serve as building blocks for the design of filters that can satisfy arbitrary specifications. They are actually sufficient for the design of all the standard types of filters, namely, Butterworth, Chebyshev, inverse-Chebyshev, and elliptic filters.

1.9.4.1 Low-Pass Section

In a *low-pass* section, we have $a_2 = a_1 = 0$ and $a_0 = K\omega_p^2$. Hence, the transfer function assumes the form

$$H_{LP}(s) = \frac{a_0}{s^2 + b_1 s + b_0} = \frac{K\omega_p^2}{s^2 + (\omega_p/Q_p)s + \omega_p^2}$$

(see Figure 1.11a)

1.9.4.2 High-Pass Section

In a *high-pass* section, we have $a_2 = K$ and $a_1 = a_0 = 0$. Hence, the transfer function assumes the form

$$H_{\text{HP}}(s) = \frac{a_2 s^2}{s^2 + b_1 s + b_0} = \frac{K s^2}{s^2 + (\omega_p/Q_p)s + \omega_p^2}$$

(see Figure 1.11b)

1.9.4.3 Bandpass Section

In a *bandpass* section, we have $a_1 = K\omega_p/Q_p$ and $a_2 = a_0 = 0$. Hence the transfer function assumes the form

$$H_{\text{BP}}(s) = \frac{a_1 s}{s^2 + b_1 s + b_0} = \frac{K(\omega_p/Q_p)s}{s^2 + (\omega_p/Q_p) + \omega_p^2}$$

(see Figure 1.11c)

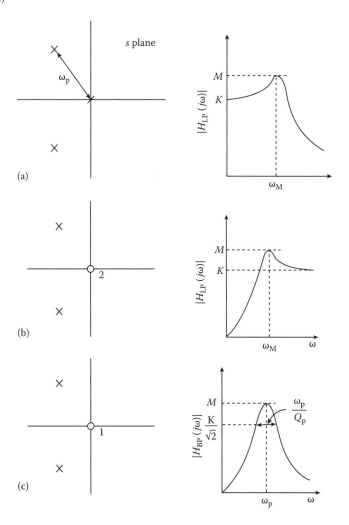

FIGURE 1.11 Basic second-order filter sections: (a) low-pass, (b) high-pass, (c) bandpass,

(continued)

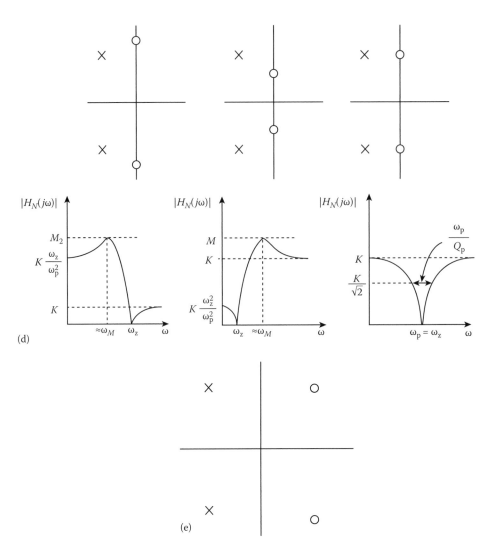

FIGURE 1.11 (continued)   (d) notch, and (e) allpass.

1.9.4.4 Notch Section

In a *notch* section, we have $a_2 = K$, $a_1 = 0$, and $a_0 = K\omega_p^2$. Hence, the transfer function assumes the form

$$H_N(s) = \frac{a_2 s^2 + a_0}{s^2 + b_1 s + b_0} = \frac{K\left(s^2 + \omega_z^2\right)}{s^2 + \left(\omega_p/Q_p\right)s + \omega_p^2}$$

(see Figure 1.11d)

1.9.4.5 Allpass Section

In an *allpass* section, we have $a_2 = K$, $a_1 = -K\omega_p/Q_p$, and $a_0 = K\omega_p^2$. Hence the transfer function assumes the form

$$H_{AP}(s) = \frac{a_2 s^2 + a_1 s + a_0}{s^2 + b_1 s + b_0} = \frac{K\left[s^2 - (\omega_p/Q_p)s + \omega_p^2\right]}{s^2 + (\omega_p/Q_p)s + \omega_p^2}$$

(see Figure 1.11e)

The design of active and switched-capacitor filters is treated in some detail in Section II.

References

1. A. Antoniou, *Digital Filters: Analysis, Design, and Applications*, 2nd ed. New York: McGraw-Hill, 1993.
2. R. J. Schwarz and B. Friedland, *Linear Systems*, New York: McGraw-Hill, 1965.
3. E. Kreyszig, *Advanced Engineering Mathematics*, 3rd ed. New York: Wiley, 1972.
4. R. Schaumann, M. S. Ghausi, and K. R. Laker, *Design of Analog Filters*, Englewood Cliffs, NJ: Prentice Hall, 1990.
5. G. Szentirmai, FILSYN—A general purpose filter synthesis program, *Proc. IEEE*, 65, 1443–1458, Oct. 1977.
6. A. S. Sedra and P. O. Brackett, *Filter Theory and Design: Active and Passive*, Portland, OR: Matrix, 1978.
7. J. K. Skwirzynski, *Design Theory and Data for Electrical Filters*, London: Van Nostrand, 1965.
8. R. Saal, *Handbook of Filter Design*, Backnang: AEG Telefunken, 1979.
9. A. I. Zverev, *Handbook of Filter Synthesis*, New York: Wiley, 1967.
10. E. Chirlian, *LC Filters: Design, Testing, and Manufacturing*, New York: Wiley, 1983.

2

Approximation

Artice M. Davis
San Jose State University

2.1 Introduction

The approximation problem for filters is illustrated in Figure 2.1. A filter is often desired to produce a given slope of gain over one or more frequency intervals, to remain constant over other intervals, and to completely reject signals having frequencies contained in still other intervals. Thus, in the example shown in the figure, the desired gain is zero for very low and very high frequencies. The centerline, shown dashed, is the nominal behavior and the shaded band shows the permissible variation in the gain characteristic. Realizable circuits must always generate smooth curves and so cannot exactly meet the piecewise linear specification represented by the centerline. Thus, the realizable behavior is shown by the smooth, dark curve that lies entirely within the shaded tolerance band.

What type of frequency response function can be postulated that will meet the required specifications and, at the same time, be realizable: constructible with a specified catalog of elements? The answer depends upon the types of elements allowed. For instance, if one allows pure delays with a common delay time, summers, and scalar multipliers, a trigonometric polynomial will work; this, however, will cause the gain function to be repeated in a periodic manner. If this is permissible, one can then realize the filter in the form of an FIR digital filter or as a commensurate transmission line filter, and in fact, it can be realized in such a fashion that the resulting phase behavior is precisely linear. If one fits the required behavior with a rational trigonometric function, a function that is the ratio of two trigonometric polynomials, an economy of hardware will result. The phase, however, will unfortunately no longer be linear. These issues are discussed at greater length in Ref. [1].

Another option would be to select an ordinary polynomial in ω as the approximating function. Polynomials, however, behave badly at infinity. They approach infinity as $\omega \to \pm\infty$, a highly undesirable solution. For this reason, one must discard polynomials. A rational function of ω, however, will work nicely for the ratio of two polynomials will approach zero as $\omega \to \pm\infty$ if the degree of the numerator polynomial is selected to be of lower degree than that of the denominator. Furthermore, by the Weierstrass theorem, such a function can approximate any continuous function arbitrarily closely over any closed interval of finite length [2]. Thus, one sees that the rational functions in ω offer a suitable approximation for analog filter design and, in fact, do not have the repetitive nature of the trigonometric rational functions.

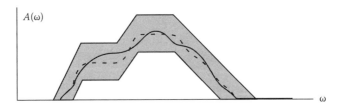

$A(\omega)$

FIGURE 2.1   General approximation problem.

Suppose, therefore, that the gain function is of the form

$$A(\omega) = \frac{N(\omega)}{D(\omega)} = \frac{a_0 + a_1\omega + a_2\omega^2 + \cdots + a_r\omega^r}{b_0 + b_1\omega + b_2\omega^2 + \cdots + b_q\omega^q} \tag{2.1}$$

where $r \leq q$ for reasons mentioned above. Assuming that the filter to be realized is constrained to be constructable with real* elements, one must require that $A(-\omega) = A(\omega)$, that is, that the gain be an event function of frequency. But then, as it is straightforward to show, one must require that all the odd coefficients of both numerator and denominator be zero. This means that the gain is a function of ω^2:

$$A(\omega) = \frac{N(\omega^2)}{D(\omega^2)} = \frac{a_0 + a_1\omega^2 + \cdots + a_m\omega^{2m}}{b_0 + b_1\omega^2 + \cdots + b_n\omega^{2n}} = A(\omega^{-2}) \tag{2.2}$$

The expression has been reindexed and the constants redefined in an obvious manner. The net result is that one must approximate the desired characteristic by the ratio of two polynomials in ω^2; the objective is to determine the numerator and denominator coefficients to meet the stated specifications. Once this is accomplished one must compute the filter transfer function $G(s)$ in order to synthesize the filter [4,5]. Assuming that $G(s)$ is real (has real coefficients), then its complex conjugate satisfies $G^*(s) = G(s^*)$, from which it follows that $G(s)$ is related to $A(\omega^2)$ by the relationship

$$[G(s)G(-s)]_{s=j\omega} = G(j\omega)G^*(j\omega) = |G(j\omega)|^2 = A^2(\omega^2) \tag{2.3}$$

In fact, it is more straightforward to simply cast the original approximation problem in terms of $A^2(\omega^2)$, rather than in terms of $A(\omega)$. In this case, Equation 2.2 becomes

$$A^2(\omega^2) = \frac{N(\omega^2)}{D(\omega^2)} = \frac{a_0 + a_1\omega^2 + \cdots + a_m\omega^{2m}}{b_0 + b_1\omega^2 + \cdots + b_n\omega^{2n}} \tag{2.4}$$

Thus, one can assume that the approximation process produces $A^2(\omega^2)$ as the ratio of two real polynomials in ω^2. Since Equation 2.3 requires the substitutions $s \to j\omega$, one also has $s^2 \to -\omega^2$, and conversely. Thus, Equation 2.3 becomes

$$G(s)G(-s) = A^2(-s^2) \tag{2.5}$$

Though this has been shown to hold only on the imaginary axis, it continues to hold for other complex values of s as well by analytic continuation.†

---

* Complex filters are quite possible to construct, as recent work [3] shows.
† A function analytic in a region is completely determined by its values along any line segment in that region—in this case, by its value along the $j\omega$ axis.

The problem now is to compute $G(s)$ from Equation 2.5, a process known as the factorization problem. The solution is not unique; in fact, the phase is arbitrary—subject only to certain realizability conditions. To see this, just let $G(j\omega) = A(\omega)e^{j\phi(\omega)}$, where $\phi\omega$ is an arbitrary phase function. Then, Equation 2.3 implies that

$$G(j\omega)G^*(j\omega) = A(\omega)e^{j\phi(\omega)} \cdot A(\omega)e^{-j\phi(\omega)} = A^2(\omega) \qquad (2.6)$$

If the resulting structure is to have the property of minimum phase [6], the phase function is determined completely by the gain function. If not, one can simply perform the factorization and accept whatever phase function results from the particular process chosen. As has been pointed out earlier in this chapter, it is often desirable that the phase be a linear function of frequency. In this case, one must follow the filter designed by the above process with a phase equalization filter, one that has constant gain and a phase characteristic that, when summed with that of the first filter, produces linear phase. As it happens, the human ear is insensitive to phase nonlinearity, so the phase is not of much importance for filters designed to operate in the audio range. For those intended for video applications, however, it is vitally important. Nonlinear phase produces, for instance, the phenomenon of multiple edges in a reproduced picture.

If completely arbitrary gain characteristics are desired, computer optimization is necessary [6]. Indeed, if phase is of great significance, computer algorithms are available for the simultaneous approximation of both gain and phase. These are complex and unwieldy to use, however, so for more modest applications the above approach relying upon gain approximation only suffices. In fact, the approach arose historically in the telephone industry in its earlier days in which voice transmission was the only concern, data and video transmission being unforeseen at the time. Furthermore, the frequency division multiplexing of voice signals was the primary concern; hence, a number of standard desired shapes of frequency response were generated: low pass, high pass, bandpass, and bandreject (or notch). Typical but stylized specification curves are shown in Figure 2.2. This figure serves to define the following parameters: the minimum passband gain A_p, the maximum stopband gain A_s, the passband cutoff frequency ω_p, the stopband cutoff frequency ω_s (the last two parameters are for low-pass and high-pass filters only), the center frequency ω_o, upper passband and stopband cutoff frequencies ω_{pu} and ω_{su}, and lower passband and stopband cutoff frequencies ω_{pl} and ω_{sl} (the last four parameters are for the bandpass

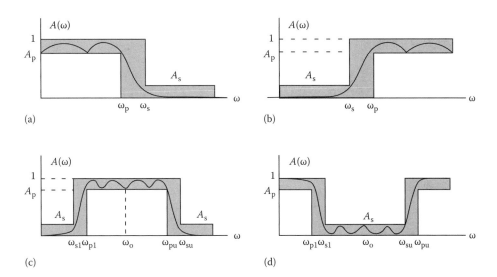

FIGURE 2.2 Catalog of basic filter types: (a) low pass, (b) high pass, (c) bandpass, and (d) bandreject.

and band-reject filters only). As shown in the figure, the maximum passband gain is usually taken to be unity. In the realization process, the transfer function scale factor is often allowed to be a free parameter that is resolved in the design procedure. The resulting "flat gain" (frequency independent) difference from unity is usually considered to be of no consequence, as long as it is not too small, thus creating signal-to-noise ratio problems. There is a fifth standard type that we have not shown: the allpass filter. It has a constant gain for all frequencies, but with a phase characteristic that can be tailored to fit a standard specification in order to compensate for phase distortion. Frequency ranges where the gain is relatively large are called passbands and those where the gain is relatively small, stopbands. Those in between—where the gain is increasing or decreasing—are termed transition bands.

In order to simplify the design even more, one bases the design of all other types of filter in terms of only one: the low pass. In this case, one says that the low-pass filter is a prototype. The specifications of the desired filter type are transformed to those of an equivalent low-pass prototype, and the transfer function for this filter is determined to meet the transformed specifications. Letting the low-pass frequency be symbolized by Ω and the original by ω, one sets

$$\Omega = f(\omega) \tag{2.7}$$

The approximation problem is then solved in terms of Ω. Letting

$$p = j\Omega \tag{2.8}$$

one then has $G(p)$, the desired transfer function.

Two approaches are now possible. One is to apply the inverse transformation, letting

$$s = j\omega = jf^{-1}(\Omega) = jf^{-1}(p/j) \tag{2.9}$$

thus obtaining the desired transfer function

$$G(s) = G(j\omega) = G[jf^{-1}(\Omega)] = G[jf^{-1}(p/j)] \tag{2.10}$$

The other consists of designing the circuit to realize the low-pass prototype filter, then transform *each of the elements* from functions of p to functions of s by means of the complex frequency transformation

$$s = j\omega = jf^{-1}(\Omega) = jf^{-1}(p/j) \tag{2.11}$$

As it happens, the transformation $p = f(s)$ has a special form. It can be shown that if f is real for real values of s, thereby having real parameters, and maps the imaginary axis of the p plane into the imaginary axis of the s plane then it must be an odd function; furthermore, if it is to map *positive real rational functions* into those of like kind (necessary for realizability with R, L, and C elements, as well as possibly ideal transformers*), it must be a *reactance function*. (See Ref. [7] for details.) Since it is always desirable from an economic standpoint to design filters of minimum order, it is desirable that the transformation be of the smallest possible degree. As a result, the following transformations are used:

$$(\text{lpp} \leftrightarrow \text{lp}) \quad p = ks \tag{2.12}$$

$$(\text{lpp} \leftrightarrow \text{hp}) \quad p = \frac{k}{s} \tag{2.13}$$

---

* For active realizations, those containing dependent sources, this condition is not necessary.

$$(\text{lpp} \leftrightarrow \text{bp}) \quad p = \frac{s^2 + \omega_o^2}{Bs} \tag{2.14}$$

$$(\text{lpp} \leftrightarrow \text{br}) \quad p = \frac{Bs}{s^2 + \omega_o^2} \tag{2.15}$$

where the parameters k, B, and ω_o are real constants to be determined by the particular set of specifications. We have used the standard abbreviations of lpp for *low-pass prototype*, lp for *low pass*, hp for *high pass*, bp for *bandpass*, and br for *bandreject*. Often the letter f is added; for example, one might use the acronym brf for *bandreject filter*. The reason for including the transformation in Equation 2.12 is to allow standardization of the lpp. For instance, one can transform from an lpp with, say, a passband cutoff frequency of 1 rad/s to a low-pass filter with a passband cutoff of perhaps 1 kHz.

As a simple example, suppose a bandreject filter were being designed and that the result of the approximation process were

$$H(p) = \frac{1}{p+1} \tag{2.16}$$

Then the br transfer function would be

$$H(s) = [H(p)]_{p=[Bs/(s^2+\omega_o^2)]} = \frac{1}{\frac{Bs}{s^2+\omega_o^2} + 1} = \frac{s^2 + \omega_o^2}{s^2 + Bs + \omega_o^2} \tag{2.17}$$

(The parameters ω_o and B would be determined by the bandreject specifications.) As one can readily see, a first-order lpp is transformed into a second-order brf. In general, for bandpass and bandreject design, the object transfer function is of twice the order of the lpp. Since the example is so simple, it can readily be seen that the circuit in Figure 2.3 realizes the lpp voltage gain function in Equation 2.16. If one applies the transformation in Equation 2.15 the 1-Ω resistor maps into a 1-Ω resistor, but the 1 F capacitor maps into a combination of elements having the admittance

$$Y(p) = p = \frac{1}{\frac{s}{B} + \frac{\omega_o^2}{Bs}} \tag{2.18}$$

But this is simply the series connection of a capacitor of value B/ω_o^2 farads and an inductor of value $1/B$ henrys. The resulting bandreject filter is shown in Figure. 2.4.

The only remaining "loose end" is the determination of the constant(s) in the appropriate transformation equation selected appropriately from Equations 2.12 through 2.15. This will be done here for the

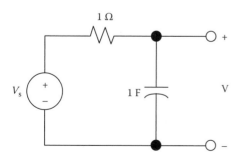

FIGURE 2.3　Low-pass prototype.

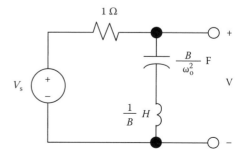

FIGURE 2.4　Resulting bandreject filter.

lpp × bp transformation (Equation 2.14). It is typical, and the reader should have no difficulty working out the other cases. Substituting $p = j\Omega$ and $s = j\omega$ in Equation 2.14, one gets

$$j\Omega = \frac{-\omega^2 + \omega_o^2}{jB\omega} \tag{2.19}$$

or

$$\Omega = \frac{\omega^2 - \omega_o^2}{B\omega} = \frac{\omega}{B} - \frac{\omega_o^2}{B\omega} \tag{2.20}$$

This clearly shows that $\omega = \pm\omega_o$ maps into $\Omega = 0$ and $\omega = \pm\infty$ into $\Omega = \pm\infty$. However, as $\omega \to 0+$, $\Omega \to -\infty$ and as $\omega \to 0-$, $\Omega \to +\infty$. Of perhaps more interest is the inverse transformation. Solving Equation 2.20 for ω in terms of Ω, one finds that*

$$\omega = \frac{B\Omega}{2} + \sqrt{\left[\frac{B\Omega}{2}\right] + \omega_o^2} \tag{2.21}$$

Now, consider pairs of values of Ω, of which one is the negative of the other. Letting ω_+ be the image of Ω with $\Omega > 0$ and ω_- be the image of $-\Omega$, one has

$$\omega_+ = \frac{B\Omega}{2} + \sqrt{\left[\frac{B\Omega}{2}\right] + \omega_o^2} \tag{2.22}$$

and

$$\omega_- = \frac{-B\Omega}{2} + \sqrt{\left[\frac{B\Omega}{2}\right] + \omega_o^2} \tag{2.23}$$

Subtracting, one obtains

$$\omega_+ - \omega_- = B\Omega \tag{2.24}$$

$$\omega_+\omega_- = \omega_o^2 \tag{2.25}$$

Thus, the geometric mean of ω_+ and ω_- is the parameter ω_o; furthermore, the lpp frequencies $\Omega = 1$ rad/s map into points whose difference is the parameter B. Recalling that $A(\Omega)$ has to be an even function of Ω, one sees that the gain magnitudes at these two points must be identical. If the lpp is designed so that $\Omega = 1$ rad/s is the "bandwidth" (single-sided), then the object bpf will have a (two-sided) bandwidth of B rad/s.

An example should clarify things. Figure 2.5 shows a set of bandpass filter gain specifications. Some slight generality has been allowed over those shown in Figure 2.2 by allowing the maximum stopband gains to be different in the two stopbands.

The graph is semilog: the vertical axis is linear with a dB scale and the horizontal axis is a log scale (base 10). The −0.1 dB minimum passband gain, by the way, is called the *passband ripple* because actual

---

* The negative sign on the radical gives $\omega < 0$ and the preceding treatment only considers positive ω.

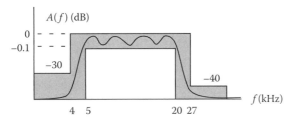

FIGURE 2.5 Bandpass filter specifications.

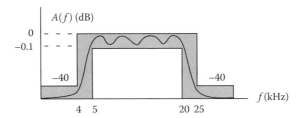

FIGURE 2.6 Bandpass filter specifications.

realized response is permitted to "ripple" back and forth between 0 and −0.1 dB. Notice that the frequency has been specified in terms of kilohertz—often a practical unit. Equation 2.20, however, can be normalized to any unit without affecting Ω. Thus, by normalizing ω to $2\pi \times 10^3$, one can substitute f in kilohertz for ω; the parameters ω_o (replaced of f_o symbolically) and B will then be in kilohertz also.

Now, notice that $5 \times 20 = 100 \neq 4 \times 27 = 108$, so the specifications are not geometrically symmetric relative to any frequency. Somewhat arbitrarily choosing $f_0 = \sqrt{5 \times 20} = 10$ kHz, one can force the specifications to have geometric symmetry by the following device: simply reduce the upper stopband cutoff frequency from 27 to 25 kHz. Then force the two stopband attenuations to be identical by decreasing the −30 dB lower stopband figure to −40 dB. This results in the modified specifications shown in Figure 2.6. If one chooses to map the upper and lower passband cutoff frequencies to $\Omega = 1$ rad/s (a quite typical choice, as many filter design catalogs are tabulated under this assumption), one then has

$$B = 20 - 5 = 15\,\text{kHz} \tag{2.26}$$

This fixes the parameters in the transformation and the lpp stopband frequency can be determined from Equation 2.20:

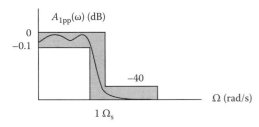

FIGURE 2.7 Bandpass filter specifications.

$$\Omega_s = \frac{25^2 - 10^2}{15 \times 25} = \frac{25}{15} - \frac{100}{375} = 1.498\,\text{rad/s} \tag{2.27}$$

The lpp specifications then assume the form shown in Figure 2.7. Once the lpp approximation problem is solved, one can then transform either the derived transfer function or the synthesized circuit back up to bandpass form since the parameters of the transformation are known.

2.2 Butterworth LPP Approximation

For performing lpp approximations, it is more convenient to work with the *characteristic function k(ω)* than with the gain function. It is defined by the equation

$$A^2(\omega) = \frac{1}{1 + K^2(\omega)} \tag{2.28}$$

Although Ω was used in Section 2.1 to denote lpp frequency, the lower case ω is used here and throughout the remainder of the section. No confusion should result because frequency will henceforth always mean lpp frequency. The main advantage in using the characteristic function is simply that it approximates zero over any frequency interval for which the gain function approximates unity. Further, it becomes infinitely large when the gain becomes zero. These ideas are illustrated in Figure 2.8. Notice that $K(\omega)$ can be either positive or negative in the passband for it is squared in the defining equation. The basic problem in lpp filter approximation is therefore to find a characteristic function that approximates zero in the passband, approximates infinity in the stopband, and makes the transition from one to the other rapidly. Ideally, it would be exactly zero in the passband, then become abruptly infinity for frequencies in the stopband.

The *n*th-order Butterworth approximation is defined by

$$K(\omega) = \omega^n \tag{2.29}$$

This characteristic function is sketched in Figure 2.9 for two values of *n*—one small and the other large. As is easily seen, the larger order provides a better approximation to the idea "brick wall" lpp response. Notice, however, that $K(1) = 1$ regardless of the order; hence $A(1) = 0.5$ (−3 dB) regardless of the order.

It is conventional to define the *loss function H(s)* to be reciprocal of the gain function:

$$H(s) = \frac{1}{G(s)} \tag{2.30}$$

Letting $s = j\omega$ and applying Equation 2.28 results in

$$|H(j\omega)|^2 - K^2(\omega) = 1 \tag{2.31}$$

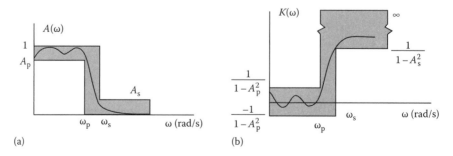

FIGURE 2.8 Filter specifications in terms of the characteristic function: (a) gain and (b) characteristic function.

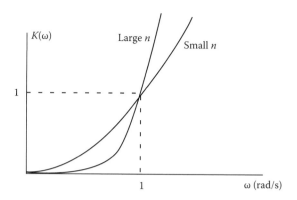

FIGURE 2.9   Butterworth characteristic function.

which is one form of *Feldtkeller's equation*, a fundamental equation in the study of filters. The loss approximates unity wherever the characteristic function approximates zero and infinity when the latter approximates infinity.

The loss function can be used to illustrate a striking property of the Butterworth approximation. Taking the *k*th derivative of Equation 2.31, one has

$$\frac{d^k |H(j\omega)|^2}{d\omega^k} = \frac{d^k K^2(\omega)}{d\omega^k} = \frac{d^k \omega^{2n}}{d\omega^k} = \frac{(2n)!}{k!} \omega^{2n-k} \tag{2.32}$$

This has the value zero at $\omega = 0$ for $k \leq 2n - 1$. It is the unique polynomial having this property among the set of all monic* polynomials of order $2n$ or lower having the value zero at the origin. But this means that the square of the Butterworth characteristic function $K(\omega)$ is the flattest of all such polynomials at $\omega = 0$. Since adding one to $K^2(\omega)$ produces the loss function $|H(j\omega)|^2$, the same is true of it relative to the set of all loss functions having the value unity at the origin. For this reason, the Butterworth approximation is often called the *maximally flat magnitude* (or MFM) approximation.

The passband ripple parameter A_p is always $1/\sqrt{2}$ for a Butterworth lpp; note that if a different ripple parameter is desired, one must treat the corresponding filter as a general low-pass filter, then apply Equation 3.8 of this book. The value of the parameter in that frequency transformation is determined by the requirement that the frequency at which the desired filter assumes the value $1/\sqrt{2}$ map into a lpp passband frequency of 1 rad/s. The required order is determined from the equation

$$\frac{1}{\sqrt{1 + \omega_s^{2n}}} \leq A_s \tag{2.33}$$

or, rearranged,

$$n \geq \frac{\log\left[\frac{1}{A_s^2} - 1\right]}{2 \log(\omega_s)} \tag{2.34}$$

The value of *n* is, of course, chosen to be the smallest integer greater than the expression on the right-hand side of Equation 2.34.

---

* A monic polynomial is one where the leading coefficient (highest power of ω) is unity.

Since only one parameter is in the Butterworth approximation, Equation 2.34 completely determines $A(\omega)$. That is, the MFM is a one-parameter approximation. The only remaining item of concern is the determination of $G(s)$, for synthesis of the actual filter requires knowledge of the transfer function. As was pointed out in Section 2.1, this is the factorization problem. In general, the solution is given by Equation 2.5, repeated here for convenience as Equation 2.35:

$$G(s)G(-s) = A^2(-s^2) \qquad (2.35)$$

In the present case, we have

$$G(s)G(-s) = \frac{1}{1 + (-s^2)^n} = \frac{1}{1 + (-1)^n s^{2n}} \qquad (2.36)$$

How does one find $G(s)$ from this equation? The solution merely lies in applying the restriction that the resulting filter is to be stable. This means that the poles of $G(s)G(-s)$ that lie in the right-half plane must be discarded and the remaining ones assigned to $G(s)$. In this connection, observe that any poles on the imaginary axis must be off even multiplicity since $G(s)G(-s)$ is an even function (or, equivalently, since $A^2(\omega^2) = |G(j\omega)|^2$ is nonnegative). Furthermore, any even-order poles of $G(s)G(-s)$ on the imaginary axis could only result from one or more poles of $G(s)$, itself, at the same location. But such a $G(s)$ represents a filter that is undesirable because, at best, it is only marginally stable. As will be shown, this situation does not occur for Butterworth filters.

The problem now is merely to find all the poles of $G(s)G(-s)$, then to sort them. These poles are located at the zeros of the denominator in Equation 2.36. Thus, one must solve

$$1 + (-s^2)^n = 0 \qquad (2.37)$$

or, equivalently,

$$s^{2n} = (-1)^{n-1} \qquad (2.38)$$

Representing s in polar coordinates by

$$s = \rho e^{j\phi} \qquad (2.39)$$

one can write Equation 2.38 in the form

$$\rho^{2n} e^{j2n\phi} = e^{j(n-1)\pi} \qquad (2.40)$$

This has the solution

$$\rho = 1 \qquad (2.41)$$

and

$$\phi = \frac{\pi}{2} + (2k - 1)\frac{\pi}{2n} \qquad (2.42)$$

where k is any integer. Of course, only those values of ϕ between 0 and 2π rad are to be considered unique.

As an example, suppose that $n = 2$; that is, one is interested in determining the transfer function of a second-order Butterworth filter. Then the unique values of ϕ determined from Equation 2.42 are $\pi/4, 3\pi/4, 5\pi/4$, and $7\pi/4$. All other values are simply these four with integer multiples of 2π added. The last two represent poles in the right-half plane, so are simply discarded. The other two correspond to poles at $s = -0.707 \pm j0.707$. Letting $D(s)$ be the numerator polynomial of $G(s)$, then, one has

$$D(s) = s^2 + \sqrt{2}s + 1 \qquad (2.43)$$

The poles of $G(s)G(-s)$ are sketched in Figure 2.10. Notice that if one assigns the left-half plane poles to $G(s)$, then those in the right-half plane will be those of $G(-s)$.

Perhaps another relatively simple example is in order. To this end, consider the third-order Butterworth transfer function $n = 3$. In this case, the polar angles of the poles of $G(s)$ are at $\pi/3, 2\pi/3, \pi, 4\pi/3, 5\pi/3$, and 2π. This pole pattern is shown in Figure 2.11. The denominator polynomial corresponds to those in the left-half plane. It is

$$D(s) = (s + 1)(s^2 + s + 1) = s^3 + 2s^2 + 2s + 1 \qquad (2.44)$$

The pattern for the general nth order case is similar—all the poles of $G(s)G(-s)$ lie on the unit circle and are equally spaced at intervals of π/n radians, but are offset by $\pi/2$ radians relative to the positive real axis. A little thought will convince one that this means that no poles ever fall on the imaginary axis.

The factorization procedure determines the denominator polynomial in $G(s)$. But what about the numerator? Since the characteristic function is a polynomial, it is clear that $G(j\omega)$, and hence $G(s)$ itself, will have a constant numerator. For this reason, the Butterworth approximation is referred to as an *all-pole* filter. Sometimes it is also called a *polynomial* filter, referring to the fact that the characteristic function is a polynomial. As was mentioned earlier, the constant is usually allowed to float freely in the synthesis process and is determined only at the conclusion of the design process. However, more can be said. Writing

$$G(s) = \frac{a}{D(s)} \qquad (2.45)$$

one can apply Equation 2.36 to show that $a^2 = 1$, provided that $|G(0)|$ is to be one, as is the case for the normalized lpp. This implies that

$$a = \pm 1 \qquad (2.46)$$

If a passive unbalanced (grounded) filter is desired the positive sign must be chosen. Otherwise, one can opt for either.

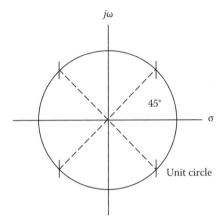

FIGURE 2.10 Poles of the second-order Butterworth transfer function.

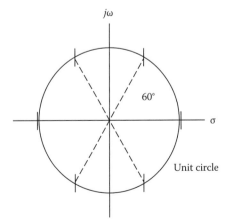

FIGURE 2.11 Poles of the third-order Butterworth transfer function.

2.3 Chebyshev LPP Approximation

The main advantage of the Butterworth approximation is that it is simple. It does not, however, draw upon the maximum approximating power of polynomials. In fact, a classical problem in mathematics is to approximate a given continuous function on a closed bounded interval with a polynomial of a specified maximum degree. One can choose to define the error of approximation in many ways, but the so-called minimax criterion seems to be the most suitable for filter design. It is the minimum value, computed over all polynomials of a specified maximum degree, of the maximum difference between the polynomial values and those of the specified function. This is illustrated in Figure 2.12. The minimax error in this case occurs at ω_x. It is the largest value of the magnitude of the difference between the function values $f(\omega)$ and those of a given candidate polynomial $p_n(\omega)$. The polynomial of best fit is the one for which this value is the smallest.

The basic lpp approximation problem is to pick the characteristic function to be that polynomial of a specified maximum degree no more than, say, n, which gives the smallest maximum error of approximation *to the constant value* 0 over the interval $0 \le \omega \le 1$ (arbitrarily assuming that the passband cutoff is to be 1 rad/s). In this special case, the solution is known in closed form: it is the Chebyshev* polynomial of degree n. Then, $K(\omega)$ is polynomial $\varepsilon T_n(\omega)$, where

$$T_n(\omega) = \cos\left[n\cos^{-1}(\omega)\right] \tag{2.47}$$

and ε is the minimax error (a constant).

It is perhaps not clear that $T_n(\omega)$ is actually a polynomial; however, upon computing the first few by applying simple trigonometric identities one has the results shown in Table 2.1. In fact, again by calling upon simple trigonometric identities, one can derive the general recurrence relation

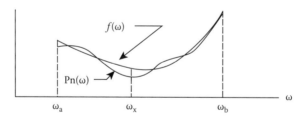

FIGURE 2.12 Minimax error criterion.

TABLE 2.1 Chebyshev Polynomial of Degree n

n	$T_n(\omega)$
0	1
1	ω
2	$2\omega^2 - 1$
3	$4\omega^3 - 3\omega$
4	$8\omega^4 - 8\omega^2 + 1$

---

* If the name of the Russian mathematician is transliterated from the French, in which the first non-Russian translations were given, it is spelled Tchebychev.

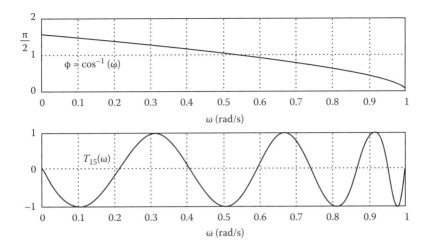

FIGURE 2.13 Behavior of the Chebyshev polynomials.

$$T_n(\omega) = 2\omega T_{n-1}(\omega) - T_{n-2}(\omega); \quad n \geq 2 \tag{2.48}$$

The Chebyshev polynomials have an enormous number of interesting properties and to explore them all would require a complete monograph. Among those of the most interest for filtering applications, however, are these. First, from the recursion relationship (Equation 2.48) one can see that $T_n(\omega)$ is indeed a polynomial of order n; furthermore, its leading coefficient is 2^{n-1}. If n is even, $T_n(\omega)$ is an even polynomial in ω and if n is odd, $T_n(\omega)$ is an odd polynomial. The basic definition in Equation 2.47 clearly shows that the extreme values of $T_n(\omega)$ over the interval $0 \leq \omega \leq 1$ are ± 1. Some insight into the behavior of the Chebyshev polynomials can be obtained by making the transformation $\phi = \cos^{-1}(\omega)$. Then, $T_n(\phi) = \cos(n\phi)$, a trigonometric function that is quite well known. The behavior of T_{15}, for example, is shown in Figure 2.13. The basic idea is this: the Chebyshev polynomial of nth order is merely a cosine of "frequency" $n/4$, which "starts" at $\omega = 1$ and "runs backward" to $\omega = 0$. Thus, it is always 1 at $\omega = 1$ and as ω goes from 1 rad/s to 0 rad/s backward, it goes through n quarter-periods (or $n/4$ full periods). Thus, at $\omega = 0$ the value of this polynomial will be either 0 or ± 1, depending upon the specific value of n. If n is even, an integral number of half-periods will have been described and the resulting value will be ± 1; if n is odd, an integral number of half-periods plus a quarter-period will have been described and the value will be zero.

Based on the foregoing theory, one sees that the best approximation to the ideal lpp characteristic over the passband is, for a given passband tolerance, ε, given by

$$A^2(\omega) = \frac{1}{1 + \varepsilon^2 T_n^2(\omega)} \tag{2.49}$$

It is, of course, known as the Chebyshev approximation and the resulting filter as the Chebyshev lpp of order n. The gain magnitude $A(\omega)$ is plotted for $n = 5$ and $\varepsilon = 0.1$ in Figure 2.14. The passband behavior looks like ripples in a container of water, and since the crests are equally spaced above and below the average value, it is called equiripple behavior. In the passband, the maximum value is 1 and the minimum value is

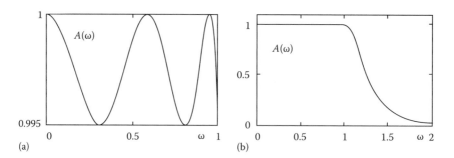

FIGURE 2.14 Frequency response of a fifth-order Chebyshev lpp: (a) passband and (b) overall.

$$A_{\min} = \frac{1}{\sqrt{1 + \varepsilon^2}} \tag{2.50}$$

The passband ripple is usually specified as the peak to peak variation in dB. Since the maximum value is one, that is 0 dB, this quantity is related to the *ripple parameter* ε through the equation

$$\text{passband ripple in dB} = 20 \log \sqrt{1 + \varepsilon^2} \tag{2.51}$$

The Chebyshev approximation is the best possible among the class of all-pole filters—*over the passband*. But what about its stopband behavior? As was pointed out previously, it is desirable that—in addition to approximating zero in the passband—the characteristic function should go to infinity as rapidly as possible in the stopband. Now it is a happy coincidence that the Chebyshev polynomial goes to infinity for $\omega > 1$ faster than any other polynomial of the same order. Thus, the Chebyshev approximation is the best possible among the class of polynomial, or all-pole, filters.

The basic definition of the Chebyshev polynomial works fine for values of ω in the passband, where $\omega \le 1$. For larger values of ω, however, $\cos^{-1}(\omega)$ is a complex number. Fortunately, there is an alternate form that avoids complex arithmetic. To derive this form, simply recognize the complex nature of $\cos^{-1}(\omega)$ explicitly and write

$$x = \cos^{-1}(\omega) \tag{2.52}$$

One then has*

$$\omega = \cos(x) = \cos[\,j(jx)] = \cosh(jx) \tag{2.53}$$

so

$$jx = \cosh^{-1}(\omega) \tag{2.54}$$

Thus, one can also write

$$T_n(\omega) = \cos[\,jn(jx)] = \cosh[n(jx)] = \cosh[n \cosh^{-1}(\omega)] \tag{2.55}$$

---

* Since $\cos(x) = \cosh(jx)$.

This result is used to compute the required filter order. Assuming as usual that A_s is the maximum allowed stopband gain, one uses the square root of Equation 2.49 to get

$$A(\omega_s) = \frac{1}{\sqrt{1 + \varepsilon^2 T_n^2(\omega_s)}} \leq A_s \tag{2.56}$$

Solving, one has

$$T_n(\omega_s) \geq \frac{1}{\varepsilon} \sqrt{\frac{1}{A_s^2} - 1} \tag{2.57}$$

Since $\omega_s \geq 1$, it is most convenient to use the hyperbolic form in Equation 2.55:

$$n \geq \frac{\cosh^{-1}\left[\frac{1}{\varepsilon}\sqrt{\frac{1}{A_s^2} - 1}\right]}{\cosh^{-1}(\omega_s)} \tag{2.58}$$

To summarize, one first determines the parameter ε from the allowed passband ripple, usually using Equation 2.51; then one determines the minimum order required using Equation 2.58. The original filter specifications must, of course, be mapped into the lpp domain through appropriate choice(s) of the constant(s) in the transformation Equations 2.12 through 2.15. Notice that the definition of passband for the Chebyshev filter differs from that of the Butterworth unless the passband ripple is 3 dB. For the Chebyshev characteristic, the passband cutoff frequency is that frequency at which the gain goes through the value $1/\sqrt{1 + \varepsilon^2}$ for the last time as ω increases.

The only remaining item is the determination of the transfer function $G(s)$ by factorization. Again, this requires the computation of the poles of $G(s)G(-s)$. Using Equation 2.49, one has

$$G(s)G(-s) = \left[A^2(\omega)\right]_{s=j\omega} = \frac{1}{1 + \varepsilon^2 \left[T_n^2(\omega)\right]_{s=j\omega}} \tag{2.59}$$

Thus, the poles are at those values of s for which

$$\left[T_n^2(\omega)\right]_{s=j\omega} = \frac{-1}{\varepsilon^2} \tag{2.60}$$

or*

$$\cos\left[n \, \cos^{-1}\left(\frac{s}{j}\right)\right] = \pm j\frac{1}{\varepsilon} \tag{2.61}$$

Letting

$$\cos^{-1}(s/j) = a + jb, \text{ there results} \tag{2.62}$$

$$\cos(na)\cosh(nb) - j\sin(na)\sinh(nb) = \pm j\frac{1}{\varepsilon} \tag{2.63}$$

---

* Since s is complex anyway, nothing is to be gained from using the hyperbolic form.

Equating real and imaginary parts, one has

$$\cos(na)\cos(nb) = 0 \tag{2.64a}$$

and

$$\sin(na)\sinh(nb) = \pm\frac{1}{\varepsilon} \tag{2.64b}$$

Since $\cosh(nb) > 0$ for any b, one must have

$$\cos(na) = 0 \tag{2.65}$$

which can hold only if

$$a = (2k+1)\frac{\pi}{2n} \quad k \text{ any integer} \tag{2.66}$$

But, in this case, $\sin[(2k+1)\pi/2] = \pm 1$, so application of Equation 2.64b gives

$$\sinh(nb) = \pm\frac{1}{\varepsilon} \tag{2.67}$$

One can now solve for b:

$$b = \pm\frac{1}{n}\sinh^{-1}\left[\frac{1}{\varepsilon}\right] \tag{2.68}$$

Equations 2.66 through 2.68 together determine a and b, hence $\cos^{-1}(s/j)$. Taking the cosine of both sides of Equation 2.62 and using Equations 2.65 and 2.68 gives

$$s = \pm\sin\left[(2k+1)\frac{\pi}{2n}\right]\sinh\left[\frac{1}{n}\sinh^{-1}\left(\frac{1}{\varepsilon}\right)\right]$$
$$+ j\cos\left[(2k+1)\frac{\pi}{2n}\right]\cosh\left[\frac{1}{n}\sinh^{-1}\left(\frac{1}{\varepsilon}\right)\right] \tag{2.69}$$

Letting $s = \sigma + j\omega$ as usual, one can rearrange Equation 2.69 into the form

$$\left[\frac{\sigma}{\sinh\left[\frac{1}{n}\sinh^{-1}\left(\frac{1}{\varepsilon}\right)\right]}\right]^2 + \left[\frac{\omega}{\cosh\left[\frac{1}{n}\sinh^{-1}\left(\frac{1}{\varepsilon}\right)\right]}\right]^2 = 1 \tag{2.70}$$

which is the equation of an ellipse in the s plane with real axis intercepts of

$$\sigma_o = \pm\sinh\left[\frac{1}{n}\sinh^{-1}\left(\frac{1}{\varepsilon}\right)\right] \tag{2.71}$$

and imaginary axis intercepts of

$$\omega_o = \pm\cosh\left[\frac{1}{n}\sinh^{-1}\left(\frac{1}{\varepsilon}\right)\right] \tag{2.72}$$

This is shown in Figure 2.15.

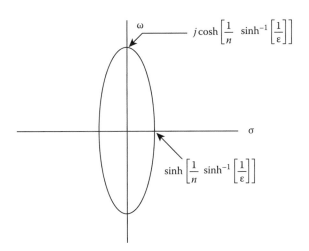

FIGURE 2.15 Pole locations for $G(s)G(-s)$ for the Chebyshev filter.

As an example, suppose that a Chebyshev lpp is to be designed that has a passband ripple of 0.1 dB, a maximum stopband gain of -20 dB, and $\omega_s = 2$ rad/s. Then, one can use Equation 2.51 to find* the ripple parameter ε:

$$\varepsilon = \sqrt{10^{0.1/10} - 1} = 0.1526 \tag{2.73}$$

Equation 2.58 gives the minimum order required:

$$n \geq \frac{\cosh^{-1}\left[\frac{1}{0.1526}\sqrt{\frac{1}{(0.1)^2} - 1}\right]}{\cosh^{-1}(2)} = \frac{\cosh^{-1}(65.20)}{\cosh^{-1}(2)} = 3.70 \tag{2.74}$$

In doing this computation by hand, one often uses the identity

$$\cosh^{-1}(x) = \ln\left[x + \sqrt{x^2 - 1}\right] \tag{2.75}$$

which can be closely approximated by

$$\cosh^{-1}(x) = \ln(2x) \tag{2.76}$$

if $x \gg 1$. In the present case, a fourth-order filter is required. The poles are shown in Table 2.2 and graphed in Figure 2.16.

By selecting the left-half plane poles and forming the corresponding factors, then multiplying them, one finds the denominator polynomial of $G(s)$ to be

$$D(s) = s^4 + 1.8040s^3 + 2.2670s^2 + 2.0257s + 0.8286 \tag{2.77}$$

---

* For practical designs, a great deal of precision is required for higher-order filters.

TABLE 2.2   Pole Locations

k	Real Part	Imaginary Part
0	0.2642	1.1226
1	0.6378	0.4650
2	0.6378	−0.4650
3	0.2642	−1.1266
4	−0.2642	−1.1266
5	−0.6378	−0.4650
6	−0.6378	0.4650
7	−0.2642	1.1226

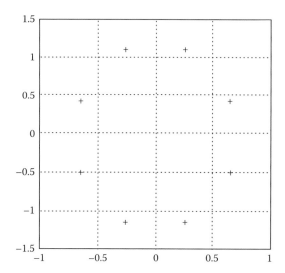

FIGURE 2.16   Pole locations for example filter.

As for the Butterworth example of Section 2.2, one finds the scale factor:

$$k = 0.8286 \tag{2.78}$$

Therefore, the complete transfer function is

$$G(s) = \frac{k}{D(s)} = \frac{0.8286}{s^4 + 1.8040s^3 + 2.2670s^2 + 2.0257s + 0.8286} \tag{2.79}$$

Of course, if completely automated algorithms are not being used to design such a filter as the one in our example, numerical assistance is required. The computation in the preceding example was performed in MATLAB®—a convenient package for many computational problems in filter design.

2.4 Bessel–Thompson LPP Approximation

Thus far, two all-pole approximations have been presented. As was pointed out in Section 2.3, the Chebyshev is better than the Butterworth—in fact, it is the best all-pole approximation available. So why bother with the Butterworth at all, other than as an item of historical interest? The answer lies in the phase. The Chebyshev filter has a phase characteristic that departs farther from linearity than that of the

Butterworth; put differently, its group delay deviates from a constant by a greater amount. Thus, the Butterworth approximation is still a viable approximation in applications where phase linearity is of some importance.

The question naturally arises as to whether there is an lpp approximation that has better phase characteristics than that of the Butterworth. The answer is yes, and that is the topic of this section, which will follow [8]—perhaps the simplest development of all.

Recall that the gain function $G(s)$ is the inversion of the loss function $H(s)$:

$$G(s) = \frac{1}{H(s)} \tag{2.80}$$

Also, recall that it was the loss function $|h(j\omega)| = 1/A(\omega)$ (or, rather, its square $|h(j\omega)|^2$) that was required to have MFM property at $\omega = 0$,

$$\left[\frac{d^k |H(j\omega)|^2}{d\omega^k}\right]_{\omega=0} = \left[\frac{d^k \omega^{2n}}{d\omega^k}\right]_{\omega=0} = \left[\frac{(2n)!}{k!}\omega^{2n-k}\right]_{\omega=0} = 0 \tag{2.81}$$

for $k = 0, 1, \ldots, 2n - 1$. The question to be asked and answered in this section is whether there exists a similar approximation for the group delay $\tau_g(\omega)$ a maximally flat delay (MFD) approximation.

To answer this question, the phase will be written in terms of the loss function $H(j\omega)$. Since the latter quantity can be written in polar form as

$$H(j\omega) = \frac{1}{G(j\omega)} = \frac{1}{A(\omega)e^{j\phi(\omega)}} = \frac{1}{A(\omega)}e^{-j\phi(\omega)} \tag{2.82}$$

the (complex) logarithm is

$$\ln H(j\omega) = -\ln A(\omega) - j\phi(\omega) \tag{2.83}$$

Thus,

$$\phi(\omega) = -\mathrm{Im}\{\ln H(j\omega)\} \tag{2.84}$$

The group delay is merely the negative of the derivative of the phase, so one has

$$\tau_g(\omega) = -\frac{d\phi}{d\omega} = \frac{d}{d\omega}\mathrm{Im}\{\ln H(j\omega)\} = \mathrm{Im}\left\{\frac{d}{d\omega}\ln H(j\omega)\right\}$$
$$= \mathrm{Im}\left\{j\frac{d}{d(j\omega)}\ln H(j\omega)\right\} = \mathrm{Re}\left\{\frac{d}{d(j\omega)}\ln H(j\omega)\right\} \tag{2.85}$$

Recalling that the even part of a complex function $F(s)$ is given by

$$\mathrm{Ev}\{F(s)\} = \frac{F(s) + F(-s)}{2} \tag{2.86}$$

one can use the symmetry property for real $F(s)$ (in the present context, $F(s)$ is assumed to be a rational function, so a real $F(s)$ is one with real coefficients) to show that

$$\text{Ev}\{F(s)\}_{s=j\omega} = \frac{F(j\omega) + F(-j\omega)}{2} = \frac{F(j\omega) + F^*(j\omega)}{2} = \text{Re}\{F(j\omega)\} \tag{2.87}$$

In this manner, one can analytically extend the group delay function $\tau_g(j\omega)$ so that it becomes a function of s:

$$\tau_g(s) = \text{Ev}\left\{\frac{d}{ds}\ln H(s)\right\} = \text{Ev}\left[\frac{H'(s)}{H(s)}\right] = \frac{1}{2}\left[\frac{H'(s)}{H(s)} + \frac{H'(-s)}{H(-s)}\right] \tag{2.88}$$

In the present case, it will be assumed that $H(s)$ is a polynomial. In this manner, an all-pole (or polynomial) filter will result. Thus, one can write

$$H(s) = \sum_{k=0}^{n} a_k^n s^k \tag{2.89}$$

The superscript on the (assumed real) coefficient matches the upper limit on the sum and is the assumed filter order. The group delay function is, thus, a real rational function:

$$\tau_g(s) = \frac{N(s)}{D(s)} \tag{2.90}$$

where $N(s)$, the numerator polynomial, and $D(s)$, the denominator polynomial, have real coefficients and are of degrees no greater than $2n-1$ and $2n$, respectively, by inspection of Equation 2.88. But, again according to Equation 2.88, $\tau_g(s)$ is an even function. Thus,

$$\tau_g(-s) = \frac{N(-s)}{D(-s)} = \tau_g(s) = \frac{N(s)}{D(s)} \tag{2.91}$$

The last equation, however, implies that

$$\frac{N(-s)}{N(s)} = \frac{D(-s)}{D(s)} = 1 \tag{2.92}$$

The last equality is arrived at by the following reasoning. $N(s)$ and $D(s)$ are assumed to have no common factors—any such have already been cancelled in the formation of $\tau_g(s)$. Thus, the two functions of s in Equation 2.92 are independent; since they must be equal for s, they must therefore equal a constant. But this constant is unity, as is easily shown by allowing s to become infinite and noting that the degrees and leading coefficients of $N(-s)$ and $N(s)$ are the same.

The implication of the preceding development is simply that $N(s)$ and $D(s)$ consist of only even powers of s. Looking at $N(s)$, for example, and letting it be written

$$N(s) = \sum_{k=0}^{2n-1} \rho_k s^k \tag{2.93}$$

one has

$$N(-s) = \sum_{k=0}^{2n-1} \rho_k(-s)^k = N(s) = \sum_{k=0}^{m} \rho_k(s)^k \tag{2.94}$$

This however implies that the ρ_k are zero for odd k. Hence $N(s)$ consists of only even powers of s. The same is true of $D(s)$ and therefore of $\tau_g(s)$. Clearly, therefore $\tau_g(\omega)$ will consist of only even powers of ω, that is, it will be a function of ω^2. Now there is to be an MFD approximation one must have by analogy with the MFM approximation in Section 2.2,

$$\frac{d^k \tau_g(\omega)}{d(\omega^2)^k} = 0 \qquad (2.95)$$

for $k = 1, 2, \ldots, n - 1$. The constraint is, of course, that τ_g must come from a polynomial loss function $H(s)$ whose zeros [poles of $G(s)$] all lie in the left-half plane.

It is convenient to normalize time so that the group delay T, say, at $\omega = 0$ is 1 s; this is equivalent to scaling the frequency variable ω to be ωT. Here, it will be assumed that this has been performed already. A slight difference exists between the MFD and MFM approximations; the latter approximates zero at $\omega = 0$, while the former approximates $T = 1s$ at $\omega = 0$. The two become the same, however, if one considers the function

$$\tau_g(s) - 1 = \frac{P(s)}{2H(s)H(-s)} \qquad (2.96)$$

where $P(s)$ has a maximum order of $2n$. The form on the right-hand side of this expression can be readily verified by consideration of Equation 2.88, the basic result for the following derivation. Furthermore, writing $P(s)$ in the form

$$P(s) = \sum_{k=0}^{n} P_k \left(s^2\right)^k \qquad (2.97)$$

it is readily seen that

$$p_0 = p_1 = \cdots = p_{n-1} = 0 \qquad (2.98)$$

The lowest-order coefficient is clearly zero because $\tau_g(0) = 1$; furthermore, all odd coefficients are zero since τ_g is even. Finally, all other coefficients in Equation 2.98 have to be zero if one imposes the MFD condition in Equation 2.95.*

At this stage, one can write the group delay function in the form

$$\tau_g(s) = \frac{N(s)}{D(s)} = 1 + \frac{p_n s^{2n}}{2H(s)H(-s)} = \frac{2H(s)H(-s) + p_n s^{2n}}{2H(s)H(-s)} \qquad (2.99)$$

It was pointed out immediately after Equation 2.90, however, that the degree of $N(s)$ is at most $2n - 1$. Hence, the coefficient of s^{2n} in the numerator of Equation 2.99 have vanished, and one therefore also has

$$2(-1)^n a_n + p_n = 0 \qquad (2.100)$$

or, equivalently,

$$p_n = 2(-1)^{n+1} a_n \qquad (2.101)$$

---

* The derivatives are easy to compute, but this is omitted here for reasons of space.

Finally, this allows one to write Equation 2.99 in the form

$$\tau_g(s) = \frac{H(s)H(-s) + (-1)^{n+1}a_n s^{2n}}{H(s)H(-s)} \tag{2.102}$$

If one now equates Equations 2.102 and 2.88 and simplifies, there results

$$H'(s)H(-s) + H'(-s)H(s) - 2H(s)H(-s) = 2(-1)^{n+1}a_n s^{2n} \tag{2.103}$$

Multiplying both sides by s^{-2n}, taking the derivative (noting that $(d/ds)H(-s) = -H'(-s)$), one obtains (after a bit of algebra)

$$\text{Ev}\{[sH''(s) - 2(s + n)H'(s) + 2nH(s)]H(-s)\} = 0 \tag{2.104}$$

Now, if s_o is a zero of this even function, the $-s_o$ will also be a zero. Further, since all of the coefficients in $H(s)$ are real, s_0^* and $-s_0^*$ must also be zeros as well; that is, the zeros must occur in a *quadrantally symmetric* manner. Each zero must belong either to $H(-s)$, or to the factor it multiplies, or to both. The degree of the entire expression in Equation 2.40 is $2n$ and $H(-s)$ has n zeros. Thus, the expression in square brackets must have n zeros. Now here is the crucial step in the logic: if the filter being designed is to be stable, then all n zeros $H(-s)$ must be in the (open) right-half plane. This implies that the factor in square brackets must have n zeros in the (open) left-half plane. Since the expression has degree n, these zeros can be found from the equation

$$sH''(s) - 2(s + n)H'(s) + 2nH(s) = 0 \tag{2.105}$$

This differential equation can be transformed into that of Bessel; here, however, the solution will be derived directly by recursion. Using Equation 2.89 for $H(s)$, computing its derivatives, reindexing, and using Equation 2.105, one obtains, for $0 \leq k \leq n - 1$,

$$(k + 1)ka_{k+1}^n - 2n(k + 1)a_{k+1}^n - 2ka_k^n + 2na_k^n = 0 \tag{2.106}$$

This produces the recursion formula

$$a_{k+1}^n = \frac{2(n - k)}{(2n - k)(k + 1)}a_k^n; \quad 0 \leq k \leq n - 1 \tag{2.107}$$

or, normalizing the one free constant so that $a_n^n = 1$ and reindexing, gives

$$a_{k+1}^n = \frac{2(n - k)}{(2n - k)(k + 1)}; \quad 0 \leq k \leq n - 1 \tag{2.108}$$

The resulting polynomials for $H(s)$ are closely allied with the Bessel polynomials. The first several are given in Table 2.3 and the corresponding gain and group delay characteristics are plotted (using MATLAB) in Figure 2.17. Notice that the higher the order, the more accurately the group delay approximates a constant and the better the gain approximates the ideal lpp; the latter behavior, however, is fairly poor. A view of the group delay behavior in the passband is shown for the third-order filter in Figure 2.18.

TABLE 2.3 Bessel Polynomials of Order n

n	$H(s)$
1	$s+1$
2	$s^2 + 3s + 3$
3	$s^3 + 6s^2 + 15s + 15$
4	$s^4 + 10s^3 + 45s^2 + 105s + 105$

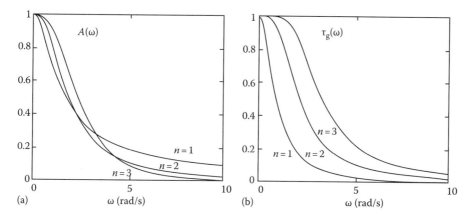

FIGURE 2.17 Gain and group delay characteristics for the Bessel–Thompson filters of orders one through three: (a) gain and (b) group delay.

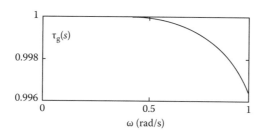

FIGURE 2.18 Group delay of third-order Bessel filter.

2.5 Elliptic Approximation

The Butterworth approximation provides a very fine approximation to the ideal "brick wall" lpp response at $\omega = 0$, but is poor for other frequencies; the Chebyshev, on the other hand, spreads out the error of approximation throughout the passband and thereby achieves a much better amplitude approximation. This is one concrete application of a general result in approximation theory known as the Weierstrass theorem [2], which asserts the possibility of uniformly approximating a continuous function on a compact set by a polynomial or by a rational function.

A compact set is the generalization of a closed and bounded interval—the setting for the Chebyshev approximation sketched in Figure 2.19. The compact set is the closed interval $[-1, 1]$,* the continuous

---

* Here, we are explicitly observing that the characteristic function to be approximated is even in ω.

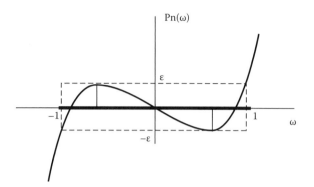

FIGURE 2.19 Chebyshev approximation.

function to be approximated is the constant 0, and the approximating function is a polynomial of degree n (in the figure n is three). A careful inspection of this figure reveals that the maximum error of approximation occurs exactly four times on a set of four discrete points within the interval and the *sign* of this maximum error alternates from one such point to its adjacent neighbor. This is no coincidence; in fact, the polynomial of best approximation of degree n is characterized by this alternation or *equiripple* property: the error function achieves its maximum value exactly $n + 1$ times on the interval of approximation and the signs at these points alternate. If one finds a polynomial by any means that has this property it is the unique polynomial of best approximation.

In the present case, one simply notes that the cosine function has the required equiripple behavior. That is, as ϕ varies between $-\pi/2$ and $\pi/2$, $\cos(n\phi)$ varies between its maximum and minimum values of ± 1 a total of $n + 1$ times. But $\cos(n\phi)$ is not a polynomial, and the problem (for reasons stated in the introduction) is to determine a *polynomial* having this property. At this point, one observes that if one makes the transformation

$$\phi = \cos^{-1}(\omega) \tag{2.109}$$

then, as ω varies from -1 to $+1$, ϕ varies from $-\pi/2$ to $+\pi/2$. This transformation is, fortunately, one-to-one over this range; furthermore, even more fortunately, the overall function

$$T_n(\omega) = \cos[n \cos^{-1}(\omega)] \tag{2.110}$$

is a polynomial as desired. Of course, the maximum error is ± 1, an impractically large value. This is easily rectified by requiring that the approximating polynomial be

$$p_n(\omega) = \varepsilon T_n(\omega) = \varepsilon \cos[n \cos^{-1}(\omega)] \tag{2.111}$$

In terms of gain and characteristic functions, therefore, one has

$$G(\omega) = \frac{1}{1 + K^2(\omega)} = \frac{1}{1 + \varepsilon^2 T_n^2(\omega)} \tag{2.112}$$

That is, $p_n(\omega) = \varepsilon T_n(\omega)$ is the best approximation to the characteristic function $K(\omega)$.

Since the gain function to be approximated is even in ω, $K(\omega)$—and, therefore, $p_n(\omega) = \varepsilon T_n(\omega)$—must be either even or odd. As one can recall, the Chebyshev polynomials have this property. For this reason, it is only necessary to discuss the situation for $\omega \geq 0$ for the extension to negative ω is then obvious.

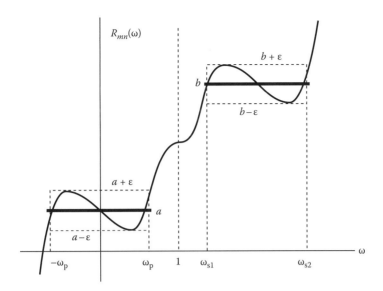

FIGURE 2.20 Chebyshev rational approximation.

The foregoing sets the stage for the more general problem, which will now be addressed. As noted previously, the Weierstrass theorem holds for more general compact sets. Thus, as a preliminary exercise, the approximation problem sketched in Figure 2.20 will be discussed. That figure shows two closed intervals as the compact set on which the approximation is to occur. The function to be approximated is assumed to be the constant a on the first interval and the constant b on the second.* In order to cast this into the filtering context, it will be assumed that the constant a is small and the constant b large. The sharpness of the transition characteristic can then be set to any desired degree by allowing ω_p to approach unity from below and ω_s^1 to approach unity from above. Notice that the frequency scaling will be different here than for the all-pole case in which the right-hand interval endpoint was allowed to be unity.

In addition to more general compact sets, the Weierstrass theorem allows the approximating function to be a general real rational function, which has been denoted by $R_{mn}(\omega)$ in the figure. This notation means that the numerator and denominator polynomials have maximum degree m and n, respectively. Now suppose that one specifies that $b = 1/a$ and suppose $R_{mn}(\omega)$ is a reciprocal function, that is, one having the property that

$$R_{mn}\left(\frac{1}{\omega}\right) = \frac{1}{R_{mn}(\omega)} \qquad (2.113)$$

Then, if one determines its coefficients such that the equiripple property holds on the interval $-\omega_p \le \omega \le \omega_p$, it will also hold on the interval $\omega_{s1} \le \omega \le \omega_{s2}$. Of course, this interval is constrained such that $\omega_{s1} = 1/\omega_p$. The other interval endpoint will then extend to $+\infty$ (and, by symmetry, there will also be an interval on the negative frequency axis with one endpoint at $-1/\omega_p$ and the other at $-\infty$). This is, of course, not a compact set but the approximation theorem continues to hold anyway because it is equivalent to approximating $a = 1/b$ on the low-pass interval $[-\omega_p, \omega_p)$. Thus, one can let $a = 0$ and $b = \infty$ and simultaneously approximate the ideal characteristic function

---

* One should note that this function is continuous on the compact set composed of the two closed intervals.

$$K(\omega) = \begin{cases} 0; & |\omega| \leq \omega_p \\ \infty; & |\omega| \geq \frac{1}{\omega_p} \end{cases} \tag{2.114}$$

with the real and even or odd (for realizability) rational function $R_{mn}(\omega)$ having the aforementioned reciprocal property. The Weierstrass theorem equiripple property for such rational functions demands that the total number of error extrema* on the compact set be $m+n+2$. (This assumes the degree of both polynomials to be relative to the variable ω^2.)

Based on the preceding discussion, one has the following form for $R_{mn}(\omega)$:

$$R_{2n,2n}(\omega) \equiv R_{2n}(\omega) = \frac{\left(\omega_1^2 - \omega^2\right)\left(\omega_3^2 - \omega^2\right)\cdots\left(\omega_{2n-1}^2 - \omega^2\right)}{(1 - \omega_1^2\omega^2)(1 - \omega_3^2\omega^2)\cdots(1 - \omega_{2n-1}^2\omega^2)} \tag{2.115}$$

$$R_{2n+1,2n}(\omega) \equiv R_{2n+1}(\omega) = \omega\frac{\left(\omega_2^2 - \omega^2\right)\left(\omega_4^2 - \omega^2\right)\cdots\left(\omega_{2n}^2 - \omega^2\right)}{(1 - \omega_2^2\omega^2)(1 - \omega_4^2\omega^2)\cdots(1 - \omega_{2n}^2\omega^2)} \tag{2.116}$$

The first is clearly an even rational function and latter odd. The problem now is to find the location of the pole and zero factors such that equiripple behavior is achieved in the passband. The even case is illustrated in Figure 2.21 for $n=2$. Notice that the upper limit of the passband frequency interval has been taken for convenience to be equal to $\sqrt{k}$; hence, the lower limit of the stopband frequency interval is $1/\sqrt{k}$. Thus,

$$k = \frac{\omega_p}{\omega_s} \tag{2.117}$$

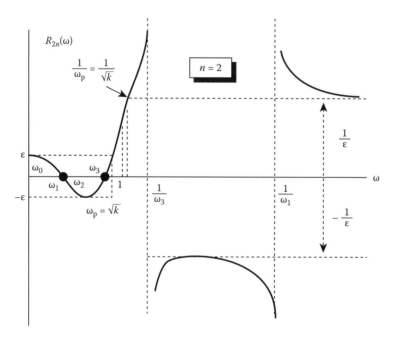

FIGURE 2.21 Typical plot of an even-order Chebyshev rational function.

---

* There can be degeneracy in the general approximation problem, but the constraints of the problem being discussed here preclude this from occurring.

is a measure of the sharpness of the transition band rolloff and is always less than unity. The closer it is to one, the sharper the rolloff. It is an arbitrarily specified parameter in the design. Notice that equiripple behavior in the passband implies equiripple behavior in the stopband—in the latter, the approximation is to the constant whose value is infinity and the minimum value (which corresponds to a maximum value of gain) is $1/\varepsilon$, while the maximum deviation from zero in the passband is ε. The zeros are all in the passband and are mirrored in poles, or infinite values, in the stopband. The notation ω_0 and ω_2 for the passband frequencies of maximum error has been introduced. In general, their indices will be even for even approximations and odd for odd approximations.

A sketch for the odd case would differ only in that the plot would go through the origin, that is, the origin would be a zero rather than a point of maximum error. Notice that the resulting filter will have finite gain unlike the Butterworth or Chebyshev, which continue to rolloff toward zero as ω becomes infinitely large. As noted in the figure, the approximating functions are called the *Chebyshev rational functions*.

The procedure [9] now is quite analogous to the Chebyshev polynomial approximation, though rather more complicated. One looks for a continuous periodic waveform that possess the desired equiripple behavior. Since the Jacobian elliptic functions are generalizations of the more ordinary sinusoids, it is only natural that they be investigated with an eye toward solving the problem under attack. With that in mind, some of the more salient properties will now be reviewed.

Consider the function

$$I(\phi, k) = -\frac{1}{\sqrt{1 - k^2 \sin^2(\phi)}} \tag{2.118}$$

which is plotted for two values of k in Figure 2.22. If $k = \pm 1$, the peak value is infinite; in this case, $I(\phi, k) = \sec(\phi)$. For smaller values of k the peak value depends upon k with smaller values of k resulting lower peak values. The peaks occur at odd multiples of $\pi/2$. Note that $I(\phi, k)$ has the constant value one for $k = 0$. Figure 2.23 shows the running integral of $I(\phi, k)$ for $k^2 = 0.99$ and for $k^2 = 0$. For $k^2 = 0$ the curve is a straight line (shown dashed); or other values of k^2, it deviates from a straight line by an amount that depends upon the size of k^2. Observe that the running integral has been plotted over one full period of $I(k, \phi)$, that is, from 0 to π. The running integral is given in analytical form by

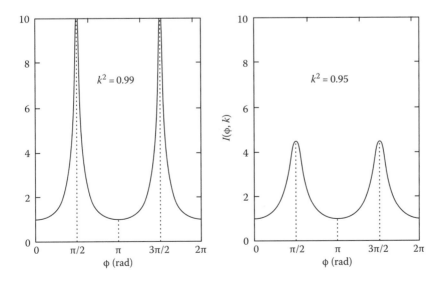

FIGURE 2.22 Plot of $I(\phi, k)$.

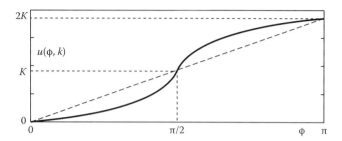

FIGURE 2.23   Running integral of I normalized to K.

$$u(\phi, k) = \int\limits_0^\phi I(\alpha, k)d\alpha = \int\limits_0^\phi \frac{1}{\sqrt{1 - k^2 \sin^2(\alpha)}}d\alpha \qquad (2.119)$$

The quantity K, shown by the lowest dashed horizontal line in Figure 2.23, is the integral of $I(\phi, k)$ from 0 to $\pi/2$; that is, it is the area beneath the $I(\phi, k)$ curve in Figure 2.22 from 0 to $\pi/2$. Thus, it is the area beneath the curve to the left of the first vertical dashed line in that figure. It is given by

$$u\left(\frac{\pi}{2}, k\right) = \int\limits_0^{\pi/2} I(\alpha, k)d\alpha = \int\limits_0^{\pi/2} \frac{1}{\sqrt{1 - k^2 \sin^2(\alpha)}}d\alpha = K \qquad (2.120)$$

and is referred to as the *complete elliptic integral of the first kind*.

The sine generalization sought can now be defined. Since the running integral $u(\phi, k)$ is monotonic, it can be inverted and thereby solved for ϕ in terms of u; for each value of u there corresponds a unique value of ϕ. The elliptic sine function is defined to be the ordinary sine function of ϕ:

$$sn(u, k) = \sin(\phi, k) \qquad (2.121)$$

Now, inspection of Figure 2.23 shows that, as ϕ progresses from 0 to 2π, $u(\phi, k)$ increases from 0 to $4K$; since angles are unique only to within multiples of 2π, therefore, it is clear that $sn(u, k)$ is periodic with period $4K$. Hence, K is a quarter-period of the elliptic sine function.

The integral in Equation 2.119 cannot be evaluated in closed form; thus, there is not a simple, compact expression for the required inverse. Therefore, the elliptic sine function can only be tabulated in numerical form or computer using numerical techniques. It is shown for $k^2 = 0.99$ in Figure 2.24 and compared with a conventional sine function having the same period ($4K$).

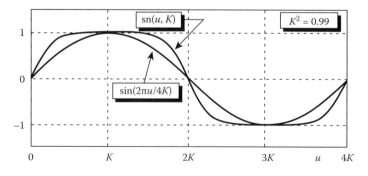

FIGURE 2.24   Plots of the elliptic and ordinary sine functions.

Recall, now, the objective of exploring the preceding generalization of the sinusoid: one is looking for a transformation that will convert the characteristic function given by Equations 2.115 and 2.116 into equivalent waveforms having the equiripple property. As one might suspect (since so much time has been spent on developing it), the elliptic sine function is precisely the transformation desired. The crucial aspect of showing this is the application of a fundamental property of $sn(u, k)$, known as an additional formula:

$$sn(u + a, k)sn(u - a, k) = \frac{sn^2(u, k) - sn^2(a, k)}{1 - k^2 sn^2(u, k)sn^2(a, k)} \qquad (2.122)$$

The right-hand side of this identity has the same form as one factor in Equations 2.115 and 2.116, that is, of one zero factor coupled with its corresponding pole factor. This suggests the transformation

$$\omega = \sqrt{k}\, sn(u, k) \qquad (2.123)$$

The passband zeros then are given by

$$\omega_i = \sqrt{k}\, sn(u_i, k) \qquad (2.124)$$

and the factors mentioned previously map into

$$\frac{\omega_i^2 - \omega^2}{1 - \omega_i^2 \omega^2} = \frac{k[sn^2(u_i, k) - sn^2(u, k)]}{1 - k^2 sn^2(u_i, k)sn^2(u, k)} = ksn(u + u_i)sn(u - u_i) \qquad (2.125)$$

For specificity, the even-order case will be discussed henceforth. The odd-order case is the same if minor notational modifications are made. Thus, one sees that

$$R_{2n}(\omega) = \prod_{i=1,3,\ldots,2n-1} \frac{\omega_i^2 - \omega^2}{1 - \omega_i^2 \omega^2} = \prod_{i=1,3,\ldots,2n-1} ksn(u + u_i)sn(u - u_i) \qquad (2.126)$$

The u_i are to be chosen. Before doing this, it helps to simplify the preceding expression by defining $u_{-i} = -u_i$ and reindexing. Calling the resulting function $G(u)$, one has

$$G(u) = \prod_{i=-1,-3,\ldots,-(2n-1)}^{i=1,3,\ldots,2n-1} ksn(u + u_i) \qquad (2.127)$$

Refer now to Figure 2.24. Each of the sn functions is periodic with period $4K$ and is completely defined by its values over one quarter-period $[0, K]$. Suppose that one defines

$$u_i = i\frac{K}{2n} \qquad (2.128)$$

Figure 2.25 shows the resulting transformation corresponding to Equations 2.123 and 2.124. As u progresses from $-K$ to $+K$, ω increases from $-\sqrt{k}$ to $+\sqrt{k}$ as desired. Furthermore, because of the symmetry of $sn(u)$, the set $\{u_i\}$ forms an additive group—adding $K/2n$ to the index of any u_i results in another u_i in the set. This means that $G(u)$ in Equation 2.127 is periodic with period $K/2n$. Thus, as ω increases from $-\sqrt{k}$ to $+\sqrt{k}$, $R_{2n}(\omega)$ achieves $2n + 2$ extrema, that is, positive and negative peak values. But this is sufficient for $R_{2n}(\omega)$ to be the Chebyshev rational function of best approximation.

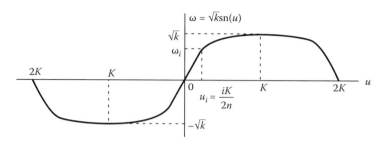

FIGURE 2.25 Elliptic sine transformation.

The symmetry of the zero distribution around the peak value of $sn(u)$, that is, around $u = K$, reveals that the peak values of $R_{2n}(\omega)$ occur between the zeros; that is, recalling that in Figure 2.21 the peak values have been defined by the symbols ω_i for even i, one has

$$\omega_i = \sqrt{k} sn\left(\frac{iK}{2n}\right); \quad i = 0, 1, 2, \ldots, 2n \tag{2.129}$$

where the odd indices correspond to the zeros and the even ones to the peak values. Note that $i = 0$ corresponds to $\omega = 0$ and $i = 2n$ to $\omega = \sqrt{k}$. This permits one to compute the minimax error:

$$\varepsilon = |R_{2n}(0)| = (\omega_1\omega_3 \cdots \omega_{2n-1})^2 \tag{2.130}$$

As pointed out previously, the analysis for odd-order filters proceeds quite analogously. The only difference lies in the computation of ε. In this case, $R_{2n+1}(0) = 0$; thus, $\omega = 0$ is a zero—not a point of maximum deviation. One can, however, note that the quantity $\varepsilon^2 - R_{2n+1}^2(\omega)$ has double zeros at the frequencies of maximum deviation (except at $\omega = \omega_{2n+1} = \sqrt{k}$) and the same denominator as $R_{2n+1}^2(\omega)$. Hence,

$$\varepsilon^2 - R_{2n+1}^2(\omega) = \frac{(\omega_1^2 - \omega^2)^2 (\omega_3^2 - \omega^2)^2 \cdots (\omega_{2n+1}^2 - \omega^2)}{(1 - \omega_2^2\omega^2)^2(1 - \omega_4^2\omega^2)^2 \cdots (1 - \omega_{2n}^2\omega^2)^2} \tag{2.131}$$

Notice here that $\omega_{2n+1} = \sqrt{k}$. Since $R_{2n+1}(0) = 0$, one can evaluate this expression at $\omega = 0$ to get

$$\varepsilon = (\omega_1\omega_3 \cdots \omega_{2n-1})^2 \sqrt{k} \tag{2.132}$$

Note that one uses the zero frequencies in the even-order case and the frequencies of maximum deviation in the odd-order case.

As an example, suppose that the object is to design an elliptic filter of order $n = 2$. Further, suppose that the passband ripple cutoff frequency is to be 0.95. Then, one has

$$\omega_p = \sqrt{k} = 0.95 \tag{2.133}$$

The quarter period of the elliptic sine function is

$$K = 2.9083 \tag{2.134}$$

TABLE 2.4   Zeros of Elliptic Rational
Functions

i	ω_i
0	0.0000
1	0.5923
2	0.8588
3	0.9352
4	0.9500

Evaluating the zeros and points of maximum deviation of the Chebyshev rational function numerically using Equation 2.129, one obtains the values shown in Table 2.4. Thus, the required elliptic rational function is

$$R_4(\omega) = \frac{(0.3508 - \omega^2)(0.8746 - \omega^2)}{(1 - 0.3508\omega^2)(1 - 0.8746\omega^2)} \tag{2.135}$$

Finally, the maximum error is given by Equation 2.130:

$$\varepsilon = (\omega_1\omega_2)^2 = (0.5923 \times 0.9352)^2 = 0.3078 \tag{2.136}$$

Figure 2.26 shows the resulting gain plot. Observe the transmission zero at $\omega = 1/0.9352 = 1.069$, corresponding to the pole of the Chebyshev rational function located at the inverse of the largest passband zero. Also, as anticipated, the minimum gain in the passband is

$$A_p = \frac{1}{\sqrt{1 + \varepsilon^2}} = \frac{1}{\sqrt{1 + (0.3078)^2}} = 0.9558 \tag{2.137}$$

and the maximum stopband gain is

$$A_s = \frac{1}{\sqrt{1 + \frac{1}{\varepsilon^2}}} = \frac{1}{\sqrt{1 + \frac{1}{(0.3078)^2}}} = 0.2942 \tag{2.138}$$

Perhaps a summary of the various filter types is in order at this point. The elliptic filter has more flexibility than the Butterworth or the Chebyshev because one can adjust its transition band rolloff independently of the passband ripple. However, as the sharpness of this rolloff increases, as a more detailed analysis shows, the stopband gain increases and the passband ripple increases. Thus, if these

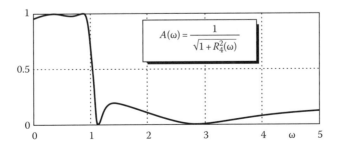

FIGURE 2.26   Gain plot for the example elliptic filter.

parameters are to be independently specified—as is often the desired approach—one must allow the order of the filter to float freely. In this case, the design becomes a bit more involved. The reader is referred to Ref. [10], which presents a simple curve for use in determining the required order. It proceeds from the result that the filter order is the integer greater than or equal to the following ratio:

$$n \geq \frac{f(M)}{f(\Omega)} \tag{2.139}$$

where

$$f(x) = \frac{K\left(\sqrt{1 - x^{-2}}\right)}{K(1/x)} \tag{2.140}$$

Lin [10] presents a curve of this function $f(x)$. K is the complete elliptic integral given in Equation 2.120. The parameters M and Ω are defined by

$$M = \sqrt{\frac{1 - 10^{-0.1A_s}}{1 - 10^{-0.1A_p}}} \tag{2.141}$$

and

$$\Omega = 1/k \tag{2.142}$$

Of course, there is still the problem of factorization. That is, now that the appropriate Chebyshev rational function is known, one must find the corresponding $G(s)$ transfer function of the filter. The overall design process is explained in some detail in Ref. [11], which develops a numerically efficient algorithm for directly computing the parameters of the transfer function.

The Butterworth and Chebyshev filters are of the all-pole variety, and this means that the synthesis of such filters is simpler than is the realization of elliptic filters, which requires the realization of transmission zeros on the finite $j\omega$ axis.

Finally, a word about phase (or group delay). Table 2.5 shows a rank ordering of the filters in terms of both gain and phase performance. As one can see, the phase behavior is inverse to the gain performance. Thus, the elliptic filter offers the very best standard approximation to be ideal lpp "brickwall" gain behavior, but its group delay deviates considerably from a constant. On the other end of the spectrum, one notes that the Bessel–Thompson filters offers excellent phase performance, but a quite high order is required to achieve a reasonable gain characteristic.

If both excellent phase and gain performance are absolutely necessary, two approaches are possible. One either uses computer optimization techniques to simultaneously approximate gain and phase, or one uses one of the filters described in this section followed by an *allpass* filter, one having unity gain over the frequency range of interest, but whose phase can be designed to have the inverse characteristic to the filter providing the desired gain. This process is known as *phase compensation*.

TABLE 2.5   A Rank Ordering of the Filters in Terms of Both Gain and Phase Performance

Filter Type	Gain Rolloff	Phase Linear
Bessel	Worst	Best
Butterworth	Poor	Better
Chebyshev	Better	Poor
Elliptic	Best	Worst

References

1. A. M. Davis, The approximation theoretic foundation of analog and digital filters, *IEEE International Symposium on Circuits and Systems,* San José, California, May, 1986.
2. T. J. Rivlin, *An Introduction to the Approximation of Functions,* New York: Dover, 1981.
3. G. R. Lang and P. O. Brackett, Complex analog filters, in *Proc. Euro. Conf. Circuit Theory, Design,* The Hague, The Netherlands, Aug. 1981, pp. 412–415.
4. W. K. Chen, *Passive and Active Filters, Theory and Implementations,* New York: Wiley, 1986.
5. L. P. Huelsman, *Theory and Design of Active RC Circuits,* New York: McGraw-Hill, 1968.
6. G. Szentirmai, *Computer-Aided Filter Design,* New York: IEEE, 1973.
7. A. M. Davis, Realizability-preserving transformations for digital and analog filters, *J. Franklin Institute*, 311(2), 111–121, Feb. 1981.
8. G. C. Temes and J. W. LaPatra, *Circuit Synthesis and Design*, New York: McGraw-Hill, 1977.
9. E. A. Guillemin, *Synthesis of Passive Networks*, New York: Wiley, 1957.
10. P. M. Lin, Single curve for determining the order of an elliptic filter, *IEEE Trans. Circuits Syst.*, 37(9), 1181–1183, Sept. 1990.
11. A. Antoniou, *Digital Filters: Analysis and Design*, New York: McGraw-Hill, 1979.

3

Frequency Transformations

Jaime Ramirez-Angulo
New Mexico State University

3.1 Low-Pass Prototype

As discussed in Chapter 2, conventional approximation techniques (Butterworth, Chebyshev, Elliptic, Bessel, etc.) lead to a normalized transfer function denoted low-pass prototype (LPP). The LPP is characterized by a passband frequency $\Omega_P = 1.0$ rad/s, a maximum passband ripple A_P (or A_{max}), a minimum stopband attenuation A_s (or A_{min}), and a stopband frequency Ω_s. A_p and A_s are usually specified in decibels. Tolerance bounds (also called box constraints) for the magnitude response of an LPP are illustrated in Figure 3.1a. The ratio Ω_s/Ω_p is called the selectivity factor and it has a value Ω_s for an LPP filter. The passband and stopband edge frequencies are defined as the maximum frequency with the maximum passband attenuation A_p and the minimum frequency with the minimum stopband attenuation A_s, respectively. The passband ripple and the minimum passband attenuation are expressed by

$$A_P = 20 \log \left| \frac{K}{H(\omega_P)} \right|, \quad A_s = 20 \log \left| \frac{K}{H(\omega_s)} \right| \tag{3.1}$$

where K is the maximum value of the magnitude response in the passband (usually unity). Figure 3.1b shows the magnitude response of a Chebyshev LPP transfer function with specifications $A_p = 2$ dB, $A_s = 45$ dB, and $\Omega_s = 1.6$.

Transformation of transfer function. Low-pass, high-pass, bandpass, and band-reject transfer functions (denoted in what follows LP, HP, BP, and BR, respectively) can be derived from an LPP transfer function through a transformation of the complex frequency variable. For convenience, the transfer function of the LPP is expressed in terms of the complex frequency variable s, where $s = u + j\Omega$ while the transfer functions obtained through the frequency transformation (low-pass, high-pass, bandpass, or band-reject) are expressed in terms of the transformed complex frequency variable $p = \sigma + j\omega$.

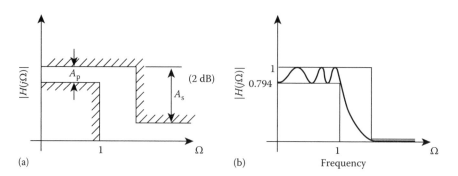

FIGURE 3.1   (a) Tolerance bounds for magnitude response of low-pass prototype and (b) Chebyshev LPP response.

The approximation of an LP, HP, BP, or BR transfer function with passband ripple (s) A_p and stopband attenuation (s) A_s involves three steps:

1. Determination of the stopband edge frequency or selectivity factor Ω_s of an LPP, which can be transformed into the desired LP, HP, BP, or BR filter.
2. Approximation of an LPP transfer function $T_{LPP}(s)$ with selectivity factor Ω_s and with same passband ripple and stopband attenuation A_p and A_s are the desired LP, HP, BP, or BR filter, respectively
3. Transformation of the LPP transfer function $T_{LPP}(s)$ into the desired transfer function (LP, HP, BP, or BR) $T(p)$ through a frequency transformation of the form

$$s = f(p) \tag{3.2}$$

Transformation of a network with LPP magnitude response into a low-pass, high-pass, bandpass, or band-rejection network can be done directly on the elements of the network. This procedure is denoted network transformation. It is very convenient in practice because element values for double terminated lossless ladder networks with LPP specifications have been extensively tabulated for some common values of A_p, A_s, and Ω_s. Also, a host of personal computer programs have become available in recent years that allow one to determine the component values of LPP ladder networks for arbitrary values A_p, A_s, and Ω_s. In what follows we study the frequency transformation $s = f(p)$ for each specific type of filter response (LP, HP, BP, and BR). We show how to calculate the selectivity factor of the equivalent LPP based on box constraint specifications for each type of filter. We then show how mapping of the imaginary frequency axis from s to p leads to LP, HP, BP, or BR magnitude responses. We analyze how poles and zeros are mapped from the s-plane to the p-plane for each transformation and finally we show the element transformations required to directly transform LPP networks into any of the filter types addressed previously.

3.2 Frequency and Impedance Scaling

3.2.1 Frequency Scaling

The simplest frequency transformation is a scaling operation expressed by

$$s = \frac{p}{\omega_o} \tag{3.3}$$

where ω_o is a frequency scaling parameter. This transformation is denoted frequency scaling and it allows one to obtain a low-pass transfer function with a nonunity passband frequency edge from an LPP transfer function.

Transformation of poles and zeros of transfer function. Consider an LPP factorized transfer function $T_{\text{LPP}}(s)$ with n poles $s_{p1}, s_{p2}, \ldots, s_{pn}$ and m zeros $s_{z1}, s_{z2}, \ldots, s_{zm}$

$$T_{\text{LPP}}(s) = K \frac{(1 - s/s_{z1})(1 - s/s_{z2}) \cdots (1 - s/s_{zm})}{(1 - s/s_{p1})(1 - s/s_{p2}) \cdots (1 - s/s_{pn})} \tag{3.4}$$

Using Equation 3.3, this transfer function becomes

$$T_{\text{LP}}(p) = K \frac{(1 - p/p_{z1})(1 - p/p_{z2}) \cdots (1 - p/p_{zm})}{(1 - p/p_{p1})(1 - p/p_{p2}) \cdots (1 - p/p_{pn})} \tag{3.5}$$

where poles and zeros (s_{zi} and s_{pj}) of the LPP transfer function $P_{\text{LPP}}(s)$ become simply poles and zeros in $T_{\text{LP}}(p)$, which are related to those of $T_{\text{LPP}}(s)$ by the scaling factor ω_o:

$$p_{pi} = \omega_o s_{pi}, \quad p_{zj} = \omega_o s_{zj} \tag{3.6}$$

To determine the magnitude (or frequency) response of the LP filter we evaluate the magnitude of the transfer function on the imaginary axis (for $s = j\Omega$). The magnitude response of the transformed transfer function $|T_{\text{LP}}(j\omega)|$ preserves a low-pass characteristic as illustrated in Figure 3.2. The frequency range from 0 to ∞ in the Ω-axis is mapped to the range 0 to ∞ in the ω-axis. A frequency and its mirror image in the negative axis $\pm\Omega$ is mapped to frequencies $\omega = \pm\omega_o\Omega$ with the same magnitude response: $|T_{\text{LPP}}(j\Omega)| = |T_{\text{LP}}(j\omega_o\Omega)|$. The passband and stopband edge frequencies $\Omega_{\text{P}} = 1$ rad/s and Ω_{s} of the LPP are mapped into passband and stopband edge frequencies $\omega_{\text{p}} = \omega_o$ and $\omega_{\text{s}} = \omega_o\Omega_{\text{s}}$, respectively. From this, it can be seen that for given low-pass filter specifications ω_{s}, ω_{p} the equivalent LPP is determined based on the relation $\Omega_{\text{s}} = \omega_{\text{s}}/\omega_{\text{p}}$, while the frequency scaling parameter ω_o corresponds to the passband edge frequency of the desired LP.

LP network transformation. Capacitors and inductors are the only elements that are frequency dependent and that can be affected by a change of frequency variable. Capacitors and inductors in an LPP network have impedances $z_c = 1/sc_n$ and $z_1 = s1_n$, respectively. Using Equation 3.3 these become $Z_c(p) = 1/pC$ and $Z_L(p) = pL$, where $C = c_n/\omega_o$ and $L = l_n/\omega_o$. The LPP to LP frequency transformation is performed directly on the network by simply dividing the values of all capacitors and inductors by the frequency scaling factor ω_o. This is illustrated in Figure 3.3a. The transformation expressed by Equation 3.3 can be applied to any type of filter and it has the effect of scaling the frequency axis without changing the shape of its magnitude response. This is illustrated in Figure 3.3, where the elements of an LPP with $A_{\text{p}} = 2$ dB, $A_{\text{s}} = 45$ dB, and selectivity $\Omega_{\text{s}} = 1.6$ rad/s are scaled (Figure 3.3b) to transform the network into an LP network with passband and stopband edge frequencies $\omega_{\text{p}} = 2\pi$ 10 krad/s and $\omega_{\text{s}} = 2\pi$ 16 krad/s (or $f_{\text{p}} = 10$ kHz and $f_{\text{s}} = 16$ kHz), respectively.

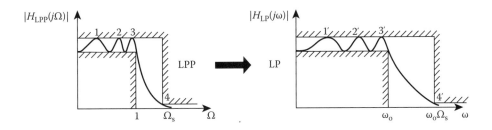

FIGURE 3.2   Derivation of low-pass response from a low-pass prototype by frequency scaling.

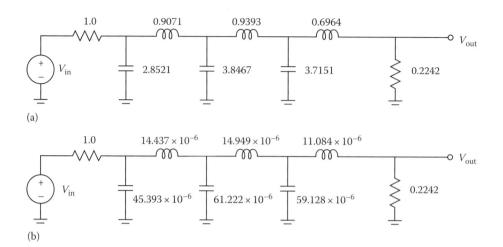

(a)

(b)

FIGURE 3.3 (a) Low-pass prototype ladder network and (b) LP network with passband frequency $f_p = 10$ kHZ derived from (a).

3.2.2 Impedance Scaling

Dimensionless transfer functions defined by ratios of voltages (V_{out}/V_{in}) or currents (I_{out}/I_{in}) remain unchanged if all impedances of a network are scaled by a common scaling factor "a." On the other hand, transfer functions of the transresistance type (V_{out}/I_{in}) or of the transconductance type (I_{out}/V_{in}) are simply modified by the impedance scaling factor a and $1/a$, respectively. If we denote a normalized impedance by z_n, then the impedance scaling operation leads to an impedance $Z = a z_n$. When applied to resistors (r_n), capacitors (c_n), inductors (l_n), transconductance gain coefficients (g_n), and transresistance gain coefficients (r_n) result in the following relations for the elements (R, C, L, g, and r) of the impedance scaled network.

$$R = a r_n$$
$$L = a l_n$$
$$C = \frac{1}{a} c_n$$
$$g = \frac{1}{a} g_n$$
$$r = a r_n$$

(3.7)

Dimensionless voltage-gain and current-gain coefficients are not affected by impedance scaling. Technologies for fabrication of microelectronic circuits (CMOS, bipolar, BiCMOS monolithic integrated circuits, thin-film and thick-film hybrid circuits) only allow elements values and time constants (or pole and zero frequencies) within certain practical ranges. Frequency and impedance scaling are very useful to scale normalized responses and network elements resulting from standard approximation procedures to values within the range achievable by the implementation technology. This is illustrated in the following example.

Example 3.1

The amplifier of Figure 3.4. is characterized by a one-pole low-pass voltage transfer function given by $H(s) = V_{out}/V_{in} = K(1 + s/\omega_p)$, where $K = g_m r_L r_1/(r_1 + r_2)$, and $\omega_p = 1/r_L C_L$. Perform frequency and impedance scaling so that the circuit pole takes a value $\omega_p = 2\pi \times 10$ Mrad/s (or $f_p = 10$ MHz) and resistance,

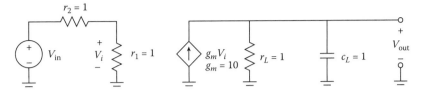

FIGURE 3.4 Normalized transconductance amplifier.

capacitance, and transconductance gain values are in the range of kΩ, pF, and μA/V, which are appropriate for the implementation of the circuit as an integrated circuit in CMOS technology.

Solution

The required location of the pole and range of values for the circuit elements can be achieved using frequency and impedance scaling factors $\omega_o = 2\pi \times 10^7$ and $a = 10^4$, respectively. These result in $R_1 = ar_1 = 10$ kΩ, $R_2 = ar_2 = 10$ kΩ, $g = g_m/a = 1000$ μA/V, $R_L = ar_L = 10$ kΩ, and $C_L = c_L/a\omega_o = 1.59$ pF.

3.3 Low-Pass to High-Pass Transformation

The LPP to high-pass transformation is defined by

$$s = \frac{\omega_o^2}{p} \tag{3.8}$$

Using this substitution in the LPP transfer function (Equation 3.4), it becomes

$$T_{\text{HP}}(P) = K \frac{p^{n-m}(p - p_{z1})(p - p_{z2}) \cdots (p - p_{zm})}{(p - p_{p1})(p - p_{p2}) \cdots (p - p_{pn})} \tag{3.9}$$

where the poles and zeros of Equation 3.8 are given by

$$p_{zi} = \frac{\omega_o^2}{s_{zi}} \quad \text{for } i \in \{1, 2, \dots, m\}$$

$$p_{pj} = \frac{\omega_o^2}{s_{pj}} \quad \text{for } j \in \{1, 2, \dots, n\} \tag{3.10}$$

It can be seen that zeros and poles of $T_{\text{HP}}(p)$ are reciprocal to those of $T_{\text{LPP}}(s)$ and scaled by the factor ω_o^2. $T_{\text{HP}}(p)$ has $n - m$ zeros at $s = 0$, which can be considered to originate from $n - m$ zeros at ∞ in $T_{\text{LPP}}(s)$.

Let us consider now the transformation of the imaginary axis in s to the imaginary axis in p. For $s = j\Omega$, p takes the form $p = j\omega$ where

$$\omega = \frac{-\omega_o^2}{\Omega} \tag{3.11}$$

From Equation 3.11, it can be seen that positive frequencies in the LPP transform to reciprocal and scaled frequencies of the HP filter. Specifically, the frequency range from 0 to ∞ in Ω maps to the frequency range $-\infty$ to 0 in ω, while the range $-\infty$ to 0 in Ω maps to 0 to ∞ in ω. The passband edge frequency

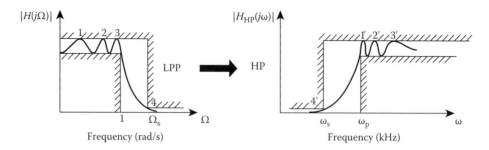

FIGURE 3.5 Transformation of a low-pass into a high-pass response.

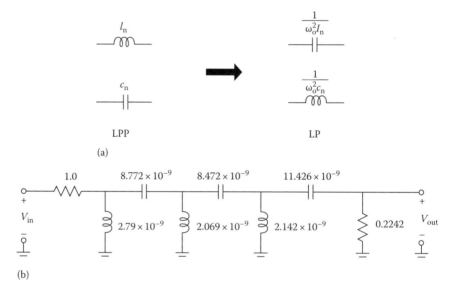

FIGURE 3.6 (a) LPP to high-pass network element transformations and (b) high-pass network derived from LPP of Figure 3.3a.

$(\Omega_p = \pm 1)$ and the stopband edge frequency $\pm \Omega_s$ of the LPP is mapped to $\omega_p = \pm \omega_o^2$ and $\omega_s = \pm \omega_o^2/\Omega_s$ in the high-pass response. This is illustrated in Figure 3.5.

The procedure to obtain the specifications of the equivalent LPP given specifications ω_p and ω_s for a HP circuit can be outlined as follows:

1. Calculate the selectivity factor of the LPP according to $\Omega_s = \omega_p/\omega_s$.
2. Approximate an LPP transfer function $T_{LPP}(s)$ with the selectivity Ω_s and the passband ripple and stopband attenuation of the desired high-pass response.
3. Perform an LPP to HP transformation either by direct substitution $p = \omega_o^2/s$ in $T_{LPP}(s)$ or by transforming poles and zeros of $T_{LPP}(s)$ using Equation 3.10.

Network transformation. Consider a capacitor c_n and an inductor l_n in an LPP network. They have impedances $z_c(s) = 1/sc_n$, $z_1(s) = sl_n$, respectively. Using Equation 3.8, these become impedances $Z_L(p) = pL$ and $Z_c = 1/pC$ in the high-pass network, where $L = 1/\omega_o^2\, c_n$ and $C = 1/\omega_o^2\, l_n$. It can be seen that an LPP to HP transformation can be done directly on an LPP network by replacing capacitors by inductors and inductors by capacitors. For illustration, Figure 3.6 shows a high-pass network with a passband edge frequency $\omega_p = 2\pi\, 20$ Mrad/s or ($fp = 20$ MHz) derived from the LPP network shown in Figure 3.6a.

3.4 Low-Pass to Bandpass Transformation

If the LPP transfer function is expressed now as a rational function

$$H_{\mathrm{LPP}}(s) = K \frac{b_0 + b_1 s + b_2 s^2 + \cdots + b_m s^m}{a_0 + a_1 s + a_2 s^2 + \cdots + a_n s^n} \tag{3.12}$$

then through the substitution

$$s = \frac{1}{\mathrm{BW}} \frac{p^2 + \omega_o^2}{p} \tag{3.13}$$

$H_{\mathrm{LPP}}(s)$ transformed into a bandpass transfer function $H_{\mathrm{BP}}(p)$ with the form

$$H_{\mathrm{BP}}(P) = K' p^{n-m} \frac{1 + B_1 p + B_2 p^2 + \cdots + B_2 p^{2m-2} + B_1 p^{2m-1} + p^{2m}}{1 + A_1 p + A_2 p^2 + \cdots + A_2 p^{2m-2} + A_1 p^{2m-1} + p^{2m}} \tag{3.14}$$

From Equation 3.14, it can be seen that the bandpass transfer function has twice as many poles and zeros as the LPP transfer function. In addition it has $n - m$ zeros at the origin. The coefficients of the numerator and denominator polynomials are symmetric and are a function of the coefficients of $H_{\mathrm{LPP}}(s)$.

In order to obtain poles and zeros of the bandpass transfer function from the poles and zeros of $T_{\mathrm{LPP}}(p)$, three points must be considered.

First, a real pole (or zero) $s_p = -u_p$ of $H_{\mathrm{LPP}}(s)$ maps into a complex conjugate pair with frequency ω_o and Q (or selectivity) factor in $H_{\mathrm{BP}}(p)$, where $Dq = \omega_o/(U_p BW)$.*

Second, a pair of complex conjugate pole (or zeros) of $H_{\mathrm{LPP}}(s)$ with frequency Ω_o and pole-quality factor q denoted by (Ω_o, q) is mapped into two pairs of complex conjugate poles (or zeros) (ω_{o1}, Q) and (ω_{o2}, Q), where the following relations apply:

$$\begin{aligned} \omega_{o1} &= \omega_o M \\ \omega_{o2} &= \frac{\omega_o}{M} \\ Q &= \frac{a}{c}\left(M + \frac{1}{M}\right) \end{aligned} \tag{3.15a}$$

and the definitions

$$\begin{aligned} a &= \frac{\omega_o}{\mathrm{BW}} \\ b &= \frac{\Omega_o}{2a} \\ c &= \frac{\Omega_o}{q} \end{aligned} \tag{3.15b}$$

$$M = \sqrt{b^2 + \sqrt{(1+b^2)^2 - \frac{c}{(2a)^2}} + \sqrt{2b}\sqrt{1 + b^2 - \frac{c}{2\Omega_o^2 +} + \sqrt{(1+b^2)^2 - \frac{c}{(2a)^2}}}}$$

apply.

---

* A complex conjugate pole pair can be expressed as $s_p, S_p{}^* = u_p \pm j\omega_p = \omega_c\, e^{\pm j\theta} = (\omega_c, Q)$, where the pole quality factor Q is given by $Q = \frac{1}{2}\cos\theta$ and $\omega_c = (G_p^2 + \omega_p^2)^{\frac{1}{2}}$, $\theta = tg^{-1}\frac{\omega_p}{G_p}$.

Narrow-band approximation. If the condition $BW/\omega_o \ll 1$ is satisfied, then following simple transformations known as the narrow band approximation* can be used to map directly poles (or zeros) from the *s*-plane to the *p*-plane

$$p_p \approx \frac{BW}{2} s_p + j\omega_o, \quad p_z \approx \frac{BW}{2} s_z + j\omega_o \tag{3.16}$$

These approximations are valid only if the transformed poles and zeros are in the vicinity of $j\omega_o$, that is, if $|s_p - \omega_o|/\omega_o \ll 1$.

Third, in order to obtain poles and zeros of the bandpass transfer function, mapping of complex zeros on the imaginary Ω axis (s_2, $s_x^* = \pm j\Omega_z$) takes place using the same mapping relations discussed next.

Mapping of imaginary frequency axis. Consider a frequency $s = j\Omega$ and its mirror image $s = -j\Omega$ in the LPP. Using Equation 3.13, these two frequencies are mapped into four frequencies: $\pm\omega_1$ and $\pm\omega_2$, where ω_1 and ω_2 are given by

$$\omega_2 = \Omega\frac{BW}{2} + \sqrt{\omega_o^2 + \left(\Omega\frac{BW}{2}\right)^2}$$
$$\omega_1 = -\Omega\frac{BW}{2} + \sqrt{\omega_o^2 + \left(\Omega\frac{BW}{2}\right)^2} \tag{3.17}$$

From Equation 3.17, the following relations can be derived:

$$\omega_2 - \omega_1 = BW\Omega$$
$$\omega_1\omega_2 = \omega_o^2 \tag{3.18}$$

It can be seen that with the LPP to bandpass transformation frequencies are mapped into bandwidths. A frequency Ω and its mirror image $-\Omega$ are mapped into two pairs of frequency points that have ω_0 as center of geometry. The interval from 0 to ∞ in the positive Ω axis maps into two intervals in the ω axis: the first from ω_o to $+\infty$ on the positive ω axis and the second from $-\omega_o$ to 0 to in the negative ω axis. The interval $-\infty$ to 0 on the negative Ω axis maps into two intervals: from $-\infty$ to $-\omega_o$ in the negative ω axis and from 0 to ω_o in the positive ω axis. The LPP passband and stopband edge frequencies $\Omega_p = \pm 1$ and $+\Omega_s$ are mapped into passband edge frequencies ω_{p1}, ω_{p2}, and into stopband edge frequencies ω_{s1}, ω_{s2} that satisfy

$$\omega_{p2} - \omega_{p1} = BW$$
$$\omega_{s2} - \omega_{s1} = BW\Omega_s \tag{3.19}$$

and

$$\omega_{p1}\omega_{p2} = \omega_{s1}\omega_{s2} = \omega_o^2 \tag{3.20}$$

Figure 3.7 shows mapping of frequency points 1, 2, 3, and 4 in the Ω axis to points $1'$, $2'$, $3'$, and $4'$ and $1''$, $2''$, $3''$, $4''$ in the ω axis of the bandpass response.

If the bandpass filter specifications do not satisfy Equation 3.20 (which is usually the case), then either one of the stopband frequencies or one of the passband frequencies has to be redefined so that they become symmetric w.r.t. ω_o and an equivalent LPP filter can be specified. For given passband and

---

* L.P. Huelsman, An algorithm for the low-pass to bandpass transformation, *IEEE Trans. Education*, EH, 72, March 1968.

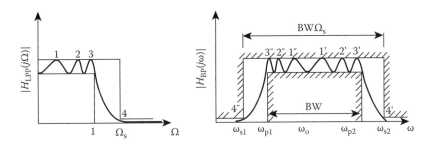

FIGURE 3.7 Low-pass to bandpass transformation.

stopband specifications ω_{p1}, ω_{p2}, ω_{s1}, ω_{s2}, A_p, A_s, the procedure to determine Ω_s for an equivalent LPP is as follows:

1. Calculate first the parameter ω_o in terms of the passband frequencies according to $\omega_o = \sqrt{\omega_{p1}\omega_{p2}}$.
2. Make the stopband frequencies geometrically symmetric with respect to ω_o determined in step 1 by redefining either ω_{s1} or ω_{s2} so that one of these frequencies becomes more constrained.* If $\omega_{s2} < \omega_o^2/\omega_{s2}$, then assign ω_{s1} the new value: $\omega_{s1} = <\omega_o^2/\omega_{s1}$. Otherwise assign ω_{s2} the new value $\omega_{s2} = \omega_o^2/\omega_{s1}$.
3. Calculate a selectivity factor of the LPP based on the redefined stopband frequency according to $\Omega_s = (\omega_{s2} - \omega_{s1})/(\omega_{p2} - \omega_{p1})$. This expression follows directly from Equations 3.19 and 3.20.
4. Calculate now the parameter ω_o in terms of the stopband frequencies, according to $\omega_o = \sqrt{\omega_{s1}\omega_{s2}}$.
5. Symmetrize the passband frequencies w.r.t. the value of ω_o determined in step 4 by constraining either ω_{p1} or ω_{p2}. If $\omega_{p1} > \omega_o^2/\omega_{p2}$, then assign ω_{p1} the new value: $\omega_{p1} > \omega_o^2/\omega_{p2}$. Otherwise assign ω_{p2} the new value $\omega_{p2} > \omega_o^2/\omega_{p1}$.
6. Calculate the selectivity factor based on the new set of passband frequencies using the same expression as in step 3.
7. Select from step 3 or step 6 the maximum selectivity factor Ω_s and determine the transformation parameters ω_o and BW from the values calculated in steps 1–3 or in 4–6, whichever sequence leads to the maximum Ω_s, which leads to the lowest order n for $T_{LPP}(s)$ and with this to the least expensive filter implementation.

Example 3.2

Consider the following nonsymmetric specifications for a bandpass filter: $\omega_{s1} = 2\pi$ 9, $\omega_{s2} = 2\pi$ 17, $\omega_{p1} = 2\pi$ 10, and $\omega_{p1} = 2\pi$ 14.4 (all frequencies specified in krad/s). Application of the above procedure leads to a new value for the upper stopband frequency $\omega_{s2} = 16$ and from this to the following parameters: BW $= 2\pi$ (14.4 $-10) = 2\pi$ 4.4, ω_o^2 $\omega_o^2 =$ 2π 9 2π 16 $= 2\pi$ 10 2π 14.4 $= (2\pi)^2$ 144, and $\Omega_s = 16 - 9/(14.4 - 10) = 1.59$.

Bandpass network transformation. Consider now the transformation of capacitors and inductors in an LPP filter. An inductor in the LPP network has an impedance $z_1(s) = sl_n$. This becomes an impedance $z_s(p) = pL_s + 1/pC_s$, where $L_s = l_n/\text{BW}$ and $C_s = \text{BW}/l_n\omega_o^2\omega_o^2$. Now consider a capacitor c_n in the LPP with admittance $Y_c(s) = sc_n$. Using the transformation (Equation 3.12), this becomes an admittance $Y_p(p) = pC_p + 1/pL_p$, where $C_p = c_n/\text{BW}$ and $L_p = \text{BW}/c_n\omega_o^2$. This indicates that to transform an LPP network into a bandpass network, inductors in the LPP network are replaced by the series connection of

---

* The term "constraint specifications" is used here in the sense of redefining either a stopband frequency or a passband frequency so that one of the transition bands becomes narrower, which corresponds to tighter design specifications.

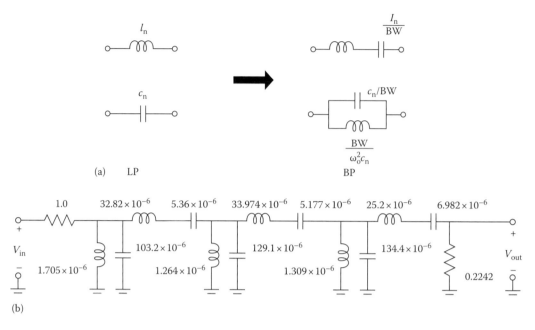

(a)   LP

(b)

FIGURE 3.8   (a) LPP to bandpass network element transformations and (b) bandpass network derived from LPP network of Figure 3.3a.

an inductor with value L_s and a capacitor with value C_s and capacitors in the LPP are replaced by the parallel combination of a capacitor C_p and inductor L_p. This is illustrated in Figure 3.8. As with other transformations, resistors remain unchanged since they are not frequency dependent.

Example 3.3

Consider the LPP network shown in Figure 3.3a with specifications $Ap = 2$ dB, $As = 45$ dB, and $\Omega_s = 1.6$. Derive a bandpass network using the parameters calculated in Example 3.2: $\omega_o^2 = (2\pi)^2\,144$, BW $= 2\pi\,4.4$ krad/s.

Solution

Straightforward application of the relations shown in Figure 3.8a leads to the network of Figure 3.8b.

3.5 Low-Pass to Band-Reject Transformation

This transformation is characterized by

$$S = \mathrm{BW}\,\frac{p}{p^2 + \omega_o^2} \tag{3.21}$$

and it can be best visualized as a sequence of two transformations through the intermediate complex frequency variables $s' = u' + j\Omega' \times 1$. A normalized LPP to high-pass transformations

$$s = \frac{1}{s'} \tag{3.22}$$

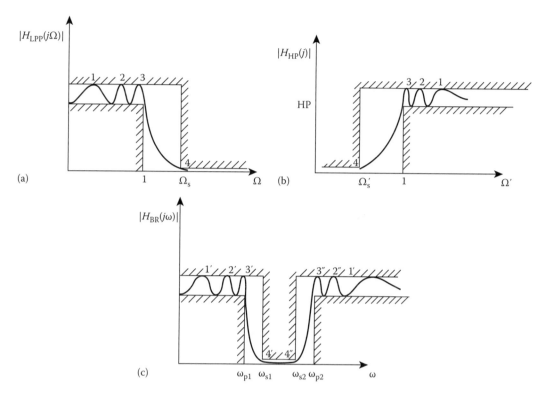

FIGURE 3.9 Low-pass to band-reject transformation: (a) Low-pass response, (b) normalized high-pass response, and (c) band-reject response derived from (a) and (b).

followed by a bandpass transformation applied to s'

$$s' = \frac{1}{\text{BW}} \frac{p^2 + \omega_0^2}{p} \tag{3.23}$$

Mapping of the imaginary ω axis to the Ω axis through this sequence of transformations leads to a band-rejection response in ω; frequency points 1, 2, 3, 4 have been singled out to better illustrate the transformation. Figure 3.9a shows the magnitude response of the LPP, Figure 3.9b shows the high-pass response obtained through Equation 3.22. This intermediate response is a normalized high-pass response with a passband extending from $\Omega'_p = 1$ to ∞ and a stopband extending from 0 to $\Omega'_s = 1/\Omega_s$. Figure 3.9c shows the magnitude response obtained by applying Equation 3.23 to the variable s'. The frequency range from $\Omega' = 0$ to ∞ in Figure 3.9b is mapped into the range from ω_o to ∞ and from $-\omega_o$ to 0. The frequency range from $-\infty$ to 0 in Ω' is mapped into the ranges from 0 to ω_o and from $-\infty$ to $-\omega_o$ in ω as indicated in Figure 3.9c. It can be seen that the bandpass transformation applied to a normalized high-pass response creates two passbands with ripple A_p from $\omega = 0$ to ω_{p1} and from ω_{p2} to ∞ and a stopband with attenuation A_s from ω_{s1} to ω_{s2}. The following conditions are satisfied:

$$\begin{aligned}
\omega_{s1}\omega_{s2} &= \omega_{p1}\omega_{p2} = \omega_o^2 \\
\omega_{s2} - \omega_{s1} &= \text{BW}\Omega'_s = \text{BW}/\Omega_s \\
\omega_{p2} - \omega_{p1} &= \text{BW}
\end{aligned} \tag{3.24}$$

From Equation 3.24, if ω_{s1}, ω_{s2}, ω_{p1}, ω_{p2} as well as A_p and A_s are specified for a band-rejection filter, then the selectivity factor of the equivalent LPP is calculated according to $\Omega_s = (\omega_{p2} - \omega_{p1})/(\omega_{s2} - \omega_{s1})$. Similar to the case of bandpass filters, the band-rejection specifications must be made symmetric with respect to ω_o so that an equivalent LPP can be specified. This is done following a similar procedure as for the bandpass transformation by constraining either one of the passband frequencies or one of the stopband frequencies so that the response becomes geometrically symmetric with respect to ω_o. The option leading to the largest selectivity factor (that corresponds in general to the least expensive network implementation) is then selected. In this case constraining design specifications refers to either increase ω_{p1} or ω_{s2} or to decrease ω_{p2} or ω_{s1}.

Example 3.4

Make following band-rejection filter specifications symmetric so that an equivalent LPP with the lowest Ω_s can be found: $\omega_{s1} = 2\pi\ 10$, $\omega_{s2} = 2\pi\ 14.4$, $\omega_{p1} = 2\pi\ 9$, $\omega_{p2} = 2\pi\ 17$.

Solution

Defining ω_o in terms of the passband frequencies the upper stopband frequency acquires the new value $\omega_{s2} = 2\pi\ 15.3$ and the selectivity factor $\Omega_s = 1.509$ is obtained. If ω_o is defined in terms of the stopband frequencies the upper passband frequency is assigned the new value $\omega_{p2} = 2\pi\ 16$ and the selectivity factor $\Omega_s = 1.59$ is obtained. Therefore, the second option with $\Omega_s = 1.59$ corresponds to the largest selectivity factor and the following transformation parameters result: $BW = \omega_{p2} - \omega_{p1} = 2\pi(16-9) = 2\pi\ 7$, $\omega_o^2 = \omega_{p1}\ \omega_{p1} = (2\pi)^2\ 144$ is made.

Transformation of poles and zeros of the LPP transfer function. To determine the poles and zeros of the band-rejection transfer function $H_{BR}(p)$ starting from those of $H_{LPP}(s)$, again a sequence of two transformations is required: poles and zeros of the LPP (denoted $s_{p1}, s_{p2}, \ldots, s_{pn}$ and $s_{z1}, s_{z2}, \ldots, s_{zn}$) are transformed into poles and zeros of $H_{HP}(s')$ $s'_{p1}, s'_{p2}, \ldots, s'_{z1}, s'_{z2}, \ldots, s'_{zm}$, which are reciprocal to those of the LPP. The high-pass also acquires $n - m$ zeros at the origin as explained in Sections 3.3 and 3.2. The transformations described in Section 3.4 are then applied to the high-pass poles zeros $s'_{p1}, s'_{p2}, \ldots, s'_{z1}, s'_{z2}, \ldots, s'_{zm}$.

Band-rejection network transformation. Using the transformation (Equation 3.21), an inductor in the LPP network with admittance $y_l(s) = 1/sl_n$ becomes an admittance $Y_p(p) = pC_p + 1/pL_p$, where $C_p = 1/BW\ l_n$ and $L_p = BW\ l_n/\omega_o^2$. A capacitor c_n in the LPP with impedance $z_c(s) = 1/sc_n$ becomes with Equation 3.20 an impedance $Z_s(p) = pL_s + 1/pC_s$, where $L_s = 1/c_n BW$ and $C_s = c_n BW/\omega_o^2$. To transform an LPP network into a bandpass network, capacitors in the LPP network are replaced by a series connection of an inductor with value L_s and a capacitor with value C_s, while inductors in the LPP are replaced by the parallel combination of a capacitor C_p and inductor L_p. This is illustrated in Figure 3.10a.

Example 3.5

Consider the LPP network shown in Figure 3.3a. It corresponds to the specifications: $A_p = 2$ dB, $A_s = 45$ dB, and $\Omega_s = 1.6$. Transform it into a bandpass network using the following parameters from Example 3.4: $\omega_o^2 = (2\pi)^2\ 144$, $BW = 2\pi\ 7$ Mrad/s (units for ω_o are microradians per second).

Solution

The circuit of Figure 3.10b is obtained applying the transformations indicated in Figure 3.10a.

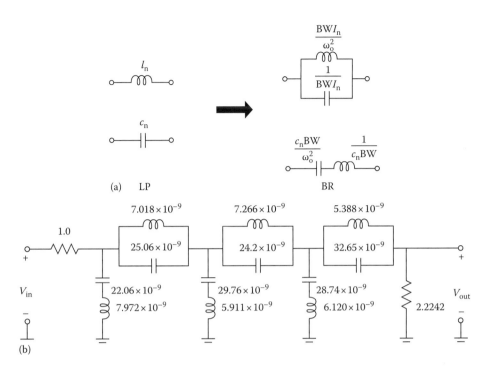

FIGURE 3.10 (a) LPP to band-reject transformations of network elements and (b) band-reject network derived from LPP of Figure 3.3a.

4

Sensitivity
and Selectivity

Igor M. Filanovsky
University of Alberta

4.1 Introduction

Using sensitivity one can evaluate the change in a filter performance characteristic (bandpass, *Q*-factor) or in a filter function (input impedance, transfer function) resulting from a change in a nominal value of one or more of the filter components. Hence, sensitivities and based on them sensitivity measures are used to compare different realizations of electric filters that meet the same specifications. Sensitivities can also be used to estimate the spread of the performance characteristic caused by the spread of the element values. In the design of filters, one is interested both in choosing realizations that have low sensitivities and in minimizing the sensitivities. This allows use of components with wider tolerances for a given variation or a given spread of the filter characteristic or function.

4.2 Definitions of Sensitivity

Let y be the filter performance characteristic and x be the value of the parameter of a filter element that is causing the characteristic change. The *relative sensitivity* is defined as follows:

$$\mathbf{S}_x^y(y,x) = \frac{\partial y}{\partial x}\frac{x}{y} = \frac{\partial y/y}{\partial x/x} = \frac{\partial(\ln y)}{\partial(\ln x)} \tag{4.1}$$

It is usually used to establish the approximate relationship between the relative changes $\delta y = \Delta y/y$ and $\delta x = \Delta x/x$. Here, Δy and Δx are absolute changes. The interpretation of relative changes δy and δx depends on the problem at hand. If these relative changes are small, one writes that

$$\delta y \approx \mathbf{S}_x^y(y,x)\delta x \tag{4.2}$$

(This relationship assumes that $\mathbf{S}_x^y$ is different from zero. If $\mathbf{S}_x^y = 0$, the relative changes δy and δx may be independent. This happens, for example, in the passband of doubly terminated LC networks [see below], where the passband attenuation always increases independently on the sign in the variation of a reactance element.) The argument in the parentheses of Equation 4.2, when does not involve any ambiguity, will usually be omitted, i.e., we will write simply $\mathbf{S}_x^y$. Some simple properties of the sensitivity determined by Equation 4.1 can be established by differentiation only. They are summarized in Table 4.1 [1].

One can also define two *semirelative sensitivities*

$$\mathbf{S}_x(y,x) = x\frac{\partial y}{\partial x} = \frac{\partial y}{\partial x/x} = \frac{\partial y}{\partial(\ln x)} \tag{4.3}$$

which is here frequently denoted by $\mathbf{S}_x(y)$, and

$$\mathbf{S}^y(y,x) = \frac{1}{y}\frac{\partial y}{\partial x} = \frac{\partial y/y}{\partial x} = \frac{\partial(\ln y)}{\partial x} \tag{4.4}$$

which is also denoted by $\mathbf{S}^y(x)$. Both these sensitivities can be used in a way similar to Equation 4.2 to establish the approximate relationships between one relative and one absolute change. Finally, the absolute sensitivity $\mathbf{S}(y,x)$ is simply a partial derivative of y with respect to x, i.e., it can be used to establish the relationship between absolute changes. The variable x represents the value for any component of the filter. The set of values for all the components will be denoted as $\mathbf{x} = \{x_i\}$, where $i = 1, 2, \ldots, n$.

TABLE 4.1   Properties of the Relative Sensitivity

Property Number	Relation	Property Number	Relation
1	$\mathbf{S}_x^{ky} = \mathbf{S}_{kx}^y = \mathbf{S}_x^y$	10	$\mathbf{S}_x^{y_1/y_2} = \mathbf{S}_x^{y_1} - \mathbf{S}_x^{y_2}$
2	$\mathbf{S}_x^x = \mathbf{S}_x^{kx} = \mathbf{S}_{kx}^{kx} = 1$	11	$\mathbf{S}_{x_1}^y = \mathbf{S}_{x_2}^y \mathbf{S}_{x_1}^{x_2}$
3	$\mathbf{S}_{1/x}^y = \mathbf{S}_x^{1/y} = -\mathbf{S}_x^y$	12^a	$\mathbf{S}_x^y = \mathbf{S}_x^{|y|} + j\arg y\mathbf{S}_x^{\arg y}$
4	$\mathbf{S}_x^{y_1 y_2} = \mathbf{S}_x^{y_1} + \mathbf{S}_x^{y_2}$	13^a	$\mathbf{S}_x^{\arg y} = \frac{1}{\arg y}\operatorname{Im}\mathbf{S}_x^y$
5	$\mathbf{S}_x^{\prod_{i=1}^n y_i} = \sum_{i=1}^n \mathbf{S}_x^{y_i}$	14^a	$\mathbf{S}_x^{|y|} = \operatorname{Re}\mathbf{S}_x^y$
6	$\mathbf{S}_x^{y^n} = n\mathbf{S}_x^y$	15	$\mathbf{S}_x^{y+z} = \frac{1}{y+z}\left(y\mathbf{S}_x^y + z\mathbf{S}_x^z\right)$
7	$\mathbf{S}_{x^n}^x = n\mathbf{S}_x^{kx^n} = n$	16	$\mathbf{S}_x^{\sum_{y=1}^n y_i} = \frac{\sum_{i=1}^n y_i\mathbf{S}_x^{y_i}}{\sum_{i=1}^n y_i}$
8	$\mathbf{S}_{x^n}^y = \frac{1}{n}\mathbf{S}_x^y$	17	$\mathbf{S}_x^{\ln y} = \frac{1}{\ln y}\mathbf{S}_x^y$
9	$\mathbf{S}_{x^n}^y = \mathbf{S}_{kx^n}^x = \frac{1}{n}$		

^{a} In this relation, y is a complex quantity and x is a real quantity.

4.3 Function Sensitivity to One Variable

Let the chosen quantity y be the filter function $F(s, x)$. When it does not involve any ambiguity, this function will be denoted as $F(s)$. The element x is a passive or active element in the circuit realization of the function. The function sensitivity is defined as

$$S_x^{F(s,x)} = \frac{\partial F(s,x)}{\partial x} \frac{x}{F(s,x)} \tag{4.5}$$

Under condition of sinusoidal steady state, when $s = j\omega$, the function $F(j\omega, x)$ can be represented as

$$F(j\omega, x) = |F(j\omega, x)|e^{j \arg F(j\omega,x)} = e^{-\alpha(\omega,x)+j\beta(\omega,x)} \tag{4.6}$$

and using the left-hand part of Equation 4.6, one finds that

$$S_x^{F(j\omega,x)} = \operatorname{Re} S_x^{F(j\omega,x)} + j \operatorname{Im} S_x^{F(j\omega,x)} = S_x^{|F(j\omega,x)|} + j \frac{\partial \arg F(j\omega, x)}{\partial x/x} \tag{4.7}$$

as follows from property 12 of Table 4.1. Thus, the real part of the function sensitivity gives the relative change in the magnitude response, and the imaginary part gives the change in the phase response, both with respect to a normalized element change. If one determines $\delta F = [F(j\omega, x) - F(j\omega, x_0)]/F(j\omega, x_0)$ and $\delta x = (x - x_0)/x_0$, where x_0 is the initial value of the element and the deflection $\Delta x = x - x_0$ is small, then Equation 4.5 is used to write

$$\delta F \approx S_x^{F(s,x)} \delta x \tag{4.8}$$

And if one determines $\delta|F| = [|F(j\omega, x) - F(j\omega, x_0)|]/|F(j\omega, x_0)|$, then using Equation 4.7, one obtains

$$\delta|F| \approx \operatorname{Re} S_x^{F(s,x)} \delta x \tag{4.9}$$

These calculations assume that the sensitivity is also calculated at $x = x_0$.

A frequently used alternate form of Equation 4.7 is obtained by using the attenuation function $\alpha(\omega, x) = \ln\left(\frac{1}{|F(j\omega,x)|}\right) = -\ln|F(j\omega, x)|$ and the phase function $\beta(\omega, x) = \arg F(j\omega, x)$ defined by the right-hand part of Equation 4.6 (this interpretation is usually used when $F(s)$ is the filter transfer function $T(s)$). In terms of these, Equation 4.7 may be rewritten as

$$S_x^{F(j\omega,x)} = -\frac{\partial \alpha(\omega, x)}{\partial x/x} + j \frac{\partial \beta(\omega, x)}{\partial x/x} = -S_x[\alpha(\omega), x] + jS_x[\beta(\omega), x] \tag{4.10}$$

From Equations 4.7 and 4.10 one concludes that $S_x^{|F(j\omega,x)|} = -S_x[\alpha(\omega), x]$. Besides, using Equations 4.7 and 4.10 one can write that

$$\Delta \arg F(j\omega, x) = \Delta\beta(\omega, x) \approx \operatorname{Im} S_x^{F(j\omega,x)} \delta x = S_x[\beta(\omega), x] \delta x \tag{4.11}$$

where $\Delta \arg F(j\omega, x) = \arg F(j\omega, x) - \arg F(j\omega, x_0)$.

Usually the filter function is a ratio of two polynomials $N(s)$ and $D(s)$, i.e.,

$$F(s) = \frac{N(s)}{D(s)} \tag{4.12}$$

Then, assuming that the coefficients of $N(s)$ and $D(s)$ depend on the element x, and using Equation 4.1, one derives the following form of Equation 4.5:

$$S_x^{F(s)} = x \left[\frac{\partial N(s)/\partial x}{N(s)} - \frac{\partial D(s)/\partial x}{D(s)} \right] \qquad (4.13)$$

which is sometimes more convenient.

4.4 Coefficient Sensitivity

In general, a network function $F(s)$ for any active or passive lumped network is a ratio of polynomials having the form

$$F(s) = \frac{N(s)}{D(s)} = \frac{a_0 + a_1 s + a_2 s^2 + \cdots + a_m s^m}{d_0 + d_1 s + d_2 s^2 + \cdots + d_n s^n} \qquad (4.14)$$

Here the coefficients a_i and d_i are real and can be functions of an arbitrary filter element x. For such an element x one may define the *relative coefficient sensitivities* as follows:

$$S_x^{a_i} = \frac{\partial a_i}{\partial x} \frac{x}{a_i}, \quad S_x^{d_i} = \frac{\partial d_i}{\partial x} \frac{x}{d_i} \qquad (4.15)$$

or the *semirelative coefficient sensitivities* (they are even more useful):

$$S_x(a_i) = x \frac{\partial a_i}{\partial x}, \quad S_x(d_i) = x \frac{\partial d_i}{\partial x} \qquad (4.16)$$

The coefficient sensitivities defined in this way are related to the function sensitivity introduced in Section 4.3. Indeed, using Equations 4.13 and 4.16 one easily obtains that

$$S_x^{F(s)} = \left[\frac{\sum_{i=0}^{m} S_x(a_i) s^i}{N(s)} - \frac{\sum_{i=0}^{n} S_x(d_i) s^i}{D(s)} \right] \qquad (4.17)$$

or, in terms of relative sensitivities, that

$$S_x^{F(s)} = \left[\frac{\sum_{i=0}^{m} S_x^{a_i} a_i s^i}{N(s)} - \frac{\sum_{i=0}^{n} S_x^{d_i} d_i s^i}{D(s)} \right] \qquad (4.18)$$

The manner in which the filter function depends on any element x is a bilinear dependence [2]. Thus, Equation 4.14 may also be written in the form

$$F(s) = \frac{N(s)}{D(s)} = \frac{N_1(s) + x N_2(s)}{D_1(s) + x D_2(s)} \qquad (4.19)$$

where $N_1(s), N_2(s), D_1(s)$, and $D_2(s)$ are polynomials with real coefficients that are not functions of the filter element x. This is true whether x is chosen to be the value of a passive resistor or capacitor, the gain of some amplifier or controlled source, etc. Only for filters with ideal transformers, ideal gyrators, and ideal negative impedance converters, the filter functions are the biquadratic functions of the ideal element parameters [2].

Because of the bilinear dependence property, there are only two ways in which a coefficient, say, a_i may depend on network element. The first of these has the form $a_i = kx$, in which case $\mathbf{S}_x^{a_i} = 1$ and $\mathbf{S}_x(a_i) = kx$; the second possible dependence for a coefficient a_i (or d_i) is $a_i = k_0 + k_1 x$, in which case $\mathbf{S}_x^{a_i} = k_1 x/(k_0 + k_1 x)$ and $\mathbf{S}_x(a_i) = k_1 x$. In this latter situation one has two cases: (1) the parities of the term are the same, and thus the magnitude of $\mathbf{S}_x^{a_i}$ is less than one and (2) the terms have opposite parities, in which case the magnitude of $\mathbf{S}_x^{a_i}$ is greater than one. In the last case the relative sensitivity $\mathbf{S}_x^{a_i}$ can have an infinite value, as a result of dividing by zero. In this case, a more meaningful measure of the change would be to use the semirelative coefficient sensitivity $\mathbf{S}_x(a_i)$.

4.5 Root Sensitivities

A filter function can also be represented as

$$F(s) = \frac{a_m \Pi_{i=0}^m (s - z_i)}{d_n \Pi_{i=0}^n (s - p_i)} \tag{4.20}$$

where z_i are zeros and p_i are poles. If $F(s)$ is also a function of the filter element x, the location of these poles and zeros will depend on this element. This dependence is described by the *semirelative root sensitivities*

$$\mathbf{S}_x(z_i) = x\frac{\partial z_i}{\partial x}, \quad \mathbf{S}_x(p_i) = x\frac{\partial p_i}{\partial x} \tag{4.21}$$

We will give calculation of the pole sensitivities only (they are used more frequently, to verify the stability) calculating the absolute change Δp_i for a given δx, the calculation of zeros follows the same pattern. Assume that p_i is a simple pole of $F(s)$, then

$$D(p_i) = D_1(p_i) + xD_2(p_i) = 0 \tag{4.22}$$

When the parameter x becomes $x + \Delta x$, the pole moves to the point $p_i + \Delta p_i$. Substituting these values in Equation 4.22 one obtains that

$$D_1(p_i + \Delta p_i) + (x + \Delta x)D_2(p_i + \Delta p_i) = 0 \tag{4.23}$$

If one uses Taylor's expansions $D_1(p_i + \Delta p_i) = D_1(p_i) + [\partial D_1(s)/\partial s]\,|_{s=p_i} \Delta p_i + \cdots$ and $D_2(p_i + \Delta p_i) = D_2(p_i) + [\partial D_2(s)/\partial s]\,|_{s=p_i} \Delta p_i + \cdots$ and substitutes them in Equation 4.23, keeping the terms of the first order of smallness, one obtains that

$$\frac{\Delta p_i}{\Delta x} = -\frac{D_2(p_i)}{D'(p_i)} \tag{4.24}$$

where $D'(p_i) = \left[\dfrac{\partial D(s)}{\partial s}\right]\bigg|_{s=p_i}$. This result allows calculation of the pole sensitivity, which becomes

$$\mathbf{S}_x(p_i) = x\frac{\partial p_i}{\partial x} = -x\frac{D_2(p_i)}{D'(p_i)} \tag{4.25}$$

One can write that $D_1(s) = b_0 + b_1 s + b_2 s^2 + \cdots$ and $D_2(s) = c_0 + c_1 s + c_2 s^2 + \cdots$. Then, taking into consideration Equation 4.19, one can write that $D(s) = d_0 + d_1 s + d_2 s^2 + \cdots = (b_0 + xc_0) + (b_1 + xc_1)s + (b_2 + xc_2)s^2 + \cdots$. Differentiating this result, one obtains that $D_2(s) = c_0 + c_1 s + c_2 s^2 + \cdots = \partial d_0/\partial x + (\partial d_1/\partial x)s + (\partial d_2/\partial x)s^2 + \cdots = (1/x)[\mathbf{S}_x(d_0) + \mathbf{S}_x(d_1)s + \mathbf{S}_x(d_2)s^2 + \cdots]$. From the other side, $D'(s) = \frac{\partial D(s)}{\partial s} = d_1 + 2d_2s + 3d_3s^2 + \cdots$. Calculating the two last expressions at $s = p_i$ and substituting them in Equation 4.25, one obtains that

$$\mathbf{S}_x(p_i) = -\frac{\sum_{j=0}^{n} p_i^j \mathbf{s}_x(d_j)}{\sum_{j=0}^{n-1}(j+1)d_{j+1}p_i^j} = -\frac{\sum_{j=0}^{n} d_j p_i^j \mathbf{s}_x^{d_j}}{\sum_{j=0}^{n-1}(j+1)d_{j+1}p_i^j} \tag{4.26}$$

The result (Equation 4.26) produces the pole sensitivity without representation of poles via coefficients of $D(s)$ (which is not possible if $n > 4$). It is convenient even for polynomials of low degree [3].

If p_i is a multiple root then the derivative $D'(p_i) = 0$ and $\mathbf{S}_x(p_i) = \infty$. But this does not mean that the variation δx causes infinitely large change of the pole location. This variation splits the multiple root into a group of simple roots. The location of these roots can be calculated in the following way. The roots are always satisfying the equation $D_1(s) + (x + \Delta x)D_2(s) = 0$, i.e., the equation $D(s) + \Delta x D_2(s) = 0$. One can rewrite the last equation as

$$1 + \delta x \left[\frac{xD_2(s)}{D(s)}\right] = 0 \tag{4.27}$$

where $\delta x = \Delta x/x$ as usual. The function $G(s) = \left[\frac{xD_2(s)}{D(s)}\right]$ can be represented as a sum of simple ratios. If the roots of $D(s)$ (i.e., the poles of $F(s)$) are simple, then

$$G(s) = \sum_{j=1}^{n} \frac{K_{ip}}{s - p_i} + K_{0p} \tag{4.28}$$

where
$$K_{0p} = G(\infty)$$
$$K_{ip} = (s - p_i)G(s)|_{s=p_i}$$

In the vicinity of $s = p_i$, Equation 4.27 can be substituted by

$$\lim_{s \to p_i}[1 + \delta x G(s)] = 1 + \delta x \frac{K_{ip}}{s - p_i} \tag{4.29}$$

Equating the right-hand side of Equation 4.29 to zero and substituting $s = p_i + \Delta p_i$ in this equation, one obtains that when $\Delta x \to 0$, the pole sensitivity can be calculated as

$$\mathbf{S}_x(p_i) = \frac{\partial p_i}{\partial x/x} = -K_{ip} \tag{4.30}$$

If a pole of $G(s)$ is not simple, but multiple, with multiplicity of k, then the limit form of Equation 4.28 will be

$$1 + \delta x \left[\frac{K_{ip}^{(1)}}{s - p_i} + \frac{K_{ip}^{(2)}}{(s - p_i)^2} + \cdots + \frac{K_{ip}^{(k)}}{(s - p_i)^k}\right] = 0 \tag{4.31}$$

If now $s = p_i + \Delta p_i$ is substituted in Equation 4.31 and only the largest term is kept, one finds that

$$\Delta p_i = \left[-\delta x K_{ip}^{(k)}\right]^{1/k} \tag{4.32}$$

Hence, these new simple roots of $D(s)$ are equiangularly spaced on a circle around p_i.

Similar calculations with similar results can be obtained for the zeros z_i of the function $F(s)$. They, together with the calculation for poles, allow one to establish the relationship between relative sensitivity of $F(s)$ and the sensitivities of its zeros and poles. Indeed, taking into consideration Equation 4.19, the result (Equation 4.13) can be rewritten as

$$\mathbf{S}_x^{F(s)} = \frac{xN_2(s)}{N(s)} - \frac{xD_2(s)}{D(s)} = H(s) - G(s) \tag{4.33}$$

Expanding both $H(s)$ and $G(s)$ into sums of simple ratios, one obtains

$$\mathbf{S}_x^{F(s)} = \sum_{i=1}^{n} \frac{\mathbf{S}_x(p_i)}{s - p_i} - \sum_{i=1}^{m} \frac{\mathbf{S}_x(z_i)}{s - z_i} + K_{0z} - K_{0p} \tag{4.34}$$

Here, $K_{0z} = H(\infty)$, $K_{0p} = G(\infty)$, and it is assumed that both zeros and poles of $F(s)$ are simple. Finally, a useful modification is obtained for the case when a coefficient d_k in the polynomial $D(s) = d_0 + d_1 s + d_2 s^2 + \cdots$ is considered as a variable parameter. If d_k is substituted by $d_k + \Delta d_k$, the polynomial $D(s)$ becomes $D(s) + \Delta d_k s^k$. For this case the function $G(s) = (\delta d_k s^k)/D(s)$ and one can write that

$$1 + \frac{\Delta d_k s^k}{D(s)} = 1 + \frac{\Delta d_k s^k}{(s - p_i)D'(s)} \tag{4.35}$$

Here $D'(s) = \partial D(s)/\partial s$ and it can be assumed that p_i is a simple root of $D(s)$. Hence, in the vicinity of $s = p_i + \Delta p_i$, the value of Δp_i can be obtained from the equation

$$1 + \frac{\Delta d_k p_i^k}{\Delta p_i D'(p_i)} = 0 \tag{4.36}$$

From Equation 4.36, one finds that

$$\frac{\Delta p_i}{\Delta d_k} = -\frac{p_i^k}{D'(p_i)} \tag{4.37}$$

and, if necessary, the pole-coefficient sensitivity

$$\mathbf{S}_{d_k}(p_i) = \frac{\partial p_i}{\partial d_k/d_k} = -d_k \frac{p_i^k}{D'(p_i)} \tag{4.38}$$

4.6 Statistical Model for One Variable

Assume that x is the value of a filter element. This x differs from the average value $\bar{x}$ in a way that cannot be controlled by the filter designer. This situation can be modeled by considering x as a random variable. Its statistical distribution depends on the manufacturing process. An approximate calculation is sufficient in most practical cases. If $F(s, x)$ is a filter function that depends on x, then the variation $F(s, x)$ around the average value $\bar{x}$ can be approximated by

$$F(s, x) \approx F(s, \bar{x}) + (x - \bar{x}) \frac{\partial F(s, x)}{\partial x}\Big|_{x=\bar{x}} \tag{4.39}$$

The statistical interpretation of Equation 4.29 follows. Due to its dependence on the random variable, it becomes a random variable as well. The values $\bar{x}$, $F(s, \bar{x}) = \bar{F}(s)$, and $\partial F(s, x)/\partial x$ calculated at $x = \bar{x}$ are constants. The last constant can be denoted as $\partial F(s, \bar{x})/\partial x$. Instead of x and $F(s, x)$ it is preferable to use their relative deviations from the average values, namely $\delta x = (x - \bar{x})/\bar{x}$ and $\delta F(s) = (F - \bar{F})/\bar{F}$. Then one obtains from Equation 4.39 that

$$\delta F(s) = \left[\frac{\bar{x}}{F(s, \bar{x})} \frac{\partial F(s, \bar{x})}{\partial x} \right] \delta x = \mathbf{S}_x^{F(s, \bar{x})} \delta x \tag{4.40}$$

Hence, in the first-order approximation the random variables $\delta F(s)$ and δx are proportional, and the proportionality factor is the same sensitivity of $F(s, x)$ with respect to x calculated at the average point $\bar{x}$. Thus, on the $j\omega$ axis the average and the variance of $\delta F(j\omega)$ and δx are related by

$$\mu_{\delta F} \approx \mathbf{S}_x^{F(s, \bar{x})} \mu_{\delta x} \tag{4.41}$$

and

$$\sigma_{\delta F}^2 = E\left\{ \left| \frac{F(j\omega, x) - F(j\omega, \bar{x})}{F(j\omega, \bar{x})} \right|^2 \right\} \approx |\mathbf{S}^{F(j\omega, \bar{x})}|^2 \sigma_{\delta x}^2 \tag{4.42}$$

where $E = \{\}$ means the expected value. Here, $\mu_{\delta x}$ is the average value of δx, $\mu_{\delta F}$ is the average value of δF, and $\sigma_{\delta x}^2$ and $\sigma_{\delta F}^2$ are the dispersions of these values. If the deviation δx is bound by the modulus $M_{\delta x}$, i.e., the probability distribution is concentrated in the interval $[-M_{\delta x}, M_{\delta x}]$, then the deviation δF is bound in the first approximation by

$$|\delta F| \leq M_{\delta F} \approx |\mathbf{S}_x^{F(j\omega, \bar{x})}| M_{\delta x} \tag{4.43}$$

Normally, the probability distribution of x should be centered around the average value $\bar{x}$, so that it can be assumed $\mu_{\delta x} = 0$. This implies that $\mu_{\delta F} = 0$ as well.

It is not difficult to see that Equations 4.8 and 4.40 are different by interpretation of δx and δF (since these were deflections from the nominal point, it was tacitly assumed that we were dealing with one sample of the filter, here they are random) and the point of calculation of sensitivity. The interpretation with a random variable is possible for Equation 4.9 as well. One has to determine $\delta|F| = [|F(j\omega, x)| - |F(j\omega, \bar{x})|]/|F(j\omega, \bar{x})|$, then using Equation 4.9 one can write

$$\mu_{\delta|F|} \approx \operatorname{Re} \mathbf{S}_x^{F(s, \bar{x})} \mu_{\delta x} \tag{4.44}$$

and

$$\sigma_{\delta|F|}^2 \approx \left(\operatorname{Re} \mathbf{S}_x^{F(j\omega, \bar{x})} \right)^2 \sigma_{\delta x}^2 \tag{4.45}$$

The result (Equation 4.11) can also be interpreted for random variables and allows one to calculate the average and the variance of the change in the filter function argument (which is hardly ever done in filter design).

4.7 Multiparameter Sensitivities and Sensitivity Measures

The multiparameter sensitivities (sometimes [4] they are called sensitivity indices) appear as an effort to introduce generalized functions that represent the influence of all filter elements. They can be used for comparison of different designs and should be minimized in the design process. The sensitivity measures appear as numbers (they are functionals of the parameter sensitivities) that should be minimized in the design.

First of all, the definition of function sensitivity given in Equation 4.5 is readily extended to determine the effect on the filter function of variation of more than one component. In this case $F(s, x_1, x_2, \ldots, x_n) = F(s, \mathbf{x})$ and one may write that

$$\frac{dF(s, \mathbf{x})}{F(s, \mathbf{x})} = d[\ln F(s, \mathbf{x})] = \sum_{i=1}^{n} \mathbf{S}_{x_i}^{F(s, \mathbf{x})} \frac{dx_i}{x_i} \tag{4.46}$$

where n is the number of components being considered. Here $\mathbf{S}_{x_i}^{F(s, \mathbf{x})} = [x_i \partial F(s, \mathbf{x})]/[F(s, \mathbf{x}) \partial x_i]$. From this result one directly (substituting $s = j\omega$ and separating real and imaginary parts) obtains that

$$\frac{d|F(j\omega, \mathbf{x})|}{|F(j\omega, \mathbf{x})|} = \sum_{i=1}^{n} \operatorname{Re} \mathbf{S}_{x_i}^{F(j\omega, \mathbf{x})} \frac{dx_i}{x_i} \tag{4.47}$$

and

$$d \arg F(j, \omega) = \sum_{i=1}^{n} \operatorname{Im} \mathbf{S}_{x_i}^{F(j\omega, \mathbf{x})} \frac{dx_i}{x_i} \tag{4.48}$$

The results (Equations 4.47 and 4.48) are used to evaluate the deviations of the magnitude and phase values of a given filter realization from their nominal values when the circuit elements have prescribed normalized deviations $\delta x_i = (x_i - x_{i0})/x_{i0}$ $(i = 1, 2, \cdots n)$. One can introduce a column vector of normalized deviation $[\delta x_1 \delta x_2 \cdots \delta x_n]^t$, where t means transpose and a sensitivity row vector

$$\mathbf{S}_{\mathbf{x}}^{F(s, \mathbf{x})} = \left[\mathbf{S}_{x_1}^{F(s, \mathbf{x})} \mathbf{S}_{x_2}^{F(s, \mathbf{x})} \cdots \mathbf{S}_{x_n}^{F(s, \mathbf{x})} \right] \tag{4.49}$$

Then defining $\delta F(s, \mathbf{x}) = [F(s, \mathbf{x}) - F(s, \mathbf{x}_0)]/F(s, \mathbf{x}_0)$, one can use Equation 4.46 to write

$$\delta F(s, \mathbf{x}) \approx \mathbf{S}_{\mathbf{x}}^{F(s, \mathbf{x})} \delta \mathbf{x} \tag{4.50}$$

which is analogous to Equation 4.8. As was mentioned before, the calculation of the filter function magnitude change is traditionally considered of primary importance in filter design. Introducing $\delta|F(s, \mathbf{x})| = [|F(j\omega, \mathbf{x})| - |F(j\omega, \mathbf{x}_0)|]/|F(j\omega, \mathbf{x}_0)|$ and using Equation 4.47 one writes

$$\delta|F(j\omega, \mathbf{x})| \approx \left[\operatorname{Re} \mathbf{S}_{\mathbf{x}}^{F(j\omega, \mathbf{x})} \right] \delta \mathbf{x} \tag{4.51}$$

where the row vector

$$\left[\operatorname{Re} \mathbf{S}_{\mathbf{x}}^{F(j\omega, \mathbf{x})} \right] = \left[\operatorname{Re} \mathbf{S}_{x_1}^{F(j\omega, \mathbf{x})} \operatorname{Re} \mathbf{S}_{x_2}^{F(j\omega, \mathbf{x})} \cdots \operatorname{Re} \mathbf{S}_{x_n}^{F(j\omega, \mathbf{x})} \right] \tag{4.52}$$

is used. This vector is determined by the function $F(s, \mathbf{x})$ and its derivatives calculated at $\mathbf{x} = \mathbf{x}_0$. To characterize and compare the vectors of this type, one can introduce different vector measures that are called *sensitivity indices*. The most frequently used ones are the *average sensitivity index*

$$\psi(F) = \sum_{i=1}^{n} \operatorname{Re} \mathbf{S}_{x_i}^{F(j\omega, \mathbf{x})} \tag{4.53}$$

then the *worst-case sensitivity index*

$$v(F) = \sum_{i=1}^{n} \left| \text{Re } S_{x_i}^{F(j\omega,\mathbf{x})} \right| \tag{4.54}$$

(sometimes it is called worst-case magnitude sensitivity), and, finally, the *quadratic sensitivity index*

$$\rho(F) = \left[\sum_{i=1}^{n} \left(\text{Re } S_{x_i}^{F(j\omega,\mathbf{x})} \right)^2 \right]^{1/2} \tag{4.55}$$

These sensitivity indices can be considered as multiparameter sensitivities.

If we let the individual nominal values of n elements be given as x_{i0}, then we may define a tolerance constant ε_i (positive number) by the requirement that

$$x_{i0}(1 - \varepsilon_i) \le x_i \le x_{i0}(1 - \varepsilon_i) \tag{4.56}$$

Then we may define a worst-case measure of sensitivity

$$M_W = \int_{\omega_1}^{\omega_2} \left(\sum_{i=1}^{n} \left| \text{Re } S_{x_i}^{F(j\omega,\mathbf{x})} \right| \varepsilon_i \right) d\omega \tag{4.57}$$

The goal of the filter design should be the search for the set of tolerance constants yielding the least expensive in the production filter. This is a difficult problem, and at the design stage it can be modeled by the minimization of the chosen sensitivity measure. In the design based on the worst-case measure of sensitivity, the usual approach [2] is to choose the tolerance constants in such a way that the contributions $\left| \text{Re } S_{x_i}^{F(j\omega,\mathbf{x})} \right| \varepsilon_i$ are approximately equal, i.e., the elements with lower sensitivities get wider tolerance constants.

For any values of the filter elements x_i satisfying Equation 4.56, the magnitude characteristic will lie within the definite bounds that are apart from the nominal characteristic by the distance less than $\varepsilon_{i\,max} v(F)$. If the tolerance constants are all equal to ε, then the maximum deviation from the nominal characteristic (when $\mathbf{x} = \mathbf{x}_0$) is thus given as $\varepsilon v(F)$. And the worst-case measure of sensitivity becomes, for this case

$$M_W = \varepsilon \int_{\omega_1}^{\omega_2} v(F) d\omega \tag{4.58}$$

Considering the imaginary parts of the sensitivity row vector one can introduce corresponding sensitivity indices and similar sensitivity measures for the filter function phase.

The element tolerances obtained using the worst-case sensitivity index and measures are extremely tight and this set of elements is frequently unfeasible. Besides, with given tolerances, the set of elements producing the worst-case sensitivity is never obtained in practice. A more feasible set of tolerances is obtained when one uses the sum of the squares of the individual functions. One may define a *quadratic measure of sensitivity* as

$$M_Q = \int_{\omega_1}^{\omega_2} \left[\sum_{i=1}^{n} \left(\text{Re } S_{x_i}^{F(j\omega,\mathbf{x})} \right)^2 \varepsilon_i^2 \right] d\omega \tag{4.59}$$

In the design using the sensitivity measure given by Equation 4.59 one also tries to get the tolerances so that the contributions of each term $\left(\operatorname{Re} S_{x_i}^{F(j\omega,\mathbf{x})}\right)^2 \varepsilon_i^2$ are approximately equal in the considered bandwidth. Again, if the tolerances are equal, then this expression is simplified to

$$M_Q = \varepsilon^2 \int\limits_{\omega_1}^{\omega_2} \rho^2(F) d\omega \tag{4.60}$$

which is useful for comparison of different filters.

As one can see, the multivariable sensitivities appear as a result of certain operations with the sensitivity row vector components. Additional multivariable sensitivities could be introduced, for example, the sum of magnitudes of vector components, the sum of their squares, etc. The multivariable sensitivities and the measures considered above represent the most frequently used filter design in the context of filter characteristic variations.

The case of random variables can also be generalized so that imprecisions of the values of several elements are simultaneously considered. Around the nominal value $\bar{\mathbf{x}} = [\bar{x}_i]$ $(i = 1, 2, \cdots n)$ the function $F(s, \bar{\mathbf{x}})$ can be approximated as

$$F(s, \mathbf{x}) \approx F(s, \bar{\mathbf{x}}) + \sum_{i=1}^{n} (x - \bar{x}_i) \frac{\partial F(s, \mathbf{x})}{\partial x_i} \tag{4.61}$$

And from this approximation one obtains

$$\delta F(s, \mathbf{x}) \approx \sum_{i=1}^{n} S_{x_i}^{F(s,\mathbf{x})} \delta x_i \tag{4.62}$$

Here, $\delta F(s, x) = [F(s, \mathbf{x}) - F(s, \bar{\mathbf{x}})]/F(s, \bar{\mathbf{x}})$ and $\delta x_i = (x_i - \bar{x}_i)/\bar{x}_i$ $(i = 1, 2, \cdots n)$. This result can be rewritten as

$$\delta F(s, \mathbf{x}) \approx S_{\mathbf{x}}^{F(s,\mathbf{x})} \delta \mathbf{x} \tag{4.63}$$

and is completely analogous to Equation 4.50. It is different in interpretation only. The components of the column vector $[\delta x_1 \delta x_2 \cdots \delta x_n]^t$ are the random variables now and the components of the row vector $S_{\mathbf{x}}^{F(s,\mathbf{x})} = \left[S_{x_1}^{F(s,\mathbf{x})} S_{x_2}^{F(s,\mathbf{x})} \cdots S_{x_n}^{F(s,\mathbf{x})}\right]$ are calculated at the point $\mathbf{x} = \bar{\mathbf{x}}$.

This interpretation allow us to obtain from Equation 4.63 that on the $j\omega$ axis

$$\mu_{\delta F} = \sum_{i=1}^{n} S_{x_i}^{F(j\omega,\mathbf{x})} \mu_i \tag{4.64}$$

Here, μ_i is the average of δx_i. If all μ_i are equal, i.e., $\mu_i = \mu_x$ $(i = 1, 2, \cdots n)$, one can introduce the average sensitivity index

$$\psi(F) = \sum_{i=1}^{n} S_{x_i}^{F(j\omega,\mathbf{x})} \tag{4.65}$$

Using Equation 4.65, the average value can be calculated as $\mu_{\delta F} = \mu_x \psi(F)$. If, in addition, the deviation of δx_i is bound by M_i, then

$$M_{\delta F} \le \sum_{i=1}^{n} \left| S_{x_i}^{F(j\omega,\mathbf{x})} \right| M_i \tag{4.66}$$

If the elements of a filter have the same precision, which means that all M_i are equal, it is reasonable to introduce the worst-case sensitivity index

$$\nu(F) = \sum_{i=1}^{n} \left| S_{x_i}^{F(j\omega,\mathbf{x})} \right| \tag{4.67}$$

so that when all M_i are equal to M_x, $M_{\delta F} = M_x \nu(F)$. Finally, one can calculate $\sigma_{\delta F}^2 = E\{[\delta F(j\omega,\mathbf{x})]^* \delta F(j\omega,\mathbf{x})\}$ (here $*$ means complex conjugate) or

$$\sigma_{\delta F}^2 = E\left\{ \left(S_{\mathbf{x}}^{F(j\omega,\mathbf{x})} \delta\mathbf{x} \right)^* \left(S_{\mathbf{x}}^{F(j\omega,\mathbf{x})} \delta\mathbf{x} \right) \right\} \tag{4.68}$$

To take into consideration possible correlation between the components δx_i one can do the following. The value of $\left(S_{\mathbf{x}}^{F(j\omega,\mathbf{x})} \delta\mathbf{x} \right)$ is a scalar. Then

$$\left(S_{\mathbf{x}}^{F(j\omega,\mathbf{x})} \delta\mathbf{x} \right) = \left(S_{\mathbf{x}}^{F(j\omega,\mathbf{x})} \delta\mathbf{x} \right)^{\mathrm{t}} = (\delta\mathbf{x})^{\mathrm{t}} \left(S_{\mathbf{x}}^{F(j\omega,\mathbf{x})} \right)^{\mathrm{t}} \tag{4.69}$$

Substituting this result into Equation 4.68 one obtains that

$$\sigma_{\delta F}^2 = E\left\{ \left(S_{\mathbf{x}}^{F(j\omega,\mathbf{x})} \right)^* (\delta\mathbf{x})^* (\delta\mathbf{x})^{\mathrm{t}} \left(S_{x_i}^{F(j\omega,\mathbf{x})} \right)^{\mathrm{t}} \right\} \tag{4.70}$$

But the components of $\delta\mathbf{x}$ are real, i.e., $(\delta\mathbf{x})^* = \delta\mathbf{x}$ and the result of multiplication in the curly brackets of Equation 4.70 is a square $n \times n$ matrix. Then $E\{(\delta\mathbf{x})^*(\delta\mathbf{x})^{\mathrm{t}}\}$ is also a square matrix

$$[\mathbf{P}] = \begin{bmatrix} \sigma_{x_1}^2 & \rho_{x_1 x_2} & \cdots & \rho_{x_1 x_n} \\ \rho_{x_2 x_1} & \sigma_{x_2}^2 & \cdots & \rho_{x_2 x_n} \\ \vdots & \vdots & \ddots & \vdots \\ \rho_{x_n x_1} & \cdots & \cdots & \sigma_{x_n}^2 \end{bmatrix} \tag{4.71}$$

the diagonal elements of which are variances of δx_i and off-diagonal terms are nonnormalized correlation coefficients. Then, Equation 4.70 can be rewritten as

$$\sigma_{\delta F}^2 = \left(S_{\mathbf{x}}^{F(j\omega,\mathbf{x})} \right)^* [\mathbf{P}] \left(S_{\mathbf{x}}^{F(j\omega,\mathbf{x})} \right)^{\mathrm{t}} \tag{4.72}$$

which is sometimes [4] called the propagation-of-variance formula. In the absence of correlation between the variations δx_i the matrix $[\mathbf{P}]$ has the diagonal terms only and Equation 4.72 becomes

$$\sigma_{\delta F}^2 = \sum_{i=1}^{n} \left| S_{x_i}^{F(j\omega,\mathbf{x})} \right|^2 \sigma_{x_i}^2 \tag{4.73}$$

If all $\sigma_{x_i}^2$ are equal to σ_x^2 one can introduce a quadratic sensitivity index

$$\rho(F) = \left[\sum_{i=1}^{n} \left| S_{x_i}^{F(j\omega,\mathbf{x})} \right|^2 \right]^{1/2} \tag{4.74}$$

and in this case $\sigma_{\delta F}^2 = \rho^2(F)\sigma_x^2$ (the value $\rho^2(F)$ is sometimes called Schoeffler multivariable sensitivity). One can also introduce two sensitivity measures, namely, the worst-case sensitivity measure

$$M_W = \int_{\omega_1}^{\omega_2} \left(\sum_{i=1}^{n} \left| \mathrm{Re}\, \mathbf{S}_{x_i}^{F(j\omega,\mathbf{x})} \right| \mu_i \right) d\omega \tag{4.75}$$

and the quadratic sensitivity measure

$$M_Q = \int_{\omega_1}^{\omega_2} \left(\mathbf{S}_{\mathbf{x}}^{F(j\omega,\mathbf{x})} \right)^* [\mathbf{P}] \left(\mathbf{S}_{\mathbf{x}}^{F(j\omega,\mathbf{x})} \right)^t d\omega \tag{4.76}$$

which, when the correlation between the elements of $\delta \mathbf{x}$ is absent, becomes

$$M_Q = \int_{\omega_1}^{\omega_2} \left[\sum_{i=1}^{n} \left| \mathbf{S}_{x_i}^{F(j\omega,\mathbf{x})} \right|^2 \sigma_i^2 \right] d\omega \tag{4.77}$$

Here, for simplicity the notation $\sigma_i = \sigma_{x_i}$ is used.

The sensitivity indices and sensitivity measures introduced for the case when $\delta \mathbf{x}$ is a random vector are cumulative; they take into consideration the variation of the amplitude and phase of the filter function. For this reason some authors prefer to use the indices as they are defined in Equations 4.65, 4.67, and 4.74 and the measures as they are defined by Equations 4.75 and 4.77 for the deterministic cases as well (the deterministic case does not assume any correlation between the variations δx_i), with the corresponding substitution of μ_i by ε_i and σ_i^2 by ε_i^2. From the other side, one can take Equation 4.51 and use it for the case of random vector $\delta \mathbf{x}$ considering, for example, the variation $\delta|F(j\omega,\mathbf{x})|$ as a random variable and calculating $\mu_{\delta|F|}$ and $\sigma_{\delta|F|}^2$, which will be characteristics of this variable. In this case one can use the results (Equations 4.75, 4.77, etc.) substituting $\mathbf{S}_{x_i}^{F(s,\mathbf{x})}$ by $\mathrm{Re}\, \mathbf{S}_{x_i}^{F(s,\mathbf{x})}$. These possibilities are responsible for many formulations of multiparameter sensitivities that represent different measures of the vector $\mathbf{S}_{\mathbf{x}}^{F(s,\mathbf{x})}$. In the design based, for example, on Equations 4.75 and 4.77, one determines the required $\sigma_{\delta F}^2$ using the reject probability [2] depending on the ratio of $\varepsilon_{\delta F}/\sigma_{\delta F}$. Here, $\varepsilon_{\delta F}$ is the tolerance of $|\delta F(j\omega,\mathbf{x})|$ and in many cases one takes $\varepsilon_{\delta F}/\sigma_{\delta F} = 2.5$ which gives the reject probability of 0.01. Then, one determines the dispersion σ_i^2 so that the contributions of each term in Equation 4.77 are equal. Finally, using the probability function that describes the distribution of δx_i within the tolerance borders one finds these borders (if, for example, the selected element has evenly distributed values $-\varepsilon_i \le \delta x_i \le \varepsilon_i$, then $\varepsilon_i = \sqrt{3}\sigma_i$; for Gaussian distribution one frequently accepts $\varepsilon_i = 2.5\sigma_i$).

The preliminary calculation of the coefficient sensitivities is useful for finding the sensitivity measures. If, for example, one calculates a multivariable statistical measure of sensitivity then one can consider that

$$F(j\omega, \mathbf{x}) = \frac{a_0 + a_1(j\omega) + \cdots + a_m(j\omega)^m}{d_0 + d(j\omega) + \cdots + d_n(j\omega)^n} = F(j\omega, \mathbf{a}, \mathbf{d}, \mathbf{x}) \tag{4.78}$$

where

$$\mathbf{a} = \begin{bmatrix} a_0 & a_1 & \cdots & a_m \end{bmatrix}^t$$
$$\mathbf{d} = \begin{bmatrix} d_0 & d_1 & \cdots & d_n \end{bmatrix}^t$$

Then, the components $\mathbf{S}_{x_i}^{F(j\omega,\mathbf{x})}$ defined earlier can be rewritten as

$$
\mathbf{S}_{x_i}^{F(j\omega,\mathbf{x})} = \frac{x_i}{F(j\omega,\mathbf{x})}\left[\frac{\partial \mathbf{a}^t}{\partial x_i}\nabla_a F(j\omega,\mathbf{x}) + \frac{\partial \mathbf{d}^t}{\partial x_i}\nabla_d F(j\omega,\mathbf{x})\right]
$$

$$
= \frac{1}{F(j\omega,\mathbf{x})}\left[\sum_{j=0}^{m}\frac{\partial F(j\omega,\mathbf{x})}{\partial a_j}\mathbf{S}_{x_i}(a_j) + \sum_{j=0}^{n}\frac{\partial F(j\omega,\mathbf{x})}{\partial d_j}\mathbf{S}_{x_i}(d_j)\right] \qquad (4.79)
$$

where

$$
\nabla_a F(j\omega,\mathbf{x}) = [\partial F(j\omega,\mathbf{x})/\partial a_0 \; \partial F(j\omega,\mathbf{x})/\partial a_1 \cdots \partial F(j\omega,\mathbf{x})/\partial a_m]^t
$$

and

$$
\nabla_d F(j\omega,\mathbf{x}) = [\partial F(j\omega,\mathbf{x})/\partial d_0 \; \partial F(j\omega,\mathbf{x})/\partial d_1 \cdots \partial F(j\omega,\mathbf{x})/\partial d_n]^t
$$

For a given transfer function, the components of the vectors $\nabla_a F(j\omega,\mathbf{x})$ and $\nabla_d F(j\omega,\mathbf{x})$ are independent of the form of the realization or the values of the elements and can be calculated in advance. If we now define a matrix $k \times (m+1)$ $\mathbf{C}_1$ as

$$
\mathbf{C}_1 = \begin{bmatrix} \mathbf{S}_{x_1}(a_0) & \mathbf{S}_{x_1}(a_1) & \cdots & \mathbf{S}_{x_1}(a_m) \\ \vdots & \vdots & \ddots & \vdots \\ \mathbf{S}_{x_k}(a_0) & \mathbf{S}_{x_k}(a_1) & \cdots & \mathbf{S}_{x_k}(a_m) \end{bmatrix} \qquad (4.80)
$$

and $k \times (n+1)$ matrix $\mathbf{C}_2$ as

$$
\mathbf{C}_2 = \begin{bmatrix} \mathbf{S}_{x_1}(d_0) & \mathbf{S}_{x_1}(d_1) & \cdots & \mathbf{S}_{x_1}(d_n) \\ \vdots & \vdots & \ddots & \vdots \\ \mathbf{S}_{x_k}(d_0) & \mathbf{S}_{x_k}(d_1) & \cdots & \mathbf{S}_{x_k}(d_n) \end{bmatrix} \qquad (4.81)
$$

then one can rewrite

$$
\left[\mathbf{S}_{\mathbf{x}}^{F(j\omega,\mathbf{x})}\right]^t = \left[\mathbf{S}_{x_1}^{F(j\omega,\mathbf{x})}\mathbf{S}_{x_2}^{F(j\omega,\mathbf{x})}\cdots \mathbf{S}_{x_k}^{F(j\omega,\mathbf{x})}\right]^t = \mathbf{C}_1\frac{\nabla_a F(j\omega,\mathbf{x})}{F(j\omega,\mathbf{x})} + \mathbf{C}_2\frac{\nabla_d F(j\omega,\mathbf{x})}{F(j\omega,\mathbf{x})} \qquad (4.82)
$$

Then, the multiparameter sensitivity measure can be written as

$$
M_Q = \int_{\omega_1}^{\omega_2}\left[\left(\frac{\nabla_a F}{F}\right)^{*t}\mathbf{C}_1^t \mathbf{P}\mathbf{C}_1\left(\frac{\nabla_a F}{F}\right) + \left(\frac{\nabla_d F}{F}\right)^{*t}\mathbf{C}_2^t \mathbf{P}\mathbf{C}_2\left(\frac{\nabla_d F}{F}\right)\right]d\omega + \int_{\omega_1}^{\omega_2}2\mathrm{Re}\left[\left(\frac{\nabla_a F}{F}\right)^{*t}\mathbf{C}_1^t \mathbf{P}\mathbf{C}_2\left(\frac{\nabla_d F}{F}\right)\right]d\omega
$$

$$
(4.83)
$$

and this definition of statistical multiparameter sensitivity measure may be directly applied to a given network realization. In a similar fashion, the matrices of unnormalized coefficient sensitivities can be used with other multiparameter sensitivity measures.

4.8 Sensitivity Invariants

When one is talking about sensitivity invariants [5–7], it is assumed that for a filter function $F(s, \mathbf{x})$ there exists the relationship

$$\sum_{i=1}^{n} \mathbf{S}_{x_i}^{F(s,\mathbf{x})} = k \tag{4.84}$$

where
  $\mathbf{x} = [x_1 x_2 \cdots x_n]^t$, as usual,
  k is a constant

These relationships are useful to check the sensitivity calculations. In the cases considered below, this constant can have one of three possible values, namely, 1, 0, and -1, and the sensitivity invariants are obtained from the homogeneity of some of the filter functions.

The function $F(s, \mathbf{x})$ is called homogeneous of order k with respect to the vector $\mathbf{x}$ if and only if it satisfies the relationship

$$F(s, \lambda \mathbf{x}) = \lambda^k F(s, \mathbf{x}) \tag{4.85}$$

where λ is an arbitrary scalar. For the homogeneous function $F(s, \mathbf{x})$ the sensitivities are related by Equation 4.84. Indeed, if one takes the logarithm of both sides of Equation 4.85, one obtains

$$\ln F(s, \lambda \mathbf{x}) = k \ln \lambda + \ln F(s, \mathbf{x}) \tag{4.86}$$

Taking the derivative of both sides of Equation 4.86 with respect to λ one obtains that

$$\frac{1}{F(s, \lambda \mathbf{x})} \left[\sum_{i=1}^{n} \frac{\partial F(s, \lambda \mathbf{x})}{\partial (\lambda x_i)} x_i \right] = \frac{K}{\lambda} \tag{4.87}$$

Substituting $\lambda = 1$ in Equation 4.87 gives Equation 4.84.

Let the filter be a passive *RLC* circuit that includes r resistors, l inductors, and c capacitors, so that $r + l + c = n$ and $\mathbf{x} = [R_1 R_2 \cdots R_r L_1 L_2 \cdots L_l D_1 D_2 \cdots D_c]$, where $D_i = 1/C_i$. One of the frequently used operations is the impedance scaling. If the scaling operation is applied to a port impedance or a transimpedance of the filter, i.e., $F(s, \mathbf{x}) = Z(s, \mathbf{x})$, then

$$Z(s, \lambda \mathbf{x}) = \lambda Z(s, \mathbf{x}) \tag{4.88}$$

Equation 4.88 is identical to Equation 4.85, with $k = 1$. Then, one can write that

$$\sum_{i=1}^{r} \mathbf{S}_{R_i}^{Z(s,\mathbf{x})} + \sum_{i=1}^{l} \mathbf{S}_{L_i}^{Z(s,\mathbf{x})} + \sum_{i=1}^{c} \mathbf{S}_{D_i}^{Z(s,\mathbf{x})} = 1 \tag{4.89}$$

Considering that $D_i = 1/C_i$ and $\mathbf{S}_{C_i}^{Z(s,\mathbf{x})} = -\mathbf{S}_{D_i}^{Z(s,\mathbf{x})}$ (see Table 4.1), this result can be rewritten as

$$\sum_{i=1}^{r} \mathbf{S}_{R_i}^{Z(s,\mathbf{x})} + \sum_{i=1}^{l} \mathbf{S}_{L_i}^{Z(s,\mathbf{x})} - \sum_{i=1}^{c} \mathbf{S}_{C_i}^{Z(s,\mathbf{x})} = 1 \tag{4.90}$$

If the same scaling operation is applied to a port admittance or a transadmittance of the filter, i.e., $F(s, \mathbf{x}) = Y(s, \mathbf{x})$, then,

$$Y(s, \lambda \mathbf{x}) = \lambda^{-1} Y(s, \mathbf{x}) \tag{4.91}$$

But Equation 4.91 is identical to Equation 4.85, with $k = -1$. Then,

$$\sum_{i=1}^{r} S_{R_i}^{Y(s,\mathbf{x})} + \sum_{i=1}^{l} S_{L_i}^{Y(s,\mathbf{x})} - \sum_{i=1}^{c} S_{C_i}^{Y(s,\mathbf{x})} = -1 \tag{4.92}$$

Finally, the transfer functions (voltage or current) do not depend on the scaling operation, i.e., if $F(s, \mathbf{x}) = T(s, \mathbf{x})$, hence,

$$T(s, \lambda \mathbf{x}) = T(s, \mathbf{x}) \tag{4.93}$$

which is identical to Equation 4.85, with $k = 0$. Then,

$$\sum_{i=1}^{r} S_{R_i}^{T(s,\mathbf{x})} + \sum_{i=1}^{l} S_{L_i}^{T(s,\mathbf{x})} - \sum_{i=1}^{c} S_{C_i}^{T(s,\mathbf{x})} = 0 \tag{4.94}$$

Additional sensitivity invariants can be obtained using the relation $S_{x_i}^{F(s,\mathbf{x})} = -S_{1/x_i}^{Z(s,\mathbf{x})}$ and using $G_i = 1/R_i$ and $\Gamma_i = 1/L_i$.

Another group of sensitivity invariants is obtained using the frequency scaling operation. The following relationship is held for a filter function:

$$F(s, R_i, \lambda L_i, \lambda C_i) = F(\lambda s, R_i, L_i, C_i) \tag{4.95}$$

Taking the logarithm of both parts, then differentiating both sides with respect to λ and substituting $\lambda = 1$ in both sides gives

$$\sum_{i=1}^{l} S_{L_i}^{F(s,\mathbf{x})} + \sum_{i=1}^{c} S_{C_i}^{F(s,\mathbf{x})} = S_s^{F(s,\mathbf{x})} \tag{4.96}$$

Substituting $s = j\omega$ in Equation 4.96 and dividing the real and imaginary parts, one obtains

$$\text{Re} \sum_{i=1}^{l} S_{L_i}^{F(j\omega,\mathbf{x})} + \text{Re} \sum_{i=1}^{c} S_{C_i}^{F(j\omega,\mathbf{x})} = \omega \frac{\partial \ln |T(\omega)|}{\partial \omega} = -\frac{\partial \alpha(\omega)}{\partial \omega} \tag{4.97}$$

and

$$\text{Im} \sum_{i=1}^{l} S_{L_i}^{F(j\omega,\mathbf{x})} + \text{Im} \sum_{i=1}^{c} S_{C_i}^{F(j\omega,\mathbf{x})} = \omega \frac{\partial \arg T(\omega)}{\partial \omega} \tag{4.98}$$

The results (Equations 4.97 and 4.98) show that in an *RLC* filter, when all inductors and capacitors (but not resistors) are subjected to the same relative change, then the resulting change in the magnitude characteristic does not depend on the circuit realization and is determined by the slope of the magnitude characteristic at a chosen frequency. A similar statement is valid for the phase characteristic.

The sensitivity invariants for passive RC circuits can be obtained from the corresponding invariants for passive RLC circuits omitting the terms for sensitivities to the inductor variations. The results can be summarized the following way. For a passive RC circuit

$$\sum_{i=1}^{r} S_{R_i}^{F(s,\mathbf{x})} - \sum_{i=1}^{c} S_{C_i}^{F(s,\mathbf{x})} = k \tag{4.99}$$

where $k = 1$ if $F(s,\mathbf{x})$ is an input impedance or transimpedance function, then $k = 0$ if $F(s,\mathbf{x})$ is a voltage- or current-transfer function, and $k = -1$ if $F(s,\mathbf{x})$ is an input admittance or transconductance function. Application of the frequency scaling gives the result

$$\sum_{i=1}^{c} S_{C_i}^{F(s,\mathbf{x})} = S_s^{F(s,\mathbf{x})} \tag{4.100}$$

and combination of Equations 4.99 and 4.100 gives

$$\sum_{i=1}^{r} S_{R_i}^{F(s,\mathbf{x})} = S_s^{F(s,\mathbf{x})} + k \tag{4.101}$$

and

$$\sum_{i=1}^{r} S_{R_i}^{F(s,\mathbf{x})} + \sum_{i=1}^{c} S_{C_i}^{F(s,\mathbf{x})} = 2S_s^{F(s,\mathbf{x})} + k \tag{4.102}$$

Considering real and imaginary parts of Equations 4.100 through 4.102, one can obtain the results that determine the limitations imposed on the sensitivity sums by the function $F(j\omega, \mathbf{x})$ when resistors and/or capacitors are subjected to the same relative change.

Finally, if the filter is not passive, then the vector of parameters
$$\mathbf{x} = \begin{bmatrix} R_1 R_2 \cdots R_r L_1 L_2 \cdots L_l C_1 C_2 \cdots C_c R_{T1} R_{T2} \cdots R_{Ta} \\ G_{T1} G_{T2} \cdots G_{Tb} A_{v1} A_{v2} \cdots A_{vp} A_{i1} A_{i2} \cdots A_{iq} \end{bmatrix}$$ includes the components of transresistors R_{Tk},
transconductances G_{Tk}, voltage amplifiers A_{vk}, and current amplifiers A_{ik}. Applying the impedance scaling one can obtain the sensitivity invariant

$$\sum_{i=1}^{r} S_{R_i}^{F(s,\mathbf{x})} + \sum_{i=1}^{l} S_{L_i}^{F(s,\mathbf{x})} - \sum_{i=1}^{c} S_{C_i}^{F(s,\mathbf{x})} + \sum_{i=1}^{a} S_{R_{Ti}}^{F(s,\mathbf{x})} - \sum_{i=1}^{b} S_{G_{Ti}}^{F(s,\mathbf{x})} = k \tag{4.103}$$

where
   $k = 1$ if $F(s,\mathbf{x})$ is an impedance function
   $k = 0$ if $F(s,\mathbf{x})$ is a transfer function
   $k = -1$ if $F(s,\mathbf{x})$ is an admittance function

The frequency scaling will give the same result as Equation 4.96.

The pole (or zero) sensitivities are also related by some invariant relationships. Indeed, the impedance scaling provides the result

$$p_k\left(\lambda R_i, \lambda L_i, \frac{C_i}{\lambda}\right) = p_k(R_i, L_i, C_i) \tag{4.104}$$

Taking the derivative of both sides of Equation 4.104 with respect to λ and substituting $\lambda = 1$, one obtains that for an arbitrary *RLC* circuit

$$\sum_{i=1}^{r} S_{R_i}(p_k) + \sum_{i=1}^{l} S_{L_i}(p_k) - \sum_{i=1}^{c} S_{C_i}(p_k) = 0 \tag{4.105}$$

This is the relationship between semirelative sensitivities. If $p_k \neq 0$ one can divide both sides of Equation 4.105 by p_k and obtain similar invariants for relative sensitivities. The frequency scaling gives

$$p_k\left(R_i, \frac{L_i}{\lambda}, \frac{C_i}{\lambda}\right) = \lambda p_k(R_i, L_i, C_i) \tag{4.106}$$

and from Equation 4.106 one obtains, for relative sensitivities only, that

$$\sum_{i=1}^{l} S_{L_i}^{p_k} + \sum_{i=1}^{c} S_{C_i}^{p_k} = -1 \tag{4.107}$$

The pole sensitivity invariants for passive *RC* circuits are obtained from Equations 4.106 and 4.107, omitting the terms corresponding to the inductor sensitivities.

4.9 Sensitivity Bounds

For some classes of filters, the worst-case magnitude sensitivity index may be shown to have a lower bound [8]. Such a bound, for example, exists for filters whose passive elements are limited to resistors, capacitors, and ideal transformers, and whose active elements are limited to gyrators characterized by two gyration resistances (realized as a series connection of transresistance amplifiers and considered as different for sensitivity calculations), current-controlled current sources (CCCSs), voltage-controlled voltage sources (VCVSs), voltage-controlled current sources (VCCSs), and current-controlled voltage sources (CCVSs). Using the sensitivity invariants, it is easy to show that for such a class of networks for any dimensionless transfer function $T(s)$

$$\sum_{i=1}^{n} S_{x_i}^{T(s)} = 2S_s^{T(s)} \tag{4.108}$$

where the x_i are taken to include only the passive elements of resistors and capacitors and the active elements of CCVSs and gyrators (if the gyrators are realized as parallel connection of transconductance amplifiers, the corresponding terms should be taken with the negative sign). Substituting $s = j\omega$ in Equation 4.108 and equating real parts, one obtains that

$$\sum_{i=1}^{n} S_{x_i}^{|T(j\omega)|} = 2S_{j\omega}^{|T(j\omega)|} \tag{4.109}$$

Applying Equation 4.54 to the above equations, one obtains for the worst-case magnitude sensitivity index

$$v(T) - \sum_{i=1}^{n} \left| S_{x_i}^{|T(j\omega)|} \right| \geq \left| \sum_{i=1}^{n} S^{|T(j\omega)|} \right| - \left| 2S^{|T(j\omega)|} \right| \tag{4.110}$$

Taking the first and the last terms of this expression, one may define a lower bound LBv(T) of the worst-case magnitude sensitivity index as

$$LBv(T) = \left| 2S_{j\omega}^{|T(j\omega)|} \right| \tag{4.111}$$

This lower bound is a function only of the transfer function $T(s)$ and is independent on the particular synthesis technique (as soon as the above-mentioned restrictions are satisfied) used to realize this transfer function. A similar lower bound may be derived (taking the imaginary parts of Equation 4.108) for worst-case phase sensitivity; but it is impossible to find the design path by which one will arrive at the circuit realizing this minimal bound.

4.10 Remarks on the Sensitivity Applications

The components of active *RC* filters have inaccuracies and parasitic components that distort the filter characteristics. The most important imperfections are the following:

1. Values of the resistors and capacitors and the values of transconductances (the gyrators can be considered usually as parallel connection of two transconductance amplifiers) are different from their nominal values. The evaluation of these effects is done using the worst-case or (more frequently) quadratic multiparameter sensitivity index, and multiparameter sensitivity measures and the tolerances are chosen so that the contributions of the passive elements' variations in the sensitivity measure are equal.

2. Operational amplifiers have finite gain and this gain is frequency dependent. The effect of finite gain is evaluated using a semirelative sensitivity of the filter function with respect to variation $1/A$, where A is the operational amplifier gain. The semirelative sensitivity is

$$-\mathbf{S}^{F(s)}\left(\frac{1}{A}\right) = -\frac{\partial F(s)}{\left[F(s)\partial\left(\frac{1}{A}\right)\right]} = \left[\frac{A^2}{F(s)}\right]\left[\frac{\partial F(s)}{\partial A}\right] = A\mathbf{S}_A^{F(s)}$$

which is called the gain-sensitivity product. In many cases, $\mathbf{S}_A^{F(s)} \to 0$ when $A \to \infty$, whereas $\mathbf{S}_{1/A}[F(s)]$ has a limit that is different from zero. The frequency dependence is difficult to take into consideration [1]. Only in case of cascade realization, as shown below, one can evaluate the effect of this frequency dependence using the sensitivity of Q-factor.

3. Temperature dependence and aging of the passive elements and operational amplifiers can be determined. The influence of temperature, aging, and other environmental factors on the values of the elements can be determined by a dependence of the probability distributions for these parameters. For example, if θ is temperature, one has to estimate $\mu_x(\theta)$ and $\sigma_x(\theta)$ before the calculation of sensitivity measures. Normally, at the nominal temperature θ_0 the average value $\mu_x(\theta_0) = 0$ and $\sigma_x(\theta_0)$ depends on the nominal precision of the elements. When the temperature changes, $\mu_x(\theta)$ increases or decreases depending on the temperature coefficients of the elements, whereas $\sigma_x(\theta)$ usually increases for any temperature variation.

4.11 Sensitivity Computations Using the Adjoint Network

The determination of the sensitivities defined in the previous sections may pose difficult computational problems. Finding the network function with the elements expressed in literal form is usually tedious and error prone and the difficulty of such a determination increases rapidly with the number of elements. Calculating the partial derivatives, which is the most important part of the sensitivity computation, provides additional tedium and increases the possibility of error still more. Thus, in general, it is advantageous to use digital computer methods to compute sensitivities. The most obvious method for doing this is to use one of

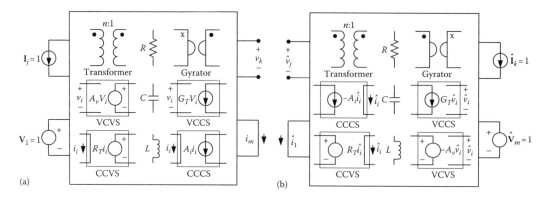

FIGURE 4.1 The components of a network (a) and its adjoint (b).

the many available computer-aided design programs (e.g., SPICE) to make an analysis of the network with nominal element values, and then repeat the analysis after having perturbed the value of one of the elements. This is not a desirable procedure, since it requires a large number of analyses. It can be justified if the network has some *a priori* known critical elements for which the analyses should be done.

The crucial part of sensitivity computation is, as was mentioned above, the calculation of network function derivatives with respect to element variations. To simplify this part of the calculation, the concept of adjoint network [9] is used. This method requires only two analyses to provide all the sensitivities of a given network immitance function.

If $\mathbf{N}$ and $\widehat{\mathbf{N}}$ are linear time invariant networks, then they are said to be adjoint (to each other) if the following hold. The two networks have the same topology and ordering of branches; thus their incidence matrices [10] are equal, namely $\mathbf{A} = \widehat{\mathbf{A}}$. If excitation with the unit current (unit voltage) at an arbitrary port j (port l) of network $\mathbf{N}$ yields a voltage (current) at an arbitrary port k (port m) of $\mathbf{N}$, excitation with the unit current (unit voltage) at port k (port m) of network $\widehat{\mathbf{N}}$ will yield the same voltage (current) as the above in port j (port l) of $\widehat{\mathbf{N}}$ (see Figure 4.1).

Figure 4.1 also shows how the adjoint network should be constructed, as follows. (a) All resistance capacitive and inductance branches and transformers in $\mathbf{N}$ are associated, respectively, with resistance, capacitive, and inductance branches and transformers in $\widehat{\mathbf{N}}$. (b) All gyrators in $\mathbf{N}$ with gyration resistance r become gyrators in $\widehat{\mathbf{N}}$ with gyration resistance $-r$. (c) All VCVSs in $\mathbf{N}$ become CCCSs in $\widehat{\mathbf{N}}$ with controlling and controlled branches reversing roles, and with the voltage amplification factor A_v becoming the current amplification factor $-A_i$. (d) All CCCSs in $\mathbf{N}$ become VCVSs in $\widehat{\mathbf{N}}$ with controlling and controlled branches reversing roles, and with the current amplification factor A_i becoming the voltage amplification factor $-A_v$. (e) All VCCCs and CCVSs have their controlling and controlled branches in $\mathbf{N}$ reversed in $\widehat{\mathbf{N}}$.

Thus, the Tellegen theorem [9] applies to the branch voltage and current variables of these two networks. If we let $\mathbf{V}$ be the vector of branch voltages and $\mathbf{I}$ be the vector of branch currents (using capital letters implies that the quantities are functions of the complex variable s), then

$$\mathbf{V}^t\widehat{\mathbf{I}} = \widehat{\mathbf{I}}^t\mathbf{V} = \widehat{\mathbf{V}}^t\mathbf{I} = \mathbf{I}^t\mathbf{V} = \mathbf{0} \tag{4.112}$$

If in both circuits all independent sources have been removed to form n external ports (as illustrated in Figure 4.2), then one can divide the variables in both circuits into two groups so that

$$\mathbf{I}^t = \begin{bmatrix} \mathbf{I}_p^t \mathbf{I}_b^t \end{bmatrix} \quad \mathbf{V}^t = \begin{bmatrix} \mathbf{V}_p^t \mathbf{V}_b^t \end{bmatrix}$$

$$\widehat{\mathbf{I}}^t = \begin{bmatrix} \widehat{\mathbf{I}}_p^t \widehat{\mathbf{I}}_b^t \end{bmatrix} \quad \widehat{\mathbf{V}}^t = \begin{bmatrix} \widehat{\mathbf{V}}_p^t \widehat{\mathbf{V}}_b^t \end{bmatrix} \tag{4.113}$$

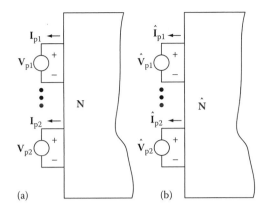

FIGURE 4.2 Separation of port variables in the network (a) and its adjoint (b).

The first are the vector of port variables $\mathbf{V}_p$ and $\mathbf{I}_p$ in $\mathbf{N}$ and $\widehat{\mathbf{V}}_p$ and $\widehat{\mathbf{I}}_p$, correspondingly, in $\widehat{\mathbf{N}}$. These variables will define n-port open-circuit impedance matrices $\mathbf{Z}_{oc}$ and $\widehat{\mathbf{Z}}_{oc}$ or n-port short-circuit admittance matrices $\mathbf{Y}_{sc}$ and $\widehat{\mathbf{Y}}_{sc}$ via the relationships

$$\mathbf{V}_p = -\mathbf{Z}_{oc}\mathbf{I}_p \quad \widehat{\mathbf{V}}_p = -\widehat{\mathbf{Z}}_{oc}\widehat{\mathbf{I}}_p$$
$$\mathbf{I}_p = -\mathbf{Y}_{sc}\mathbf{V}_p \quad \widehat{\mathbf{I}}_p = -\widehat{\mathbf{Y}}_{sc}\widehat{\mathbf{V}}_p \tag{4.114}$$

Then, the rest of the variables will be nonport variables (including the variables of dependent source branches) $\mathbf{V}_b$ and $\mathbf{I}_b$ for $\mathbf{N}$ and $\widehat{\mathbf{V}}_b$ and $\widehat{\mathbf{I}}_b$ for $\widehat{\mathbf{N}}$. These variables may define branch impedance matrices $\mathbf{Z}_b$ and $\widehat{\mathbf{Z}}_b$ and branch admittance matrices $\mathbf{Y}_b$ and $\widehat{\mathbf{Y}}_b$ by the relationships

$$\mathbf{V}_b = \mathbf{Z}_b\mathbf{I}_b \quad \widehat{\mathbf{V}}_b = \widehat{\mathbf{Z}}_b\widehat{\mathbf{I}}_b$$
$$\mathbf{I}_b = \mathbf{Y}_b\mathbf{V}_b \quad \widehat{\mathbf{I}}_b = \widehat{\mathbf{Y}}_b\widehat{\mathbf{V}}_b \tag{4.115}$$

If the branch impedance and branch admittance matrices do not exist, a hybrid matrix may be used to relate the branch variables. For $\mathbf{N}$, this may be put in the form

$$\begin{bmatrix} \mathbf{V}_{b1} \\ \mathbf{I}_{b2} \end{bmatrix} = \begin{bmatrix} \mathbf{H}_{11} & \mathbf{H}_{12} \\ \mathbf{H}_{21} & \mathbf{H}_{22} \end{bmatrix} \begin{bmatrix} \mathbf{I}_{b1} \\ \mathbf{V}_{b2} \end{bmatrix} \tag{4.116}$$

Similarly, for $\widehat{\mathbf{N}}$ one may write

$$\begin{bmatrix} \widehat{\mathbf{V}}_{b1} \\ \widehat{\mathbf{I}}_{b2} \end{bmatrix} = \begin{bmatrix} \widehat{\mathbf{H}}_{11} & \widehat{\mathbf{H}}_{12} \\ \widehat{\mathbf{H}}_{21} & \widehat{\mathbf{H}}_{22} \end{bmatrix} \begin{bmatrix} \widehat{\mathbf{I}}_{b1} \\ \widehat{\mathbf{V}}_{b2} \end{bmatrix} \tag{4.117}$$

For the adjoint networks, the branch impedance and branch admittance matrices (if they exist) are transposed, namely

$$\mathbf{Z}_b^t = \widehat{\mathbf{Z}}_b \quad \mathbf{Y}_b^t = \widehat{\mathbf{Y}}_b \tag{4.118}$$

and, if a hybrid representation is used, the matrices are connected by the relationship

$$\begin{bmatrix} \mathbf{H}_{11}^t & -\mathbf{H}_{21}^t \\ -\mathbf{H}_{12}^t & \mathbf{H}_{22}^t \end{bmatrix} = \begin{bmatrix} \widehat{\mathbf{H}}_{11} & \widehat{\mathbf{H}}_{12} \\ \widehat{\mathbf{H}}_{21} & \widehat{\mathbf{H}}_{22} \end{bmatrix} \tag{4.119}$$

As a result of these relationships, if no controlled sources are present in two networks, they are identical. In the general case it may be shown that

$$\mathbf{Z}_{oc}^t = \widehat{\mathbf{Z}}_{oc} \quad \mathbf{Y}_{sc}^t = \widehat{\mathbf{Y}}_{sc} \tag{4.120}$$

The application of adjoint circuits for sensitivity calculations requires that, first of all, using the port variables as independent ones, one finds the branch variables, and this is done for both circuits. Assume, for example, that the branch impedance matrices exist. Then, using unity-valued excitation currents as the components of $\mathbf{I}_p$ and $\hat{\mathbf{I}}_p$ one has, first, to find $\mathbf{I}_b$ and $\hat{\mathbf{I}}_b$. Now, in the original network let the elements be perturbed. The resulting vector of currents may thus be written as $\mathbf{I} + \Delta\mathbf{I}$ and the resulting vector of voltages as $\mathbf{V} + \Delta\mathbf{V}$. From Kirchhoff's current law, we have $\mathbf{A}(\mathbf{I} + \Delta\mathbf{I}) = \mathbf{0}$, and since $\mathbf{AI} = \mathbf{0}$ one also has $\mathbf{A}\Delta\mathbf{I} = \mathbf{0}$. Thus, $\Delta\mathbf{I} = \Delta\mathbf{I}_p + \Delta\mathbf{V}_b$ may be substituted in any of the relations in Equation 4.112. By similar reasoning one can conclude that it is possible as well to substitute the perturbation vector $\Delta\mathbf{V} = \Delta\mathbf{V}_p + \Delta\mathbf{V}_b$ instead of $\mathbf{V}$ in these relations. Making these substitutions, one obtains

$$\hat{\mathbf{V}}_p^t \Delta\mathbf{I}_p + \hat{\mathbf{V}}_b^t \Delta\mathbf{I}_p = 0$$
$$\mathbf{I}_p^t \Delta\mathbf{V}_p + \hat{\mathbf{I}}_b^t \Delta\mathbf{V}_b = 0 \tag{4.121}$$

Subtracting these two equations, one has

$$\hat{\mathbf{V}}_p^t \Delta\mathbf{I}_p - \mathbf{I}_p^t \Delta\mathbf{V}_p + \hat{\mathbf{V}}_b^t \Delta\mathbf{I}_p - \hat{\mathbf{I}}_b^t \Delta\mathbf{V}_b = 0 \tag{4.122}$$

To a first-order approximation, we have

$$\Delta\mathbf{V}_p = -\Delta(\mathbf{Z}_{oc}\mathbf{I}_p) \approx -\Delta\mathbf{Z}_{oc}\mathbf{I}_p - \mathbf{Z}_{oc}\Delta\mathbf{I}_p$$
$$\Delta\mathbf{V}_b = -\Delta(\mathbf{Z}_b\mathbf{I}_b) \approx -\Delta\mathbf{Z}_b\mathbf{I}_b - \mathbf{Z}_b\Delta\mathbf{I}_b \tag{4.123}$$

Substituting Equation 4.123 into Equation 4.122 and taking into consideration that $\hat{\mathbf{V}}_p^t = -\hat{\mathbf{I}}_p^t \hat{\mathbf{Z}}_{oc}^t = -\hat{\mathbf{I}}_p^t \mathbf{Z}_{oc}$ and that $\hat{\mathbf{V}}_b^t = \hat{\mathbf{I}}_b^t \mathbf{Z}_b$, one can simplify the result (Equation 4.122) to

$$\hat{\mathbf{I}}_p^t \Delta\mathbf{Z}_{oc}\mathbf{I}_p = \hat{\mathbf{I}}_b^t \Delta\mathbf{Z}_b\mathbf{I}_b \tag{4.124}$$

Equation 4.124 clearly shows that if all currents in the original network and its adjoint are known, one can easily calculate the absolute sensitivities $\mathbf{S}(Z_{ij}, \mathbf{Z}_b) \approx \Delta Z_{ij}/\Delta\mathbf{Z}_b$, which can be used for calculation of the corresponding relative sensitivities. Here, Z_{ij} is an element of the n-port open circuit impedance matrix. Then, if necessary, these sensitivities can be used for evaluation of the transfer function sensitivity or the sensitivities of other functions derived via the n-port open circuit impedance matrix.

Usually, the transfer function calculation can be easily reduced to the calculation of a particular Z_{ij}. In this case one can choose $\hat{I}_j = 1, \hat{I}_k = 0$ (for all $k \neq j$) as an excitation in the adjoint network. Then, Equation 4.124 becomes

$$\Delta Z_{ij} = \hat{\mathbf{I}}_b^t \Delta\mathbf{Z}_b\mathbf{I}_b \tag{4.125}$$

where $\mathbf{I}_b$ and $\hat{\mathbf{I}}_b$ are the branch currents in $\mathbf{N}$ and $\hat{\mathbf{N}}$ corresponding to the indicated excitation. The relations for other types of matrices are obtained in the same manner.

4.12 General Methods of Reducing Sensitivity

It is very desirable from the start of the realization procedure to concentrate on circuits that give lower sensitivity in comparison to other circuits. The practice of active filter realization allows formulation of some general suggestions ensuring that filter realizations will have low sensitivities to component variations.

It is possible to transfer the problem of lower sensitivity at the approximation stage, i.e., before any realization. It is obvious that a low-order transfer function $T(s)$ that just satisfies the specifications will require tighter tolerances in comparison to a higher order transfer function that easily satisfies the specifications. Figure 4.3 shows an example of such an approach for a low-pass filter. Hence, increasing the order of approximation and introducing a redundancy one achieves a set of wider element tolerances.

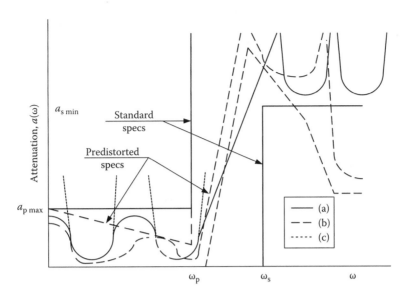

FIGURE 4.3 Attenuation requirements, their predistortion, and attenuation of different realizations: (a) attenuation of a standard circuit of a higher order; (b) attenuation of the nonstandard circuit; and (c) passband attenuation in the stages of cascade realization.

Usually the most critical region where it is difficult to satisfy the specifications is the edge of the passband. Two approaches can be used to find the function that will have less sensitivity in this frequency region. One way is to introduce predistortion in the transfer function specifications. This is also shown in Figure 4.3. The transfer function should satisfy the predistorted (tapered) specifications. It can be obtained directly if the numerical packages solving the approximation problem are available. One can also take a standard table higher order transfer function satisfying the modified specifications and then modify it to more uniformly use the tapered specifications (this allows the increase of the component tolerances even more).

Another way [11] is to preserve the initial transfer function specifications and to use transfer functions with a limited value of the maximum Q of the transfer function poles. In Ref. [11] one can find such transfer functions corresponding to the Cauer approximation. The nonstandard LC circuits corresponding to these approaches cannot be tabulated and simulated; hence, neither of them is widely used. In addition, they imply the cascaded (building-block) realization that intrinsically has worse sensitivity than the realizations using simulation of doubly terminated lossless matched filters.

4.13 Cascaded Realization of Active Filters

The cascaded (building-block) realization is based on the assumption that the transfer function will have low sensitivity if the realization provides tight control of the transfer function poles and zeros. The relationship between the element value and the transfer function poles and zeros can be established relatively easily if the transfer function is not more complicated than biquadratic (i.e., the ratio of two second-order polynomials). It is difficult (or even impossible) to establish such correspondence if, say, the denominator polynomial degree is higher than two. For a high-degree polynomial, a small variation in the polynomial coefficient can result in a large or undesirable migration of the root (it can move to the right-half plane). This justifies the cascaded approach: one hopes to have a low sensitivity of the transfer function under realization if one chooses a method allowing tight control of the poles' and zeros' locations; hence, cascade realization with $T(s) = T_1(s)T_2(s) \cdots T_i(s) \cdots T_k(s)$. Also, if one chooses for realization of each function $T_i(s)$ the method of lowest sensitivity (discussed below), then it will be possible to obtain the largest element tolerances.

If the realization by cascade connection is chosen it is still a choice of optimum factorization of the transfer function $T(s)$ into low-order factors. This optimum depends on the filter application, the chosen method of the factor realization, and the transfer function itself. It is recommended [1] that, for realization of lower sensitivity, the poles and zeros in the partial functions $T_i(s)$ are located as far apart as possible. This statement is not always true and such a choice of poles and zeros in $T_i(s)$ is in contradiction with the requirement of high dynamic range of the stage. In general, CAD methods should be used.

The methods that are most popular for the realization of the partial transfer functions are mostly limited by the filters providing the output voltage at the output of an operational amplifier. The state–space relations satisfy this requirement and provide the direct realizations of the polynomial coefficients. It is not occasionally that such an approach is used by a series of manufacturers. This is not the best method from the sensitivity point of view; the methods of realization using gyrators usually give better results [3]. But the cascade realization of gyrator filters require buffers between the blocks, and is better to use this approach if the filter is not realized in cascade form. Other realization methods [1] are also occasionally used, mostly because of their simplicity.

If the transfer function is realized in a cascade form and $T(s) = T_1(s)T_2(s) \cdots T_i(s) \cdots T_k(s)$, the element x is located only in one stage. If this is the stage realizing $T_i(s)$, then

$$\mathbf{S}_x^{T(s)} = \mathbf{S}_x^{T_i(s)} \tag{4.126}$$

Assume that this $T_i(s)$ has the form

$$T_i(s) = K \frac{s^2 + \left(\frac{\omega_z}{Q_z}\right) + \omega_z^2}{s^2 + \left(\frac{\omega_p}{Q_p}\right) + \omega_p^2} \tag{4.127}$$

Then one can write

$$\mathbf{S}_x^{T_i(s)} = \mathbf{S}_K^{T_i(s)}\mathbf{S}_x^K + \mathbf{S}_{\omega_z}^{T_i(s)}\mathbf{S}_x^{\omega_z} + \mathbf{S}_{1/Q_z}^{T_i(s)}\mathbf{S}_x^{1/Q_z} + \mathbf{S}_{\omega_p}^{T_i(s)}\mathbf{S}_x^{\omega_p} + \mathbf{S}_{1/Q_p}^{T_i(s)}\mathbf{S}_x^{1/Q_p} \tag{4.128}$$

The second multiplier in each term of this sum depends on the stage realization method. In the first term $\mathbf{S}_K^{T_i(s)} = 1$, the first multipliers in other terms depend on Q-factors of zeros and poles. It is enough to consider the influence of the terms related to the poles. One can notice that $\mathbf{S}_{1/Q_p}^{T_i(s)}\mathbf{S}_x^{1/Q_p} = \mathbf{S}^{T_i(s)}\left[T_i(s), \frac{1}{Q_p}\right]\mathbf{S}_x\left(\frac{1}{Q_p}, x\right)$, then, calculating $\mathbf{S}^{T_i(s)}\left[T_i(s), \frac{1}{Q_p}\right]$ (the calculation of the semirelative sensitivity is done for convenience of graphic representation) and $\mathbf{S}_{\omega_p}^{T_i(s)}$, one obtains

$$\mathbf{S}^{T_i(s)}\left[T_i(s), \frac{1}{Q_p}\right] = \frac{1}{T_i(s)}\frac{\partial T_i(s)}{\partial\left(\frac{1}{Q_p}\right)} = -\frac{1}{\left(\frac{s}{\omega_p}\right) + \left(\frac{\omega_p}{s}\right) + \left(\frac{1}{Q_p}\right)} \tag{4.129}$$

and

$$\mathbf{S}_{\omega_p}^{T_i(s)} = \frac{\omega_p}{T_i(s)}\frac{\partial T_i(s)}{\partial\omega_p} = -\frac{\left(\frac{1}{Q_p}\right) + 2\left(\frac{\omega_p}{s}\right)}{\left(\frac{s}{\omega_p}\right) + \left(\frac{\omega_p}{s}\right) + \left(\frac{1}{Q_p}\right)} \tag{4.130}$$

Introducing the normalized frequency $\Omega = \left(\frac{\omega}{\omega_p} - \frac{\omega_p}{\omega}\right)$, one can find that

$$\text{Re}\,\mathbf{S}^{T_i(s)}\left[T_i(s), \frac{1}{Q_p}\right] = -\frac{\frac{1}{Q_p}}{\Omega^2 + \left(\frac{1}{Q_p}\right)^2} \tag{4.131}$$

$$\text{Im}\,\mathbf{S}^{T_i(s)}\left[T_i(s), \frac{1}{Q_p}\right] = -\frac{\Omega}{\Omega^2 + \left(\frac{1}{Q_p}\right)^2} \tag{4.132}$$

$$\text{Re}\,\mathbf{S}^{T_i(s)}_{\omega_p} = \frac{\Omega\sqrt{\Omega^2 + 4} - \Omega^2 - \left(\frac{1}{Q_p}\right)^2}{\Omega^2 + \left(\frac{1}{Q_p}\right)^2} \tag{4.133}$$

$$\text{Im}\,\mathbf{S}^{T_i(s)}_{\omega_p} = \frac{\left(\frac{1}{Q_p}\right)\sqrt{\Omega^2 + 4}}{\Omega^2 + \left(\frac{1}{Q_p}\right)^2} \tag{4.134}$$

Figure 4.4 shows the graphs of these four functions. They allow the following conclusions [4]. The functions reach high values in the vicinity of $\Omega = 0$, i.e., when $\omega \approx \omega_p$. This means that in the filter

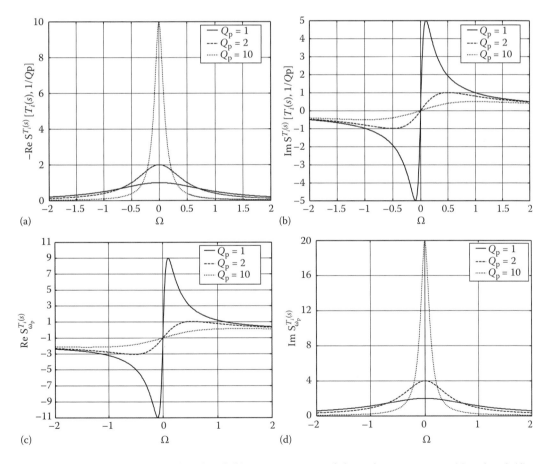

FIGURE 4.4 Stage sensitivities: (a) real and (b) imaginary parts of the Q-factor sensitivity; (c) real and (d) imaginary parts of the pole frequency sensitivity.

passband, especially if the poles and zeros are sufficiently divided (which is the condition of optimal cascading), one can neglect the contribution of zeros in modification of the transfer function $T_i(s)$. When Q_p becomes higher, this vicinity of $\omega \approx \omega_p$ with a rapid change of the sensitivity functions becomes relatively smaller in the case of $\operatorname{Re} \mathbf{S}^{T_i(s)}\left[T_i(s), \frac{1}{Q_p}\right]$ (Figure 4.4a) and $\operatorname{Im} \mathbf{S}^{T_i(s)}_{\omega_p}$ (Figure 4.4d) and not so small in the case of $\operatorname{Im} \mathbf{S}^{T_i(s)}\left[T_i(s), \frac{1}{Q_p}\right]$ (Figure 4.4b) and $\operatorname{Re} \mathbf{S}^{T_i(s)}_{\omega_p}$ (Figure 4.4c). Normally, the second multipliers in the terms of the sum in Equation 4.128 are all real; this means that the function $\operatorname{Re} \mathbf{S}^{T_i(s)}_{\omega_p}$ is the most important one in estimation of the sensitivity to variations of passive elements. In many realization methods [1,3] one obtains that $\omega_p \propto (R_1 R_2 C_1 C_2)^{-1/2}$. This implies that $S^{\omega_p}_{R_1} = S^{\omega_p}_{R_2} = S^{\omega_p}_{C_1} = S^{\omega_p}_{C_2} = -1/2$. Thus, finally, the function $\operatorname{Re} \mathbf{S}^{T_i(s)}_{\omega_p}$ (which can be called the main passive sensitivity term) will determine the maximum realizable Q_p for given tolerances of passive elements (or for the elements that are simulated as passive elements).

If a stage is realized using, for example, a state–space approach, it includes operational amplifiers. The stage is usually designed assuming ideal operational amplifiers, then the realization errors are analyzed considering that the operational amplifiers can be described by a model

$$A(s) = \frac{A_0}{1 + \left(\frac{s}{\omega_0}\right)} = \frac{\text{GBW}}{s + \omega_0} \tag{4.135}$$

where

A_0 is the dc gain, is the amplifier bandwidth
$\text{GBW} = A_0\omega_0$ is the gain-bandwidth product

If the stage transfer function is derived anew, with amplifiers described by the model equation (Equation 4.135), then $T_i(s)$ will no longer be a biquadratic. It will be a ratio of two higher degree polynomials, and the error analysis becomes very complicated [1]. To do an approximate analysis, one can pretend that the amplifier gain is simply a real constant A. Then, the transfer function $T_i(s)$ will preserve its biquadratic appearance, and the term $1/Q_p$ will be possible to represent as

$$\frac{1}{Q_p} = \frac{1}{Q} + \frac{k}{A} \tag{4.136}$$

The first term in Equation 4.136 is determined by the ratio of passive elements, the second term (k is the design constant) can be considered as an *absolute* change $\Delta(1/Q_p)$, which on the $j\omega$ axis becomes

$$\Delta\left(\frac{1}{Q_p}\right) \approx \frac{k}{A(j\omega)} = \frac{1}{A_0} + j\frac{k\omega}{\text{GBW}} \approx j\frac{k\omega}{\text{GBW}} \tag{4.137}$$

Then, in calculation of $\delta|T_i| = \operatorname{Re} \mathbf{S}^{T_i(s)}_K$, the function $\operatorname{Im} \mathbf{S}^{T_i(s)}_{\omega_p}$ becomes important (it can be called the main active sensitivity term) and the product $\left[\operatorname{Im} \mathbf{S}^{T_i(s)}_{\omega_p}\left(\frac{k\omega}{\text{GBW}}\right)\right]$ allows evaluation of the limitations on the Q-factor caused by the stage operational amplifiers.

The relationships given next are useful for any arbitrary realization, but pertain more to the cascade realization, where one can better control the pole's location when the circuit parameters are changing. If the filter transfer function is represented as

$$T(s) = \frac{a_m \Pi^m_{i=1}(s - z_i)}{d_n \Pi^n_{i=1}(s - p_i)} \tag{4.138}$$

then taking the logarithm of Equation 4.138 and its derivatives gives

$$\frac{dT(s)}{T(s)} = \frac{d(a_m/d_n)}{a_m/d_n} - \sum_{i=1}^{m} \frac{dz_i}{s - z_i} + \sum_{i=1}^{n} \frac{dp_i}{s - p_i} \tag{4.139}$$

Multiplying both sides of Equation 4.139 by x and expressing the differentials via partial derivatives one obtains

$$\mathbf{S}_x^{T(s)} = \mathbf{S}_x^{a_m/d_n} - \sum_{i=1}^{m} \frac{\mathbf{S}_x(z_i)}{s - z_i} + \sum_{i=1}^{n} \frac{\mathbf{S}_x(p_i)}{s - p_i} \tag{4.140}$$

In the vicinity of the pole $p_i = -\sigma_i + j\omega_i$, the sensitivity is determined by the term $\mathbf{S}_x(p_i)/(s - p_i)$. Besides $\mathbf{S}_x(p_i) = -\mathbf{S}_x(\sigma_i) + j\mathbf{S}_x(\omega_i)$ and on the $j\omega$ axis in this region one has

$$\frac{\mathbf{S}_x(p_i)}{j\omega - p_i} = -\frac{\mathbf{S}_x(\sigma_i)\sigma_i - \mathbf{S}_x(\omega_i)(\omega - \omega_i)}{\sigma_i^2 + (\omega - \omega_i)^2} + j\frac{\mathbf{S}_x(\omega_i)\sigma_i + \mathbf{S}_x(\sigma_i)(\omega - \omega_i)}{\sigma_i^2 + (\omega - \omega_i)^2} \tag{4.141}$$

Hence, when $\omega = \omega_i$

$$\mathbf{S}_x^{|T(j\omega)|} \approx -\frac{\mathbf{S}_x(\sigma_i)\sigma_i - \mathbf{S}_x(\omega_i)(\omega - \omega_i)}{\sigma_i^2 + (\omega - \omega_i)^2} \tag{4.142}$$

and

$$\mathbf{S}_x[\arg T(j\omega)] \approx \frac{\mathbf{S}_x(\omega_i)\sigma_i + \mathbf{S}_x(\sigma_i)(\omega - \omega_i)}{\sigma_i^2 + (\omega - \omega_i)^2} \tag{4.143}$$

Usually Equations 4.142 and 4.143 are considered at the point $\omega - \omega_i$, where $\mathbf{S}_x^{|T(j\omega)|} = -\mathbf{S}_x(\sigma_i)/\sigma_i$ and $\mathbf{S}_x[\arg T(j\omega)] = \mathbf{S}_x(\omega_i)/\sigma_i$. The frequent conclusion that follows is that the pole's movement toward the $j\omega$ axis is more dangerous (it introduces transfer function magnitude change) than the movement parallel to the $j\omega$ axis. But it is not difficult to see that in the immediate vicinity of this point, at $\omega = \omega_i \pm \sigma_i$, one has $\mathbf{S}_x^{|T(j\omega)|} = [-\mathbf{S}_x(\sigma_i) \pm \mathbf{S}_x(\omega_i)]/(2\sigma_i)$ and $\mathbf{S}_x[\arg T(j\omega)] = [\mathbf{S}_x(\omega_i) \pm j\mathbf{S}_x(\sigma_i)]/(2\sigma_i)$; i.e., one has to reduce both components of the pole movement. If care is taken to get $\mathbf{S}_x(\sigma_i) = 0$, then, indeed, $\mathbf{S}_x^{|T(j\omega)|} = 0$ at $\omega = \omega_i$, but at $\omega = \omega_i + \sigma_i$ (closer to the edge of the passband) $\mathbf{S}_x^{|T(j\omega)|} = [\mathbf{S}_x(\omega_i)/\sigma_i]$ and this can result in an essential $\delta|T(j\omega)|$.

Finally, some additional relationships between different sensitivities can be obtained from the definition of $Q_p = \omega_p/(2\sigma_i) = \left[\sqrt{\sigma_i^2 + \omega_i^2}\right]/(2\sigma_i) \approx \omega_i/(2\sigma_i)$. One can find that

$$\mathbf{S}_x(\omega_p) = \frac{\mathbf{S}_x(\sigma_i)}{2Q_p} + \sqrt{1 - \left(\frac{1}{4Q_p^2}\right)}\mathbf{S}_x(\omega_i) \tag{4.144}$$

and

$$\mathbf{S}_x(Q_p) = \frac{\sqrt{4Q_p^2 - 1}\mathbf{S}_x(\omega_i) - (4Q_p^2 - 1)\mathbf{S}_x(\sigma_i)}{\omega_p} \tag{4.145}$$

for semirelative sensitivities. From this basic definition of the Q-factor, one can also derive the relationships

$$\mathbf{S}_x^{Q_p} = \frac{\mathbf{S}_x^{\sigma_i}}{4Q_p^2} + \sqrt{1 - \left(\frac{1}{4Q_p^2}\right)} \mathbf{S}_x^{\omega_i} \tag{4.146}$$

and

$$\mathbf{S}_x^{Q_p} \approx \mathbf{S}_x^{\omega_i} - \mathbf{S}_x^{\sigma_i} \tag{4.147}$$

involving relative sensitivities. Another group of results can be obtained considering relative sensitivity of the pole $p_i = -\sigma_i + j\omega_i$. For example, one can find that

$$\mathbf{S}_x^{Q_p} = -\sqrt{4Q_p^2 - 1}\,\mathrm{Im}\,\mathbf{S}_x^{p_i} = \frac{x}{\omega_p^2}\left(\frac{\omega_i}{\sigma_i}\right)\left(\sigma_i \frac{\partial\omega_i}{\partial x} - \omega_i \frac{\partial\sigma_i}{\partial x}\right) \tag{4.148}$$

If $\dfrac{\partial\omega_i}{\partial x} = 0$ and $\dfrac{\partial\sigma_i}{\partial x} = \text{constant}$, then

$$\mathbf{S}_x^{Q_p} \approx kQ_p \tag{4.149}$$

which shows that in this case, $\mathbf{S}_x^{Q_p}$ increases proportionally to the Q-factor independently of the cause of this high sensitivity.

4.14 Simulation of Doubly Terminated Matched Lossless Filters

By cascading the first- and second-order sections (occasionally a third-order section is realized in odd-order filters instead of the cascade connection of a first- and a second-order section) any high-order transfer function $T(s)$ can be realized. In practice, however, the resulting circuit is difficult to fabricate for high-order and/or highly selective filters. The transfer function of such filters usually contains a pair of complex–conjugate poles very close to the $j\omega$ axis. The sensitivity of this section that realizes high-Q poles is high and the element tolerances for this section can be very tight. The section can be unacceptable for fabrication.

For filters that have such high-Q transfer function poles, other design techniques are often used. The most successful and widely used of these alternative strategies are based on simulating the low-sensitivity transfer function of a doubly-terminated lossless (reactance) two-port.

Assume that the two-port shown in Figure 4.5 is lossless and the transfer function $T(s) = V_2(s)/E(s)$ is realized. Considering power relations, one can show [12] that for steady-state sinusoidal operation the equation

$$|\rho(j\omega)|^2 + \frac{4R_1}{R_2}|T(j\omega)|^2 = 1 \tag{4.150}$$

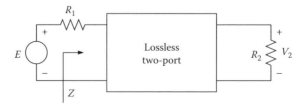

FIGURE 4.5   Lossless two-port with resistive loadings.

is valid for this circuit. Here, $\rho(j\omega) = [R_1 - Z(j\omega)]/[R_1 + Z(j\omega)]$ is the reflection coefficient and $Z(j\omega)$ is the input impedance of the loaded two-port. In many cases, the filter requirements are formulated for the transducer function:

$$H(s) = \sqrt{\frac{R_2}{4R_1}} \frac{E(s)}{V_2(s)} = \sqrt{\frac{R_2}{4R_1}} \frac{1}{T(s)} \tag{4.151}$$

For this function

$$\ln H(j\omega) = \alpha(\omega) + j\varphi(\omega) = -\ln|T(j\omega)| + \ln\left[\sqrt{(R_2/4R_1)}\right] - j\arg T(j\omega) \tag{4.152}$$

Here, $\alpha(\omega)$ is attenuation (it is different from the previously used only by the value of $\ln\left[\sqrt{(R_2/4R_1)}\right]$) and $\varphi(\omega) = -\beta(\omega) = -\arg T(j\omega)$ is phase. The impedance $Z(j\omega)$ satisfies the condition $\mathrm{Re}\, Z(j\omega) \geq 0$ (as for any passive circuit input impedance), which means that $|\rho(j\omega)|^2 < 1$. Then, as it follows from Equation 4.150 $|H(j\omega)|^2 \geq 1$ and $\alpha(\omega) \geq 0$.

When a filter is designed using $\alpha(\omega)$, the attenuation is optimized so that it is zero in one or more passband points (Figure 4.3 shows, for example, the attenuation characteristics with two zeros in the passband). Then the attenuation partial derivative with respect to the value of each of the two-port elements is equal to zero at the attenuation zeros. Indeed, let x_i be any element of the two-port and $\partial\alpha/\partial x_i$ be the partial derivative of $\alpha(\omega, x_i)$ with respect to that element. Suppose that x_i does not have its nominal value and differs from it by a small Δx_i variation. Expanding $\alpha(\omega, x_i)$ in the Taylor series one obtains that

$$\alpha(\omega, x_i + \Delta x_i) \cong \alpha(\omega, x_i) + \Delta x_i \frac{\partial\alpha(\omega, x_i)}{\partial x_i} \tag{4.153}$$

If ω_k is an attenuation zero, then $\alpha(\omega_k, x_i) = 0$; but, as was mentioned before, $\alpha(\omega) \geq 0$, and from Equation 4.153, one obtains that at the point $\omega = \omega_k$ one has

$$\Delta x_i \frac{\partial\alpha(\omega_k, x_i)}{\partial x_i} \geq 0 \tag{4.154}$$

Now, the variation Δx_i was of unspecified sign. Therefore, Equation 4.154 can only be satisfied with equality sign, which means that $\frac{\partial\alpha(\omega_k, x_i)}{\partial x_i} = 0$.

This result is called the *Fettweis-Orchard theorem* [4] and it explains why the preference is always given to the filter realized as a nondissipative two-port between resistive terminations or to the simulation of such a filter if the filter should be realized as an active circuit. First, considering the real parts of Equation 4.152, one obtains that

$$\frac{\partial\alpha(\omega, x_i)}{\partial x_i} = \mathbf{S}_{x_i}[\alpha(\omega)] = -\frac{x_i}{|T(j\omega)|} \frac{\partial|T(j\omega)|}{\partial x_i} = -\mathbf{S}_{x_i}^{|T(j\omega)|} \tag{4.155}$$

when x_i is any element inside the two-port. Hence, the points where $\alpha(\omega) = 0$ and, simultaneously, $\frac{\partial\alpha(\omega, x_i)}{\partial x_i} = 0$ are the points of zero sensitivity not only for attenuation but for transfer function magnitude as well (as a result of Equation 4.155 and this discussion, one cannot use at these points the relationships $d|T(j\omega)| \approx \mathbf{S}_{x_i}^{|T(j\omega)|} \delta x_i$; the relative change $\delta|T(j\omega)|$ is always negative and different from zero with δx_i of unspecified sign). Moreover, if α is small, $\partial\alpha/\partial x_i$ will also remain small [13], which means that the sensitivities are small in the whole passband. If $x_i = R_1$ or $x_i = R_2$, one obtains from Equation 4.152 that

$$R_1 \frac{\partial \alpha(\omega, R_1)}{\partial R_1} = -\mathbf{S}_{R_1}^{|T(j\omega)|} - \frac{1}{2} \tag{4.156}$$

and

$$R_2 \frac{\partial \alpha(\omega, R_2)}{\partial R_2} = -\mathbf{S}_{R_2}^{|T(j\omega)|} + \frac{1}{2} \tag{4.157}$$

The derivatives $\frac{\partial \alpha}{\partial R_i}$ $(i = 1, 2)$ are also zero at the points where $\alpha = 0$ and they are small when α remains small [13]. This means that in the passband $\mathbf{S}_{R_1}^{|T(j\omega)|} \approx -1/2$ and $\mathbf{S}_{R_2}^{|T(j\omega)|} \approx 1/2$. Thus, $|T(j\omega)|$ will share the zero sensitivity of α with respect to all the elements inside the two-port, but due to the terms $\pm 1/2$ in Equations 4.156 and 4.157 a change either in R_1 or R_2 will produce a frequency-independent shift (which can usually be tolerated) in $|T(j\omega)|$ in addition to the small effects proportional to $\frac{\partial \alpha}{\partial R_i}$. This is the basis of the low sensitivity of conventional *LC*-ladder filters and of those active, switched-capacitor, or digital filters that are based on *LC* filter model. This is valid with the condition that the transfer functions of the active filter and *LC* prototype are the same and the parameters of the two filters enter their respective transfer functions the same way.

The Fettweis-Orchard theorem explains why the filtering characteristics sought are those with the maximum number of attenuation zeros (for a given order of the transfer function $T(s)$). It also helps to understand why it is difficult to design a filter that simultaneously meets the requirements of α and $\varphi(\omega)$ (or to $|T(j\omega)|$ and $\beta(\omega)$); the degrees of freedom used for optimizing $\varphi(\omega)$ will not be available to attain the maximum number of attenuation zeros. It also explains why a cascade realization is more sensitive than the realization based on *LC* lossless model. Indeed, assume that, say, one of the characteristics of Figure 4.3 is realized by two cascaded sections (with the attenuation of each section shown by the dash-and-dotted line) with each actual section realized in doubly terminated matched lossless form. Each such section of the cascaded filter will be matched at one frequency and the sensitivities to the elements that are in unmatched sections will be different from zero. In addition, the attenuation ripple in each section is usually much larger than the total ripple, and the derivative $\frac{\partial \alpha}{\partial x_i}$, which is, in the first approximation, proportional to the attenuation ripple, will not be small. Indeed, practice shows [4] that there is, in fact, a substantial increase in sensitivity in the factored realization.

4.15 Sensitivity of Active *RC* Filters

The required component tolerances are very important factors determining the cost of filters. They are especially important with integrated realizations (where the tolerances are usually higher than in discrete technology). Also, the active filter realizations commonly require tighter tolerances than *LC* realizations. Yet two classes of active *RC* filters have tolerances comparable with those of passive *LC* filters. These are analog-computer and gyrator filters that simulate doubly terminated passive *LC* filters. The tolerance comparison [4] shows the tolerance advantages (sometimes by an order of magnitude) of the doubly terminated lossless structure as compared to any cascade realization. These are the only methods that are now used [14] for high-order high-Q sharp cutoff filters with tight tolerances. For less demanding requirements, cascaded realizations could be used. The main advantages that are put forth in this case are the ease of design and simplicity of tuning. But even here the tolerance comparison [4] shows that the stages have better tolerances if they are realized using gyrators and computer simulation methods.

4.16 Errors in Sensitivity Comparisons

In conclusion we briefly outline some common errors in sensitivity comparison. More detailed treatment can be found in Ref. [4].

1. Calculating the wrong sensitivities. The calculated sensitivities should have as close a relation to the filter specification as possible. In general, for a filter specified in the frequency domain, the sensitivities of amplitude and phase should be calculated along the $j\omega$ axis. Sensitivities of poles, zeros, Q's, resonant frequencies, etc. should be carefully interpreted in the context of their connection with amplitude and phase sensitivities.

2. Sensitivities of optimized designs. The optimization should use a criterion as closely related as possible to filter specifications. The use of a criterion that is not closely related to the filter specifications (e.g., pole sensitivity) can lead to valid conclusions if the filters being compared differ by an order of magnitude in sensitivity [4]. A sensitivity comparison is valid only if all the circuits have been optimized using the criterion on which they will be compared. The optimized circuit should not be compared with a nonoptimized one. Another error is to optimize one part of the transfer function (usually the denominator) and forgetting about the modifying effect of the numerator.

3. Comparing the incomparable. A frequent error occurs when comparing sensitivities with respect to different types of elements. In general, different types of elements can be realized with different tolerances, and the comparison is valid only if sensitivities are weighted proportionally. Besides, there are basic differences in variability between circuit parameters with physical dimensions and those without. The latter are often determined in the circuit as the ratio of dimensional quantities (as a result, the tolerance of the ratio will be about double the tolerances of the two-dimensional quantities determining them). In integrated technologies the dimensioned quantities usually have worse tolerances but better matching and tracking ability, especially with temperature. Hence, any conclusion involving sensitivities to different types of components, in addition, is technologically dependent.

4. Correlations between component values. The correlations between components are neglected when they are essential (this is usually done for simplification of the statistical analysis). From the other side, an unwarranted correlation is introduced when it does not exist. This is a frequent case where the realization involves cancellation of terms that are equal when the elements have their nominal values (e.g., a cancellation of a pole of one section of a filter by a zero of another section, cancellation of a positive conductance by a negative conductance).

5. Incomplete analysis. Very often, only sensitivities to variations of a single component (usually an amplifier gain) are considered. This is satisfactory only if it is the most critical component, which is seldom the case. Another form of incomplete analysis is to calculate only one coordinate of a complex sensitivity measure (S_x^Q is calculated while $S_x^{\omega_0}$ is ignored). Also, frequency-dependent sensitivities are calculated and compared at one discrete frequency instead of being calculated in frequency intervals.

6. First-order differential sensitivities are the most commonly calculated. But the fact is that $\frac{\partial y}{\partial x} = 0$ implies that the variation of y with x is quadratic at the point considered. A consequence of this is that zero sensitivities do not imply infinitely wide tolerances for the components in question. Similarly, infinite sensitivities do not imply infinitely narrow tolerances. Infinite values arise if the nominal value of y is zero, and the finite variations of x will almost always give finite variations of y.

References

1. L.P. Huelsman and P.E. Allen, *Introduction to the Theory and Design of Active Filters*, New York: McGraw-Hill, 1980.
2. K. Geher, *Theory of Network Tolerances*, Budapest, Hungary: Akademiai Kiado, 1971.
3. W.E. Heinlein and W.H. Holmes, *Active Filters for Integrated Circuits*, London: Prentice Hall, 1974.
4. M. Hasler and J. Neirynck, *Electric Filters*, Dedham, MA: Artech House, 1986.
5. S.K. Mitra, Ed., *Active Inductorless Filters*, New York: IEEE, 1971.

6. L.P. Huelsman, Ed., Active RC-filters: Theory and application, in *Benchmark Papers in Electrical Engineering and Computer Science*, vol. 15, Stroudsburg, PA: Dowden, Hutchinson and Ross, Inc., 1976.

7. A.F. Schwarz, *Computer-Aided Design of Microelectronic Circuits and Systems*, vol. 1. Orlando, FL: Academic, 1987.

8. M.L. Blostein, Some bounds on the sensitivity in RLC networks, in *Proc. 1st Allerton Conf. Circuits Syst. Theory*, Monticello, IL, November 15–17, 1963, pp. 488–501.

9. R.K. Brayton and R. Spence, Sensitivity and optimization, in *Computer-Aided Design of Electronic Circuits*, vol. 2, Amsterdam: Elsevier, 1980.

10. C. Desoer and E.S. Kuh, *Basic Circuit Theory*, New York: McGraw-Hill, 1969.

11. M. Biey and A. Premoli, *Cauer and MCPER Functions for Low-Q Filter Design*, St. Saphorin, Switzerland: Georgi, 1980.

12. N. Balabanian and T.A. Bickart, *Electrical Network Theory*, New York: Wiley, 1969.

13. H.J. Orchard, Loss sensitivities in singly and doubly terminated filters, *IEEE Trans. Circuits Syst.*, CAS-26, 293–297, 1979.

14. B. Nauta, *Analog CMOS Filters for Very High Frequencies*, Boston, MA: Kluwer Academic, 1993.

5

Passive Immittances and Positive-Real Functions

Wai-Kai Chen
University of Illinois at Chicago

In this chapter on passive filters, we deal with the design of one-port networks composed exclusively of passive elements such as resistors R, inductors L, capacitors C, and coupled inductors M. The one-ports are specified by their driving-point *immittances*, *impedances*, or *admittances*. Our basic problem is that given an immittance function, is it possible to find a one-port composed only of R, L, C, and M elements called the *RLCM one-port network* that realizes the given immittance function? This is known as the realizability problem, and its complete solution was first given by Brune [1].

Consider a linear RLCM one-port network of Figure 5.1 excited by a voltage source $V_1(s)$. For our purposes, we assume that there are b branches and the branch corresponding to the voltage source $V_1(s)$ is numbered branch 1 and all other branches are numbered from 2 to b. The Laplace transformed Kirchhoff current law equation can be written as

$$\mathbf{AI}(s) = \mathbf{0} \tag{5.1}$$

where
   $\mathbf{A}$ is the basis incidence matrix
   $\mathbf{I}(s)$ is the branch-current vector of the network

If $\mathbf{V}_n(s)$ is the nodal voltage vector, then the branch-voltage vector $\mathbf{V}(s)$ can be expressed in terms of $\mathbf{V}_n(s)$ by

$$\mathbf{V}(s) = \mathbf{A}'\mathbf{V}_n(s) \tag{5.2}$$

where the prime denotes the matrix transpose. Taking the complex conjugate of Equation 5.1 in conjunction with Equation 5.2 gives

$$\mathbf{V}'(s)\bar{\mathbf{I}}(s) = \mathbf{V}'_n(s)\mathbf{A}\bar{\mathbf{I}}(s) = \mathbf{V}'_n(s)\mathbf{0} = 0 \tag{5.3}$$

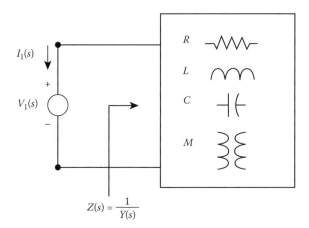

FIGURE 5.1   General linear *RLCM* one-port network.

or

$$\sum_{k=1}^{b} V_k(s)\bar{I}_k(s) = 0 \tag{5.4}$$

where $V_k(s)$ and $I_k(s)$ are the branch voltages and the branch currents, respectively.

From Figure 5.1, the driving-point impedance of the one-port is defined to be the ratio of $V_1(s)$ to $-I_1(s)$, or

$$Z(s) \equiv \frac{V_1(s)}{-I_1(s)} = \frac{V_1(s)\bar{I}_1(s)}{-I_1(s)\bar{I}_1(s)} = -\frac{V_1(s)\bar{I}_1(s)}{|I_1(s)|^2} \tag{5.5}$$

Equation 5.4 can be rewritten as

$$-V_1(s)\bar{I}_1(s) = \sum_{k=2}^{b} V_k(s)\bar{I}_k(s) \tag{5.6}$$

Substituting this in Equation 5.5 yields

$$Z(s) = \frac{1}{|I_1(s)|^2} \sum_{k=2}^{b} V_k(s)\bar{I}_k(s) \tag{5.7}$$

Likewise, the dual relation of the input admittance

$$Y(s) \equiv \frac{-I_1(s)}{V_1(s)} = \frac{1}{|V_1(s)|^2} \sum_{k=2}^{b} \bar{V}_k(s)I_k(s) \tag{5.8}$$

holds.

We know consider individual types of elements inside the one-port. For a resistive branch k of resistance R_k, we have

$$V_k(s) = R_k I_k(s) \tag{5.9}$$

For a capacitive branch of capacitance C,

$$V_k(s) = \frac{1}{sC_k} I_k(s) \tag{5.10}$$

Finally, for an inductive branch of self-inductance L_k and mutual inductances M_{kj},

$$V_k(s) = sL_k I_k(s) + \sum_{\text{all } j, j \neq k} sM_{kj} I_j(s) \tag{5.11}$$

Substituting these in Equation 5.7 and grouping the summation as sums of all resistors R, all capacitors C, and all inductors LM, we obtain

$$Z(s) = \frac{1}{|I_1(s)|^2} \left[\sum_R R_k |I_k(s)|^2 + \sum_C \frac{1}{sC_k} |I_k(s)|^2 + \sum_{LM} \left(sL_k |I_k(s)|^2 + \sum_{\text{all } j, j \neq k} sM_{kj} I_j(s) \bar{I}_k(s) \right) \right]$$

$$= \frac{1}{|I_1(s)|^2} \left[F_0(s) + \frac{1}{s} V_0(s) + sM_0(s) \right] \tag{5.12}$$

where

$$F_0(s) \equiv \sum_R R_k |I_k(s)|^2 \geq 0 \tag{5.13a}$$

$$V_0(s) \equiv \sum_C \frac{1}{C_k} |I_k(s)|^2 \geq 0 \tag{5.13b}$$

$$M_0(s) \equiv \sum_{LM} \left(L_k |I_k(s)|^2 + \sum_{\text{all } j, j \neq k} M_{kj} I_j(s) \bar{I}_k(s) \right) \tag{5.13c}$$

These quantities are closely related to the average power and stored energies of the one-port under steady-state sinusoidal conditions. The average power dissipated in the resistors is

$$P_{\text{ave}} = \frac{1}{2} \sum_R R_k |I_k(j\omega)|^2 = \frac{1}{2} F_0(j\omega) \tag{5.14}$$

showing that $F_0(j\omega)$ represents twice the average power dissipated in the resistors of the one-port. The average electric energy stored in the capacitors is

$$E_C = \frac{1}{4\omega^2} \sum_C \frac{1}{C_k} |I_k(j\omega)|^2 = \frac{1}{4\omega^2} V_0(j\omega) \tag{5.15}$$

Thus, $V_0(j\omega)$ denotes $4\omega^2$ times the average electric energy stored in the capacitors. Similarly, the average magnetic energy stored in the inductors is

$$E_M = \frac{1}{4} \sum_{LM} \left[L_k |I_k(j\omega)|^2 + \sum_{\text{all } q, q \neq k} M_{kq} I_q(j\omega) \bar{I}_k(j\omega) \right] = \frac{1}{4} M_0(j\omega) \tag{5.16}$$

indicating that $M_0(j\omega)$ represents four times the average magnetic energy stored in the inductors.

Therefore, all the three quantities $F_0(j\omega)$, $V_0(j\omega)$, and $M_0(j\omega)$ are real and nonnegative, and Equation 5.12 can be rewritten as

$$Z(s) = \frac{1}{|I_1(s)|^2} \left(F_0 + \frac{1}{s} V_0 + sM_0 \right) \tag{5.17}$$

Likewise, the dual result for $Y(s)$ is found to be

$$Y(s) = \frac{1}{|V_1(s)|^2} \left(F_0 + \frac{1}{s} V_0 + \bar{s} M_0 \right) \tag{5.18}$$

Now, we set $s = \sigma + j\omega$ and compute the real part and imaginary part of $Z(s)$ and obtain

$$\text{Re } Z(s) = \frac{1}{|I_1(s)|^2} \left(F_0 + \frac{\sigma}{\sigma^2 + \omega^2} V_0 + \sigma M_0 \right) \tag{5.19}$$

$$\text{Im } Z(s) = \frac{\omega}{|I_1(s)|^2} \left(M_0 - \frac{1}{\sigma^2 + \omega^2} V_0 \right) \tag{5.20}$$

where

Re stands for "real part of"

Im for "imaginary part of"

These equations are valid irrespective of the value of s, except at the zeros of $I_1(s)$. They are extremely important in that many analytic properties of passive impedances can be obtained from them. The following is one of such consequences:

THEOREM 5.1

If $Z(s)$ is the driving-point impedance of a linear, passive, lumped, reciprocal, and time-invariant one-port network N, then

1. *Whenever $\sigma \geq 0$, Re $Z(s) \geq 0$.*
2. *If N contains no resistors, then*
   $\sigma > 0$ implies Re $Z(s) > 0$
   $\sigma = 0$ implies Re $Z(s) = 0$
   $\sigma < 0$ implies Re $Z(s) < 0$
3. *If N contains no capacitors, then*
   $\omega > 0$ implies Im $Z(s) > 0$
   $\omega = 0$ implies Im $Z(s) = 0$
   $\omega < 0$ implies Im $Z(s) < 0$
4. *If N contains no self- and mutual-inductors, then*
   $\omega > 0$ implies Im $Z(s) < 0$
   $\omega = 0$ implies Im $Z(s) = 0$
   $\omega < 0$ implies Im $Z(s) > 0$

Similar results can be stated for the admittance function $Y(s)$ simply by replacing $Z(s)$, Re $Z(s)$, and Im $Z(s)$ by $Y(s)$, Re $Y(s)$, and Im $Y(s)$, respectively.

The theorem states that the driving-point impedance $Z(s)$ of a passive LMC, RLM, or RC one-port network maps different regions of the complex-frequency s-plane into various regions of the Z-plane. Now, we assert that the driving-point immittance of a passive one-port is a positive-real function, and every positive-real function can be realized as the input immittance of an RLCM one-port network.

Definition 5.1: *Positive-real function. A positive-real function F(s)*, abbreviated as a PR function, is an analytic function of the complex variable $s = \sigma + j\omega$ satisfying the following three conditions:

1. $F(s)$ is analytic in the open RHS (*right-half* of the *s*-plane), i.e., $\sigma > 0$.
2. $F(\bar{s}) = \overline{F}(s)$ for all *s* in the open RHS.
3. Re $F(s) \geq 0$ whenever Re $s \geq 0$.

The concept of a positive-real function, as well as many of its properties, is credited to Otto Brune [1]. Our objective is to show that positive realness is a necessary and sufficient condition for a passive one-port immittance. The above definition holds for both rational and transcendental functions. A rational function is defined as a ratio of two polynomials. Network functions associated with any linear lumped system, with which we deal exclusively in this section, are rational. In the case of rational functions, not all three conditions in the definition are independent. For example, the analyticity requirement is implied by the other two. The second condition is equivalent to stating the $F(s)$ is real when *s* is real, and for a rational $F(s)$ it is always satisfied if all the coefficients of the polynomial are real.

Some important properties of a positive-real function can be stated as follows:

1. If $F_1(s)$ and $F_2(s)$ are positive real, so is $F_1[F_2(s)]$.
2. If $F(s)$ is positive real, so are $1/F(s)$ and $F(1/s)$.
3. A positive-real function is devoid of poles and zeros in the open RHS.
4. If a positive-real function has any poles or zeros on the $j\omega$-axis (0 and ∞ included), such poles and zeros must be simple. At a simple pole on the $j\omega$-axis, the residue is real positive.

Property 1 states that a positive-real function of a positive-real function is itself positive real, and property 2 shows that the reciprocal of a positive-real function is positive real. The real significance of the positive-real functions is its use in the characterization of the passive one-port immittances. This characterization is one of the most penetrating results in network theory, and is stated as

THEOREM 5.2

A real rational function is the driving-point immittance of a linear, passive, lumped, reciprocal, and time-invariant one-port network if and only if it is positive real.

The necessity of the theorem follows directly from Equation 5.19. The sufficiency was first established by Brune in 1930 by showing that any given positive-real rational function can be realized as the input immittance of a passive one-port network using only the passive elements such as resistors, capacitors, and self- and mutual inductors. A formal constructive proof will be presented in the following section.

Example 5.1

Consider the passive one-port of Figure 5.2, the driving-point impedance of which is found to be

$$Z(s) = \frac{3s^2 + s + 2}{2s^2 + s + 3} \tag{5.21}$$

To verify that the function $Z(s)$ is positive real, we compute its real part by substituting $s = \sigma + j\omega$ and obtain

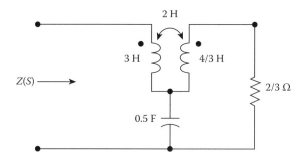

FIGURE 5.2   A passive one-port network.

$$\text{Re } Z(s) = \text{Re } Z(\sigma + j\omega) = \text{Re}\frac{3(\sigma + j\omega)^2 + (\sigma + j\omega) + 2}{2(\sigma + j\omega)^2 + (\sigma + j\omega) + 3}$$

$$= \frac{6(\omega^2 - 1)^2 + (12\omega^2\sigma + 5\omega^2 + 6\sigma^3 + 5\sigma^2 + 14\sigma + 5)\sigma}{(2\sigma^2 - 2\omega^2 + \sigma + 3)^2 + \omega^2(4\sigma + 1)^2} \geq 0, \quad \sigma \geq 0 \qquad (5.22)$$

This, in conjunction with the facts that $Z(s)$ is analytic in the open RHS and that all the coefficients of $Z(s)$ are real, shows that $Z(s)$ is positive real.

Observe that if the function $Z(s)$ is of high order, the task of ascertaining its positive realness is difficult if condition 3 of Definition 5.1 is employed for checking. Hence, it is desirable to have alternate but much simpler conditions for testing. For this reason, we introduce the following equivalent conditions that are relatively easy to apply:

THEOREM 5.3

A rational function $F(s)$ is positive real if and only if the following conditions are satisfied:

1. *$F(s)$ is real when s is real.*
2. *$F(s)$ has no poles in the open RHS.*
3. *Poles of $F(s)$ on the $j\omega$-axis, if they exist, are simple, and residues evaluated at these poles are real and positive.*
4. *Re $F(j\omega) \geq 0$ for all ω, except at the poles.*

PROOF: From the definition of a PR function, we see immediately that all the conditions are necessary. To prove sufficiency, we expand $F(s)$ in a partial fraction as

$$F(s) = \left[k_\infty s + \frac{k_0}{s} + \sum_x \left(\frac{k_x}{s + j\omega_x} + \frac{k_x}{s - j\omega_x} \right) \right] + F_1(s)$$

$$= \left(k_\infty s + \frac{k_0}{s} + \sum_x \frac{2k_x s}{s^2 + \omega_x^2} \right) + F_1(s) \qquad (5.23)$$

where k_∞, k_0, and k_x are residues evaluated at the $j\omega$-axis poles $j\infty$, 0, and $\pm j\omega_x$, respectively, and are real and positive. $F_1(s)$ is the function formed by the terms corresponding to the open LHS (left-half of the s-plane) poles of $F(s)$, and therefore is analytic in the RHS and the entire $j\omega$-axis including the point at infinity. For such a function, the minimum value of the real part throughout the region where the function is analytic lies on the boundary, namely, the $j\omega$-axis. (See, for example, Churchill [2]. This shows that the minimum value of Re $F_1(s)$ for all Re $s \geq 0$ occurs on the $j\omega$-axis; but according to Equation 5.23, this value is nonnegative:

$$\text{Re } F_1(j\omega) = \text{Re } F(j\omega) \geq 0 \tag{5.24}$$

Thus, the real part of $F_1(s)$ is nonnegative everywhere in the closed RHS or

$$\text{Re } F_1(s) \geq 0 \quad \text{for Re } s \geq 0 \tag{5.25}$$

This, together with the fact that $F_1(s)$ is real whenever s is real, shows that $F_1(s)$ is positive real.

Since each term inside the parentheses of Equation 5.23 is positive real, and since the sum of two or more positive-real functions is positive real, $F(s)$ is positive real. This completes the proof of the theorem.

In testing for positive realness, we may eliminate some functions from consideration by inspection because they violate certain simple necessary conditions. For example, a function cannot be PR if it has a pole or zero in the open RHS. Another simple test is that the highest powers of s in numerator and denominator not differ by more than unity, because a PR function can have at most a simple pole or zero at the origin or infinity, both of which lie on the $j\omega$-axis.

A Hurwitz polynomial is a polynomial devoid of zeros in the open RHS. Thus, it may have zeros on the $j\omega$-axis. To distinguish such a polynomial from the one that has zeros neither in the open RHS nor on the $j\omega$-axis, the latter is referred to as a strictly Hurwitz polynomial. For computational purposes, Theorem 5.3 can be reformulated and put in a much more convenient form.

THEOREM 5.4

A rational function represented in the form

$$F(s) = \frac{P(s)}{Q(s)} = \frac{m_1(s) + n_1(s)}{m_2(s) + n_2(s)} \tag{5.26}$$

where $m_1(s)$, $m_2(s)$, and $n_1(s)$, $n_2(s)$ are the even and odd parts of the polynomials $P(s)$ and $Q(s)$, respectively, is positive real if and only if the following conditions are satisfied:

1. *$F(s)$ is real when s is real.*
2. *$P(s) + Q(s)$ is strictly Hurwitz.*
3. *$m_1(j\omega)m_2(j\omega) - n_1(j\omega)n_2(j\omega) \geq 0$ for all ω.*

A real polynomial is strictly Hurwitz if and only if the continued-fraction expansion of the ratio of the even part to the odd part or the odd part to the even part of the polynomial yields only real and positive coefficients, and does not terminate prematurely. For $P(s) + Q(s)$ to be strictly Hurwitz, it is necessary and sufficient that the continued-fraction expansion

$$\left[\frac{m_1(s) + m_2(s)}{n_1(s) + n_2(s)} \right]^{\pm 1} = \alpha_1 s + \cfrac{1}{\alpha_2 s + \cfrac{1}{\ddots + \cfrac{1}{\alpha_k s}}} \tag{5.27}$$

yields only real and positive α's, and does not terminate prematurely, i.e., k must equal the degree $m_1(s) + m_2(s)$ or $n_1(s) + n_2(s)$, whichever is larger. It can be shown that the third condition of the theorem is satisfied if and only if its left-hand-side polynomial does not have real positive roots of odd multiplicity. This may be determined by factoring it or by the use of the Sturm's theorem, which can be found in most texts on elementary theory of equations. We illustrate the above procedure by the following examples.

Example 5.2

Test the following function to see if it is PR:

$$F(s) = \frac{2s^4 + 4s^3 + 5s^2 + 5s + 2}{s^3 + s^2 + s + 1} \tag{5.28}$$

For illustrative purposes, we follow the three steps outlined in the theorem, as follows:

$$F(s) = \frac{2s^4 + 4s^3 + 5s^2 + 5s + 2}{s^3 + s^2 + s + 1} = \frac{P(s)}{Q(s)} = \frac{m_1(s) + n_1(s)}{m_2(s) + n_2(s)} \tag{5.29}$$

where

$$m_1(s) = 2s^4 + 5s^2 + 2, \quad n_1(s) = 4s^3 + 5s \tag{5.30a}$$

$$m_2(s) = s^2 + 1, \quad n_2(s) = s^3 + s \tag{5.30b}$$

Condition 1 is clearly satisfied. To test condition 2, we perform the Hurwitz test, which gives

$$\frac{m_1(s) + m_2(s)}{n_1(s) + n_2(s)} = \frac{2s^4 + 6s^2 + 3}{5s^3 + 6s} = \frac{2}{5}s + \cfrac{1}{\frac{25}{18}s + \cfrac{1}{\frac{324}{165}s + \cfrac{1}{\frac{33}{54}s}}} \tag{5.31}$$

Since all the coefficients are real and positive and since the continued-fraction expansion does not terminate prematurely, the polynomial $P(s) + Q(s)$ is strictly Hurwitz. Thus, condition 2 is satisfied.

To test condition 3, we compute

$$m_1(j\omega)m_2(j\omega) - n_1(j\omega)n_2(j\omega) = 2\omega^6 - 2\omega^4 - 2\omega^2 + 2$$

$$= 2(\omega^2 + 1)(\omega^2 - 1)^2 \geq 0 \tag{5.32}$$

which is nonnegative for all ω, or, equivalently, which does not possess any real positive roots of odd multiplicity. Therefore $F(s)$ is positive real.

References

1. O. Brune, Synthesis of a finite two-terminal network whose driving-point impedance is a prescribed function of frequency, *J. Math. Phys.*, 10, 191–236, 1931.
2. R.V. Churchill, *Introduction to Complex Variables and Applications*, New York: McGraw-Hill, 1960.

6

Passive Cascade Synthesis

Wai-Kai Chen
University of Illinois at Chicago

6.1 Introduction

In this chapter, we demonstrate that any rational positive-real function can be realized as the input immittance of a passive one-port network terminated in a resistor, thereby also proving the sufficiency of Theorem 5.2.

Consider the even part

$$\text{Ev } Z(s) = r(s) = \frac{1}{2}[Z(s) + Z(-s)] \tag{6.1}$$

of a given rational positive-real impedance $Z(s)$. As in Equation 5.26, we first separate the numerator and denominator polynomials of $Z(s)$ into even and odd parts, and write

$$Z(s) = \frac{m_1 + n_1}{m_2 + n_2} \tag{6.2}$$

Then, we have

$$r(s) = \frac{m_1 m_2 - n_1 n_2}{m_2^2 + n_2^2} \tag{6.3}$$

showing that if s_0 is a zero or pole of $r(s)$, so is $-s_0$. Thus, the zeros and poles of $r(s)$ possess quadrantal symmetry with respect to both the real and imaginary axes. They may appear in pairs on the real axis, in pairs on the $j\omega$-axis, or in the form of sets of quadruplets in the complex-frequency plane. Furthermore, for a positive-real $Z(s)$, the $j\omega$-axis zeros of $r(j\omega)$ are required to be of even multiplicity in order that Re Z $(j\omega) = r(j\omega)$ never be negative.

Suppose that we can extract from $Z(s)$ a set of open-circuit impedance parameters $z_{ij}(s)$ characterizing a component two-port network, as depicted in Figure 6.1, which produces one pair of real axis zeros, one pair of $j\omega$-axis zeros, or one set of quadruplet of zeros of $r(s)$, and leaves a rational positive-real impedance $Z_1(s)$ of lower degree, the even part of which $r_1(s)$ is devoid of these zeros but contains all

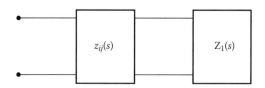

FIGURE 6.1 Two-port network terminated in $Z_1(s)$.

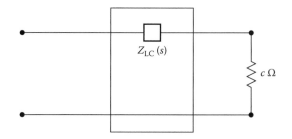

FIGURE 6.2 Two-port network terminated in a resistor.

other zeros of $r(s)$. After a finite q steps, we arrive at a rational positive-real impedance $Z_q(s)$, the even part $r_q(s)$ of which is devoid of zeros in the entire complex-frequency plane, meaning that its even part must be a nonnegative constant c:

$$r_q(s) = \frac{1}{2}\left[Z_q(s) + Z_q(-s)\right] = c \tag{6.4}$$

Therefore, $Z_q(s)$ is expressible as the sum of a reactance function* $Z_{LC}(s)$ and a resistance c:

$$Z_q(s) = Z_{LC}(s) + c \tag{6.5}$$

which can be realized as the input impedance of a lossless two-port network terminated in a c-ohm resistor, as shown in Figure 6.2.

To motivate our discussion, we first present a theorem credited to Richards [1,2], which is intimately tied up with the famous Bott–Duffin technique [3].

THEOREM 6.1

Let $Z(s)$ be a positive-real function that is neither of the form Ls nor $1/Cs$. Let k be an arbitrary positive-real constant. Then, the Richards function

$$W(s) = \frac{kZ(s) - sZ(k)}{kZ(k) - sZ(s)} \tag{6.6}$$

is also positive real.

The degree of a rational function is defined as the sum of the degrees of its relatively prime numerator and denominator polynomials. Thus, the Richards function $W(s)$ is also rational, the degree of which is

---

* A formal definition will be given in Chapter 7.

not greater than that of $Z(s)$. It was first pointed out by Richards that if k can be chosen so that the even part of $Z(s)$ vanishes at k, then the degree of $W(s)$ is at least two less than that of $Z(s)$. Let

$$s_0 = \sigma_0 + j\omega_0 \tag{6.7}$$

be a point in the closed RHS. Then, according to the preceding theorem, the function

$$\widehat{W}_1(s) = \frac{s_0 Z(s) - s Z(s_0)}{s_0 Z(s_0) - s Z(s)} \tag{6.8}$$

is positive real if s_0 is positive real; and the function

$$W_1(s) = Z(s_0)\widehat{W}_1(\bar{s}_0)\frac{\bar{s}_0 \widehat{W}_1(s) - s\widehat{W}_1(\bar{s}_0)}{\bar{s}_0 \widehat{W}_1(\bar{s}_0) - s\widehat{W}_1(s)} \tag{6.9}$$

is positive real if s_0 is a positive-real constant and $\widehat{W}_1(s)$ is a positive-real function. Substituting Equation 6.8 in Equation 6.9 yields

$$W_1(s) = \frac{D_1(s)Z(s) - B_1(s)}{-C_1(s)Z(s) + A_1(s)} \tag{6.10}$$

where

$$A_1(s) = q_4 s^2 + |s_0|^2 \tag{6.11a}$$

$$B_1(s) = q_2 s \tag{6.11b}$$

$$C_1(s) = q_3 s \tag{6.11c}$$

$$D_1(s) = q_1 s^2 + |s_0|^2 \tag{6.11d}$$

$$q_1 = \frac{R_0/\sigma_0 - X_0/\omega_0}{R_0/\sigma_0 + X_0/\omega_0} \tag{6.12a}$$

$$q_2 = \frac{2|Z_0|^2}{R_0/\sigma_0 + X_0/\omega_0} \tag{6.12b}$$

$$q_3 = \frac{2}{R_0/\sigma_0 - X_0/\omega_0} \tag{6.12c}$$

$$q_4 = \frac{R_0/\sigma_0 + X_0/\omega_0}{R_0/\sigma_0 - X_0/\omega_0} = \frac{1}{q_1} \tag{6.12d}$$

in which

$$Z(s_0) = R_0 + jX_0 \equiv Z_0 \tag{6.13}$$

In the case $\omega_0 = 0$, then X_0/ω_0 must be replaced by $Z'(\sigma_0)$:

$$\frac{X_0}{\omega_0} \rightarrow Z'(\sigma_0) = \left.\frac{dZ(s)}{ds}\right|_{s=\sigma_0} \tag{6.14a}$$

For $\sigma = 0$ and $R_0 = 0$, R_0/σ_0 is replaced by $X'(\omega_0)$:

$$\frac{R_0}{\sigma_0} \rightarrow X'(\omega_0) = \left.\frac{dZ(s)}{ds}\right|_{s=j\omega_0} \tag{6.14b}$$

Definition 6.1:   *Index set*

For a given positive-real function $Z(s)$, let s_0 be any point in the open RHS or any finite nonzero point on the $j\omega$-axis where $Z(s)$ is analytic. Then, the set of four real numbers q_1, q_2, q_3, and q_4, as defined in Equations 6.12 through 6.14, is called the index set assigned to the point s_0 by the positive-real function $Z(s)$.

We illustrate this concept by the following example.

Example 6.1

Determine the index set assigned to the point $s_0 = 0.4551 + j1.099$ by the positive-real function

$$Z(s) = \frac{s^2 + s + 1}{s^2 + s + 2} \tag{6.15}$$

From definition and Equation 6.15, we have

$$s_0 = 0.4551 + j1.099 = \sigma_0 + j\omega_0 \tag{6.16a}$$

$$Z(s_0) = Z(0.4551 + j1.099)$$

$$= 0.7770 + j0.3218 = 0.8410e^{j22.5°}$$

$$= R_0 + jX_0 \equiv Z_0 \tag{6.16b}$$

$$|Z_0|^2 = 0.7073 \tag{6.17}$$

obtaining from Equation 6.12

$$q_1 = 0.707, \quad q_2 = 0.707, \quad q_3 = 1.414, \quad q_4 = 1.414 \tag{6.18}$$

With these preliminaries, we now state the following theorem, which forms the cornerstone of the method of cascade synthesis of a rational positive-real impedance according to the Darlington theory [4].

THEOREM 6.2

Let $Z(s)$ be a positive-real function, which is neither of the form Ls nor $1/Cs$, L and C being real nonnegative constants. Let $s_0 = \sigma_0 + j\omega_0$ be a finite nonzero point in the closed RHS where $Z(s)$ is analytic, then the function

$$W_1(s) = \frac{D_1(s)Z(s) - B_1(s)}{-C_1(s)Z(s) + A_1(s)} \tag{6.19}$$

is positive real; where A_1, B_1, C_1, and D_1 are defined in Equation 6.11 and $\{q_1, q_2, q_3, q_4\}$ is the index set assigned to the point s_0 by $Z(s)$. Furthermore, $W_1(s)$ possesses the following attributes:

(1) If $Z(s)$ is rational, $W_1(s)$ is rational, the degree of which is not greater than that of $Z(s)$, or

$$\text{degree } W_1(s) \le \text{degree } Z(s) \tag{6.20}$$

(2) If $Z(s)$ rational and if s_0 is a zero of its even part $r(s)$, then

$$\text{degree } W_1(s) \le \text{degree } Z(s) - 4, \quad \omega_0 \ne 0 \tag{6.21a}$$

$$\text{degree } W_1(s) \leq \text{degree } Z(s) - 2, \quad \omega_0 = 0 \tag{6.21b}$$

(3) *If $s_0 = \sigma_0 > 0$ is a real zero of $r(s)$ of at least multiplicity 2 and if $Z(s)$ is rational, then*

$$\text{degree } W_1(s) \leq \text{degree } Z(s) - 4 \tag{6.22}$$

We remark that since $Z(s)$ is positive real, all the points in the open RHS are admissible. Any point on the $j\omega$-axis, exclusive of the origin and infinity, where $Z(s)$ is analytic is admissible as s_0.

We are now in a position to show that any positive-real function can be realized as the input impedance of a lossless one-port network terminated in a resistor. Our starting point is Equation 6.19, which after solving $Z(s)$ in terms of $W_1(s)$ yields

$$Z(s) = \frac{A_1(s)W_1(s) + B_1(s)}{C_1(s)W_1(s) + D_1(s)} \tag{6.23}$$

It can be shown that $Z(s)$ can be realized as the input impedance of a two-port network N_1, which is characterized by its transmission matrix

$$\mathbf{T}_1(s) = \begin{bmatrix} A_1(s) & B_1(s) \\ C_1(s) & D_1(s) \end{bmatrix} \tag{6.24}$$

terminated in $W_1(s)$, as depicted in Figure 6.3. To see this, we first compute the corresponding impedance matrix $\mathbf{Z}_1(s)$ of N_1 from $\mathbf{T}_1(s)$ and obtain

$$\mathbf{Z}_1(s) = \begin{bmatrix} z_{11}(s) & z_{12}(s) \\ z_{21}(s) & z_{22}(s) \end{bmatrix} = \frac{1}{C_1(s)} \begin{bmatrix} A_1(s) & A_1(s)D_1(s) - B_1(s)C_1(s) \\ 1 & D_1(s) \end{bmatrix} \tag{6.25}$$

The input impedance $Z_{11}(s)$ of N_1 with the output port terminating in $W_1(s)$ is found to be

$$Z_{11}(s) = z_{11}(s) - \frac{z_{12}(s)z_{21}(s)}{z_{22}(s) + W_1(s)} = \frac{A_1(s)W_1(s) + B_1(s)}{C_1(s)W_1(s) + D_1(s)} = Z(s) \tag{6.26}$$

The determinant of the transmission matrix $\mathbf{T}_1(s)$ is computed as

$$\det \mathbf{T}_1(s) = A_1(s)D_1(s) - B_1(s)C_1(s) = s^4 + 2\left(\omega_0^2 - \sigma_0^2\right)s^2 + |s_0|^4 \tag{6.27}$$

Observe that $\det \mathbf{T}_1(s)$ depends only upon the point s_0 and not on $Z(s)$, and that the input impedance $Z_{11}(s)$ remains unaltered if each element of $\mathbf{T}_1(s)$ is multiplied or divided by a nonzero

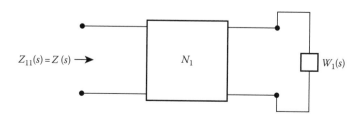

FIGURE 6.3   Two-port network N_1 terminated in the impedance $W_1(s)$.

finite quality. To complete the realization, we must now demonstrate that the two-port network N_1 is physically realizable.

6.2 Type-E Section

Consider the lossless nonreciprocal two-port network of Figure 6.4 known as the type-E section. Our objective is to show that this two-port realizes N_1. To this end, we first compute its impedance matrix $\mathbf{Z}_E(s)$ as

$$\mathbf{Z}_E(s) = \begin{bmatrix} L_1(s) + 1/Cs & Ms + 1/Cs + \zeta \\ Ms + 1/Cs - \zeta & L_2s + 1/Cs \end{bmatrix} \tag{6.28}$$

where $M^2 = L_1 L_2$, the determinant of which is given by

$$\det \mathbf{Z}_E(s) = \frac{L_1 + L_2 - 2M + \zeta^2 C}{C} \tag{6.29}$$

a constant independent of s due to perfect coupling. From the impedance matrix $\mathbf{Z}_E(s)$, its corresponding transmission matrix $\mathbf{T}_E(s)$ is found to be

$$\mathbf{T}_E(s) = \frac{1}{MCs^2 - \zeta Cs + 1} \begin{bmatrix} L_1 Cs^2 + 1 & (L_1 + L_2 - 2M + \zeta^2 C)s \\ Cs & L_2 Cs^2 + 1 \end{bmatrix} \tag{6.30}$$

To show that the type-E section realizes N_1, we divide each element of $\mathbf{T}_1(s)$ of Equation 6.24 by $|s_0|^2$ $(MCs^2 - \zeta Cs + 1)$. This manipulation will not affect the input impedance $Z(s)$ but it will result in a transmission matrix having the form of $\mathbf{T}_E(s)$. Comparing this new matrix with Equation 6.30 in conjunction with Equation 6.11 yields the following identifications:

$$L_1 C = \frac{q_4}{|s_0|^2} \tag{6.31a}$$

$$L_1 + L_2 - 2M + \zeta^2 C = \frac{q_2}{|s_0|^2} \tag{6.31b}$$

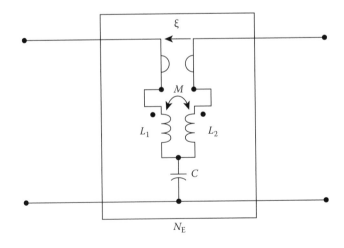

FIGURE 6.4 Type-E section.

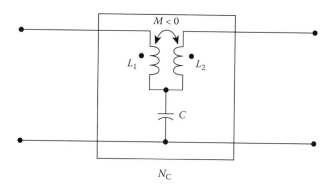

FIGURE 6.5 Darlington type-C section.

$$C = \frac{q_3}{|s_0|^2} \tag{6.31c}$$

$$L_2 C = \frac{q_1}{|s_0|^2} \tag{6.31d}$$

Solving these for the element values of the type-E section, we obtain

$$L_1 = \frac{q_4}{q_3} = \frac{1}{q_1 q_3}, \quad L_2 = \frac{q_1}{q_3} \tag{6.32a}$$

$$C = \frac{q_3}{|s_0|^2}, \quad M = \frac{1}{q_3}, \quad \zeta = \pm \frac{2\sigma_0}{q_3} \tag{6.32b}$$

Since the elements of the index set assigned to the point s_0 by $Z(s)$ are positive and finited for all admissible points s_0 in the closed RHS except at those admissible points $s_0 = j\omega_0$ on the $j\omega$-axis where $R_0 \neq 0$, all the elements in Equation 6.32 are physical. But at those admissible points $s_0 = j\omega_0$ where $R_0 \neq 0$, R_0/σ_0 becomes infinity and $q_1 = q_4 = 1$ and $q_2 = q_3 = 0$. Under this situation, $W_1(s) = Z(s)$ and the corresponding two-port network N_1 degenerates into a pair of wires.

Appealing to Theorem 6.2 shows that if s_0 is chosen to be a complex open RHS zero of the even part $r(s)$ of $Z(s)$, the type-E section is capable of extracting a set of quadrantal zeros of $r(s)$ and leads to at least a four-degree reduction. For zeros of $r(s)$ on the $j\omega$-axis or the σ-axis, the type-E section degenerates into other types of sections, as follows:

Case 1. $s_0 = \sigma_0 > 0$. Then, we have $\zeta = 0$ and

$$L_1 = \frac{q_4}{q_3} = \frac{1}{q_1 q_3}, \quad L_2 = \frac{q_1}{q_3}, \quad C = \frac{q_3}{\sigma_0^2}, \quad M = -\frac{1}{q_3} < 0 \tag{6.33}$$

The type-E section degenerates to the Darlington type-C section, as shown in Figure 6.5.

Case 2. $s_0 = j\omega_0$ and $R_0 = 0$. In this case, we replace R_0/σ_0 by $X'(\omega_0)$ and the gyrator in the type-E section can be avoided because $\zeta = 0$. The type-E section degenerates into the Brune section of Figure 6.6, the element values of which are given by

$$L_1 = \frac{q_4}{q_3} = \frac{1}{q_1 q_3} = \frac{\omega_0 X'(\omega_0) + X_0}{2\omega_0} \tag{6.34a}$$

$$L_2 = \frac{q_1}{q_3} = \frac{[\omega_0 X'(\omega_0) - X_0]^2}{2\omega_0 [\omega_0 X'(\omega_0) + X_0]} \tag{6.34b}$$

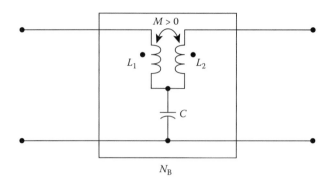

FIGURE 6.6   Brune section.

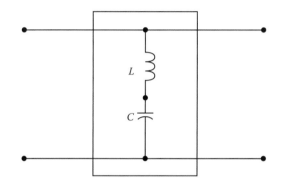

FIGURE 6.7   Degenerate Brune section.

$$C = \frac{q_3}{|s_0|^2} = \frac{2}{\omega_0[\omega_0 X'(\omega_0) - X_0]} \tag{6.34c}$$

$$M = \frac{1}{q_3} = \frac{\omega_0 X'(\omega_0) - X_0}{2\omega_0} > 0 \tag{6.34d}$$

In particular, if $X_0 = 0$ or $Z(j\omega_0) = 0$, the Brune section degenerates into the two-port network of Figure 6.7 with element values

$$L = \frac{1}{2} X'(\omega_0), \quad C = \frac{2}{\omega_0^2 X'(\omega_0)} \tag{6.35}$$

As ω_0 approaches zeros, this degenerate Brune section goes into the type-A section of Figure 6.8 with

$$L = \frac{1}{2} Z'(0) \tag{6.36}$$

When ω_0 approaches infinity, the degenerate Brune section collapses into the type-B section of Figure 6.9 with

$$\frac{2}{C} = \lim_{s \to \infty} sZ(s) \tag{6.37}$$

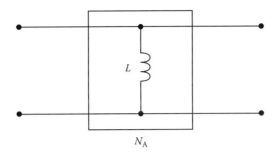

FIGURE 6.8   Type-A section.

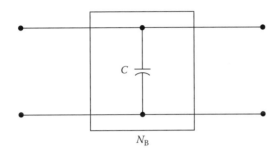

FIGURE 6.9   Type-B section.

Case 3. $s_0 = j\omega_0$ and $R_0 \neq 0$. In this case, R_0/σ_0 is infinity and

$$q_1 = q_4 = 1, \quad q_2 = q_3 = 0 \tag{6.38}$$

The type-E section degenerates into a pair of wires.

Therefore, the Brune section is capable of extracting any $j\omega$-axis zero of the even part of a positive-real impedance, and leads to at least a four-degree reduction if $j\omega_0$ is nonzero and finite, a two-degree reduction otherwise. The latter corresponds to the type-A or type-B section.

Example 6.2

Consider the positive-real impedance

$$Z(s) = \frac{8s^2 + 9s + 10}{2s^2 + 4s + 4} \tag{6.39}$$

The zeros of its even part $r(s)$ are found from the polynomial

$$m_1 m_2 - n_1 n_2 = 16\left(s^4 + s^2 + 2.5\right) \tag{6.40}$$

obtaining

$$s_0 = \sigma_0 + j\omega_0 = 0.735 + j1.020 \tag{6.41a}$$

$$Z(s_0) \equiv R_0 + jX_0 = 2.633 + j0.4279 = 2.667e^{j9.23} \equiv Z_0 \tag{6.41b}$$

The elements of the index set assigned to s_0 by $Z(s)$ are computed as

$$q_1 = 0.7904, \quad q_2 = 3.556, \quad q_3 = 0.6323, \quad q_4 = 1.265 \tag{6.42}$$

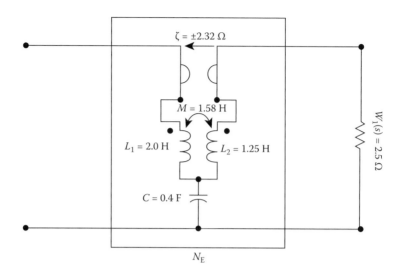

FIGURE 6.10   Realization of the impedance function of (6.39).

Substituting these in Equation 6.32 yields the element values of the type-E section as shown in Figure 6.10.

$$L_1 = 2 \text{ H}, \quad L_2 = 1.25 \text{ H}, \quad C = 0.40 \text{ F} \tag{6.43a}$$

$$M = 1.58 \text{ H}, \quad \zeta = \pm 2.32 \ \Omega \tag{6.43b}$$

The terminating impedance $W_1(s)$ is a resistance of value

$$W_1(s) = Z(0) = 2.5 \ \Omega \tag{6.44}$$

as shown in Figure 6.10.

6.3 Richards Section

In this part, we show that any positive real zero $s_0 = \sigma_0$ of the even part $r(s)$ of a positive-real impedance $Z(s)$, in addition to being realized by the reciprocal Darlington type-C section, can also be realized by a nonreciprocal section called the Richards section of Figure 6.11.

Let $Z(s)$ be a rational positive-real function. Then according to Theorem 6.1, for any positive real σ_0, the function

$$W_1(s) = Z(\sigma_0) \frac{\sigma_0 Z(s) - s Z(\sigma_0)}{\sigma_0 Z(\sigma_0) - s Z(s)} \tag{6.45}$$

is also rational and positive real, the degree of which is not greater than that of $Z(s)$. As pointed out by Richards [1], if σ_0 is a zero of $r(s)$, then

$$\text{degree } W_1(s) \leq \text{degree } Z(s) - 2 \tag{6.46}$$

Inverting Equation 6.45 for $Z(s)$ yields

$$Z(s) = \frac{\sigma_0 W_1(s) + s Z(\sigma_0)}{s W_1(s)/Z(\sigma_0) + \sigma_0} \tag{6.47}$$

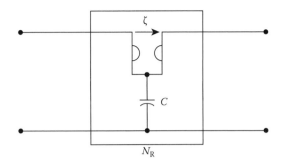

FIGURE 6.11 Richards section.

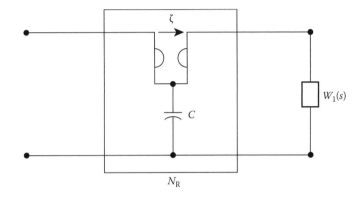

FIGURE 6.12 Realization of $Z(s)$ by Richards section.

This impedance can be realized by the Richards section terminated in the impedance $W_1(s)$ as indicated in Figure 6.12 with the element values

$$C = \frac{1}{\sigma_0 Z(\sigma_0)}, \quad \zeta = \pm Z(\sigma_0) \tag{6.48}$$

6.4 Darlington Type-D Section

In the foregoing, we have demonstrated that the lossless two-port network N_1 can be realized by the lossless nonreciprocal type-E section, which degenerates into the classical type-A, type-B, type-C, and the Brune sections when the even part zero s_0 of the positive-real impedance is restricted to the $j\omega$-axis or the positive σ-axis. In the present section, we show that N_1 can also be realized by a lossless reciprocal two-port network by the application of Theorem 6.1 twice.

Let s_0 be a zero of the even part $r(s)$ of a rational positive-real impedance $Z(s)$. By Theorem 6.2, the function $W_1(s)$ of Equation 6.19 is also rational positive real, and its degree is at least four or two less that that of $Z(s)$, depending on whether $\omega_0 \neq 0$ or $\omega_0 = 0$. Now, apply Theorem 6.2 to $W_1(s)$ at the same point s_0. Then, the function

$$W_2(s) = \frac{D_2(s)W_1(s) - B_2(S)}{-C_2(s)W_1(s) + A_2(s)} \tag{6.49}$$

is rational positive real, the degree of which cannot exceed that of $W_1(s)$, being at least two or four degrees less than that of $Z(s)$, where

$$A_2(s) = p_4 s^2 + |s_0|^2 \tag{6.50a}$$

$$B_2(s) = p_2 s \tag{6.50b}$$

$$C_2(s) = p_3 s \tag{6.50c}$$

$$D_2(s) = p_1 s^2 + |s_0|^2 \tag{6.50d}$$

and $\{p_1, p_2, p_3, p_4\}$ is the index set assigned to the point s_0 by the positive-real function $W_1(s)$. Solving for $W_1(s)$ in Equation 6.49 gives

$$W_1(s) = \frac{A_2(s)W_2(s) + B_2(s)}{C_2(s)W_2(s) + D_2(s)} \tag{6.51}$$

which can be realized as the input impedance of a two-port network N_2 characterized by the transmission matrix

$$\mathbf{T}_2(s) = \begin{bmatrix} A_2(s) & B_2(s) \\ C_2(s) & D_2(s) \end{bmatrix} \tag{6.52}$$

terminated in $W_2(s)$, as depicted in Figure 6.13.

Consider the cascade connection of the two-port N_1 of Figure 6.3 and N_2 of Figure 6.13 terminated in $W_2(s)$, as shown in Figure 6.14. The transmission matrix $\mathbf{T}(s)$ of the overall two-port network N is simply the product of the transmission matrices of the individual two-ports:

$$\mathbf{T}(s) = \mathbf{T}_1(s)\mathbf{T}_2(s) \tag{6.53}$$

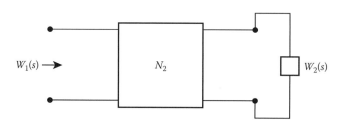

FIGURE 6.13 Realization of the impedance function $W_1(s)$.

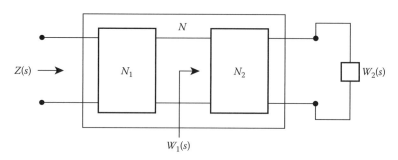

FIGURE 6.14 Cascade connection of two-port networks N_1 and N_2.

the determinant of which is found to be

$$\det \ \mathbf{T}(s) = [\det \ \mathbf{T}_1(s)][\det \ \mathbf{T}_2(s)] = \left[s^4 + 2(\omega_0^2 - \sigma_0^2)s^2 + |s_0|^4\right]^2 \tag{6.54}$$

Thus, when N is terminated in $W_2(s)$, the input impedance of Figure 6.14 is $Z(s)$. This impedance remains unaltered if each element of $\mathbf{T}_1(s)$ is divided by a nonzero finite quantity. For our purposes, we stipulate that the two-port N_1 be characterized by the transmission matrix

$$\widehat{\mathbf{T}}_1(s) = \frac{1}{\Delta(s)}\mathbf{T}_1(s) \tag{6.55}$$

where

$$\Delta(s) = s^4 + 2(\omega_0^2 - \sigma_0^2)s^2 + |s_0|^4 \tag{6.56}$$

Using this matrix $\widehat{\mathbf{T}}_1(s)$ for N_1, the transmission matrix $\widehat{\mathbf{T}}_1(s)$ of the overall two-port network N becomes

$$
\begin{aligned}
\widehat{\mathbf{T}}(s) &= \frac{1}{\Delta(s)}\mathbf{T}_1(s)\mathbf{T}_2(s) \equiv \frac{1}{\Delta(s)}\begin{bmatrix} A(s) & B(s) \\ C(s) & D(s) \end{bmatrix} \\
&= \frac{1}{\Delta(s)}\begin{bmatrix} \begin{array}{l} p_4 q_4 s^4 + |s_0|^4 \\ + \left[(p_4 + q_4)|s_0|^2 + p_3 q_2\right]s^2 \end{array} & \begin{array}{l} (p_2 q_4 + p_1 q_2)s^3 \\ +(p_2 + q_2)|s_0|^2 s \end{array} \\ \begin{array}{l} (p_4 q_3 + p_3 q_1)s^3 \\ +(p_3 + q_3)|s_0|^2 s \end{array} & \begin{array}{l} p_1 q_1 s^4 + |s_0|^4 + \left[(p_1 + q_1)|s_0|^2 \\ + p_2 q_3\right]s^2 \end{array} \end{bmatrix}
\end{aligned} \tag{6.57}
$$

The corresponding impedance matrix $\mathbf{Z}(s)$ of the overall two-port network is found to be

$$\mathbf{Z}(s) = \frac{1}{C(s)}\begin{bmatrix} A(s) & \Delta(s) \\ \Delta(s) & D(s) \end{bmatrix} \tag{6.58}$$

showing that N is reciprocal because $\mathbf{Z}(s)$ is symmetric.

Now consider the reciprocal lossless Darlington type-D section N_D of Figure 6.15 with two perfectly coupled transformers

$$L_1 L_2 = M_1^2, \quad L_3 L_4 = M_2^2 \tag{6.59}$$

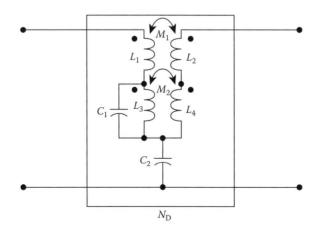

FIGURE 6.15 Darlington type-D section N_D.

The impedance matrix $\mathbf{Z}_D(s)$ of N_D is found to be

$$\mathbf{Z}_D(s) = \begin{bmatrix} L_1 s + \frac{1}{C_2 s} + \dfrac{s/C_1}{s^2 + \omega_a^2} & M_1 s + \frac{1}{C_2 s} + \dfrac{\omega_a^2 M_2 s}{s^2 + \omega_a^2} \\[2mm] M_1 s + \frac{1}{C_2 s} + \dfrac{\omega_a^2 M_2 s}{s^2 + \omega_a^2} & L_2 s + \frac{1}{C_2 s} + \dfrac{\omega_a^2 L_4 s}{s^2 + \omega_a^2} \end{bmatrix} \tag{6.60}$$

where $\omega_a^2 = 1/C_1 L_3$.

Setting $\mathbf{Z}_D(s) = \mathbf{Z}(s)$ in conjunction with Equation 6.57 and after considerable algebraic manipulations, we can make the following identifications:

$$L_1 = \frac{p_4 q_4}{p_4 q_3 + p_3 q_1} \tag{6.61a}$$

$$L_2 = \frac{p_1 q_1}{p_4 q_3 + p_3 q_1} = \frac{M_1^2}{L_1} \tag{6.61b}$$

$$M_1 = \frac{1}{p_4 q_3 + p_3 q_1} = \sqrt{L_1 L_2} \tag{6.61c}$$

$$C_2 = \frac{p_3 + q_3}{|s_0|^2} \tag{6.61d}$$

$$\omega_a^2 = \omega_1^2 = \frac{|s_0|^2 (p_3 + q_3)}{p_4 q_3 + p_3 q_1} \tag{6.61e}$$

$$M_2 = -\frac{\omega_1^4 - 2(\omega_0^2 - \sigma_0^2)\omega_1^2 + |s_0|^4}{\omega_1^4 (p_4 q_3 + p_3 q_1)}$$

$$= -\frac{p_3^2 q_3^2 |\overline{W}_1(s_0) + Z(s_0) q_1|^2}{|s_0|^2 (p_4 q_3 + p_3 q_1)(p_3 + q_3)^2} \le 0 \tag{6.61f}$$

$$L_4 = \frac{[(p_1 + q_1)|s_0|^2 + p_2 q_3]\omega_1^2 - p_1 q_1 \omega_1^4 - |s_0|^4}{\omega_1^4 (p_4 q_3 + p_3 q_1)} = -\frac{q_3 M_2}{p_3} \tag{6.61g}$$

$$L_3 = \frac{M_2^2}{L_4} = -\frac{p_3 M_2}{q_3} \tag{6.61h}$$

$$C_1 = \frac{1}{\omega_1^2 L_3} = -\frac{q_3}{\omega_1^2 p_3 M_2} \tag{6.61i}$$

Thus, all the element values except M_2 are nonnegative, and the lossless reciprocal Darlington type-D section is equivalent to the two type-E sections in cascade.

Example 6.3

Consider the positive-real impedance

$$Z(s) = \frac{6s^2 + 5s + 6}{2s^2 + 4s + 4} \tag{6.62}$$

the even part of which has a zero at

$$s_0 = \sigma_0 + j\omega_0 = 0.61139 + j1.02005 \tag{6.63}$$

The elements of the index set assigned to s_0 by $Z(s)$ are given by

$$q_1 = 0.70711, \quad q_2 = 1.76784, \quad q_3 = 0.94283, \quad q_4 = 1.41421 \tag{6.64}$$

The terminating impedance $W_1(s)$ is determined to be

$$W_1(s) = W_1(0) = Z(0) = 1.5 \ \Omega \tag{6.65}$$

The elements of the index set assigned to the point s_0 by $W_1(s)$ are found to be

$$p_1 = 1, \quad p_2 = 1.83417, \quad p_3 = 0.81519, \quad p_4 = 1 \tag{6.66}$$

Substituting these in Equation 6.61 yields the desired element values of the type-D section, as follows:

$$L_1 = \frac{p_4 q_4}{p_4 q_3 + p_3 q_1} = 0.93086 \text{ H} \tag{6.67a}$$

$$L_2 = \frac{p_1 q_1}{p_4 q_3 + p_3 q_1} = 0.46543 \text{ H} \tag{6.67b}$$

$$M_1 = \frac{1}{p_3 q_3 + p_3 q_1} = 0.65822 \text{ H} \tag{6.67c}$$

$$C_2 = \frac{p_3 + q_3}{|s_0|^2} = 1.24303 \text{ F} \tag{6.67d}$$

$$\omega_1^2 = \frac{|s_0^2|(p_3 + q_3)}{p_4 q_3 + p_3 q_1} = 1.63656 \tag{6.67e}$$

$$M_2 = -\frac{\omega_1^4 - 2(\omega_0^2 - \sigma_0^2)\omega_1^2 + |s_0|^4}{\omega_1^4(p_4 q_3 + p_3 q_1)} = -0.61350 \text{ H} \tag{6.67f}$$

$$L_4 = -\frac{q_3 M_2}{p_3} = 0.70956 \text{ H} \tag{6.67g}$$

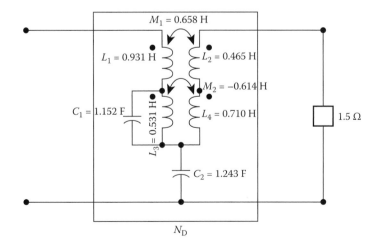

FIGURE 6.16 Darlington type-D section terminated in a resistor.

$$L_3 = \frac{M_2^2}{L_4} = -\frac{p_3 M_2}{q_3} = 0.53044 \text{ H} \tag{6.67h}$$

$$C_1 = \frac{1}{\omega_1^2 L_3} = -\frac{q_3}{\omega_1^2 p_3 M_2} = 1.15193 \text{ F} \tag{6.67i}$$

The complete network together with its termination is presented in Figure 6.16.

References

1. P. I. Richards, A special class of functions with positive real part in half-plane, *Duke Math. J.*, 14, 777–786, 1947.
2. P. I. Richards, General impedance-function theory, *Quart. Appl. Math.*, 6, 21–29, 1948.
3. R. Bott and R. J. Duffin, Impedance synthesis without the use of transformers, *J. Appl. Phys.*, 20, 816, 1949.
4. W. K. Chen, *Passive and Active Filters: Theory and Implementations,* New York: John Wiley & Sons, 1986.

7

Synthesis of LCM and RC One-Port Networks

Wai-Kai Chen
University of Illinois at Chicago

7.1 Introduction

In Chapter 6, we showed that any positive-real function can be realized as the input immittance of a passive one-port network, which is describable as a lossless two-port network terminated in a resistor. Therefore, insofar as the input immittance is concerned, any passive network is equivalent to one containing at most one resistor. In this section, we consider the synthesis of a one-port network composed only of self and mutual inductors and capacitors called the LCM one-port, or a one-port composed only of resistors and capacitors called the RC one-port.

7.2 LCM One-Port Networks

Consider the input impedance $Z(s)$ of an LCM one-port network written in the form

$$Z(s) = \frac{m_1 + n_1}{m_2 + n_2} \tag{7.1}$$

the even part of which is given by

$$r(s) = \frac{m_1 m_2 - n_1 n_2}{m_2^2 - n_2^2} \tag{7.2}$$

Since the one-port is lossless, we have

$$r(j\omega) = \operatorname{Re} Z(j\omega) = 0 \quad \text{for all } \omega \tag{7.3}$$

To make Re $Z(j\omega) = 0$, there are three nontrivial ways: (1) $m_1 = 0$ and $n_2 = 0$, (2) $m_2 = 0$ and $n_1 = 0$, and (3) $m_1 m_2 - n_1 n_2 = 0$. The first possibility leads $Z(s)$ to n_1/m_2, the second to m_1/n_2. For the third possibility, we require that $m_1 m_2 = n_1 n_2$ or

$$(m_1 + n_1)m_2 = (m_2 + n_2)n_1 \tag{7.4}$$

which is equivalent to

$$Z(s) = \frac{m_1 + n_1}{m_2 + n_2} = \frac{n_1}{m_2} \tag{7.5}$$

Therefore, the driving-point immittance of a lossless network is always the quotient of even to odd or odd to even polynomials. Its zeros and poles must occur in quadrantal symmetry, being symmetric with respect to both axes. As a result, they are simple and purely imaginary from stability considerations, or $Z(s)$ can be explicitly written as

$$Z(s) = H \frac{\left(s^2 + \omega_{z1}^2\right)\left(s^2 + \omega_{z2}^2\right)\left(s^2 + \omega_{z3}^2\right)\cdots}{s\left(s^2 + \omega_{p1}^2\right)\left(s^2 + \omega_{p2}^2\right)\cdots} \tag{7.6}$$

where $\omega_{z1} \geq 0$. This equation can be expanded in partial fraction as

$$Z(s) = Hs + \frac{K_0}{s} + \sum_{i=1}^{n} \frac{2K_i s}{s^2 + \omega_i^2} \tag{7.7}$$

where $\omega_{pi} = \omega_i$, and the residues H, K_0, and K_i are all real and positive.

Substituting $s = j\omega$ and writing $Z(j\omega) = \operatorname{Re} Z(j\omega) + j \operatorname{Im} Z(j\omega)$ results in an odd function known as the reactance function $X(\omega)$:

$$X(\omega) = \operatorname{Im} Z(j\omega) = H\omega - \frac{K_0}{\omega} + \sum_{i=1}^{n} \frac{2K_i \omega}{-\omega^2 + \omega_i^2} \tag{7.8}$$

Taking the derivatives on both sides yields

$$\frac{dX(\omega)}{d\omega} = H + \frac{K_0}{\omega^2} + \sum_{i=1}^{n} \frac{2K_i\left(\omega^2 + \omega_i^2\right)}{\left(-\omega^2 + \omega_i^2\right)^2} \tag{7.9}$$

Since every factor in this equation is positive for all positive and negative values of ω, we conclude that

$$\frac{dX(\omega)}{d\omega} > 0 \quad \text{for } -\infty < \omega < \infty \tag{7.10}$$

It states that the slope of the reactance function versus frequency curve is always positive, as depicted in Figure 7.1. Consequently, the poles and zeros of $Z(s)$ alternate along the $j\omega$-axis. This is known as the separation property for reactance function credited to Foster [1]. Because of this, the pole and zero frequencies of Equation 7.6 are related by

$$0 \leq \omega_{z1} < \omega_{p1} < \omega_{z2} < \omega_{p2} < \cdots \tag{7.11}$$

We now consider the realization of $Z(s)$. If each term on the right-hand side of Equation 7.7 can be identified as the input impedance of the LC one-port, the series connection of these one-ports would yield the desired realization. The first term is the impedance of an inductor of inductance H, and the second term corresponds to a capacitor of capacitance $1/K_0$. Each of the remaining term can be realized

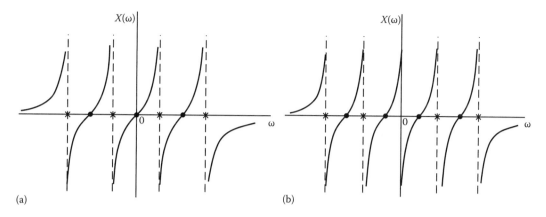

FIGURE 7.1 Plots of reactance function $X(\omega)$ versus ω. (a) The origin is a zero. (b) The origin is a pole.

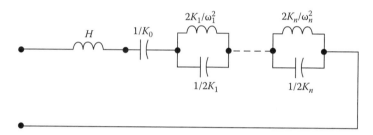

FIGURE 7.2 First Foster canonical form.

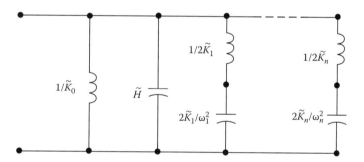

FIGURE 7.3 Second Foster canonical form.

as a parallel combination of an inductor of inductance $2K_i/\omega_i^2$ and a capacitor of capacitance $1/2K_i$. The resulting realization is shown in Figure 7.2 known as the first Foster canonical form. Likewise, if we consider the admittance function $Y(s) = 1/Z(s)$ and expanded it in partial fraction, we obtain

$$Y(s) = \tilde{H}s + \frac{\tilde{K}_0}{s} + \sum_{i=1}^{n} \frac{2\tilde{K}_i s}{s^2 + \omega_i^2} \tag{7.12}$$

which can be realized by the one-port of Figure 7.3 known as the second Foster canonical form. The term canonical form refers to a network containing the minimum number of elements to meet given specifications.

We summarize the preceding results by stating the following theorem:

THEOREM 7.1

A real rational function is the input immittance function of an LCM one-port network if and only if all of its zeros and poles are simple, lie on the jω-axis, and alternate with each other.

In addition to the two Foster canonical forms, there is another synthesis procedure, that gives rise to one-ports known as the Cauer canonical form [2]. Let us expand, $Z(s)$ in a continued fraction

$$Z(s) = \frac{m(s)}{n(s)} = L_1 s + \cfrac{1}{C_2 s + \cfrac{1}{L_3 s + \cfrac{1}{C_4 s + \cfrac{1}{\ddots}}}} \tag{7.13}$$

where $m(s)$ is assumed to be of higher degree than $n(s)$. Otherwise, we expand $Y(s) = 1/Z(s)$ instead of $Z(s)$. Equation 7.13 can be realized as the input impedance of the LC ladder network of Figure 7.4 and is known as the first Cauer canonical form.

Suppose now that we rearrange the numerator and denominator polynomials $m(s)$ and $n(s)$ in ascending order of s, and expand the resulting function in a continued fraction. Such an expansion yields

$$
\begin{aligned}
Z(s) &= \frac{m(s)}{n(s)} = \frac{a_0 + a_2 s^2 + \cdots + a_{k-2} s^{k-2} + a_k s^k}{b_1 s + b_3 s^3 + \cdots + b_{k-1} s^{k-1}} \\
&= \frac{1}{C_1 s} + \cfrac{1}{L_2 s + \cfrac{1}{C_3 s + \cfrac{1}{L_4 s + \cfrac{1}{\ddots}}}}
\end{aligned}
\tag{7.14}
$$

which can be realized by the LC ladder of Figure 7.5 known as the second Cauer canonical form.

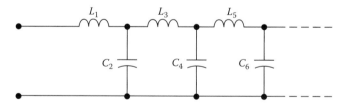

FIGURE 7.4 First Cauer canonical form.

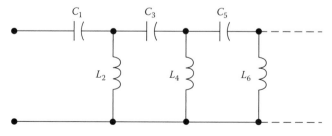

FIGURE 7.5 Second Cauer canonical form.

Example 7.1

Consider the reactance function

$$Z(s) = \frac{s(s^2 + 4)(s^2 + 36)}{(s^2 + 1)(s^2 + 25)(s^2 + 81)} \tag{7.15}$$

For the first Foster canonical form, we expand $Z(s)$ in a partial fraction

$$Z(s) = \frac{7s/128}{s^2 + 1} + \frac{11s/64}{s^2 + 25} + \frac{99s/128}{s^2 + 81} \tag{7.16}$$

and obtain the one-port network of Figure 7.6.

For the second Foster canonical form, we expand $Y(s) = 1/Z(s)$ in a partial fraction

$$Y(s) = s + \frac{225/16}{s} + \frac{4851s/128}{s^2 + 4} + \frac{1925s/128}{s^2 + 36} \tag{7.17}$$

and obtain the one-port network of Figure 7.7.

For the Cauer canonical form, we expand the function in a continued fraction

$$Z(s) = \cfrac{1}{s + \cfrac{1}{0.015s + \cfrac{1}{6.48s + \cfrac{1}{8.28 \times 10^{-3}s + \cfrac{1}{12.88s + \cfrac{1}{0.048s}}}}}} \tag{7.18}$$

and obtain the one-port network of Figure 7.8.

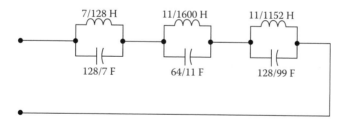

FIGURE 7.6 First Foster canonical form.

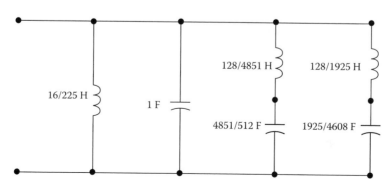

FIGURE 7.7 Second Foster canonical form.

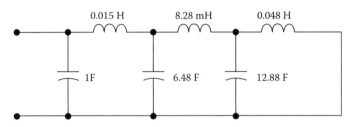

FIGURE 7.8   First Cauer canonical form.

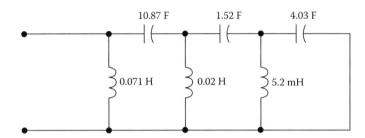

FIGURE 7.9   Second Cauer canonical form.

For the second Cauer canonical form, we rearrange the polynomials in ascending order of s, then expand the resulting function in a continued fraction, and obtain

$$Z(s) = \cfrac{1}{\cfrac{14.06}{s} + \cfrac{1}{\cfrac{0.092}{s} + \cfrac{1}{\cfrac{49.84}{s} + \cfrac{1}{\cfrac{0.66}{s} + \cfrac{1}{\cfrac{192.26}{s} + \cfrac{1}{\cfrac{0.248}{s}}}}}}} \tag{7.19}$$

The desired LC ladder is shown in Figure 7.9.

7.3 RC One-Port Networks

In this part, we exploit the properties of impedance functions of the RC one-ports from the known properties of the LCM one-ports of Section 7.2.

From a given RC one-port N_{RC}, we construct an LC one-port N_{LC} by replacing each resistor of resistance R_i by an inductor of inductance $L_i = R_i$. Suppose that we use loop analysis for both N_{RC} and N_{LC}, and choose the same set of loop currents. In addition, assume that the voltage source at the input port is traversed only by loop current ①. Then the input impedance $Z_{LC}(s)$ of N_{LC} is determined by the equation

$$Z_{LC}(s) = \frac{\tilde{\Delta}(s)}{\tilde{\Delta}_{11}(s)} \tag{7.20}$$

where
   $\tilde{\Delta}$ is the loop determinant
   $\tilde{\Delta}_{11}$ is the cofactor corresponding to loop current 1 in N_{LC}

Similarly, the input impedance $Z_{RC}(s)$ of N_{RC} can be written as

$$Z_{RC}(s) = \frac{\Delta(s)}{\Delta_{11}(s)} \tag{7.21}$$

where Δ is the loop determinant and Δ_{11} is the cofactor corresponding to loop current 1 in N_{RC}. It is not difficult to see that these loop determinants and cofactors are related by

$$\Delta(s) = \left. \frac{\tilde{\Delta}(p)}{p^r} \right|_{p^2 = s} \tag{7.22a}$$

$$\Delta_{11}(s) = \left. \frac{\tilde{\Delta}_{11}(p)}{p^{r-1}} \right|_{p^2 = s} \tag{7.22b}$$

where r is the order of the loop determinants $\tilde{\Delta}$ and Δ. Combining Equations 7.20 through 7.22 yields

$$Z_{RC}(s) = \left[\frac{1}{p} Z_{LC}(p) \right]_{p^2 = s} \tag{7.23}$$

This relation allows us to deduce the properties of RC networks from those of the LC networks. Substituting Equations 7.6 and 7.7 in Equation 7.23, we obtain the general forms of the RC impedance function as

$$Z_{RC}(s) = H \frac{(s + \sigma_{z1})(s + \sigma_{z2})(s + \sigma_{z3}) \cdots}{s(s + \sigma_{p1})(s + \sigma_{p2}) \cdots} = H + \frac{K_0}{s} + \sum_{i=1}^{n} \frac{\hat{K}_i}{s + \sigma_i} \tag{7.24}$$

where $\sigma_{zj} = \omega_{zj}^2$, $\sigma_{pi} = \omega_{pi}^2$, $\sigma_i = \omega_i^2$ and $\hat{K}_i = 2K_i$, and from Equation 7.11

$$0 \leq \sigma_{z1} < \sigma_{p1} < \sigma_{z2} < \sigma_{p2} < \cdots \tag{7.25}$$

Thus, the zeros and poles of an RC impedance alternate along the nonpositive real axis. This property turns out also to be sufficient to characterize the RC impedances.

THEOREM 7.2

A real rational function is the driving-point impedance of an RC one-port network if and only if all the poles and zeros are simple, lie on the nonpositive real axis, and alternate with each other, the first critical frequency (pole or zero) being a pole.
    The slope of $Z_{RC}(\sigma)$ is found from Equation 7.24 to be

$$\frac{dZ_{RC}(\sigma)}{d\sigma} = -\frac{K_0}{\sigma^2} - \sum_{i=1}^{n} \frac{\hat{K}_i}{(\sigma + \sigma_i)^2} \tag{7.26}$$

which is negative for all values of σ, since K_0, and $\hat{k}_i$ are positive. Thus, we have

$$\frac{dZ_{RC}(\sigma)}{d\sigma} < 0 \tag{7.27}$$

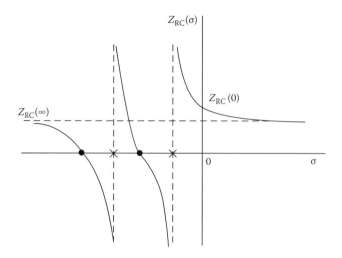

FIGURE 7.10   Plot of $Z_{RC}(\sigma)$ as a function of σ.

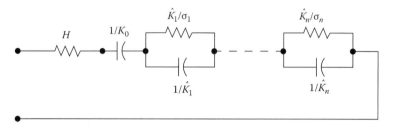

FIGURE 7.11   First Foster Canonical Form.

A plot of $Z_{RC}(\sigma)$ as a function of σ is shown in Figure 7.10. Since no poles and zeros exist along the positive real axis, we have

$$Z_{RC}(\infty) \leq Z_{RC}(0) \tag{7.28}$$

We now proceed to the realization of the RC one-port networks. Suppose that we are given $Z_{RC}(\sigma)$ as in Equation 7.24. By analogy to the LC case, this impedance can be realized by the one-port network of Figure 7.11 called the first Foster canonical form for the RC impedance.

To obtain the second Foster canonical form, we expand $Y_{RC}(s)/s$ in a partial fraction, where $Y_{RC}(s) = 1/Z_{RC}(s)$, and then multiply the resulting equation by s. The reason is that a direct partial-fraction expansion of $Y_{RC}(s)$ will result in negative residues. Proceeding in this way, we obtain

$$Y_{RC}(s) = K_0 + K_\infty s + \sum_{i=1}^{n} \frac{K_i s}{s + \sigma_i} \tag{7.29}$$

yielding the one-port network of Figure 7.12.

As before, in addition to the two Foster forms, RC ladder realizations are also possible. Following the LC case, we perform a continued-fraction expansion of $Z_{RC}(s)$ and obtain

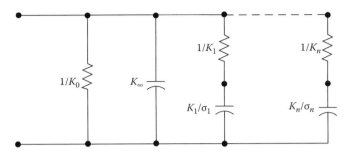

FIGURE 7.12 Second Foster canonical form.

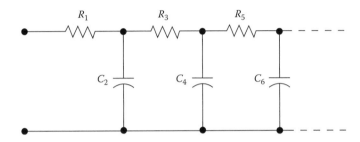

FIGURE 7.13 First Cauer canonical form.

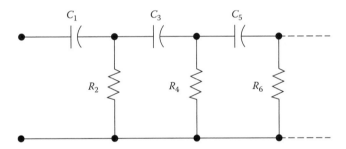

FIGURE 7.14 Second Cauer canonical form.

$$Z_{RC}(s) = R_1 + \cfrac{1}{C_2 s + \cfrac{1}{R_3 + \cfrac{1}{C_4 s + \cfrac{1}{\ddots}}}} \tag{7.30}$$

This expansion can be realized by the ladder network of Figure 7.13 known as the first Cauer canonical form for RC impedance. If we rearrange the terms of $Z_{RC}(s)$ so that the numerator and denominator polynomials appear in ascending order of s, the resulting continued-fraction expansion takes the general form

$$Z_{RC}(s) = \cfrac{1}{C_1 s} + \cfrac{1}{\cfrac{1}{R_2} + \cfrac{1}{\cfrac{1}{C_3 s} + \cfrac{1}{\cfrac{1}{R_4} + \cfrac{1}{\cfrac{1}{C_5 s} + \cfrac{1}{\ddots}}}}} \tag{7.31}$$

yielding the second Cauer canonical form of Figure 7.14.

Example 7.2

Consider the impedance function

$$Z(s) = \frac{s^2 + 12s + 35}{s^2 + 10s + 24} \tag{7.32}$$

To obtain the first Foster canonical form, we expand $Z(s)$ in partial fraction as

$$Z(s) = 1 + \frac{3/2}{s+4} + \frac{1/2}{s+6} \tag{7.33}$$

and obtain the one-port of Figure 7.15.
    For the second Foster canonical form, the proper function to expand is $Y(s)/s$, yielding

$$Y(s) = \frac{24}{35} + \frac{s/10}{s+5} + \frac{3s/14}{s+7} \tag{7.34}$$

The corresponding realization is shown in Figure 7.16.
    For the first Cauer canonical form, we expand $Z(s)$ in a continued fraction as

$$Z(s) = 1 + \cfrac{1}{\cfrac{s}{2} + \cfrac{1}{\cfrac{4}{9} + \cfrac{1}{27s + \cfrac{1}{\cfrac{1}{2} + \cfrac{1}{\cfrac{1}{72}}}}}} \tag{7.35}$$

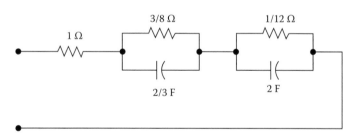

FIGURE 7.15   First Foster canonical form.

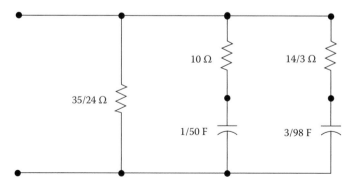

FIGURE 7.16   Second Foster canonical form.

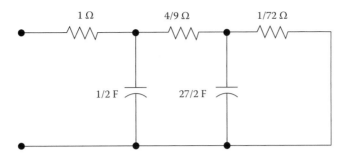

FIGURE 7.17 First Cauer canonical form.

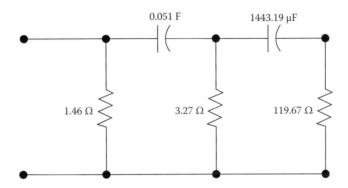

FIGURE 7.18 Second Cauer canonical form.

which can be realized by the ladder of Figure 7.17. To obtain the second Cauer canonical form, we rearrange the numerator and denominator polynomials in ascending order of s, and then expand in a continued fraction

$$Z(s) = \frac{35 + 12s + s^2}{24 + 10s + s^2} = \cfrac{1}{0.69 + \cfrac{1}{\cfrac{19.76}{s} + \cfrac{1}{0.306 + \cfrac{1}{\cfrac{692.91}{s} + \cfrac{1}{8.36 \times 10^{-3}}}}}} \tag{7.36}$$

yielding the ladder of Figure 7.18.

References

1. R. M. Foster, A reactance theorem, *Bell System Tech. J.*, 3, 259–267, 1924.
2. W. Cauer, *Synthesis of Linear Communication Networks,* New York: McGraw-Hill, Chapters 5 and 6, 1958.

8

Two-Part Synthesis by Ladder Development

Wai-Kai Chen

University of Illinois at Chicago

8.1 Introduction

In two-port synthesis, specifications are often given in terms of the transfer functions such as the transfer voltage ratio, transfer current ratio, transfer impedance, or transfer admittance. The actual realization, however, is accomplished by means of the y- or z-parameters. Figure 8.1 shows a two-port network driven by a voltage source with output terminating in an impedance $Z_2(s)$. It is straightforward to show that the transfer voltage ratio function $G_{12}(s)$ can be expressed in terms of its y-parameters $y_{ij}(s)$ or z-parameters $z_{ij}(s)$ by the equation

$$G_{12}(s) = \frac{V_2}{V_1} = \frac{-y_{21}}{y_{22} + Y_2} \tag{8.1}$$

where $Y_2(s) = 1/Z_2(s)$. When the output is open-circuited, Equation 8.1 becomes

$$G_{12}(s) = \frac{V_2}{V_1} = \frac{-y_{21}}{y_{22}} = \frac{z_{21}}{z_{11}} \tag{8.2}$$

Likewise, the transfer current ratio $\alpha_{12}(s)$ can be expressed as

$$\alpha_{12}(s) = -\frac{I_2}{I_1} = \frac{z_{21}}{z_{22} + Z_2} \tag{8.3}$$

The zeros of transmission of a two-port network are defined as the frequencies at which the two-port results in zero output for a finite input. They play an important role in ladder development. There are many ways of producing zeros of transmission. One possibility to prevent the input signal from reaching the output is by shorting together all transmission paths or by opening all transmission paths by means of a series or parallel resonance. Another possibility is that signals transmitted by different paths cancel at the output.

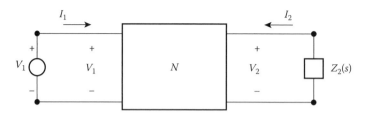

FIGURE 8.1   Terminated two-port network.

Observe from Equations 8.1 and 8.2 that zero output, $V_2 = 0$, implies a zero for each of these functions. Therefore, zeros of transmission are zeros of $-y_{21}$ or z_{21} provided y_{21} and y_{22} or z_{21} and z_{11} have the same poles. For the ladder network, the transmission can be interrupted only by the short circuit in a shunt arm or an open circuit in a series arm. The short circuit of a shunt arm corresponds to the pole frequencies of its admittances, whereas the open circuit of a series arm corresponds to the pole frequencies of its impedances. Therefore, the zeros of transmission of a ladder network can be identified directly with the zeros of the impedances of the shunt arms and the poles of the impedances of the series arms. For the LC ladder, all zeros of transmission lie on the $j\omega$-axis, and for the RC ladder they are on the nonpositive real axis of the complex-frequency s-plane.

8.2 LC Ladder

For LC ladders, the conditions imposed on $-y_{21}$ and y_{22} or z_{21} and z_{22} are that the driving-point functions y_{22} and z_{22} be positive real with poles and zeros interlaced on the $j\omega$-axis. The transfer functions $-y_{21}$ and z_{21}, assuming to have the same poles as y_{22} or z_{22}, must have all of its zeros on the $j\omega$-axis. However, these zeros need not be interlaced with the poles and they may not be simple. Our strategy in realization is that of carrying out the driving-point synthesis of y_{22} or z_{22}, using Cauer ladder development method, in such a way that the zeros of transmission are realized at the same time. The procedure consists of two steps: a zero-shifting step and a zero-producing step, as described below.

Zero shifting by partial removal. Consider an impedance of Equation 7.6, the partial-fraction expansion of which is given in Equation 7.7. The first term on the right-hand side of Equation 7.7 is due to the contribution of the pole at the infinity. If this term Hs is subtracted from $Z(s)$, the resulting function $Z(s) - Hs$ is devoid of the pole at the infinity. We say that the pole at infinity has been removed completely. Instead of complete removal of this pole, suppose that we subtract a fraction of the terms Hs from $Z(s)$ by introducing a constant k_p such that

$$Z_1(s) = Z(s) - k_p Hs, \quad k_p < 1 \tag{8.4}$$

We say that the pole at infinity has been partially removed or weakened. The function $Z_1(s)$ that results from the partial removal of the pole at infinity still possesses the pole at infinity. Since all the zeros of $Z_1(s)$ are again located on the $j\omega$-axis, these zeros are found by substituting $s = j\omega$ in Equation 8.4,

$$X_1(\omega) = X(\omega) - k_p H\omega \tag{8.5}$$

where $Z_1(j\omega) = jX_1(\omega)$. The zeros of $X_1(\omega)$ are values of ω satisfying the equation

$$X(\omega) = k_p H\omega \tag{8.6}$$

Solutions to this equation are found graphically from the intersections of the curves $X(\omega)$ and $k_p H\omega$, as depicted in Figure 8.2. Observe that all the zeros in the resulting function are shifted toward the pole

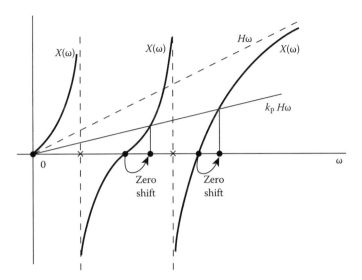

FIGURE 8.2 Zero shifting by weakening the pole at the infinity.

being weakened. The amount of shift of the zeros from their original positions depends on the value of k_p and the proximity of a zero to the pole being weakened.

We next consider the term K_0/s in Equation 7.7 due to the pole at the origin. The partial removal of this pole is equivalent to the operation

$$Z_2(s) = Z(s) - k_p \frac{K_0}{s} \tag{8.7}$$

As before, the zeros of $Z_2(s)$ are defined by the intersections of the curves $X(\omega)$ and $-k_p(K_0/\omega)$ with ω,

$$X(\omega) = -k_p \frac{K_0}{\omega} \tag{8.8}$$

as illustrated in Figure 8.3. Observe again that the zeros are shifted toward the pole being weakened, which is at the origin.

Finally, for the finite nonzero poles, the corresponding factors take the general form $2K_i s/(s^2 + \omega_i^2)$. The partial removal of this pair of complex conjugate poles results in the new function

$$Z_3(s) = Z(s) - k_p \frac{2K_i s}{s^2 + \omega_i^2}, \quad k_p < 1 \tag{8.9}$$

The zeros of this function are defined by the intersections of the plots of $X(\omega)$ and $-k_p(2K_i\omega/(\omega^2 - \omega_i^2))$ with ω,

$$X(\omega) = -k_p \frac{2K_i\omega}{\omega^2 - \omega_i^2}, \quad k_p < 1 \tag{8.10}$$

as illustrated in Figure 8.4.

Our conclusion is that the partial removal of a pole shifts the zeros toward that pole. The amount of shift depends on the value of k_p and the proximity of a zero to that pole, but in no case can a zero be shifted beyond an adjacent pole.

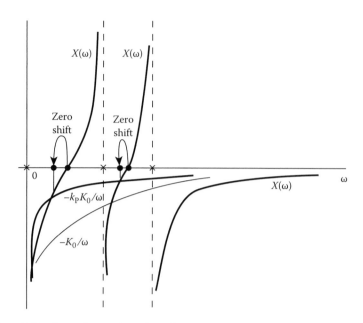

FIGURE 8.3   Zero shifting by weakening the pole at the origin.

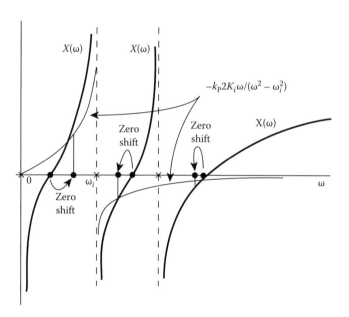

FIGURE 8.4   Zero shifting by weakening a finite nonzero pole.

Zero producing by complete pole removal. After a zero of transmission has been shifted to a desired location by the partial removal of an appropriate pole, the realization of this zero of transmission is accomplished by the complete removal of the pole of the reciprocal function corresponding to the shifted zero. For the LC ladder two-ports, a series combination of an inductor L and a capacitor C produces a zero at its resonant frequency $\omega = 1/\sqrt{LC}$ and this network is used in the shunt arm in the ladder to produce the desired zero of transmission. Likewise, the parallel connection of L and C yields

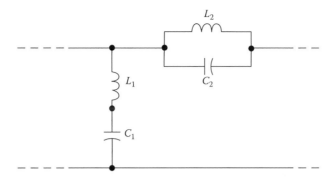

FIGURE 8.5   Zero producing sections in a ladder network.

an infinite impedance at its resonant frequency, and is used in the series arm of the ladder. They are shown in Figure 8.5.

Example 8.1

We wish to design a lossless two-port network terminated in a 100-Ω resistor to meet the specifications for the transfer voltage-ratio function

$$G_{12}(s) = \frac{V_2}{V_1} = K\frac{s^2 + 4}{s^3 + 4s^2 + 9s + 4} \tag{8.11}$$

within a multiplicative constant.

Since magnitude scaling does not affect the voltage transfer ratio, without loss of generality, we first assume that the terminating resistor is 1 Ω. Equation 8.11 can be rewritten as

$$G_{12}(s) = K\frac{\dfrac{s^2 + 4}{s(s^2 + 9)}}{\dfrac{4(s^2 + 1)}{s(s^2 + 9)} + 1} = \frac{-y_{21}}{y_{22} + Y_2} = \frac{-y_{21}}{y_{22} + 1} \tag{8.12}$$

We can make the following identifications:

$$-y_{21}(s) = K\frac{s^2 + 4}{s(s^2 + 9)}, \quad y_{22}(s) = 4\frac{s^2 + 1}{s(s^2 + 9)} \tag{8.13}$$

Both functions have the same poles at $s = 0$ and $s = \pm j3$ and the zeros of transmission of the ladder are located at $s = \pm j2$ and $s = \infty$. These poles and zeros are shown in Figure 8.6. To realize the zero of transmission at $s = \infty$, we remove the pole of $z_1(s) = 1/y_{22}(s)$ at $s = \infty$.

$$z_1(s) = \frac{1}{y_{22}(s)} = \frac{s(s^2 + 9)}{4(s^2 + 1)} = \frac{2s}{s^2 + 1} + \frac{s}{4} \tag{8.14}$$

After subtracting the term $s/4$ corresponding to an inductor of inductance 1/4 H shown in Figure 8.7 from $z_1(s)$, the remaining impedance $z_2(s)$ is found to be

$$z_2(s) = \frac{2s}{s^2 + 1} \tag{8.15}$$

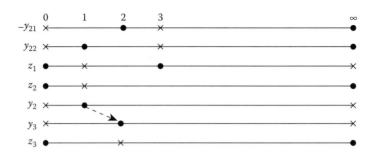

FIGURE 8.6   Poles and zeros of LC immittances.

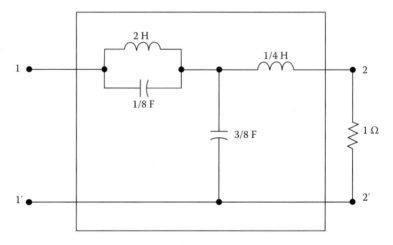

FIGURE 8.7   LC ladder realization of the transfer voltage ratio (Equation 8.11).

To realize the zero of transmission at $s = \pm j2$, we partially weaken the pole of the admittance

$$y_2(s) \equiv \frac{1}{z_2(s)} = \frac{s^2 + 1}{2s} \tag{8.16}$$

at $s = \infty$ in order to shift the zero at $s = \pm j$ to $s = \pm j2$, or

$$y_3(\pm j2) = y_2(s) - k_p \frac{1}{2} s \bigg|_{s = \pm j2} = \frac{-4 + 1}{\pm j2 \times 2} \mp k_p \frac{1}{2} j2 = 0 \tag{8.17}$$

yielding $k_p = 3/4$. The new function becomes

$$y_3(s) = \frac{s^2 + 1}{2s} - \frac{3}{8}s = \frac{s^2 + 4}{8s} \tag{8.18}$$

after the removal of a shunt capacitor of capacitance 3/8 F, as shown in Figure 8.7. The factor $(s^2 + 4)$ in the numerator was anticipated because our objective was to produce a zero in the driving-point admittance $y_3(s)$ at $s = \pm j2$. To realize this zero, we consider the reciprocal function $z_3(s) = 1/y_3(s)$ by complete removal of its pole at $s = \pm j2$. This yields a parallel connection of $L = 2$ H and $C = 1/8$ F shown in Figure 8.7. The final realization is obtained by magnitude-scaling by a factor of 100. The realized constant K is found from the network to be $K = 1$.

8.3 RC Ladder

We now consider the realization of RC ladder with prescribed $-y_{21}(s)$ and $y_{22}(s)$ or $z_{21}(s)$ and $z_{22}(s)$. Following the LC case, the zero shifting for the RC driving-point functions is accomplished by one or any combination of the following three operations:

1. Partial removal of a constant $Z(\infty)$ from $Z(s)$
2. Partial removal of a constant $Y(0)$ from $Y(s)$
3. Partial removal of a pole from $Z(s)$ or $Y(s)$

The first operation permits a series resistance to be removed so that the resulting impedance is still positive real and possesses a desired zero of transmission:

$$Z_1(s) = Z(s) - k_p Z(\infty), \quad k_p \le 1 \tag{8.19}$$

the zeros of which occur at those values of σ satisfying

$$Z(\sigma) = k_p Z(\infty), \quad k_p \le 1 \tag{8.20}$$

Observe that from Figure 8.8 all zeros are shifted toward $s = \sigma = -\infty$ by the partial removal of $Z(\infty)$ because the slope is negative for the RC impedances.

The second operation corresponds to the removal of a shunt resistance, and the remaining admittance

$$Y_1(s) = Y(s) - k_p Y(0), \quad k_p \le 1 \tag{8.21}$$

is still positive real, the zeros of which occur at those values of σ satisfying

$$Y(\sigma) = k_p Y(0), \quad k_p \le 1 \tag{8.22}$$

Observe from Figure 8.9 that these zeros again are shifted toward $s = 0$ in relation to those of $Y(s)$.

Finally, the partial removal of a pole of $Z(s)$ results in a parallel connection of a resistor and a capacitor, and the remaining impedance becomes

$$Z_2(s) = Z(s) - k_p \frac{K_i}{s + \sigma_i}, \quad k_p < 1 \tag{8.23}$$

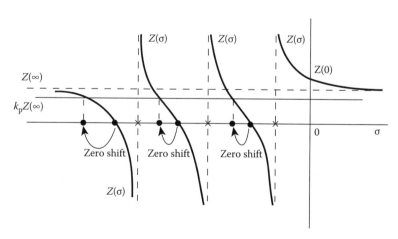

FIGURE 8.8 Zero shifting by the partial removal of $Z(\infty)$.

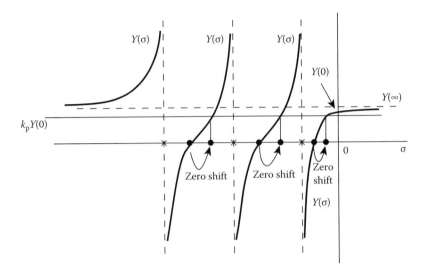

FIGURE 8.9 Zero shifting by partial removal of $Y(0)$.

the zeros of which occur at those values of σ satisfying

$$Z(\sigma) = k_p \frac{K_i}{\sigma + \sigma_i}, \quad k_p < 1 \tag{8.24}$$

As indicated in Figure 8.10, these zeros are again shifted toward the pole at $s = -\sigma_i$ being partially weakened.

Therefore, the partial removal of a constant or a pole shifts the zeros of the remaining function toward the quantity being weakened. The amount of shift depends on the value of k_p and the proximity of a zero to that quantity. Once a zero is shifted to an appropriate location, its realization is accomplished by the complete removal of the pole of the reciprocal function corresponding to the shifted zero.

Example 8.2

We wish to realize the transfer impedance

$$Z_{12}(s) = K \frac{s(s+1)}{2s^2 + 18s + 34} \tag{8.25}$$

of an RC two-port network terminated in a $1\,\Omega$ resistor.

From Equation 8.3, the transfer impedance can be written as

$$Z_{12}(s) = \frac{V_2}{I_1} = \frac{-I_2 \times 1}{I_1} = \frac{-I_2}{I_1} = \frac{z_{21}}{z_{22} + 1} = \frac{K \dfrac{s(s+1)}{(s+2)(s+5)}}{\dfrac{(s+3)(s+8)}{(s+2)(s+5)} + 1} \tag{8.26}$$

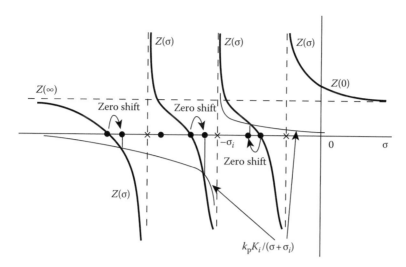

FIGURE 8.10 Zero shifting by partial removal of a pole of an RC impedance.

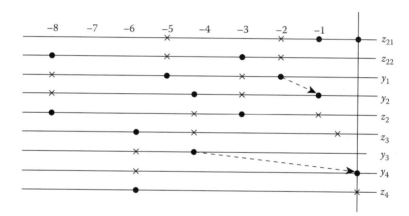

FIGURE 8.11 Zeros and poles of RC immittances.

identifying

$$z_{21}(s) = K\frac{s(s+1)}{(s+2)(s+5)}, \quad z_{22}(s) = \frac{(s+3)(s+8)}{(s+2)(s+5)} \tag{8.27}$$

The zeros of transmission are located at $s = 0$ and $s = -1$. Suppose that we wish to realize these zeros in the order of -1 and 0. For this we consider the reciprocal function $y_1(s) = 1/z_{22}(s)$, the zeros. and poles of which are shown in Figure 8.11. Clearly, the zero at $s = -2$ can be shifted to $s = -1$ by partial removal of the constant $y_1(0) = 5/12$,

$$y_2(-1) = y_1(-1) - k_p y_1(0) = 0 \tag{8.28}$$

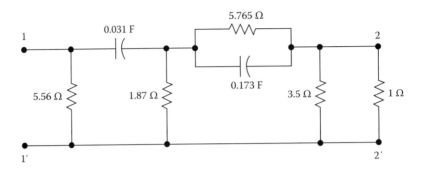

FIGURE 8.12   RC ladder realization of the transfer impedance $Z_{12}(s)$.

obtaining $k_p = 24/35$ and

$$y_2(s) = y_1(s) - \frac{24}{35}y_1(0) = \frac{(s+1)(5s+22)}{7(s^2 + 11s + 24)} \tag{8.29}$$

This admittance has a zero at $s = -1$, as expected. The partially removed constant $k_p y_1(0)$ corresponds to a shunt resistor of resistance $3.5\,\Omega$. To realize the zero of transmission at $s = -1$, we consider the reciprocal function $z_2(s) = 1/y_2(s)$ by the complete removal of its pole at $s = -1$. The remaining impedance $z_3(s)$ becomes

$$z_3(s) = z_2(s) - \frac{\frac{98}{17}}{s+1} = \frac{7}{5} \times \frac{s + \frac{100}{17}}{s + \frac{22}{5}} \tag{8.30}$$

The removed pole corresponds to a parallel connection of a resistor of resistance $98/17\,\Omega$ and a capacitor of capacitance $17/98$ F, as illustrated in Figure 8.12.

For the zero of transmission at $s = 0$, we consider the reciprocal function $y_3(s) = 1/z_3(s)$. The zero of $y_3(s)$ at $s = -22/5$ can be shifted to 0 by partial removal of the constant $y_3(0) = 187/350$, or

$$y_4(0) = y_3(0) - k_p y_3(0) = 0 \tag{8.31}$$

yielding $k_p = 1$. The remaining admittance $y_4(s)$ becomes

$$y_4(s) = y_3(s) - y_3(0) = \frac{\frac{9}{50}s}{s + \frac{100}{17}} \tag{8.32}$$

showing a zero at the origin, as anticipated. This zero is realized by the complete removal of the pole at the origin of its reciprocal $z_4(s) = 1/y_4(s)$, yielding

$$z_5(s) = z_4(s) - \frac{32.68}{s} = 5.56 \tag{8.33}$$

The complete realization is presented in Figure 8.12, from which the constant K is found to be $K = 1$.

8.4 Parallel or Series Ladders

The zeros of transmission of the LC ladders are restricted to the $j\omega$-axis and those of the RC ladders to the negative real axis of the s-plane. For complex zeros of transmission such as those needed for certain phase-correction applications, they cannot be realized by a single LC or RC ladder because there is only a single transmission path from the input to the output. The use of parallel or series ladders, on the other

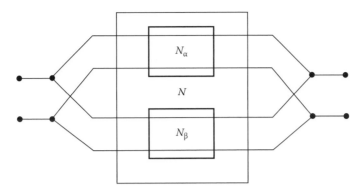

FIGURE 8.13 Parallel connection of two ladder networks.

hand, provides a conduit for multiple path signal transmission, so that the signals arriving at the output through the various paths may cancel one another, resulting in the zero output for a finite input. Therefore, they are capable of producing complex zeros of transmission. This structure was first suggested by Guillemin [1].

Figure 8.13 is the parallel connection of the ladder networks N_α and N_β. The y-parameters y_{ij} of the composite two-port N can be expressed in terms of those $y'_{ij\alpha}$ and $y'_{ij\beta}$ of the component two-ports N_α and N_β by the equation

$$y_{ij} = y'_{ij\alpha} + y'_{ij\beta}, \quad i, j = 1, 2 \tag{8.34}$$

Thus, to realize $-y_{21}(s)$ and $y_{22}(s)$, we may separate them into pairs like $-y'_{21\alpha}, y'_{22\alpha}$ and $-y'_{21\beta}, y'_{22\beta}$ realize an individual pair as an LC or RC ladder. Then connect these individual ladders in parallel to realize $-y_{21}(s)$ and $y_{22}(s)$. In order for the procedure to succeed, we must resolve the following problem. Recall than in the Cauer development of LC and RC ladders, $-y_{21}(s)$ is realized only within the multiplicative constant k. Thus, the transfer admittances realized by the component two-ports actually will be $-k_\alpha y'_{21\alpha}$ and $-k_\beta y'_{21\beta}$. The sum of these two functions will not result in the desired $-k y_{21}$ unless $k = k_\alpha = k_\beta$. To circumvent this difficulty, we introduce an additional degree of freedom by adjusting the admittance level of the α-ladder N_α by a factor b_α and the β-ladder N_β by b_β. Then the functions of the resulting realizations become

$$-y'_{21\alpha} = -b_\alpha k_\alpha y_{21\alpha}, \quad y'_{22\alpha} = b_\alpha y_{22\alpha} \tag{8.35a}$$

$$-y'_{21\beta} = -b_\beta k_\beta y_{21\beta}, \quad y'_{22\beta} = b_\beta y_{22\beta} \tag{8.35b}$$

where $y_{ij} = y_{ij\alpha} + y_{ij\beta}$, $i, j = 1, 2$. Substituting these in Equation 8.34 gives

$$y_{21} = b_\alpha k_\alpha y_{21\alpha} + b_\beta k_\beta y_{21\beta} \tag{8.36a}$$

$$y_{22} = b_\alpha y_{22\alpha} + b_\beta y_{22\beta} \tag{8.36b}$$

Our objective is to choose b_α and b_β to satisfy the above equations, once k_α and k_β are known. One way to meet these requirements is to let $y'_{22\alpha}$ and $y'_{22\beta}$ have the same zeros and poles as y_{22} but different scale factors such that $y'_{22\alpha} = b_\alpha y_{22}$ and $y'_{22\beta} = b_\beta y_{22}$, obtaining from Equation 8.36b

$$b_\alpha + b_\beta = 1 \tag{8.37}$$

Since we can only realize y_{21} within a multiplicative constant, we replace y_{21} by ky_{21} in Equation 8.36a, and set

$$b_\alpha k_\alpha = b_\beta k_\beta = k \tag{8.38}$$

These two equations can be solved to yield the desired scale factors b_α and b_β.

In general, for m ladders in parallel, we require

$$b_1 + b_2 + \cdots + b_m = 1 \tag{8.39a}$$

$$b_1 k_1 = b_2 k_2 = \cdots = b_m k_m = k \tag{8.39b}$$

where
$\quad b_i$ is the admittance scale factor for the ith ladder
$\quad k_i$ is the realized multiplicative constant of the transfer admittance of the ith ladder

Once k_i are known, these m simultaneous equations can be solved for the m unknowns b_i, and the admittance level of each ladder can be scaled accordingly. The scaled ladders are connected in parallel to realize the $-y_{21}(s)$ specifications within the multiplicative constant k, and the $y_{22}(s)$ specifications exactly.

Similar results are obtained by using the z-parameters $z_{ij}(s)$ and the series connection of the component two-port networks, the details of which are omitted. However, a design example will be presented below.

Example 8.3

We wish to realize an RC two-port network to meet the following specifications:

$$-y_{21}(s) = k\frac{s^2 + 1}{(s + 5)(s + 9)} \tag{8.40a}$$

$$y_{22}(s) = \frac{(s + 3)(s + 7)}{(s + 5)(s + 9)} \tag{8.40b}$$

Since the zeros of transmission are located at $s = \pm j1$, they cannot be realized by a single RC ladder. For our purposes, we choose the y-parameters of the component two-ports as

$$-y_{21\alpha}(s) = k_\alpha \frac{s^2}{(s + 5)(s + 9)}, \quad y_{22\alpha}(s) = y_{22}(s) \tag{8.41a}$$

$$-y_{21\beta}(s) = \frac{k_\beta}{(s + 5)(s + 9)}, \quad y_{22\beta}(s) = y_{22}(s) \tag{8.41b}$$

For the α-ladder, since the zeros of transmission are all at the origin, they can be realized by the second Cauer canonical form shown in Figure 8.14 with $k_\alpha = 0.043$. For the β-ladder, since the zeros of transmission are all at the infinity, they can be realized by the first Cauer canonical form of Figure 8.15 with $k_\beta = 21$.

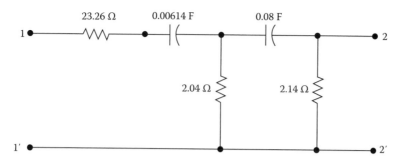

FIGURE 8.14 RC α-ladder realization.

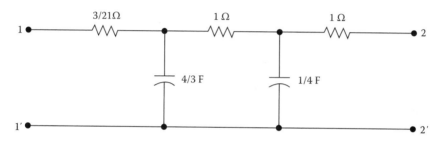

FIGURE 8.15 RC β-ladder realization.

Our next task is to adjust the admittance level of the individual ladders so that when they are connected in parallel, the $y_{22}(s)$ specifications are realized exactly, and the $-y_{21}(s)$ specifications are realized to within a multiplicative constant k. Appealing to Equations 8.37 and 8.38, we have

$$b_\alpha + b_\beta = 1 \qquad (8.42a)$$

$$0.043b_\alpha = 21b_\beta = k \qquad (8.42b)$$

yielding

$$b_\alpha = 0.998 \cong 1, \quad b_\beta = 2.043 \times 10^{-3} \qquad (8.43)$$

We now adjust the admittance level of the β-ladder by a factor $b_\beta = 2.043 \times 10^{-3}$, leaving the α-ladder intact. The final realization is achieved by the parallel connection of the α-ladder and the resulting β-ladder shown in Figure 8.16. The realized multiplicative constant for the overall transfer admittance $-y_{21}(s)$ is found to be $k = 0.043$.

Example 8.4

We wish to realize the open-circuit transfer voltage ratio

$$G_{12}(s) = \frac{z_{21}(s)}{z_{11}(s)} = k\frac{s^2 - 2s + 10}{s^2 + 8s + 15} \qquad (8.44)$$

by an RC two-port network.

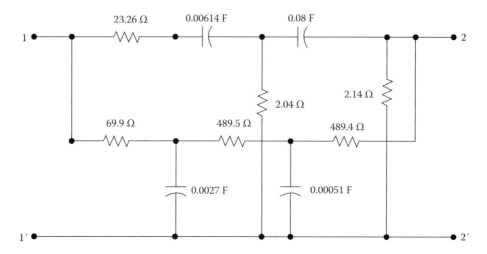

FIGURE 8.16   Parallel RC ladder realization of $-y_{21}(s)$ and $y_{22}(s)$.

First, we multiply the numerator and denominator of $G_{12}(s)$ by the factor $(s+2)$ to yield

$$G_{12}(s) = \frac{z_{21}(s)}{z_{11}(s)} = k\frac{s^3 + 6s + 20}{(s+2)(s+3)(s+5)} \tag{8.45}$$

and then divide by $(s+1)(s+2.5)(s+4)$ in order to make the following identifications:

$$z_{21}(s) = k\frac{s^3 + 6s + 20}{(s+1)(s+2.5)(s+4)} \tag{8.46a}$$

$$z_{11}(s) = \frac{(s+2)(s+3)(s+5)}{(s+1)(s+2.5)(s+4)} \tag{8.46b}$$

We next decompose the pairs into three pairs as follows:

$$z_{21\alpha}(s) = k_\alpha \frac{s^3}{(s+1)(s+2.5)(s+4)}, \quad z_{11\alpha}(s) = z_{11}(s) \tag{8.47a}$$

$$z_{21\beta}(s) = k_\beta \frac{6s}{(s+1)(s+2.5)(s+4)}, \quad z_{11\beta}(s) = z_{11}(s) \tag{8.47b}$$

$$z_{21\gamma}(s) = k_\gamma \frac{20}{(s+1)(s+2.5)(s+4)}, \quad z_{11\gamma}(s) = z_{11}(s) \tag{8.47c}$$

1. *The α-ladder.* Since all the zeros of transmission are located at the origin, it can be realized by the second Cauer canonical form of Figure 8.17 with $k_\alpha = 1$.
2. *The β-ladder.* Since one of the zeros of transmission is located at the origin, and two others at the infinity, the first half of the β-ladder can be realized as the second Cauer canonical form and the second half of the β-ladder by the first Cauer canonical forms as illustrated in Figure 8.18 with $k_\beta = 0.0117$.
3. *The γ-ladder.* Since all of its zeros of transmission are located at the infinity, it can be realized by the first Cauer canonical form of Figure 8.19 with $k_\gamma = 0.0145$.

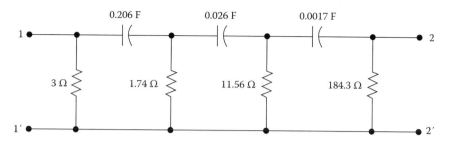

FIGURE 8.17 α-Ladder.

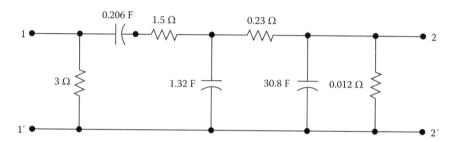

FIGURE 8.18 β-Ladder.

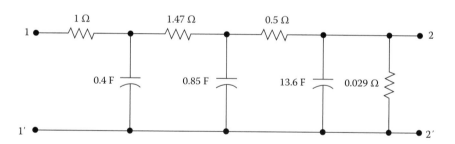

FIGURE 8.19 γ-Ladder.

We next adjust the admittance level of the ladders so that when they are connected in parallel, the desired specifications are realized. From Equation 8.39 we require that

$$b_\alpha + b_\beta + b_\gamma = 1 \tag{8.48a}$$

$$b_\alpha k_\alpha = b_\beta k_\beta = b_\gamma k_\gamma = k \tag{8.48b}$$

These equations can be solved to yield

$$b_\alpha = 0.0064, \quad b_\beta = 0.552, \quad b_\gamma = 0.4416 \tag{8.49}$$

with $k = 0.0064$. The final two-port network is shown in Figure 8.20.

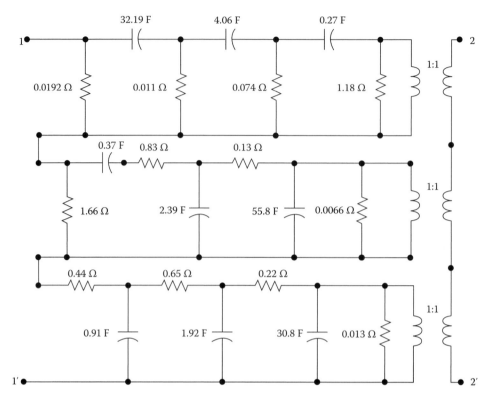

FIGURE 8.20 RC two-port realization of the open-circuit voltage ratio $G_{12}(s)$.

Reference

1. E. A. Guillemin, Synthesis of RC networks, *J. Math. Phys.*, 28, 22–42, 1949.

9

Design of Resistively Terminated Networks

Wai-Kai Chen
University of Illinois at Chicago

9.1 Introduction

In the design of communication systems, it is frequently required to synthesize a coupling network that will transform a given frequency-dependent load impedance into another specified one. We refer to this operation as impedance matching or equalization, and the resulting coupling network as a matching network or equalizer.

Refer to the network configuration of Figure 9.1 where the source is represented either by its Thévenin equivalent or by its Norton equivalent. Our objective here is to design a lossless two-port network or equalizer N, which when inserted between a resistive source and a resistive load will yield a preassigned transducer power-gain characteristic over the entire sinusoidal frequency spectrum. Explicit formulas for the design of Butterworth and Chebyshev LC ladder networks will be given. The more complicated situation where the load is frequency dependent will be discussed in Chapter 10.

In the networks of Figure 9.1, let $Z_{11}(s)$ and $Z_{22}(s)$ be the impedances looking into the input and output ports when the output and input ports are terminated in $z_2(s)$ and $z_1(s)$, respectively. The input and output reflection coefficients are defined by

$$\rho_{11}(s) = \frac{Z_{11}(s) - z_1(-s)}{Z_{11}(s) + z_1(s)} \tag{9.1a}$$

$$\rho_{22}(s) = \frac{Z_{22}(s) - z_2(-s)}{Z_{22}(s) + z_2(s)} \tag{9.1b}$$

respectively. We now demonstrate that the transducer power gain $G(\omega^2)$ defined as the ratio of average power delivered to the load to the maximum available average power at the source is given by

$$G(\omega^2) = 1 - |\rho_{11}(j\omega)|^2 \equiv |\rho_{21}(j\omega)|^2 \tag{9.2}$$

where $\rho_{21}(s)$ is known as the transmission coefficient. To prove this, we first compute

$$1 - \rho_{11}(s)\rho_{11}(-s) = \frac{4r_1(s)R_{11}(s)}{[Z_{11}(s) + z_1(s)][Z_{11}(-s) + z_1(-s)]} \tag{9.3}$$

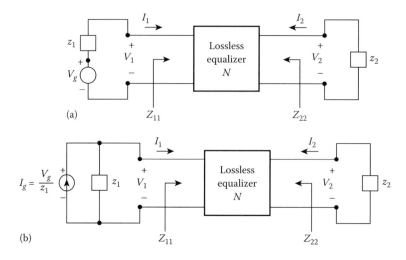

FIGURE 9.1 General broadband matching configuration. (a) The Thévenin equivalent. (b) The Norton equivalent.

where

$$r_1(s) = \text{Ev } z_1(s) = \frac{1}{2}[z_1(s) + z_1(-s)] \tag{9.4a}$$

$$R_{11}(s) = \text{Ev } Z_{11}(s) = \frac{1}{2}[Z_{11}(s) + Z_{11}(-s)] \tag{9.4b}$$

are the even parts of $z_1(s)$ and $Z_{11}(s)$, respectively. Thus, the impedance facing the voltage source V_g is given by

$$\frac{V_g(s)}{I_1(s)} = Z_{11}(s) + z_1(s) \tag{9.5}$$

On the $j\omega$-axis, Equation 9.3 reduces to

$$1 - |\rho_{11}(j\omega)|^2 = \frac{4r_1(j\omega)R_{11}(j\omega)}{|Z_{11}(j\omega) + z_1(j\omega)|^2} = \frac{|I_1(j\omega)|^2 R_{11}(j\omega)}{|V_g(j\omega)|^2/4r_1(j\omega)} \tag{9.6}$$

The power input to the network of Figure 9.1a under sinusoidal steady state is

$$P_{\text{in}} = |I_1(j\omega)|^2 R_{11}(j\omega) \tag{9.7}$$

while the power output to the load is

$$P_{\text{out}} = |I_2(j\omega)|^2 r_2(j\omega) \tag{9.8}$$

where

$$r_2(s) = \text{Ev } z_2(s) = \frac{1}{2}[z_2(s) + z_2(-s)] \tag{9.9}$$

is the even part of $z_2(s)$. Since the two-port N is lossless, the power input must be equal to power output or

$$|I_1(j\omega)|^2 R_{11}(j\omega) = |I_2(j\omega)|^2 r_2(j\omega) \tag{9.10}$$

The maximum average power that the source combination is capable of delivering to the network occurs when the input port is conjugately matched or $Z_{11}(j\omega) = -z_1(j\omega)$. Under this condition, the maximum available average power from the source combination is

$$P_{ava} = \frac{|V_g(j\omega)|^2}{4r_1(j\omega)} \tag{9.11}$$

Substituting Equations 9.8 and 9.11 in Equation 9.6 yields

$$1 - |\rho_{11}(j\omega)|^2 = \frac{|I_2(j\omega)|^2 r_2(j\omega)}{|V_g(j\omega)|^2/4r_1(j\omega)} = \frac{P_{out}}{P_{ava}} = G(\omega^2) \tag{9.12}$$

Using Equation 9.2 shows that

$$|\rho_{11}(j\omega)|^2 + |\rho_{21}(j\omega)|^2 = 1 \tag{9.13}$$

or

$$
\begin{aligned}
|\rho_{11}(j\omega)|^2 &= 1 - \frac{\text{average power to load}}{\text{average power available}} \\
&= \frac{\text{average power available} - \text{average power to load}}{\text{average power available}} \\
&= \frac{\text{``average reflected'' power}}{\text{average power available}}
\end{aligned}
\tag{9.14}
$$

$$|\rho_{21}(j\omega)|^2 = \frac{\text{average power to load}}{\text{average power available}} \tag{9.15}$$

Therefore, the magnitude squared of the reflection coefficient $|\rho_{11}(j\omega)|^2$ denotes the fraction of the maximum available average power that is reflected back to the source, and the magnitude squared of the transmission coefficient $|\rho_{21}(j\omega)|^2$ represents the fraction of the maximum available average power that is transmitted to the load from the source. In fact, their names are suggested by these interpretations. We remark that since the transducer power gain G is a function of ω^2, it is written as $G(\omega^2)$ to emphasize this.

We next express the transmission coefficient in terms of other specifications. From Equation 9.12, we have

$$|\rho_{21}[j\omega]|^2 = 4r_1(j\omega)r_2(j\omega)\left|\frac{I_2(j\omega)}{V_g(j\omega)}\right|^2 \tag{9.16}$$

Substituting $V_g(j\omega) = z_1(j\omega)I_g(j\omega)$ in Equation 9.16 gives

$$|\rho_{21}(j\omega)|^2 = \frac{4r_1(j\omega)r_2(j\omega)}{|z_1(j\omega)|^2}\left|\frac{I_2(j\omega)}{I_g(j\omega)}\right|^2 \tag{9.17}$$

In terms of the transfer voltage ratio and transfer impedance, we apply the relation $V_2(j\omega) = -I_2(j\omega)$ $z_2(j\omega)$ and obtain

$$|\rho_{21}(j\omega)|^2 = \frac{4r_1(j\omega)r_2(j\omega)}{|z_2(j\omega)|^2}\left|\frac{V_2(j\omega)}{V_g(j\omega)}\right|^2 \tag{9.18}$$

$$|\rho_{21}(j\omega)|^2 = \frac{4r_1(j\omega)r_2(j\omega)}{|z_1(j\omega)z_2(j\omega)|^2}\left|\frac{V_2(j\omega)}{I_g(j\omega)}\right|^2 \tag{9.19}$$

Similarly, we can derive a relation between the output reflection coefficient $\rho_{22}(j\omega)$ and the transmission coefficient $\rho_{12}(j\omega)$ magnitude squared as

$$|\rho_{12}(j\omega)|^2 \equiv 1 - |\rho_{22}(j\omega)|^2 \tag{9.20}$$

In fact, for the lossless reciprocal two-port network N we have

$$|\rho_{21}(j\omega)|^2 = |\rho_{12}(j\omega)|^2 = 1 - |\rho_{11}(j\omega)|^2 = 1 - |\rho_{22}(j\omega)|^2 \tag{9.21}$$

9.2 Double-Terminated Butterworth Networks

In this part, we show how to design a lossless two-port network operating between a resistive generator with internal resistance R_1 and a resistive load with resistance R_2 to yield the nth-order Butterworth transducer power-gain characteristic

$$G(\omega^2) = |\rho_{21}(j\omega)|^2 = \frac{K_n}{1 + (\omega/\omega_c)^{2n}} \tag{9.22}$$

Since for a passive network $G(\omega^2)$ is bounded between 0 and 1, the DC gain K_n is restricted by

$$0 \leq K_n \leq 1 \tag{9.23}$$

Substituting Equation 9.22 in Equation 9.12 yields the squared magnitude of the input reflection coefficient as

$$|\rho_{11}(j\omega)|^2 = 1 - G(\omega^2) = 1 - |\rho_{21}(j\omega)|^2 = \frac{1 - K_n + (\omega/\omega_c)^{2n}}{1 + (\omega/\omega_c)^{2n}} \tag{9.24}$$

or

$$\rho_{11}(j\omega)\rho_{11}(-j\omega) = \alpha^{2n}\frac{1 + (\omega/\alpha\omega_c)^{2n}}{1 + (\omega/\omega_c)^{2n}} \tag{9.25}$$

where

$$\alpha = (1 - K_n)^{1/2n} \tag{9.26}$$

Appealing to analytic continuation by substituting ω by $-js$ results in

$$\rho_{11}(s)\rho_{11}(-s) = \alpha^{2n}\frac{1+(-1)^n x^{2n}}{1+(-1)^n y^{2n}} \tag{9.27}$$

where

$$y = \frac{s}{\omega_c}, \quad x = \frac{y}{\alpha} \tag{9.28}$$

To obtain the input reflection coefficient $\rho_{11}(s)$ from $\rho_{11}(s)\,\rho_{11}(-s)$, we need to assign the zeros and poles of Equation 9.27. Since $\rho_{11}(s)$ is devoid of poles in the closed RHS, we must assign all the LHS poles to $\rho_{11}(s)$. The zeros of $\rho_{11}(s)$, however, may lie in the RHS, so that in general a number of different numerators are possible. For our purposes, we choose only the LHS zeros for $\rho_{11}(s)$. Define a minimum-phase reflection coefficient to be one that is devoid of zeros in the open RHS. Then, the minimum-phase solution of Equation 9.27 can be written as

$$\rho_{11}(s) = \pm\alpha^n\frac{q(x)}{q(y)} \tag{9.29}$$

where $q(x)$ is the Hurwitz polynomial with unity leading coefficient formed by the LHS roots of the equation $1+(-1)^n x^{2n} = 0$. From Equation 9.1a, the input impedance is found to be

$$Z_{11}(s) = R_1\frac{1+\rho_{11}(s)}{1-\rho_{11}(s)} \tag{9.30}$$

Combining this with Equation 9.29 yields

$$Z_{11}(s) = R_1\frac{q(y)\pm\alpha^n q(x)}{q(y)\mp\alpha^n q(x)} \tag{9.31}$$

If both R_1 and R_2 are specified, then the DC gain K_n cannot be chosen independently. In fact, by substituting $s=0$ in Equation 9.31 and assuming that $K_n \neq 0$ we obtain

$$\frac{R_2}{R_1} = \left(\frac{1+\alpha^n}{1-\alpha^n}\right)^{\pm 1} \tag{9.32}$$

where the $\pm$ signs are determined, respectively, according to $R_2 \geq R_1$ and $R_2 \leq R_1$. Therefore, if any two of the three quantities R_1, R_2, and K_n are specified, the third one is fixed.

We now show that the input impedance $Z_{11}(s)$ can be realized by an LC ladder terminated in a resistor. In fact, explicit formulas for their element values will be given, thereby reducing the design problem to simple arithmetic. Depending upon the choice of the plus and minus signs in Equation 9.32, two cases are distinguished.

Case 1. $\rho_{11}(0) \geq 0$. With the choice of the plus sign, the input impedance becomes

$$Z_{11}(s) = R_1\frac{q(y)+\alpha^n q(x)}{q(y)-\alpha^n q(x)} \tag{9.33}$$

which can be expanded in a continued fraction about infinity, as in the first Cauer canonical form, and results in an LC ladder terminated in a resistor:

$$Z_{11}(s) = L_1 s + \cfrac{1}{C_2 S + \cfrac{1}{L_3 s + \cfrac{1}{\ddots + \cfrac{1}{W}}}} \tag{9.34}$$

where W is a constant representing either a resistance or conductance. Depending upon whether n is odd or even, the LC ladder has the configuration of Figure 9.2. The element values can be computed by the following recurrence formulas:

$$L_1 = \frac{2R_1 \sin \pi/2n}{(1-\alpha)\omega_c} \tag{9.35}$$

$$L_{2m-1}C_{2m} = \frac{4 \sin \gamma_{4m-3} \sin \gamma_{4m-1}}{\omega_c^2(1 - 2\alpha \cos \gamma_{4m-2} + \alpha^2)} \tag{9.36a}$$

$$L_{2m+1}C_{2m} = \frac{4 \sin \gamma_{4m-1} \sin \gamma_{4m+1}}{\omega_c^2(1 - 2\alpha \cos \gamma_{4m} + \alpha^2)} \tag{9.36b}$$

for $m = 1, 2, \ldots, \lceil n/2 \rceil$, the largest integer not greater than $n/2$; where

$$\gamma_m = \frac{m\pi}{2n} \tag{9.37}$$

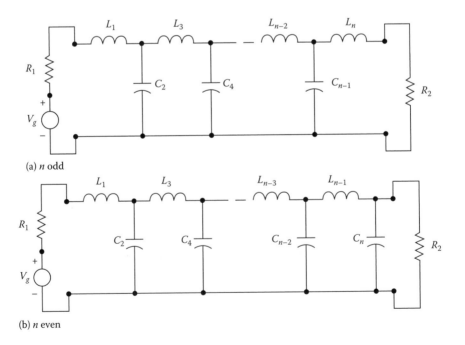

(a) *n* odd

(b) *n* even

FIGURE 9.2   Butterworth LC ladder networks for $\rho_{11}(0) \geq 0$.

The values of the final elements can also be calculated directly by

$$L_n = \frac{2R_2 \sin \pi/2n}{(1+\alpha)\omega_c}, \quad n \text{ odd} \tag{9.38a}$$

$$C_n = \frac{2 \sin \pi/2n}{R_2(1+\alpha)\omega_c}, \quad n \text{ even} \tag{9.38b}$$

A complete derivation of these formulas was first given by Bossé [1]. Hence we can calculate the element values starting from either the first or the last element. When $R_1 = R_2$, formulas Equation 9.36 reduce to

$$L_{2m-1} = \frac{2R_1 \sin \gamma_{4m-3}}{\omega_c} \tag{9.39a}$$

$$C_{2m} = \frac{2 \sin \gamma_{4m-1}}{R_1 \omega_c} \tag{9.39b}$$

Example 9.1

Given

$$R_1 = 70 \ \Omega, \quad R_2 = 200 \ \Omega, \quad \omega_c = 10^5 \text{ rad/s}, \quad n = 4 \tag{9.40}$$

obtain a Butterworth LC ladder to meet these specifications.

Since $R_2 > R_1$, we choose the plus sign in Equation 9.32 and obtain $\rho_{11}(0) \geq 0$, $\alpha = 0.833$, and $\gamma_m = 22.5m$. Thus, from Equations 9.35 and 9.36 the element values are found to be

$$L_1 = \frac{2 \times 70 \sin 22.5°}{(1-0.833) \times 10^5} = 3.2081 \text{ mH} \tag{9.41a}$$

$$C_2 = \frac{4 \sin 22.5° \sin 67.5°}{L_1(1.6939 - 1.666 \cos 45°) \times 10^{10}} = 0.085456 \ \mu\text{F} \tag{9.41b}$$

$$L_3 = \frac{4 \sin 67.5° \sin 112.5°}{C_2(1.6939 - 1.666 \cos 90°) \times 10^{10}} = 2.3587 \text{ mH} \tag{9.41c}$$

$$C_4 = \frac{4 \sin 112.5° \sin 157.5°}{L_3(1.6939 - 1.666 \cos 135°) \times 10^{10}} = 0.020877 \ \mu\text{F} \tag{9.41d}$$

Alternatively, C_4 can be computed directly from Equation 9.39b as

$$C_4 = \frac{2 \sin 22.5°}{200 \times (1+0.833) \times 10^5} = 0.020877 \ \mu\text{F} \tag{9.42}$$

The ladder network together with its termination is presented in Figure 9.3. This network possesses the fourth-order Butterworth transducer power-gain response with a DC gain

$$K_4 = 1 - \alpha^8 = 0.7682 \tag{9.43}$$

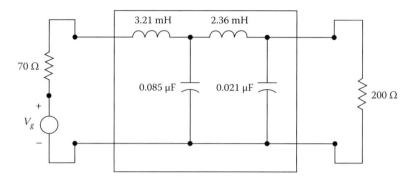

FIGURE 9.3   Fourth-order Butterworth LC ladder network.

Case 2. $\rho_{11}(0) < 0$. With the choice of the minus sign, the input impedance can be expanded in a continued fraction as

$$\frac{1}{Z_{11}(s)} = C_1 s + \cfrac{1}{L_2 s + \cfrac{1}{C_3 s + \cfrac{1}{\ddots + \cfrac{1}{W}}}} \tag{9.44}$$

which can be realized by the LC ladder networks of Figure 9.4, depending on whether W is even or odd, where W is the terminating resistance or conductance. Formulas for the element values are similar to

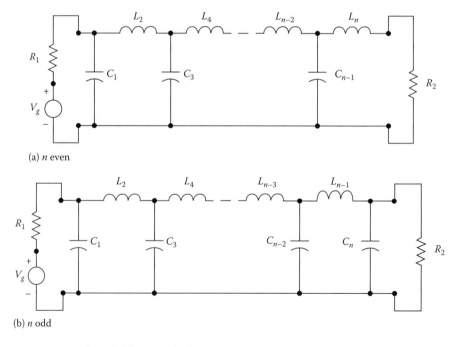

FIGURE 9.4   Butterworth LC ladder networks for $\rho_{11}(0) < 0$.

those given in Equations 9.35 through 9.39 except that the roles of C's and L's are interchanged and R_1 and R_2 are replaced by their reciprocals.

$$C_1 = \frac{2\sin\pi/2n}{R_1(1-\alpha)\omega_c} \tag{9.45a}$$

$$C_{2m-1}L_{2m} = \frac{4\sin\gamma_{4m-3}\sin\gamma_{4m-1}}{\omega_c^2(1-2\alpha\cos\gamma_{4m-2}+\alpha^2)} \tag{9.45b}$$

$$C_{2m+1}L_{2m} = \frac{4\sin\gamma_{4m-1}\sin\gamma_{4m+1}}{\omega_c^2(1-2\alpha\cos\gamma_{4m}+\alpha^2)} \tag{9.45c}$$

for $m = 1, 2, \ldots, \lceil n/2 \rceil$. The values of the final elements can also be calculated directly by

$$C_n = \frac{2\sin\pi/2n}{R_2(1+\alpha)\omega_c}, \quad n \text{ odd} \tag{9.46a}$$

$$L_n = \frac{2R_2\ \sin\pi/2n}{(1+\alpha)\omega_c}, \quad n \text{ even} \tag{9.46b}$$

9.3 Double-Terminated Chebyshev Networks

Now, we consider the problem of synthesizing an LC ladder which when connected between a resistive source of internal resistance R_1 and a resistive load of resistance R_2 will yield a preassigned Chebyshev transducer power-gain characteristic

$$G(\omega^2) = |\rho_{21}(j\omega)|^2 = \frac{K_n}{1+\varepsilon^2 C_n^2(\omega/\omega_c)} \tag{9.47}$$

with K_n bounded between 0 and 1. Following Equation 9.24, the squared magnitude of the input reflection coefficient can be written as

$$|\rho_{11}(j\omega)|^2 = 1 - G(\omega^2) = 1 - |\rho_{21}(j\omega)|^2 = \frac{1-K_n+\varepsilon^2 C_n^2(\omega/\omega_c)}{1+\varepsilon^2 C_n^2(\omega/\omega_c)} \tag{9.48}$$

Appealing to analytic continuation, we obtain

$$\rho_{11}(s)\rho_{11}(-s) = (1-K_n)\frac{1+\hat\varepsilon^2 C_n^2(-jy)}{1+\varepsilon^2 C_n^2(-jy)} \tag{9.49}$$

where

$$\hat\varepsilon = \frac{\varepsilon}{\sqrt{1-K_n}} \tag{9.50}$$

As in the Butterworth case, we assign LHS poles to $\rho_{11}(s)$ and the minimum-phase solution of Equation 9.49 becomes

$$\rho_{11}(s) = \pm\frac{\hat p(y)}{p(y)} \tag{9.51}$$

where $p(y)$ and $\hat{p}(y)$ are the Hurwitz polynomials with unity leading coefficient formed by the LHS roots of the equations $1 + \varepsilon^2 C_n^2(-jy) = 0$ and $1 + \hat{\varepsilon}^2 C_n^2(-jy) = 0$, respectively. From Equation 9.1a, the input impedance of the LC ladder when the output port is terminated in R_2 is found to be

$$Z_{11}(s) = R_1 \frac{p(y) \pm \hat{p}(y)}{p(y) \mp \hat{p}(y)} \tag{9.52}$$

A relationship among the quantities R_1, R_2, and K_n is given by

$$\frac{R_2}{R_1} = \left(\frac{1 + \sqrt{1 - K_n}}{1 - \sqrt{1 - K_n}} \right)^{\pm 1}, \quad n \text{ odd} \tag{9.53a}$$

$$= \left(\frac{\sqrt{1 + \varepsilon^2} + \sqrt{1 + \varepsilon^2 - K_n}}{\sqrt{1 + \varepsilon^2} - \sqrt{1 + \varepsilon^2 - K_n}} \right)^{\pm 1}, \quad n \text{ even} \tag{9.53b}$$

where the $\pm$ signs are determined, respectively, according to $R_2 \geq R_1$ and $R_2 \leq R_1$. Therefore, if n is odd and the DC gain is specified, the ratio of the terminating resistances is fixed by Equation 9.53a. On the other hand, if n is even and the peak-to-peak ripple in the passband and K_n or the DC gain is specified, the ratio of the resistances is given by Equation 9.53b.

We now show that the input impedance $Z_{11}(s)$ can be realized by an LC ladder terminated in a resistor. Again, explicit formulas for their element values will be given, thereby reducing the design problem to simple arithmetic. Depending upon the choice of the plus and minus signs in Equation 9.51, two cases are distinguished.

Case 1. $\rho_{11}(0) \geq 0$. With the choice of the plus sign, the input impedance becomes

$$Z_{11}(s) = R_1 \frac{p(y) + \hat{p}(y)}{p(y) - \hat{p}(y)} \tag{9.54}$$

which can be expanded in a continued fraction as in Equation 9.34. Depending on whether n is odd or even, the corresponding LC ladder network has the configurations of Figure 9.2. The element values can be computed by the following recurrence formulas:

$$L_1 = \frac{2R_1 \sin \pi / 2n}{(\sinh a - \sinh \hat{a})\dot{\omega}_c} \tag{9.55}$$

$$L_{2m-1}C_{2m} = \frac{4 \sin \gamma_{4m-3} \sin \gamma_{4m-1}}{\omega_c^2 f_{2m-1}(\sinh a, \sinh \hat{a})} \tag{9.56a}$$

$$L_{2m+1}C_{2m} = \frac{4 \sin \gamma_{4m-1} \sin \gamma_{4m+1}}{\omega_c^2 f_{2m}(\sinh a, \sinh \hat{a})} \tag{9.56b}$$

for $m = 1, 2, \ldots, \lceil n/2 \rceil$, where

$$\gamma_m = \frac{m\pi}{2n} \tag{9.57a}$$

$$\hat{a} = \frac{1}{n} \sinh^{-1} \left(\frac{\sqrt{1 - K_n}}{\varepsilon} \right) \tag{9.57b}$$

$$f_m(u, v) = u^2 + v^2 + \sin^2 \gamma_{2m} - 2uv \cos \gamma_{2m} \tag{9.57c}$$

In addition, the values of the last elements can also be computed directly by the equations

$$L_n = \frac{2R_2 \sin \pi/2n}{(\sinh a + \sinh \hat{a})\omega_c}, \quad n \text{ odd} \tag{9.58a}$$

$$C_n = \frac{2 \sin \pi/2n}{R_2(\sinh a + \sinh \hat{a})\omega_c}, \quad n \text{ even} \tag{9.58b}$$

A formal proof of these formulas was first given by Takahasi [2]. Hence, we can calculate the element values starting from either the first or the last element.

Example 9.2

Given

$$R_1 = 150 \ \Omega, \quad R_2 = 470 \ \Omega, \quad \omega_c = 10^8\pi \text{ rad/s}, \quad n = 4 \tag{9.59}$$

find a Chebyshev LC ladder network to meet these specifications with peak-to-peak ripple in the passband not exceeding 1.5 dB.

Since R_1 and R_2 are both specified, the minimum passband gain $G_{\min}$ is fixed by Equation 9.53b as

$$G_{\min} \equiv \frac{K_n}{1 + \varepsilon^2} = 1 - \left(\frac{\dfrac{470}{150} - 1}{\dfrac{470}{150} + 1}\right)^2 = 0.7336 \tag{9.60}$$

For the 1.5 dB ripple in the passband, the corresponding ripple factor is given by

$$\varepsilon = \sqrt{10^{0.15} - 1} = 0.64229 \tag{9.61}$$

obtaining $K_4 = 1.036$, which is too large for the network to be physically realizable. Thus, let $K_4 = 1$, the maximum permissible value, and the corresponding ripple factor becomes

$$\varepsilon = \sqrt{\frac{1}{G_{\min}} - 1} = \sqrt{\frac{1}{0.7336} - 1} = 0.6026 \quad \text{or} \quad 1.345 \text{ dB} \leq 1.5 \text{ dB} \tag{9.62}$$

We next compute the quantities

$$\gamma_m = \frac{m\pi}{2n} = 22.5m \tag{9.63a}$$

$$a = \frac{1}{4} \sinh^{-1} \frac{1}{0.6026} = 0.32, \quad \hat{a} = 0 \tag{9.63b}$$

$$f_m(\sinh 0.32, 0) = \sinh^2 0.32 + \sin^2 \gamma_{2m} = 0.1059 + \sin^2 \gamma_{2m} \tag{9.63c}$$

Appealing to formulas Equations 9.55 and 9.56, the element values are calculated as follows:

$$L_1 = \frac{2 \times 150 \ \sin 22.5°}{(\sinh 0.32 - \sinh 0) \times 10^8\pi} = 1.123 \ \mu\text{H} \tag{9.64a}$$

$$C_2 = \frac{4 \sin 22.5° \sin 67.5°}{L_1 10^{16} \pi^2 \left(\sinh^2 0.32 + \sin^2 45°\right)} = 21.062 \text{ pF} \tag{9.64b}$$

$$L_3 = \frac{4 \sin 67.5° \sin 112.5°}{C_2 10^{16} \pi^2 \left(\sinh^2 0.32 + \sin^2 90°\right)} = 1.485 \text{ μH} \tag{9.64c}$$

$$C_4 = \frac{4 \sin 112.5° \sin 157.5°}{L_3 10^{16} \pi^2 \left(\sinh^2 0.32 + \sin^2 135°\right)} = 15.924 \text{ pF} \tag{9.64d}$$

Alternatively, the last capacitance can also be computed directly from Equation 9.58b as

$$C_4 = \frac{2 \sin 22.5°}{470 \times (\sinh 0.32 + \sinh 0) \times 10^8 \pi} = 15.925 \text{ pF} \tag{9.65}$$

The LC ladder together with its terminations is presented in Figure 9.5.

Case 2. $\rho_{11}(0) < 0$. With the choice of the minus sign in Equation 9.51, the input impedance, aside from the constant R_1, becomes the reciprocal of Equation 9.54

$$Z_{11}(s) = R_1 \frac{p(y) - \hat{p}(y)}{p(y) + \hat{p}(y)} \tag{9.66}$$

and can be expanded in a continued fraction as that shown in Equation 9.44. Depending on whether n is even or odd, the LC ladder network has the configurations of Figure 9.4. Formulas for the element values are similar to those given in Equations 9.55 through 9.58 except that the roles of C's and L's are interchanged and R_1 and R_2 are replaced by their reciprocals:

$$C_1 = \frac{2 \sin \pi/2n}{R_1 (\sinh a - \sinh \hat{a}) \omega_c} \tag{9.67a}$$

$$C_{2m-1} L_{2m} = \frac{4 \sin \gamma_{4m-3} \sin \gamma_{4m-1}}{\omega_c^2 f_{2m-1} (\sinh a, \sinh \hat{a})} \tag{9.67b}$$

$$C_{2m+1} L_{2m} = \frac{4 \sin \gamma_{4m-1} \sin \gamma_{4m+1}}{\omega_c^2 f_{2m} (\sinh a, \sinh \hat{a})} \tag{9.67c}$$

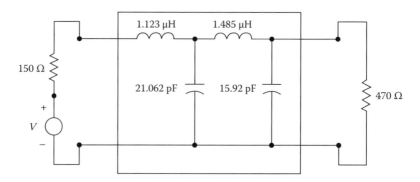

FIGURE 9.5 Fourth-order Chebyshev LC ladder network for $\rho_{11}(0) \geq 0$.

for $m = 1, 2, \ldots, \lceil n/2 \rceil$, where γ_m and $f_m(\sinh a, \sinh \hat{a})$ are defined in Equation 9.57. In addition, the values of the last elements can also be computed directly by the formulas

$$C_n = \frac{2 \sin \pi/2n}{R_2(\sinh a + \sinh \hat{a})\omega_c}, \quad n \text{ odd} \tag{9.68a}$$

$$L_n = \frac{2R_2 \sin \pi/2n}{(\sinh a + \sinh \hat{a})\omega_c}, \quad n \text{ even} \tag{9.68b}$$

References

1. G. Bossé, Siebketten ohne Dämpfungsschwankungen im Durchlassbereich (Potenzketten), *Frequenz*, 5, 279–284, 1951.
2. H. Takahasi, On the ladder-type filter network with Tchebysheff response, *J. Inst. Elec. Commun. Engrs. Japan*, 34, 65–74, 1951.

10

Design of Broadband Matching Networks

Wai-Kai Chen
University of Illinois at Chicago

10.1 Introduction

Refer to the network configuration of Figure 10.1 where the source is represented either by its Thévenin equivalent or by its Norton equivalent. The load impedance $z_2(s)$ is assumed to be strictly passive over a frequency band of interest, because the matching problem cannot be meaningfully defined if the load is purely reactive. Our objective is to design an "optimum" lossless two-port network or equalizer N to match out the load impedance $z_2(s)$ to the resistive source impedance $z_1(s) = R_1$, and to achieve a preassigned transducer power-gain characteristic $G(\omega^2)$ over the entire sinusoidal frequency spectrum.

As stated in Chapter 9, the output reflection coefficient is given by

$$\rho_{22}(s) = \frac{Z_{22}(s) - z_2(-s)}{Z_{22}(s) + z_2(s)} \tag{10.1}$$

where $Z_{22}(s)$ is the impedance looking into the output port when the input port is terminated in the source resistance R_1. As shown in Chapter 9, the transducer power gain $G(\omega^2)$ is related to the transmission and reflection coefficients by the equation

$$G(\omega^2) = |\rho_{21}(j\omega)|^2 = |\rho_{12}(j\omega)|^2 = 1 - |\rho_{11}(j\omega)|^2 = 1 - |\rho_{22}(j\omega)|^2 \tag{10.2}$$

Recall that in computing $\rho_{11}(s)$ from $\rho_{11}(s)\rho_{11}(-s)$ we assign all of the LHS poles of $\rho_{11}(s)\rho_{11}(-s)$ to $\rho_{11}(s)$ because with resistive load $z_2(s) = R_2$, $\rho_{11}(s)$ is devoid of poles in the RHS. For the complex load, the poles of $\rho_{22}(s)$ include those of $z_2(-s)$, which may lie in the open RHS. As a result, the assignment of poles of $\rho_{22}(s)\rho_{22}(-s)$ is not unique. Furthermore, the nonanalyticity of $\rho_{22}(s)$ leaves much to be desired in terms of our ability to manipulate. For these reasons, we consider the normalized reflection coefficient defined by

$$\rho(s) = A(s)\rho_{22}(s) = A(s)\frac{Z_{22}(s) - z_2(-s)}{Z_{22}(s) + z_2(s)} \tag{10.3}$$

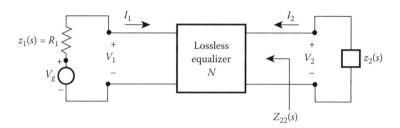

FIGURE 10.1   Schematic of broadband matching.

where

$$A(s) = \prod_{i=1}^{q} \frac{s - s_i}{s + s_i}, \quad \mathrm{Re}\; s_i > 0 \tag{10.4}$$

is the real all-pass function defined by the open RHS poles s_i ($i = 1, 2, \ldots, q$) of $z_2(-s)$. An all-pass function is a function whose zeros are all located in the open RHS and whose poles are located at the LHS mirror image of the zeros. Therefore, it is analytic in the closed RHS and such that

$$A(s)A(-s) = 1 \tag{10.5}$$

On the $j\omega$-axis, the magnitude of $A(j\omega)$ is unity, being flat for all sinusoidal frequencies, and we have

$$|\rho(j\omega)| = |A(j\omega)\rho_{22}(j\omega)| = |\rho_{22}(j\omega)| \tag{10.6}$$

and Equation 10.2 becomes

$$G(\omega^2) = 1 - |\rho(j\omega)|^2 \tag{10.7}$$

This equation together with the normalized reflection coefficient $\rho(s)$ of Equation 10.3 forms the cornerstone of Youla's theory of broadband matching [1].

10.2 Basic Coefficient Constraints

In Chapter 8, we define the zeros of transmission for a terminate two-port network as the frequencies at which a zero output results for a finite input. We extend this concept by defining the zeros of transmission for a one-port impedance.

Definition 10.1:   Zero of transmission. For a given impedance $z_2(s)$, a closed RHS zero of multiplicity k of the function

$$w(s) \equiv \frac{r_2(s)}{z_2(s)} \tag{10.8}$$

where $r_2(s)$ is the even part of $z_2(s)$, is called a zero of transmission of order k of $z_2(s)$.

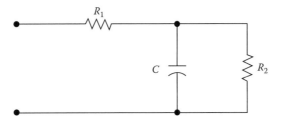

FIGURE 10.2   RC one-port network.

The reason for this name is that if we realize the impedance $z_2(s)$ as the input impedance of a lossless two-port network terminated in a $1 - \Omega$ resistor, the magnitude squared of the transfer impedance function $Z_{12}(j\omega)$ between the $1 - \Omega$ resistor and the input equals the real part of the input impedance,

$$|Z_{12}(j\omega)|^2 = \operatorname{Re} z_2(j\omega) = r_2(j\omega) \tag{10.9}$$

After appealing to analytic continuation by substituting ω by $-js$, the zeros of $r_2(s)$ are seen to be the zeros of transmission of the lossless two-port.

Consider, for example, the RC impedance $z_2(s)$ of Figure 10.2,

$$z_2(s) = R_1 + \frac{R_2}{R_2 Cs + 1} \tag{10.10}$$

the even part of which is given by

$$r_2(s) = \frac{1}{2}[z_2(s) + z_2(-s)] = \frac{R_1 + R_2 - R_1 R_2^2 C^2 s^2}{1 - R_2^2 C^2 s^2} \tag{10.11}$$

obtaining

$$w(s) = \frac{r_2(s)}{z_2(s)} = \frac{R_2 Cs^2 - (R_1 + R_2)/R_1 R_2 C}{(R_2 Cs - 1)[s + (R_1 + R_2)/R_1 R_2 C]} \tag{10.12}$$

Thus, the impedance $z_2(s)$ has a zero of transmission of order 1 located at

$$s = \sigma_0 = \frac{1}{R_2 C}\sqrt{1 + \frac{R_2}{R_1}} \tag{10.13}$$

For our purposes, the zeros of transmission are divided into four mutually exclusive classes.

Definition 10.2:   *Classification of zeros of transmission.* Let $s_0 = \sigma_0 + j\omega_0$ be a zero of transmission of an impedance $z_2(s)$. Then s_0 belongs to one of the following four mutually exclusive classes depending on σ_0 and $z_2(s_0)$, as follows:

Class I: $\sigma_0 > 0$, which includes all the open RHS zeros of transmission.
Class II: $\sigma_0 = 0$ and $z_2(\omega_0) = 0$.

Class III: $\sigma_0 = 0$ and $0 < |z_2(j\omega 0)| < \infty$.
Class IV: $\sigma_0 = 0$ and $|z_2(j\omega 0)| = \infty$.

For the impedance $z_2(s)$ of Equation 10.10, its zero of transmission given in Equation 10.13 belongs to Class I of order 1. If $z_2(s)$ is the load of the network of Figure 10.1, it imposes the basic constraints on the normalized reflection coefficient $\rho(s)$. These constraints are important in that they are necessary and sufficient for $\rho(s)$ to be physically realizable, and are most conveniently formulated in terms of the coefficients of the Laurent series expansions of the following quantities about a zero of transmission $s_0 = \sigma_0 + j\omega_0$ of order k of $z_2(s)$:

$$\rho(s) = \rho_0 + \rho_1(s - s_0) + \rho_2(s - s_0)^2 + \cdots = \sum_{m=0}^{\infty} \rho_m(s - s_0)^m \tag{10.14a}$$

$$A(s) = A_0 + A_1(s - s_0) + A_2(s - s_0)^2 + \cdots = \sum_{m=0}^{\infty} A_m(s - s_0)^m \tag{10.14b}$$

$$F(s) \equiv 2r_2(s)A(s) = F_0 + F_1(s - s_0) + F_2(s - s_0)^2 + \cdots = \sum_{m=0}^{\infty} F_m(s - s_0)^m \tag{10.14c}$$

We remark that the expansions of the Laurent type can be found by any method because it is unique, and the resulting expansion is *the* Laurent series expansion. For the zero of transmission at infinity, the expansions take the form

$$\rho(s) = \rho_0 + \frac{\rho_1}{s} + \frac{\rho_2}{s^2} + \frac{\rho_3}{s^3} + \cdots = \sum_{m=0}^{\infty} \frac{\rho_m}{s^m} \tag{10.15a}$$

$$A(s) = A_0 + \frac{A_1}{s} + \frac{A_2}{s^2} + \frac{A_3}{s^3} + \cdots = \sum_{m=0}^{\infty} \frac{A_m}{s^m} \tag{10.15b}$$

$$F(s) = F_0 + \frac{F_1}{s} + \frac{F_2}{s^2} + \frac{F_3}{s^3} + \cdots = \sum_{m=0}^{\infty} \frac{F_m}{s^m} \tag{10.15c}$$

In fact, they can be obtained by means of the binomial expansion formula

$$(s + c)^n = s^n + ns^{n-1}c + \frac{n(n-1)}{2!}s^{n-2}c^2 + \cdots \tag{10.16}$$

which is valid for all values of n if $|s| > |c|$, and is valid only for nonnegative integers n if $|s| \le |c|$.

Example 10.1

Assume that the network of Figure 10.1 is terminated in the passive impedance

$$z_2(s) = \frac{s}{s^2 + 2s + 1} \tag{10.17}$$

and possesses the transducer power-gain characteristic

$$G(\omega^2) = \frac{K\omega^2}{\omega^4 - \omega^2 + 1}, \quad 0 \le K \le 1 \tag{10.18}$$

We first compute the even part

$$r_2(s) = \frac{1}{2}[z_2(s) + z_2(-s)] = \frac{-2s^2}{(s^2 + 2s + 1)(s^2 - 2s + 1)} \tag{10.19}$$

of $z_2(s)$, and obtain the function

$$w(s) = \frac{r_2(s)}{z_2(s)} = \frac{-2s}{s^2 - 2s + 1} \tag{10.20}$$

showing that $z_2(s)$ possesses two Class II zeros of transmission at $s = 0$ and ∞. Since the pole of $z_2(-s)$ is located at $s = 1$ of order 2, the all-pass function $A(s)$ takes the form

$$A(s) = \frac{s^2 - 2s + 1}{s^2 + 2s + 1} \tag{10.21}$$

The other required functions are found to be

$$F(s) = 2A(s)r_2(s) = -\frac{4s^2}{(s^2 + 2s + 1)^2} \tag{10.22}$$

$$\rho(s)\rho(-s) = 1 - G(-s^2) = \frac{s^4 + (1 + K)s^2 + 1}{s^4 + s^2 + 1} \tag{10.23}$$

The minimum-phase solution of Equation 10.23 is determined as

$$\pm\hat{\rho}(s) = \frac{s^2 + \sqrt{1 - K}s + 1}{s^2 + s + 1} \tag{10.24}$$

Now, we expand the functions $A(s)$, $F(s)$, and $\hat{\rho}(s)$ in Laurent series about the zeros of transmission at the origin and at infinity, and obtain

$$A(s) = 1 - 4s + \cdots = 1 - \frac{4}{s} + \cdots \tag{10.25}$$

$$F(s) = 0 + 0 - 4s^2 + \cdots = 0 + 0 - \frac{4}{s^2} + \cdots \tag{10.26}$$

$$\pm\hat{\rho}(s) = 1 + (\sqrt{1 - K} - 1)s + \cdots = 1 + \frac{\sqrt{1 - K} - 1}{s} + \cdots \tag{10.27}$$

In both expansions, we can make the following identifications:

$$A_0 = 1, \quad F_0 = 0, \quad \rho_0 = 1 \tag{10.28a}$$

$$A_1 = -4, \quad F_1 = 0, \quad \rho_1 = \sqrt{1 - K} - 1 \tag{10.28b}$$

$$F_2 = -4 \tag{10.28c}$$

10.2.1 Basic Coefficient Constraints on $\rho(S)$

The basic constraints imposed on the normalized reflection coefficient $\rho(s)$ by a load impedance $z_2(s)$ are most succinctly expressed in terms of the coefficients of the Laurent series expansions (Equation 10.14) of the functions $\rho(s)$, $A(s)$, and $F(s)$ about each zero of transmission $s_0 = \sigma_0 + j\omega_0$. Depending on the classification of the zero of transmission, one of the following four sets of coefficient conditions must be satisfied:

Class I: For $x = 0, 1, 2, \ldots, k - 1$

$$A_x = \rho_x \tag{10.29a}$$

Class II: $A_x = \rho_x$ for $x = 0, 1, 2, \ldots, k - 1$, and

$$\frac{A_k - \rho_k}{F_{k+1}} \geq 0 \tag{10.29b}$$

Class III: $A_x = \rho_x$ for $x = 0, 1, 2, \ldots, k - 2, k \geq 2$, and

$$\frac{A_{k-1} - \rho_{k-1}}{F_k} \geq 0 \tag{10.29c}$$

Class IV: $A_x = \rho_x$ for $x = 0, 1, 2, \ldots, k - 1$, and

$$\frac{F_{k-1}}{A_k - \rho_k} \geq a_{-1}, \text{ the residue of } z_2(s) \text{ evaluated at the poles} = j\omega_0 \tag{10.29d}$$

To determine the normalized reflection coefficient ρ (s) from a preassigned transducer power-gain characteristic $G(\omega^2)$, we appeal to Equation 10.7 and analytic continuation by replacing ω by $-js$ and obtain

$$\rho(s)\rho(-s) = 1 - G(-s^2) \tag{10.30}$$

Since the zeros and poles of $\rho(s)\rho(-s)$ must appear in quadrantal symmetry, being symmetric with respect to both the real and imaginary axes of the s-plane, and since $\rho(s)$ is analytic in the closed RHS, the open LHS poles of $\rho(s)\rho(-s)$ belong to $\rho(s)$ whereas those in the open RHS belong to $\rho(-s)$. For a lumped system, $\rho(s)$ is devoid of poles on the $j\omega$-axis. For the zeros, no unique ways are available to assign them. The only requirement is that the complex–conjugate pair of zeros must be assigned together. However, if we specify that $\rho(s)$ be made a minimum-phase function, then all the open LHS zeros of $\rho(s)\rho(-s)$ are assigned to $\rho(s)$. The $j\omega$-axis zeros of $\rho(s)\rho(-s)$ are of even multiplicity, and thus they are divided equally between $\rho(s)$ and $\rho(-s)$. Therefore, $\rho(s)$ is uniquely determined by the zeros and poles of $\rho(s)\rho(-s)$ only if $\rho(s)$ is required to be minimum-phase.

Let $\hat{\rho}(s)$ be the minimum-phase solution of Equation 10.30. Then, any solution of the form

$$\rho(s) = \pm\eta(s)\hat{\rho}(s) \tag{10.31}$$

is admissible, where $\acute{\eta}(s)$ is an arbitrary real all-pass function possessing the property that

$$\eta(s)\eta(-s) = 1 \tag{10.32}$$

The significance of these coefficient constraints is that they are both necessary and sufficient for the physical realizability of $\rho(s)$, and is summarized in the following theorem. The proof of this result can be found in Ref. [2].

THEOREM 10.1

Given a strictly passive impedance $z_2(s)$, the function defined by the equation

$$Z_{22}(s) \equiv \frac{F(s)}{A(s) - \rho(s)} - z_2(s) \tag{10.33}$$

is positive real if and only if $|\rho(j\omega)| \leq 1$ for all ω and the coefficient conditions Equation 10.29 are satisfied.

The function $Z_{22}(s)$ defined in Equation 10.33 is actually the back-end impedance of a desired equalizer. To see this, we solve for $Z_{22}(s)$ in Equation 10.3 and obtain

$$Z_{22}(s) = \frac{A(s)[z_2(s) + z_2(-s)]}{A(s) - \rho(s)} - z_2(s) = \frac{F(s)}{A(s) - \rho(s)} - z_2(s) \tag{10.34}$$

which is guaranteed to be positive real by Theorem 10.1. This impedance can be realized as the input impedance of a lossless two-port network terminated in a resistor. The removal of this resistor gives the desired matching network. An ideal transformer may be needed at the input port to compensate for the actual level of the generator resistance R_1.

Example 10.2

Design a lossless matching network to equalize the load impedance

$$z_2(s) = \frac{s}{s^2 + 2s + 1} \tag{10.35}$$

to a resistive generator of internal resistance of 0.5 Ω and to achieve the transducer power-gain characteristic

$$G(\omega^2) = \frac{K\omega^2}{\omega^4 - \omega^2 + 1}, \quad 0 \leq K \leq 1 \tag{10.36}$$

From Example 10.1, the load possesses two Class II zeros of transmission of order 1 at $s = 0$ and $s = \infty$. The coefficients of the Laurent series expansions of the functions $A(s)$, $F(s)$, and $\rho(s)$ about the zeros of transmission at $s = 0$ and $s = \infty$ were computed in Example 10.1 as

$$A_0 = 1, \quad F_0 = 0, \quad \rho_0 = 1 \tag{10.37a}$$

$$A_1 = -4, \quad F_1 = 0, \quad \rho_1 = \sqrt{1 - K} - 1 \tag{10.37b}$$

$$F_2 = -4 \tag{10.37c}$$

The coefficient constraints (Equation 10.29b) for the Class II zeros of transmission of order 1 become

$$A_0 = \rho_0, \quad \frac{A_1 - \rho_1}{F_2} \geq 0 \tag{10.38}$$

Clearly, the first condition is always satisfied. To meet the second requirement, we set

$$\frac{-4 - (\sqrt{1 - K} - 1)}{-4} \geq 0 \tag{10.39}$$

or $0 \leq K \leq 1$, showing that the maximum realizable K is 1. For our purposes, set $K = 1$ and choose the plus sign in Equation 10.24. From Equation 10.34, the equalizer back-end impedance is computed as

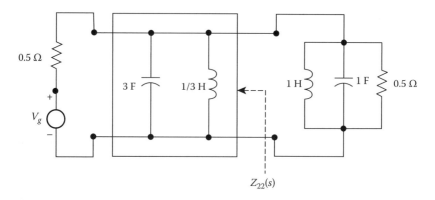

FIGURE 10.3   Equalizer having transducer power-gain characteristics Equation 10.36.

$$Z_{22}(s) = \frac{F(s)}{A(s) - \hat{\rho}(s)} - z_2(s) = \frac{\frac{-4s^2}{(s^2+2s+1)^2}}{\frac{s^2-2s+1}{s^2+2s+1} - \frac{s^2+1}{s^2+s+1}} - \frac{s}{s^2+2s+1}$$

$$= \frac{s}{3s^2 + 2s + 3} \tag{10.40}$$

This impedance can be realized as the input impedance of the parallel connection of an inductor $L = 1/3$ H, a capacitor $C = 3$F, and a resistor $R = 0.5\ \Omega$. The resulting equalizer together with the load is presented in Figure 10.3.

10.3 Design Procedure

We now outline an eight-step procedure for the design of an optimum lossless matching network that equalizes a frequency-dependent load impedance $z_2(s)$ to a resistive generator of internal resistance R_1 and achieves a preassigned transducer power-gain characteristic $G(\omega^2)$ over the entire sinusoidal frequency spectrum.

Step 1. From a preassigned transducer power-gain characteristic $G(\omega^2)$, verify that $G(\omega^2)$ is an even rational real function and satisfies the inequality

$$0 \leq G(\omega^2) \leq 1 \quad \text{for all } \omega \tag{10.41}$$

The gain level is usually not specified to allow some flexibility.

Step 2. From a prescribed strictly passive load impedance $z_2(s)$, compute

$$r_2(s) \equiv \text{Ev } z_2(s) = \frac{1}{2}[z_2(s) + z_2(-s)] \tag{10.42}$$

$$A(s) = \prod_{i=1}^{q} \frac{s - s_i}{s + s_i}, \quad \text{Re } s_i > 0 \tag{10.43}$$

where s_i ($i = 1, 2, \ldots, q$) are the open RHS poles of $z_2(-s)$, and

$$F(s) = 2A(s)r_2(s) \tag{10.44}$$

Step 3. Determine the locations and the orders of the zeros of transmission of $z_2(s)$, which are defined as the closed RHS zeros of the function

$$w(s) = \frac{r_2(s)}{z_2(s)} \tag{10.45}$$

and divide them into respective classes according to Definition 10.2.

Step 4. Perform the unique factorization of the function

$$\hat{\rho}(s)\hat{\rho}(-s) = 1 - G(-s^2) \tag{10.46}$$

in which the numerator of the minimum-phase solution $\hat{\rho}(s)$ is a Hurwitz polynomial and the denominator $\hat{\rho}(s)$ is a strictly Hurwitz polynomial.

Step 5. Obtain the Laurent series expansions of the functions $A(s)$, $F(s)$, and $\hat{\rho}(s)$ about each zero of transmission s_0 of $z_2(s)$, as follows:

$$A(s) = \sum_{m=0}^{\infty} A_m (s - s_0)^m \tag{10.47a}$$

$$F(s) = \sum_{m=0}^{\infty} F_m (s - s_0)^m \tag{10.47b}$$

$$\hat{\rho}(s) = \sum_{m=0}^{\infty} \rho_m (s - s_0)^m \tag{10.47c}$$

They may be obtained by any available methods.

Step 6. According to the classes of zeros of transmission, list the basic constraints (Equation 10.29) imposed on the coefficients of Equation 10.47. The gain level is ascertained from these constraints. If not all the constraints are satisfied, consider the more general solution

$$\rho(s) = \pm \acute{\eta}(s)\hat{\rho}(s) \tag{10.48}$$

where $\acute{\eta}(s)$ is an arbitrary real all-pass function. Then repeat Step 5 for $\rho(s)$, starting with lower-order $\acute{\eta}(s)$. If the constraints still cannot be satisfied, modify the preassigned transducer power-gain characteristics $G(\omega^2)$. Otherwise, no match exists.

Step 7. Having successfully carried out Step 6, the equalizer back-end impedance is determined by the equation

$$Z_{22}(s) \equiv \frac{F(s)}{A(s) - \rho(s)} - z_2(s) \tag{10.49}$$

where $\rho(s)$ may be $\hat{\rho}(s)$ and $Z_{22}(s)$ is guaranteed to be positive real.

Step 8. Realize $Z_{22}(s)$ as the input impedance of a lossless two-port network terminated in a resistor. An ideal transformer may be required at the input port to compensate for the actual level of the generator resistance R_1. This completes the design of an equalizer.

Example 10.3

Design a lossless matching network to equalize the RLC load as shown in Figure 10.4 to a resistive generator and to achieve the fifth-order Butterworth transducer power-gain characteristic with a maximal DC gain. The cutoff frequency is 10^8 rad/s.

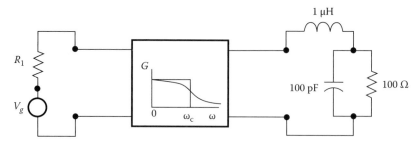

FIGURE 10.4 Broadband matching of an RLC load to a resistive generator.

To simplify the computation, we magnitude-scale the network by a factor of 10^{-2} and frequency-scale it by a factor of 10^{-8}. Thus, s denotes the normalized complex frequency and ω the normalized real frequency. The load impedance becomes

$$z_2(s) = \frac{s^2 + s + 1}{s + 1} \tag{10.50}$$

We now follow the eight steps outlined below to design a lossless equalizer to meet the desired specifications.

Step 1. The fifth-order Butterworth transducer power-gain characteristic is given by

$$G(\omega^2) = \frac{K_5}{1 + \omega^{10}}, \quad 0 \le K_5 \le 1 \tag{10.51}$$

Our objective is to maximize the DC gain K_5.

Step 2. From the load impedance $z_2(s)$, we compute the functions

$$r_2(s) = \frac{1}{2}[z_2(s) + z_2(-s)] = \frac{1}{1 - s^2} \tag{10.52}$$

$$A(s) = \frac{s - 1}{s + 1} \tag{10.53}$$

where $s = 1$ is the open RHS pole of $z_2(-s)$, and

$$F(s) = 2A(s)r_2(s) = \frac{-2}{(s + 1)^2} \tag{10.54}$$

Step 3. The zeros of transmission $z_2(s)$ are defined by the closed RHS zero of the function.

$$w(s) = \frac{r_2(s)}{z_2(s)} = \frac{1}{(s^2 + s + 1)(1 - s)} \tag{10.55}$$

indicating that $s = \infty$ is a Class IV zero of transmission of order 3.

Step 4. Substituting Equation 10.51 in Equation 10.46 with $-js$ replacing ω gives

$$\rho(s)\rho(-s) = 1 - G(-s^2) = 1 - \frac{K_5}{1 - s^{10}} = \alpha^{10}\frac{1 - x^{10}}{1 - s^{10}} \tag{10.56}$$

where

$$\alpha = (1 - K_5)^{1/10}, \quad x = \frac{s}{\alpha} \tag{10.57}$$

The minimum-phase solution of Equation 10.56 is found to be

$$\hat{\rho}(s) = \frac{s^5 + 3.23607\alpha s^4 + 5.23607\alpha^2 s^3 + 5.23607\alpha^3 s^2 + 3.23607\alpha^4 s + \alpha^5}{s^5 + 3.23607 s^4 + 5.23607 s^3 + 5.23607 s^2 + 3.23607 s + 1} \tag{10.58}$$

The maximum attainable DC gain will be ascertained later from the coefficient conditions.

Step 5. The Laurent series expansions of the functions $A(s)$, $F(s)$, and $\hat{\rho}(s)$ about the zero of transmission $s_0 = \infty$ of $z_2(s)$ are obtained as follows:

$$A(s) = \frac{s-1}{s+1} = 1 - \frac{2}{s} + \frac{2}{s^2} - \frac{2}{s^3} + \cdots \tag{10.59a}$$

$$F(s) = \frac{-2}{(s+1)^2} = 0 + 0 - \frac{2}{s^2} + \frac{4}{s^3} + \cdots \tag{10.59b}$$

$$\hat{\rho}(s) = 1 + \frac{3.23607(\alpha - 1)}{s} + \frac{5.23607(\alpha - 1)^2}{s^2}$$
$$+ \frac{5.23607(\alpha^3 - 3.23607\alpha^2 + 3.23607\alpha - 1)}{s^3} + \cdots \tag{10.59c}$$

Step 6. For a Class IV zero of transmission of order 3, the coefficient conditions are, from Equation 10.29d with $k = 3$,

$$A_m = \rho_m, \quad m = 0, 1, 2 \tag{10.60a}$$

$$\frac{F_2}{A_3 - \rho_3} \geq a_{-1}(\infty) = 1 \tag{10.60b}$$

where $a_{-1}(\infty)$ is the residue of $z_2(s)$ evaluated at the pole $s = \infty$, which is also the zero of transmission of $z_2(s)$. Substituting the coefficients of Equation 10.59 in Equation 10.60 yields the constraints imposed on K_5 as

$$A_0 = 1 = \rho_0 \tag{10.61a}$$

$$A_1 = -2 = \rho_1 = 3.23607(\alpha - 1) \tag{10.61b}$$

$$A_2 = 2 = \rho_2 = 5.23607(\alpha - 1)^2 \tag{10.61c}$$

yielding $\alpha = 0.3819664$, and

$$\frac{F_2}{A_3 - p_3} = \frac{-2}{-2 - 5.23607(\alpha^3 - 3.23607\alpha^2 + 3.23607\alpha - 1)}$$
$$\geq a_{-1}(\infty) = 1 \tag{10.62}$$

This inequality is satisfied for $\alpha = 0.3819664$. Hence, we choose $\alpha = 0.3819664$ and obtain from Equation 10.57 the maximum realizable DC gain K_5 as

$$K_5 = 1 - \alpha^{10} = 0.99993 \tag{10.63}$$

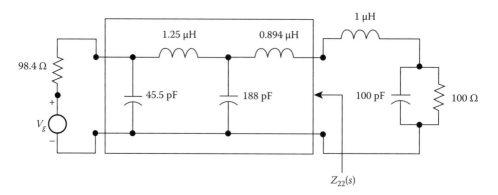

FIGURE 10.5   Fifth-order Butterworth broadband matching equalizer.

With this value of K_5, the minimum-phase reflection coefficient becomes

$$\hat{\rho}(s) = \frac{s^5 + 1.23607s^4 + 0.76393s^3 + 0.2918s^2 + 0.068884s + 0.0081307}{s^5 + 3.23607s^4 + 5.23607s^3 + 5.23607s^2 + 3.23607s + 1} \tag{10.64}$$

Step 7. The equalizer back-end impedance is determined by

$$
\begin{aligned}
Z_{22}(s) &\equiv \frac{F(s)}{A(s) - \hat{\rho}(s)} - z_2(s) = \frac{\frac{-2}{(s+1)^2}}{\frac{s-1}{s+1} - \hat{\rho}(s)} - \frac{s^2 + s + 1}{s+1} \\
&= \frac{0.94427s^4 + 2.1115s^3 + 2.6312s^2 + 2.1591s + 0.9919}{1.0557s^3 + 2.3607s^2 + 2.3131s + 1.0081}
\end{aligned} \tag{10.65}
$$

Step 8. Expanding $Z_{22}(s)$ in a continued fraction results in

$$Z_{22}(s) = 0.894s + \cfrac{1}{1.88s + \cfrac{1}{1.25s + \cfrac{1}{0.455s + \frac{1}{0.984}}}} \tag{10.66}$$

which can be identified as an LC ladder network terminated in a resistor. Denormalizing the element values with regard to magnitude-scaling by a factor of 100 and frequency-scaling by a factor 10^8 gives the final design of the equalizer of Figure 10.5.

Example 10.4

Design a lossless equalizer to match the load

$$z_2(s) = \frac{s^2 + 9s + 8}{s^2 + 2s + 2} \tag{10.67}$$

to a resistive generator and to achieve the largest flat transducer power gain over the entire sinusoidal frequency spectrum.

Step 1. For truly-flat transducer power gain, let

$$G(\omega^2) = K, \quad 0 \le K \le 1 \tag{10.68}$$

Step 2. The following functions are computed from $z_2(s)$:

$$r_2(s) \frac{(s^2 - 4)^2}{(s^2 + 2s + 2)(s^2 - 2s + 2)} \tag{10.69a}$$

$$A(s) = \frac{s^2 - 2s + 2}{s^2 + 2s + 2} \tag{10.69b}$$

$$F(s) = \frac{2(s^4 - 8s^2 + 16)}{(s^2 + 2s + 2)^2} \tag{10.69c}$$

Step 3. Since

$$w(s) = \frac{r_2(s)}{z_2(s)} = \frac{(s^2 - 4)^2}{(s^2 - 2s + 2)(s^2 + 9s + 8)} \tag{10.70}$$

the load impedance $z_2(s)$ possesses a Class I zero of transmission of order 2 at $s = 2$.

Step 4. Substituting Equation 10.68 in Equation 10.46 yields

$$\rho(s)\rho(-s) = 1 - K \tag{10.71a}$$

the minimum-phase solution of which is found to be

$$\hat{\rho}(s) = \pm\sqrt{1 - K} \tag{10.71b}$$

Step 5. For a Class I zero of transmission of order 2, the coefficient conditions (Equation 10.29a) become

$$A(2) = \hat{\rho}(2) \tag{10.72a}$$

$$A_1 = \left. \frac{dA(s)}{ds} \right|_{s=2} = \rho_1 = \left. \frac{d\rho(s)}{ds} \right|_{s=2} \tag{10.72b}$$

The Laurent series expansions of the functions $A(s)$, $F(s)$, and $\hat{\rho}(s)$ about the zero of transmission at $s = 2$ are not needed.

Step 6. Substituting Equations 10.69b and 10.71b in Equation 10.72 gives

$$A_0 = A(2) = 0.2 = \rho_0 = \pm\sqrt{1 - K} \tag{10.73a}$$

$$A_1 = \left. \frac{dA(s)}{ds} \right|_{s=2} = 0.08 \neq \rho_1 = 0 \tag{10.73b}$$

Since the coefficient conditions cannot all be satisfied without the insertion of a real all-pass function, let

$$\rho(s) = \eta(s)\hat{\rho}(s) = \frac{s - \sigma_1}{s + \sigma_1}\hat{\rho}(s) \tag{10.74}$$

Using this $\rho(s)$ in Equation 10.72 results in the new constraints.

$$A_0 = 0.2 = \rho_0 = \pm\frac{2 - \sigma_1}{2 + \sigma_1}\sqrt{1 - K} \tag{10.75a}$$

$$A_1 = 0.08 = \rho_1 = \pm\frac{2\sigma_1\sqrt{1 - K}}{(2 + \sigma_1)^2} \tag{10.75b}$$

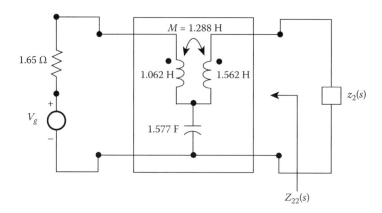

FIGURE 10.6   Lossless equalizer having a truly flat transducer power gain.

which can be combined to yield

$$\sigma_1^2 + 5\sigma_1 - 4 = 0 \tag{10.76}$$

obtaining $\sigma_1 = 0.70156$ or -5.7016. Choosing $\sigma_1 = 0.70156$ and the plus sign for $\rho(s)$, the maximum permissible flat transducer power gain is found to be

$$K_{max} = 0.82684 \tag{10.77}$$

Step 7. The equalizer back-end impedance is determined as

$$Z_{22}(s) \equiv \frac{F(s)}{A(s) - \rho(s)} - z_2(s) = \frac{1.4161s + 0.8192}{0.58388s + 0.49675} \tag{10.78}$$

Step 8. The positive-real impedance $Z_{22}(s)$ can be realized as the input impedance of a lossless two-port network terminated in a resistor. The overall network is presented in Figure 10.6.

10.4 Explicit Formulas for the RLC Load

In many practical cases, the source can usually be represented by an ideal voltage source in series with a pure resistor, which may be the Thévenin equivalent of some other network, and the load is composed of the parallel combination of a resistor and a capacitor and then in series with an inductor, as shown in Figure 10.7, which may include the parasitic effects of a physical device. The problem is to match out this load and source over a preassigned frequency band to within a given tolerance, and to achieve a prescribed transducer power-gain characteristic $G(\omega^2)$. In the case that $G(\omega^2)$ is of Butterworth or Chebyshev type of response, explicit formulas for the design of such optimum matching networks for any RLC load of the type shown in Figure 10.7 are available, thereby avoiding the necessity of applying the coefficient constraints and solving the nonlinear equations for selecting the optimum design parameters. As a result, we reduce the design of these equalizers to simple arithmetic.

10.4.1 Butterworth Networks

Refer to Figure 10.7. We wish to match out the load impedance

$$z_2(s) = Ls + \frac{R}{RCs + 1} \tag{10.79}$$

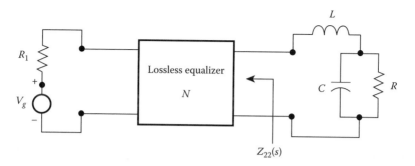

FIGURE 10.7 Broadband matching of an RLC load to a resistive source.

to a resistive generator and to achieve the *n*th-order Butterworth transducer power-gain characteristic

$$G(\omega^2) = \frac{K_n}{1 + (\omega/\omega_c)^{2n}}, \quad 0 \le K_n \le 1 \tag{10.80}$$

with maximum attainable DC gain K_n, where ω_c is the 3 dB bandwidth or the radian cutoff frequency. The even part $r_2(s)$ of $z_2(s)$ is found to be

$$r_2(s) = \frac{-R}{R^2 C^2 s^2 - 1} \tag{10.81}$$

Since $z_2(-s)$ has an open RHS pole at $s = 1/RC$, the all-pass real function defined by this pole is given by

$$A(s) = \frac{s - 1/RC}{s + 1/RC} = \frac{RCs - 1}{RCs + 1} \tag{10.82}$$

yielding

$$F(s) = 2A(s)r_2(s) = \frac{-2R}{(RCs + 1)^2} \tag{10.83}$$

We next replace ω by $-js$ in Equation 10.80 and substitute the resulting equation in Equation 10.46 to obtain

$$\rho(s)\rho(-s) = \alpha^{2n} \frac{1 + (-1)^n x^{2n}}{1 + (-1)^n y^{2n}} \tag{10.84}$$

where

$$y = \frac{s}{\omega_c}, \quad x = \frac{y}{\alpha} \tag{10.85a}$$

$$\alpha = (1 - K_n)^{1/2n} \tag{10.85b}$$

As previously shown in Equation 9.29, the minimum-phase solution $\hat{\rho}(s)$ of Equation 10.84 is found to be

$$\hat{\rho}(s) = \alpha^n \frac{q(x)}{q(y)} \tag{10.86}$$

For our purposes, we consider the more general solution

$$\rho(s) = \pm \acute{\eta}(s)\hat{\rho}(s) \tag{10.87}$$

where $\acute{\eta}(s)$ is an arbitrary first-order real all-pass function of the form

$$\eta(s) = \frac{s - \sigma_1}{s + \sigma_1}, \quad \sigma_1 \geq 0 \tag{10.88}$$

Since the load impedance $z_2(s)$ possesses a Class IV zero of transmission at infinity of order 3, the coefficient constraints become

$$A_m = \rho_m, \quad m = 0, 1, 2 \tag{10.89a}$$

$$L_a \equiv \frac{F_2}{A_3 - \rho_3} \geq L \tag{10.89b}$$

After substituting the coefficients F_2, A_3, and ρ_3 from the Laurent series expansions of $F(s)$, $A(s)$, and $\rho(s)$ in Equation 10.89b, Equation 10.89a can all be satisfied by requiring that the DC gain be

$$K_n = 1 - \left[1 - \frac{2(1 - RC\sigma_1)\sin\gamma_1}{RC\omega_c}\right]^{2n} \tag{10.90}$$

where γ_m is defined in Equation 9.37, and after considerable mathematical manipulations the constraint (Equation 10.89b) becomes

$$L_a = \frac{4R\sin\gamma_1 \sin\gamma_3}{(1 - RC\sigma_1)\left[RC\omega_c^2(\alpha^2 - 2\alpha\cos\gamma_2 + 1) + 4\sigma_1 \sin\gamma_1 \sin\gamma_3\right]} \geq L \tag{10.91}$$

The details of these derivations can be found in Ref. [3]. Thus, with K_n as specified in Equation 10.90, a match is possible if and only if the series inductance L does not exceed a critical inductance L_a. To show that any RLC load can be matched, we must demonstrate that there exists a nonnegative real σ_1 such that L_a can be made at least as large as the given inductance L and satisfies the constraint (Equation 10.90) with $0 \leq K_n \leq 1$. To this end, four cases are distinguished. Let

$$L_{a1} = \frac{R^2 C\omega_c \sin\gamma_3}{\left[(RC\omega_c - \sin\gamma_1)^2 + \cos^2\gamma_1\right]\omega_c \sin\gamma_1} > 0 \tag{10.92}$$

$$L_{a2} = \frac{8R\sin^2\gamma_1 \sin\gamma_3}{\left[(RC\omega_c - \sin\gamma_3)^2 + (1 + 4\sin^2\gamma_1)\sin\gamma_1 \sin\gamma_3\right]\omega_c} > 0 \tag{10.93}$$

Case 1. $RC\omega_c \geq 2\sin\gamma_1$ and $L_{a1} \geq L$. Under this situation, $\sigma_1 = 0$ and the maximum attainable K_n is given by Equation 10.90. The equalizer back-end impedance $Z_{22}(s)$ can be expanded in a continued fraction as

$$Z_{22}(s) = (L_{a1} - L)s + \cfrac{1}{C_2 s + \cfrac{1}{L_3 s + \cfrac{1}{\ddots + \frac{1}{W}}}} \tag{10.94}$$

where
W is a constant representing either a resistance or a conductance, and

$$L_1 = L_{a1} \tag{10.95a}$$

$$C_{2m}L_{2m-1} = \frac{4\sin\gamma_{4m-1}\sin\gamma_{4m+1}}{\omega_c^2(1 - 2\alpha\cos\gamma_{4m} + \alpha^2)}, \quad m \le \frac{1}{2}(n-1) \tag{10.95b}$$

$$C_{2m}L_{2m+1} = \frac{4\sin\gamma_{4m+1}\sin\gamma_{4m+3}}{\omega_c^2(1 - 2\alpha\cos\gamma_{4m+2} + \alpha^2)}, \quad m < \frac{1}{2}(n-1) \tag{10.95c}$$

where $m = 1, 2, \ldots, \lfloor\frac{1}{2}(n-1)\rfloor$, $n > 1$. In addition, the final reactive element can also be computed directly by the formulas

$$C_{n-1} = \frac{2(1+\alpha^n)\sin\gamma_1}{R(1-\alpha^n)(1+\alpha)\omega_c}, \quad n \text{ odd} \tag{10.96a}$$

$$L_{n-1} = \frac{2R(1-\alpha^n)\sin\gamma_1}{(1+\alpha^n)(1+\alpha)\omega_c}, \quad n \text{ even} \tag{10.96b}$$

Equation 10.94 can be identified as an LC ladder terminated in a resistor, as depicted in Figure 10.8. The terminating resistance is determined by

$$R_{22} = R\frac{1-\alpha^n}{1+\alpha^n} \tag{10.97}$$

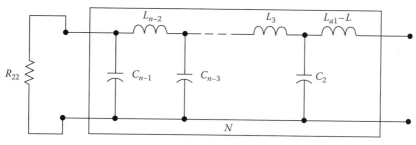

(a) *n* odd

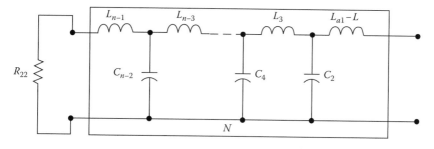

(b) *n* even

FIGURE 10.8 *n*th-order Butterworth ladder network *N*.

Case 2. $RC\omega_c \geq 2\sin\gamma_1$ and $L_{a1} < L$. Under this situation, σ_1 is nonzero and can be determined by the formula

$$\sigma_1 = \frac{1}{RC}\left(1 + 2\sqrt{p}\sinh\frac{\varphi}{3} - \frac{2RC\omega_c\sin^2\gamma_1 + \sin\gamma_3}{3\sin\gamma_1}\right) \tag{10.98}$$

where

$$p = \frac{(RC\omega_c - 2\sin\gamma_1)^2\sin\gamma_3}{9\sin\gamma_1} > 0 \tag{10.99a}$$

$$w = \frac{(2RC\omega_c\sin^2\gamma_1 + \sin\gamma_3)}{54\sin^3\gamma_1}\left[3(RC\omega_c - 2\sin\gamma_1)^2\sin\gamma_1\sin\gamma_3\right.$$
$$\left. + \left(2RC\omega_c\sin^2\gamma_1 + \sin\gamma_3\right)^2\right] - \frac{R^2C\sin\gamma_3}{2L\sin\gamma_1} \tag{10.99b}$$

$$\varphi = \sinh^{-1}\frac{w}{(\sqrt{p})^3} \tag{10.99c}$$

Using this value of σ_1, the DC gain K_n is computed by Equation 10.90.

Case 3. $RC\omega_c < 2\sin\gamma_1$ and $L_{a2} \geq L$. Then, we have

$$K_n = 1 \tag{10.100a}$$

$$\sigma_1 = \frac{1}{RC}\left(1 - \frac{RC\omega_c}{2\sin\gamma_1}\right) > 0 \tag{10.100b}$$

Case 4. $RC\omega_c < 2\sin\gamma_1$ and $L_{a2} < L$. Then the desired value of σ_1 can be computed by Equation 10.98. Using this value of σ_1, the DC gain K_n is computed by Equation 10.90.

Example 10.5

Let

$$R = 100\,\Omega, \quad C = 100\,\text{pF}, \quad L = 0.5\,\mu\text{F} \tag{10.101a}$$

$$n = 6, \quad \omega_c = 10^8\,\text{rad/s} \tag{10.101b}$$

From Equation 10.92, we first compute

$$L_{a1} = \frac{100\sin 45°}{[(1 - \sin 15°)^2 + \cos^2 15°] \times 10^8\sin 15°} = 1.84304\,\mu\text{H} \tag{10.102}$$

Since $L_{a1} > L$ and

$$RC\omega_c = 1 > 2\sin 15° = 0.517638 \tag{10.103}$$

Case 1 applies and the matching network can be realized as an LC ladder terminating in a resistor as shown in Figure 10.8b. With $\sigma_1 = 0$, the maximum attainable DC gain K_6 is from Equation 10.90

$$K_6 = 1 - \left(1 - \frac{2\sin 15°}{RC\omega_c}\right)^{12} = 0.999841 \tag{10.104}$$

giving from Equation 10.85b

$$\alpha = (1 - K_6)^{1/12} = 0.482362 \tag{10.105}$$

Applying Equation 10.95 yields the element values of the LC ladder network, as follows:

$$L_1 = L_{a1} = 1.84304 \, \mu H \tag{10.106a}$$

$$C_2 = \frac{4\sin 45° \sin 75°}{1.84304 \times 10^{-6} \times 10^{16}(1 - 2 \times 0.482362 \cos 60° + 0.482362^2)}$$
$$= 197.566 \, pF \tag{10.106b}$$

$$L_3 = \frac{4\sin 75° \sin 105°}{197.566 \times 10^{-12} \times 10^{16}(1 - 2 \times 0.482362 \cos 90° + 0.482362^2)}$$
$$= 1.53245 \, \mu H \tag{10.106c}$$

$$C_4 = \frac{4\sin 105° \sin 135°}{1.53245 \times 10^{-6} \times 10^{16}(1 - 2 \times 0.482362 \cos 120° + 0.482362^2)}$$
$$= 103.951 \, pF \tag{10.106d}$$

$$L_5 = \frac{4\sin 135° \sin 165°}{103.951 \times 10^{-12} \times 10^{16}(1 - 2 \times 0.482362 \cos 150° + 0.482362^2)}$$
$$= 0.34051 \, \mu H \tag{10.106e}$$

The last reactive elements L_5 can also be calculated directly from Equation 10.96b as

$$L_5 = \frac{2 \times 100(1 - 0.482362^6)\sin 15°}{(1 + 0.482362^6)(1 + 0.482362)10^8} = 0.34051 \, \mu H \tag{10.107}$$

Finally, the terminating resistance is determined from Equation 10.97 as

$$R_{22} = 100\frac{1 - 0.482362^6}{1 + 0.482362^6} = 97.512 \, \Omega \tag{10.108}$$

The matching network together with its terminations is presented in Figure 10.9. We remark that for computational accuracy we retain five significant figures in all the calculations. In practice, one or two significant digits are sufficient, as indicated in the figure.

Example 10.6

Let

$$R = 100 \, \Omega, \quad C = 50 \, pF, \quad L = 0.5 \, \mu F \tag{10.109a}$$

$$n = 5, \quad \omega_c = 10^8 \, rad/s \tag{10.109b}$$

Since

$$RC\omega_c = 0.5 < 2\sin 18° = 0.618 \tag{10.110}$$

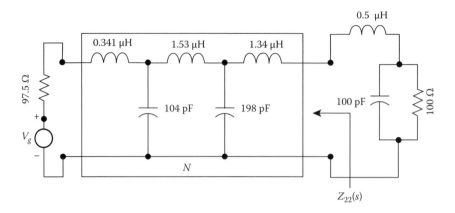

FIGURE 10.9   Sixth-order Butterworth matching network.

and from Equation 10.93,

$$L_{a2} = 1.401 \ \mu H > L = 0.5 \ \mu H \tag{10.111}$$

Case 3 applies, and we have $K_5 = 1$ and from Equation 10.100b

$$\sigma_1 = 0.381966 \times 10^8 \tag{10.112}$$

The normalized reflection coefficient is found to be

$$\rho(y) = \frac{(y - 0.381966)y^5}{(y + 0.381966)(y^5 + 3.23607y^4 + 5.23607y^3 + 5.23607y^2 + 3.23607y + 1)} \tag{10.113}$$

where $y = s/10^8$. Finally, we compute the equalizer back-end impedance as

$$
\begin{aligned}
\frac{Z_{22}(s)}{100} &= \frac{F(y)}{A(y) - \rho(y)} - z_2(y) \\[2mm]
&= \frac{\frac{-2}{(0.5\,y+1)^2}}{\frac{0.5\,y-1}{0.5\,y+1} - \rho(s)} - 0.5\,y - \frac{1}{0.5\,y+1} \\[2mm]
&= \frac{2.573\,y^5 + 4.1631\,y^4 + 5.177\,y^3 + 4.2136\,y^2 + 2.045\,y + 0.38197}{2.8541\,y^4 + 4.618\,y^3 + 4.118\,y^2 + 2.045\,y + 0.38197} \\[2mm]
&= 0.9015\,y + \cfrac{1}{1.949\,y + \cfrac{1}{1.821\,y + \cfrac{1}{0.8002\,y + \cfrac{1}{\cfrac{1.005\,y + 0.3822}{0.9944\,y + 0.3822}}}}}
\end{aligned}
\tag{10.114}
$$

The final matching network together with its terminations is presented in Figure 10.10.

Example 10.7

Let

$$R = 100 \ \Omega, \quad C = 50 \ pF, \quad L = 3 \ \mu F \tag{10.115a}$$

$$n = 4, \quad \omega_c = 10^8 \ rad/s \tag{10.115b}$$

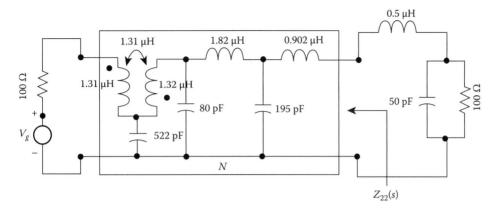

FIGURE 10.10   Fifth-order Butterworth matching network.

Since

$$RC\omega_c = 0.5 < 2\sin 22.5° = 0.76537 \tag{10.116}$$

and from Equation 10.93

$$L_{a2} = 1.462 \ \mu\text{H} < L = 3 \ \mu\text{H} \tag{10.117}$$

Case 4 applies, and from Equation 10.98

$$\sigma_1 = 1.63129 \times 10^8 \tag{10.118}$$

where Equation 10.99

$$p = 0.0188898, \quad w = 0.2304, \quad \varphi = 5.17894 \tag{10.119}$$

From Equation 10.90, the maximum attainable DC gain is obtained as

$$K_4 = 1 - \left[1 - \frac{2(1 - 100 \times 50 \times 10^{-12} \times 1.63129 \times 10^8)\sin 22.5°}{100 \times 50 \times 10^{-12} \times 10^8}\right]^8$$
$$= 0.929525 \tag{10.120}$$

giving from Equation 10.85b

$$\alpha = (1 - K_4)^{1/8} = 0.717802 \tag{10.121}$$

Finally, the normalized reflection coefficient $\rho(s)$ is obtained as

$$\rho(s) = \frac{(y - 1.63129)(y^4 + 1.87571\,y^3 + 1.75913\,y^2 + 0.966439\,y + 0.265471)}{(y + 1.63129)(y^4 + 2.61313\,y^3 + 3.41421\,y^2 + 2.61313\,y + 1)} \tag{10.122}$$

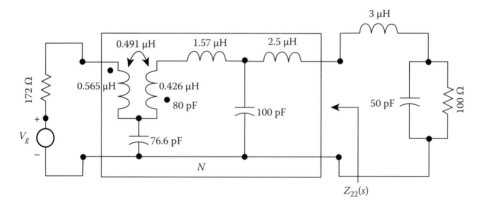

FIGURE 10.11   Fourth-order Butterworth matching network.

where $y = s/10^8$. Finally, we compute the equalizer back-end impedance as

$$
\begin{aligned}
\frac{Z_{22}(s)}{100} &= \frac{F(y)}{A(y) - \rho(y)} - z_2(y) = \frac{\frac{-2}{(0.5y+1)^2}}{\frac{0.5y-1}{0.5y+1} - \rho(s)} - 0.5\,y - \frac{1}{0.5\,y + 1} \\[2mm]
&= \frac{3.3333\,y^4 + 7.4814\,y^3 + 8.6261\,y^2 + 5.9748\,y + 2.0642}{1.3333\,y^3 + 2.9926\,y^2 + 2.9195\,y + 1.1982} \\[2mm]
&= 2.5\,y + \cfrac{1}{1.0046\,y + \cfrac{1}{1.5691\,y + \cfrac{1}{\cfrac{1.0991\,y + 2.0642}{0.84591\,y + 1.1982}}}}
\end{aligned}
\tag{10.123}
$$

The final matching network together with its terminations is presented in Figure 10.11.

10.4.2 Chebyshev Networks

Refer again to Figure 10.7. We wish to match out the load impedance

$$
z_2(s) = Ls + \frac{R}{RCs + 1}
\tag{10.124}
$$

to a resistive generator and to achieve the nth-order Chebyshev transducer power-gain characteristic

$$
G(\omega^2) = \frac{K_n}{1 + \varepsilon^2 C_n^2(\omega/\omega_c)}, \quad 0 \le K_n \le 1
\tag{10.125}
$$

with maximum attainable constant K_n. Following Equation 10.84, we obtain

$$
\rho(s)\rho(-s) = (1 - K_n)\frac{1 + \hat{\varepsilon}^2 C_n^2(-jy)}{1 + \varepsilon^2 C_n^2(-jy)}
\tag{10.126}
$$

where $y = s/\omega_c$ and

$$
\hat{\varepsilon} = \frac{\varepsilon}{\sqrt{1 - K_n}}
\tag{10.127}
$$

As in Equation 10.86, let $\hat{\rho}(s)$ be the minimum-phase solution of Equation 10.126. For our purposes, we consider the more general solution

$$\rho(s) = \pm \eta(s)\hat{\rho}(s) \tag{10.128}$$

where $\acute{\eta}(s)$ is an arbitrary first-order real all-pass function of the form

$$\eta(s) = \frac{s - \sigma_1}{s + \sigma_1}, \quad \sigma_1 \geq 0 \tag{10.129}$$

Since the load impedance $z_2(s)$ possesses a Class IV zero of transmission at infinity of order 3, the coefficient constraints become

$$A_m = \rho_m, \quad m = 0, 1, 2 \tag{10.130a}$$

$$\frac{F_2}{A_3 - \rho_3} \geq L \tag{10.130b}$$

After substituting the coefficients F_2, A_3, and ρ_3 from the Laurent series expansions of $F(s)$, $A(s)$, and $\rho(s)$ in Equation 10.130a, they lead to the constraints on the constant K_n as

$$K_n = 1 - \varepsilon^2 \sinh^2 \left\{ n \sinh^{-1} \left[\sinh a - \frac{2(1 - RC\sigma_1) \sin \gamma_1}{RC\omega_c} \right] \right\} \tag{10.131}$$

where γ_m is defined in Equation 9.37, and

$$a = \frac{1}{n} \sinh^{-1} \frac{1}{\varepsilon} \tag{10.132}$$

To apply the constraint (Equation 10.130b), we rewrite it as

$$L_b \equiv \frac{F_2}{A_3 - \rho_3} \geq L \tag{10.133}$$

After substituting the coefficients F_2, A_3, and ρ_3 from the Laurent series expansions of $F(s)$, $A(s)$, and $\rho(s)$ in Equation 10.133 and after considerable mathematical manipulations, we obtain

$$L_b = \frac{4R \sin \gamma_1 \sin \gamma_3}{(1 - RC\sigma_1)\left[RC\omega_c^2 f_1 (\sin a, \sinh \hat{a}) + 4\sigma_1 \sin \gamma_1 \sin \gamma_3 \right]} \geq L \tag{10.134}$$

where

$$\hat{a} = \frac{1}{n} \sinh^{-1} \frac{1}{\hat{\varepsilon}} = \frac{1}{n} \sinh^{-1} \frac{\sqrt{1 - K_n}}{\varepsilon} \tag{10.135}$$

$$f_m(x, y) = x^2 + y^2 - 2xy \cos \gamma_{2m} + \sin^2 \gamma_{2m},$$
$$m = 1, 2, \ldots, \left| \frac{1}{2} n \right| \tag{10.136}$$

The details of these derivations can be found in Ref. [3]. Thus, with K_n as specified in Equation 10.131, a match is possible if and only if the series inductance L does not exceed a critical inductance L_b. To show that any RLC load can be matched, we must demonstrate that there exists a nonnegative real σ_1 such that L_b can be made at least as large as the given inductance L and satisfies the constraint (Equation 10.131) with $0 \leq K_n \leq 1$. To this end, four cases are distinguished. Let

$$L_{b1} = \frac{R^2 C \omega_c \sin \gamma_3}{\left[(1 - RC\omega_c \sinh a \sin \gamma_1)^2 + R^2 C^2 \omega_c^2 \cosh^2 a \cos^2 \gamma_1\right] \omega_c \sin \gamma_1} > 0 \tag{10.137}$$

$$L_{b2} = \frac{8R \sin^2 \gamma_1 \sin \gamma_3}{\left[(RC\omega_c \sinh a - \sin \gamma_3)^2 + (1 + 4\sin^2 \gamma_1) \sin \gamma_1 \sin \gamma_3 + R^2 C^2 \omega_c^2 \sin^2 \gamma_2\right] \omega_c \sinh a} > 0 \tag{10.138}$$

Observe that both L_{b1} and L_{b2} are positive.

Case 1. $RC\omega_c \sinh a \geq 2 \sin \gamma_1$ and $L_{b1} \geq L$. Under this situation, $\sigma_1 = 0$ and the maximum attainable K_n is given by

$$K_n = 1 - \varepsilon^2 \sinh^2 \left[n \; \sinh^{-1} \left(\sinh a - \frac{2 \sin \gamma_1}{RC\omega_c} \right) \right] \tag{10.139}$$

The equalizer back-end impedance $Z_{22}(s)$ can be expanded in a continued fraction as in Equation 10.94 with L_{b1} replacing L_{a1} and realized by the LC ladders Figure 10.8 with

$$L_1 = L_{b1} \tag{10.140a}$$

$$C_{2m} L_{2m-1} = \frac{4 \sin \gamma_{4m-1} \sin \gamma_{4m+1}}{\omega_c^2 f_{2m}(\sinh a, \sinh \hat{a})}, \quad m < \frac{1}{2}(n-1) \tag{10.140b}$$

$$C_{2m} L_{2m+1} = \frac{4 \sin \gamma_{4m+1} \sin \gamma_{4m+3}}{\omega_c^2 f_{2m+1}(\sinh a, \sinh \hat{a})}, \quad m < \frac{1}{2}(n-1) \tag{10.140c}$$

where $m = 1, 2, \ldots, \left[\frac{1}{2}(n-1)\right], n > 1$. In addition, the final reactive element can also be computed directly by the formulas

$$C_{n-1} = \frac{2(\sinh na + \sinh n\hat{a}) \sin \gamma_1}{R\omega_c(\sinh a + \sinh \hat{a})(\sinh na - \sinh n\hat{a})}, \quad n \text{ odd} \tag{10.141a}$$

$$L_{n-1} = \frac{2R(\cosh na - \cosh n\hat{a}) \sin \gamma_1}{\omega_c(\sinh a + \sinh \hat{a})(\cosh na + \cosh n\hat{a})}, \quad n \text{ even} \tag{10.141b}$$

The terminating resistance is determined by

$$R_{22} = R \frac{\sinh na - \sinh n\hat{a}}{\sinh na + \sinh n\hat{a}}, \quad n \text{ odd} \tag{10.142a}$$

$$R_{22} = R \frac{\cosh na - \cosh n\hat{a}}{\cosh na + \cosh n\hat{a}}, \quad n \text{ even} \tag{10.142b}$$

Case 2. $RC\omega_c \sinh a \geq 2 \sin \gamma_1$ and $L_{b1} < L$. Under this situation, σ_1 is nonzero and can be determined by the formula

$$\sigma_1 = \frac{1}{RC}\left(1 + 2\sqrt{q}\sinh\frac{\varphi}{3} - \frac{2RC\omega_c\sin^2\gamma_1\sinh a + \sin\gamma_3}{3\sin\gamma_1}\right) \tag{10.143}$$

where

$$q = \frac{(RC\omega_c\sinh a - 2\sin\gamma_1)^2\sin\gamma_3 + 3R^2C^2\omega_c^2\sin\gamma_1\cos^2\gamma_1}{9\sin\gamma_1} > 0 \tag{10.144a}$$

$$\zeta = \frac{(2RC\omega_c\sin^2\gamma_1\sinh a + \sin\gamma_3)}{54\sin^3\gamma_1}$$

$$\times\left[3(RC\omega_c\sinh a - 2\sin\gamma_1)^2\sin\gamma_1\sin\gamma_3\right.$$

$$+ 2.25R^2C^2\omega_c^2\sin^2\gamma_2 + \left(2RC\omega_c\sin^2\gamma_1\sinh a + \sin\gamma_3\right)^2\right]$$

$$-\frac{R^2C\sin\gamma_3}{2L\sin\gamma_1} \tag{10.144b}$$

$$\varphi = \sinh^{-1}\frac{\zeta}{\left(\sqrt{q}\right)^3} \tag{10.144c}$$

Using this value of σ_1, the constant K_n is computed by Equation 10.131.
Case 3. $RC\omega_c\sinh a < 2\sin\gamma_1$ and $L_{b2} \geq L$. Then, we have

$$K_n = 1 \tag{10.145a}$$

$$\sigma_1 = \frac{1}{RC}\left(1 - \frac{RC\omega_c\sinh a}{2\sin\gamma_1}\right) > 0 \tag{10.145b}$$

Case 4. $RC\omega_c\sinh a < 2\sin\gamma_1$ and $L_{b2} < L$. Then the desired value of σ_1 can be computed by formula given by Equation 10.143. Using this value of σ_1, the constant K_n is computed by Equation 10.131.

Example 10.8

Let

$$R = 100\ \Omega, \quad C = 500\ \text{pF}, \quad L = 0.5\ \mu\text{F} \tag{10.146a}$$

$$n = 6, \quad \varepsilon = 0.50885(1\ \text{dB ripple}), \quad \omega_c = 10^8\ \text{rad/s} \tag{10.146b}$$

From Equation 10.137, we first compute

$$L_{b1} = \frac{500\sin 45°}{\left[(1 - 5\sinh 0.237996\sin 15°)^2 + 25\cosh^2 0.237996\cos^2 15°\right]10^8\sin 15°}$$

$$= 0.54323\ \mu\text{H} \tag{10.147}$$

Since $L_{b1} > L$ and

$$RC\omega_c\sinh a = 5\sinh 0.237996 = 1.20125 > 2\sin 15° = 0.517638 \tag{10.148}$$

Case 1 applies and the matching network can be realized as an LC ladder terminating in a resistor as shown in Figure 10.8. With $\sigma_1 = 0$, the maximum attainable DC gain K_6 is from Equation 10.131

$$K_6 = 1 - 0.50885^2 \sinh^2 \left[6 \sinh^{-1} \left(\sinh 0.237996 - \frac{2 \sin 15°}{5} \right) \right] = 0.78462 \qquad (10.149)$$

giving from Equations 10.127 and 10.135

$$\hat{\varepsilon} = \frac{0.50885}{\sqrt{1 - 0.78462}} = 1.0964 \qquad (10.150a)$$

$$\hat{\alpha} = \frac{1}{6} \sinh^{-1} \frac{1}{1.0964} = 0.13630 \qquad (10.150b)$$

Applying formulas given by Equation 10.140 yields the element values of the LC ladder network, as follows:

$$L_1 = L_{b1} = 0.54323\ \mu H, \quad C_2 = 634\ pF, \quad L_3 = 0.547\ \mu H \qquad (10.151a)$$

$$C_4 = 581\ pF, \quad L_5 = 0.329\ \mu H \qquad (10.151b)$$

The last reactive element L_5 can be calculated directly from Equation 10.141b as

$$L_5 = \frac{2 \times 100[\cosh(6 \times 0.237996) - \cosh(6 \times 0.13630)]\sin 15°}{10^8 (\sinh 0.237996 + \sinh 0.13630)[\cosh(6 \times 0.237996) + \cosh(6 \times 0.13630)]}$$
$$= 0.328603\ \mu H \qquad (10.152)$$

Finally, the terminating resistance is determined from Equation 10.142b as

$$R_{22} = 100 \frac{\cosh(6 \times 0.237996) - \cosh(6 \times 0.13630)}{\cosh(6 \times 0.237996) + \cosh(6 \times 0.13630)} = 23.93062\ \Omega \qquad (10.153)$$

The matching network together with its terminations is presented in Figure 10.12. We remark that for computational accuracy we retain five significant figures in all the calculations. In practice, one or two significant digits are sufficient, as indicated in the figure.

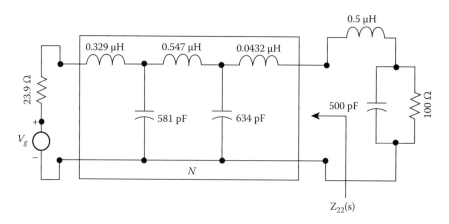

FIGURE 10.12 Sixth-order Chebyshev matching network.

Example 10.9

Let

$$R = 100\,\Omega, \quad C = 500\,\text{pF}, \quad L = 1\,\mu\text{H} \tag{10.154a}$$

$$n = 5, \quad \varepsilon = 0.50885\,(1\,\text{dB ripple}), \quad \omega_c = 10^8\,\text{rad/s} \tag{10.154b}$$

From Equation 10.137, we first compute

$$L_{b1} = 0.52755\,\mu\text{H} \tag{10.155}$$

Since $L_{b1} < L$ and

$$RC\omega_c \sinh a = 5\sinh 0.28560 = 1.44747 > 2\sin 18° = 0.61803 \tag{10.156}$$

Case 2 applies. From Equation 10.144, we obtain

$$q = 7.73769, \quad \zeta = 7.84729, \quad \varphi = 0.356959 \tag{10.157}$$

Substituting these in Equation 10.143 gives

$$\sigma_1 = 0.0985305 \times 10^8 \tag{10.158}$$

From Equation 10.131, the maximum attainable constant K_5 is found to be

$$K_5 = 0.509206 \tag{10.159}$$

The rest of the calculations proceed exactly as in the previous example, and the details are omitted.

Example 10.10

Let

$$R = 100\,\Omega, \quad C = 100\,\text{pF}, \quad L = 0.5\,\mu\text{F} \tag{10.160a}$$

$$n = 5, \quad \varepsilon = 0.76478\,(2\,\text{dB ripple}), \quad \omega_c = 10^8\,\text{rad/s} \tag{10.160b}$$

We first compute

$$a = \frac{1}{5}\sinh^{-1}\frac{1}{\varepsilon} = 0.2166104 \tag{10.161}$$

and from Equation 10.138

$$L_{b2} = 2.72234\,\mu\text{H} \tag{10.162}$$

Since $L_{b2} \ge L$ and

$$RC\omega_c \sinh a = \sinh 0.2166104 = 0.218308 < 2\sin 18° = 0.618034 \tag{10.163}$$

Case 3 applies. Then $K_5 = 1$ and from Equation 10.145b

$$\sigma_1 = \frac{1}{10^{-8}}\left(1 - \frac{\sinh 0.2166104}{2 \sin 18°}\right) = 0.64677 \times 10^8 \tag{10.164}$$

For $K_5 = 1$, Equation 10.126 degenerates into

$$\rho(y)\rho(-y) = \frac{\varepsilon^2 C_5^2(-jy)}{1 + \varepsilon^2 C_5^2(-jy)} \tag{10.165}$$

where $y = s/10^8$, the minimum-phase solution of which is found to be

$$\hat{\rho}(y) = \frac{y^5 + 1.25 y^3 + 0.3125 y}{y^5 + 0.706461 y^4 + 1.49954 y^3 + 0.693477 y^2 + 0.459349 y + 0.0817225} \tag{10.166}$$

A more general solution is given by

$$\rho(y) = \frac{y - 0.64677 \times 10^8}{y + 0.64677 \times 10^8}\hat{\rho}(y) \tag{10.167}$$

Finally, we compute the equalizer back-end impedance as

$$
\begin{aligned}
\frac{Z_{22}(y)}{100} &= \frac{F(y)}{A(y) - \rho(y)} - z_2(y) \\
&= \frac{\frac{-2}{(y+1)^2}}{\frac{y-1}{y+1} - \rho(x)} - 0.5y - \frac{1}{y+1} \\
&= \frac{1.63267 y^5 + 0.576709 y^4 + 2.15208 y^3 + 0.533447 y^2 + 0.554503 y + 0.0528557}{0.734663 y^4 + 0.259505 y^3 + 0.639438 y^2 + 0.123844 y + 0.0528557} \\
&= 2.222 y + \cfrac{1}{1.005 y + \cfrac{1}{3.651 y + \cfrac{1}{0.8204 y + \cfrac{1}{\cfrac{0.2441 y + 0.05286}{0.02736 y + 0.05286}}}}}
\end{aligned}
\tag{10.168}
$$

The final matching network together with its terminations is presented in Figure 10.13.

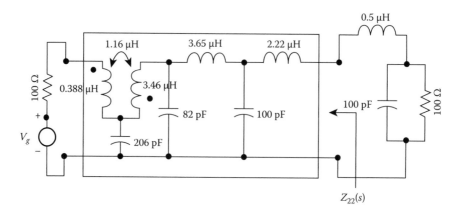

FIGURE 10.13 Fifth-order Chebyshev matching network.

Example 10.11

Let

$$R = 100\,\Omega, \quad C = 100\,\text{pF}, \quad L = 3\,\mu\text{F} \tag{10.169a}$$

$$n = 4, \quad \varepsilon = 0.76478\,(2\text{ dB ripple}), \quad \omega_c = 10^8\,\text{rad/s} \tag{10.169b}$$

We first compute

$$a = \frac{1}{4}\sinh^{-1}\frac{1}{0.76478} = 0.27076 \tag{10.170}$$

and from Equation 10.138

$$L_{b2} = 2.66312\,\mu\text{H} \tag{10.171}$$

Since $L_{b2} < L$ and

$$RC\omega_c\sinh a = \sinh 0.27076 = 0.27408 < 2\sin 22.5° = 0.765367 \tag{10.172}$$

Case 4 applies. From Equation 10.144, we obtain

$$q = 0.349261, \quad \zeta = 0.390434, \quad \varphi = 1.39406 \tag{10.173}$$

Substituting these in Equation 10.143 gives

$$\sigma_1 = 0.694564 \times 10^8 \tag{10.174}$$

obtaining from Equations 10.127 and 10.131

$$K_4 = 0.984668, \quad \hat{\varepsilon} = 6.17635 \tag{10.175}$$

From Equation 10.126,

$$\rho(y)\rho(-y) = (1 - K_4)\frac{1 + \hat{\varepsilon}^2 C_4^2(-jy)}{1 + \varepsilon^2 C_4^2(-jy)} \tag{10.176}$$

where $y = s/10^8$, the minimum-phase solution of which is found to be

$$\hat{\rho}(y) = \frac{y^4 + 0.105343\,y^3 + 1.00555\,y^2 + 0.0682693\,y + 0.126628}{y^4 + 0.716215\,y^3 + 1.25648\,y^2 + 0.516798\,y + 0.205765} \tag{10.177}$$

A more general solution is given by

$$\rho(y) = \frac{y - 0.694564 \times 10^8}{y + 0.694564 \times 10^8}\hat{\rho}(s) \tag{10.178}$$

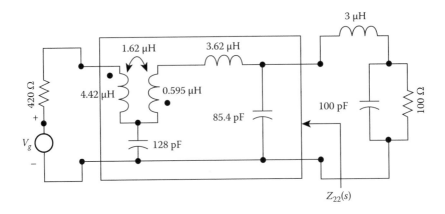

FIGURE 10.14   A fourth-order Chebyshev matching network.

Finally, we compute the equalizer back-end impedance as

$$
\begin{aligned}
\frac{Z_{22}(y)}{100} &= \frac{F(y)}{A(y) - \rho(y)} - z_2(y) = \frac{\frac{-2}{(y+1)^2}}{\frac{y-1}{y+1} - \rho(s)} - \frac{3y^2 + 3y + 1}{y + 1} \\
&= \frac{0.780486\, y^2 + 0.320604\, y + 0.23087}{0.666665\, y^3 + 0.273851\, y^2 + 0.413057\, y + 0.0549659} \\
&= \cfrac{1}{0.8542\, y + \cfrac{1}{3.616\, y + \cfrac{1}{\cfrac{0.1219\, y + 0.2309}{0.2159\, y + 0.05497}}}}
\end{aligned}
\tag{10.179}
$$

The final matching network together with its terminations is presented in Figure 10.14.

References

1. D. C. Youla, A new theory of broad-band matching, *IEEE Trans. Circuit Theory*, CT-11, 30–50, 1964.
2. W. K. Chen, *Theory and Design of Broadband Matching Networks*, Cambridge, UK: Pergamon Press, 1976.
3. W. K. Chen, Explicit formulas for the synthesis of optimum broadband impedance-matching networks, *IEEE Trans. Circuits Syst.*, CAS-24, 157–169, 1977.

Active Filters

Wai-Kai Chen
University of Illinois at Chicago

11

Low-Gain Active Filters

Phillip E. Allen
Georgia Institute of Technology

Benjamin J. Blalock
The University of Tennessee

Stephen W. Milam
RF Micro-Devices

11.1 Introduction

Active filters consist of only amplifiers, resistors, and capacitors. Complex roots are achieved by the use of feedback eliminating the need for inductors. The gain of the amplifier can be finite or infinite (an op-amp). This section describes active filters using low-gain or finite-gain amplifiers. Filter design equations and examples will be given along with the performance limits of low-gain amplifier filters.

11.2 First- and Second-Order Transfer Functions

Before discussing the characteristics and the synthesis of filters, it is important to understand the transfer functions of first- and second-order filters. Later we will explain the implementations of these filters and show how to construct higher order filters from first- and second-order sections. The transfer functions of most first- and second-order filters are examined in the following.

11.2.1 First-Order Transfer Functions

A standard form of the transfer function of a first-order low-pass filter is

$$T_{\text{LP}}(s) = \frac{T_{\text{LP}}(j0)\omega_{\text{o}}}{s + \omega_{\text{o}}}$$

(11.1)

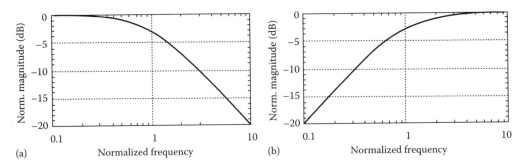

FIGURE 11.1   Normalized magnitude response of (a) first-order low-pass and (b) first-order high-pass filters.

where
 $T_{LP}(j0)$ is the value of $T_{LP}(s)$ at dc
 ω_o is the pole frequency

It is common practice to normalize both the magnitude and frequency. Normalizing Equation 11.1 yields

$$T_{LPn}(s_n) = \frac{T_{LP}(s_n\omega_o)}{|T_{LP}(j0)|} = \frac{1}{s_n + 1} \tag{11.2}$$

where $s_n = s/\omega_o$ and amplitude has been normalized as

$$T_{LPn}(s) = \frac{T_{LP}(s)}{|T_{LP}(j0)|} \tag{11.3}$$

The equivalent normalized forms of a first-order high-pass filter are

$$T_{HP}(s) = \frac{T_{HP}(j\infty)s}{s + \omega_o} \tag{11.4}$$

and

$$T_{HPn}(s_n) = \frac{T_{HP}(s_n\omega_o)}{|T_{HP}(j\infty)|} = \frac{s_n}{s_n + 1} \tag{11.5}$$

where $T_{HP}(j\infty) = T_{HP}(s) \mid$ at $\omega = \infty$. The normalized magnitude responses of these functions are shown in Figure 11.1.

11.2.2 Second-Order Transfer Functions

The standard form of a second-order low-pass filter is given as

$$T_{LP}(s) = \frac{T_{LP}(j0)\omega_o^2}{s^2 + \left(\frac{\omega_o}{Q}\right)s + \omega_o^2} \tag{11.6}$$

where
 $T_{LP}(j0)$ is the value of $T_{LP}(s)$ at dc
 ω_o is the pole frequency
 Q is the pole Q or the pole quality factor

The damping factor ζ, which may be better known to the reader, is given as

$$\zeta = \frac{1}{2Q} \tag{11.7}$$

The poles of the transfer function of Equation 11.7 are

$$p_1, p_2 = \frac{-\omega_o}{2Q} \pm j\left(\frac{\omega_o}{2Q}\right)\sqrt{4Q^2 - 1} \tag{11.8}$$

Normalization of Equation 11.6 in both amplitude and frequency gives

$$T_{LPn}(s_n) = \frac{T_{LP}(s_n\omega_o)}{|T_{LP}(j0)|} = \frac{1}{s_n^2 + \frac{s_n}{Q} + 1} \tag{11.9}$$

where $s_n = s/\omega_o$. The standard second-order, high-pass and bandpass transfer functions are

$$T_{HP}(s) = \frac{T_{HP}(j\infty)s^2}{s^2 + \left(\frac{\omega_o}{Q}\right)s + \omega_o^2} \tag{11.10}$$

and

$$T_{BP}(s) = \frac{T_{BP}(j\omega_o)\left(\frac{\omega_o}{Q}\right)s}{s^2 + \left(\frac{\omega_o}{Q}\right)s + \omega_o^2} \tag{11.11}$$

where $T_{BP}(j\omega_o) = T_{BP}(s)$ at $s = j\omega = j\omega_o$. The poles of the second-order high-pass and bandpass transfer functions are given by Equation 11.8.

We can normalize these equations as we did for $T_{LP}(s)$ to get

$$T_{HPn}(s_n) = \frac{T_{HP}(s_n\omega_o)}{|T_{HP}(j\infty)|} = \frac{s_n^2}{s_n^2 + \frac{s_n}{Q} + 1} \tag{11.12}$$

$$T_{BPn}(s_n) = \frac{T_{BP}(s_n\omega_o)}{|T_{BP}(j\omega_o)|} = \frac{\frac{s_n}{Q}}{s_n^2 + \frac{s_n}{Q} + 1} \tag{11.13}$$

where

$$T_{HPn}(s) = \frac{T_{HP}(s)}{|T_{HP}(j\infty)|} \tag{11.14}$$

and

$$T_{BPn}(s) = \frac{T_{BP}(s)}{|T_{BP}(j\omega_o)|} \tag{11.15}$$

Two other types of second-order transfer function filters that we have not covered here are the bandstop and the all-pass. These transfer functions have the same poles as the previous ones. However, the zeros of the bandstop transfer function are on the $j\omega$ axis while the zeros of the all-pass transfer function are quadratically symmetric to the poles (they are mirror images of the poles in the right-half plane). Both of these transfer functions can be implemented by a second-order biquadratic transfer function whose transfer function is given as

$$T_{BQ}(s) = \frac{K\left[s^2 \pm \left(\frac{\omega_z}{Q_z}\right)s + \omega_z^2\right]}{s^2 + \left(\frac{\omega_p}{Q_p}\right)s + \omega_p^2} \qquad (11.16)$$

where

K is a constant
ω_z is the zero frequency
Q_z the zero Q
ω_p is the pole frequency
Q_p the pole Q

11.2.3 Frequency Response (Magnitude and Phase)

The magnitude and phase response of the normalized second-order low-pass transfer function is shown in Figure 11.2, where Q is a parameter. In this figure we see that Q influences the frequency response near ω_o. If Q is greater than 0.707, then the normalized magnitude response has a peak value of

$$|T_n[\omega_n(\max)]| = \frac{Q}{\sqrt{1 - (1/4Q^2)}} \qquad (11.17)$$

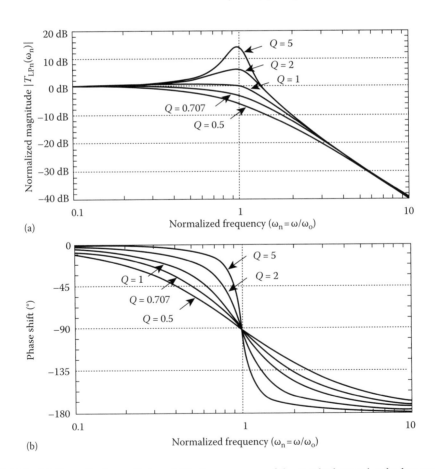

FIGURE 11.2 (a) Normalized magnitude and (b) phase response of the standard second-order low-pass transfer function with Q as a parameter.

at a frequency of

$$\omega_n(\max) = \sqrt{1 - \frac{1}{2Q^2}} \tag{11.18}$$

If the transfer function is multiplied by -1, the phase shift is shifted vertically by $\pm 180°$.

The magnitude and phase response of the normalized second-order high-pass transfer function is shown in Figure 11.3. For Q greater than 0.707 the peak value of the normalized magnitude response is as

$$\omega_n(\max) = \frac{1}{\sqrt{1 - \frac{1}{2Q^2}}} \tag{11.19}$$

The normalized frequency response of the standard second-order bandpass transfer function is shown in Figure 11.4. The slopes of the normalized magnitude curves at frequencies much greater or much less than ω_o are ± 20 dB/decade rather than ± 40 dB/decade of the second-order high and low-pass transfer functions. This difference is because one pole is causing the high-frequency roll-off while the other pole is causing the low-frequency roll-off. The peak of the magnitude occurs at $\omega = \omega_o$ or $\omega_n = 1$.

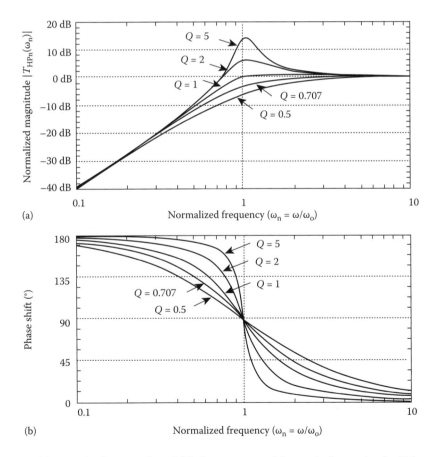

FIGURE 11.3 (a) Normalized magnitude and (b) phase response of the standard second-order high-pass transfer function with Q as a parameter.

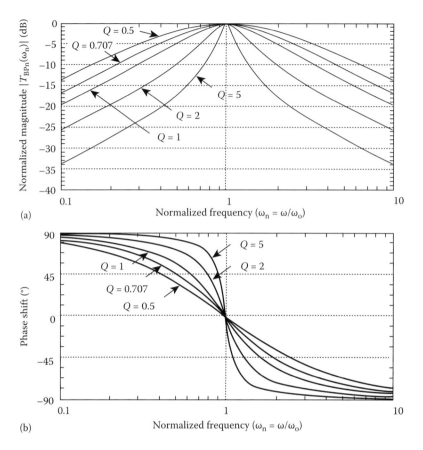

FIGURE 11.4 (a) Normalized magnitude and (b) phase response of the standard second-order bandpass transfer function with Q as a parameter.

11.2.4 Tuning Active Filters

A general tuning procedure for most second-order active filters is outlined below. This method is illustrated for adjusting the magnitude of the frequency response of a low-pass filter. The filter parameters are assumed to be the pole frequency f_o, the pole Q, and the gain $T(j0)$.

1. The component(s) which set(s) the parameter f_o is (are) tuned by adjusting the magnitude of the filter response to be $T(j0)/10$ or $T(j0)$ (dB) -20 dB at $10f_o$.
2. The component(s) that set(s) the parameter $T(j0)$ is (are) tuned by adjusting the magnitude to $T(j0)$ at $f_o/10$.
3. The component(s) that set(s) the parameter Q is (are) tuned by adjusting the magnitude of the peak (if there is one) to the value given by Figure 11.2. If there is no peaking, then adjust so that the magnitude at f_o is correct (i.e., -3 dB for $Q = 0.707$).

The tuning procedure should follow in the order of steps 1 through 3 and may be repeated if necessary. One could also use the phase shift to help in the tuning of the filter. The concept of the above tuning procedure is easily adaptable to other types of second-order filters.

11.3 First-Order Filter Realizations

A first-order filter has only one pole and zero. For stability, the pole must be on the negative real axis but the zero may be on the negative or positive real axis. A single-amplifier low-gain realization of a first-order low-pass filter is shown in Figure 11.5a. The transfer function for this filter is

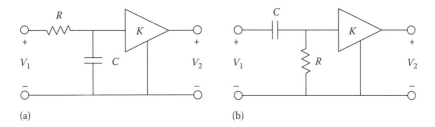

FIGURE 11.5 (a) Low-pass and (b) high-pass first-order filters.

$$T(s) = \frac{V_2(s)}{V_1(s)} = \frac{K/RC}{s + 1/RC} = \frac{K\omega_o}{s + \omega_o} = \frac{T_{LP}(0)\omega_o}{s + \omega_o} \tag{11.20}$$

The low-frequency ($\omega \ll \omega_o$) gain magnitude and polarity are set by K.

First-order high-pass filters can also be realized in a single-amplifier low-gain circuit (see Figure 11.5b). The transfer function for this filter is given in Equation 11.4, and in this case $T_{HP}(j\infty) = K$. Both low-pass and high-pass magnitude responses are shown in Figure 11.1. Note that the first-order filters exhibit no peaking and the frequency-dependent part of the magnitude response approaches ± 20 dB/decade.

11.4 Second-Order Positive-Gain Filters

Practical realizations for second-order filters using positive gain amplifiers are presented here. These filters are easy to design and have been extensively used in many applications. The first realization is a low-pass Sallen and Key filter and is shown in Figure 11.6 [8].

The transfer function of Figure 11.6 is

$$\frac{V_2(s)}{V_1(s)} = \frac{\frac{K}{R_1 R_3 C_2 C_4}}{s^2 + s\left(\frac{1}{R_3 C_4} + \frac{1}{R_1 C_2} + \frac{1}{R_3 C_2} - \frac{K}{R_3 C_4}\right) + \frac{1}{R_1 R_3 C_2 C_4}} \tag{11.21}$$

Equating this transfer function with the standard form of a second-order low-pass filter given in Equation 11.6 gives

$$\omega_o = \frac{1}{\sqrt{R_1 R_3 C_2 C_4}} \tag{11.22}$$

$$\frac{1}{Q} = \sqrt{\frac{R_3 C_4}{R_1 C_2}} + \sqrt{\frac{R_1 C_4}{R_3 C_2}} + (1 - K)\sqrt{\frac{R_1 C_2}{R_3 C_4}} \tag{11.23}$$

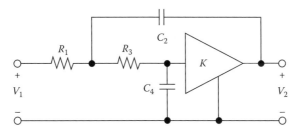

FIGURE 11.6 Low-pass Sallen and Key filter.

and

$$T_{LP}(j0) = K \tag{11.24}$$

These three equations have five unknowns, giving the designer some freedom in selecting component values. Two examples showing different techniques for defining these values of the circuit in Figure 11.6 are given below.

Example 11.1

An equal-resistance equal-capacitance low-pass filter. For this example, $R = R_1 = R_3$ and $C = C_2 = C_4$. A Butterworth low-pass filter characteristic is needed with $\omega_o = 6283$ rad/s (1 kHz) and $Q = 0.7071$. With these constraints,

$$\omega_o = \frac{1}{RC} \quad \frac{1}{Q} = 3 - K \tag{11.25}$$

and $RC = 159 \ \mu s$, $T_{LP}(j0) = K = 1.586$. Selecting $C = 0.1 \ \mu F$ yields $R = 1.59 \ k\Omega$.

Example 11.2

A unity-gain low-pass filter. For this example let $K = 1$. Therefore, Equation 11.23 becomes

$$\frac{1}{Q} = \sqrt{\frac{R_3 C_4}{R_1 C_2}} + \sqrt{\frac{R_1 C_4}{R_3 C_2}} = \sqrt{\frac{C_4}{C_2}} \left(\sqrt{\frac{R_3}{R_1}} + \sqrt{\frac{R_1}{R_3}} \right) \tag{11.26}$$

The desired transfer function, with the complex frequency normalized by a factor of 10^4 rad/s, is

$$\frac{V_2(s)}{V_1(s)} = \frac{0.988}{s^2 + 0.179s + 0.988} = \frac{T_{LP}(j0)\omega_o^2}{s^2 + \frac{\omega_o}{Q}s + \omega_o^2} \tag{11.27}$$

and $\omega_o = 0.994$ rad/s, $Q = 5.553$. To obtain a real-valued resistor ratio, pick

$$\frac{C_4}{C_2} \le \frac{1}{4Q^2} = 0.00811 \tag{11.28}$$

or $C_4/C_2 = 0.001$ in this case. Equation 11.26 yields two solutions for the ratio R_3/R_1, 30.3977 and 0.0329. From Equation 11.22 with $\omega_o = 9.94$ krad/s, $R_1 C_2 = 577 \ \mu s$. If C_2 is selected as 0.1 μF, then $C_4 = 100$ pF, $R_1 = 5.77 \ k\Omega$, and $R_3 = 175.4 \ k\Omega$.

A Sallen and Key bandpass circuit is shown in Figure 11.7, and its voltage transfer function is

$$\frac{V_2(s)}{V_1(s)} = \frac{\frac{sK}{R_1 C_5}}{s^2 + s\left(\frac{1}{R_1 C_5} + \frac{1}{R_2 C_5} + \frac{1}{R_4 C_5} + \frac{1}{R_4 C_3} - \frac{K}{R_2 C_5} \right) + \frac{1}{R_4 C_3 C_5}\left(\frac{1}{R_1} + \frac{1}{R_2} \right)} \tag{11.29}$$

Equating this transfer function to that of Equation 11.11 results in

$$\omega_o = \sqrt{\frac{1 + \frac{R_1}{R_2}}{R_1 R_4 C_3 C_5}} \tag{11.30}$$

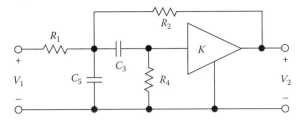

FIGURE 11.7 Bandpass Sallen and Key filter.

$$\frac{1}{Q} = \frac{\left[1 + \frac{R_1}{R_2}(1 - K)\right]\sqrt{\frac{R_4 C_3}{R_1 C_5}} + \sqrt{\frac{R_1 C_3}{R_4 C_5}} + \sqrt{\frac{R_1 C_5}{R_4 C_3}}}{\sqrt{1 + \sqrt{\frac{R_1}{R_2}}}} \tag{11.31}$$

and

$$T_{BP}(j0) = \frac{\frac{K}{R_1 C_5}}{\frac{1}{R_1 C_5} + \frac{1}{R_2 C_5} + \frac{1}{R_4 C_5} + \frac{1}{R_4 C_3} + \frac{K}{R_2 C_5}} \tag{11.32}$$

These three equations contain six unknowns. A common constraint is to set $K = 2$, requiring the gain block to have two equal-valued resistors connected around an op-amp.

Example 11.3

An equal-capacitance gain-of-2 bandpass filter. Design a bandpass filter with $\omega_n = 1$ and $Q = 2$. Arbitrarily select $C_3 = C_5 = 1$ F and $K = 2$. If R_1 is selected to be 1 Ω, then Equations 11.11, 11.29, and 11.30 give

$$\omega_o^2 = \left(1 + \frac{R_1}{R_2}\right)\frac{1}{R_4} = 1 \tag{11.33}$$

and

$$\frac{\omega_o}{Q} = 1 - \frac{1}{R_2} + \frac{2}{R_4} = \frac{1}{2} \tag{11.34}$$

Solving these equations gives $R_2 = 0.7403$ Ω and $R_4 = 2.3508$ Ω. Practical values are achieved after frequency and impedance denormalizations are made.

The high-pass filter, the third type of filter to be discussed, is shown in Figure 11.8. Its transfer function is

$$\frac{V_2(s)}{V_1(s)} = \frac{s^2 K}{s^2 + s\left(\frac{1}{R_2 C_1} + \frac{1}{R_4 C_3} + \frac{1}{R_4 C_1} - \frac{K}{R_2 C_1}\right) + \frac{1}{R_2 R_4 C_1 C_3}} \tag{11.35}$$

Equating Equation 11.35 to Equation 11.10 results in

$$\omega_o = \frac{1}{\sqrt{R_2 R_4 C_1 C_3}} \tag{11.36}$$

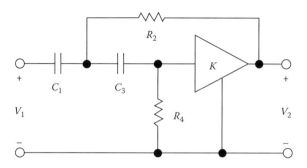

FIGURE 11.8 High-pass Sallen and Key filter.

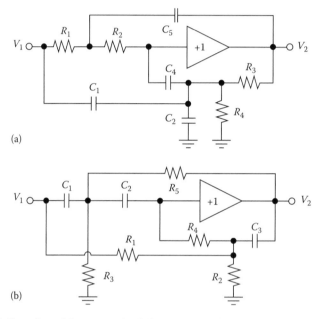

(a)

(b)

FIGURE 11.9 Notch filters derived from center-loaded twin-T networks: (a) $\omega_z \leq \omega_p$ and (b) $\omega_z < \omega_p$. (From Sedra, A. S. and Brackett, P. O., *Filter Theory and Design: Active and Passive*, Matrix, Portland, OR, 1978. With permission.)

$$\frac{1}{Q} = \sqrt{\frac{R_4 C_3}{R_2 C_1}} + \sqrt{\frac{R_2 C_1}{R_4 C_3}} + \sqrt{\frac{R_2 C_3}{R_4 C_1}} - K\sqrt{\frac{R_4 C_3}{R_2 C_1}} \tag{11.37}$$

and

$$T_{HP}(j\infty) = K \tag{11.38}$$

The design procedure using these equations is similar to that for the low-pass and bandpass filters.

The last type of second-order filter in this section is the notch filter, shown in Figure 11.9. The general transfer function for the notch filter is

$$\frac{V_2(s)}{V_1(s)} = T_N(j0)\frac{s^2 + \omega_z^2}{s^2 + \frac{\omega_p}{Q_p}s + \omega_p^2} \tag{11.39}$$

where

ω_p and ω_z are the pole and zero frequencies (ω_o), respectively
Q_p is the pole Q

For the circuit of Figure 11.9a,

$$T_N(j0) = \frac{\omega_z^2}{\omega_p^2}, \quad R_2 C_4 = \frac{\sqrt{1+a}}{\omega_p} \tag{11.40}$$

and

$$R_2 = (1+a)R_1 = 2R_3 = aR_4, \quad C_4 = \frac{\omega_z^2}{\omega_p^2}C_1 = C_2 + C_1 = \frac{C_5}{2} \tag{11.41}$$

For the circuit of Figure 11.9b.

$$T_N(j0) = 1, \quad R_2 C_4 = \frac{\sqrt{1+a}}{\omega_p} \tag{11.42}$$

and

$$R_4 = (1+a)\frac{\omega_z^2}{\omega_p^2}R_1 = (1+a)(R_1 + R_2) = aR_3 = 2R_5, \quad C_1 = C_2 = \frac{C_3}{2} \tag{11.43}$$

The Q for both notch circuits is

$$Q = \frac{\sqrt{1+a}}{a} \tag{11.44}$$

so a is a helpful design parameter for these circuits. For typical values of Q, say 0.5, 1, and 5, the corresponding values of a are 4.828, 1.618, and 0.2210, respectively.

11.5 Second-Order Biquadratic Filters

Next, second-order biquadratic functions will be described. These functions are general realizations of the second-order transfer function. Filters implementing biquadratic functions, often referred to as biquads, are found in many signal processing applications.

11.5.1 Biquadratic Transfer Function

The general form of the second-order biquadratic transfer function is

$$T(s) = H\frac{s^2 + b_1 s + b_0}{s^2 + a_1 s + a_0} = H\frac{s^2 + \frac{\omega_z}{Q_z}s + \omega_z^2}{s^2 + \frac{\omega_p}{Q_p}s + \omega_p^2} \tag{11.45}$$

with the pole locations given by Equation 11.8 and the zero locations by

$$z_1, z_2 = -\frac{\omega_z}{2Q_z} \pm j\omega_z\sqrt{1 - \frac{1}{4Q_z^2}} \tag{11.46}$$

where Q_p and Q_z are the pole and zero Q, respectively. Filters capable of implementing this voltage transfer function are called biquads since both the numerator and denominator of their transfer functions contain biquadratic expressions. The zeros described by the numerator of Equation 11.45 strongly influence the magnitude response of the biquadratic transfer function and determine the filter type (low-pass, high-pass, etc.).

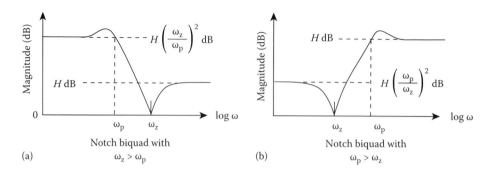

FIGURE 11.10 Frequency response of second-order notch filters. (a) $\omega_z > \omega_p$ and (b) $\omega_p > \omega_z$.

The notch filter form of the biquadratic transfer function is

$$T_{NF}(s) = H\frac{s^2 + \omega_z^2}{s^2 + \frac{\omega_p}{Q_p}s + \omega_p^2} \tag{11.47}$$

with zeros located at $s = \pm j\omega_z$. Attenuation of high or low frequencies is determined by selection of ω_z relative to ω_p. The low-pass notch filter (LPN) requires $\omega_z > \omega_p$, shown in Figure 11.10a and the high-pass notch filter (HPN) requires $\omega_p > \omega_z$, shown in Figure 11.10b.

An all-pass filter implemented using the biquadratic transfer function has the general form

$$T_{AP}(s) = H\frac{s^2 - \frac{\omega_z}{Q_z}s + \omega_z^2}{s^2 + \frac{\omega_p}{Q_p}s + \omega_p^2} \tag{11.48}$$

The all-pass magnitude response is independent of frequency; i.e., its magnitude is constant. The all-pass filter finds use in shaping the phase response of a system. To accomplish this, the all-pass has right-half plane zeros that are mirror images around the imaginary axis of left-half plane poles.

11.5.2 Biquad Implementations

A single-amplifier low-gain realization [4] of the biquadratic transfer function with transmission zeros on the $j\omega$ axis is shown in Figure 11.11. Zeros are generated by the circuit's input "twin-T" RC network. This filter is well suited for implementing elliptic functions. Its transfer function has the form

$$T(s) = H\frac{s^2 + \omega_z^2}{s^2 + \frac{\omega_p}{Q_p}s + \omega_p^2} = H\frac{s^2 + b_0^2}{s^2 + a_1 s + a_0^2} \tag{11.49}$$

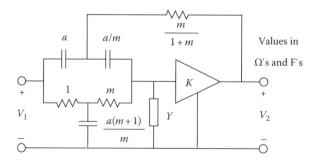

FIGURE 11.11 Low-gain realization for biquadratic network functions. (From Huelsman, L. P. and Allen, P. E., *Introduction to the Theory and Design of Active Filters*, McGraw-Hill, New York, 1980. With permission.)

The selection of element $Y(s)$ in the filter circuit can be a resistor or capacitor. If the condition $\omega_p > \omega_z$ is desired, $Y = 1/R$ is chosen. For this choice the resulting transfer function has the form

$$T(s) = \frac{V_2(s)}{V_1(s)} = \frac{K(s^2 + 1/a^2)}{s^2 + (m + 1/a)[1/R + (2 - K)/m]s + [1 + (m + 1)/R]/a^2} \tag{11.50}$$

where, K, m, and a are defined in Figure 11.11. The design parameter m is used to control the spread of circuit element values. The design equations for this filter resulting from equating Equation 11.49 and 11.50 are

$$a = \frac{1}{\sqrt{b_o}} \tag{11.51}$$

$$R = \frac{m + 1}{a_o/b_o - 1} \tag{11.52}$$

$$K = 2 + \frac{m}{m + 1}\left(\frac{a_o}{b_o} - 1 - \frac{a_1}{\sqrt{b_o}}\right) \tag{11.53}$$

and

$$H = K \tag{11.54}$$

If the condition $\omega_z > \omega_p$ is desired, then $Y = s(aC)$ is chosen. This choice has the following transfer function:

$$\frac{V_2(s)}{V_1(s)} = \frac{\dfrac{K(s^2 + 1/a^2)}{(m + 1)C + 1}}{s^2 + \left\{\dfrac{(m + 1)[C + (2 - K)/m]}{a[(m + 1)C + 1]}\right\}s + \dfrac{1}{a^2[(m + 1)C + 1]}} \tag{11.55}$$

The design equations follow from equating Equations 11.49 and 11.55 to get

$$a = \frac{1}{\sqrt{b_o}} \tag{11.56}$$

$$C = \frac{b_o/a_o - 1}{m + 1} \tag{11.57}$$

$$K = 2 + \frac{m}{m + 1}\left(\frac{b_o}{a_o} - 1 - \frac{a_1\sqrt{b_o}}{a_o}\right) \tag{11.58}$$

and

$$H = \frac{a_o}{b_o}K \tag{11.59}$$

As before, the factor m is chosen arbitrarily to control the spread of element values.

Example 11.4

A low-pass elliptic filter. It is desired to realize the elliptic voltage transfer function given as

$$\frac{V_2(s)}{V_1(s)} = \frac{H(s^2 + 2.235990)}{s^2 + 0.641131s + 1.235820}$$

This transfer function will have a 1 dB ripple in the passband, and at least 6 dB of attenuation for all frequencies greater than 1.2 rad/s [12]. Obviously, the second choice described above applies. From Equations 11.56 through 11.59, with $m = 0.2$, we find that $a = 0.66875$, $C = 0.67443$, $K = 2.0056$, and $H = 1.1085$.

Biquadratic transfer functions can also be implemented using the two-amplifier low-gain configuration in Figure 11.12 [5]. When zeros located off the $j\omega$ axis are desired, the design equations for the two-amplifier low-gain configuration are simpler than those required by the single-amplifier low-gain configuration. Also note that the required gain blocks of -1 and $+2$ are readily implemented using operational amplifiers. The transfer function for this configuration is

$$\frac{V_2(s)}{V_1(s)} = \frac{2(Y_1 - Y_2)}{Y_3 - Y_4} \tag{11.60}$$

The values of the admittances are determined by separately dividing the numerator and denominator by $s + c$, where c is a convenient value greater than zero. Partial fraction expansion results in expressions that can be implemented using RC networks. From the partial fraction expansion of the numerator divided by $s + c$, the pole residue at $s = -c$ is

$$k_b = \frac{H(c^2 - cb_1 + b_o)}{-c} \tag{11.61}$$

Depending on c, b_1, and b_o, the quality k_b can be positive or negative. If positive, Y_1 is

$$Y_1 = \frac{Hs}{2} + \frac{Hb_o}{2c} + \frac{1}{2/k_b + 2c/k_b s} \tag{11.62}$$

with $Y_2 = 0$, removing the inverting gain amplifier from the circuit. The RC network used to realize Y_1 when k_b is positive is shown in Figure 11.13a. If k_b is negative, Y_1 and Y_2 become

$$Y_1 = \frac{Hs}{2} + \frac{Hb_o}{2c} \quad \text{and} \quad Y_2 = \frac{1}{2/|k_b| + 2c/|k_b|s} \tag{11.63}$$

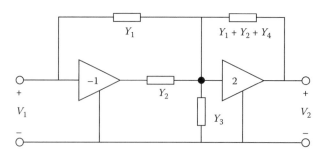

FIGURE 11.12   Two-amplifier low-gain biquad configuration.

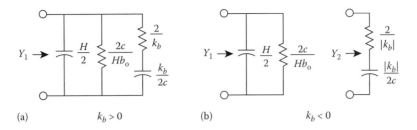

FIGURE 11.13 Realizations for (a) $Y_1(s)$ and (b) $Y_2(s)$ in Figure 11.12.

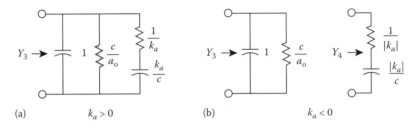

FIGURE 11.14 Realizations for (a) $Y_3(s)$ and (b) $Y_4(s)$ in Figure 11.12.

Realizations for Y_1 and Y_2 for this case are shown in Figure 11.13b. In determining Y_3 and Y_4, the partial fraction expansion of the denominator divided by $s + c$ yields a pole residue at $s = -c$,

$$k_a = \frac{c_2 - ca_1 + a_o}{-c} \tag{11.64}$$

If k_a is positive, then $Y_4 = 0$ and Y_3 is

$$Y_3 = s + \frac{a_o}{c} + \frac{1}{1/k_a + c/k_a s} \tag{11.65}$$

The realization of Y_3 is shown in Figure 11.14a. For a negative k_a, Y_3 and Y_4 become

$$Y_3 = s + \frac{a_o}{c} \tag{11.66}$$

and

$$Y_4 = \frac{1}{1/|k_a| + c/|k_a|s} \tag{11.67}$$

RC network realizations for Y_3 and Y_4 are shown in Figure 11.14b.

Example 11.5

An all-pass function [5]. It is desired to use the configuration of Figure 11.12 to realize the following all-pass function:

$$\frac{V_2(s)}{V_1(s)} = \frac{s^2 - 4s + 4}{s^2 + 4s + 4}$$

The choice of $c = +2$ will simplify the network. For this choice the numerator partial-fraction expansion is

$$\frac{s^2 = 4s + 4}{s + 2} = s + 2 - \frac{8s}{s + 2}$$

Hence $Y_1(s) = (s + 2)/2$ and $Y_2(s) = 4s/(s + 2)$. The denominator partial-fraction expansion is

$$\frac{s^2 + 4s + 4}{s + 2} = s + 2$$

Thus, $Y_3(s) = (s + 2) + [4s/(s + 2)]$ and $Y_4(s) = 0$.

11.6 Higher Order Filters

Many applications require filters of order greater than two. One way to realize these filters is simply to cascade second-order filters to implement the higher order filter. If the order is odd, then one first-order (or third-order) section will be required. The advantage of the cascade approach is that it builds on the precious techniques described in this section. The desired high-order transfer function of the filter $T(s)$ will be broken into second-order functions $T_k(s)$ so that

$$T(s) = T_1(s)T_2(s)T_3(s) \cdots T_{n/2}(s) \tag{11.68}$$

Since the output impedance of the second-order sections is low, the sections can be cascaded without significant interaction.

If $T(s)$ is an odd-order function then a first-order passive network (like those shown in Figure 11.15) can be added to the cascade of second-order functions. Both of the sections shown have nonzero output impedances, so it is advisable to use them only in the last stage of a cascade filter. Alternatively, Figure 11.5 could be used.

Several considerations should be taken into account when cascading second-order sections together. Dynamic range is one of these considerations. To maximize the dynamic range of the filter, the peak gain of each of the individual transfer functions should be equal to the maximum gain of the overall transfer function [10].

To maximize the signal-to-noise of the cascade filter, the magnitude curve in the passband of each of the individual transfer functions should be flat when possible. Otherwise, signals with frequencies in a minimum gain region of an individual transfer function will have a lower signal-to-noise ratio than other signals [10]. Another consideration is minimum noise which is achieved by placing high gain stages first.

Designing a higher order filter as a cascade of second-order sections allows the designer many options. For instance, a fourth-order bandpass filter could result from the cascade of two second-order bandpass sections or one second-order low-pass and a second-order high-pass. Since the bandpass section is the easiest of the three types to tune [10] it might be selected instead of the low-pass–high-pass combination.

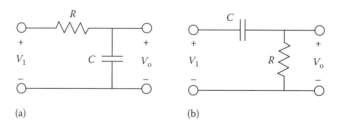

FIGURE 11.15 First-order filter sections: (a) low-pass and (b) high-pass.

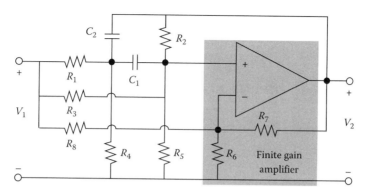

FIGURE 11.16   Biquad multiloop feedback stage.

Other guidelines for the cascade of second-order sections include putting the low-pass or bandpass sections at the input of the cascade in order to decrease the high-frequency content of the signal, which could avoid slewing in following sections. Using a bandpass or high-pass section at the output of the cascade can reduce dc offsets or low-frequency ripple.

One option that is now available is to use CAD tools in the design of filters. When high-order filters are needed, CAD tools can save the designer time. The examples that are worked later in this section require the use of filter tables interspersed with calculations. A CAD tool such as MicroSim Filter Designer [6] can hasten this process since it eliminates the need to refer to tables. The designer can also compare filters of different orders and different approximation methods (Butterworth, Chebyshev, etc.) to determine which one is the most practical for a specific application. The Filter Designer is able to design using the cascade method as discussed in this section, and after proposing a circuit it can generate a netlist so that a program such as SPICE can simulate the filter along with additional circuitry, if necessary.

The Filter Designer allows the user to specify a filter with either a passband/stopband description, center frequency/bandwidth description, or a combination of filter order and type (low-pass, high-pass, etc.). Then an approximation type is chosen, with Butterworth, Chebyshev, inverse Chebyshev, Bessel, and elliptic types available. A Bode plot, pole-zero plot, step or impulse response plot of the transfer function can be inspected. The designer can then select a specific circuit that will be used to realize the filter. The Sallen and Key low-pass and high-pass stages and the biquad multiloop feedback (Figure 11.16) stages are included. The s-coefficients of the filter can also be displayed.

The Filter Designer allows modifications to the specifications, s-coefficients, or actual components of the circuit in order to examine their effect on the transfer function. There is also a rounding function that will round components to 1%, 5%, or 10% values for resistors and 5% for capacitors and then recompute the transfer function based on these new component values. Finally, the package offers a resize function that will rescale components without changing the transfer function.

Example 11.6

A fifth-order low-pass filter. Design a low-pass filter using the Chebyshev approximation with a 1 dB equiripple passband, a 500 Hz passband, 45 dB attenuation at 1 kHz, and a gain of 10. Using a nomograph for Chebyshev magnitude functions, the order of this function is found to be five [5]. Using this information and a chart of quadratic factors of 1.0 dB equal-ripple magnitude low-pass functions, the transfer function is found to be

$$T(s_n) = \frac{0.9883K_1}{s_n^2 + 0.1789s_n + 0.9883} \cdot \frac{0.4293K_2}{s_n^2 + 0.4684s_n + 0.4293} \cdot \frac{0.2895K_3}{s_n + 0.2895}$$

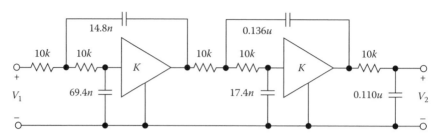

FIGURE 11.17   Fifth-order low-pass filter.

Denormalizing the frequency with $s_n = s/500(2\pi)$ gives

$$T_1(s) = \frac{9,751,000K_1}{(s^2 + 562s + 9,751,000)}$$

$$T_2(s) = \frac{4,240,000K_2}{(s^2 + 1470s + 4,240,000)}$$

and

$$T_3(s) = \frac{909K_3}{s + 909}$$

Stages 1 and 2 will be Sallen and Key low-pass filters with $R = 10$ kΩ. Following the guidelines given above, stage 1 will have the gain of 10 and successive stages will have unity gain. Using Equation 11.23 for stage 1 ($R_2 = R_4 = R$),

$$\frac{1}{Q} = 2\sqrt{\frac{C_4}{C_2}} + (1 - K)\sqrt{\frac{C_2}{C_4}} = 0.1800$$

So $C_4 = 4.696C_2$. Substituting this into Equation 11.22,

$$\omega_n = \frac{1}{2.167RC_2} = 3123 \, \text{rad/s}$$

Thus, $C_2 = 14.8$ nF and $C_4 = 69.4$ μF. Analysis for stage 2 with $K_2 = 1$ yields $C_2 = 0.135$ μF and $C_4 = 17.4$ nF. Stage 3 is simply an RC low-pass section with $R = 10$ kΩ and $C = 0.110$ μF. The schematic for this filter is shown in Figure 11.17.

Example 11.7

A fourth-order bandpass filter. Design a bandpass filter centered at 4 kHz with a 1 kHz bandwidth. The time delay of the filter should be 0.5 ms (with less than 5% error at the center frequency).

The group of filters with constant time delays are called Thomson filters. For this design, begin with a low-pass filter with the normalized bandwidth of 1/4 kHz, or 0.25. Consulting a table for delay error in Thomson filter [5] shows that a fourth-order filter is needed. From tables for Thomson filters, the quadratic factors for the fourth-order low-pass filter give the overall transfer function

$$T_{LP}(s_n) = \frac{1}{\left(s_n^2 + 4.208s_n + 11.488\right)\left(s_n^2 + 5.792s_n + 9.140\right)}$$

To convert $T_{LP}(s_n)$ to a bandpass function $T_{BP}(s_n)$, the transformation

$$s_n = \frac{\omega_R}{BW} \frac{s_{bn}^2 + 1}{s_{bn}} = 4\frac{s_{bn}^2 + 1}{s_{bn}}$$

is used, giving

$$T_{BPn}(s) = \frac{s^4/256}{(s^4 + 1.05s^3 + 2.72s^2 + 1.05s + 1)(s^4 + 1.45s^3 + 2.57s^2 + 1.45s + 1)}$$

ω_R and BW are the geometric center frequency and bandwidth of the bandpass filter. The polynomials in the denominator of the expression are difficult to factor, so an alternate method can be used to find the poles of the normalized bandpass function. Each pole p_{ln} of the normalized low-pass transfer function generates two poles in the normalized bandpass transfer function [5–10] given by

$$p_{bn} = \frac{BW\, p_{1n}}{2\omega_R} \pm \sqrt{\left(\frac{BW\, p_{1n}}{2\omega_R}\right)^2 - 1} \tag{11.69}$$

Using Equation 11.69 with each of the poles of $T_{LP}(s_n)$ gives the following pairs of poles (which are also given in tables for Thomson filters):

$$p_{b1n} = -0.1777 \pm j0.6919 \quad p_{b2n} = -0.3483 \pm j1.356$$
$$p_{b3n} = -0.3202 \pm j0.8310 \quad p_{b4n} = -0.4038 \pm j1.0478$$

Note that each low-pass pole does not generate a conjugate pair, but when taken together, the bandpass poles generated by the low-pass poles will be conjugates. The unnormalized transfer functions generated by the normalized bandpass poles are

$$T_{BP1}(s) = \frac{6283s}{(s^2 + 8932s + 322.3 \times 10^6)}$$

$$T_{BP2}(s) = \frac{6283s}{(s^2 + 1751s + 1238 \times 10^6)}$$

$$T_{BP3}(s) = \frac{6283s}{(s^2 + 16100s + 501.0 \times 10^6)}$$

and

$$T_{BP4}(s) = \frac{6283s}{(s^2 + 20300s + 796.5 \times 10^6)}$$

The overall transfer function realized by these four intermediate transfer functions could also be grouped into two low-pass stages followed by two high-pass stages, however, this example will use four bandpass filters. Using the equal-capacitance gain of two-design strategy for the Sallen and Key bandpass stage,

$$\omega_n^2 = \left(1 + \frac{R_1}{R_2}\right)\frac{1}{R_1 R_4 C^2} \tag{11.70}$$

$$\frac{1}{Q} = \left[\left(1 - \frac{R_1}{R_2}\right)\sqrt{\frac{R_4}{R_1}} + 2\sqrt{\frac{R_1}{R_4}}\right]\sqrt{\frac{R_2}{R_1 + R_2}} \tag{11.71}$$

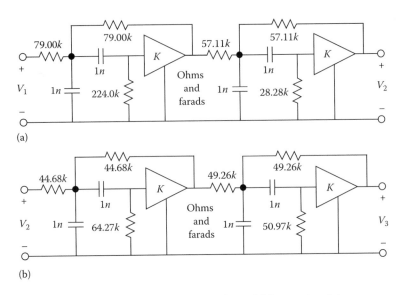

(a)

(b)

FIGURE 11.18 Fourth-order bandpass filter. (a) Stages 1 and 2 and (b) stages 3 and 4.

and

$$H_o = \left(\dfrac{\dfrac{K}{R_1}}{\dfrac{1}{R_1} + \dfrac{1}{R_2} + \dfrac{2}{R_4} - \dfrac{K}{R_2}} \right) \tag{11.72}$$

Making the further assignment $R_2 = R_1$ simplifies the equations to

$$\omega_n^2 = \dfrac{2}{R_1 R_4 C^2}, \quad \dfrac{1}{Q} = 2\sqrt{\dfrac{R_1}{R_4}}, \quad \text{and} \quad H_o = 2Q^2$$

Selecting $C = 1$ nF gives the component values shown in Figure 11.18, the schematic diagram for the fourth-order bandpass filter. While the gain of the filter is greater than unity, the output can be attenuated if unity gain is desired.

Example 11.8

A sixth-order high-pass filter. Design a Butterworth high-pass filter with a -3 dB frequency of 100 kHz and a stopband attenuation of 30 dB at 55 kHz. A nomograph for Butterworth (maximally flat) functions indicates that this filter must have an order of at least six. A table for quadratic factors of Butterworth functions [5] gives

$$T(s_n) = \dfrac{K_1}{\left(s_n^2 + 0.518 s_n + 1\right)} \dfrac{K_2}{\left(s_n^2 + 1.414 s_n + 1\right)} \dfrac{K_3}{\left(s_n^2 + 1.932 s_n + 1\right)}$$

Using $s_{nH} = 1/s_n$ for low-pass to high-pass transformation and $s_n = s/2\pi \times 10^5$ to denormalize the frequency,

$$T_1(s) = \dfrac{s^2 K_1}{s^2 + 325,500 s + 394.8 \times 10^9}$$

$$T_2(s) = \dfrac{s^2 K_2}{s^2 + 888,600 s + 394.8 \times s 10^9}$$

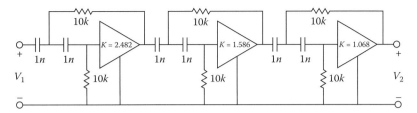

FIGURE 11.19   A sixth-order high-pass filter (all component values are Ω or F).

and

$$T_3(s) = \frac{s^2 K_3}{s^2 + 1,214,000s + 394.8 \times 10^9}$$

The Sallen and Key high-pass filter can be used to implement all three of these stages. Picking a constant R, constant C design style with $\omega_n = 10^5$ rad/s gives $R = 10$ kΩ and $C = 1$ nF for each stage. Equation 11.23 then simplifies to

$$\frac{1}{Q} = 3 - K$$

for each stage. Thus $K_1 = 2.482$, $K_2 = 1.586$, and $K_3 = 1.068$, and the resulting high-pass filter is shown in Figure 11.19.

Example 11.9

A fifth-order low-pass filter. Design a low-pass filter with an elliptic characteristic such that it has a 1 dB ripple in the passband (which is 500 Hz) and 65 dB attenuation at 1 kHz. From a nomograph for elliptic magnitude functions, the required order of this filter is five. A chart for a fifth-order elliptic filter with 1 dB passband ripple gives

$$T(s_n) = \frac{H\left(s_n^2 + 4.365\right)\left(s_n^2 + 10.568\right)}{(s_n + 0.3126)\left(s_n^2 + 0.4647s_n + 0.4719\right)\left(s_n^2 + 0.1552s_n + 0.9919\right)}$$

This transfer function can be realized with two notch filters like Figure 11.9a and a first-order low-pass section. Denormalizing the frequency with $s_n = s/1000\pi$ gives

$$T_1(s) = \frac{H_1(s^2 + 43,080,000)}{s^2 + 1,460s + 4,657,000}$$

$$T_2(s) = \frac{H_2(s^2 + 104,300,000)}{s^2 + 487.6s + 9,790,000}$$

and

$$T_3(s) = \frac{982.1}{s + 982.1}$$

For the first stage, use Equations 11.40 and 11.44, $H_1 = 0.1081$, and $a = 0.9429$. Setting $C_1 = 1$ nF and applying Equations 11.40 and 11.41,

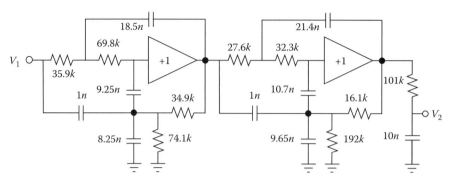

FIGURE 11.20   Fifth-order low-pass filter (all component values are Ω or F).

$$C_1 = 1\,\text{nF}, \qquad C_4 = 9.25\,\text{nF}, \qquad C_2 = 8.25\,\text{nF}, \qquad C_5 = 18.5\,\text{nF}$$
$$R_2 = 69.8\,\text{k}\Omega, \qquad R_1 = 35.9\,\text{k}\Omega, \qquad R_3 = 34.9\,\text{k}\Omega, \qquad R_1 = 74.1\,\text{k}\Omega$$

Similarly, in the second stage $H_2 = 0.0939$, $a = 0.1684$, and

$$C_1 = 1\,\text{nF}, \qquad C_4 = 10.7\,\text{nF}, \qquad C_2 = 9.65\,\text{nF}, \qquad C_5 = 21.4\,\text{nF}$$
$$R_2 = 32.3\,\text{k}\Omega, \qquad R_1 = 27.6\,\text{k}\Omega, \qquad R_3 = 16.1\,\text{k}\Omega, \qquad R_4 = 192\,\text{k}\Omega$$

The third stage can be a simple RC low-pass with $C = 10$ nF and $R = 101$ kΩ. The resulting cascade filter is shown in Figure 11.20.

These examples illustrate both the use of the Filter Designer and the method of cascading first- and second-order stages to achieve higher order filters.

11.7 Influence of Nonidealities

Effective filter design requires an understanding of how circuit nonidealities influence performance. Important nonidealities to consider include passive component tolerance (i.e., resistor and capacitor accuracy), amplifier gain accuracy, finite amplifier gain-bandwidth, amplifier slew rate, and noise.

11.7.1 Sensitivity Analysis

Classical sensitivity functions [5] are valuable tools for analyzing the influence of nonidealities on filter performance. The sensitivity function, s_x^Y [2], by definition, describes the change in a performance characteristic of interest, say y, due to the change in nominal value of some element x,

$$S_x^y = \left(\frac{\partial y}{\partial x}\right)\frac{x}{y} = \frac{\partial y/y}{\partial y/x} = \frac{\partial (l_n\, y)}{\partial (l_n\, x)} \tag{11.73}$$

The mathematical properties of the sensitivity function are given in Table 11.1 [5] for convenient reference.

A valuable result of the sensitivity function is that it enables us to estimate the percentage change in a performance characteristic due to variation in passive circuit elements from their nominal values. Consider, for example, a low-pass filter's cutoff frequency ω_o, which is a function of R_1, R_2, C_1, and C_2. Using sensitivity functions, the percentage change in ω_o is estimated as [3]

$$\frac{\Delta\omega_o}{\omega_o} \approx S_{R_1}^{\omega_o}\left(\frac{\Delta R_1}{R_1}\right) + S_{R_2}^{\omega_o}\left(\frac{\Delta R_2}{R_2}\right) + S_{C_1}^{\omega_o}\left(\frac{\Delta C_1}{C_1}\right) + S_{C_2}^{\omega_o}\left(\frac{\Delta C_2}{C_2}\right) \tag{11.74}$$

TABLE 11.1 Properties of Sensitivity Function

$$S_x^{ky} = S_{kx}^y = S_x^y \qquad\qquad S_x^{y_1/y_2} = S_x^{y_1} - S_x^{y_2}$$

$$S_x^x = S_x^{kx} = S_{kx}^{kx} = 1 \qquad\qquad S_{x_1}^y = S_{x_2}^y S_{x_1}^{y_2}$$

$$S_{1/x}^y = S_x^{1/y} = -S_x^y \qquad\qquad S_x^y = S_x^{|y|} + j \arg y S_x^{\arg y}$$

$$S_x^{y_1 y_2} = S_x^{y_1} + S_x^{y_2} \qquad\qquad S_x^{\arg y} = \frac{1}{\arg y} \operatorname{Im} S_x^y(*)$$

$$S_x^{\prod_{i=1}^n y_i} = \sum_{i=1}^n S_x^{y_i} \qquad\qquad S_x^{|y|} = \operatorname{Re} S_x^y(*)$$

$$S_x^{y^n} = n S_x^y \qquad\qquad S_x^{y+x} = \frac{1}{y+z}\left(y S_x^y + z S_x^z \right)$$

$$S_x^{x^n} = S_x^{kx^n} = n \qquad\qquad S_x^{\sum_{i=1}^n y_i} = \frac{\sum_{i=1}^n y_i S_x^{y_i}}{\sum_{i=1}^n y_i}$$

$$S_{x^n}^y = \frac{1}{n} S_x^y \qquad\qquad S_x^{\ln y} = \frac{1}{\ln y} S_x^y$$

$$S_{x^n}^x = S_{kx^n}^x = \frac{1}{n}$$

Source: Huelsman, L.P. and Allen, P.E., *Introduction to the Theory and Design of Active Filters*, McGraw-Hill, New York, 1980.

Note: Relations denoted by "(*)" use *y* to indicate a complex quantity and *x* to indicate a real quantity.

Not that the quantity $\Delta\omega_o/\omega_o$ represents a small differential change in the cutoff frequency ω_o. When considering the root locus of a filter design, the change in pole location due to a change in gain K, for example, could be described by [3]

$$S_K^{p_1} = \left(\frac{\partial \sigma_o}{\partial K} \right) + j \frac{\partial \omega_o}{\partial K} \left(\frac{K}{\omega_o} \right) \tag{11.75}$$

where the pole $p_1 = \sigma_o + j\omega_o$. Furthermore, the filter transfer function's magnitude and phase sensitivity is defined by [3]

$$S_x^{T(j\omega)} = S_x^{|T(j\omega)|} + S_x^{\theta(j\omega)} = \operatorname{Re} S_x^{T(j\omega)} + \frac{1}{\theta(\omega)} \operatorname{Im} S_x^{T(j\omega)} \tag{11.76}$$

and

$$S_x^{T(j\omega)} = \frac{x}{|T(j\omega)|} \frac{\partial}{\partial x} |T(j\omega)| + jx \frac{\partial \theta(\omega)}{\partial x} \tag{11.77}$$

where $T(j\omega) = |T(j\omega)| \exp[j\theta(\omega)]$. The interested reader is directed to Ref. [5, Chapter 3] for a detailed discussion of sensitivity.

11.7.2 Gain-Bandwidth

Now let us consider some of the amplifier nonidealities that influence the performance of our low-gain amplifier filter realizations. There are two amplifier implementations of interest each using an operational amplifier. They are the noninverting configuration and the inverting configuration shown in Figure 11.21.

The noninverting amplifier is described by

$$K = \frac{V_2}{V_1} = \frac{A_d(s)}{1 + A_d(s)/K_o} \tag{11.78}$$

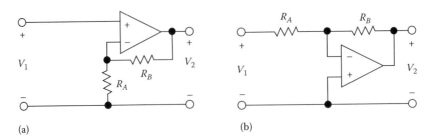

FIGURE 11.21   (a) Noninverting amplifier and (b) inverting amplifier.

where
$K_o = 1 + R_B/R_A$ is the ideal gain
$A_d(s)$ is the operational amplifier's differential gain

Using the dominant-pole model for the op-amp,

$$A_d(s) = \frac{GB}{s + \omega_a} = \frac{A_o \omega_a}{s + \omega_a} \tag{11.79}$$

where
A_o is the dc gain
ω_a is the dominant pole
GB is the gain-bandwidth product of the op-amp

Inserting Equation 11.79 into Equation 11.78 gives

$$K = \frac{GB}{s + \omega_a(1 + A_o/K_o)} \approx \frac{GB}{s + GB/K_o} \tag{11.80}$$

The approximation of the K expression is valid provided $A_o \gg K_o$. The magnitude and phase of Equation 11.80 are

$$|K(j\omega)| = \frac{GB}{\sqrt{\omega^2 + (GB/K_o)^2}} \tag{11.81}$$

and

$$\arg[K(j\omega)] = -\tan^{-1}\left(\frac{\omega}{GB}K_o\right) \tag{11.82}$$

Note that for $\omega > GB/K_o$, Equations 11.81 and 11.82 may be approximated by

$$|K(j\omega)| \approx K_o \tag{11.83}$$

and

$$\arg[K(j\omega)] \approx -\frac{\omega}{GB}K_o \tag{11.84}$$

The phase expression (Equation 11.84) describes the phase lag introduced to an active filter realization by the op-amp's finite gain-bandwidth product. To illustrate the importance of taking this into account, consider a positive-gain filter with $K_o = 3$ and operation frequency $\omega = GB/30$ [5]. According to Equation 11.84, a phase lag of 5.73° is introduced to the filter by the op-amp. Such phase lag might significantly impair the performance of some filter realizations.

The inverting amplifier of Figure 11.21b is described by

$$K = \frac{V_2}{V_1} = \frac{-A_d(s)[R_B/(R_A + R_B)]}{1 + A_d(s)[R_A/(R_A + R_B)]} \tag{11.85}$$

Inserting Equation 11.79 into Equation 11.85 gives

$$K = \frac{-[R_B/(R_A + R_B)]GB}{s + \omega_a[1 + A_oR_A/(R_A + R_B)]} \approx \frac{-[R_B/(R_A + R_B)]GB}{s + GB[R_A/(R_A + R_B)]} \tag{11.86}$$

where as before the approximation is valid if $A_o \gg (R_A + R_B)/R_A$. The magnitude and phase expressions for Equation 11.86 are

$$|K(j\omega)| = \frac{GBR_B/(R_A + R_B)}{\sqrt{\omega^2 + [GBR_A/(R_A + R_B)]^2}} \tag{11.87}$$

and

$$\arg[K(j\omega)] = \pi - \tan^{-1}\left[\frac{\omega}{GB}\left(1 + \frac{R_B}{R_A}\right)\right] \tag{11.88}$$

If $\omega < GB\,R_A/(R_A + R_B)$, then Equations 11.87 and 11.88 are approximated by

$$|K(j\omega)| \approx \frac{R_B}{R_A} \tag{11.89}$$

and

$$\arg[K(j\omega)] \approx \pi - \left[\frac{\omega}{GB}\left(1 + \frac{R_B}{R_A}\right)\right] \tag{11.90}$$

An important limitation of the inverting configuration is its bandwidth [5], compared to that of the noninverting configuration, when small values of gain K are desired. Take, for example, the case where a unity-gain amplifier is needed. Using the amplifier of Figure 11.21a, the resistor values would be $R_B = 0$ and $R_A = \infty$. From Equation 11.80 we see that this places a pole at $s = -GB$. Using the amplifier of Figure 11.21b, the resistors would be selected such that $R_B = R_A$. From Equation 11.86 we see that a pole located at $s = -GB/2$ results. Hence, for unity-gain realizations, the inverting configuration has half the bandwidth of the noninverting case. However, for higher gains, as the ratio R_B/R_A becomes large, both configurations yield a pole location that approaches $s = -(GB)R_A/R_B$.

When using the inverting configuration described above to implement a negative-gain realization, the input impedance to the amplifier is approximately R_A. R_A must then be carefully selected to avoid loading the RC network. Another problem is related to maintaining low phase lag when a high Q is desired [5]. A Q of 10 requires that the ratio R_B/R_A be approximately 900. However, to avoid a phase lag of 6°, Equation 11.90 indicates that for $Q = 10$ the maximum filter design frequency cannot exceed approximately GB/8500. If GB = 1 MHz, say if the 741 op-amp is being used, then the filter cannot function properly above a frequency of about 100 Hz.

We have seen that a practical amplifier implementation contributes a pole to the gain K of a filter due to the op-amp's finite gain-bandwidth product GB. This means that an active filter's transfer function will have an additional pole that is attributed to the amplifier. The poles then of an active filter will be influenced by this additional pole. Consider the positive-gain amplifier used in the Sallen–Key low-pass filter (Figure 11.6). The filter's ideal transfer function is expressed in Equation 11.21. For the equal R, equal C case ($R_1 = R_3$, $C_2 = C_4$) we have $R_B = R_A[2 - (1/Q)]$. By inserting Equation 11.80 into Equation 11.21, the transfer function becomes [3]

$$T(s_n) = \frac{V_2}{V_1} = \frac{GB_n}{s_n^3 + s_n^2\left(3 + \frac{GB_n}{1-\frac{1}{Q}}\right) + s_n\left(1 + \frac{GB_n}{3Q-1}\right) + \frac{GB_n}{3-\frac{1}{Q}}} \tag{11.91}$$

where the normalized terms are defined by $s_n = s/\omega_o$ and GB_n/ω_o. The manner in which the poles of Equation 11.91 are influenced by the amplifier's finite gain-bandwidth product is illustrated in Figure 11.22a. This plot is the upper left-half s_n plane. The solid lines correspond to constant values of Q and the circles correspond to discrete values of GB_n, which were the same for each Q. For $Q = 0.5$ the GB_n values are labeled. Note that the $GB_n = \infty$ case shown is the ideal case where amplifier GB is infinite. The poles shift toward the origin as ω_o approaches a value near GB. Furthermore, for large design values of Q, i.e., greater than 3, the actual Q value will decrease as ω_o approaches GB.

As another example, consider the Sallen–Key low-pass again but for the case where positive unity-gain and equal resistance ($R_1 = R_3$) are desired. When the amplifier's gain-bandwidth product is taken into account, the filter's transfer function becomes [3]

$$T(s) = \frac{V_2}{V_1} = \frac{GB_n}{s_n^3 + s_n^2\left(2Q + \frac{1}{Q} + GB_n\right) + s_n\left(1 + \frac{GB_n}{Q}\right) + GB_n} \tag{11.92}$$

where again $s_n = s/\omega_o$ and $GB_n = GB/\omega_o$. The poles of Equation 11.92 are shown in Figure 11.22b. The plots are similar but the Figure 11.22b) plot constant GB_n values for the range of Q shown are closer

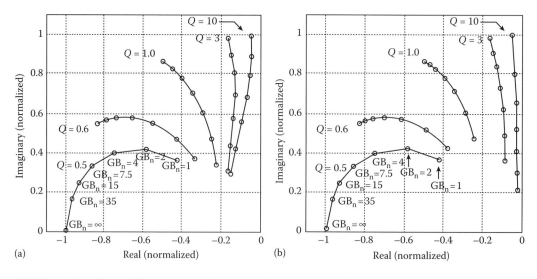

FIGURE 11.22 Effect of GB on the poles of the Sallen–Key second-order low-pass case. (a) Equal R, equal C case and (b) unity gain, equal R case.

TABLE 11.2   Summary of Finite GB on Filter Parameters

Sallen–Key Low-Pass Realizations

Equal R, equal C	$\omega_0(\text{actual}) \approx \omega_0(\text{design})\left[1 - \frac{1}{2}\left(3 - \frac{1}{Q}\right)^2 \frac{\omega_0}{GB}\right]$
	$Q(\text{actual}) \approx Q(\text{design})\left[1 - \frac{1}{2}\left(3 - \frac{1}{Q}\right)^2 \frac{\omega_0}{GB}\right]$
	$GB \geq 45\left(1 - \frac{1}{3Q}\right)^2 \omega_0^*$
Unity-gain, equal R	$\omega_0(\text{actual}) \approx \omega_0(\text{design})\left(1 - \frac{\omega_0 Q}{GB}\right)$
	$Q(\text{actual}) \approx Q(\text{design})\left(1 + \frac{\omega_0 Q}{GB}\right)$
	$GB \geq 10\omega_0^*$

"*" denotes the condition for $\Delta\omega_0/\omega_0 = \Delta Q/Q \leq 10\%$, i.e., for ω_0 and Q to change less than 10% [3].

together than the Figure 11.22a plot. This means that the poles of the unity-gain, equal R design are less sensitive to amplifier gain-bandwidth product than the equal R, equal C case. The influence GB has on Sallen–Key low-pass filter parameters is summarized in Table 11.2.

11.7.3 Slew Rate

Slew rate is another operational amplifier parameter that warrants consideration in active filter design. When a filter signal level is large and/or the frequency is high, op-amp slew rate (SR) limitations can affect the small-signal behavior of amplifiers used in low-gain filter realizations. The amplifier's amplitude and phase response can be distorted by large-signal amplitudes. An analysis of slew rate induced distortion is presented in Ref. [1]. For the inverting amplifier (Figure 11.21b), slewing increases the small-signal phase lag. This distortion is described by Ref. [5]

$$\arg[K(j\omega)] = \pi - \tan^{-1}\left(\frac{(\omega/GB(1 + R_B/R_A))}{N(A)}\right) \tag{11.93}$$

where

   $N(A)$ is a describing function
   A is the peak amplitude of the sinusoidal voltage input to the amplifier

$N(A) = 1$ while the op-amp is not slewing, yielding the same phase lag described by Equation 11.88. However, for conditions where slewing is present, $N(A)$ becomes less than 1. Under such conditions the filter may become unstable. Refer to Ref. [1] and Ref. [5, Chapter 4, Section 6] for a detailed discussion of slew rate distortion.

Selecting an operational amplifier with higher slew rate can reduce large-signal distortion. Given the choice of FET input or BJT input op-amps, FET input op-amps tend to have higher slew rates and higher input signal thresholds before slewing occurs. When selecting an amplifier for a filter it is important to remember that slewing occurs when the magnitude of the output signal slope exceeds SR and when the input level exceeds a given voltage threshold. Often the slew rate of a filter's amplifier sets the final limit of performance that can be attained in a given active filter realization at high frequencies.

11.7.4 Noise

The noise performance of a given filter design will ultimately determine the circuit's signal-to-noise ratio and dynamic range [7]. For these reasons, a strong understanding of noise analysis is required for effective filter design. In this section noise analysis of the noninverting and inverting op-amp

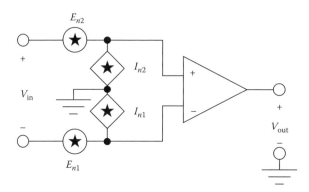

FIGURE 11.23   Operational amplifier noise model with four noise sources.

configurations are discussed for the implementation of the gain block K. An example noise analysis is illustrated for a simple filter design.

Proceeding with noise analysis, consider the four noise sources associated with operational amplifier. Shown in Figure 11.23, E_{n1} and E_{n2} represent the op-amp's voltage noise referred to the negative and positive inputs, respectively, and I_{n1} and I_{n2} represent the op-amp's current noise referred to the negative and positive inputs, respectively. E_{n1} and E_{n2} are related to E_n, from an op-amp manufacture's data sheet, through the relationship

$$E_n = \sqrt{E_{n1}^2 + E_{n2}^2} \ \left[V/\sqrt{Hz}\right] \tag{11.94}$$

I_n is also usually reported in an op-amp manufacturer's data sheet. The value given for I_n is used for I_{n1} and I_{n2}.

$$I_n = I_{n1} = I_{n2} \ \left[A/\sqrt{Hz}\right] \tag{11.95}$$

The analysis of the filter noise performance begins with the identification of all noise sources, which include the op-amp noise sources and all resistor thermal noise ($E_t = \sqrt{4\,kt\,R\Delta f}$ where 4kT is approximately 1.65×10^{-20} V²/Ω; Hz at room temperature and Δf is the noise bandwidth). To begin the analysis, uncorrelated independent signal generators can be substituted for each noise source, voltage sources for noise voltage sources, and current sources for current noise sources. By applying superposition and using the rms value of the result, the noise contribution of each element to the total output noise E_{no} can be determined. Only squared rms values can be summed to determine E_{no}^2. The resultant expression then for E_{no}^2 should never involve subtraction of noise sources, only addition. Sources of noise in a circuit always add to the total noise at the circuit's output. The total noise referred to the input E_{ni} is simply E_{no} divided by the circuit's transfer function gain. Capacitors ideally only influence noise performance through their effect on the circuit's transfer function. That is, the transfer gain seen by a noise source to the circuit's output can be influenced by capacitors depending on the circuit. As a result, of course, a circuit's noise performance can be a strong function of frequency.

Consider again the noninverting op-amp configuration, which can be used to implement positive K. The circuit is shown in Figure 11.24a with appropriate noise sources included. Resistor R_s has been included to represent the driving source's (V_1) output resistance. The simplified circuit lumps together the amplifier's noise sources into E_{na}, shown in Figure 11.24b. E_{na} is described by Equation 11.96. The circuit in Figure 11.24b can be used in place of the positive gain block K to simplify the noise analysis of a larger circuit. R_s obviously will be determined by the circuit seen looking back from the amplifier's input:

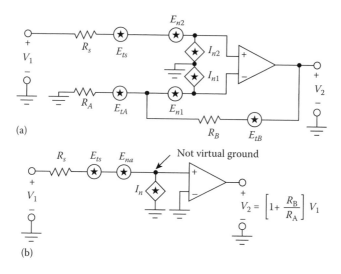

(a)

(b)

FIGURE 11.24   (a) Noninverting op-amp configuration with noise sources and (b) simplified noninverting amplifier circuit with noise sources.

$$E_{na}^2 = E_n^2 + \left(\frac{R_A}{R_A + R_B}\right)^2 (E_{tB}^2) + \left(\frac{R_B}{R_A + R_B}\right)^2 (E_{tA}^2) + I_{n1}^2 (R_A \| R_B)^2 \tag{11.96}$$

The inverting op-amp configuration for implementing a negative gain is shown in Figure 11.25a with noise sources included. Here again, the op-amp's noise sources can be lumped together and referred to the input as noise source E_{nb} shown in Figure 11.25b. For simplicity, R_s has been lumped together with R_A.

E_{nb} is described by Equation 11.97. The simplified circuit in Figure 11.25b is convenient for simplifying the noise analysis of larger circuit that uses a negative gain block.

$$E_{nb}^2 = \left(1 + \frac{R_A}{R_B}\right)^2 (E_n^2) + \left(\frac{R_A}{R_B}\right)^2 (E_{tB}^2) \tag{11.97}$$

Example 11.10

First-order low-pass filter noise analysis. Noise analysis of a low-pass filter with noninverting midband gain of 6 dB and -20 dB/decade gain roll-off after a 20 kHz cutoff frequency is desired. Recall from Section 11.3 that $\omega_o = RC$. Let us designate R_1 for R and C_2 for C. To obtain $\omega_o = 2\pi(20$ kHz), commercially available component values $R_1 = 16.9$ kΩ and $C_2 = 470$ pF (such as a 1% metal-film resistor for R_1 and a ceramic capacitor for C_2) can be used. Implement K with the noninverting op-amp configuration and use Figure 11.24b as a starting point of the noise analysis. The midband gain of 6 dB can be achieved by selecting $R_A = R_B$. For this example let $R_A = R_B = 10.0$ kΩ, also available in 1% metal film. Including the thermal noise source for R_1, the circuit shown in Figure 11.26 results.

From this circuit we can easily derive the output noise of the low-pass filter.

$$E_{no}^2 = K^2 \left[E_{na}^2 + \left|\frac{Z_2}{R_1 + Z_2}\right|^2 (E_{t1}^2) + |R_1\|Z_2|^2 (I_n^2) \right] \tag{11.98}$$

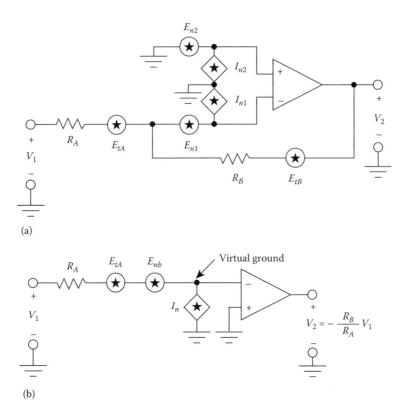

(a)

(b)

FIGURE 11.25   (a) Inverting op-amp configuration with noise sources and (b) simplified inverting amplifier circuit with noise sources.

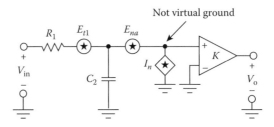

FIGURE 11.26   Positive gain first-order low-pass filter with noise sources.

where

$Z_2 = 1/sC_2$, the impedance of the capacitor

E_{na} is defined by Equation 11.96

The commercial op-amp OP-27 can be used to implement K and thus determines the values of E_n and I_n. For OP-27, $E_n = 3\,\text{nV}/\sqrt{\text{Hz}}$ and $I_n = 0.4\,\text{pA}/\sqrt{\text{Hz}}$. A comparison of the hand analysis to a PSpice simulation of this low-pass filter is shown in Figure 11.27. The frequency dependence of K was not included in the hand analysis since the filter required a low midband gain of two and low cutoff frequency relative to the OP-27's 8 MHz gain-bandwidth product. The simulated frequency response is shown in Figure 11.28. The computer simulation did include a complete model for the OP-27.

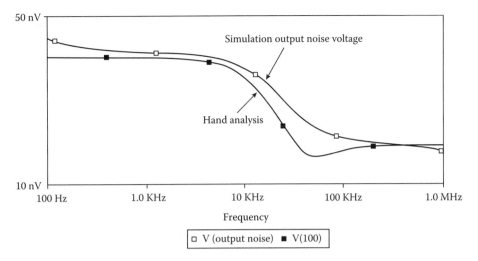

FIGURE 11.27   Comparison of calculated output noise with simulation.

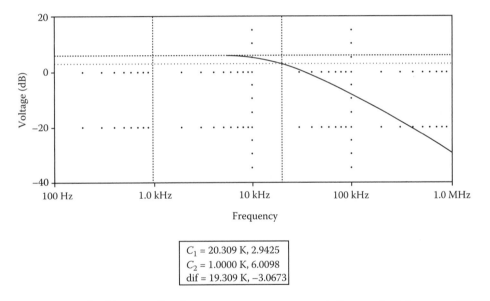

$$\boxed{\begin{array}{l} C_1 = 20.309 \text{ K}, 2.9425 \\ C_2 = 1.0000 \text{ K}, 6.0098 \\ \text{dif} = 19.309 \text{ K}, -3.0673 \end{array}}$$

FIGURE 11.28   Simulated low-pass filter frequency response. The cursors indicate a -3 dB frequency of 20 kHz.

As indicated by Figure 11.27, the output noise predicted by Equation 11.98 agrees well with the simulation. Analyzing the noise contribution of each circuit element to the filter's total output noise can provide insight for improving the design. To this end, Table 11.3 was generated for this filter.

Note that at 1 kHz R_1 is the dominant source of noise. A smaller value of R_1 could be chosen at the expense of increasing C_2. Care must be taken in doing so because changing C_2 also changes the gain multiplier for I_n and R_1's noise source, as described in Equation 11.98. The noise bandwidth, using a single-pole response approximation, is 20 kHz $(\pi/2) = 31.4$ kHz. Hence, $E_{ni} = (20.2 \text{ nV}/\sqrt{\text{Hz}})$. $(31.4$ kHz$)^{1/2} = 3.58$ μV. To achieve then a signal-to-noise ratio of 10, the input signal level must be 35.8 μV. This would be more difficult to attain if 741 op-amp has been used, with its $E_n = 20.2$ nV$/\sqrt{\text{Hz}}$, rather than the OP-27.

TABLE 11.3   Noise Contributions of Each Circuit Element at 1 kHz

Noise Source	Noise Value	Gain Multiplier	Output Noise Contribution	Input Noise Contribution
E_n	$3\,\text{nV}/\sqrt{\text{Hz}}$	2	$6\,\text{nV}/\sqrt{\text{Hz}}$	$3\,\text{nV}/\sqrt{\text{Hz}}$
I_n	$0.4\,\text{pA}/\sqrt{\text{Hz}}$	35.2 k	$14.1\,\text{nV}/\sqrt{\text{Hz}}$	$7.04\,\text{nV}/\sqrt{\text{Hz}}$
R_A	$12.65\,\text{nV}/\sqrt{\text{Hz}}$	1	$12.65\,\text{nV}/\sqrt{\text{Hz}}$	$6.325\,\text{nV}/\sqrt{\text{Hz}}$
R_B	$12.65\,\text{nV}/\sqrt{\text{Hz}}$	1	$12.65\,\text{nV}/\sqrt{\text{Hz}}$	$6.325\,\text{nV}/\sqrt{\text{Hz}}$
R_1	$16.4\,\text{nV}/\sqrt{\text{Hz}}$	1.998	$32.8\,\text{nV}/\sqrt{\text{Hz}}$	$16.4\,\text{nV}/\sqrt{\text{Hz}}$
Total noise contributions			$40.4\,\text{nV}/\sqrt{\text{Hz}}$	$20.2\,\text{nV}/\sqrt{\text{Hz}}$

Often in signal processing circuits several stages are cascaded to shape the response of the system. In doing so, the noise contributed by each stage is an important consideration. Placement of each stage within the cascade ultimately determines the E_{ni} of the entire cascade. The minimum E_{ni} is achieved by placing the highest gain first. By doing so the noise of all the following stages is divided by the single largest gain stage of the entire cascade. This comes, however, at the sacrifice of dynamic range. With the highest gain first in the cascade, each following stage has a larger input signal than in the case where the highest gain stage is last or somewhere between first and last in the cascade. Dynamic range is lost since there is a finite limit in the input signal level to any circuit before distortion results.

11.8 Summary

This chapter has examined active filters using low-gain amplifiers. These filters are capable of realizing any second-order transfer function. Higher order transfer functions are realized using cascaded first- and second-order stages. The nonideal behavior of finite gain-bandwidth, slew rate, and noise was examined. More information on this category of filters can be found in the references.

References

1. P. E. Allen, Slew induced distortion in operational amplifiers, *IEEE J. Solid-State Circuits,* SC1,2, 39–44, Feb. 1977.
2. H. W. Bode, *Network Analysis and Feedback Amplifier Design*, Princeton, NJ: Van Nostrand, 1945.
3. E. J. Kennedy, *Operational Amplifier Circuits—Theory and Applications,* New York: Holt, Rinehart & Winston, 1988.
4. W. J. Kerwin and L. P. Huelsman, The design of high-performance active *RC* bandpass filters, In *Proc. IEEE Int. Conv. Rec.,* pt. 10, Mar. 1966, pp. 74–80.
5. L. P. Huelsman and P. E. Allen, *Introduction to the Theory and Design of Active Filters*, New York: McGraw-Hill, 1980.
6. MicroSim Corporation, *PSpice Circuit Synthesis, version 4.05,* Irvine, CA: MicroSim, 1991.
7. C. D. Motchenbacher and J. A. Connelly, *Low Noise Electronic System Design*, New York: Wiley, 1993.
8. R. P. Sallen and E. L. Key, A practical method of designing *RC* active filters, *IRE Trans. Circuit Theory,* CT-2, 74–85, 1955.
9. R. Schaumann, M. S. Ghausi, and K. R. Laker, *Design of Analog Filters: Passive, Active RC, and Switched Capacitor*, Englewood Cliffs, NJ: Prentice-Hall, 1990.
10. A. S. Sedra and P. O. Brackett, *Filter Theory and Design: Active and Passive*, Portland, OR: Matrix, 1978.
11. A. S. Sedra and K. C. Smith, *Microelectronic Circuits*, New York: Holt, Rinehart & Winston, 1987.
12. A. Zverev, *Handbook of Filter Synthesis*, New York: Wiley, 1967.

12

Single-Amplifier Multiple-Feedback Filters

F. William Stephenson
Virginia Polytechnic Institute
and State University

12.1 Introduction

In this section we will consider the design of second-order sections that incorporate a single operational amplifier. Such designs are based upon one of the earliest approaches to *RC* active filter synthesis, which has proven to be a fundamentally sound technique for over 30 years. Furthermore, this basic topology has formed the basis for designs as technology has evolved from discrete component assemblies to monolithic realizations. Hence, the circuits presented here truly represent reliable well-tested building blocks for sections of modest selectivity.

12.2 General Structure for Single-Amplifier Filters

The general structure of Figure 12.1 forms the basis for the development of infinite-gain single-amplifier configurations. Simple circuit analysis may be invoked to obtain the open-circuit voltage transfer function. For the passive network

$$I_1 = y_{11}V_1 + y_{12}V_2 + y_{13}V_3 \tag{12.1}$$

$$I_2 = y_{21}V_1 + y_{22}V_2 + y_{23}V_3 \tag{12.2}$$

$$I_3 = y_{31}V_1 + y_{32}V_2 + y_{33}V_3 \tag{12.3}$$

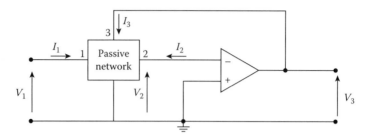

FIGURE 12.1   General infinite-gain single-amplifier structure.

For the amplifier, ideal except for finite gain A

$$V_3 = -AV_2 \tag{12.4}$$

Noting that $I_2 = 0$, the above equations reduce to the following expression for the voltage transfer function:

$$\frac{V_3}{V_1} = \frac{-y_{31}}{y_{32} + \frac{y_{33}}{A}} \tag{12.5}$$

As $A \rightarrow \infty$, which we can expect at low frequencies, the above expression reduces to the more familiar

$$\frac{V_3}{V_1} = -\frac{y_{31}}{y_{32}} \tag{12.6}$$

Theoretically, a wide range of transfer characteristics can be realized by appropriate synthesis of the passive network [1]. However, it is not advisable to extend synthesis beyond second-order functions for structures containing only one operational amplifier due to the ensuing problems of sensitivity and tuning. Furthermore, notch functions require double-element replacements [2] or parallel ladder arrangements [3], which are nontrivial to design, and whose performance is inferior to that resulting from other topologies such as those discussed in Chapters 13 and 14.

While formal synthesis techniques could be used to meet particular requirements, the most common approach is to use a double-ladder realization of the passive network, as shown in Figure 12.2. This arrangement, commonly referred to as the multiple-loop feedback (MFB) structure, is described by the following voltage transfer ratio:

$$\frac{V_3}{V_1} = \frac{-Y_1 Y_3}{Y_5(Y_1 + Y_2 + Y_3 + Y_4) + Y_3 Y_4} \tag{12.7}$$

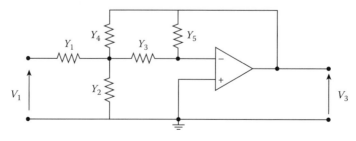

FIGURE 12.2   General double ladder multiple-feedback network.

This negative feedback arrangement yields highly stable realizations. The basic all-pole (low-pass, bandpass, high-pass) functions can be realized by single-element replacements for the admittances $Y_1, \ldots, Y_5$, as described in the following section.

12.3 All-Pole Realizations of the MFB Structure

12.3.1 Low-Pass Structure

The general form of the second-order all-pole low-pass structure is described by the following transfer ratio:

$$\frac{V_3}{V_1} = \frac{H}{s^2 + \frac{\omega_p s}{Q_p} + \omega_p^2} \tag{12.8}$$

By comparing the above requirement with Equation 12.7 it is clear that both Y_1 and Y_3 must represent conductances. Furthermore, by reviewing the requirements for the denominator, Y_5 and Y_2 must be capacitors, while Y_4 is a conductance.

12.3.2 High-Pass Structure

The general form of the second-order all-pole high-pass transfer function is

$$\frac{V_3}{V_1} = \frac{H s^2}{s^2 + \frac{\omega_p s}{Q_p} + \omega_p^2} \tag{12.9}$$

With reference to Equation 12.7, it is seen that both Y_1 and Y_3 must represent capacitors. There is a need for a third capacitor ($Y_4 = sC_4$) to yield the s^2 term in the denominator function. The remaining two elements, Y_2 and Y_5, represent conductances.

12.3.3 Bandpass Structure

The general form of the second-order all-pole bandpass transfer function is

$$\frac{V_3}{V_1} = \frac{H s}{s^2 + \frac{\omega_p s}{Q_p} + \omega_p^2} \tag{12.10}$$

Two solutions exist since Y_1 and Y_3 can be either capacitive or conductive. Choosing $Y_1 = G_1$ and $Y_3 = sC_3$ yields $Y_4 = sC_4$ and Y_2, Y_5 are both conductances.

The general forms of the above realizations are summarized in Table 12.1 [4].

12.4 MFB All-Pole Designs

MFB designs are typically reserved for sections having a pole-Q of 10 or less. One of the reasons for this constraint is the reliance upon component ratios for achieving Q. This can be illustrated by consideration of the low-pass structure for which

$$\frac{V_3}{V_1} = \frac{-G_1 G_3}{s^2 C_2 C_5 + s C_5 (G_1 + G_3 + G_4) + G_3 G_4} \tag{12.11}$$

TABLE 12.1   MFB All-Pole Realizations

Filter Type	Network	Voltage Transfer Function
Low-pass		$\dfrac{-G_1G_3}{s^2C_2C_5 + sC_5(G_1 + G_3 + G_4) + G_3G_4}$
High-pass		$\dfrac{-s^2C_1C_3}{s^2C_3C_4 + sG_5(C_1 + C_3 + C_4) + G_2G_5}$
Bandpass		$\dfrac{-sG_1C_3}{s^2C_3C_4 + sG_5(C_3 + C_4) + G_5(G_1 + G_2)}$

By comparison with Equation 12.8,

$$Q_p = \frac{\sqrt{C_2C_5G_3G_4}}{C_5(G_1 + G_3 + G_4)}$$

or, in terms of component ratios:

$$Q_p = \frac{\sqrt{C_2}}{\sqrt{C_5}}\left(\frac{1}{\frac{G_1}{\sqrt{G_3G_4}} + \sqrt{\frac{G_3}{G_4}} + \sqrt{\frac{G_4}{G_3}}}\right) \tag{12.12}$$

Hence, high Q_p can only be achieved by means of high component spreads. In general terms, a Q_p of value n requires a component spread proportional to n^2.

Filter design is effected by means of coefficient matching. Thus, for the low-pass case, comparison of like coefficients in Equation 12.8 and the transfer ratio in Table 12.1 yields

$$G_1G_3 = H \tag{12.13}$$

TABLE 12.2 Element Values for the MFB All-Pole Realizations

Element (Table 12.1)	Low-pass	Bandpass	High-Pass
Y_1	$G_1 = \frac{H}{\omega_p}$	$G_1 = H$	$C_1 = H$
Y_2	$C_2 = \dfrac{Q_p\left(2\omega_p^2 + H\right)}{\omega_p^2}$	$G_2 = 2\omega_p Q_p - H$	$G_2 = \omega_p(2 + H)Q_p$
Y_3	$G_3 = \omega_p$	$C_3 = 1$	$C_3 = 1$
Y_4	$G_4 = G_3$	$C_4 = C_3$	$C_4 = C_3$
Y_5	$C_5 = \dfrac{\omega_p^2}{Q_p\left(2\omega_p^2 + H\right)}$	$G_5 = \dfrac{\omega_p}{2Q_p}$	$G_5 = \dfrac{\omega_p}{Q_p(2 + H)}$

$$C_2 C_5 = 1 \tag{12.14}$$

$$C_5(G_1 + G_3 + G_4) = \frac{\omega_p}{Q_p} \tag{12.15}$$

$$G_3 G_4 = \omega_p^2 \tag{12.16}$$

These equations do not yield an equal-capacitor solution but can be solved for equal-resistor pairs. Hence, if $G_1 = G_3$,

$$G_1 = G_3 = \sqrt{H} \quad \text{(From Equation 12.13)}$$

$$G_4 = \frac{\omega_p^2}{\sqrt{H}} \quad \text{(From Equation 12.16)}$$

Then,

$$C_5 = \frac{\omega_p \sqrt{H}}{Q_p\left(2H + \omega_p^2\right)} = \frac{1}{C_2}$$

An alternative solution for which $G_3 = G_4$ is shown in Table 12.2, together with equal-capacitor designs for the bandpass and high-pass cases.

The conditions [4] for maximum Q_p in the bandpass realization require $C_3 = C_4$ and $G_1 = G_2 = nG_5$, where n is a real number. This yields a maximum Q_p of $\sqrt{n/2}$, and requires that $H = \omega_p Q_p$.

Example 12.1

Using the cascade approach, design a four-pole Butterworth bandpass filter having a Q of 5, a center frequency of 1.5 kHz, and midband gain of 20 dB. Assume that only 6800 pF capacitors are available.

Solution

The low-pass prototype is the second-order Butterworth characteristic having a dc gain of 10 (i.e., 20 dB). Thus,

$$T(s) = \frac{10}{s^2 + \sqrt{2}s + 1} \tag{i}$$

The low-pass-to-bandpass frequency transformation for a Q of 5 entails replacing s in (i) by $5(s+1/s)$. This yields the following bandpass function for realization:

$$\frac{V_o}{V_i} = \frac{0.4s^2}{s^4 + 0.28284s^3 + 2.04s^2 + 0.28284s + 1}$$

$$= \frac{-sH_1}{(s^2 + 0.15142s + 1.15218)} \cdot \frac{-sH_2}{(s^2 + 0.13142s + 0.86792)} \qquad \text{(ii)}$$

$$\text{(section 1)} \qquad\qquad\qquad \text{(section 2)}$$

$$Q_1 = Q_2 = 7.089$$
$$\omega_{p1} = 1.0734; \quad \omega_{p2} = 0.9316$$

As expected, the Q-factors of the cascaded sections are equal in the transformed bandpass characteristic. However, the order of cascade is still important. So as to reduce the noise output of the filter, it is necessary to apportion most of the gain to section 1 of the cascade. Section 2 then filters out the noise without introducing excessive passband gain. In the calculation that follows, it is important to note that the peak gain of a bandpass section is given by HQ/ω_p.

Since the overall peak gain of the cascade is to be 10, let this also be the peak gain of section 1. Hence,

$$\frac{H_1 Q_1}{\omega_{p1}} = 6.6041 H_1 = 10$$

giving $H_1 = 1.514$.
    Furthermore, from (ii)

$$H_1 H_2 = 0.4$$

so that $H_2 = 0.264$.

The design of each bandpass section proceeds by coefficient matching, conveniently simplified by Table 12.2. Setting $C_3 = C_4 = 1$ F, the normalized resistor values for section 1 may be determined as

$$R_1 = 0.661 \; \Omega; \quad R_2 = 0.073 \; \Omega; \quad R_5 = 13.208 \; \Omega$$

The impedance denormalization factor is determined as

$$z_n = \frac{10^{12}}{2\pi \times 1500 \times 6800} = 15{,}603$$

Thus, the final component values for section 1 are

$$C_1 = C_2 = 6800 \text{ pF}$$
$$\left.\begin{array}{l} R_1 = 10.2 \text{ k}\Omega \\ R_2 = 1.13 \text{ k}\Omega \\ R_5 = 205 \text{ k}\Omega \end{array}\right\} \text{ Standard 1\% values}$$

Note the large spread in resistance values ($R_5/R_2 \simeq 4Q^2$) and the fact that this circuit is only suitable for low-Q realizations. It should also be noted that the amplifier open-loop gain at ω_p must be much greater than $4Q^2$ if it is not to cause significant differences between the design and measured values of Q.

    The component values for section 2 are determined in an identical fashion.

Example 12.2

Design the MFB bandpass filter characterized in Example 12.1 as a high-pass/low-pass cascade of second-order sections. Use the design equations of Table 12.2 and, where possible, set capacitors equal to 5600 pF. It is suggested that you use the same impedance denormalization factor in each stage. Select the nearest preferred 1% resistor values.

Solution

Since the peak gain of the overall cascade is to be 10, let this also be the gain of stage 1 (this solution yields the best noise performance). The peak gain of the low-pass section is given by

$$\frac{H_1 Q}{\sqrt{1 - 1/2Q^2}} = 7.16 H_1 = 10 \quad \therefore H_1 = 1.397$$

The overall transfer function (from Example 12.1) is

$$\frac{V_o}{V_i} = \frac{0.4 s^2}{s^4 + 0.2824 s^3 + 2.04 s^2 + 0.2824 s + 1}$$

$$\therefore H_1 H_2 = 0.4 \quad \text{so that } H_2 = 0.286$$

Thus, assuming a low-pass/high-pass cascade, we have

$$\frac{V_o}{V_i} = \underbrace{\frac{1.397}{s^2 + 0.15145 + 1.1522}}_{\text{section 1}} \cdot \underbrace{\frac{0.286 s^2}{s^2 + 0.1314 s + 0.8679}}_{\text{section 2}}$$

Design the low-pass section (section 1) using Table 12.2 to yield

$$G_1 = \frac{H_1}{\sqrt{1.15218}} = 1.301, \quad \text{so that } R_1 = 0.7684\,\Omega$$

$$C_2 = \frac{\left(2\omega_p^2 + H_1\right)}{\omega_p^2} \quad Q_p = 22.77\,\text{F}$$

$$G_3 = \omega_p = 1.0733, \quad \text{so that } R_3 = 0.9316\,\Omega = R_4$$

$$C_5 = 1/C_2 = 0.0439\,\text{F}$$

Now, design the high-pass section (section 2) to yield

$$C_1' = 0.286\,\text{F}$$

$$G_2' = \omega_p(2 + H_2)Q = 15.099, \quad \text{so that } R_2' = 0.0662\,\Omega$$

$$C_3' = C_4' = 1\,\text{F}$$

$$G_5' = \frac{\omega_p}{Q_p(2 + H)} = 0.0574, \quad \text{so that } R_5' = 17.397\,\Omega$$

To obtain as many 5600 pF capacitors as possible, the two sections should be denormalized separately. However, in this example, a single impedance denormalization will be used. Setting $C_3' = C_4' = 5600$ pF yields $z_n = 18,947$.

This leads to the following component values:

Low-Pass Stage	High-Pass Stage
$R_1 = 2.204$ kΩ (2.21 kΩ)	$C_1' = 1602$ pF
$C_2 = 0.128$ μF	$R_2' = 1.254$ kΩ (1.24 kΩ)
$R_3 = 17.651$ kΩ (17.8 kΩ) $= R_4$	$C_3' = C_4' = 5600$ pF
$C_5 = 246$ pF	$R_5' = 329.62$ kΩ (332 kΩ)

Note: 1% values in parentheses.

12.5 Practical Considerations in the Design of MFB Filters

Sensitivity, the effects of finite amplifier gain, and tuning are all of importance in practical designs. The following discussion is based upon the bandpass case.

12.5.1 Sensitivity

Taking account of finite amplifier gain A, but assuming $R_{in} = \infty$ and $R_o = 0$ for the amplifier, the bandpass transfer function becomes

$$\frac{V_3}{V_1} = \frac{-sG_1/C_4\left(1 + \frac{1}{A}\right)}{s^2 + s\left\{\frac{G_5(C_3 + C_4)}{C_3 C_4} + \frac{G_1 + G_2}{C_4(1 + A)}\right\} + \left\{\frac{G_5(G_1 + G_2)}{C_3 C_4}\right\}} \tag{12.17}$$

which is identical to the expression in Table 12.1 if $A = \infty$.

Assuming a maximum Q design,

$$Q = \frac{Q_p}{1 + \frac{2Q_p^2}{(1 + A)}} \tag{12.18}$$

where

Q_p is the desired selectively

Q is the actual Q-factor in the presence of finite amplifier gain

If $A \gg 2Q - 1$, the classical Q-sensitivity may be derived as

$$S_A^Q = \frac{2Q^2}{A} \tag{12.19}$$

which is uncomfortably high. By contrast, the passive sensitivities are relatively low:

$$S_{C_3, C_4}^Q = 0; \quad S_{G_5}^Q = -0.5; \quad S_{G_1, G_2}^Q = 0.25$$

while the ω_p sensitivities are all ± 0.5.

12.5.2 Effect of Finite Amplifier Gain

The effect of finite amplifier gain can be further illustrated by plotting Equation 12.18 for various Q-factors, and for two commercial operational amplifiers. Assuming a single-pole roll-off model, the frequency dependence of open-loop gain for μA741 and LF351 amplifiers is as follows:

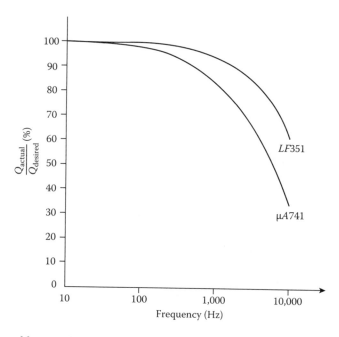

FIGURE 12.3 Effect of finite amplifier gain on the design Q for MFB bandpass realizations using two commercial operational amplifiers.

Frequency (Hz)	Gain	
	$\mu A741$	*LF351*
10	1.3×10^5	3.16×10^5
100	10^4	3.16×10^4
1,000	10^3	3.16×10^3
10,000	10^2	3.16×10^2

Figure 12.3 shows the rather dramatic fall-off in actual Q as frequency increases (and hence gain decreases). Thus, for designs of modest Q (note that $A \gg 2Q - 1$), a very high quality amplifier is needed if the center frequency is more than a few kilohertz. For example, the *LF351* with a unity gain frequency of 4 MHz will yield 6% error in Q at a frequency of only 1 kHz.

12.5.3 Tuning

Limited functional tuning of the bandpass section is possible. For example, the midband (peak) gain

$$K_o = \frac{G_1 C_3}{G_5(C_3 + G_4)} \tag{12.20}$$

may be adjusted by means of either G_1 or G_5.

Subsequent adjustment of either Q_p or ω_p is possible via G_2. In view of the discussion above, it is most likely that adjustment of Q_p will be desired. However, since the expressions for Q_p and ω_p are so similar, any adjustment of Q_p is likely to require an iterative procedure to ensure that ω_p does not change undesirably.

A more desirable functional tuning result is obtained in circumstances where it is necessary to preserve a constant bandwidth, i.e., in a spectrum analyzer. Since

$$\omega_p = \sqrt{G_5(G_1 + G_2)/C_3 C_4} \tag{12.21}$$

and bandwidth, B, is

$$B = \frac{\omega_p}{Q_p} = G_5(C_3 + C_4) \tag{12.22}$$

adjustment of G_2 will allow for a frequency sweep without affecting K_o or B.

An alternative to functional tuning may be found by adopting deterministic [5] or automatic [6] tuning procedures. These are particularly applicable to hybrid microelectronic or monolithic realizations.

12.6 Modified Multiple-Loop Feedback Structure

In negative feedback topologies such as the MFB, "high" values of Q_p are obtained at the expense of large spreads in element values. By contrast, in positive feedback topologies such as those attributed to Sallen and Key, Q_p is enhanced by subtracting a term from the s^1 (damping) coefficient in the denominator. The two techniques are combined in the MMFB (Deliyannis) arrangement [7] of Figure 12.4.

Analysis of the circuit yields the bandpass transfer function as

$$\frac{V_o}{V_i} = \frac{-sC_3 G_1 (1 + k)}{s^2 C_3 C_4 + s\{G_5(C_3 + C_4) - kC_3 G_1\} + G_1 G_5} \tag{12.23}$$

where $k = G_b/G_a$, and the Q-enhancement term "$-kC_3 G_1$" signifies the presence of positive feedback. This latter term is also evident in the expression for Q_p:

$$Q_p = \frac{\sqrt{G_1/G_5}}{\left\{\sqrt{\frac{C_4}{C_3}} + \sqrt{\frac{C_3}{C_4}} - k\frac{G_1}{G_5}\sqrt{\frac{C_1}{C_2}}\right\}} \tag{12.24}$$

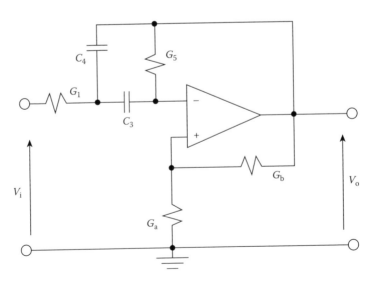

FIGURE 12.4 MMFB structure.

The design process consists of matching coefficients in Equation 12.23 with those of the standard bandpass expression of Equation 12.10. The design steps have been conveniently summarized by Huelsman [8] for the equal-capacitor solution and the following procedure is essentially the same as that described by him.

Example 12.3

Design a second-order bandpass filter with a center frequency of 1 kHz, a pole-Q of 8, and a maximum resistance spread of 50. Assume that the only available capacitors are of value 6800 pF.

1. The above constraint suggests an equal-valued capacitor solution. Thus, set $C_3 = C_4 = C$.
2. Determine the resistance ratio parameter n_o that would be required if there were no positive feedback. From Section 12.4, $n_o = 4Q_p^2 = 256$.
3. Select the desired ratio n (where n is greater than 1 but less than 256) and use it to determine the amount of positive feedback k. From Equation 12.24,

$$Q_p = \frac{\sqrt{n}}{2 - kn}$$

so that

$$k = \frac{1}{\sqrt{n}} \left\{ \frac{2}{\sqrt{n}} - \frac{1}{Q_p} \right\}$$

Since $n = 50$ and $Q_p = 8$, $k = 0.0316$.
4. A convenient value may now be selected for R_B. If $R_B = 110\,\text{k}\Omega$, then $R_A = R_B (0.0316) = 3.48\,\text{k}\Omega$.
5. Since, from Equation 12.23,

$$\omega_p = \sqrt{\frac{G_1 G_5}{C_3 C_4}}$$

and $G_1/G_5 = n$, we may determine G_5 as

$$G_5 = \frac{\omega_p C}{\sqrt{n}}$$

Since $C = 6800\,\text{pF}$, $n = 50$, and $G_5 = 1/R_5$:

$$R_5 = \frac{\sqrt{50}}{2\pi 10^3 \times 6.8 \times 10^{-9}} = 165.5\,\text{k}\Omega$$

Hence $R_1 = R_5/n = 3.31\,\text{k}\Omega$.
6. Using 1% preferred resistor values we have

$$R_B = 110\,\text{k}\Omega; \quad R_A = 3.16\,\text{k}\Omega$$
$$R_5 = 165\,\text{k}\Omega; \quad R_1 = 3.48\,\text{k}\Omega$$

Judicious use of positive feedback in the Deliyannis circuit can yield bandpass filters with Q values as high as 15–20 at modest center frequencies. A more detailed discussion of the optimization of this structure may be found elsewhere [9].

12.7 Biquadratic MMFB Structure

A generalization of the MMFB arrangement, yielding a fully biquadratic transfer ratio is shown in Figure 12.5. If the gain functions K_1, K_2, K_3 are realized by resistive potential dividers, the circuit reduces to the more familiar Friend biquad of Figure 12.6, for which

$$\frac{V_o}{V_i} = \frac{cs^2 + ds + e}{s^2 + as + b} \tag{12.25}$$

where

$$K_1 = \frac{R_5}{R_4 + R_5}; \quad K_2 = \frac{R_D}{R_c + R_D}; \quad K_3 = \frac{R_7}{R_6 + R_7}$$

$$R_1 = \frac{R_4 R_5}{R_4 + R_5}; \quad R_A = \frac{R_c R_D}{R_c + R_D}; \quad R_3 = \frac{R_6 R_7}{R_6 + R_7} \tag{12.26}$$

This structure is capable of yielding a full range of biquads of modest pole Q, including notch functions derived as elliptic characteristics of low modular angle. It was used extensively in the Bell System, where the benefits of large-scale manufacture were possible. Using the standard tantalum thin film process, and deterministic tuning by means of laser trimming, quite exacting realizations were possible [10].

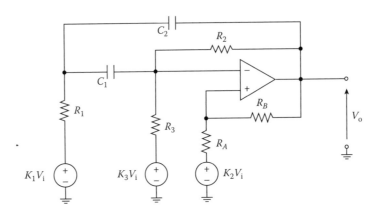

FIGURE 12.5 Generalization of the MMFB circuit.

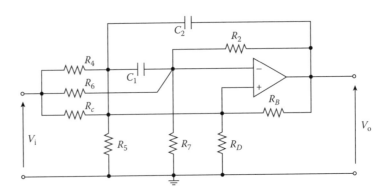

FIGURE 12.6 Friend biquad.

The structure is less suited to discrete component realizations. However, design is possible by coefficient matching. The reader is referred to an excellent step-by-step procedure developed by Huelsman [11].

12.8 Conclusions

The multiple-feedback structure is one of the most basic active filter building blocks. It is extremely reliable when used to realize low-Q (<10), low frequency (up to 15 kHz) second-order sections of the low-pass, bandpass, and high-pass forms. Stability is ensured by the negative feedback topology, though component spreads are proportional to Q^2.

The disadvantage of larger component spread may be reduced by the judicious use of positive feedback. This approach may be extended to yield the widely used Friend biquad, which allows the realization of notch and other approximations requiring a pair of imaginary zeros.

All networks described in this section readily lend themselves to the cascade method for realizing higher-order filters.

References

1. G. K. Aggarwal, On *n*th order simulation by one operational amplifier, *Proc. IEEE*, 52, 969, 1969.
2. P. L. Taylor, Flexible design method for active *RC* two-ports, *Proc. IEE*, 110, 1607–1616, 1963.
3. A. G. J. Holt and J. I. Sewell, Active *RC* filters employing a single operational amplifier to obtain biquadratic responses, *Proc. IEE*, 112, 2227–2234, 1965.
4. P. Bowron and F. W. Stephenson, *Active Filters for Communications and Instrumentation*, Berkshire, England: McGraw-Hill, 1979, p. 170.
5. R. A. Friedenson, R. W. Daniels, R. J. Dow, and P. H. McDonald, *RC* active filters for the *D3* channel bank filter, *Bell Syst. Tech. J.*, 54(3), 507–529, 1975.
6. A. B. Grebene and H. R. Camenzind, Frequency-selective integrated circuits using phase-locked techniques, *IEEE J. Solid-State Circuits*, SC-4, 216–225, Aug. 1969.
7. T. Deliyannis, High-*Q* factor circuit with reduced sensitivity, *Electron. Lett.*, 4(26), 577–579, 1968.
8. L. P. Huelsman, *Active and Passive Filter Design*, New York: McGraw-Hill, 1993, ch. 5, pp. 277–278.
9. M. S. Ghausi and K. R. Laker, *Modern Filter Design*, Englewood Cliffs, NJ: Prentice Hall, 1981, ch. 4, pp. 197–201.
10. J. J. Friend, C. A. Harris, and D. Hilberman, STAR: An active biquadratic filter section, *IEEE Trans. Circuits Syst.*, CAS-22, 115–121, Feb. 1975.
11. L. P. Huelsman, Multiple-loop feedback filters, in *RC Active Filter Design Handbook* (F. W. Stephenson, Ed.), New York: Wiley, 1985, ch. 7, pp. 201–203.

13

Multiple-Amplifier Biquads

Norbert J. Fliege
University of Mannheim

13.1 Introduction

The step from single-amplifier to multiple-amplifier biquadratic filter sections provides several benefits. The most important benefits are as follows:

- Reduced passive element spread, i.e., the ratio between the largest and the smallest values of resistors and/or capacitors can be reduced compared to the single-amplifier case.
- The required amplifier gains in some circuits grow linearly or less with the Q-factor of the complex pole pairs.
- Multiple-amplifier biquads often provide lower sensitivities to both passive and active components.
- Most of the multiple-amplifier biquads are more universal filter structures realizing the general biquadratic transfer function.
- Most of the filter parameters such as pole and zero Q-factors, pole and zero frequencies, and the gain factor of the transfer function can be tuned independently.
- Designing multiple-amplifier filters, often the values of the capacitors can be chosen freely. The filter parameters are then determined by resistors, which is less costly than by capacitors.

On the other hand, these benefits must be paid for by increased space requirements and increased power dissipation. However, today there are low-cost and low-power integrated-circuit op-amps available with up to four op-amps on one chip. Therefore, size and power dissipation are often no longer the main problem.

In the following, we will first consider biquadratic filter sections and dual-amplifier twin-*T* biquads. Both circuit families are directly derived from single-amplifier circuits. Next we will derive filter circuits having a quite different origin: filters that are derived from the generalized impedance converter (GIC) and filters derived from state-variable representation of linear systems on the analog computer. Finally, we will briefly consider filter circuits based on first-order all-pass sections.

13.2 Biquads with Decoupled Time Constants

One of the simplest methods for improving the performance of a biquad is demonstrated in Figure 13.1.

Figure 13.1a shows a well-known Sallen and Key bandpass circuit [1], which offers moderate pole sensitivities with respect to the passive elements. It can easily be shown that this circuit has the following transfer voltage ratio:

$$H(s) = \frac{V_o}{V_i} = \frac{-K \cdot sT_2}{s^2 T_1 T_2 (1 + K) + s(T_{12} + T_1 + T_2) + 1} \tag{13.1}$$

with the amplifier gain K and the time constants $T_1 = R_1 C_1$, $T_2 = R_2 C_2$, and $T_{12} = R_1 C_2$. The gain requirement is more than $4Q_p^2 - 1$ with Q_p being the Q-factor of the pole pair. This relationship limits the circuit to low or medium-Q applications.

If we insert another amplifier between the two RC networks $R_1 C_1$ and $R_2 C_2$, we obtain the biquad in Figure 13.1b [2,3], which possesses a transfer voltage ratio

$$H(s) = \frac{V_o}{V_i} = \frac{K_1 K_2 \cdot sT_2}{s^2 T_1 T_2 (1 - K_1 K_2) + s(T_1 + T_2) + 1} \tag{13.2}$$

Here the product $K_1 K_2$ of the gain factors plays the role of the gain $-K$ in Figure 13.1a. One of the gain factors must be positive, the other negative. A comparison of Equations 13.1 and 13.2 shows that after isolating both RC networks the "cross time constant" T_{12} disappears. The two time constants T_1 and T_2 are decoupled. From this change we can derive two benefits. Both factors K_1 and K_2 require only a gain of approximately $2Q_p$ and, as will be shown later, this circuit can be designed with zero Q-sensitivity.

Next, we generalize the structure in Figure 13.1b by replacing the passive elements by general admittances; see Figure 13.2. In this circuit, the two subnetworks Y_{1a}, Y_{1b} and Y_{2a}, Y_{2b} are decoupled.

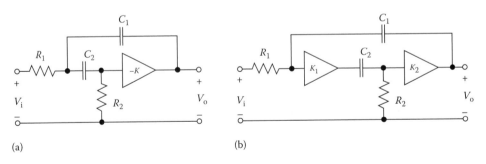

(a)   (b)

FIGURE 13.1   (a) Sallen and Key bandpass filter and (b) dual-amplifier bandpass filter with decoupled time constants.

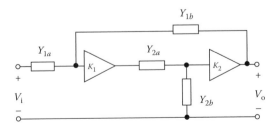

FIGURE 13.2   General biquad with decoupled networks.

By a simple analysis we obtain the transfer voltage ratio of the general circuit in Figure 13.2:

$$H(s) = \frac{V_o}{V_i} = \frac{K_1 K_2 Y_{1a} Y_{2a}}{(Y_{2a} + Y_{2b})(Y_{1a} + Y_{1b}) - K_1 K_2 Y_{1b} Y_{2a}} \tag{13.3}$$

By prespecifying the types of passive elements in Figure 13.2 and Equation 13.3, respectively, we next will derive a low-pass, a bandpass, and high-pass filter section.

With the prespecified elements $Y_{1a} = G_1$, $Y_{1b} = sC_1$, $Y_{2a} = G_2$, and $Y_{2b} = sC_2$ we obtain from Equation 13.3 the low-pass transfer function:

$$H_{LP}(s) = \frac{V_o}{V_i} = \frac{K_1 K_2 G_1 G_2}{s^2 C_1 C_2 + s[G_1 C_2 + G_2 C_1(1 - K_1 K_2)] + G_1 G_2}$$

$$= \frac{K_1 K_2}{s^2 T_1 T_2 + s[T_2 + T_1(1 - K_1 K_2)] + 1} \tag{13.4}$$

If we predefine $K_1 = K_2 = 1$ we simply get voltage followers in the filter circuit and additionally a simple design procedure resulting in a low-sensitivity filter section. Figure 13.3 shows the low-pass filter section with two op-amps and four passive elements.

For $K_1 = K_2 = 1$ we will have dc gain $H_0 = 1$, a pole frequency

$$\omega_p = 1/\sqrt{T_1 T_2} \tag{13.5}$$

and a Q-factor

$$Q_p = \sqrt{T_2/T_1} \tag{13.6}$$

Thus, designing the filter section, from the predefined parameters ω_p and Q_p we determine the time constants

$$T_2 = \frac{Q_p}{\omega_p}, \quad T_1 = \frac{1}{Q_p \omega_p} \tag{13.7}$$

Finally, we choose the values of the capacitors and calculate the resistors from the time constants.

From Equations 13.5 and 13.6, we can immediately read the pole sensitivities:

$$S_{R_1}^{\omega_p} = S_{C_1}^{\omega_p} = S_{R_2}^{\omega_p} = S_{C_2}^{\omega_p} = -\frac{1}{2} \tag{13.8}$$

$$S_{R_1}^{Q_p} = S_{C_1}^{Q_p} = -S_{R_2}^{Q_p} = -S_{C_2}^{Q_p} = -\frac{1}{2} \tag{13.9}$$

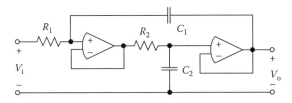

FIGURE 13.3 Dual op-amp low-pass filter section.

This result is comparable with the pole sensitivities of passive second-order *RLC* networks.

Next we will develop a high-pass filter section by prespecifying $Y_{1a} = sC_1$, $Y_{1b} = G_1$, $Y_{2a} = sC_2$, and $Y_{2b} = G_2$ resulting in a transfer voltage ratio:

$$H_{\text{HP}}(s) = \frac{V_o}{V_i} = \frac{s^2 K_1 K_2 C_1 C_2}{s^2 C_1 C_2 + s[G_2 C_1 + G_1 C_2 (1 - K_1 K_2)] + G_1 G_2}$$

$$= \frac{s^2 K_1 K_2 T_1 T_2}{s^2 T_1 T_2 + s[T_1 + T_2 (1 - K_1 K_2)] + 1} \tag{13.10}$$

The only difference of this transfer function compared with the low-pass transfer function in Equation 13.4 is the high-pass term (with s^2) in the numerator and the fact that the time constants T_1 and T_2 are interchanged in the denominator polynomial. Thus we can transfer the results in Equation 13.5 through 13.9 to the high-pass filter by only replacing the index 1 by 2 and vice versa. Figure 13.4 shows the high-pass filter section.

The bandpass filter circuit mentioned at the beginning of this section is defined by the following types of elements: $Y_{1a} = G_1$, $Y_{1b} = sC_1$, $Y_{2a} = sC_2$, and $Y_{2b} = G_2$. Its transfer function is written in Equation 13.2. The bandpass design is somewhat different from that of the low-pass and the high-pass filters described above. If we set the design values

$$G_1 = G_2 = 1, \quad C_1 = C_2 = \frac{1}{2Q_p} \tag{13.11}$$

and

$$-K_1 K_2 = 4Q_p^2 - 1 \tag{13.12}$$

we obtain a special design where all passive Q-sensitivities are zero and all other sensitivities are very low [2,3]:

$$S_{R_1}^{Q_p} = S_{C_1}^{Q_p} = S_{R_2}^{Q_p} = S_{C_2}^{Q_p} = 0 \tag{13.13}$$

$$S_{R_1}^{\omega_p} = S_{C_1}^{\omega_p} = S_{R_2}^{\omega_p} = S_{C_2}^{\omega_p} = -\frac{1}{2} \tag{13.14}$$

$$S_{K_1}^{Q_p} = S_{K_2}^{Q_p} = \frac{-K_1 K_2}{2(1 - K_1 K_2)} < \frac{1}{2} \tag{13.15}$$

$$S_{K_1}^{\omega_p} = S_{K_2}^{\omega_p} = -\frac{1}{2}\left(1 - \frac{1}{4Q_p^2}\right) \tag{13.16}$$

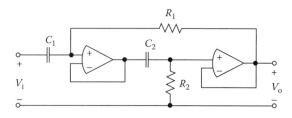

FIGURE 13.4   Dual op-amp high-pass filter section.

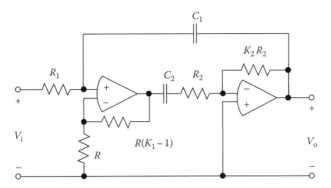

FIGURE 13.5 Dual op-amp bandpass filter section.

Figure 13.5 shows the bandpass filter section realized with two op-amps. In this circuit K_1 is a noninverting and K_2 an inverting amplifier. The resistor R_2 of the passive network R_2, C_2 is realized by the input resistor of the inverting amplifier.

13.3 Dual-Amplifier Twin-*T* Biquads

Twin-*T* feedback networks are easily tunable and provide relatively favorable sensitivity properties. In active filters, a twin-*T* network is connected between the input and the output of an inverting amplifier. It has been shown [4] that the sensitivity to the active element can be substantially reduced when a symmetrical feedback network is used, i.e., when the dc gain of the feedback network is equal to the high-frequency gain.

In a single-amplifier twin-*T* biquad, the output port of the feedback network is loaded by the input resistance of the inverting amplifier. As a consequence, the feedback network is no longer symmetrical. Here again we can significantly improve the behavior of the twin-*T* biquad by introducing a second amplifier: inserting a voltage follower between the twin-*T* network and the inverting amplifier maintains the symmetry of the feedback network. Figure 13.6 shows the corresponding dual-amp twin-*T* resonator* with the usually chosen passive element relations.

From the resonator in Figure 31.6 several filter circuits can be derived by inserting the input voltage source in one of the grounded branches and by taking one of the two amplifier outputs as output terminal

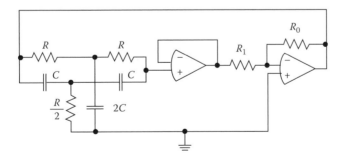

FIGURE 13.6 Dual-amp twin-*T* resonator.

---

* A resonator is a circuit without input and output terminals that only serves to show the feedback mechanism.

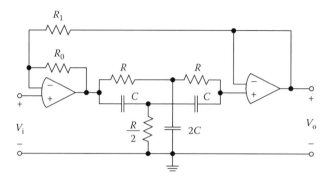

FIGURE 13.7   Dual-amp twin-T band-rejection filter.

of the filter section. The voltage transfer ratios of all these circuits will have different numerator polynomials but the same denominator polynomial. As an example, Figure 31.7 shows a band-rejection filter derived from the dual-amp twin-T resonator in Figure 13.6.

The voltage transfer ratio of the band-rejection filter in Figure 13.7 can be calculated to be

$$H_{BR}(s) = \frac{V_o}{V_i} = \frac{s^2 T^2 + 1}{s^2 T^2 + s4T/K + 1} \tag{13.17}$$

with the time constant $T = RC$ and the gain $K = (R_0 + R_1)/R_1$ of the series connection of the two op-amp circuits. Designing this filter, we determine the time constant T from the notch frequency ω_z or pole frequency ω_p, respectively,

$$T = \frac{1}{\omega_z} = \frac{1}{\omega_p} \tag{13.18}$$

and the amplifier gain from the Q-factor:

$$K = 4 \cdot Q_p \tag{13.19}$$

In Ref. [4], it is shown that the filter has favorably low gain-sensitivity products.

In order to obtain further filter variants, in Ref. [4] a complementary circuit is derived from the resonator in Figure 13.6; see Figure 13.8.

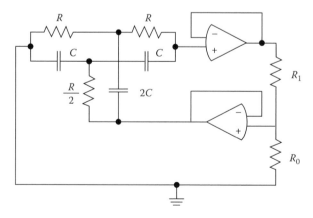

FIGURE 13.8   Complementary circuit of the resonator in Figure 13.6.

This resonator is especially useful for deriving filter sections with frequently applied transfer functions. In Figures 13.9 through 13.13 some of these filter sections are depicted. Their voltage transfer ratios are given by

$$H(s) = \frac{V_o}{V_i} = \frac{N(s)}{s^2 T^2 + s4T/K + 1} \tag{13.20}$$

The band-rejection filter in Figure 13.9 has a numerator polynomial $N(S) = s^2 T^2 + 1$. The filter section in Figure 13.10 is low-pass with a numerator $N(s) = 1$, the filter section in Figure 13.11 is high-pass

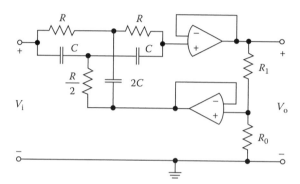

FIGURE 13.9 Twin-T band-rejection filter.

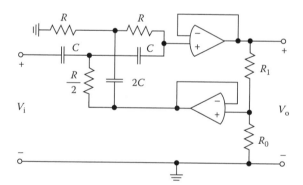

FIGURE 13.10 Twin-T low-pass filter.

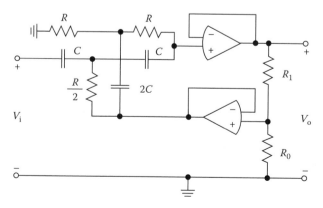

FIGURE 13.11 Twin-T high-pass filter.

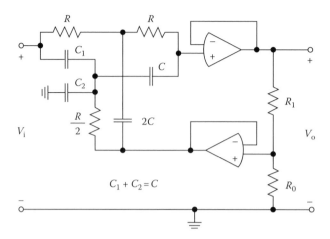

FIGURE 13.12   Twin-*T* Cauer low-pass filter section.

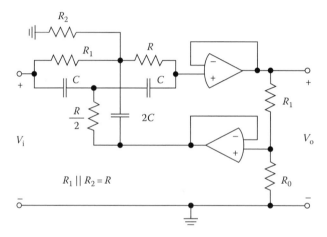

FIGURE 13.13   Twin-*T* Cauer high-pass filter section.

with $N(s) = s^2 T^2$. The design of these sections is exactly the same as that of the filter in Figure 13.7. It also provides the same pole sensitivities.

The two structures in Figure 13.12 and 13.13 are particularly suitable for realizing elliptical filter sections (also called Cauer filters). The Cauer low-pass filter section in Figure 13.12 has a numerator polynomial:

$$N(s) = s^2 R^2 CC_1 + 1, \quad C_1 + C_2 = C \tag{13.21}$$

In this circuit, the first capacitor C is split into a parallel connection of two capacitors C_1 and C_2, where C_2 is grounded. The zeros are on the $j\omega$ axis at positions $\pm j\omega_z$ with $\omega_z = 1/\sqrt{R^2 CC_1}$. The notch frequency ω_z is greater than the pole frequency ω_p. Designing the filter section, we determine the parameters T and K as described above. The splitting ratio of the input capacitor is determined by the poles and zeros:

$$\frac{C_2}{C_1} = \frac{\omega_z^2}{\omega_p^2} - 1 \tag{13.22}$$

Finally, in Figure 13.13 we have a Cauer high-pass filter section with a numerator polynomial:

$$N(s) = s^2 R R_1 C^2 + 1, \quad R_1 \| R_2 = R \tag{13.23}$$

Here the zero frequency ω_z is less than the pole frequency ω_p. As in the low-pass case, the splitting ratio of the input resistors is determined by these two frequencies:

$$\frac{R_1}{R_2} = \frac{\omega_p^2}{\omega_z^2} - 1 \tag{13.24}$$

The circuit in Figure 13.6 is actually of third order. By matching the passive components, as shown in Figure 13.6, we obtain a second-order transfer function. The question may arise as to what happens if we have a small mismatch due to the tolerances of practical components. In this case, a third pole and a third zero appear on the negative real axis of the s-plane. They do not cancel each other exactly, but approximately. In general, the existence of the third pole and the third zero does not affect the frequency response of the filter significantly. However, there is a second effect due to a small mismatch of passive components that is more severe: the position of the pole pair desired by design is changed. This change can be estimated by the pole sensitivities with respect to the passive elements [4].

13.4 GIC-Derived Dual-Amplifier Biquads

In this section, we consider a class of biquadratic building blocks with two op-amps that are derived from the generalized impedance converter (GIC). A catalog of such building blocks realizing a wide variety of network functions, including elliptic and all-pass ones, was published by Fliege [5].

Figure 13.14b shows the general filter structure, which is based on the resonator with two nullors in Figure 13.14a. Each nullor consists of one nullator and one norator and constitutes a model for the ideal op-amp. Combining the norator between the node between the admittances Y_2 and Y_3 and ground and the nullator across Y_4 and Y_6 yields the op-amp μ_1 in Figure 13.14b. It can readily be verified that the voltage transfer ratio of the circuit in Figure 13.14b with $\mu_i = \infty$, $i = 1, 2$ is given by

$$H(s) = \frac{V_o}{V_i} = \frac{Y_{6b}(Y_2 Y_4 + Y_{1a} Y_4) + Y_{1b}(Y_3 Y_5 - Y_{6a} Y_4)}{Y_1 Y_3 Y_5 + Y_2 Y_4 Y_6} \tag{13.25}$$

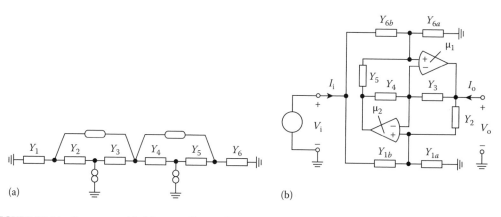

FIGURE 13.14   Resonator with (a) two nullors and (b) the corresponding general dual-amplifier filter structure.

with

$$Y_1 = Y_{1a} + Y_{1b}, \quad Y_6 = Y_{6a} + Y_{6b} \tag{13.26}$$

If we choose the node between Y_4 and Y_5 as the output of the building block we will obtain a similar transfer function. We only have to interchange the admittances Y_1 by Y_6, Y_2 by Y_5, and Y_3 by Y_4.

First, we will derive a second-order low-pass building block from the general structure and we will take this low-pass filter as a prototype for the whole circuit family to explain their advantageous characteristics. If we choose the passive elements of the building block as $Y_{1a} = G_1$, $Y_{1b} = 0$, $Y_2 = G_2$, $Y_3 = sC_3$, $Y_4 = G_4$, $Y_5 = G_5 + sC_5$, $Y_{6a} = 0$, and $Y_{6b} = G_6$ we obtain the low-pass building block in Figure 13.15. Its voltage transfer ratio reads

$$H_{LP}(s) = \frac{V_o}{V_i} = \frac{G_6 G_4 (G_2 + G_1)}{s^2 C_3 C_5 G_1 + s C_3 G_5 G_1 + G_2 G_4 G_6}$$

$$= H_0 \cdot \frac{\omega_p^2}{s^2 + s\omega_p/Q_p + \omega_p^2} \tag{13.27}$$

with dc gain

$$H_0 = \frac{1 + \alpha_{21}}{\alpha_{21}} \tag{13.28}$$

pole frequency

$$\omega_p = \sqrt{\frac{\alpha_{21}}{T_{34} T_{56}}} \tag{13.29}$$

and Q-factor

$$Q_p = \omega_p \cdot T_{55} \tag{13.30}$$

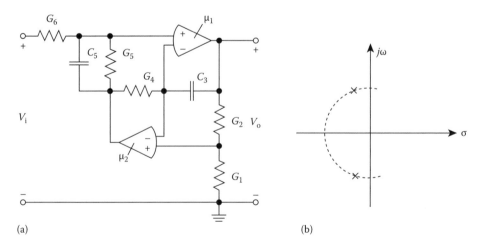

FIGURE 13.15 Second-order low-pass building block. (a) The circuitry. (b) The pole locations.

Here, the time constants are defined as

$$T_{\mu\nu} = \frac{C_\mu}{G_\nu} \tag{13.31}$$

and the conductance ratios as

$$\alpha_{\mu\nu} = \frac{G_\mu}{G_\nu} \tag{13.32}$$

Figure 13.15b shows the location of the finite poles and zeros in the s-plane.

In the following, we predefine $G_1 = G_2$, i.e., $\alpha_{21} = 1$, and $T_{34} = T_{56} = T$. This provides optimum pole sensitivities with respect to the passive elements and with respect to the gain-bandwidth products of the operational amplifiers. Hence, the gain factor H_0 always has the value 2.

Based on these parameter restrictions, we can establish an easy design procedure. First, we can choose the capacitors $C_3 = C_5 = C$. Then, given the pole parameters ω_p and Q_p, we determine the two resistors

$$R_4 = R_6 = R = \frac{1}{\omega_p \cdot C} \tag{13.33}$$

and from Equation 13.30 the resistor

$$R_5 = Q_p \cdot R \tag{13.34}$$

There are three key features that make these dual-amplifier biquads particularly favorable for practical applications.

- Biquad building blocks have a low spread of elements. The resistors $R_1 = R_2$ can be chosen freely. The same holds for the two capacitors $C_3 = C_5$. The frequency-determining resistors $R_4 = R_6$ have equal values, too. There is only the Q-determining resistor R_5, which differs by a factor of Q_p.
- Relative sensitivities of the gain and the pole parameters with respect to the passive elements are of the same order of magnitude as in case of a second-order passive *RLC* network; see Equation 13.35.
- Impact of the op-amp gain and gain-bandwidth product on the pole parameter is extremely low.

A sensitivity analysis of the biquad circuit in Figure 13.15 gives the following results:

$$
\begin{aligned}
S_{R_1}^{H_0} &= -S_{R_2}^{H_0} = -\frac{1}{2}s \\
S_{R_1}^{\omega_p} &= -S_{R_2}^{\omega_p} = -S_{C_3}^{\omega_p} = -S_{R_4}^{\omega_p} = -S_{C_5}^{\omega_p} = -S_{R_6}^{\omega_p} = \frac{1}{2} \\
S_{R_1}^{Q_p} &= -S_{R_2}^{Q_p} = -S_{C_3}^{Q_p} = -S_{R_4}^{Q_p} = +S_{C_5}^{Q_p} = -S_{R_6}^{Q_p} = \frac{1}{2} \\
S_{R_5}^{Q_p} &= 1.
\end{aligned}
\tag{13.35}
$$

In most practical applications, we can describe the op-amp dynamics by a one-pole model:

$$\mu(s) = \frac{\mu_0}{1 + (s/\omega_c)} = \frac{1}{(1/\mu_0) + (s/\omega_T)} \tag{13.36}$$

with the dc gain μ_0, the 3-dB cutoff frequency ω_c, and the gain-bandwidth product $\omega_T = \mu_0\omega_c = 1/T_T$.

If the circuit elements are chosen properly, e.g., choosing network parameters $\alpha_{21} = 1$ and $T_{34} = T_{56} = T = 1/\omega_p$, an analysis of the impact of the parameters μ_{0i} and $T_{Ti} = 1/\omega_{Ti}$ of the two op-amps μ_i, $i = 1, 2$, on the pole parameters of the transfer function yields

$$\frac{\Delta\omega_p}{\omega_p} \approx \frac{1}{\mu_{01}} - \frac{1}{\mu_{02}} - \frac{T_{T1} + T_{T2}}{T} \tag{13.37}$$

$$\frac{\Delta Q_p}{Q_p} \approx -2Q_p \left(\frac{1}{\mu_{01}} + \frac{1}{\mu_{02}} + \frac{T_{T1} - T_{T2}}{T} \right) \tag{13.38}$$

In case of high pole frequencies the impact of the dc gains μ_{01} and μ_{02} can be neglected against that of the gain-bandwidth products. From Equation 13.37, after some intermediate steps, we obtain

$$\omega_p + \Delta\omega_p \approx \frac{1}{T + T_{T1} + T_{T2}} \tag{13.39}$$

In case of an ideal op-amp we have $\omega_p = 1/T$. The nonideal op-amp causes a pole frequency change $\Delta\omega_p$ due to the time constants T_{Ti}. The frequency determining time constant T of the passive network is increased by the two time constants T_{Ti} of the op-amps, which are reciprocals of the gain-bandwidth products ω_{Ti}.

The most interesting result is the Q-factor change in Equation 13.38. If the frequency responses of the two op-amps are matched, as is usually the case with dual packages, the impact of the two gain-bandwidth products cancels out. Thus, we have nearly no Q enhancement at higher pole frequencies.

To make the GIC-derived biquad almost independent of op-amp parameters, the above-mentioned conditions ($\alpha_{21} = 1$, $T_{34} = T_{56}$) must be met by the passive elements. In a more general view, this result holds for the whole family of GIC-derived building blocks. Each of these building blocks has two frequency-determining time constants and one resistive voltage divider (G_1 and G_2 or G_3 and G_4). In any case, to obtain independence of op-amp parameters we have to choose equal time constants and a voltage divider with equal resistors. If the time constants or the resistors, respectively, do not match exactly, the independence of op-amp parameters remains nearly unchanged.

Next we will derive a bandpass building block from the general biquad in Figure 13.14b. If we choose the passive elements as $Y_{1a} = G_1$, $Y_{1b} = 0$, $Y_2 = G_2$, $Y_3 = G_3$, $Y_4 = sC_4$, $Y_5 = G_5$, $Y_{6a} = sC_6$, and $Y_{6b} = G_6$, we obtain the bandpass building block in Figure 13.16, which has a voltage transfer ratio:

$$H_{BP}(s) = \frac{V_o}{V_i} = \frac{G_6(sC_4G_2 + sC_4G_1)}{s^2C_4C_6G_2 + sC_4G_2G_6 + G_1G_3G_5}$$

$$= H_0 \cdot \frac{s}{s^2 + s\omega_p/Q_p + \omega_p^2} \tag{13.40}$$

with pole frequency

$$\omega_p = \sqrt{\alpha_{12}/(T_{43}T_{65})} \tag{13.41}$$

Q-factor

$$Q_p = \omega_p \cdot T_{66} \tag{13.42}$$

and midband gain

$$H_0 Q_p = (1 + \alpha_{12}) \tag{13.43}$$

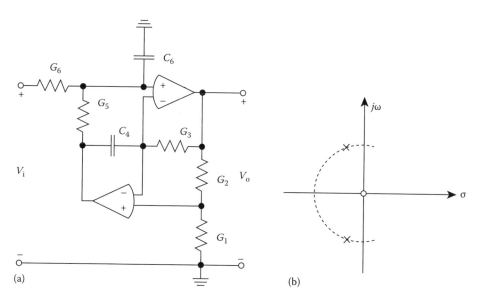

FIGURE 13.16 Second-order bandpass building block. (a) The circuitry. (b) The pole and zero locations.

The filter design is similar to that of the low-pass building block. First we choose the two resistors $R_1 = R_2$ and the two capacitors $C_4 = C_6 = C$. Then, from the pole frequency ω_p we determine the two resistors:

$$R_3 = R_5 = R = \frac{1}{\omega_p \cdot C} \tag{13.44}$$

and from the Q-factor Q_p the resistor

$$R_6 = Q_p \cdot R \tag{13.45}$$

The pole sensitivities with respect to the passive elements and the relationship between the gain and the gain-bandwidth product of the op-amps are the same as in case of the low-pass filters. It should be mentioned that the maximum of the magnitude response $|H(j\omega)|$ occurs at $\omega = \omega_p$ and has exactly the value 2, independent of the Q-factor and the time constants of the circuit*; see Equation 13.43.

We can derive a high-pass building block by choosing the following elements: $Y_{1a} = G_1$, $Y_{1b} = 0$, $Y_2 = G_2$, $Y_3 = G_3$, $Y_4 = sC_4$, $Y_5 = G_5$, $Y_{6a} = G_6$, and $Y_{6b} = sC_6$; see Figure 13.17.

This filter circuit has a voltage transfer ratio

$$H_{HP}(s) = \frac{V_o}{V_i} = \frac{sC_6(sC_4G_2 + sC_4G_1)}{s^2C_4C_6G_2 + sC_4G_2G_6 + G_1G_3G_5}$$

$$= (1 + \alpha_{12}) \cdot \frac{s^2}{s^2 + s\omega_p/Q_p + \omega_p^2} \tag{13.46}$$

with ω_p as in Equation 13.41 and Q_p as in Equation 13.42.

The design of the high-pass building block in Figure 13.17 is identical to that of the bandpass described above. Both building blocks also have the same pole sensitivities and the same impact of the op-amps on the pole parameters.

---

* Only assuming $\alpha_{12} = 1$.

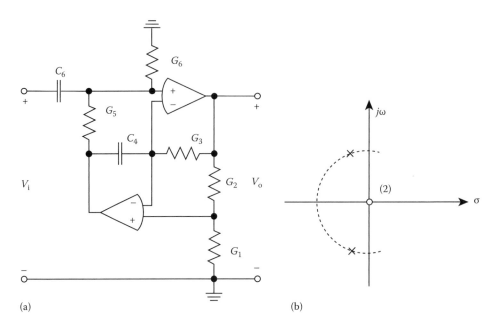

FIGURE 13.17   Second-order high-pass building block. (a) The circuitry. (b) The pole and zero locations.

If we feed the input signal not only through the capacitor C_6, but additionally through the element G_1, and if we set again $\alpha_{12} = 1$ we obtain the all-pass building block in Figure 13.18.

Its voltage transfer ratio reads

$$H_{AP}(s) = \frac{V_o}{V_i} = \frac{s^2 C_6 C_4 G_2 - s C_4 G_1 G_6 + G_1 G_3 G_5}{s^2 C_4 C_6 G_2 + s C_4 G_2 G_6 + G_1 G_3 G_5}$$

$$= \frac{s^2 - \dfrac{s}{T_{66}} + \dfrac{1}{T_{43} T_{65}}}{s^2 + \dfrac{s}{T_{66}} + \dfrac{1}{T_{43} T_{65}}} \tag{13.47}$$

The all-pass building block is designed exactly as the high-pass or bandpass circuit.

From the all-pass building block, we can derive a second-order notch filter by adding a conductor G_{6b} in parallel with the capacitor G_6; see Figure 13.19.

The circuit in Figure 13.19 has the general voltage transfer ratio

$$H(s) = \frac{V_o}{V_i} = \frac{s C_6 C_4 G_2 + s C_4 (G_2 G_{6b} - G_1 G_{6a}) + G_1 G_3 G_5}{s^2 C_4 C_6 G_2 + s C_4 G_2 G_6 + G_1 G_3 G_5} \tag{13.48}$$

with $G_6 = G_{6a} + G_{6b}$. By setting $G_2 G_{6b} = G_1 G_{6a}$ (normally $G_1 = G_2$ and $G_{6a} = G_{6b}$) we obtain a second-order notch filter with a voltage transfer ratio:

$$H(s) = \frac{s^2 + \dfrac{\alpha_{12}}{T_{43} T_{65}}}{s^2 + \dfrac{s}{T_{66}} + \dfrac{\alpha_{12}}{T_{43} T_{65}}} \tag{13.49}$$

where $\alpha_{12} = 1$ if $G_1 = G_2$.

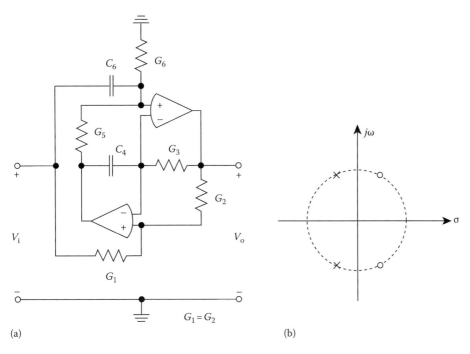

FIGURE 13.18 Second-order all-pass building block. (a) The circuitry. (b) The pole and zero locations.

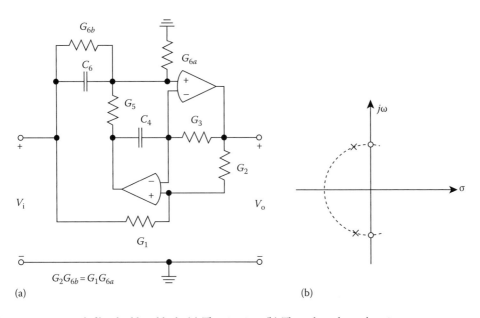

FIGURE 13.19 Notch filter building block. (a) The circuitry. (b) The pole and zero locations.

Finally we will derive two second-order elliptic or Cauer-type building blocks from the general dual-amplifier structure in Figure 13.14b. To get a Cauer low-pass building block we take $Y_{1a} = G_1$, $Y_{1b} = C_1$, $Y_2 = G_2$, $Y_3 = G_3$, $Y_4 = G_4$, $Y_5 = sC_5$, $Y_{6a} = 0$, and $Y_{6b} = G_6$; see Figure 13.20.

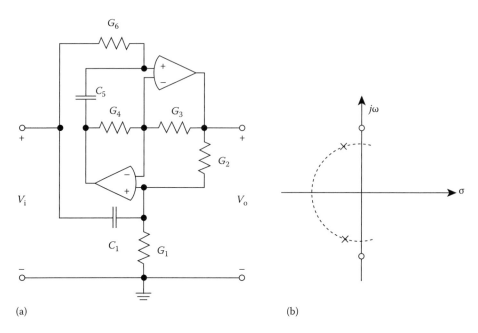

FIGURE 13.20 Second-order Cauer low-pass building block. (a) The circuitry. (b) The pole and zero locations.

This filter circuit has a voltage transfer ratio

$$H(s) = \frac{V_o}{V_i} = \frac{s^2 C_1 G_3 C_5 + G_6(G_2 G_4 + G_1 G_4)}{s^2 C_1 G_3 C_5 + s G_1 G_3 C_5 + G_2 G_4 G_6}$$

$$= \frac{s^2 + \dfrac{\alpha_{43}(1 + \alpha_{12})}{T_{12} T_{56}}}{s^2 + \dfrac{s}{T_{11}} + \dfrac{\alpha_{43}}{T_{12} T_{56}}}$$

$$= \frac{s^2 + \omega_z^2}{s^2 + s\omega_p/Q_p + \omega_p^2} \tag{13.50}$$

The magnitude ω_z of the zeros is always greater than the magnitude ω_p of the poles. Additionally, without any matching of elements, the real part of the zeros is always zero.

The design can proceed in the following steps. First we predefine $\alpha_{43} = 1$ and choose due to practical considerations the values of the resistors $R_3 = R_4$. We also select the two capacitors C_1 and C_5. Then, comparing the first coefficient in the denominator of $H(s)$ in Equation 13.50, namely, $T_{11} = Q_p/\omega_p$ with $T_{11} = C_1 R_1$, we determine the resistor

$$R_1 = \frac{Q_p}{\omega_p C_1} \tag{13.51}$$

Further, comparing the last coefficients in the numerator and denominator yields

$$\frac{\omega_z^2}{\omega_p^2} = 1 + \alpha_{12} \tag{13.52}$$

With $\alpha_{12} = R_2/R_1$ we can solve Equation 13.52 for

$$R_2 = R_1 \cdot \left(\frac{\omega_z^2}{\omega_p^2} - 1 \right) \tag{13.53}$$

Finally, from $\omega_p^2 = 1/(T_{12}T_{56})$ we can determine the resistor

$$R_6 = \frac{1}{\omega_p^2 C_1 C_5 R_2} \tag{13.54}$$

In order to get the three resistors R_1, R_2, and R_6 in the same order of magnitude it might be advisable to predefine different values for the capacitors C_1 and C_5.

A Cauer high-pass building block is depicted in Figure 13.21. Its voltage transfer ratio reads

$$H(s) = \frac{V_o}{V_i} = \frac{s^2 C_2 G_4 C_6 + G_1 (G_3 G_5 - G_6 G_4)}{s^2 C_2 G_4 C_6 + s G_6 G_4 C_2 + G_1 G_3 G_5}$$

$$= \frac{s^2 + \dfrac{\alpha_{34} - \alpha_{65}}{T_{21}T_{65}}}{s^2 + s\dfrac{1}{T_{11}} + \dfrac{\alpha_{43}}{T_{12}T_{56}}}$$

$$= \frac{s^2 + \omega_z^2}{s^2 + s\omega_p/Q_p + \omega_p^2} \tag{13.55}$$

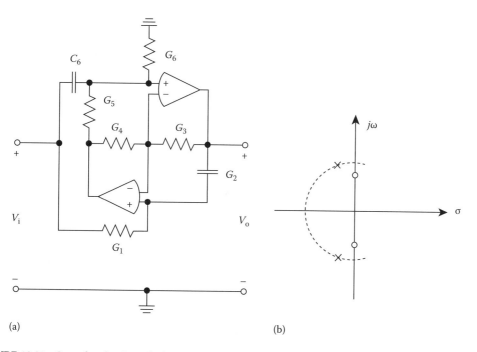

(a)　　　　　　　　　　　　　　(b)

FIGURE 13.21 Second-order Cauer high-pass building block. (a) The circuitry. (b) The pole and zero locations.

The design of this building block is similar to that of the low-pass circuit. First we choose the resistors $R_3 = R_4$ and the capacitors C_2 and C_6 and then, with predefined parameters Q_p, ω_p, and ω_z, we determine the remaining three resistors:

$$R_6 = \frac{Q_p}{\omega_p C_6} \tag{13.56}$$

$$R_5 = R_6 \cdot \left(1 - \frac{\omega_z^2}{\omega_p^2}\right) \tag{13.57}$$

$$R_1 = \frac{1}{\omega_p^2 C_6 C_2 R_5} \tag{13.58}$$

13.5 GIC-Derived Three-Amplifier Biquads

In order to get more flexibility for realizing arbitrary second-order transfer functions and to obtain less resistor spread for realizing a given pole-Q, we can extend the resonator in Figure 13.14a to a resonator with three nullors; see Figure 13.22.

There are two different biquads known from the literature that are based on the resonator in Figure 13.22. The first one is proposed by Mikhael and Bhattacharyya [6] and is shown in Figure 13.23.

This filter circuit requires a small resistor spread to realize high pole-Q. The zeros of the transfer function are formed with a resistive feed-forward network providing a flexible design with arbitrary numerator coefficients. The voltage transfer ratio V_3/V_i is

$$H_3(s) = \frac{V_3}{V_i} = \frac{N_3(s)}{D(s)} \tag{13.59}$$

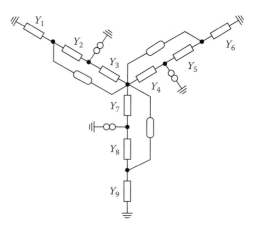

FIGURE 13.22   Resonator with three nullors.

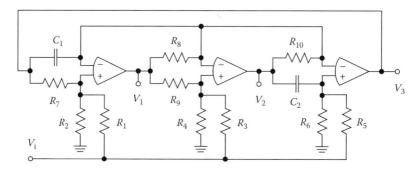

FIGURE 13.23   Mikhael–Bhattacharyya biquad.

with

$$N_3(s) = s^2 C_1 C_2 G_1 G_9 + s C_2 (G_7 G_8 G_3 + G_8 G_2 G_3 - G_8 G_1 G_4)$$
$$+ (G_9 G_7 G_{10} G_5 + G_9 G_{10} G_2 G_5 - G_9 G_{10} G_4 G_6) \qquad (13.60)$$

and

$$D(s) = s^2 C_1 C_2 (G_1 G_9 + G_2 G_9) + s C_2 (G_7 G_8 G_3 + G_7 G_8 G_4)$$
$$+ (G_9 G_7 G_{10} G_5 + G_9 G_{10} G_7 G_6) \qquad (13.61)$$

The two other output nodes lead to similar transfer expressions.

The second biquad, which is based on the resonator in Figure 13.22, was proposed by Padukone et al. [7] and is depicted in Figure 13.24. Assuming ideal op-amps and choosing V_3 as output voltage, we obtain the following transfer function:

$$H_3(s) = \frac{V_3}{V_i} = \frac{N_3(s)}{D(s)} \qquad (13.62)$$

with

$$N_3(s) = s^2 [C_2 C_3 (G_2 G_6 - G_1 G_3) + C_1 C_2 G_1 G_8]$$
$$+ s[C_1 G_2 G_5 G_9 - C_3 G_2 G_5 G_7 + C_2 G_1 G_4 G_8] + (G_2 G_4 G_5 G_9) \qquad (13.63)$$

and

$$D(s) = s^2 (C_1 + C_3) C_2 G_2 G_6 + s C_2 G_1 G_4 (G_3 + G_8)$$
$$+ G_2 G_4 G_5 (G_7 + G_9) \qquad (13.64)$$

It has been shown [7] that the pole sensitivities to all passive components are not greater than unity. The filter section has the main advantage of being particularly insensitive to gain-bandwidth variations even when the op-amps are mismatched.

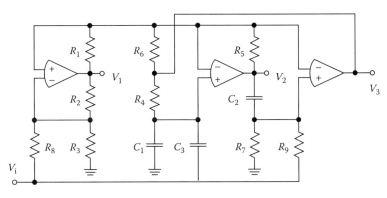

FIGURE 13.24 Padukone–Mulawka–Ghausi biquad.

13.6 State-Variable-Based Biquads

A frequently-used multiple-amplifier biquad is the circuit proposed by Kerwin et al. [8]. This filter circuit has extreme flexibility, good performance, and low sensitivities to the passive components. The filter is based on analog computer structures [9], which are derived from the state-variable representation of linear continuous systems. Therefore, these filters are also referred to as *state-variable filters*.

Figure 13.25a shows the basic analog computer structure consisting of one summing amplifier and two integrators. We assume both integrators to have the same transfer function $-1/(sT)$, where T is called the integrator time constant. Analyzing this structure yields

$$V_1 = -K_1 \cdot V_3 + K_2 \cdot V_i + K_3 \cdot V_2 \tag{13.65}$$

$$V_2 = -\frac{1}{sT} \cdot V_1 \tag{13.66}$$

$$V_3 = -\frac{1}{sT} \cdot V_2 \tag{13.67}$$

which results in

$$H_{HP}(s) = \frac{V_1}{V_i} = \frac{s^2 T^2 K_2}{s^2 T^2 + sTK_3 + K_1} \tag{13.68}$$

Using Equation 13.66, we can immediately derive the voltage transfer ratio V_2/V_i from Equation 13.68:

$$H_{BP}(s) = \frac{V_2}{V_i} = \frac{-sTK_2}{s^2 T^2 + sTK_3 + K_1} \tag{13.69}$$

Finally, with Equation 13.67 we obtain from Equation 13.69

$$H_{LP}(s) = \frac{V_3}{V_i} = \frac{K_2}{s^2 T^2 + sTK_3 + K_1} \tag{13.70}$$

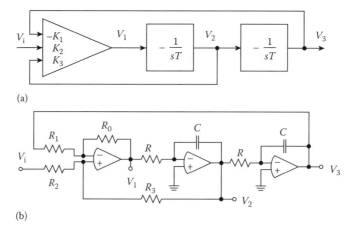

(a)

(b)

FIGURE 13.25   (a) Second-order analog computer structure and (b) state-variable filter section proposed by Kerwin, Huelsman, and Newcomb.

Thus, the structure in Figure 13.25a simultaneously realizes a high-pass filter, a bandpass filter, and a low-pass filter. The corresponding filter circuit proposed by Kerwin, Huelsman, and Newcomb is depicted in Figure 13.25b. The integrators consist of one op-amp, one resistor R, and one capacitor C. The time constant is given by $T = RC$. The three gain factors of the summing amplifier are determined by the four resistors $R_0 - R_3$:

$$K_1 = \frac{R_0}{R_1} \tag{13.71}$$

$$K_2 = \frac{R_3}{R_2 + R_3}\left(1 + \frac{R_0}{R_1}\right) \tag{13.72}$$

$$K_3 = \frac{R_2}{R_2 + R_3}\left(1 + \frac{R_0}{R_1}\right) \tag{13.73}$$

Obviously, the integrator time constant T plays the role of a reciprocal normalization frequency. Thus, if we refer the frequency variable s to $1/T$, we obtain from Equation 13.70 the normalized low-pass transfer function:

$$H_{LP}(s) = \frac{V_3}{V_i} = \frac{K_2}{s^2 + sK_3 + K_1} \tag{13.74}$$

Given the normalized pole frequency ω_p and the pole Q-factor Q_p, we can design the filter section by equating

$$K_1 = \omega_p^2 \tag{13.75}$$

and

$$K_3 = \frac{\omega_p}{Q_p} \tag{13.76}$$

Then K_2 is fixed by the dc gain of the transfer function:

$$K_2 = H_0 \tag{13.77}$$

The state-variable filter circuit can be extended to a general biquad by adding an output amplifier that sums the three voltages V_1, V_2, and V_3. Figure 13.26a shows a state-variable filter with an output amplifier summing the voltages V_1, V_2, and V_3 of the circuit in Figure 13.25b. This biquad has been proposed also by Kerwin et al. [8]. As alternative circuits, the amplifiers in Figure 13.26b and c can be used. Figure 13.26b shows an output amplifier that realizes the following sum:

$$V_o = \alpha_1 V_1 + \alpha_2 V_2 + \alpha_3 V_3 \tag{13.78}$$

with

$$\alpha_1 = -\frac{R_{10}}{R_{11}}, \quad \alpha_2 = \frac{R_{14}}{R_{12} + R_{14}}\left(1 + \frac{R_{10}}{R_{11}\|R_{13}}\right), \quad \alpha_3 = -\frac{R_{10}}{R_{13}} \tag{13.79}$$

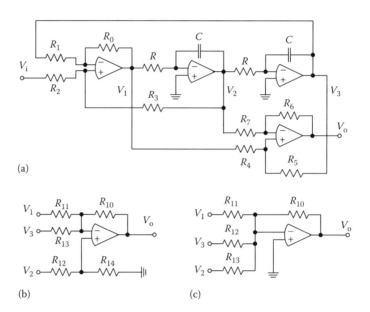

(a)

(b)   (c)

FIGURE 13.26   (a) State-variable filter with output amplifier, (b) output amplifier with inverting and noninverting inputs, and (c) output amplifier with three inverting inputs.

If we solve Equations 13.68 through 13.70 for the three output voltages and substitute them in Equation 13.79 with normalized variables s we obtain

$$H(s) = \frac{V_o}{V_i} = -K_2 \frac{|\alpha_1|s^2 + |\alpha_2|s + |\alpha_3|}{D(s)} \tag{13.80}$$

All numerator coefficients have the same sign. Therefore, the zeros of the transfer function are in the left-s-half plane. If we set $\alpha_2 = 0$, i.e., if we delete the voltage divider R_{12}, R_{14} and ground the noninverting input terminal of the op-amp, we obtain zeros on the $j\omega$ axis.

When designing the biquad, R_{14} and R_{10} may be used to scale the impedance level of the two resistive subnetworks. Then from the three numerator coefficients or from the overall gain constant of the transfer function, the zero frequency ω_z, and the zero Q-factor Q_z we can easily determine the remaining resistors R_{11}, R_{12}, and R_{13}.

The output amplifier in Figure 13.26c has three inverting inputs. Summing the voltages V_1, V_2, and V_3 leads to a numerator polynomial where the sign of the middle coefficient is different from the sign of the other two. Thus, the zeros are in the right-s-half plane. Again, we can delete the resistor R_{12} to realize zero on the $j\omega$ axis.

A second state-variable biquad circuit proposed by Tow and Thomas [10–12] yields similar performance to that of the Kerwin–Huelsman–Newcomb circuit. It uses a feedback loop with one damped integrator, one integrator, and one inverting amplifier; see Figure 13.27a. Figure 13.27b shows the Tow–Thomas circuit with three op-amps.

The damped integrator has a transfer function

$$\frac{V_1}{V_3} = \frac{-1}{sT + \alpha}\bigg|_{V_i = 0} \tag{13.81}$$

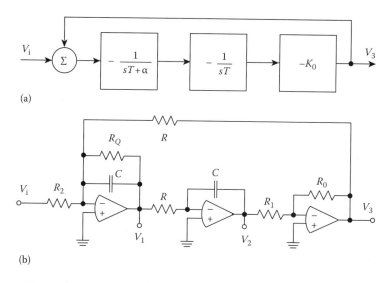

FIGURE 13.27 (a) Principle and (b) three op-amp realization of Tow–Thomas filter.

with $T = RC$ and $\alpha = R/R_Q$. The inverting amplifier has a gain

$$\frac{V_3}{V_2} = -\frac{R_0}{R_1} = -K_0 \tag{13.82}$$

An analysis of the circuit in Figure 13.27b with V_i being the input voltage and V_3 the output voltage yields a transfer function

$$H_{\mathrm{LP}}(s) = \frac{V_3}{V_i} = \frac{-K_0 \alpha_2}{s^2 T^2 + sTK_Q + K_0} \tag{13.83}$$

with

$$K_0 = \frac{R_0}{R_1}, \quad \alpha_2 = \frac{R}{R_2}, \quad K_Q = \frac{R}{R_Q} \tag{13.84}$$

For the design of the filter, the integrator time constant T serves as a reciprocal normalization frequency. Then, from the predefined normalized pole frequency we can determine the resistor ratio K_0, from the pole Q-factor the ratio K_Q, and from the dc gain of the filter section the ratio α_2. Choosing convenient values for C and R_0, we finally determine the resistors R, R_1, R_Q, and R_2 from the parameters T, K_0, K_Q, and α_2, respectively.

The filter circuit in Figure 13.26 requires an additional op-amp to realize a transfer function with a general second-degree numerator polynomial. An alternative method is to feed fractions of the input signal forward into the input of each op-amp. This is realized in the multiple-input Tow–Thomas biquad [10]; see Figure 13.28. The transfer function of this circuit can be calculated to be

$$H(s) = \frac{V_o}{V_i} = -\frac{s^2 T^2 \alpha_4 + sT(K_Q \alpha_4 - K_0 \alpha_3) + [K_0(\alpha_2 - K_Q \alpha_3)]}{s^2 T^2 + sTK_Q + K_0} \tag{13.85}$$

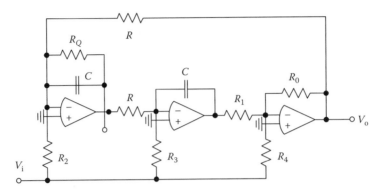

FIGURE 13.28   Generalized Tow–Thomas biquad.

with K_0, α_2, and K_Q as defined in Equation 13.84 and

$$\alpha_3 = \frac{R}{R_3}, \quad \alpha_4 = \frac{R_0}{R_4} \tag{13.86}$$

Thus, arbitrary numerator coefficients can be predescribed. In particular, if we choose $\alpha_3 = \alpha_4 = 0$ we obtain the low-pass filter circuit in Figure 13.27b and the transfer function in Equation 13.83.

When the state-variable filters described above are used to realize high-Q filter functions, the Q practically obtained is usually higher than that desired in the design. This effect is called Q enhancement and is caused by the phase lag introduced by the nonideal op-amps. One way to solve this problem is to use integrators with phase compensation.

Figure 13.29 shows a noninverting integrator with an additional op-amp for phase lag compensation. A detailed description of this circuit can be found in Ref. [13]. Putting this noninverting integrator together with an inverting integrator in a feedback loop results in a resonator with a Q-factor that is almost independent of the gain-bandwidth product of the op-amps. Thus, nearly no Q enhancement occurs.

Exactly this feedback loop is used in the Åkerberg–Mossberg biquad [14]; see Figure 13.30. In this circuit, a noninverting integrator with phase lag compensation together with an inverting damped integrator is connected as a feedback loop. More details about this filter section can be found in Refs. [4,13,14].

Finally, let us consider the general biquad proposed by Berka and Herpy [15]; see Figure 13.31. This biquad is also based on a state-variable representation and requires a second-order differentiator and a damped integrator. One of the main advantages of this circuit is extremely low sensitivities. A detailed description of the filter circuit and its design can be found in Ref. [4].

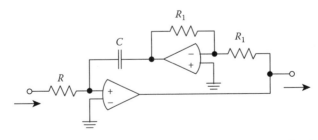

FIGURE 13.29   Noninverting integrator using an additional op-amp to compensate for phase lag.

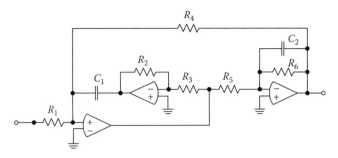

FIGURE 13.30 Åkerberg–Mossberg biquad.

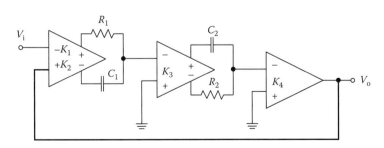

FIGURE 13.31 Berka–Herpy biquad.

13.7 All-Pass-Based Biquads

Finally, we will briefly consider two circuits that are based on first-order all-pass sections. Figure 13.32 shows the classical filter circuit introduced by Tarmy and Ghausi [16].

In this circuit, amplifiers with differential inputs and differential outputs are used. In the feedback loop, two all-pass sections with voltage transfer ratios

$$H_i(s) = \frac{sT_i - 1}{sT_i + 1} \tag{13.87}$$

$T_i = R_i C_i$, $i = 1, 2$, and one inverting amplifier are cascaded. The overall transfer function is

$$H(s) = \frac{V_o}{V_i} = \alpha_1 \frac{s^2 T_1 T_2 - s(T_1 + T_2) + 1}{s^2 T_1 T_2 + s\alpha_2(T_1 + T_2) + 1} \tag{13.88}$$

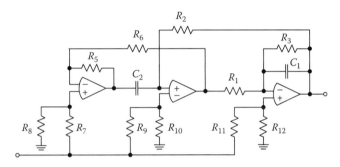

FIGURE 13.32 Tarmy–Ghausi circuit.

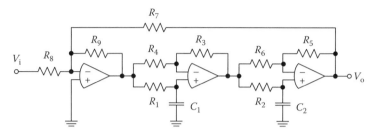

FIGURE 13.33 Tarmy–Ghausi circuit modified by Moschytz.

with

$$\alpha_1 = \frac{K_1 K_3 K_4}{1 + K_2 K_3 K_4}, \quad \alpha_2 = \frac{1 - K_2 K_3 K_4}{1 + K_2 K_3 K_4} \tag{13.89}$$

The main advantages of the Tarmy–Ghausi circuit are its low sensitivities to the gain-bandwidth products of the op-amps and thus its favorable performance at high pole frequencies and high pole Q-factors.

Most of the popular low-cost op-amps are not configured for differential output operations. This problem is bypassed by a circuit proposed by Moschytz [17]; see Figure 13.33.

The circuit in Figure 13.33 is also based on first-order all-pass sections. But the all-pass circuits are realized by means of only single-ended op-amps. Therefore, this circuit is more convenient for practical realizations.

13.8 Summary

In this chapter, we have shown that a lot of different multiple-amplifier biquads are known from the literature. Hence, a design engineer who aims to realize a high-performance active filter is in the favorable situation to find a rich variety of alternative circuits and design methods. On the other hand, this variety is also confusing and the design engineer may need assistance in deciding among all these circuits and methods. Unfortunately, the multiple-amplifier biquads mentioned in this chapter cannot be classified in a simple way. Therefore, only a rough guidance can be given. In a practical case, it is advisable to compare two or three solutions next to the predefined demands by a thorough analysis and then to find the final solution.

The biquads with decoupled time constants offer some of the benefits mentioned at the beginning of this section at low costs, namely, small passive element spread and low sensitivities. The dual-amplifier twin-T biquads and the GIC-derived dual-amplifier biquads offer a trade-off between costs and performance. The twin-T biquads are particularly suitable for the design of elliptic filters. The special merits of the GIC-derived biquads are the small number of passive components and the independence of the op-amp parameters. Thus, this biquad represents a robust filter solution for many different applications. Additionally, these filters can be easily designed.

The GIC-derived three-amplifier biquads and the state-variable-based filters offer additional flexibility with respect to an independent choice and an independent tuning of the filter parameters. Typically, these circuits are used in applications with switched parameters, e.g., filters with switched cutoff frequencies. This flexibility is paid for by a higher number of op-amps and passive components. It should be mentioned that the original state-variable circuits, i.e., two integrators in a loop, are rather sensitive to the phase lag introduced by the nonideal op-amps. This is always a problem when simultaneously high-frequency and high-Q performance shall be achieved. In such applications, biquads with compensated phase lag should be applied, e.g., the biquads proposed by Åkerberg and Mossberg, Berka and Herpy, or Tarmy and Ghausi.

References

1. R. P. Sallen and E. L. Key, A practical method of designing *RC* active filters, *IRE Trans. Circuits Theory*, CT-2, 75–85, 1955.
2. P. R. Geffe, A Q-invariant active resonator, *Proc. IEEE* (Lett.), 57, 1442, 1969.
3. M. A. Soderstrand and S. K. Mitra, Extremely low sensitivity active *RC* filter, *Proc. IEEE (Lett.)*, 57, 2175, 1969.
4. M. Herpy and J.-C. Berka, *Active RC Filter Design*, Amsterdam: Elsevier, 1986.
5. N. J. Fliege, A new class of second-order rc-active filters with two operational amplifiers, *Nachrichtentech. Z.*, 26, 279–282, 1973.
6. W. B. Mikhael and B. B. Bhattacharyya, A practical design for insensitive *RC*-active filters, *IEEE Trans. Circuits Syst.*, CAS-22, 407–415, 1975.
7. P. Padukone, J. Mulawka, and M. S. Ghausi, An active biquadratic section with reduced sensitivity to operational amplifier imperfections, *J. Franklin Inst.*, 30(1), 27–40, 1980.
8. W. J. Kerwin, L. P. Huelsman, and R. W. Newcomb, State-variable synthesis for insensitive integrated circuit transfer functions, *IEEE J. Solid-State Circuits*, SC-2, 87–92, 1967.
9. W. Schüßler, Schaltung und Messung von Übertragungsfunktionen an einem Analogrechner, *Archiv der Elektrischen Übertragung*, 13, 405–419, 1959.
10. P. E. Fleischer and J. Tow, Design formulas for biquad active filters using three operational amplifiers, *Proc. IEEE*, 61, 662–663, 1973.
11. L. C. Thomas, The biquad: Part I—Some practical design considerations, *IEEE Trans. Circuits Theory*, CT-18, 350–357; —, The biquad: Part II—A multipurpose active filtering system, *IEEE Trans. Circuits Theory*, CT-18, 358–361, 1971.
12. J. Tow, Design formulas for active *RC* filters using operational-amplifier biquad, *Electron. Lett.*, 5, 339–341, 1969.
13. L. P. Huelsman and P. E. Allen, *Introduction to the Theory and Design of Active Filters*, New York: McGraw-Hill, 1980.
14. D. Åkerberg and K. Mossberg, A versatile active *RC* building block with inherent compensation for the finite bandwidth of the amplifier, *IEEE Trans. Circuit Syst.*, CAS-21, 75–78, 1974.
15. J. C. Berka and M. Herpy, Novel active rc building block with optimal sensitivity, *Electron. Lett.*, 17, 887–888, 1981.
16. R. Tarmy and M. S. Ghausi, Very high-Q insensitive *RC* networks, *IEEE Trans. Circuits Theory*, CT-17, 358–366, 1970.
17. G. S. Moschytz, High-Q factor insensitive active *RC* networks, similar to the Tarmy–Ghausi circuit but using single-ended operational amplifiers, *Electron. Lett.*, 8, 458–459, 1972.

14

The Current Generalized Immittance Converter Biquads*

Wasfy B. Mikhael
University of Central Florida

14.1 Introduction

Current generalized immittance convertors (CGICs) have been used to realize high-performance active biquads with 2 OAs, 3 OAs, or *n* OAs per sections [1,5].

In this chapter two- and three-amplifier biquads are presented that are based on CGICs. Although several biquads have been reported, the ones presented here have proved to be clearly superior

---

* Wasfy B. Mikhael, "Chapter 9: Biquad II," in *RC Active Filter Design Handbook*, Stephenson, New York: Wiley, 1985.

to single-amplifier biquads. This is because the design carried out [1] was constrained by stringent performance criteria satisfying important properties and features such as stability, versatility, insensitivity to component tolerances and drift, low dependence on the op-amp (OA) frequency limitations, finite gain, tunability, small spread in component values, and minimum total capacitance.

In addition, the performance of the CGIC-based biquads presented here are comparable to multiple OA biquads. On the other hand, the 3-OA CGIC biquad is shown to yield additional performance improvements over the 2-OA CGIC biquad. Also, the CGIC biquads use the OAs in the differential mode.

In the following discussion the generalized structure of the 2-OA and 3-OA CGIC biquads are presented. Illustrative examples of the element identification to realize the most commonly used biquads are tabulated. Stability and sensitivity properties are discussed. A design procedure for each biquad is described that minimizes the active sensitivities while maintaining the filter's stability. Several second-order design examples are given. A sixth-order Chebyshev LPF and a sixth-order elliptic BPF are designed using the design values and tuning procedure suggested. The excellent performance of the resulting realizations is experimentally verified. A universal 2-OA GIC hybrid implementation using thick film is also described. 2-OA CGIC biquadratic active filter realizations for extended high-frequency applications employing the composite operational amplifiers technique [2,3] are given.

14.2 Biquadratic Structure Using the Antoniou CGIC (2-OA CGIC Biquad)

Consider the network of Figure 14.1, which is simply Antoniou's CGIC [4], with two new ports created across 3-G and 4-G. This is represented symbolically in Figure 14.2.

A new configuration is now obtained, as shown in Figure 14.3. A synthesis procedure is now described that uses this configuration. The transfer functions between the input and output terminals 2, 3, and 4, assuming ideal OAs are readily obtained as

$$\frac{V_3}{V_1} = T_1 = \{Y_5 + h(s)[Y_7(1 + Y_6/Y_2) - Y_5Y_8/Y_2]\}/D(s) \tag{14.1a}$$

$$\frac{V_4}{V_i} = T_2 = [Y_5(1 + Y_8/Y_4) - Y_6Y_7/Y_4 + h(s)Y_7]/D(s) \tag{14.1b}$$

$$\frac{V_2}{V_i} = T_3 = [Y_5 + h(s)Y_7]/D(s) \tag{14.1c}$$

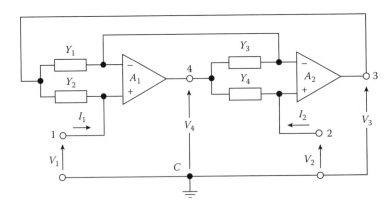

FIGURE 14.1 Antoniou's CGIC with additional ports 3G and 4G.

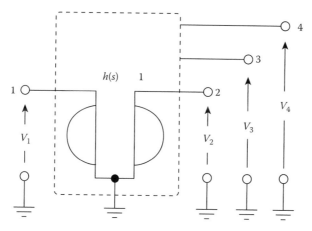

FIGURE 14.2 Symbolic representation of the CGIC in Figure 14.1.

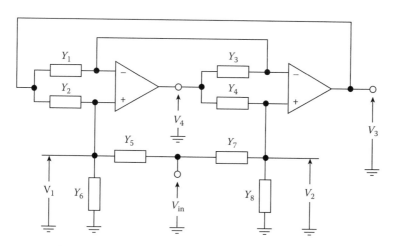

FIGURE 14.3 Basic configuration.

where

$$h(s) = Y_2 Y_3 / Y_1 Y_4$$
$$D(s) = (Y_5 + Y_6) + h(s)(Y_7 + Y_8) \tag{14.1d}$$

The conversion function $h(s)$ and the admittances $Y_5 - Y_8$ can be selected in many different ways, and it is clear that any stable second-order transfer function with any desired zero and pole locations can be realized.

Letting

$$Y_i = sC_i + G_i \tag{14.2}$$

where $i = 1$–4, we have from Equation 14.1

$$h(s) = \frac{(sC_2 + G_2)(sC_3 + G_3)}{(sC_1 + G_1)(sC_4 + G_4)} \tag{14.3}$$

Clearly, by omitting one or more conductances and/or capacitances, a number of specific conversion functions can be generated.

14.3 Realization of the Most Commonly Used Biquad Functions Using the 2-OA CGIC

The most frequently used second-order transfer functions have already been described in Chapter 11. Realizations of these functions are given in Table 14.1. The LP, HP, BP, and BS sections are produced by choosing $h(s)$ in a simple manner such as $K_1 s$, $K_2 s^2$, $K_3 s + K_4 s^2$ or their reciprocals. Circuits 3, 4, and 7 can be regarded as realizations of simple RLC networks [4,5]. Second-order AP sections can be obtained from circuits 11 and 12 of Table 14.1.

Figures 14.1 and 14.2 and Table 14.1 show that, with the exception of circuit 10, the response is obtained from the output of an OA. Owing to the low output resistance of the amplifier, any number of sections can be cascaded without using isolation amplifiers.

14.4 Stability during Activation and Sensitivity Using the 2-OA CGIC Biquad

14.4.1 Stability Properties

It has been shown [6] that some networks using CGICs can be conditionally stable, that is, a circuit can lock in an unstable mode during activation (just after switching on the power supply). For amplifiers with a finite open loop gain A, the circuit of Figure 14.3 gives

$$\frac{V_k}{V_i} = \frac{N_k(s)}{D(s)}$$

where $k = 2$, 3, 4 and

$$D(s) = M_1 Y_1 + M_2 Y_3$$
$$+ (1 + M_1)(1 + M_2)[Y_1/A_1 + Y_3/A_2 + Y_1/A_1 A_2 + Y_3/A_1 A_2] \tag{14.4}$$

$$M_1 = \frac{Y_5 + Y_6}{Y_2}$$

$$M_2 = \frac{Y_7 + Y_8}{Y_4}$$

It can be easily shown that for the circuit in Figure 14.3 low-frequency unstable modes cannot arise during activation. This is due to the absence of differences and changes of sign in the denominator coefficients. Thus any combination of OA gains that occur during transients and power supply switching does not result in saturation or low-frequency instability before reaching the steady state [7].

14.4.2 Sensitivity Analysis

The pole Q-factor Q_p, the undamped frequency of oscillation ω_p, the notch frequency ω_n, and the multiplier constants H_{BS}, H_{LP}, H_{HP}, and H_{BP}, as well as Q_z, ω_z, and H_{AP}, have been previously defined in

TABLE 14.1 Element Identification for Realizing the Most Commonly Used Transfer Functions

Circuit Number	$H(s)$	Y_1	Y_2	Y_3	Y_4	Y_5	Y_6	Y_7	Y_8	Transfer Function	Remarks
1	$\dfrac{sC_2(sC_3+G_3)}{G_1G_4}$	G_1	sC_2	sC_3+G_3	G_4	G_5	0	0	G_8	$T_2 = \dfrac{G_1G_5(G_4+G_8)}{G_1G_5G_4+sC_2G_5G_8+s^2C_2C_3C_8}$	LP
2	$\dfrac{G_2G_3}{sC_1(sC_4+G_4)}$	sC_1	G_2	G_3	$sC_4+G_4^a$	0	G_6	G_7	$sC_8^a+G_8^a$	$T_1 = \dfrac{G_2G_3G_7\left(1+\frac{G_8}{G_7}\right)}{G_2G_3(G_7+G_8)+s(C_2G_4G_6+C_8G_2G_3)+s^2C_1C_4C_6}$	LP
3	$\dfrac{sC_3G_2}{G_1G_4}$	G_1	G_2	sC_3	G_4	0	G_6	sC_7	$sC_8^a+G_8$	$T_1 = \dfrac{s^2C_3C_7(G_2+G_6)}{G_6G_1G_4+sC_3G_2G_8+s^2(C_7+C_8)G_2C_3}$	HP
4	$\dfrac{sC_3G_2}{G_1G_4}$	G_1	G_2	sC_3	G_4	0	G_6	$\dfrac{sC_7}{1+sC_7R_7}$	0	$T_1 = \dfrac{s^2C_3C_7(G_2+G_6)}{G_6G_1G_4+sC_7G_1G_4G_6R_7+s^2C_3C_7G_2}$	HP
5	$\dfrac{s^2C_2C_3}{G_1G_4}$	G_1	G_2	sC_3	G_4	0	G_6	0	G_8	$T_2 = \dfrac{sC_5\left(1+\frac{G_6}{G_5}\right)G_1G_4}{G_1G_4G_6+sC_5G_1G_4+s^2C_2C_3G_8}$	BP
6	$\dfrac{G_2G_3}{sC_1(sC_4+G_4)}$	sC_1	G_2	G_3	$sC_4+G_4^a$	0	G_6	sC_7	G_8	$T_1 = \dfrac{sC_7G_2G_3\left(1+\frac{G_6}{G_7}\right)}{G_2G_3G_8+s(C_7G_2G_3+C_1G_4G_8)+s^2C_1C_4C_6}$	BP
7	$\dfrac{sC_3G_2}{G_1G_4}$	G_1	G_2	sC_3	G_4	0	G_6	sC_7	$sC_8+G_8^a$	$T_1 = \dfrac{sC_3G_7(G_2+G_6)}{G_1G_4G_6+sC_3G_2(G_7+G_8)+s^2C_3C_8G_2}$	BP
8	$\dfrac{G_2G_3}{s^2C_1C_4}$	sC_1	G_2	G_3	sC_4	G_6	0	G_7	sC_8	$T_2 = \dfrac{s^2C_1G_3(C_4+C_8)+G_2G_3G_7}{s^2C_1C_5C_4+sC_8G_2G_3+G_2G_3G_7}$	N
9	$\dfrac{sG_2C_3}{G_1G_4}$	G_1	G_2	sC_3	G_4	G_5	0	sC_7	G_8	$T_2 = \dfrac{s^2C_3C_7G_2+G_1G_5(G_4+G_8)}{s^2C_3C_7G_2+sC_3G_2G_8+G_1G_4G_5}$	N
10	$\dfrac{sG_2C_3}{G_1G_4}$	G_1	G_2	sC_3	G_4	G_5	G_6^a	sC_7	sC_8+G_8	$T_1 = \dfrac{\left(s^2+\frac{G_1G_4G_5}{C_3C_7G_2}\right)\left(\frac{C_8}{C_7+C_8}\right)}{s^2+\frac{G_8}{C_7+C_8}+\frac{G_1G_4(G_5+G_6)}{C_3C_7G_2(C_7+C_8)}}$	N
11	$\dfrac{G_2G_3}{s^2C_1C_4}$	sC_1	G_2	G_3	sC_4	G_5	G_6^a	G_7	sC_8	$T_1 = \dfrac{s^2C_1G_4G_5-sG_5C_8G_3+G_3G_7(G_2+G_6)}{s^2C_1C_4(G_5+G_6)+sC_8G_2G_3+G_2G_3G_7}$	Nonminimum phase For all pass: $G_6=0$, $G_5=G_2$
12[b]	$\dfrac{sG_2G_3}{G_1G_4}$	G_1	G_2	sC_3	G_4	G_5	G_6	sC_7	G_8	$T_1 = \dfrac{s^2C_3C_7(G_2+G_6)-sC_3G_5G_8+G_1G_4G_5}{s^2C_3C_7G_2+sC_3G_2G_8+(G_5+G_6)G_1G_4}$	Nonminimum phase For all pass: $G_6=0$, $G_2=G_5$

[a] These elements can be set equal to zero.

[b] A special case of Circuit 12—namely, the all-pass case—has also been independently proposed by J. T. Lim, Bell Northern Research, Canada, as communicated privately by him to Dr. A. Antoniou.

Chapter 7. The sensitivity of a quantity x with respect to variations in an element e is denoted by S_e^x. For ideal amplifiers, the use of Table 14.1 leads to

$$0 \leq S_e^x \leq 1 \tag{14.5}$$

where

  x represents any one of these H, ω, or Q qualities
  e represents any capacitance or conductance

14.5 Design and Tuning Procedure of the 2-OA CGIC Biquad

A design procedure is now described. Table 14.1 shows that there are several degrees of freedom in the choice of element values. These may be used to minimize $S_A^{Q_p}$ or the spread of element values. By using minimum sensitivity (to the OA parameters) constraints in circuits 1, 3, 7, 10, and 12, possible sets of element values for LP, HP, BP, BS, and AP sections have been obtained, as shown in Table 14.2. In addition, this choice of elements guarantees stable operation with real OA's. It is seen that the notch frequency ω_n, the undamped frequency of oscillation ω_p, and Q-factor Q_p can be easily adjusted by sequentially trimming three distinct resistors. A tuning sequence is also given in Table 14.2.

It is to be noted from Table 14.2 for the design of the BS sections, that

$$\omega_n^2 = \frac{\omega_p G_s}{C_s} \tag{14.6}$$

where

$$G_5 + G_6 = G_1, \quad C_7 + C_8 + C, \quad \text{and} \quad \omega_p = \frac{G}{C}$$

TABLE 14.2 Design Values and Tuning Procedure

Circuit Number (from Table 14.1)	Design Values	Transfer Function Realized	Tuning Sequence		
			ω_o	ω_p	Q_p
1	$G_1 = G_4 = G_5 = G_8 = G$, $G_3 = G/Q_p$, $C_2 = C_3 = C$, where $C = \frac{G}{\omega_p}$	$T_2 = \frac{2\omega_p^2}{D(s)}$	—	G_8	G_3
3	$G_1 = G_2 = G_4 = G_6 = G$, $G_8 = G/Q_p$, $C_8 = 0$, $C_3 = C_7 = C$, where $C = \frac{G}{\omega_p}$	$T_1 = \frac{2s^2}{D(s)}$	—	G_4	G_8
7	$G_1 = G_2 = G_4 = G_6 = G$, $G_7 = G/Q_p$, $G_8 = 0$, $C_3 = C_8 = C$, where $C = \frac{G}{\omega_p}$	$T_1 = \frac{\left(\frac{2\omega_p}{Q_p}\right)s}{D(s)}$	—	G_2	G_7
10	$G_1 = G_2 = G_4 = G_5 + G_6 = G$, $G_8 = G/Q_p$, $C_3 = C_7 + C_8 = C$, where $\omega_p = \frac{G}{C}$ and $\omega_n^2 = \omega_p \frac{G_5}{C_7}$	$T_3 = \frac{C_7}{C}\frac{(s^2+\omega_n^2)}{D(s)}$	G_2	G_6	G_8
12	$G_1 = G_2 = G_4 = G_5 = G$, $G_6 = 0$, $C_3 = C_7 = C$, $C_8 = G/Q_p$, where $\omega_p = \frac{G}{C}$	$T_1 = \frac{D(-s)}{D(s)}$	G_4		G_8

Note: $D(s) = s^2 + (\omega_p/Q_p)s + \omega_p^2$.

Hence

$$\frac{w_n^2}{w_p^2} = \frac{G_5}{G} \cdot \frac{C}{C_7} \tag{14.7}$$

where

$$\frac{G_5}{G} \text{ is always} \leq 1 \quad \text{and} \quad \frac{C}{C_7} \text{ is always} \geq 1$$

It is seen from Equation 14.7 that for $\omega_n \leq \omega_p$, C_8 can be set to zero ($C_7 = C$), and only two capacitors per section are used. For $\omega_n > \omega_p$, three capacitors are required. In all cases, regardless of the choice of C_7 and C_8, the total capacitance per section is equal to $2C$.

In most applications where notch sections are used, ω_n/ω_p is close to unity and care should be taken when choosing G_5 and C_7 in Equation 14.7. A suitable choice of G_5, G_6, G_7, and C_8 values may be obtained by letting

$$\frac{G_5}{G} = \frac{1}{K} \quad \text{and} \quad \frac{C}{C_7} = 2$$

Thus K is approximately equal to 2 ($K < 2$ for $\omega_n > \omega_p$ and $K > 2$ for $\omega_n \leq \omega_p$). The suggested choice yields a capacitor spread of 2 and both R_5 and R_6 are approximately equal to $2R$.

14.6 Design Examples Using the 2-OA CGIC Biquad

Example 14.1

Design a second-order Butterworth ($Q_p = 0.707$) LP filter having a cutoff frequency $f_p = 20{,}000/2\pi$ Hz.

Procedure

1. Circuit 1 in Table 14.2 realizes an LPF. The design equations are also given in Table 14.2.
2. First we choose an appropriate value for C, say 10 nF. Thus $C = C_2 = C_3 = 10$ nF.
3. Now, $R = \frac{1}{C\omega_p} = \frac{1}{20{,}000 \times 10^{-8}}$
   Therefore, $R = 5$ kΩ
4. Consequently, $R = R_1 = R_4 = R_5 = R_8 = 5$ kΩ and $R_3 = RQ_p = 3.535$ kΩ.
5. The circuit is shown in Figure 14.4a. It is noted that the low-frequency gain of the LP filter H_{LP} is 2.

A simple procedure can be followed to scale H_{LP} by a factor x less than unity, that is, effectively multiplying the transfer function realized by x. This is done by replacing the resistance R_5 by two resistors R_A, and R_B (in series with the input V_{in}) in the manner shown in Figure 14.4b, where

$$R_5 = R = 5 \text{ k}\Omega$$
$$= R_A \parallel R_B$$

The desired gain and scale factor $x = R_B/(R_A + R_B)$. Thus for $x = \frac{1}{2}$, resulting in a dc gain of the LP filter of unity, the choice of resistors R_A and R_B is $R_A = R_B = 10$ kΩ.

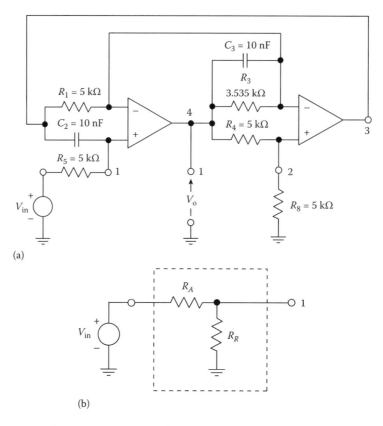

(a)

(b)

FIGURE 14.4 (a) Second-order Butterworth LPF design example and (b) controlling the gain factor of the LPF.

If functional tuning of the filter is desired, the tuning sequence of circuit 1 in Table 14.2 can be followed. First, ω_p is adjusted by applying a sinusoidal input at the desired ω_p frequency. Then R_B is tuned until ω_p realized equals the desired value. This can be detected by monitoring the phase angle of the output relative to the input. When the proper ω_p is reached, the output lags the input by 90°. Next, to adjust the Q_p, the filter gain H_{dc} of the LPF at a frequency much lower than ω_p is determined. Then an input at ω_p is applied. R_3 is adjusted until the gain of the LPF at ω_P is Q_P desired H_{dc}.

Example 14.2

Design a second-order BP filter with $Q_p = 10$ and $f_p = 10,000/2\pi$ Hz.

Procedure

1. Circuit 7 in Table 14.2 realizes a BP filter. The design equations are also given in Table 14.2.
2. First we choose a suitable value for C, say 10 nF.
3. Thus $C_3 = C_8 = C = 10$ nF.
4. Hence $R = (1/C\omega_p) = 10$ kΩ. Consequently, $R = R_1 = R_2 = R_4$, $R_6 = 10$ kΩ, and $R_7 = RQ_p = 100$ kΩ.
5. The circuit is shown in Figure 14.5. The gain at resonance, that is, at $\omega = \omega_p$, is equal to 2. To scale the gain by a factor x less than 2, the resistor R_7 is split into two resistors in a manner similar to that in Figure 14.4b and explained in Example 14.1.

Again, if functional tuning is desired, the sequence in Table 14.2 can be followed.

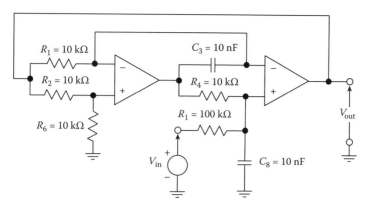

FIGURE 14.5 Design of a second-order BPF.

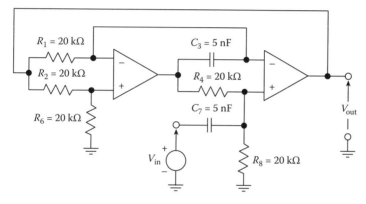

FIGURE 14.6 Design of a second-order HPF.

Example 14.3

Design a second-order HP filter with $Q_p = 1$ and $f_p = 10,000/2\pi$ Hz.

Procedure

1. Circuit 3 in Table 14.2 realizes an HP filter.
2. Let us choose $C_3 = C_7 = C = 5$ nF. Hence R can be computed as $R = \dfrac{1}{\omega_p C} = \dfrac{1}{10,000 \times 50 \times 10^{-9}}$
   Therefore, $R = 20$ kΩ.
3. Consequently, $R_1 = R_2 = R_4 = R_6 = R = 20$ kΩ and $R_8 = RQ = 20$ kΩ.
4. The realization is shown in Figure 14.6. The gain at high-frequency H_{HP} is equal to 2.

14.7 Practical High-Order Design Examples Using the 2-OA CGIC Biquad

Using Table 14.2, 1% metal-film resistors, 2% polystyrene capacitors and μ A741 OA's, a sixth-order Chebyshev LP filter, and a sixth-order elliptic BP filter were designed and constructed. The LP filter has a maximum passband attenuation of 1.0 dB; bandwidth $= 3979$ Hz. The BP filter has the following specifications:

Center frequency $= 1500$ Hz
Passband $= 60$ Hz
Maximum passband attenuation $= 0.3$ dB
Minimum stopband attenuation outside the frequency range $1408 \rightarrow 1595$ Hz $= 38$ dB

14.7.1 Low-Pass Filter

The realization uses cascaded section of type 1, in Table 14.2, as shown in Figure 14.7a. The measured frequency response (input level $= 50$ mV), shown in Figure 14.7b and c, agrees with the theoretical response. The effect of dc-supply variations is illustrated in Figure 14.7d. The deviation in the passband ripple is about 0.1 dB for supply voltage in the range 5–15 V. The effect of temperature variations is illustrated in Figure 14.7e, which shows the frequency response at $-10°$C (right-hand curve), 20°C, and 70°C (left-hand curve). The last peak has been displaced horizontally by 42 Hz, which corresponds to a

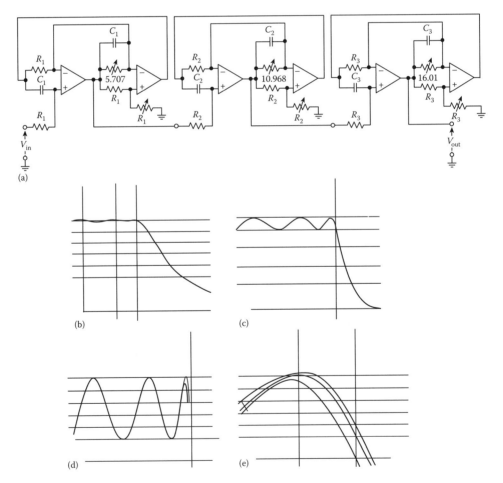

FIGURE 14.7 (a) Realization of the sixth-order Chebyshev low-pass filter. Frequency responses: (b) logarithmic gain scale and linear frequency scale; (c) linear gain and frequency scales; (d) for supply voltages ± 5 V (lower curve) and ± 15 V, input level $= 0.05$ V; (e) at temperatures $-10°$C (right-hand curve), 20°C, and 70°C (left-hand curve).

change of 133 ppm/°C. The frequency displacement is due to passive element variations and is within the predicted value.

14.7.2 Bandpass Filter

The realization uses cascaded sections of the types 7 and 10, in Table 14.2, and is shown in Figure 14.8a. The measured frequency response is shown in Figure 14.8b and c, and it is in agreement with the theoretical response. Figure 14.8d shows the frequency response for supply voltages of 7.5 V (lower curve) and 15 V; the input is 0.3 V. The passband ripple remains less than 0.39 dB and the deviation in the stopband is negligible. Figure 14.8e and f illustrate the effect of temperature variations. The passband ripple remains less than 0.35 dB in the temperature range −10°C to 70°C. A center frequency displacement of 15 Hz has been measured that corresponds to a change of 125 ppm/°C.

14.8 Universal 2-OA CGIC Biquad

Study of Table 14.2 suggests that several circuits may be combined to form a universal biquad. This can be achieved on a single substrate using thick-film technology.

Upon examining the element identification and design values in Table 14.2 it is easy to see that one common thick-film substrate can be made to realize circuits 1, 3, 7, and 10 in Table 14.2 (other circuits from Table 14.1 can be included if desired) with no duplication in OAs and chip capacitors and minimal duplication in resistors. The superposition of circuits 1, 3, 7, and 10 is shown in Figure 14.9. The following items should be noted.

1. Each resistor having the same subscript represents one resistor only and needs to appear once in a given biquad realization and thus once on the substrate. As an example, for $R_J = RQ_p$, only one R_J is needed with connection to several nodes. The unwanted connections may be opened during the trimming process according to the type of circuit required.
2. Three capacitor pads are needed; they are marked 1, 2, and 3 in Figure 14.9. To obtain capacitor 4, either capacitor 2 or 3 connections are made common with capacitor 4 terminals. The capacitor pad terminals are available on the external terminals of the substrate.* The chip capacitors are reflow-soldered in the appropriate locations based on the circuit realized.

A dual CGIC universal biquad implemented using thick-film resistors, chip NPO capacitors, and one quad OA is shown in Figure 14.10. Note that this hybrid array is capable of realizing gyrators and simulating inductors and super capacitors. Sample results are given in Figure 14.11 for realizing different biquadratic functions using this implementation.

14.9 3-OA CGIC Biquadratic Structure

Consider the circuit shown in Figure 14.12 where the output can be taken to be V_1, V_2, or V_3. Assuming ideal OAs, the transfer functions between the input and output terminals 1, 2, and 3 can be readily obtained as

$$V_1/V_i = T_1$$
$$= \{Y_{11}Y_{12}[Y_1(1 + Y_4/Y_9) - Y_2Y_3/Y_9]$$
$$+ (Y_{12}Y_3Y_7Y_8)/Y_9 + [(1 + Y_4/Y_9)Y_5 - Y_3Y_6/Y_9]Y_7Y_{10}\}/D(s) \tag{14.8a}$$

---

* This can readily be understood by examining the different realizations to be obtained from this layout.

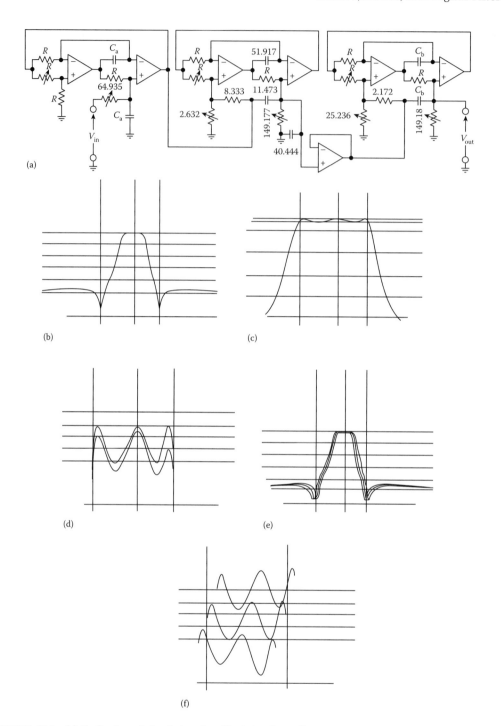

FIGURE 14.8 (a) Realization of the sixth-order elliptic bandpass filter. Frequency responses of bandpass filter: (b) logarithmic gain and linear frequency scales; (c) linear gain and frequency scales; (d) for supply voltages of ±7.5 V (lower curve) and ±15 V, input level of 0.3 V; (e) at temperatures of 10°C (right-hand curve), 20°C, and 70°C (left-handed curve); and (f) expanded passband of Figure 14.8e.

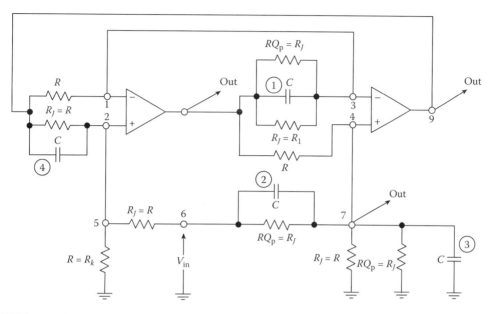

FIGURE 14.9 Superposition of circuits 1, 3, 7, and 10 from Table 14.2. Note: Out = output terminals.

FIGURE 14.10 Dual CGIC universal biquad implemented using thick-film resistors; chip NPO capacitors and one quad OA.

$$V_2/V_i = T_2$$
$$= \{Y_{11}Y_{12}Y_1 + Y_{12}[Y_1Y_6 + (Y_3Y_7Y_8)/Y_9 - Y_5Y_6]$$
$$+ [Y_3Y_6 - Y_4Y_5](Y_7Y_8Y_{12})/Y_9Y_{11} + Y_5Y_7Y_{10}\}/D(s) \qquad (14.8b)$$

$$V_3/V_i = T_3$$
$$= \{Y_1Y_{11}Y_{12} + Y_{12}[(1 + Y_2/Y_7)Y_3 - Y_1Y_4/Y_7]Y_8/Y_9$$
$$+ [(1 + Y_2/Y_7)Y_5 - Y_1Y_6/Y_7]Y_7Y_{10}\}D(s) \qquad (14.8c)$$

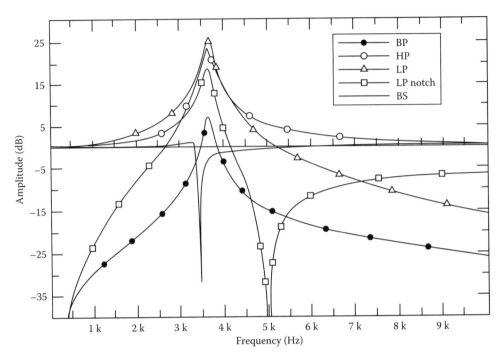

FIGURE 14.11   Different second-order realizations obtained from the universal CGIC biquad.

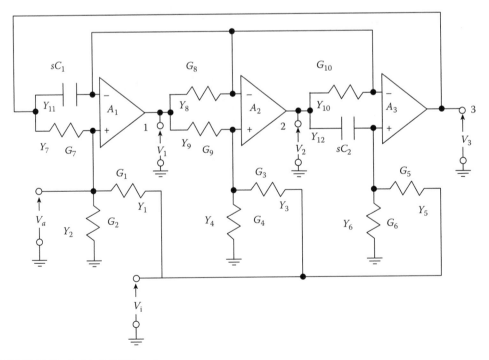

FIGURE 14.12   3-OA CGIC biquad.

where

$$D(s) = Y_{11}Y_{12}(Y_1 + Y_2) + Y_{12}(Y_3 + Y_4)Y_7Y_8/Y_9 + (Y_5 + Y_6)Y_7Y_{10} \qquad (14.9)$$

The admittances Y_1–Y_{12} can be selected in many ways and any stable second-order transfer function can be realized. For the purposes of this chapter, Y_1–Y_{10} are taken to be purely conductive, while Y_{11} and Y_{12} are purely capacitive, that is,

$$\begin{aligned}
Y_1 - Y_{10} &= G_1 - G_{10} \\
Y_{11} &= sC_1, \quad Y_{12} = sC_2
\end{aligned} \qquad (14.10)$$

Any rational and stable transfer function can be expressed as a product of second-order transfer functions of the form

$$T(s) = \frac{a_2 s^2 + a_1 s + a_0}{b_2 s^2 + b_1 s + b_0} \qquad (14.11)$$

where $a_1 = a_2 = 0$, $a_0 = a_1 = 0$, $a_0 = a_2 = 0$, or $a_1 = 0$, for an LP, HP, BP, or N section, respectively. These section can be realized by choosing the G_i's ($i = 1$–10) properly in Equation 14.8. By comparing Equations 14.8 through 14.11, circuits 1–4 in Table 14.3 can be obtained.

All-pass transfer functions can be realized by setting $a_2 = b_2$, $a_1 = -b_1$, $a_0 = b_0$; these can be obtained from circuit 5 of Table 14.3.

It can be easily shown that this biquad possesses similar excellent low sensitivity properties and stability during activation as those of the 2-OA CGIC biquad, given in Section 14.4.

14.10 Design and Tuning Procedure of the 3-OA CGIC Biquad

Several degrees of freedom exist in choosing element values, as shown in Table 14.3. These are used to satisfy the constraints of the given design, namely, those of low sensitivity, reduced dependence on the

TABLE 14.3 Element Identification for Realizing the Most Commonly Used Transfer Functions

Circuits Number	G_1	G_2	G_3	G_4	G_5	G_6	Transfer Function	Remarks
1	0		0			a	$T_3 = \dfrac{\left(1 + \frac{G_2}{G_7}\right)\frac{G_5 G_7 G_{10}}{C_1 C_2}}{s^2 G_2 + s\frac{G_4 G_7 G_8}{C_1 G_9} + \frac{(G_5 + G_6)G_7 G_{10}}{C_1 C_2}}$	LP
2		a	0		0		$T_1 = \dfrac{s^2 G_1\left(1 + \frac{G_4}{G_9}\right)}{s^2(G_1 + G_2) + s\frac{G_4 G_7 G_8}{C_1 G_9} + \frac{G_6 G_7 G_{10}}{C_1 C_2}}$	HP
3	0			a		0	$T_3 = \dfrac{s\left(1 + \frac{G_2}{G_7}\right)\frac{G_3 G_7 G_8}{C_1 G_9}}{s^2 G_2 + s\frac{(G_3 + G_4)G_7 G_8}{C_1 G_9} + \frac{G_6 G_7 G_{10}}{C_1 C_2}}$	BP
4		a	0		a		$T_1 = G_1\left(1 + \frac{G_4}{G_9}\right)\dfrac{s^2 + \frac{G_5 G_7 G_{10}}{C_1 C_2 G_1}}{s^2(G_1 + G_2) + s\frac{G_4 G_7 G_8}{C_1 G_9} + \frac{(G_5 + G_6)G_7 G_{10}}{C_1 C_2}}$	N
5		a	0		0		$T_3 = \dfrac{s^2 G_1 - s\frac{G_1 G_4 G_8}{C_1 G_9} + \left(1 + \frac{G_2}{G_7}\right)\frac{G_5 G_7 G_{10}}{C_1 C_2}}{s^2(G_1 + G_2) + s\frac{G_4 G_7 G_8}{C_1 G_9} + \frac{G_5 G_7 G_{10}}{C_1 C_2}}$	Nonminimum phase[b]

Notes: $Y_7 - Y_{10} = G_7 - G_{10}$ always.

$Y_{11} - sC_1$, $Y_{12} = sC_2$.

[a] These elements can be set to zero.

[b] For all-pass $G_2 = 0$, $G_7 = G_1$.

TABLE 14.4 Design Values and Tuning Procedures

Circuit Number	Design Values	Transfer Function Realized	Tuning Sequence			
			ω_n	ω_p	Q_p	Q_z
1	$R_2 = R$, $R_4 = RQ_p^{1/2}$, $R_5 = 2R/(\alpha H_{LP})$, $R_6 = R/[\alpha(1 - H_{LP}/2)]$, $H_{LP} < 2$	$T_3 = H_{LP}\dfrac{\omega_p^2}{D(s)}$	—	R_5	R_4	—
2	$R_1 = R(1 + \bar{Q}_p^{1/2})/H_{HP}$, $R_2 = R(1 + \bar{Q}_p^{1/2})\left[1 + \bar{Q}_p^{1/2} - H_{HP}\right]$, $R_4 = RQ_p^{1/2}$, $R_6 = R/\alpha$, $H_{HP} < 1 + \bar{Q}_p^{1/2}$	$T_1 = H_{HP}\dfrac{s^2}{D(s)}$	—	R_6	R_4	—
3	$R_2 = R$, $R_3 = 2RQ_p^{1/2}H_{BP}$, $R_4 = RQ_p^{1/2}/1 - H_{BP}/2$, $R_6 = R/\alpha$, $-H_{BP} < 2$	$T_3 = H_{BP}\dfrac{\frac{s\omega_p}{Q_p}}{D(s)}$	—	R_6	R_3	—
4	$R_1 = R(1 + \bar{Q}_p^{1/2})/H_N$, $R_2 = 1/(G - G_1)$, $R_4 = R\bar{Q}_p^{1/2}$, $R_6 = 1/(\alpha G - G_5)$, $R_5 = R\omega_p^2(1 + \bar{Q}_p^{1/2})/(\alpha H_N\omega_n^2)$, $H_N < (1 + \bar{Q}_p^{1/2})$, $H_N =< \omega_p^2(1 + \bar{Q}_p^{1/2})/\omega_n^2$ for $\omega_n > \omega_p$	$T_1 = H_N\dfrac{(s^2 + \omega_n^2)}{D(s)}$	R_5	R_6	R_4	—
5	For all-pass: $R_1 = R$, $R_4 = RQ_p^{1/2}$ $R_5 = R/\alpha$, $R_1 = R_7$, $R_2 = \infty$, $H_{AP} = 1$	$T_1 = H_{AP}\dfrac{s^2 - \frac{\omega_x}{Q_z}s + \omega_z^2}{D(s)}$	R_5	R_4		

Notes: $D(s) = s^2 + (\omega_p/Q_p)s + \omega_p^2$, $\alpha = 2Q_p^{1/2}/(1 + Q_p^{1/2})$.
$C_1 = C_2 = C = 1/(\omega_p R)$, $R_{10} = \alpha R$, $R_8 = RQ_p^{1/2}$, $R_7 = R_8 = R$.

OA finite gain and bandwidth, and low element-spread design, in circuits 1–5 in Table 14.3. Using these constraints, a possible design for LP, HP, BP, AP, and N sections is obtained, as indicated in Table 14.4. It is seen that Q_p, ω_p, ω_n, Q_z, and ω_z, can be independently adjusted by trimming at most three resistors. A trimming sequence is also given in Table 14.4.

14.11 Practical Sixth-Order Elliptic BP Filter Design Using the 3-OA CGIC Biquad

The sixth-order elliptic bandpass filter specified in Section 14.7 was designed using the 3-OA CGIC biquad and similar components to those in Section 14.7.

The realization shown in Figure 14.13a uses cascaded sections of the types 3 and 4 in Table 14.4. The element design values are also given in Figure 14.13a. The measured frequency response is shown in Figure 14.13b and c; it is in agreement with the theoretical response. Figure 14.13d shows the frequency response for supply voltages of ±7.5 V (lower curve), and ±15 V (upper curve); the input voltage is 0.3 V. The passband ripple remains less than 0.34 dB and the deviation in the stopband is negligible. Figure 14.13e and f illustrate the effect of temperature variations. The passband ripple remains less than 0.5 dB in the temperature range from $-10°C$ (right-hand curve) to $70°C$ (left-hand curve). A center frequency displacement of 9 Hz has been measured, which corresponds to a change of 75 ppm/°C.

These results illustrate the additional performance improvements compared with the results in Section 14.7 using the 2-OA CGIC biquad.

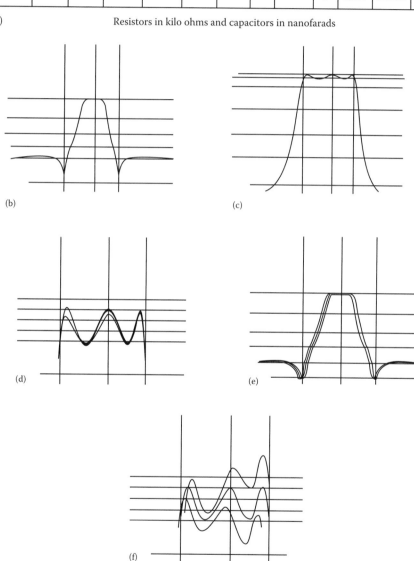

Section No	R_1	R_2	R_3	R_4	R_5	R_6	R_7	R_8	R_9	R_{10}	C_1	C_2
1	∞	2	11.396	∞	∞	1.175	2	11.396	2	3.403	53.052	53.052
2	8.939	2.576	∞	17.273	4.592	1.426	2	17.273	2	3.585	51.917	51.917
3	4.309	3.732	∞	17.273	2.611	1.948	2	17.273	2	3.585	54.211	54.211

(a) Resistors in kilo ohms and capacitors in nanofarads

(b)

(c)

(d)

(e)

(f)

FIGURE 14.13 (a) Sixth-order elliptic bandpass filter. Resistors in kiloohms and capacitors in nanofarads. (b)–(f) Frequency response using 3-OA CGIC. (b) Logarithmic gain and linear frequency scales; (c) linear gain and frequency scales; (d) frequency response for supply voltages of ±7.5 V (lower curve) and ±15 V, input level of 0.3 V; (e) frequency response; and (f) at temperatures of −10°C (right-hand curve), and 20°C and 70°C (left-hand curve).

14.12 Composite GIC Biquad: 2 C 2-OA CGIC

To obtain the composite CGIC biquad [3], each single OA in the original 2-OA CGIC biquad is replaced by a composite amplifier each constructed using two regular OAs [2] and denoted C2OA. All possible combinations of the four C2OA structures in Ref. [3] were used to replace the two OAs in the CIC network. Although several useful combinations were obtained, it was found that the best combination is shown in Figure 14.14, where A_1 is replaced by C2OA-4 and A_2 is replaced by C2OA-3 in the CGIC of Figure 14.1.

Computer simulation plots and experimental results of Figure 14.15 show clearly the considerable improvements of the new CGIC filter responses over those of a 2-OA CGIC implemented using regular OAs.

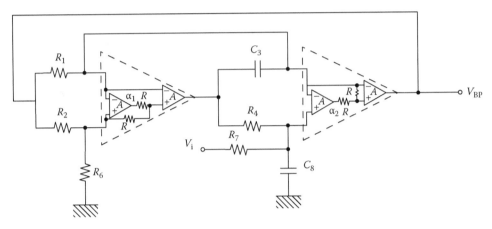

FIGURE 14.14   Practical BP filter realization of the composite GIC using C2OA-4 and C2OA-3.

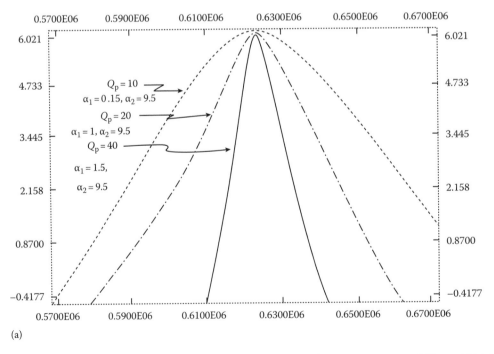

(a)

FIGURE 14.15   (a) Computer plots of the composite GIC BP filter frequency responses of Figure 14.14 for $f_0 = 100$ kHz and $Q_p = 10, 20, 40$.

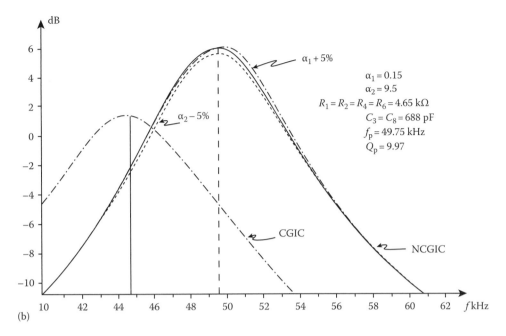

FIGURE 14.15 (continued)   (b) Experimental results of single and composite GIC BP filters of Figure 6.3, and the response sensitivity for a_1 and a_2 variations ($f_0 = 50$ kHz, $Q_p = 10$).

Similar improvements in inductance simulation applications, employing the 2C2-OA CGIC, have been reported.

14.13 Summary

Two configurations have been presented for the synthesis of *RC*-active networks. These have been used to design a number of universal second-order sections such as low-pass, high-pass, bandpass, and bandstop sections. By using these sections, most of the practical filter specifications can be realized. Each section employs a CGIC, which can be implemented by using 2 OAs in the 2-OA CGIC biquad and 3 OAs in the 3-OA CGIC design. The sensitivities of Q_p, ω_p, Q_z, ω_z, and also the multiplier constant of the realization, have been found to be low with respect to the passive- and active-element variations. A simple functional tuning sequence in which only resistors are adjusted has been described. With the exception of one of the notch sections using the 2-OA CGIC, of all the circuits considered, the output can be located at an OA output. Consequently, these sections can be cascaded without the need for isolating amplifiers. Although exact matching of the gain bandwidth products of the OAs in the CGIC biquad is not essential, optimum results are obtained when they are matched within practical ranges. This is achieved easily by using dual or quad OAs. Also, the choice of the element values for optimum performance is also given. The additional performance improvements using the 3-OA CGICiquads are illustrated experimentally. Useful operating frequencies extension employing the composite operational amplifier technique in the 2-OA CGIC biquad is also given.

References

1. W. B. Mikhael and B. B. Bhattacharyya, A practical design for insensitive *RC*-active filters, *IEEE Trans. Circuits Syst.*, CAS-21, 75–78, Jan. 1974.
2. W. B. Mikhael and S. Michael, Composite operational amplifiers: Generations and finite gain applications, *IEEE Trans. Circuit Syst.*, CAS-34, 449–460, May 1987.

3. S. Micheal and W. B. Mikhael, High frequency filtering and inductance simulation using new composite generalized immittance converters, in *Proc. IEEE ISCAS*, Kyoto, Japan, June 1985, pp. 299–300.

4. A. Antoniou, Realization of gyrators using operational amplifiers and their use in *RC* active network synthesis, *Proc. IEEE*, 116, 1838–1850, Nov. 1969.

5. J. Valihora, Modern technology applied to network implementation, in *Proc. IEEE Int. Symp. Circuit Theory*, Los Angeles, CA, Apr. 1972, pp. 169–173.

6. A. Antoniou, Stability properties of some gyrator circuits, *Electron. Lett.*, 4, 510–512, 1968.

7. B. B. Bhattacharyya, W. B. Mikhael, and A. Antoniou, Design of *RC*-active networks by using generalized immittance converters, in *Proc. IEEE Int. Symp. Circuit Theory*, Toronto, Canada, Apr. 1973, pp. 290–294.

15

High-Order Filters

Rolf Schaumann
Portland State University

15.1 Introduction

With the realization of second-order filters discussed in the previous chapters of this section, we will now treat methods for practical filter implementations of order higher than two. Specifically, we will investigate how to realize efficiently, with low sensitivities to component tolerances, the input-to-output voltage transfer function

$$H(s) = \frac{V_{\text{out}}}{V_{\text{in}}} = \frac{N(s)}{D(s)} = \frac{a_m s^m + a_{m-1} s^{m-1} + \cdots + a_1 s + a_0}{s^n + b_{n-1} s^{n-1} + \cdots + b_1 s + b_0} \tag{15.1}$$

where $n \geq m$ and $n > 2$. The sensitivity behavior of high-order filter realizations shows that, in general, it is not advisable to realize the transfer function $H(s)$ in the so-called direct form [5, Chapter 3] (see also Chapter 5). By direct form we mean an implementation of Equation 15.1 that uses only one or maybe two active devices, such as operational amplifiers (op-amps) or operational transconductance amplifiers (OTAs), embedded in a high-order passive RC network. Although it is possible in principle to realize Equation 15.1 in direct form, the resulting circuits are normally so sensitive to component tolerances as to be impractical. Since the direct form for the realization of high-order functions is ruled out, in this section we present those methods that result in designs of practical manufacturable active filters with acceptably low sensitivity, the *cascade* approach, the *multiple-loop feedback topology*, and *ladder simulations*. Both cascade and multiple-loop feedback techniques are modular, with active biquads used as the fundamental building blocks. The ladder simulation method seeks active realizations that inherit the low passband sensitivity properties of passive doubly terminated LC ladder filters (see Chapter 9).

In the cascade approach, a high-order function $H(s)$ is factored into low (first or second) order subnetworks, which are realized as discussed in the previous chapters of this section and connected in cascade such that their product implements the prescribed function $H(s)$. The method is widely employed in industry; it is well understood, very easy to design, and efficient in its use of active devices. It uses a modular approach and results in filters that, for the most part, show satisfactory performance in practice. The main advantage of cascade filters is their generality, i.e., any arbitrary stable transfer function can be realized as a cascade circuit, and tuning is very easy because each biquad is responsible for the realization of only one pole pair (and zero pair): the realizations of the individual critical frequencies of the filter are decoupled from each other. The disadvantage of this decoupling is that for filters of high order, say $n > 8$, with stringent requirements and tight tolerances, the passband sensitivity of cascade designs to component variations is often found to remain still too sensitive. In these cases, the following approaches lead to more reliable circuits.

The multiple-loop feedback or coupled-biquad methods also split the high-order transfer function into second-order subnetworks. These are interconnected in some type of feedback configuration that introduces coupling chosen to reduce the transfer function sensitivities. The multiple-loop feedback approach retains the modularity of cascade designs but at the same time yields high-order filter realizations with noticeably better passband sensitivities. Of the numerous topologies that have been proposed in the literature, see, e.g., Ref. [5, Chapter 6], we discuss only the follow-the-leader feedback (FLF) and the LF (leapfrog) methods. Both are particularly well suited for all-pole characteristics but can be extended to realizations of general high-order transfer functions. Although based on coupling of biquads in a feedback configuration, the LF procedure is actually derived from an *LC* ladder simulation and will, therefore, be treated as part of that method.

As the name implies, the ladder simulation approach uses an active circuit to simulate the behavior of doubly terminated *LC* ladders in an attempt to inherit their excellent low passband-sensitivity properties. The methods fall into two groups. One is based on element replacement or substitution, where the inductors are simulated via electronic circuits whose input impedance is inductive over the necessary frequency range; the resulting active "components" are then inserted into the *LC* filter topology. The second group may be labeled operational simulation of the *LC* ladder, where the active circuit is configured to realize the internal operation, i.e., the equations, of the *LC* prototype. Active filters simulating the structure or the behavior of *LC* ladders have been found to have the lowest passband sensitivities among active filters and, consequently, to be the most appropriate if filter requirements are stringent. They can draw on the wealth of information available for lossless filters, e.g., Refs. [5, Chapter 2], [9, Chapter 13], [7] that can be used directly in the design of active ladder simulations. A disadvantage of this design method is that a passive *LC* prototype must, of course, exist* before an active simulation can be attempted.

15.2 Cascade Realizations

Without loss of generality we may assume in our discussion of active filters that the polynomials $N(s)$ and $D(s)$ of Equation 15.1 are even, i.e., both n and m are even. An odd function can always be factored into the product of an even function and a first-order function, where the latter can easily be realized by a passive *RC* network and can be appended to the high-order active filter as an additional section. Thus, we can factor Equation 15.1 into the product of second-order pole-zero pairs, so that the high-order transfer function $H(s)$ is factored into the product of the second-order functions

$$T_i(s) = k_i \frac{\alpha_{2i}s^2 + \alpha_{1i}s + \alpha_{0i}}{s^2 + s\omega_{0i}/Q_i + \omega_{0i}^2} = k_i t_i(s) \tag{15.2}$$

---

* The realizability conditions for passive *LC* filters are more restrictive than those for active *RC* filters; specifically, the numerator $N(s)$ of Equation 16.1 must be strictly even or odd so that only $j\omega$-axis transmission zeros can be realized.

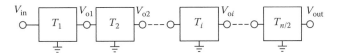

FIGURE 15.1 Cascade realization of an nth-order transfer function; n is assumed even.

such that

$$H(s) = \prod_{i=1}^{n/2} T_i(s) = \prod_{i=1}^{n/2} k_i \frac{\alpha_{2i}s^2 + \alpha_{1i}s + \alpha_{0i}}{s^2 + s\omega_{0i}/Q_i + \omega_{0i}^2} = \prod_{i=1}^{n/2} k_i t_i(s) \tag{15.3}$$

In Equation 15.2, ω_{0i} is the pole frequency and Q_i the pole quality factor; the coefficients α_{2i}, α_{1i}, and α_{0i} determine the type of second-order function $T_i(s)$ that can be realized by an appropriate choice of biquad from the literature, e.g., Refs. [5, Chapter 5], [9, Chapters 4 and 5] (see also Chapters 12 through 14). The transfer functions $T_i(s)$ of the individual biquads are scaled by a suitably defined gain constant k_i, e.g., such that the leading coefficient in the numerator of the gain-scaled transfer function $t_i(s)$ is unity or such that $|t_i(j\omega)| = 1$ at some desired frequency. If we assume now that the output impedances of the biquads are small compared to their input impedances, all second-order blocks can be connected in cascade as in Figure 15.1 without causing mutual interactions due to loading. The product of the biquadratic functions is then realized as required by Equation 15.3.

The process is straightforward and leads to many possibilities for cascading the identified circuit blocks, but several questions must still be answered:

1. Which pair of zeros of Equation 15.1 should be assigned to which pole-pair when the biquadratic functions $T_i(s)$ are formed? Since we have $n/2$ pole pairs and $n/2$ zero pairs (counting zeros at 0 and at ∞) we can select from $(n/2)!$ possible pole-zero pairings.
2. In which order should the biquads be cascaded? For $n/2$ biquads, we have $(n/2)!$ possible ways of section ordering.
3. How should the gain constants k_i in Equation 15.2 be chosen to determine the signal level for each biquad? In other words, what is the optimum gain assignment?

Because the total prescribed transfer function $H(s)$ is simply the product of $T_i(s)$, the choices in 1, 2, and 3 are quite arbitrary as far as $H(s)$ is concerned. However, they determine significantly the dynamic range,[*] i.e., the distance between the maximum possible undistorted signal (limited by the active devices) and the noise floor, because the maximum and minimum signal levels throughout the cascade filter depend on the pole-zero pairings, cascading sequence, and gain assignment.

There exist well-developed techniques and algorithms for answering the questions of pole-zero pairing, section ordering, and gain assignment exactly [3, Chapter 1], [5, Chapter 6]; they rely heavily on computer routines and are too lengthy to be treated in this chapter. We will only give a few rules-of-thumb, which are based on the intuitive observation that in the passband the magnitude of each biquad should vary as little as possible. This keeps signal levels as equal as possible versus frequency and avoids in-band attenuation and the need for subsequent amplification that will at the same time raise the noise floor and thereby further limit the dynamic range. Note also that in-band signal amplification may overdrive the amplifier stages and cause distortion. The simple rules below provide generally adequate designs.

---

[*] Pole-zero pairing also affects to some extent the sensitivity performance, but the effect usually is not very strong and will not be discussed in this chapter. For a detailed treatment see Ref. [3].

1. Assign each zero or zero-pair to the closest pole-pair.
2. Sequence the sections in the order of increasing values of Q_i, i.e., $Q_1 < Q_2 < \cdots < Q_{n/2}$, so that the section with the flattest transfer function magnitude comes first, the next flattest one follows, and so on. If the position of any section is predetermined,* use the sequencing rule for the remaining sections.

After performing steps (1) and (2), assign the gain-scaling constants k_i such that the maximum output signals of all sections in the cascade are the same, i.e.,

$$\max\left| V_{oi}(j\omega)\right| = \max\left| V_{o,n/2}(j\omega)\right| = \max\left| V_{\text{out}}(j\omega)\right| \quad i = 1,\ldots,n/2 - 1 \tag{15.4}$$

This can be done readily with the help of a network analysis routine or simulator, such as SPICE, by computing the output signals at each biquad: using the notation, from Equation 15.3,

$$H(s) = \prod_{i=1}^{n/2} T_i(s) = \prod_{i=1}^{n/2} k_i t_i(s) = \prod_{i=1}^{n/2} k_i \prod_{i=1}^{n/2} t_i(s) \tag{15.5}$$

and

$$H_i(s) = \prod_{j=1}^{i} T_j(s) = \prod_{j=1}^{i} k_j t_j(s) = \prod_{j=1}^{i} k_j \prod_{j=1}^{i} t_j(s) \tag{15.6}$$

we label the total prescribed gain constant

$$K = \prod_{i=1}^{n/2} k_i \tag{15.7}$$

such that

$$\max\left| \prod_{i=1}^{n/2} t_i(j\omega)\right| = M_{n/2} \tag{15.8a}$$

is some given value. Further, let us denote the maxima of the intermediate gain-scaled transfer functions by M_i, i.e.,

$$\max\left| \prod_{k=1}^{i} t_k(j\omega)\right| = M_i \quad i = 1,\ldots,n/2 - 1 \tag{15.8b}$$

then we obtain [5, Chapter 6]

$$k_1 = K\frac{M_{n/2}}{M_1} \quad \text{and} \quad k_j = \frac{M_{j-1}}{M_j} \quad j = 2,\ldots,n/2 \tag{15.9}$$

---

* Such as, e.g., placing a lowpass at the input may be preferred to avoid unnecessary high-frequency signals from entering the filter.

Choosing the gain constants as in Equation 15.9 guarantees that the same maximum voltage appears at all biquad outputs to assure that the largest possible signal can be processed without distortion.

To illustrate the process, let us realize the sixth-order transfer function

$$H(s) = \frac{0.7560s^3}{(s^2 + 0.5704s + 1)(s^2 + 0.4216s + 2.9224)(s^2 + 0.1443s + 0.3422)} \tag{15.10}$$

where the frequency parameter s is normalized with respect to $\omega_n = 130{,}590 \text{ s}^{-1}$. It defines a sixth-order bandpass filter with a 1 dB equiripple passband in $12 \text{ kHz} \leq f \leq 36 \text{ kHz}$ and at least 25 dB attenuation in $f \leq 4.8 \text{ kHz}$ and $f \geq 72 \text{ kHz}$ [5, Chapter 1]. $H(s)$ can be factored into the product of

$$T_1(s) = k_1 t_1(s) = \frac{k_1 s}{s^2 + 0.5704s + 1} \quad \text{with}$$

$$Q_1 = \frac{\sqrt{1}}{0.5704} \approx 1.75 \tag{15.11a}$$

$$T_2(s) = k_2 t_2(s) = \frac{k_2 s}{s^2 + 0.4216s + 2.9224} \quad \text{with}$$

$$Q_2 = \frac{\sqrt{2.9224}}{0.4216} \approx 4.06 \tag{15.11b}$$

$$T_3(s) = k_3 t_3(s) = \frac{k_3 s}{s^2 + 0.1443s + 0.3422} \quad \text{with}$$

$$Q_3 = \frac{\sqrt{0.3422}}{0.1443} \approx 4.06 \tag{15.11c}$$

which are cascaded in the order of increasing values of Q, i.e., $T_1 T_2 T_3$. We can compute readily the maximum values at the output of the sections in the cascade as $M_1 = |t_1|_{\max} \approx 1.75$, $M_2 = |t_1 t_2|_{\max} \approx 1.92$, and $M_3 = |t_1 t_2 t_3|_{\max} \approx 1.32$ to yield by Equation 15.9

$$k_1 = 0.7560 \frac{1.32}{1.75} \approx 0.57 \quad k_2 = \frac{1.75}{1.92} \approx 0.91 \quad k_3 = \frac{1.92}{1.32} \approx 1.45 \tag{15.12}$$

Let us build the sections with the Åckerberg–Mossberg circuit [9, Chapter 4.4] shown in Figure 15.2. It realizes

$$\frac{V_o}{V_i} = -\frac{k\omega_0^2}{s^2 + s\omega_0/Q + \omega_0^2} \quad \text{with } \omega_0 = \frac{1}{RC} \tag{15.13}$$

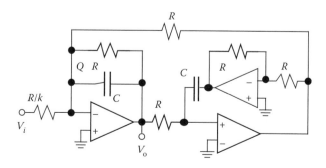

FIGURE 15.2 Åckerberg–Mossberg bandpass circuit.

We obtain readily from Equations 15.11 through 15.13 the following (rounded) component values:

$$\text{for } T_1: \quad R = 1.5 \text{ k}\Omega \quad QR = 2.7 \text{ k}\Omega \quad R/k = 2.7 \text{ k}\Omega$$

$$\text{for } T_2: \quad R = 0.9 \text{ k}\Omega \quad QR = 3.6 \text{ k}\Omega \quad R/k = 1.7 \text{ k}\Omega$$

$$\text{for } T_3: \quad R = 2.6 \text{ k}\Omega \quad QR = 10 \text{ k}\Omega \quad R/k = 1 \text{ k}\Omega$$

The three sections are then interconnected in cascade in the order $T_1 T_2 T_3$.

15.3 Multiple-Loop Feedback Realizations

These topologies are also based on biquad building blocks, which are then embedded, as the name implies, into multiple-loop resistive feedback configurations. The resulting coupling between sections is selected such that transfer function sensitivities are reduced below those of cascade circuits. It has been shown that the sensitivity behavior of the different available configurations is comparable. We shall, therefore, concentrate our discussion only on the FLF and, as part of the ladder simulation techniques, on the leapfrog (LF) topologies, which have the advantage of being relatively easy to derive without any sacrifice in performance. Our derivation will reflect the fact that both configurations* are particularly convenient for geometrically symmetrical bandpass functions and that the LF topology is obtained from a direct simulation of an *LC* lowpass ladder.

15.3.1 Follow-the-Leader Feedback Topology

The FLF topology consists of a cascade of biquads whose outputs are fed back into a summer at the filter's input. At the same time, the biquad outputs may be fed forward into a second summer at the filter's output to permit an easy realization of arbitrary transmission zeros. The actual implementation of the summers and the feedback factors is shown in Figure 15.3; if there are n noninteracting biquads, the order of the realized transfer function $H(s)$ is $2n$. Assuming that the two summer op-amps are ideal, routine analysis yields

$$-V_0 = \frac{R_{F0}}{R_{\text{in}}} V_{\text{in}} + \sum_{i=1}^{n} \frac{R_{F0}}{R_{Fi}} V_i = \alpha V_{\text{in}} + \sum_{i=1}^{n} F_i V_i \tag{15.14}$$

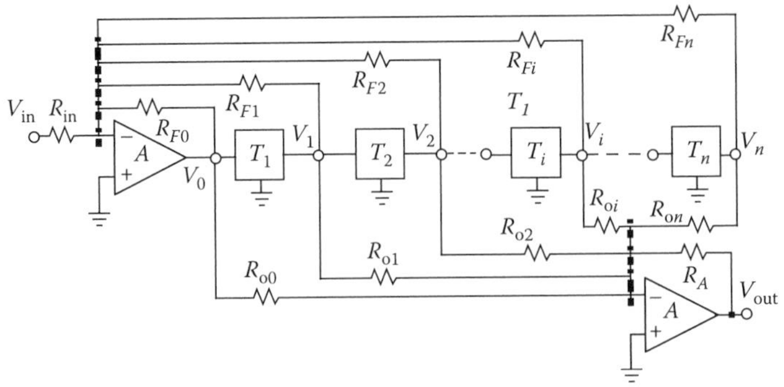

FIGURE 15.3   FLF circuit built from second-order sections $T_i(s)$ and a feedback network consisting of an op-amp summer with resistors R_{Fi}. Also shown is an output summer with resistors R_{oi} to facilitate the realization of arbitrary transmission zeros.

---

* As are all other multiple-loop feedback circuits.

where we defined $\alpha = R_{F0}/R_{in}$ and the feedback factors $F_i = R_{F0}/R_{Fi}$. Similarly, we find for the output summer

$$V_{out} = -\sum_{i=0}^{n} K_i V_i = -\sum_{i=0}^{n} \frac{R_A}{R_{oi}} V_i \tag{15.15}$$

from which the definition of the resistor ratios K_i is apparent. Any of the parameters F_i and K_i may, of course, be reduced to zero by replacing the corresponding resistor, R_{Fi} or R_{oi}, respectively, by an open circuit. Finally, the internal voltages V_i can be computed from

$$V_i = V_0 \prod_{j=1}^{i} T_j(s) \quad i = 1, \ldots, n \tag{15.16}$$

so that with Equation 15.14

$$H_0(s) = \frac{V_0}{V_{in}} = -\frac{\alpha}{1 + \sum_{k=1}^{n} \left[F_k \prod_{j=1}^{k} T_j(s) \right]} \tag{15.17}$$

which with Equation 15.16 yields

$$H_i(s) = \frac{V_i}{V_{in}} = -\frac{N_i(s)}{D(s)} = -\frac{\alpha \prod_{j=1}^{i} T_j(s)}{1 + \sum_{k=1}^{n} \left[F_k \prod_{j=1}^{k} T_j(s) \right]} \quad i = 1, \ldots, n \tag{15.18}$$

Note that

$$H_n(s) = \frac{V_n}{V_{in}} = -\frac{N_n(s)}{D(s)} = -\frac{\alpha \prod_{j=1}^{n} T_j(s)}{1 + \sum_{k=1}^{n} \left[F_k \prod_{j=1}^{k} T_j(s) \right]} \tag{15.19}$$

is the transfer function of the FLF network without the output summer, i.e., with $R_{oi} = \infty$ for all i.

By Equation 15.19, the transmission zeros of $H_n(s)$ are set by the zeros of $T_j(s)$, i.e., by the feedforward path, whereas the poles of $H_n(s)$ are determined by the feedback network and involve both the poles and zeros of the biquads $T_j(s)$ and the feedback factors F_k. Typically, an FLF network is designed with second-order bandpass biquads

$$T_i(s) = A_i \frac{s/Q_i}{s^2 + s/Q_i + 1} = A_i t_i(s) \tag{15.20}$$

so that $H_n(s)$ has all transmission zeros at the origin, i.e., it is an all-pole bandpass function. Note that in Equation 15.20 the frequency parameter s is normalized to the pole frequency ω_0 ($\omega_{0i} = \omega_0$ for all i is assumed), and Q_i and A_i are the section's pole quality factor and midband gain, respectively. Designing an FLF network with arbitrary zeros requires second-order sections with finite transmission zeros, which leads to quite difficult design procedures. It is much simpler to use Equation 15.15 with Equation 15.18 to yield

$$H(s) = \frac{V_{\text{out}}}{V_{\text{in}}} = \frac{N_{2n}(s)}{D_{2n}(s)} = \alpha \frac{K_0 + \sum_{k=1}^{n}\left[K_k \prod_{j=1}^{k} T_j(s)\right]}{1 + \sum_{k=1}^{n}\left[F_k \prod_{j=1}^{k} T_j(s)\right]} \tag{15.21}$$

which realizes the transfer function of the complete circuit in Figure 15.3 with an arbitrary numerator polynomial $N_{2n}(s)$, even for second-order bandpass functions $T_i(s)$ as in Equation 15.20. It is a ratio of two polynomials whose roots can be set by an appropriate choice of the functions $T_i(s)$, the parameters K_i for the transmission zeros, and the feedback factors F_i for the poles.

We illustrate the design procedure by considering a specific case. For $n = 3$, Equation 15.21 becomes

$$H(s) = \alpha \frac{K_0 + K_1 T_1 + K_2 T_1 T_2 + K_3 T_1 T_2 T_3}{1 + F_1 T_1 + F_2 T_1 T_2 + F_3 T_1 T_2 T_3} \tag{15.22}$$

Next we transform the bandpass functions $T_i(s)$ in Equation 15.20 into lowpass functions by the lowpass-to-bandpass transformation (see Chapter 3)

$$p = Q\frac{s^2 + 1}{s} \tag{15.23}$$

where
$Q = \omega_0 / B$ is the "quality factor" of the high-order bandpass with bandcenter ω_0 and bandwidth B
p is the normalized lowpass frequency

This step transforms the bandpass functions (Equation 15.20) with all identical pole frequencies ω_0 into the first-order lowpass functions

$$T_{i\text{LP}}(p) = \frac{A_i Q/Q_i}{p + Q/Q_i} = \frac{q_i}{p + q_i} \tag{15.24}$$

where
$q_i = Q/Q_i$
A_i is the dc gain of the lowpass section

Applying Equation 15.23 to the prescribed function $H(s)$ of order $2n$ in Equation 15.22 converts it into a prototype lowpass function $H_{\text{LP}}(p)$ of order n. Substituting Equation 15.24 into the numerator and denominator expressions of that function, of order $n = 3$ in our case, shows that the zeros and poles, respectively, are determined by

$$N_3(p) = \alpha K_0 \prod_{j=1}^{3}(p + q_j) + \sum_{j=1}^{2}\left[k_j \prod_{i=j+1}^{3}(p + q_i)\right] + k_3 \tag{15.25a}$$

and

$$D_3(p) = \prod_{j=1}^{3}(p + q_j) + \sum_{k=1}^{2}\left[f_k \prod_{i=k+1}^{3}(p + q_i)\right] + f_3 \tag{15.25b}$$

where we introduced the abbreviations

$$f_i = F_i \prod_{j=1}^{i} A_j q_j \quad \text{and} \quad k_i = \alpha K_i \prod_{j=1}^{i} A_j q_j \tag{15.26}$$

To realize the prescribed third-order function

$$H_{LP}(p) = \frac{V_o}{V_i} = \frac{a_3 p^3 + a_2 p^2 + a_1 p + a_0}{p^3 + b_2 p^2 + b_1 p + b_0} \tag{15.27}$$

we compare coefficients between Equations 15.25 and 15.27. For the denominator terms we obtain

$$b_2 = q_1 + q_2 + q_3 + f_1$$
$$b_1 = q_1 q_2 + q_1 q_3 + q_2 q_3 + f_1(q_2 + q_3) + f_2$$
$$b_0 = q_1 q_2 q_3 + f_1 q_2 q_3 + f_2 q_3 + f_3$$

These are three equations in six unknowns, f_i and q_i, $i = 1, \ldots, 3$, which can be written more conveniently in matrix form:

$$\begin{pmatrix} 1 & 0 & 0 \\ q_2 + q_3 & 1 & 0 \\ q_2 q_3 & q_3 & 1 \end{pmatrix} \begin{pmatrix} f_1 \\ f_2 \\ f_3 \end{pmatrix} = \begin{pmatrix} b_2 - (q_1 + q_2 + q_3) \\ b_1 - (q_1 q_2 + q_1 q_3 + q_2 q_3) \\ b_0 - q_1 q_2 q_3 \end{pmatrix} \tag{15.28}$$

The transmission zeros are found via an identical process: the unknown parameters k_i are computed from an equation of the form (Equation 15.28) with f_i replaced by $k_i/(\alpha K_0)$ and b_i replaced by a_i/a_3, $i = 1, \ldots, 3$. Also, $K_0 = a_3/\alpha$.

The unknown parameters f_i can be solved from the matrix expression (Equation 15.28). It is a set of linear equations whose coefficients are functions of the prescribed coefficients b_i and of the numbers q_i which for given Q are determined by the quality factors Q_i of the second-order sections $T_i(s)$. Thus, the Q_i are free parameters that may be selected to satisfy any criteria that may lead to a better-working circuit. The free design parameters may be chosen, for example, to reduce a circuit's sensitivity to element variations. This leads to a multiparameter (i.e., the n Q_i-values) optimization problem whose solution requires the availability of the appropriate computer algorithms. If such software is not available, specific values of Q_i can be chosen. The design becomes particularly simple if all the Q_i-factors are equal, a choice that has the additional practical advantage of resulting in all identical second-order building blocks, $T_i(s) = T(s)$. For this reason, this approach has been referred to as the "Primary Resonator Block" (PRB) technique. The passband sensitivity performance of PRB circuits is almost as good as that of fully optimized FLF structures. The relevant equations are derived in the following:

With $q_i = q$ for all i we find from Equation 15.28

$$\begin{pmatrix} 1 & 0 & 0 \\ 2q & 1 & 0 \\ q^2 & q & 1 \end{pmatrix} \begin{pmatrix} f_1 \\ f_2 \\ f_3 \end{pmatrix} = \begin{pmatrix} b_2 - 3q \\ b_1 - 3q^2 \\ b_0 - q^3 \end{pmatrix} \tag{15.29}$$

It shows that

$$F_1 A_1 q = f_1 = b_2 - 3q$$
$$F_2 A_1 A_2 q^2 = f_2 = b_1 - 3q^2 - 2q f_1$$
$$F_3 A_1 A_2 A_3 q^3 = f_3 = b_0 - q^3 - q^2 f_1 - q f_2 \tag{15.30}$$

The system (Equation 15.30) represents three equations for the four unknowns, q, f_i, $i = 1, \ldots, 3$, (in general, one obtains n equations for $n+1$ unknowns) so that one parameter, q, can still be used for optimization purposes. This single degree of freedom is often eliminated by choosing

$$q = \frac{b_{n-1}}{nb_n} \tag{15.31a}$$

i.e., $q = b_2/3$ in our example, which means $f_1 = 0$. The remaining feedback factors can then be computed recursively from Equation 15.30. The systematic nature of the equations makes it apparent how to proceed for $n > 3$. As a matter of fact, it is not difficult to show that, with $f_1 = 0$, in general

$$f_2 = b_{n-2} - \frac{n(n-1)}{2!} q^2 b_n \tag{15.31b}$$

$$f_i = b_{n-i} - \frac{q^i}{(n-i)!} \left[\frac{n!}{i!} b_n + \sum_{j=1}^{i-1} \frac{f_j}{q^j} \frac{(n-j)!}{(i-j)!} \right] \quad i = 3, \ldots, n \tag{15.31c}$$

Note that b_n usually equals unity. As mentioned earlier, equations of identical form, with f_i replaced by $k_i/(\alpha K_0)$ and b_i by a_i/a_n with $K_0 = a_n/\alpha$, are used to determine the summing coefficients K_i of the output summer, which establishes the transmission zeros. Thus, given a geometrically symmetrical bandpass function with center frequency ω_0, quality factor Q, and bandwidth B, Equation 15.31 can be used to calculate the parameter q and all feedback and summing coefficients required for a PRB design. All second-order bandpass sections, Equation 15.20, are tuned to the same pole-frequency, $\omega = \omega_0$, and have the same pole quality factor, $Q_p = Q/q$, where $Q = \omega_0/B$.

As discussed, the design procedure computes only the *products* f_i and k_i; the actual values of F_i, K_i, and A_i are not uniquely determined. As a matter of fact, the gain constants A_i are *free* parameters that are selected to maximize the circuit's dynamic range in much the same way as for cascade designs.* With a few simple approximations, it can be shown [5, Chapter 6] that the appropriate choice of gain constants in the FLF circuit is

$$A_i \approx \sqrt{1 + (Q_i/Q)^2} = \sqrt{1 + q_i^{-2}} \quad i = 1, \ldots, n \tag{15.32a}$$

The same equation holds for the PRB case where $Q_i = Q_p$ for all i so that

$$A_i = A \approx \sqrt{1 + (Q_p/Q)^2} = \sqrt{1 + q^{-2}} \tag{15.32b}$$

The following example demonstrates the complete FLF (PRB) design process.

Let us illustrate the multiple-loop feedback procedure by realizing again the bandpass function (Equation 15.10), but now as an FLF (PRB) circuit. The previous data specify that the bandpass function should be converted into a prototype lowpass function with bandcenter $\omega_0/(2\pi) = \sqrt{12 \cdot 36}$ kHz and bandwidth $B/(2\pi) = (36 - 12)$ kHz by the transformation Equation 15.23,

$$p = Q \frac{s^2 + 1}{s} = \frac{\omega_0}{B} \frac{s^2 + 1}{s} = \frac{\sqrt{12.36}}{36 - 12} \frac{s^2 + 1}{s} = 0.866 \frac{s^2 + 1}{s} \tag{15.33}$$

---

* For practical FLF (PRB) designs, this step of scaling the signal levels is very important because an inadvertently poor choice of gain constants can result in very large internal signals and, consequently, very poor dynamic range.

where s is normalized by ω_0 as before. Substituting Equation 15.33 into Equation 15.10 results in

$$H_{LP}(s) = \frac{0.491}{p^3 + 0.984p^2 + 1.236p + 0.491} \tag{15.34}$$

which corresponds to Equation 15.27 with $a_i = 0$, $i = 1, 2, 3$. To realize this function, we need three first-order lowpass sections of the form (Equation 15.24), i.e.,

$$T_{LP}(p) = A\frac{q}{p+q} \tag{15.35}$$

where $A = \sqrt{1 + 1/q^2}$ from Equation 15.32b. We proceed with Equation 15.31a to choose $q = 0.984/3 = 0.328$, so that $A = 3.209$. With these numbers, we can apply Equation 15.33 to Equation 15.35 to obtain the second-order bandpass that must be realized:

$$T_{BP}(p) = 3.209\frac{0.328}{0.866\frac{s^2+1}{s} + 0.328} = \frac{1.2153s}{s^2 + 0.379s + 1} \tag{15.36}$$

The feedback resistors we obtain by Equation 15.30:

$$1.052F_1 = f_1 = 0 \rightarrow F_1 = 0$$
$$1.108F_2 = f_2 = b_1 - 3q^2 = 0.913 \rightarrow F_2 = 0.777 \tag{15.37}$$
$$1.166F_3 = f_3 = b_0 - q^3 - qf_2 = 0.371 \rightarrow F_3 = 0.318$$

Choosing, e.g., $R_{F0} = 10\ \text{k}\Omega$ results in $R_{Fi} = R_{F0}/F_i$, that is, $R_{F1} = \infty$, $R_{F2} = 12.9\ \text{k}\Omega$, and $R_{F3} = 31.5\ \text{k}\Omega$. Also, $R_{F0}/R_{in} = \alpha = 0.491\ (Aq)^{-3} = 0.421 \rightarrow R_{in} = 23.7\ \text{k}\Omega$.

There remains the choice of second-order bandpass sections. Notice from Equation 15.19 that the sections must have positive, i.e., noninverting, gain to keep the feedback loops stable. We choose GIC (general impedance converter) sections [5, Chapter 4], [9, Chapter 4.5], one of which is shown explicitly in Figure 15.4. They realize

$$T(s) = \frac{V_o}{V_i} = \frac{\left(1 + \frac{G_3}{G_2}\right)\frac{G_1}{C}s}{s^2 + \frac{G_1}{C}s + \left(\frac{G}{C}\right)^2\frac{G_3}{G_2}} = \frac{K\frac{\omega_0}{Q}s}{s^2 + \frac{\omega_0}{Q}s + \omega_0^2} \tag{15.38}$$

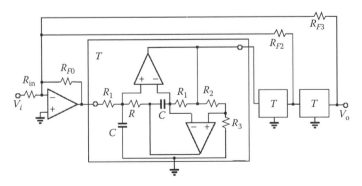

FIGURE 15.4 PRB topology with three identical GIC bandpass sections. The circuit for one section is shown explicitly, as is the input summer with the feedback resistors.

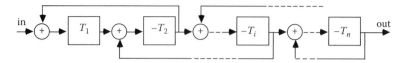

FIGURE 15.5   Leapfrog topology.

Comparing Equation 15.38 with Equation 15.36, choosing $R_3 = 5$ kΩ, $C = 2$ nF, and remembering that s is normalized with respect to $\omega_0 = 130{,}590$ s^{-1} results in the rounded values

$$R = 5.5 \text{ k}\Omega \quad R_1 = 10 \text{ k}\Omega \quad R_2 = 10 \text{ k}\Omega$$

15.3.2 Leapfrog Topology

The LF configuration is pictured in Figure 15.5. Each of the boxes labeled T_i realizes a second-order transfer function. The feedback loops always comprise two sections; thus, inverting and noninverting sections must alternate to keep the loop gains negative and the loops stable. If the circuit is derived from a resistively terminated lossless ladder filter as is normally the case, T_1 and T_n are lossy and all the internal sections are lossless. A lossless block implies a function T_i with infinite Q, which may not be stable by itself, but the overall feedback connection guarantees stability.

An LF circuit can be derived from the configuration in Figure 15.5 by direct analysis with, e.g., bandpass transfer functions as in Equation 15.20 assumed for the blocks T_i. Comparing the resulting equation with that of a prescribed filter yields expressions that permit determining all circuit parameters in a similar way as for FLF filters [1]. Because the topology is identical to that derived from a signal-flow graph representation of an *LC* ladder filter, we do not consider the details of the LF approach here, but instead proceed directly to the ladder simulation techniques.

15.4 Ladder Simulations

Although transfer functions of *LC* ladders are more restrictive than those realizable by cascade circuits (see footnote in Section 15.2), lossless ladder filters designed for maximum power transfer have received considerable attention in the active filter literature because of their significant advantage of having the lowest possible sensitivities to component tolerances in the passband. Indeed, the majority of high-order active filters with demanding specifications are being designed as simulated *LC* ladders. Many active circuit structures have been developed, which simulate the performance of passive *LC* ladders and inherit their good sensitivity performance. The ladder simulations can be classified into two main groups: *operational simulation* and *element substitution*. Starting from an existing *LC* prototype ladder, the operational simulation models the internal operation of the ladder by simulating the circuit equations, i.e., Kirchhoff's voltage and current laws and the *I–V* relationships of the ladder arms. Fundamentally, this procedure simulates the *signal-flow graph* (SFG) of the ladder where all voltages and all currents are considered signals, which are integrated on the inductors and capacitors, respectively. The SFG method is developed in Section 15.4.1. The element substitution procedure replaces all inductors or inductive branches by active networks whose input impedance is inductive over the frequency range of interest. A practical approach to this method is presented in Section 15.4.3. For more detailed discussions of these important and practical procedures, we refer the reader to a modern text on active filters, such as Refs. [5,6,9] and, for approaches using operational transconductance amplifiers, to Section 16.3.

15.4.1 Signal-Flow Graph Methods

To derive the SFG method, consider the ladder structure in Figure 15.6, whose branches may contain arbitrary combinations of capacitors and inductors. In general, resistors are permitted to allow for lossy components. Labeling the combination of elements in the series arms as admittances Y_i, $i = 2, 4$, and those in the shunt branches as impedances Z_j, $j = 1, 3, 5$, we can readily analyze the ladder by writing Kirchhoff's laws and the I–V relationships for the ladder arms as follows:

$$I_i = G_i(V_i - V_1) \quad V_1 = Z_1 I_1 = Z_1(I_i - I_2)$$
$$I_2 = Y_2 V_2 = Y_2(V_1 - V_3) \quad V_3 = Z_3 I_3 = Z_3(I_2 - I_4) \tag{15.39}$$
$$I_4 = Y_4 V_4 = Y_4(V_3 - V_5) \quad V_5 = V_o = Z_5 I_5 = Z_5(I_4 - I_6) \quad I_6 = G_o V_o$$

In the active simulation of this circuit, all currents and voltages are to be represented as *voltage* signals. To reflect this in the expressions, we use a resistive scaling factor R as shown in one of these equations as an example:

$$V_3 = \frac{Z_3}{R} I_3 R = \frac{Z_3}{R}(I_2 R - I_4 R) \tag{15.40}$$

and introduce the notation

$$I_k R = i_k \quad V_k = v_k \quad G_i R = g_i \quad Z_k/R = z_k \quad Y_k R = y_k \tag{15.41}$$

The lower-case symbols are used to represent the *scaled* quantities; notice that z_k and y_k are dimensionless voltage transfer functions, also called *transmittances*, and that both i_k and v_k are voltages. We shall retain the symbol i_k to remind ourselves of the origin of that signal as a current in the original ladder. Equation 15.39 then takes on the form

$$i_i = g_i[v_i + (-v_1)] \quad -v_1 = -z_1 i_1 = -z_1 [i_i + (-i_2)]$$
$$-i_2 = y_2(-v_2) = y_2[(-v_1) + v_3] \quad v_3 = -z_3(-i_3) = -z_3[(-i_2) + i_4] \tag{15.42}$$
$$i_4 = y_4 v_4 = y_4[v_3 + (-v_5)] \quad -v_5 = -v_o = -z_5 i_5 = -z_5[i_4 + (-i_6)] \quad -i_6 = g_o(-v_o)$$

where we have made all the transmittances z_i inverting and assigned signs to the signals in a consistent fashion, such that only *signal addition* is required in the circuit to be derived from these equations. We made this choice because addition can be performed at op-amp summing nodes with no additional circuitry (see below), whereas subtraction would require either differential amplifiers or inverters. The price to be paid for this convenience is that in some cases the overall transfer function suffers a sign inversion (a 180° phase shift), which is of no consequence in most cases. The signal-flow block diagram

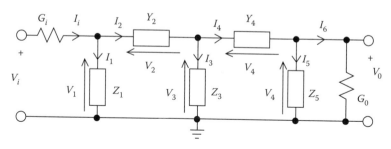

FIGURE 15.6 Ladder network.

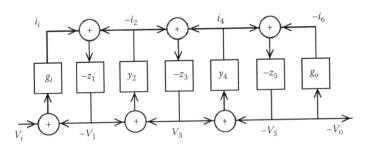

FIGURE 15.7   Signal-flow graph block diagram realizing Equation 15.42.

implementing Equation 15.42 is shown in Figure 15.7. As is customary, all voltage signals are drawn at the bottom of the diagram, and those derived from currents at the top. We observe that the circuit consists of a number of interconnected loops of two transmittances each and that all loop-gains are negative as required for stability. Notice that redrawing this figure in the form of Figure 15.5 results in an identical configuration, i.e., as mentioned earlier, the leapfrog method is derived from a ladder simulation technique.

To determine how the transmittances are to be implemented, we need to know which elements are in the ladder arms. Consider first the simple case of an all-pole lowpass ladder where $Z_i = 1/(sC_i)$ and $Y_j = 1/(sL_j)$, i.e., $z_i = 1/(sC_iR)$ and $y_i = 1/(sL_i/R)$ (see Figure 15.11). Evidently then, for this case all transmittances are integrators. Suitable circuits are shown in Figure 15.8 where for each integrator we have used two inputs in anticipation of the final realization, which has to sum two signals as indicated in Figure 15.7. The circuits realize

$$V_o = \pm \frac{G_1 V_1 + G_2 V_2}{sC + G_3} \tag{15.43}$$

where the plus sign is valid for the series transmittances $y_i(s)$ in Figure 15.8b and the minus sign for the shunt transmittances $z_i(s)^*$ in Figure 15.8a. G_3 is zero if the integration is lossless as required in an internal branch of the ladder; in the two end branches, G_3 is used to implement the source and load resistors of the ladder.

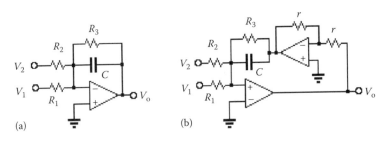

(a)   (b)

FIGURE 15.8   (a) Inverting lossy Miller integrator; (b) noninverting lossy phase-lead integrator.

---

* These two circuits are good candidates for building two-integrator loops as required in Figure 15.7 because the op-amp in the Miller integrator causes a phase lag, whereas the op-amps in the noninverting integrator cause a phase lead of the same magnitude. In the loop, these two phase shifts just cancel and cause no errors in circuit performance. Notice that the Åckerberg–Mossberg biquad in Figure 15.2 is also constructed as a loop of these two integrators.

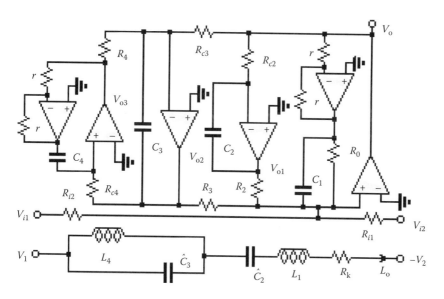

FIGURE 15.9 Active realization of a series ladder branch (plus sign in Equation 15.46).

The next question to be answered is what circuitry will realize more general ladder arms which may contain both series and parallel *LC* elements, and resistors to handle the source or load terminations (or losses if required). Such general series and shunt branches are shown at the bottom of Figures 15.9 and 15.10, respectively, where we have labeled the capacitors in the *passive* network as "$\hat{C}$" to be able to distinguish them from the capacitors in the active circuit (labeled *C* without circumflex). We have chosen the signs of the relevant voltages and currents in Figures 15.9 and 15.10 appropriately to obtain

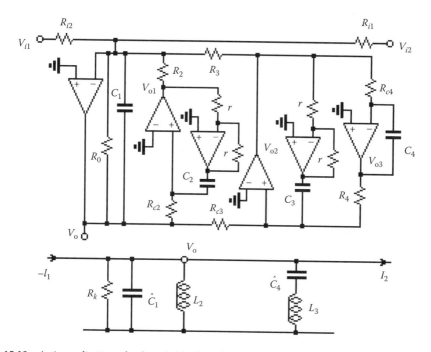

FIGURE 15.10 Active realization of a shunt ladder branch minus sign in Equation 15.46.

noninverting transmittances $y(s)$ for the series arms and *inverting transmittances* $z(s)$ for the shunt arms as requested in Figure 15.7. Recall that this choice permits the flowgraph to be realized with only *summing* functions. For easy reference, the active RC circuits realizing these branches are shown directly above the passive LC arms [5, Chapter 6], [9, Chapter 15].

The passive series branch in Figure 15.9 realizes the current I_o

$$I_o R_p = +Y(s)R_p(aV_1 + bV_2)$$

$$= +\frac{1}{R_k/R_p + sL_1/R_p + \frac{1}{s\hat{C}_2 R_p} + \frac{1}{s\hat{C}_3 R_p + \frac{1}{sL_4/R_p}}}(aV_1 + bV_2) \tag{15.44a}$$

which was converted into a voltage through multiplication with a scaling resistor R_p (p stands for *passive*; R_p is the resistor used to scale the *passive* circuit). Also, we have multiplied the input signals by two constants, a and b, in anticipation of future scaling possibilities in the *active* circuit. Using, as before, *lower-case* symbols for the *normalized* variables, we obtain for the series branch

$$i_o = +y(s)(av_1 + bv_2) = +\frac{1}{r_k + sl_1 + \frac{1}{sc_2} + \frac{1}{sc_3 + \frac{1}{sl_4}}}(av_1 + bv_2) \tag{15.44b}$$

In an analogous fashion we find for the voltage V_o in the passive shunt branch in Figure 15.10, after impedance-level scaling with R_p and signal-level scaling with a and b, the expression

$$V_o = -\frac{Z(s)}{R_p}(aI_1 R_p + bI_2 R_p)$$

$$= -\frac{1}{G_k R_p + s\hat{C}_1 R_p + \frac{1}{sL_2/R_p} + \frac{1}{sL_3/R_p + \frac{1}{s\hat{C}_4 R_p}}}(aI_1 R_p + bI_2 R_p) \tag{15.45a}$$

which with lower-case notation gives

$$v_o = -z(s)(ai_1 + bi_2) = -\frac{1}{g_k + sc_1 + \frac{1}{sl_2} + \frac{1}{sl_3 + \frac{1}{sc_4}}}(ai_1 + bi_2) \tag{15.45b}$$

Turning now to the active RC branches in Figures 15.9 and 15.10, elementary analysis of the two circuits, assuming ideal op-amps,* results in

$$V_o = \pm\frac{R_a G_{i1} V_{i1} + R_a G_{i2} V_{i2}}{R_a G_0 + sC_1 R_a + \frac{R_a G_2}{sC_2 R_{c2}} + \frac{R_a G_3}{sC_3 R_{c3} + \frac{G_4 R_{c3}}{sC_4 R_{c4}}}} \tag{15.46}$$

where we used a normalizing resistor R_a (a stands for *active*; R_a is the resistor used to scale the *active* circuit). In Equation 15.46 the plus sign is valid for the series arm, Figure 15.9, and the minus sign for the shunt arm, Figure 15.10.

---

* Using a more realistic op-amp model, $A(s) \approx \omega_t/s$, one can show [5, Chapter 6] that Equation 16.46 to a first-order approximation is multiplied by $(1 + j\omega/\omega_t) \approx \exp(j\omega/\omega_t)$ for the series arm (plus sign in Equation 16.46) and by $(1 - j\omega/\omega_t) \approx \exp(-j\omega/\omega_t)$ for the shunt arm [minus sign in Equation 16.46]. Thus, in the loop gains determined by the product of an inverting and a noninverting branch, op-amp effects cancel to a first order, justifying the assumption of ideal op-amps. See footnote in Section 15.4.1.

Notice that Equation 15.46 is of the same form as Equations 15.44 and 15.45 so that we can compare the expressions term-by-term to obtain the component values required for the active circuit (s) to realize the prescribed passive ladder arms. Thus, we find for the *series* branch, Figure 15.9, by comparing coefficients between Equations 15.46 and 15.44a, and assuming all equal capacitors C in the active circuit:

$$R_{i1} = \frac{R_a}{a} \quad R_{i2} = \frac{R_a}{b} \quad R_0 = \frac{R_a R_p}{R_k} \quad C = \frac{L_1}{R_a R_p}$$

$$R_{c2} R_2 = \frac{\hat{C}_2}{C} R_a R_p \quad R_{c3} R_3 = \frac{\hat{C}_3}{C} R_a R_p \quad R_{c4} R_4 = \frac{L_4 \hat{C}_3}{C^2} \tag{15.47a}$$

In an identical fashion we obtain from Equations 15.46 and 15.45a for the components of the *shunt* arm

$$R_{i1} = \frac{R_a}{a} \quad R_{i2} = \frac{R_a}{b} \quad R_0 = R_k \frac{R_a}{R_p} \quad C = \hat{C}_1 \frac{R_p}{R_a}$$

$$R_{c2} R_2 = \frac{L_2}{C} \frac{R_a}{R_p} \quad R_{c3} R_3 = \frac{L_3}{C} \frac{R_a}{R_p} \quad R_{c4} R_4 = \frac{L_3 \hat{C}_4}{C^2} \tag{15.47b}$$

The scaling resistors R_a and R_p are chosen to obtain convenient element values. Note that each of the last three equations for both circuits determines only the *product* of two resistors. This leaves three degrees of freedom that are normally chosen to maximize dynamic range by equalizing the maximum signal levels at all op-amp outputs. We provide some discussion of these matters in Section 15.4.2.

We have displayed the active and passive branches in Figures 15.9 and 15.10 together to illustrate the one-to-one correspondence of the circuits. For example, if we wish to design an all-pole lowpass filter, such as the one in Figure 15.11, where the internal series arms consist of a single inductor and each internal shunt arm of a single capacitor, the corresponding active realizations reduce to those of Figure 15.8b and a, respectively, with $R_3 = \infty$. For the end branches, R_s in series with L_1 and R_l in parallel with $\hat{C}_4$, we obtain the circuits in Figure 15.8b and a, respectively, with R_3 *finite* to account for the source and load resistors. The remaining branches in the active circuits are absent. Assembling the resulting circuits as prescribed in Figure 15.7 leads to the active SFG filter in Figure 15.12, where we have for convenience chosen all identical capacitors, C, and have multiplied the input by an arbitrary constant K because the active circuit may realize a gain scaling factor. The component values are computed from a set of equations similar to Equation 15.47. To show in some detail how they are arrived at, we derive the equations for each integrator in Figure 15.12 and compare them with the corresponding arm in the passive ladder. Recalling that signals with lower-case symbols in the active circuit are voltages, we obtain

$$\frac{G_1 R_a V_{in} + G_2 R_a(-v_2)}{sCR_a + G_3 R_a} \rightarrow \frac{K V_{in} - V_2}{sL_1/R_p + R_s/R_p} \tag{15.48a}$$

$$\frac{G_4 R_a i_1 + G_5 R_a(-i_3)}{sCR_a} \rightarrow \frac{I_1 - I_3}{s\hat{C}_2 R_p} \tag{15.48b}$$

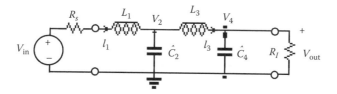

FIGURE 15.11 Fourth-order all-pole lowpass ladder.

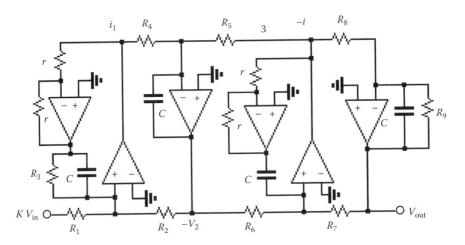

FIGURE 15.12 Active realization of the ladder in Figure 15.11.

$$\frac{G_6 R_a(-v_2) + G_7 R_a V_{out}}{sCR_a} \rightarrow \frac{-V_2 + V_4}{sL_3/R_p} \qquad (15.48c)$$

$$\frac{G_8 R_a(-i_3)}{sCR_a + G_9 R_a} \rightarrow \frac{-i_3}{s\hat{C}_4 R_p + R_p/R_l} \qquad (15.48d)$$

where we used scaling resistors for the active (R_a) and the passive (R_p) circuits as before. Choosing a convenient value for C in the active circuit and equating the time constants and the dc gains in Equation 15.48a results in the following expressions:

$$R_3 = \frac{L_1}{C}\frac{1}{R_s} \qquad R_1 = R_3\frac{R_s}{KR_p} \qquad R_2 = R_3\frac{R_s}{R_p} \qquad (15.49a)$$

Similarly,

$$R_4 = R_5 = \frac{\hat{C}_2}{C} \qquad R_6 = R_7 = \frac{L_3}{C}\frac{1}{R_p} \qquad R_8 = \frac{\hat{C}_4}{C} \qquad R_9 = \frac{\hat{C}_4}{C} \qquad R_l = \frac{R_l}{R_p}R_8 \qquad (15.49b)$$

15.4.2 Maximization of the Dynamic Range

The remaining task in a signal-flow graph simulation of an *LC* ladder is that of voltage-level scaling for dynamic range maximization. As in cascade design, we need to achieve that all op-amps in $0 \le \omega \le \infty$ see the same maximum signal level so that no op-amp becomes overdriven sooner than any other one. It may be accomplished by noting that a scale factor can be inserted into any signal line in a signal-flow graph *as long as the loop gains are not changed.* Such signal-level scaling will not affect the transfer function except for an overall gain factor. The procedure can be illustrated in the flow diagram in Figure 15.7. If we simplify the circuit by combining the self-loops at input and output and employ the scale factors α, β, γ, δ and their inverses to the loops in Figure 15.7, we obtain the modified flow diagram in Figure 15.13. Simple analysis shows that the transfer function has not changed except for a multiplication by the factor $\alpha\beta\gamma\delta$, which is canceled by the multiplier $(\alpha\beta\gamma\delta)^{-1}$ at the input. To understand how the scale factors are

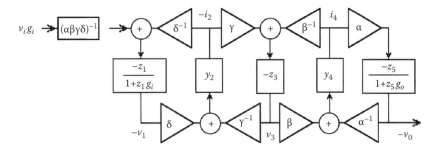

FIGURE 15.13   Using scale factor for signal-level equalization.

computed, assume that in Figure 15.7, i.e., before scaling, the maximum of $|i_4(j\omega)|$ is α times as large as the maximum of $|v_o(j\omega)|$, where α may be less than or larger than unity. Since

$$-v_o = \frac{-z_5}{1 + z_5 g_o} i_4$$

the maxima can be equalized, i.e., the level of v_o can be increased by α, if we apply a gain scale factor α as indicated in Figure 15.13. To keep the loop gain unchanged, a second scale factor α^{-1} is inserted as shown. Continuing, if in Figure 15.7 max $|v_3(j\omega)|$ is β times as large as max $|i_4(j\omega)|$, we raise the level of $|i_4|$ by a gain factor β and correct the loop gain by a second factor β^{-1} as shown in Figure 15.13. In a similar fashion we obtain the maxima of the remaining voltages and apply the appropriate scale factors γ, δ, γ^{-1}, and δ^{-1} in Figure 15.13.

It is easy to determine the relevant maxima needed for calculating the gain factors. Recall that the node voltages v_i, v_1, i_2, v_3, i_4, and $v_5 = v_o$ in the signal-flow graph of Figure 15.13 correspond directly to the actual currents and voltages in the original ladder, Figure 15.6, and that their maxima in $0 \leq \omega \leq \infty$ can be evaluated readily with any network analysis program. For the circuit in Figure 15.13, the scale factors that ensure that the currents (normalized by R_p) in all series ladder arms and the voltages in all shunt ladder arms have the same maxima are then obtained as

$$\alpha = \frac{\max|i_4|}{\max|v_o|} \qquad \beta = \frac{\max|v_3|}{\max|i_4|} \qquad \gamma = \frac{\max|i_2|}{\max|v_3|} \qquad \delta = \frac{\max|v_1|}{\max|i_2|} \tag{15.50}$$

The procedure thus far takes care of ladders with only one element in each branch, such as all-pole lowpass filters (Figure 15.11). In the more general case, we must also equalize the maxima of the magnitudes of V_{o1}, V_{o2}, and V_{o3} at the *internal* op-amp outputs in Figures 15.9 and 15.10. This is achieved easily if we remember that the "external" voltage maxima of $|V_{i1}|$, $|V_{i2}|$, and $|V_o|$ are already equalized by the previous steps leading to Equation 15.50, and that in the passive *series* branch V_{o1} represents the voltage on $\hat{C}_2$, V_{o2} stands for the voltage on $\hat{C}_3$, and V_{o3} corresponds to the current (times R_p) through the inductor L_4. After finding the maxima of these signals with the help of a network analysis program and computing the scale factors

$$m_1 = \frac{\max|V_{o1}|}{\max|V_o|} \qquad m_2 = \frac{\max|V_{o2}|}{\max|V_o|} \qquad m_3 = \frac{\max|V_{o3}|}{\max|V_o|} \tag{15.51}$$

we can equalize all internal branch voltages of the *series* arm (Figure 15.9) by modifying the design Equations 15.47a as follows [5, Chapter 6]:

$$R_{i1} = \frac{R_a}{a} \quad R_{i2} = \frac{R_a}{b} \quad R_0 = \frac{R_a R_p}{R_k} \quad C = \frac{L_1}{R_a R_p}$$

$$m_1 R_{c2} \frac{R_2}{m_1} = \frac{\hat{C}_2}{C} R_a R_p \quad m_2 R_{c3} \frac{R_3}{m_2} = \frac{\hat{C}_3}{C} R_a R_p \quad \frac{m_3}{m_2} R_{c4} \frac{R_4}{m_3/m_2} = \frac{L_4 \hat{C}_3}{C^2} \tag{15.52}$$

A possible choice for the element values given by products is

$$m_1 R_{c2} = \frac{R_2}{m_1} = \sqrt{\frac{\hat{C}_2}{C} R_a R_p} \quad m_2 R_{c3} = \frac{R_3}{m_2} = \sqrt{\frac{\hat{C}_3}{C} R_a R_p}$$

$$\frac{m_3}{m_2} R_{c4} = \frac{m_2}{m_3} R_4 = \frac{\sqrt{L_4 \hat{C}_3}}{C} \tag{15.53a}$$

In the passive *shunt* branch, V_{o1} and V_{o2} represent the currents (times R_p) through the inductors L_2 and L_3, respectively, and V_{o3} stands for the voltage across $\hat{C}_4$. In an identical fashion we obtain then with Equation 15.50 from Equation 15.47b for the components of the *shunt* arm

$$R_{i1} = \frac{R_a}{a} \quad R_{i2} = \frac{R_a}{b} \quad R_0 = R_k \frac{R_a}{R_p} \quad C = \hat{C}_1 \frac{R_p}{R_a}$$

$$m_1 R_{c2} = \frac{R_2}{m_1} = \sqrt{\frac{L_2}{C} \frac{R_a}{R_p}} \quad m_2 R_{c3} = \frac{R_3}{m_2} = \sqrt{\frac{L_3}{C} \frac{R_a}{R_p}} \tag{15.53b}$$

$$\frac{m_3}{m_2} R_{c4} = \frac{m_2}{m_3} R_4 = \sqrt{\frac{L_3 \hat{C}_4}{C}}$$

Next we present an example [2], [5, Chapter 6] in which the reader may follow the different steps discussed.

To simulate the fourth-order elliptic lowpass ladder filter in Figure 15.14a by the signal-flow graph technique, we first reduce the loop count by a source transformation. The resulting circuit, after

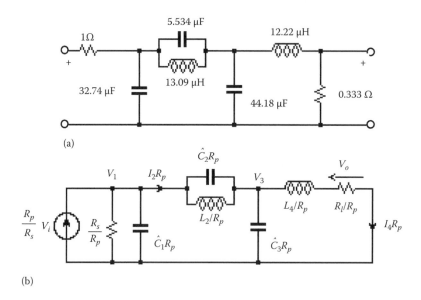

(a)

(b)

FIGURE 15.14   (a) Fourth-order elliptic *LC* lowpass ladder; (b) the circuit after source transformation and impedance scaling.

TABLE 15.1 Voltage and Current Maxima for Figure 15.14a

Voltage or Current	Maximum of Voltage or Current
Voltage across $\hat{C}_1$	0.699 V
Current through L_2	1.550 A
Voltage across $\hat{C}_2$	1.262 V
Current through $L_1 \| \hat{C}_2$	1.125 A
Voltage across $\hat{C}_2$	0.690 V
Current through L_4	0.866 A

impedance scaling, is shown in Figure 15.14b. To use signal-level scaling for optimizing the dynamic range, the relevant maxima of the LC ladder currents and voltages are needed. Table 15.1 lists the relevant data obtained with the help of network analysis software. From these numbers we find

$$\alpha = \frac{\max|v_3|}{\max|i_4 R_p|} = \frac{0.69}{0.866} = 0.797 \quad \beta = \frac{\max|i_2 R_p|}{\max|v_3|} = \frac{1.125}{0.690} = 1.630$$

$$\gamma = \frac{\max|v_1|}{\max|i_2 R_p|} = \frac{0.699}{1.125} = 0.621 \tag{15.54}$$

For ease of reference, the signal-flow graph with scaling factors is shown in Figure 15.15. Let $R_p = 1\ \Omega$. Note that the input signal voltage $i_{in} = R_p V_i / R_s$ has been multiplied by a factor K to permit realizing an arbitrary signal gain. If the desired gain is unity, and since the *dc* gain in the passive LC ladder equals

$$\frac{V_o}{V_i} = \frac{R_l}{R_s + R_l} = \frac{0.333}{1.333} = 0.25$$

we need to choose $K = 1/0.25 = 4$. Thus, the input signal i_i is multiplied by

$$\frac{K}{\alpha\beta\gamma} = \frac{4}{0.797 \cdot 1.630 \cdot 0.621} = 4.958 \tag{15.55}$$

The transmittances are defined as

$$z_1 = \frac{1}{s\hat{C}_1 R_p + G_s R_p} \quad y_2 = \frac{1}{\frac{1}{s\hat{C}_2 R_p + R_p/(sL_p)}}$$

$$z_3 = \frac{1}{s\hat{C}_3} \quad y_4 = \frac{1}{sL_4/R_p + R_l/R_p} \tag{15.56}$$

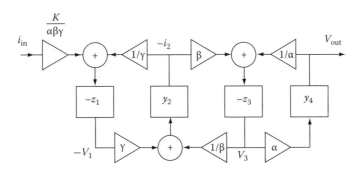

FIGURE 15.15 Signal-flow graph for the filter in Figure 15.14.

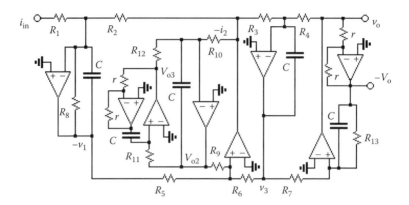

FIGURE 15.16   Active realization of the circuit in Figure 15.14.

With the numbers in Equations 15.54 and 15.55 we obtain from the signal-flow graph

$$-v_1 = -z_1\left(4.958i_{\text{in}} - \frac{1}{0.621}i_2\right) \quad -i_2 = y_2\left(-0.621v_1 + \frac{1}{1.630}v_3\right)$$

$$v_3 = -z_3\left(-1.630i_2 + \frac{1}{0.797}v_{\text{out}}\right) \quad v_{\text{out}} = y_4 0.792v_3 \tag{15.57}$$

The reader is encouraged to verify from Equations 15.56 and 15.57 and Figures 15.9 and 15.10 that the active SFG realization of the ladder is as shown in Figure 15.16. For instance, the circuitry between the nodes v_1, v_3, and i_2 implements the parallel LC series branch to realize the finite transmission zero, obtained from Figure 15.9 by setting the components G_0, C_1, G_2, G_{c2}, and C_2 to zero. The element values in the active circuit are now determined readily by comparing the circuit equations for the active circuit,

$$-v_1 = -\frac{G_1 R_{\text{a1}} i_{\text{in}} + G_2 R_{\text{a1}}(-i_2)}{sCR_{\text{a1}} + G_8 R_{\text{a1}}} \quad -i_2 = \frac{G_5 R_{\text{a2}}(-v_1) + G_6 R_{\text{a2}} v_3}{\frac{G_9 R_{\text{a2}}}{sCR_{10} + \frac{G_{12}R_{10}}{sCR_{11}}}}$$

$$v_3 = -\frac{G_3 R_{\text{a3}}(-i_2) + G_4 R_{\text{a3}} v_o}{sCR_{\text{a3}}} \quad v_o = \frac{G_7 R_{\text{a4}} v_3}{sCR_{\text{a4}} + G_{13} R_{\text{a4}}} \tag{15.58}$$

to Equation 15.57 and using Equation 15.56.* Comparing the coefficients results in

$$G_1 R_{\text{a1}} = 4.958 \quad G_2 R_{\text{a1}} = \frac{1}{0.621} \quad G_5 R_{\text{a2}} = 0.621 \quad G_6 R_{\text{a2}} = \frac{1}{1.630}$$

$$G_3 R_{\text{a3}} = 1.630 \quad G_4 R_{\text{a3}} = \frac{1}{0.797} \quad G_7 R_{\text{a4}} = 0.797$$

Further,

$$\hat{C}_1 R_{\text{p}} = CR_{\text{a1}} \quad \hat{C}_2 R_{\text{p}} = C\frac{R_9 R_{10}}{R_{\text{a2}}} \quad \hat{C}_3 R_{\text{p}} = CR_{\text{a3}} \quad \frac{L_4}{R_{\text{p}}} = CR_{\text{a4}}$$

Thus, with $R_{\text{p}} = 1\ \Omega$ and choosing $C = 5$ nF,

---

* Observe that for greater flexibility we permitted a different scaling resistor $R_{\text{a}i}$, $i = 1, \ldots, 4$, in each branch.

$$R_{a1} = \frac{\hat{C}_1}{C} R_p = 6.55 \text{ k}\Omega \quad R_{a3} = \frac{\hat{C}_3}{C} R_p = 8.84 \text{ k}\Omega \quad R_{a4} = \frac{L_4}{CR_p} = 2.44 \text{ k}\Omega$$

R_{a2} is undetermined; choosing $R_{a2} = 5 \text{ k}\Omega$ leads to the feed-in resistors for each branch

$$R_1 = \frac{R_{a1}}{4.958} = 0.800 \text{ k}\Omega \quad R_2 = 0.621 R_{a1} = 4.07 \text{ k}\Omega \quad R_3 = \frac{R_{a3}}{1.630} = 5.42 \text{ k}\Omega$$

$$R_4 = 0.797 R_{a3} = 7.05 \text{ k}\Omega \quad R_5 = \frac{R_{a2}}{0.621} = 8.05 \text{ k}\Omega$$

$$R_6 = 1.630 R_{a3} = 8.15 \text{ k}\Omega \quad R_7 = \frac{R_{a4}}{0.797} = 3.06 \text{ k}\Omega$$

The remaining components are determined from the equations

$$R_8 = R_s \frac{R_{a1}}{R_p} = 6.55 \text{ k}\Omega \quad R_{13} = \frac{R_{a4} R_p}{R_1} = 7.33 \text{ k}\Omega$$

and

$$R_9 R_{10} = R_p R_{a2} \frac{\hat{C}_2}{C} = (2.35 \text{ k}\Omega)^2 \quad R_{11} R_{12} = \frac{L_2}{C} \frac{R_9 R_{10}}{R_{a2} R_p} = (1.70 \text{ k}\Omega)^2$$

Since only the products of these resistors are given, we select their values uniquely for dynamic range maximization. According to Equation 15.51 we compute from Table 15.1

$$m_2 = \frac{\max|v_{c2}|}{\max|i_2|} = \frac{1.262}{1.125} = 1.12 \quad \text{and} \quad m_3 = \frac{\max|i_{L2}|}{\max|i_2|} = \frac{1.550}{1.125} = 1.38$$

to yield from Equation 15.53a:

$$R_9 = m_2 \, 2.35 \text{ k}\Omega = 2.64 \text{ k}\Omega \quad R_{10} = \frac{2.35 \text{ k}\Omega}{m_2} = 2.10 \text{ k}\Omega$$

$$R_{11} = \frac{m_2}{m_3} 1.70 \text{ k}\Omega = 1.38 \text{ k}\Omega \quad R_{12} = \frac{m_3}{m_2} 1.70 \text{ k}\Omega = 2.09 \text{ k}\Omega$$

15.4.3 Element Substitution

LC ladders are known to yield low-sensitivity filters with excellent performance, but high-quality inductors cannot be implemented in microelectronic form. An appealing solution to the filter design problem is, therefore, to retain the ladder structure and to simulate the behavior of the inductors by circuits consisting of resistors, capacitors, and op-amps. A proven technique uses *impedance converters*, electronic circuits whose input impedance is proportional to frequency when loaded by the appropriate element at the output. The best-known impedance converter is the gyrator, a two-port circuit whose input impedance is inversely proportional to the load impedance, i.e.,

$$Z_{in}(s) = \frac{r^2}{Z_L(s)} \tag{15.59}$$

The parameter r is called the *gyration resistance*. Clearly, when the load is a capacitor, $Z_L(s) = 1/(sC)$, $Z_{in}(s) = sr^2 C$ is the impedance of an inductor of value $L = r^2 C$. Gyrators are very easy to realize with

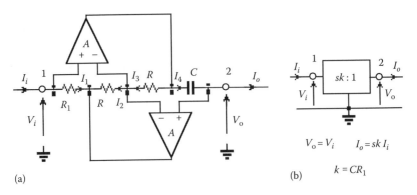

FIGURE 15.17   General impedance converter: (a) circuit; (b) symbolic representation.

transconductors (voltage-to-current converters) and are widely used in transconductance-C filters; see Section 16.3. However, no high-quality gyrators with good performance beyond the audio range have been designed to date with op-amps. If op-amps are to be used, a different kind of impedance converter is employed, one that converts a load resistor R_L into an inductive impedance, such that

$$Z_{in}(s) = (sk)R_L \qquad (15.60)$$

A good circuit that performs this function is Antoniou's general impedance converter (GIC) shown in Figure 15.17a. The circuit, with elements slightly rearranged, was encountered in Figure 15.4, where we used the GIC to realize a second-order bandpass function. The circuit is readily analyzed if we recall that the voltage measured between the op-amp input terminals and the currents flowing into the op-amp input terminals are zero. Thus, we obtain from Figure 15.17a the set of equations

$$V_o = V_i \quad \frac{I_4}{sC} = I_3R \quad I_2R = I_1R_1 \quad I_2 = I_3 \quad I_i = I_1 \quad I_4 = I_o \qquad (15.61)$$

These equations indicate that the terminal behavior of the general impedance converter is described by

$$V_o = V_i \quad I_o = sCR_1L_i = skI_i \qquad (15.62)$$

that is

$$\frac{V_i}{I_i} = Z_{in}(s) = sk\frac{V_o}{I_o} = sk\, Z_L(s) \qquad (15.63)$$

Notice that the input impedance is inductive as prescribed by Equation 15.60 if the load is a resistor.* Figure 15.17b also shows the circuit symbol we will use for the GIC in the following to keep the circuit diagrams simple. This impedance converter and its function of converting a resistive load into an inductive input impedance is the basis for Gorski-Popiel's embedding technique [5, Chapter 6], [9, Chapter 14.4], which permits replacing the inductors in an LC filter by resistors.

---

* To optimize the performance of the GIC, i.e., to make it optimally independent of the finite gain-bandwidth product of the op-amps, the GIC elements should be chosen as follows: For an arbitrary load $Z_L(s)$ one chooses $\omega_c C = 1/*Z_L(j\omega_c)*$. ω_c is some critical frequency, normally chosen at the upper passband corner. If the load is resistive, $Z_L = R_L$, select $C = 1/(\omega_c R_L)$ [6].

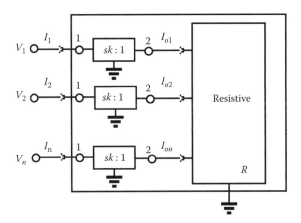

FIGURE 15.18 Simulation of an inductance network.

15.4.3.1 Inductor Replacement: Gorski-Popiel's Method

To understand the behavior of the circuit in Figure 15.18, recall from Equation 15.62 that the voltages at terminals 1 and 2 of the GICs are the same. Then, by superposition the typical input current I_{oi} into the resistive network R can be written as

$$I_{oi} = \frac{1}{R_{i1}}V_1 + \frac{1}{R_{i2}}V_2 + \cdots + \frac{1}{R_{in}}V_n = \sum_{j=1}^{n}\frac{1}{R_{ij}}V_j \quad i = 1, \ldots, n \tag{15.64}$$

where the parameters R_{ij} are given by the resistors and the configuration of R. Using the current relationship $I_{oi} = skI_i$ of the impedance converters in Equation 15.64 results in

$$I_i = \sum_{j=1}^{n}\frac{1}{skR_{ij}}V_j \quad i = 1, \ldots, n \tag{15.65}$$

which makes the combined network consisting of the n GICs and the network R look purely inductive, i.e., each resistor R_r in the network R appears replaced by an inductor of value

$$L_r = skR_r = sCR_1R_r \tag{15.66}$$

Examples of this process are contained in Figure 15.19. Figure 15.19a and b illustrate the conventional realizations of a grounded and a floating inductor requiring one and two converters, respectively. This is

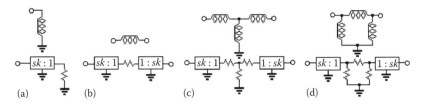

FIGURE 15.19 Elementary inductance networks and their GIC-R equivalents: (a) grounded inductor (b) floating inductor (c) inductive T (d) inductive Π.

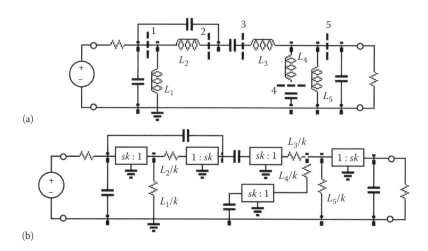

FIGURE 15.20 *LC* ladder realization with GICs: (a) a choice of cuts; (b) the cuts replaced by GICs.

completely analogous to the use of gyrators, where a grounded inductor requires one gyrator, but a floating inductor is realized with two gyrators connected back-to-back. Figure 15.19c and d show how the process is extended to complete inductive subnetworks: the GICs are used to isolate the subnetworks from the remainder of the circuit; there is no need to convert each inductor separately. The method is further illustrated with the filter in Figure 15.20. The *LC* ladder is designed to specifications by use of the appropriate design tables [4,7] or filter design software [8]. In the resulting ladder, the inductive subnetworks are separated as shown in Figure 15.20a by the five dashed cuts. The cuts are repaired by inserting GICs with conversion factor k and the correct orientation (output, terminal 2 in Figure 15.17, toward the inductors). Note that all GICs must have the same conversion factor k. Finally, the inductors L_i are replaced by resistors of value L_i/k as shown in Figure 15.20b. A numerical example will illustrate the design steps:

> Assume we wish to implement a sixth-order bandpass filter with the following specifications:
> Maximally flat passband with ≤ 3 dB attenuation in 900 Hz $\leq f \leq 1200$ Hz
> Transmission zero at $f_z = 1582.5$ Hz; source and load resistors $R = 3$ kΩ

Using the appropriate software or tables, the *LC* filter in Figure 15.21a is found with element values in kiloohm, megahertz, and nanofarad. Figure 15.21b shows the circuit redrawn to help identify the inductive subnetwork and the locations of the cuts. Note that only three impedance converters are used rather than six (two each for the two floating inductors and one each for the grounded ones), if a direct conversion of the individual inductors had been attempted.

For the final active realization, we assumed a conversion parameter $k = CR_1 = 30$ μs. Finally, to design the GICs we compute the impedances seen into the nodes where the GICs see the (now) resistive subnetworks. Using analysis software to compute $|V_i/I_i|$ at these nodes, we find

$$\text{for GIC}_a, \ |Z_a| \approx 18 \text{ k}\Omega, \text{ and for GIC}_b \text{ and GIC}_c, |Z_b| \approx |Z_c| \approx 4.7 \text{ k}\Omega$$

so that with $\omega_c = 2\pi \cdot 1.2$ kHz and $k = 30$ μs the design elements are

$$C_a = \frac{1}{\omega_c |z_a|} \approx 7.4 \text{ nF} \quad R_{1a} = \frac{k}{C_a} \approx 4 \text{ k}\Omega \quad C_b = C_c \approx 28 \text{ nF} \quad R_b = R_c \approx 1 \text{ k}\Omega$$

To complete the design, we choose the resistors R in Figure 15.17a as $R = 1$ kΩ.

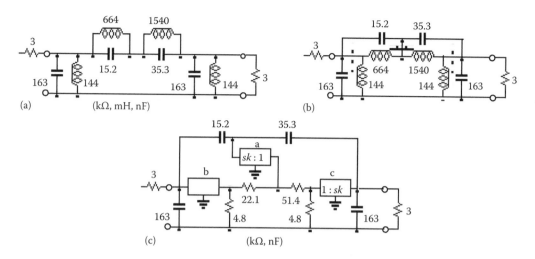

FIGURE 15.21 Sixth-order *LC* bandpass filter and active realization using impedance converters. (a) The LC filter with element values in kiloohm, megahertz, and nanoforad. (b) The circuit redrawn to help identify the inductive subnetwork and the locations of the cuts. (c) The final active realization using a conversion parameter $k = CR_i = 30\mu s$.

15.5 Summary

In this chapter, we discussed the more practical techniques for the design of active filters of order higher than two: *cascade* design, *multiple-loop feedback* approaches, and methods that *simulate the behavior of LC ladder* filters. We pointed out that direct realization methods are impractical because they result in high sensitivities to component values. In many applications, a cascade design leads to satisfactory results. The practical advantages of cascade circuits are modularity, ease of design, flexibility, very simple tuning procedures, and economical use of op-amps with as few as one op-amp per pole pair. Also, we pointed out again that an arbitrary transfer function can be realized with the cascade design method, i.e., no restrictions are placed on the permitted locations of poles and zeros.

For challenging filter requirements, the cascade topologies may still be too sensitive to parameter changes. In those cases the designer can use the FLF configuration or, for best performance, a ladder simulation, provided that a passive prototype ladder exists. FLF circuits retain the advantage of modularity, but if optimal performance is desired, computer-aided optimization routines must generally be used to adjust the available free design parameters. However, excellent performance with very simple design procedures, no optimization, and high modularity can be obtained by use of the *primary-resonator-block* (PRB) method, where all biquad building blocks are identical.

From the point of view of minimum passband sensitivity to component tolerances, the best active filters are obtained by simulating *LC ladders*. If the prescribed transfer characteristic can at all be realized as a passive *LC* ladder, the designer can make use of the wealth of available information about the design of such circuits, and simple procedures are available for "translating" the passive *LC* circuit into its active counterpart. We may either take the passive circuit and replace the inductors by active networks, or we imitate the mathematical behavior of the whole *LC* circuit by realizing the integrating action of inductors and capacitors via active *RC* integrators. Both approaches result in active circuits of high quality; a disadvantage is that more op-amps may be required than in cascade or FLF methods. This drawback is offset, however, by the fact that the sensitivities to component tolerances are very nearly as low as those of the originating ladder.

An alternative method for the active realization of a passive *LC* ladder is obtained by scaling each impedance of the passive circuit by $1/(ks)$ (Bruton's transformation) [5, Chapter 6], [9, Chapter 14.5].

This impedance transformation converts resistors into capacitors, inductors into resistors, and capacitors into "*frequency-dependent negative resistors* (FDNRs)," which can readily be realized with Antoniou's GIC circuit. The procedure is especially useful for passive prototype circuits that have only *grounded* capacitors because capacitors are converted to FDNRs and floating FDNRs are very difficult to implement in practice. The method results in biasing difficulties for active elements, and because the *entire* ladder must be transformed, the active circuit no longer contains source and load resistors. If these two components are prescribed and must be maintained as in the original passive circuit, additional buffers are required. The method was not discussed in this text because it shows no advantages over the Gorski-Popiel procedure.

An important practical aspect is the *limited dynamic range* of active filters, restricted by noise and by the finite linear signal swing of op-amps. Dynamic range maximization should be addressed whenever possible. Apart from designing low-noise circuits, the procedures always proceed to equalize the op-amp output voltages by exploiting available free gain constants or impedance scaling factors. It is a disadvantage of the element substitution method that no general dynamic range scaling method appears to be available.

The methods discussed in this chapter dealt with the design of filters in *discrete* form, that is, separate passive components and operational or transconductance amplifiers are assembled on, e.g., a printed circuit board to make up the desired filter. Chapter 16 addresses modifications in the design approach that lead to filters realizable in fully integrated form.

References

1. Laker, K. R. and M. S. Ghausi, A comparison of active multiple-loop feedback techniques for realizing high-order bandpass filters, *IEEE Trans. Circuits Syst.*, CAS-21, 774–783, November 1974.
2. Martin, K. and A. S. Sedra, Design of signal-flow graph (SFG) active filters, *IEEE Trans. Circuits Syst.*, CAS-25, 185–195, September 1978.
3. Moschytz, G. S., *Linear Integrated Networks—Design*, New York: Van Nostrand Reinhold, 1975.
4. Saal, R., *Handbook of Filter Design*, Berlin: AEG-Telefunken, 1979.
5. Schaumann, R., M. S. Ghausi, and K. R. Laker, *Design of Analog Filters: Passive, Active RC and Switched Capacitor*, Englewood Cliffs, NJ: Prentice-Hall, 1990.
6. Sedra, A. S. and P. O. Brackett, *Filter Theory and Design: Active and Passive*, Portland, OR: Matrix, 1978.
7. Zverev, A. I., *Handbook of Filter Synthesis*, New York: Wiley, 1967.
8. Silveira, B. M., C. Ouslis, and A. S. Sedra, Passive ladder synthesis in filtorX, *IEEE Int. Symp. Circuits Syst.*, 1431–1434, 1993.
9. Schaumann, R. and M. Van Valkenburg, *Design of Analog Filters*, New York: Oxford University Press, 2001.

16

Continuous-Time Integrated Filters

Rolf Schaumann
Portland State University

16.1 Introduction

All modern signal-processing systems include various types of electrical filters that the designer has to realize in an appropriate technology. The literature contains many well-defined filter design techniques [1–3], and computer programs are available, which help the designer find the appropriate transfer function that describes the required filter characteristics mathematically. The reader may also refer to Section I in this book and the other chapters of this section (Section II).

Once the filter's transfer function is obtained, implementation methods must be found that are compatible with the technology selected for the design of the total system. In some situations, considerations of power consumption, frequency range, signal level, or production numbers may dictate *discrete* (*passive* or *active*) filter realizations. Often, however, as much as possible of the total system must be *fully integrated* in microelectronic form, so that the filters can be implemented in the same technology.

Often, digital (Section III) or sampled-data (Chapter 18) implementations are suitable for realizing the filter requirements. However, in modern communications applications, the required frequency range is so high that digital or sampled-data circuitry is inappropriate or too expensive so that *continuous-time* (*c-t*) filters are necessary. In addition, filters in many signal-processing situations must interface with the "real world," where the input and output signals take on continuous values as functions of the continuous variable time, i.e., they are *c-t* signals. In these situations *c-t* antialiasing and reconstruction filters are often required. Because the performance of the total filter system is of relevance and not just the performance of the intrinsic filter, the designer may have to consider if it might not be preferable to implement the entire system in the *c-t* domain rather than as a digital or sampled-data system. At least at low frequencies the latter methods have the advantages of very high accuracy, better signal-to-noise ratio, and little or no parameter drifts, but they entail a number of problems connected with analog-to-digital (A/D) and digital-to-analog (D/A) conversion (see Chapter 10 of *Analog and VLSI Circuits*), sample-and-hold, switching, antialiasing, and reconstruction circuitry.

Traditionally, *c-t* filters were implemented as discrete designs. Well-understood procedures exist for deriving passive *LC* filters (Section I) from a given transfer function with prescribed complex natural frequencies, e.g., [1, Chapter 2], [2], [3, Chapter 13]. To date no practical methods exist for building high-quality, i.e., low-loss, inductors on an integrated circuit (IC) chip.* The required complex natural frequencies must, therefore, be realized by using *gain*, i.e., as we saw earlier in this section, by embedding an operational amplifier (op-amp; see Chapter 16 of *Fundamentals of Circuits and Filters*) in an *RC* feedback network [1,3]. Since op-amps, resistors, and capacitors can be implemented on an integrated circuit, it appears that with active *RC* networks the problem of monolithic filter design is solved in principle: all active devices and any necessary capacitors and resistors can be integrated together on one silicon chip. Although this conclusion is correct, the designer needs to consider four other factors that are important in integrated *c-t* filter design and perhaps are not immediately obvious.

The first item concerns the most important design task for achieving commercially practical designs: integrated filters must be *electronically tunable*, preferably by an *automatic* tuning scheme. Because of its importance, we shall devote a separate section, Section 16.4, to this topic. The second item deals with the economics of practical implementations of active filters: in *discrete* designs, the cost of components and stocking them usually necessitate designing the filter with a minimum number of active devices. One, two, or possibly three op-amps per pole pair are used and the smallest number of different (if possible, all identical) capacitors. In *integrated* realizations, capacitors are determined by processing mask dimensions and the number of different capacitor values is unimportant, as long as the *element spread* is not excessive. Further, active devices frequently occupy less chip area than passive elements so that it is often preferable to use active elements instead of passive ones.[†] Also, the designer should remember that in IC technology it is not *easy* to generate accurate absolute component values, but that *ratios* of like components, such as capacitor ratios, can be realized very precisely. The third observation pertains to the fact that filters usually have to share an integrated circuit with other, possibly switched or digital, systems so that the ac ground lines (power supply and ground wires) are likely to contain switching transients and generally are noisy. Measuring the analog signals relative to ac ground, therefore, may result in designs with poor signal-to-noise ratio and low power-supply rejection. The situation is remedied in practice by building continuous-time filters in fully differential, balanced form, where the signals are referred to each other as $V = V^+ - V^-$ as shown in Figure 16.4b through d. An additional advantage of this arrangement is that the signal range is doubled (for an added 6 dB of signal-to-noise ratio) and that the even-order harmonics in the nonlinear operation of the active devices cancel. All filters in this chapter are understood, therefore, to be designed in fully differential form. Finally, we point out that communication circuitry is often required to operate at hundreds of megahertz or higher, where op-amp–based active *RC* filters will not function because of the op-amps' limited bandwidth.

Today, *c-t* filters integrated in bipolar, CMOS, or BiCMOS technology are no longer academic curiosities but a commercial reality (see Ref. [5] for some recent advances in the field). In the following we discuss the main methods that have proven to be reliable. First, we present *MOSFET-C filters*, whose design methods resemble most closely the standard active *RC* procedures discussed in Chapters 11–14; they can, therefore, be most readily understood by the reader without requiring further background. Next, we introduce the *transconductance-C* (also referred to as *g_m-C*) technique, which is currently the predominant method for

---

* Spiral inductors of a few nanohenry or microhenry in size, however, can be used at gigahertz frequencies. Typically, they are extremely lossy, with quality factors of the order of only 10. Since such low values of *Q* are unacceptable for the design of high-quality selective filters, the inductor losses must be reduced. This is accomplished by placing a simulated negative resistor (Figure 16.17c) in series or parallel with the inductor. Similarly, since the spiral inductor itself cannot be tuned, variable current-shunting circuitry is employed to change the effective inductor value; see Ref. [4].

† However, keeping the number of active devices small remains important because active devices consume power and generate noise.

c-t integrated filters. Designs based on transconductors lead to filters for the higher operating frequencies that are important for modern communication systems.

16.2 MOSFET-*C* Filters

As mentioned in the introduction, the MOSFET-*C* method follows standard op-amp–based active filter techniques [6], which, as we saw in previous chapters, rely heavily on the availability of integrators and summers. The only difference is that the method replaces the resistors used in the conventional active *RC* integrating and summing circuitry by MOSFET devices (Figure 16.1) based in the triode (ohmic) region ($V_C - V_T > V_D$, see Section 1.2 of *Analog and VLSI Circuits*). Defining the source (V_S), drain (V_D), gate (V_C), and substrate (V_B) voltages as shown in Figure 16.1, the resulting nonlinear drain current *I* becomes

$$
\begin{aligned}
I &= \frac{W}{L}\mu C_{ox}\left[(V_C - V_T)(V_D - V_S) + a_2\left(V_D^2 - V_S^2\right) + a_3\left(V_D^3 - V_S^3\right) + \cdots\right] \\
&= \left[\frac{W}{L}\mu C_{ox}(V_C - V_T)\right](V_D - V_S) + b_2\left(V_D^2 - V_S^2\right) + b_3\left(V_D^3 - V_S^3\right) + \cdots
\end{aligned}
\tag{16.1}
$$

where
W and L are the channel width and length, respectively
μ is the effective mobility
C_{ox} is the gate capacitance per unit area
V_T is the threshold voltage of the device

The term in brackets is a tunable conductor,

$$
G(V_C) = \frac{W}{L}\mu C_{ox}(V_C - V_t)
\tag{16.2}
$$

where the gate voltage V_C is used for control or *tuning*. Thus, the drain current of a MOSFET in the triode region is proportional to the drain-to-source voltage, but contains nonlinear second- and higher order terms. Third- and higher order terms can be shown to be small and will be neglected in the following. Now consider placing two of these MOS conductances in parallel as shown in Figure 16.2a. Note that the two devices are driven in balanced form by $+V_1$ and $-V_1$ at one pair of terminals and that the other terminals are at the same voltage *V*. Applying these conditions to Equation 16.1 results in

$$
I_1 = G(V_C)(+V_1 - V) + b_2\left[(+V_1)^2 - V^2\right] \tag{16.3a}
$$

$$
I_2 = G(V_C)(-V_1 - V) + b_2\left[(-V_1)^2 - V^2\right] \tag{16.3b}
$$

Consequently, apart from the neglected high-order odd terms, the difference current $I_1 - I_2$ is perfectly linear in the applied signal voltage V_1:

$$
I_1 - I_2 = 2G(V_C)V_1 \tag{16.4}
$$

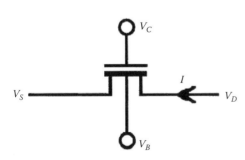

FIGURE 16.1 MOS transistor biased in the triode region.

Thus, in the MOSFET-*C* method, the even-order non-linearities can be shown to be eliminated by carefully balanced circuit design, where all signals are measured

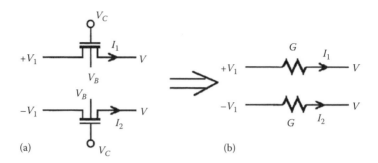

FIGURE 16.2   (a) Balanced linear MOSFET conductance; (b) equivalent resistor circuit.

strictly differentially [6]. Note that the expression for $I_1 - I_2$ in Equation 16.4 is the same as that for the linear resistive equivalent circuit in Figure 16.2b. This means that the MOSFET circuit in Figure 16.2a can be substituted for a pair of resistors in any *appropriate* active RC circuit; appropriate means that the resistor pair is driven by balanced signals at one end and that the voltages at the other two terminals are the same (in practice, usually virtual ground). In addition to these conditions, the substitution is valid if the MOSFETs are operated in the triode region and if $v_1(t)$ and $v(t)$ are safely within the range $|V_C - V_B^*|$. Of course, in practice there remains some small distortion due to mismatches and the neglected odd-order terms. Odd-order nonlinearities are usually small enough to be negligible. Typically, the remaining nonlinearities, arising from odd harmonics, device mismatch, and body effects, are found to be of the order of 0.1% for 1 V signals.

In Section 16.2.1, we discuss the MOSFET-C integrator that is the fundamental building block used in *all* MOSFET-C active filter designs. The integrator will initially be used to construct first- and second-order MOSFET-C sections from which higher order filters can be assembled by the cascade method (Section 16.2.2). In Section 16.2.3, we show how simulated LC ladder filters are designed by use of integrators.

16.2.1 Basic Building Blocks

Substituting MOSFETs for resistors in an active RC prototype works correctly if the active RC circuit is of a form where all resistors come in balanced pairs, with one end of the resistor pair voltage-driven and the two other terminals seeing the same voltage (see Figure 16.2a). Active RC circuits do not normally satisfy these conditions, but many can readily be converted into that form if a balanced symmetrical op-amp as pictured in Figure 16.3 is available. It is important to note that this structure is not simply a differential op-amp, but that the input and output voltages are symmetrical with respect to a ground reference. Using a balanced op-amp, the conversion of many active RC prototypes into balanced form proceeds simply by taking the single-ended circuits and mirroring them at ground as shown below.

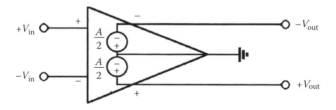

FIGURE 16.3   Balanced operational amplifier configuration realizing $V_{out} = -AV_{in}$.

16.2.1.1 Integrators

Mirroring the active *RC* integrator structure in Figure 16.4a at ground and using the balanced op-amp in Figure 16.3 leads to the balanced integrator in Figure 16.4b. Note that the two resistors are connected in the configuration prescribed in Figure 16.2 with $V = 0$ (virtual ground) so that they may be replaced by the MOSFET equivalent in Figure 16.4c. Analysis of the circuit in the time domain results in

$$-v_{\text{out}} = v^+ - \frac{1}{C}\int_{-\infty}^{t} i_2(t)dt; \quad +v_{\text{out}} = v^- - \frac{1}{C}\int_{-\infty}^{t} i_1(t)dt \tag{16.5a}$$

$$v_{\text{out}} - (-v_{\text{out}}) = v^- - v^+ - \frac{1}{C}\int_{-\infty}^{t} [i_1(t) - i_2(t)]dt \tag{16.5b}$$

$$2v_{\text{out}} = 0 - \frac{1}{C}\int_{-\infty}^{t} 2G(V_C)v_1(t)dt = -\frac{1}{C}\int_{-\infty}^{t} \frac{2v_1(t)}{R(V_C)}dt \tag{16.5c}$$

or, after simplifying and using the Laplace transform, in the frequency domain,

$$v_{\text{out}} = -\frac{1}{CR(V_C)}\int_{-\infty}^{t} v_1(t)dt \rightarrow V_{\text{out}}(s) = -\frac{1}{sCR(V_C)}V_1(s) \tag{16.5d}$$

We see that the MOSFET-C integrator realizes exactly the same transfer function as the active *RC* prototype and that the integration time constant $CR(V_C)$ is tunable by the control voltage V_C that is applied to all gates. Note that the circuit in Figure 16.4c is not a differential integrator. To build a

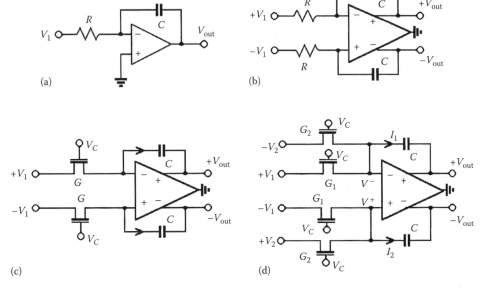

FIGURE 16.4 (a) Active *RC* integrator; (b) fully balanced equivalent; (c) MOSFET-*C* equivalent; and (d) differential MOSFET-*C* integrator.

differential integrator, one must connect a second pair of balanced resistors (MOSFETs) with inputs $-V_2$ and $+V_2$ as in Figure 16.4d to yield

$$v_{out} = -\frac{1}{C} \int_{-\infty}^{t} \left[\frac{v_1(t)}{R_1} - \frac{v_2(t)}{R_2} \right] dt \rightarrow V_{out}(s) = -\frac{1}{sC} \left[\frac{V_1(s)}{R_1} - \frac{V_2(s)}{R_2} \right] \tag{16.6}$$

The same principle can also be used to build *programmable* integrators and, therefore, programmable filters: consider in Figure 16.4d the terminals for V_1 and $-V_2$ connected, i.e., $V_1 = -V_2$. The two resistors are then connected in parallel to give a resistive path $G_1 + G_2$, and the two resistor values can be controlled by different gate voltages V_{C1} and V_{C2}. Similarly, additional *balanced* MOSFET-resistor paths may be connected from V_1 or other signals to the integrator inputs (summing nodes). These paths can be turned on or off by an appropriate choice of gate voltages, so that transfer functions with different parameters (such as gains, quality factors, or pole-frequencies) or even transfer functions of different types (such as bandpass, low-pass, etc.) are obtainable.

If better linearity is required than is obtainable with the two-transistor MOSFET circuit in Figure 16.4c that replaces each resistor in the balanced active *RC* structure by one MOSFET, a four-MOSFET cross-coupled modification for each resistor pair can be used instead [7, paper 2-B.7]. The configuration is illustrated in Figure 16.5. The current difference can be shown to equal

$$\Delta I = I_1 - I_2 = K(V_{C1} - V_{C2})[V_1 - (-V_1)] \tag{16.7}$$

that is, ΔI is proportional to the product of the difference of the input signals and the difference of the applied gate voltages V_{Ci}, $i = 1$, 2. This indicates that one may interchange the input and gate-control voltages and reduce the drive requirements of the previous op-amp stage because no resistive current flows, only the small gate-capacitor current flows with the input applied at the control gates (now at $\pm V_1$). The price paid for this advantage is that the requirements on the control voltage source become more difficult and may necessitate an appropriate common-mode voltage on the signal lines.

Additional resistive or capacitive inputs may, of course, be used to design more general lossy integrators as illustrated in Figure 16.6. By writing the node equation at the summing node, this circuit may be analyzed to realize

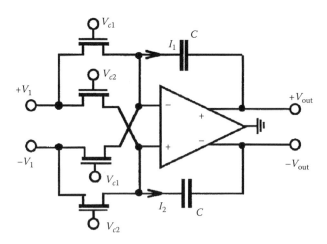

FIGURE 16.5 Integrator using a four-MOSFET configuration for greater linearity.

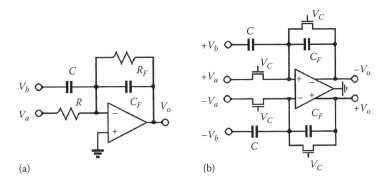

FIGURE 16.6 (a) Lossy integrator with capacitive feed-in branch; (b) MOSFET-C equivalent.

$$V_o = -\frac{1}{sC_F + G_F}(GV_a + sCV_b) \tag{16.8}$$

where the conductors are implemented in Figure 16.6b as

$$R(V_C) = \frac{1}{\mu C_{ox}(W/L)(V_C - V_T)}; \quad R_F(V_C) = \frac{1}{\mu C_{ox}(W_F/L_F)(V_C - V_T)} \tag{16.9}$$

as was derived in Equation 16.2. The controlling gate voltages V_C for the two resistors R and R_F may, in general, be different so that the resistor values may be tuned (or turned on or off) independently.

With lossy and lossless MOSFET-C integrators available, we can obtain not only first- and second-order filters as shown in Section 16.2.1.2, but also simulations of LC ladders in much the same way as was done in Chapter 15 for active RC circuits. We discuss this procedure in some detail in Section 16.2.3.

16.2.1.2 First- and Second-Order Sections

Based on the principle of balancing a single-ended structure, appropriate* standard classical active RC filters from the literature can be converted into balanced form and resistors can be replaced by MOSFETs. As an illustration, consider the single-ended prototype (Tow–Thomas) biquad in Figure 16.7a. As is typical for second-order active RC filter sections (see Chapter 13), the circuit is a two-integrator loop consisting of inverting lossy and lossless integrators (in addition to an inverter to keep the loop gain negative for stability reasons). The realized transfer function is not important to our discussion, but we point out that all resistors are voltage-driven (by the signal source or by an op-amp) and at their other ends all resistors are connected to an op-amp input, i.e., they are at virtual ground. Thus, this circuit satisfies our earlier condition for conversion to a MOSFET-C structure. Figure 16.7b shows the balanced active RC equivalent that is necessary to eliminate the nonlinear performance. Replacing the resistors by MOSFETs biased in the triode region leads to the final MOSFET-C version in Figure 16.7c. The W/L ratios are chosen appropriately to realize the prescribed resistor values

$$G_i = \frac{W_i}{L_i}\mu C_{ox}(V_C - V_T) \tag{16.10}$$

with excellent resistor matching given by aspect ratios

---

* All resistors must be voltage-driven from one side and see the same voltage, usually virtual ground, on the other.

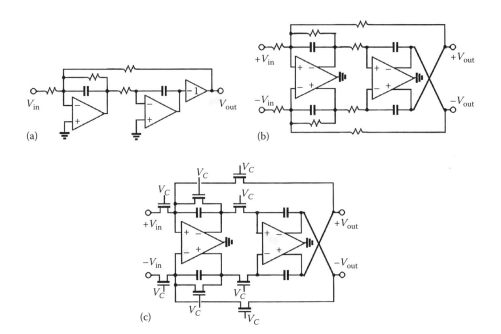

FIGURE 16.7   (a) Active *RC* prototype (Tow–Thomas biquad) for conversion into a MOSFET-C structure; (b) fully balanced version of the biquad with resistors; and (c) fully balanced version of the biquad with MOSFETs.

$$\frac{G_i}{G_k} = \frac{(W/L)_i}{(W/L)_k} \tag{16.11}$$

The voltage-variable resistors given by Equation 16.10 permit loss and time constants to be electronically tuned. Note that the inverter in the original circuit is not needed in Figure 16.7b and c because inversion in the differential topology is obtained by crossing wires.

The circuit may be made *programmable* by connecting additional MOSFET resistors with appropriate W/L ratios in parallel with the fundamental ones shown in Figure 16.7c and then switching them on or off as required by the desired values of the filter coefficients.

The above method indicates how first- or second-order sections can be obtained by choosing any suitable configuration from the active *RC* filter literature [1, Chapter 5], [3, Chapters 4 and 5] and converting it to balanced MOSFET-C form. Next we show a couple of entirely general first- and second-order sections that can be developed from the integrator in Figure 16.6b. The resulting circuits are shown in Figure 16.8 [7, paper 2-A.2]. If we combine the inputs $V_a = V_b = V_i$ of the integrator in Figure 16.6b and add two further cross-coupled feed-in capacitors, we obtain the first-order circuit in Figure 16.8a. Writing the node equation at the (inverting or noninverting)* op-amp input results in the transfer function

$$T_1(s) = \frac{V_o}{V_i} = \frac{s(C_1 - C_2) + G_1}{sC_F + G_2} \tag{16.12}$$

Note that this first-order circuit can realize zeros anywhere on the real axis. Similarly, if we combine two such integrators in a loop, with the individual signal paths and signs selected to assure negative feedback,

---

* Because the circuit is completely symmetrical, it is only necessary to derive the equations for one side, e.g., for V_o; the expressions for the other side, i.e., $-V_o$, are the same.

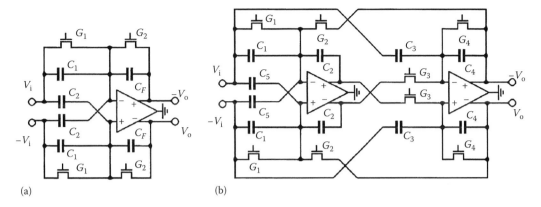

FIGURE 16.8 General (a) first- and (b) second-order MOSFET-C sections. The MOSFETs are labeled by the conductance values they are to implement.

TABLE 16.1 Functions Realizable with Figure 16.8b

Function	Choice of Elements
Bandpass	$G_1 = C_3 = 0;\ C_5 = 0$ if noninverting, $C_1 = 0$ if inverting
Low-pass	$C_1 = C_3 = C_5 = 0$
High-pass	$C_1 = C_5 = G_1 = 0$
Notch	$C_1 = C_5$
All-pass	$(C_1 - C_5)G_3 = -C_2 G_4$

we obtain the general second-order section in Figure 16.8b. Writing again the node equations at the op-amp input nodes leads to the transfer function

$$T_2(s) = \frac{V_o}{V_i} = \frac{s^2 C_2 C_3 + s(C_1 - C_5)G_3 + G_1 G_3}{s^2 C_2 C_4 + s C_2 G_4 + G_2 G_3} \tag{16.13}$$

Observe that this circuit can realize zeros anywhere in the s-plane, depending on the choice of element values. Consequently, the sections in Figure 16.8 can be used to implement arbitrary high-order transfer functions in a cascade topology. Specifically, for the indicated choice of elements the general biquad function in Equation 16.13 realizes the different transfer functions in Table 16.1.

16.2.2 Cascade Realizations

The realization of high-order transfer functions as a connection of low-order sections, including the *cascade*, *multiple-loop feedback*, and *coupled-biquad* approaches, is identical to that discussed for discrete active *RC* filters. The difference lies only in the implementation of the sections in fully integrated form. We do not repeat the process here but only discuss the most prevalent method, cascade design, in terms of an example and encourage the reader to refer to the earlier discussion in Chapter 15 for details. To repeat briefly, if a high-order function

$$H(s) = \frac{V_{out}}{V_{in}} = \frac{N(s)}{D(s)} = \frac{a_m s^m + a_{m-1} s^{m-1} + \cdots + a_1 s + a_0}{s^n + b_{n-1} s^{n-1} + \cdots + b_1 s + b_0} \tag{16.14}$$

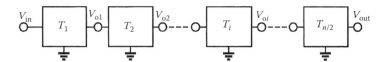

FIGURE 16.9 Cascade realization of an nth-order transfer function.

with $m \leq n$ and $n > 2$ is given, it is factored into second-order sections (and one first-order section if n is odd),

$$H(s) = \prod_{i=1}^{n/2} T_i(s) = \prod_{i=1}^{n/2} k_i \frac{a_{2i}s^2 + a_{1i}s + a_{0i}}{s^2 + s\omega_{0i}/Q_i + \omega_{0i}^2} = \prod_{i=1}^{n/2} k_i t_i(s) \tag{16.15}$$

where each of the second-order functions

$$T_i(s) = \frac{V_{oi}}{V_{oi-1}} = k_i \frac{a_{2i}s^2 + a_{1i}s + a_{0i}}{s^2 + s\omega_{0i}/Q_i + \omega_{0i}^2} = k_i t_i(s) \tag{16.16}$$

is implemented as a suitable biquad with specified pole quality factor Q_i and pole frequency ω_{0i}, such as the one in Figure 16.8b realizing Equation 16.13. As was explained in Chapter 15, k_i is a suitable gain constant, chosen to equalize the signal levels in order to optimize the dynamic range, and $t_i(s)$ is a gain-scaled transfer function. Note that in writing Equation 16.15 we assumed that n is even. Figure 16.9 shows the general (single-ended) structure of the filter.

To provide an example, assume we wish to realize a delay of $\tau_D = 0.187$ μs via a fifth-order Bessel approximation, but with a transmission zero at $f/f_n = 4.67$ to improve the attenuation in the stopband.* The transfer function is found to be

$$H_5(s) = \frac{43.3315(s^2 + 21.809)}{(s + 3.6467)(s^2 + 6.7040s + 14.2729)(s^2 + 4.6494s + 18.1563)} \tag{16.17}$$

The normalizing frequency is $f_n = 1/(2\pi\tau_D) = 850$ kHz [1, Chapter 1], [3, Chapter 10]. Let us choose a cascade implementation with the circuits in Figure 16.8. Factoring Equation 16.17 leads to the first-order and the two second-order functions on the left-hand side of Equation 16.18 to be realized by the functions on the right, which are obtained from Equations 16.12 and 16.13:

$$T_1(s) = \frac{3.6467}{s + 3.6467} \rightarrow T_1(s) = \frac{V_{o1}}{V_{in}} = \frac{G_1}{sC_F + G_2} \tag{16.18a}$$

$$T_2(s) = \frac{14.2729}{s^2 + 6.7040s + 14.2729} \rightarrow T_2(s) = \frac{V_{o2}}{V_{o1}} = \frac{G_1 G_3}{s^2 C_2 C_4 + s C_2 G_4 + G_2 G_3} \tag{16.18b}$$

$$T_3(s) = \frac{0.8325(s^2 + 21.809)}{s^2 + 4.6494s + 18.1563} \rightarrow T_3(s) = \frac{V_{o3}}{V_{o2}} = \frac{s^2 C_2 C_3 + G_1 G_3}{s^2 C_2 C_4 + s C_2 G_4 + G_2 G_3} \tag{16.18c}$$

The components of the first- and second-order filters in Figure 16.8 are to be determined from these equations. The gain constants in the function were chosen to result in unity gain at dc. Comparing coefficients, we find with $\omega_n = 1/\tau_D \approx 2\pi \cdot 850$ krad s$^{-1} \approx 5.341$ Mrad s^{-1}:

---

* Note that transmission zeros on the $j\omega$-axis can be added [3, Chapter 10.8] to a linear-phase network, such as a Bessel filter, without changing the phase because a zero factor, $(\omega_z^2 - \omega^2)$, is a purely real number on the $j\omega$-axis.

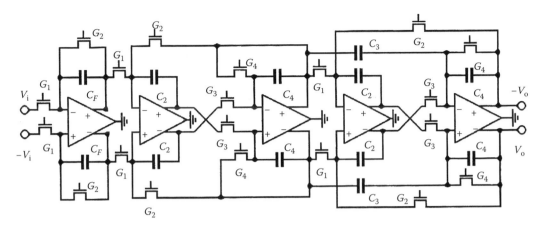

FIGURE 16.10 Circuit to realize the filter described by Equation 16.17. Note that for easy reference we have kept the subscripts on the elements in each section the same as in Figure 16.8.

From Equation 16.18a

$$G_1 = G_2 = 3.6467\omega_n C_F \qquad (16.19a)$$

Choosing $C_F = 2$ pF gives $G_1 = G_2 = 38.952$ μS; also $C_1 = C_2 = 0$.
From Equation 16.18b

$$G_1 G_3 = G_2 G_3 = 14.2729\omega_n^2 C_2 C_4; \quad G_4 = 6.7040\omega_n C_4 \qquad (16.19b)$$

Choosing $C_2 = C_4 = 2$ pF results in $G_1 = G_2 = G_3 = 40.354$ μS and $G_4 = 71.6$ μS; also, $C_1 = C_3 = C_5 = 0$.
From Equation 16.18c, $G_1 G_3 = 0.8325 \times 21.809\omega_n^2 C_2 C_4 = 18.1563\omega_n^2 C_2 C_4$;

$$G_2 G_3 = 18.1563\omega_n^2 C_2 C_4; \quad G_4 = 4.6494\omega_n C_4; \quad C_3 = 0.8325 C_4 \qquad (16.19c)$$

Choosing $C_2 = 2$ pF and $C_4 = 10$ pF yields $C_3 = 8.325$ pF, $G_1 = G_2 = G_3 = 101.77$ μS and $G_4 = 248.32$ μS; also, $C_1 = C_5 = 0$. The remaining task is to convert the resistors into MOSFET devices. Assume the process provides transistors with $\mu C_{ox} = 120$ μA/V^2 and $V_T = 0.9$ V. For the choice of $V_C = 2$ V, the aspect ratios are then calculated from the above conductance values and from Equation 16.10 via

$$\frac{W_i}{L_i} = \frac{G_i(V_C)}{\mu C_{ox}(V_C - V_T)} = \frac{G_i(V_C)}{120 \ \mu A/V^2 \times 1.1V} = \frac{G_i/\mu S}{132} \qquad (16.20)$$

For instance, in the first-order section we find $W_1/L_1 = W_2/L_2 = 38.952/132 = 1/3.389$. The resulting circuit is shown in Figure 16.10.

16.2.3 Ladder Simulations

Using MOSFET-C integrators, the ladder simulation method for the MOSFET-C approach is entirely analogous to the active RC procedures discussed earlier in this book (see Chapter 15). The process is illustrated by a step-by-step generic example, which should guide the reader when implementing a specific design.

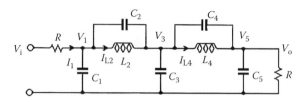

FIGURE 16.11 Fifth-order elliptic *LC* low-pass filter.

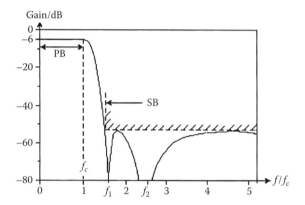

FIGURE 16.12 Transfer function magnitude of a fifth-order elliptic low-pass function.

Assume a fifth-order elliptic low-pass filter with the transfer function

$$H_{\text{ell}}(s) = \frac{\left(s^2 + \omega_1^2\right)\left(s^2 + \omega_2^2\right)}{(s + a)(s^2 + bs + c)(s^2 + ds + e)} \qquad (16.21)$$

is prescribed, which with the help of readily available ladder synthesis software is realized by the *LC* ladder structure in Figure 16.11 with known component values. A plot of such a function is shown in Figure 16.12. The two transmission zeros f_1 and f_2 in the figure are obtained when L_2, C_2, and L_4, C_4, respectively, resonate. The *LC* active simulation proceeds by deriving the signal-flow graph equations [1, Chapter 6], [3, Chapter 15] or by writing the loop and node equations of the ladder along with the *V–I* relationships describing the functions of the elements:

$$I_1 = \frac{V_i - V_1}{R}, \quad I_{L2} = \frac{V_1 - V_3}{sL_2}, \quad I_{L4} = \frac{V_3 - V_5}{sL_4} \qquad (16.22a)$$

$$V_1 = \frac{I_1 - [I_{L2} + sC_2(V_1 - V_3)]}{sC_1} \qquad (16.22b)$$

$$V_3 = \frac{I_{L2} - I_{L4} + sC_2(V_1 - V_3) - sC_4(V_3 - V_5)}{sC_1} \qquad (16.22c)$$

$$V_5 = V_o = \frac{I_{L4} + sC_4(V_3 - V_5)}{sC_5 + G} \qquad (16.22d)$$

We recognize that these equations represent *integrations* of voltages into currents and currents into voltages. We also note that the currents through the capacitors C_2 and C_4 can be taken care of efficiently without resorting to integration: by connecting C_2 and C_4 directly to the voltage nodes V_1, V_3, and V_3, V_5, respectively, they conduct the currents as prescribed in Equation 16.22b through d. Next we reformat Equation 16.22a through d by eliminating I_1 from Equation 16.22a and b and rewriting Equation 16.22b through d such that V_1, V_3, and V_5, respectively, appear only on the left-hand side. The result is the new set of equations

$$I_{L2} = \frac{V_1 - V_3}{sL_2}; \quad I_{L4} = \frac{V_3 - V_5}{sL_4} \tag{16.23a}$$

$$V_1 = \frac{V_iG - I_{L2} + sC_2V_3}{s(C_1 + C_2) + G} \tag{16.23b}$$

$$V_3 = \frac{I_{L2} - I_{L4} + sC_2V_1 + sC_4V_5}{s(C_2 + C_3 + C_4)} \tag{16.23c}$$

$$V_5 = V_o = \frac{I_{L4} + sC_4V_3}{s(C_4 + C_5) + G} \tag{16.23d}$$

Recall that all signals in the MOSFET-C circuit are voltages, which in turn produce currents summed at the op-amp inputs, and that the integration constant must be *time* rather than *capacitance*. This is achieved by scaling the equations by a resistor R. We illustrate the process on Equation 16.23a and b:

$$I_{L4} = \frac{V_3 - V_5}{sL_4} \rightarrow \frac{I_{L4}}{G} = \frac{GV_3 - GV_5}{s(L_4G)G} \tag{16.24a}$$

$$V_1 = \frac{V_iG - I_{L2} + sC_2V_3}{s(C_1 + C_2) + G} \rightarrow V_1 = \frac{V_iG - (I_{L2}/G)G + sC_2V_3}{s(C_1 + C_2) + G} \tag{16.24b}$$

Notice that L_4G^2 has the unit of *farad*, i.e., it is a capacitor that, however, in Figures 16.13 and 16.14 will be labeled L_4 (and L_2, respectively) to help keep track of its origin. Similarly, I_{L2}/G and I_{L4}/G will be labeled V_{I2} and V_{I4}, respectively. The integrators can now be realized as in Figure 16.6b and then interconnected as the equations prescribe. Figure 16.13 illustrates the process for Equation 16.24. The MOSFET-C implementation of all appropriately scaled equations (Equation 16.23) leads to the

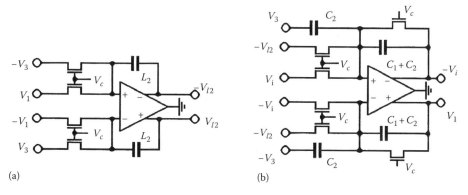

(a)   (b)

FIGURE 16.13 MOSFET-C implementation of (a) Equation 16.24a and (b) Equation 16.24b. All MOSFETs realize the value G.

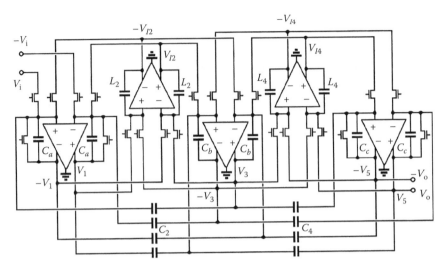

FIGURE 16.14 Operational simulation of the *LC* ladder of Figure 16.11 with MOSFET-*C* integrators; $C_a = C_1 + C_2$, $C_b = C_2 + C_3 + C_4$, $C_c = C_4 + C_5$.

fifth-order elliptic filter shown in Figure 16.14. As given in Equation 16.10, the aspect ratio of each MOSFET and the control voltage V_C are adjusted to realize the corresponding resistor values of the standard active *RC* implementation and all MOSFET gates are controlled by the same V_C for tuning purposes. Arrays of MOSFETs controlled by different values of V_C can be used to achieve programmable filter coefficients.

An often cited advantage of the MOSFET-*C* technique is the reduced sensitivity to parasitic capacitors, whereas the g_m-*C* approach discussed next must carefully account for parasitics by predistortion. Note from Figures 16.6b, 16.8, 16.10, and 16.14 that all capacitors and the MOSFET resistors are connected to voltage-driven nodes or to virtual ground so that parasitic capacitors to ground are of no consequence as long as amplifiers with sufficiently high gain and wide bandwidth are used. Fortunately, such amplifiers are being developed [7, paper 2-B.5] so that MOSFET-*C* circuits promise to become increasingly attractive in the future.

16.3 g_m-*C* Filters

At the time of this writing, the dominant active device used in the design of integrated continuous-time filters is the *transconductor* (g_m) or the *operational transconductance amplifier* (OTA) [8, Chapter 5], [3, Chapter 16]. Both names, g_m-*C* filters and OTA-*C* filters, are used in the literature; we will use the term g_m-*C* filter in this text. The main reasons for this prevalence appear to be the simple systematic design methods for g_m-*C* filters and, especially, the higher range of frequencies over which g_m-based filters can operate. Also, OTAs often have simpler circuitry (fewer elements) than op-amps. A transconductor is a voltage-to-current converter described by

$$I_{\text{out}} = g_m(s)V_{\text{in}} \tag{16.25}$$

where $g_m(s)$ is the frequency-dependent transconductance parameter with units of ampere/volt or siemens, abbreviated S. Typical values for g_m are tens to hundreds of microsiemens in CMOS and up to millisiemens in bipolar technology. A simplified small-signal equivalent circuit is shown in Figure 16.15. The dashed components in Figure 16.15 are parasitics that in an ideal OTA should be

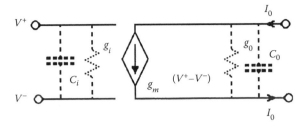

FIGURE 16.15   Small-signal equivalent circuit.

zero but that in practice must be accounted for. For common designs in CMOS technology, the input conductance g_i is zero; the input and output capacitances, c_i and c_o, are typically of the order of 0.05 pF or less and the output resistance $r_o = 1/g_o$ is in the range of 50 kΩ to 1 MΩ. The bandwidth of well-designed transconductors is so large that g_m in many cases can be regarded as constant, $g_m(s) = g_{m0}$, but for critical applications in high-frequency designs the transconductance pole and the resulting phase errors must be considered. A good model for these situations is

$$g_m(s) \approx g_{m0}e^{-s\tau} \approx \frac{g_{m0}}{1 + s\tau} \approx g_{m0}(1 - s\tau) \qquad (16.26)$$

where $f = 1/(2\pi\tau)$ is the pole location (typically at several 100 MHz to 10 GHz) and the phase error $\Delta\phi = -\omega\tau$ is considered small, i.e., $\omega\tau \ll 1$. The three different approximations in Equation 16.26 for representing the frequency dependence are equivalent; the pole is used most often, the zero frequently results in simpler equations and algebra, and the phase may give better insight into the behavior of feedback loops.

The most commonly used circuit symbols are shown in Figure 16.16. Note that OTA designs with multiple differential inputs as in Figure 16.16c are readily available. They often lead to simpler filter designs with less silicon area and power consumption because only the OTA input stages must be duplicated, whereas the remaining parts of the OTA, such as output and common-mode feedback circuitry, can be shared. Essentially, if two OTAs with the same g_m value in a filter have a common output node (a frequent situation), the two OTAs can be merged into the circuit of Figure 16.16c, thus saving circuitry and power.

Filter design methods discussed in this section use only OTAs and capacitors: OTAs to provide gain and capacitors to provide integration. To establish time constants, resistors may also be required, but their function can be obtained from OTAs: "resistors" of value $1/g_m$ can be simulated by connecting the OTA output to its input with the polarities shown in Figure 16.17a and b. Inverting one pair of

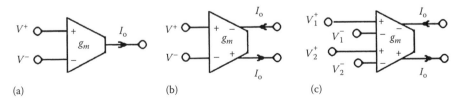

FIGURE 16.16   OTA symbols (a) differential input–single-ended output: $I_o = g_m(V^+ - V^-)$; (b) fully differential; and (c) with multiple inputs: $I_o = g_m[(V_1^+ - V_1^-) + (V_2^+ - V_2^-)]$.

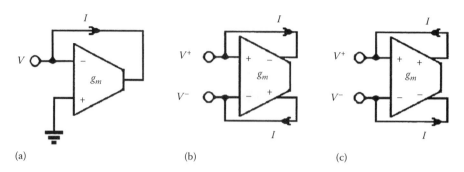

(a)   (b)   (c)

FIGURE 16.17   Simulated resistors: (a) positive single-ended of value $V/I = 1/g_m$; (b) positive differential resistor of value $(V^+ - V^-)/I = 1/g_m$; and (c) negative differential resistor $-1/g_m$.

terminals results in a negative resistor as in Figure 16.17c.* Since transconductors and capacitors can be used to build all components necessary for designing the filters, they are called transconductance-C or g_m-C filters. We discuss the remaining "composite" building blocks, integrators and gyrators[†] subsequently in Section 16.3.1.

In this section we will not go into the electronic circuit design methods for OTAs, but refer the reader to the literature, which contains a great number of useful transconductance designs in all current technologies. References [5,7] contain numerous papers that discuss practical transconductance circuits. The most popular designs currently use CMOS, but bipolar and BiCMOS are also widely employed, and GaAs has been proposed for applications at the highest frequencies or under unusually severe environmental conditions. Since transconductors are almost always used in open loop without local feedback, their input stages must handle the full amplitude of the signal to be processed. Typically, the OTA input stage is a differential pair with quite limited signal swing before nonlinearities become unacceptable. Thus, much design expertise has gone into developing linearization schemes for transconductance circuits. They have resulted in designs that can handle signals of the order of volts with nonlinearities of a fraction of 1%. Apart from simple source-degeneration techniques, the most commonly employed approaches use variations of the principle of taking the difference between the drain currents of two MOS devices in the saturation region but driven differentially, so that the difference current is linear in V_{gs}:

$$I_d^+ = k(V_{gs} - V_T)^2 = k\left(V_{gs}^2 + V_T^2 - 2V_{gs}V_T\right)$$
$$I_d^- = k(-V_{gs} - V_T)^2 = k\left(V_{gs}^2 + V_T^2 + 2V_{gs}V_T\right) \qquad (16.27)$$
$$\Delta I_d = I_d^+ - I_d^- = -4V_{gs}V_T$$

Another approach reasons that the most linear (trans)conductance behavior should be obtainable from the current through a resistor. Thus, operating an MOS device in the resistive (triode) region,

$$I_d = k\left[(V_{gs} - V_T)V_{ds} - 0.5V_{ds}^2\right]$$

and taking the derivative with respect to V_{gs} for constant $V_{ds} = V_{DS}$ results in a perfectly linear transconductance,

---

* Such "negative resistors" are often used to cancel losses, for example to increase the dc gain of transconductors or the quality factors of filter stages. Specifically, a negative resistor can be employed to increase the quality factor of the inductor in Figures 16.21 and 16.22.

† A gyrator is a two-port circuit whose input impedance is inversely proportional to the load impedance: $Z_{in}(s) = r^2/Z_{load}(s)$. If $Z_{load} = 1/(sC)$, the input is inductive, $Z_{in}(s) = sr^2C = sL$. r is called the gyration resistance (see the discussion to follow).

$$g_m = \frac{dI_d}{dV_{gs}} = kV_{DS} \qquad (16.28)$$

that furthermore can be adjusted (tuned) by varying a dc bias voltage (V_{DS}) as long as V_{DS} stays small enough for the transistor to remain in the triode region. The circuitry surrounding the triode-region MOS device must assure that V_{DS} remains constant and independent of the signal.

As mentioned, the literature contains numerous practical CMOS, bipolar, or biCMOS transconductance designs that require low power supply voltages (± 1.5 V, or 0 to 3 V, or even less for low-power applications), and have acceptable signal swing (of the order of volts), with low nonlinearities (as low as a small fraction of 1%) and wide bandwidth (up to several hundred megahertz and even into the gigahertz range). Two further aspects of OTA design should be stressed at this point. First, since g_m-C filters often contain many transconductors, the designer ought to strive for simple OTA circuitry. It saves silicon real estate and at the same time often results in better frequency performance because of reduced parasitics at internal nodes. We point out though that there exists a trade-off between simple circuitry and large voltage swing: a wide linear signal range always requires special linearizing circuit techniques and, therefore, additional components. The second issue pertains to tuning. We mentioned earlier that continuous-time filters always require tuning steps to eliminate the effects of fabrication tolerances and component drifts. In IC technologies this implies that the circuit components must be electronically adjustable. Since (MOS) capacitors are generally fixed,* all tuning must be handled via the transconductance cells by changing one or more bias points. Usually two adjustments are needed: the magnitude of g_m must be varied to permit tuning the frequency parameters set by g_m/C-ratios and, as explained later, the phase $\Delta\phi$ must be varied to permit tuning the filters' quality factors. We discuss tuning methods in some detail in Section 16.4.

In the Section 16.3.1 we introduce the central building blocks from which g_m-C filters are constructed, the integrator and the gyrator. Just as we saw in the discussion of MOSFET-C circuits and in the earlier treatment of active RC filters, integrators are fundamental to the development of active filter structures, both for second-order sections and cascade designs, as well as for higher order LC ladder simulations. Gyrators along with capacitors are used to replace inductors in passive RLC filters so that many passive filter structures, such as LC ladders, can be directly translated into g_m-C form.

16.3.1 Basic Building Blocks

16.3.1.1 Integrators

Integrators are obtained readily by loading an OTA with a floating or a grounded capacitor as shown in Figure 16.18. Observe that the simpler technology of grounded capacitors requires four times the capacitor value and silicon area. Ideally, the integrator realizes the transfer function

$$\frac{V_o}{V_i} = -\frac{g_m}{sC} \qquad (16.29a)$$

but notice that the function is sensitive to unavoidable parasitic capacitors as well as to the OTA output conductance g_o. Observe from Figure 16.19 that the output conductance is in parallel with the integrating capacitor C, and that the output capacitances C_o from the positive and negative output nodes of the OTA circuitry to ground add to the value of C. Furthermore, the designer should bear in mind that in IC technology floating capacitors have a substantial parasitic capacitance C_s (about 10% of the value of C) from the bottom plate to the substrate, i.e., to ac ground. To maintain symmetry, the designer may wish

---

* To implement small values of variable capacitor, MOS transistors connected as varactors can be used. The capacitor can be varied, e.g., via substrate bias.

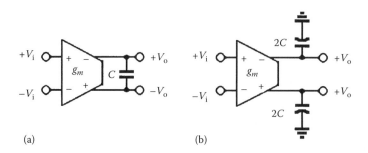

FIGURE 16.18   Integrator. The integrator capacitor may be floating (a) or grounded (b). Note that the grounded-capacitor realization requires four times the area.

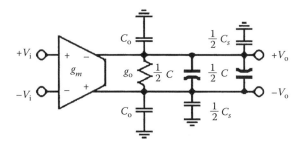

FIGURE 16.19   Parasitic capacitors associated with a g_m-C integrator.

to split the integrating capacitor into two halves connected such that the parasitic bottom plate capacitors $0.5C_s$ appear at the two OTA outputs. The situation is illustrated in Figure 16.19. Taking the parasitics into consideration, evidently, the integrator realizes

$$\frac{V_o}{V_i} = -\frac{g_m}{sC_{int} + g_o}\bigg|_{s=j\omega} = -\frac{g_m}{j\omega C_{int}(1 - j/Q_{int})} \tag{16.29b}$$

that is, it becomes lossy with a finite integrator quality factor

$$Q_{int} = \frac{\omega C_{int}}{g_o} \tag{16.30a}$$

and an effective integrating capacitor equal to

$$C_{int} = C + \frac{1}{2}\left(\frac{C_s}{2} + C_o\right) \tag{16.30b}$$

To maintain the correct integration constant as nominally designed, the circuit capacitor C should be predistorted to reflect the parasitics appearing at the integration nodes. The parasitics should be estimated as best as possible, for example, from a layout process file, and their values subtracted from the nominal value of C in the final layout. If grounded capacitors are used as in Figure 16.18b, the bottom plate should, of course, be connected to ground so that the substrate capacitances are connected between ground and the power supply. Thus, they are shorted out for the signals and play no role. Observe that the presence of parasitic capacitors tends to limit the high-frequency performance of these

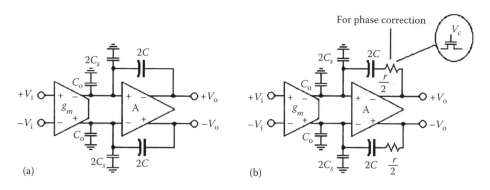

FIGURE 16.20   (a) g_m-C–op-amp integrator; (b) nominal correction for phase errors.

filters because high-frequency filters require large time constants, g_m/C, i.e., small capacitors.* The smallest capacitor C, however, must obviously be larger than the sum of all parasitics connected at the integrator output nodes to be able to absorb these parasitics.[†] Because the values of the parasitic capacitors can generally only be estimated, one typically chooses C to be three to five times larger than the expected parasitics to maintain some predictability in the design. The reader will notice that integrators with grounded capacitors, Figure 16.18b, have a small advantage in high-frequency circuits where parasitic capacitors become large relative to C. Because of the problem with parasitic capacitors, an alternative "g_m-C–op-amp" integrator is also used. It employs an op-amp to minimize their effects for the price of a second active device, increased noise, silicon area, and power consumption. Figure 16.20a shows the configuration. Notice that now the parasitic capacitors play no role because they are connected between ground and virtual ground (the op-amp inputs) so that they are never charged. A more careful analysis shows that the integrator realizes

$$\frac{V_o}{V_i} = \frac{g_m}{sC} \frac{1}{1+\frac{1}{A(s)}\left(1+\frac{C_p}{C}\right)} \approx \frac{g_m}{sC} \frac{1}{1+\frac{s}{\omega_t}\left(1+\frac{C_p}{C}\right)} \tag{16.31}$$

where $C_p = 0.5(C_o + 2C_s)$ represents the total parasitic capacitance at each of the op-amp input terminals. Evidently, the integrator has acquired a parasitic pole at

$$s = -\omega_t \frac{C}{C+C_p} \tag{16.32}$$

where we have modeled the amplifier gain as $A(s) \approx \omega_t/s$. The high-frequency performance is now limited by the op-amp behavior. It has been shown, though, that a low-gain wideband amplifier, essentially a second OTA with dc gain g_m/g_o, can be used for this application. Nevertheless, the second active device introduces parasitic poles (and zeros) whose effects must be carefully evaluated in practice. The dominant pole introduced by the op-amp may be canceled *nominally* by an rC phase lead as shown in Figure 16.20b. The circuit realizes

---

* Increasing g_m is generally not a satisfactory solution because it also increases the parasitics.
[†] Obviously, the highest operating frequencies are obtained for $C=0$, that is, when a parasitic capacitor C_p is used as the integrating capacitor C. In that case, operating frequencies in the gigahertz range can be realized: $g_m/C \approx 2\,\pi \times 1$ GHz for $g_m = 100\ \mu S$ and $C_p \approx 15$ fF. Naturally, because of the uncertainty in the design values, an automatic tuning scheme is unavoidable.

$$\frac{V_o}{V_i} = \frac{g_m}{sC} \frac{1 + sCr}{1 - \frac{\omega}{\omega_t}\omega C_p r + \frac{s}{\omega_t}\left(1 + \frac{C_p}{C}\right)} \approx \frac{g_m}{sC} \frac{1 + sCr}{1 + \frac{s}{\omega_t}\left(1 + \frac{C_p}{C}\right)} \tag{16.33}$$

so that r should be chosen as

$$r = \frac{1}{\omega_t C}\left(1 + \frac{C_p}{C}\right) \tag{16.34}$$

to cancel the pole. The small resistor r may be a MOSFET in the triode region as indicated in Figure 16.20b so that $r(V_C)$ becomes variable for any necessary fine adjustments. Notice that the cancellation is only *nominal* because Equation 16.34 can never be satisfied exactly because of the uncertain values of ω_t and C_p.

16.3.1.2 Gyrators

A gyrator is defined by the equations

$$I_i = \frac{1}{r} V_o; \quad I_o = -\frac{1}{r} V_i \tag{16.35a}$$

where r is the so-called gyration resistance and the currents are defined as positive when flowing into the gyrator. Thus, the input impedance $Z_{in}(s)$ is inversely proportional to the load impedance $Z_{load}(s)$:

$$Z_{in}(s) = \frac{V_i}{I_i} = r^2 \frac{-I_o}{V_o} = r^2 \frac{1}{Z_{load}(s)} \tag{16.35b}$$

If a gyrator is loaded by a capacitor, $Z_{load}(s) = 1/(sC)$, the input impedance is proportional to frequency, i.e., it behaves like an inductor of value r^2C:

$$Z_{in}(s) = sr^2C = sL \tag{16.36}$$

Equation 16.35a indicates that a gyrator can be interpreted as a connection of an inverting and a noninverting transconductor of value $g_m = 1/r$ as shown in Figure 16.21a. This fact makes it very easy to build excellent gyrators with OTAs (see Figure 16.21b and c), whereas it has been found quite difficult to obtain good gyrators with op-amps. An exception is Antoniou's general impedance converter (GIC) [1, Chapter 5], [3, Chapter 14.1]. However, it is useful only at relatively moderate frequencies (up to about 5%–10% of the op-amp's gain-bandwidth product);[*] also, the circuit contains resistors that are not voltage-driven and, therefore, cannot readily be translated into a MOSFET-C equivalent as was discussed previously. The availability of good gyrators provides us with a convenient method for building high-frequency integrated g_m-C ladder filters, which is based on inductor-replacement to be discussed in Section 16.3.3.1.

Notice that the comments made earlier about the effects of parasitic capacitors also apply to inductor simulation: the parasitic input and output capacitors of the OTAs and the bottom-plate-to-substrate capacitances (Figure 16.19) add to the capacitor used to set the value of the simulated inductor. For instance, using the same notation as in Figure 16.19, the effective capacitor in Figure 16.21b equals

---

[*] As stated, Antoniou's GIC is really an *impedance converter*, not a gyrator (see Equation 15.63 and Figure 15.17). The GIC converts a load resistor into an inductive input impedance, rather than inverting a capacitor as described in Equation 16.35b.

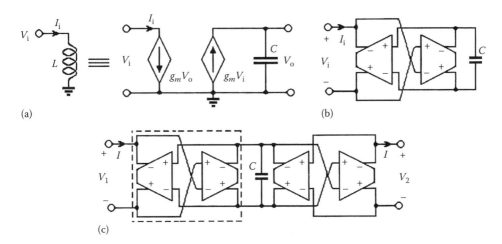

FIGURE 16.21 (a) Controlled-source realization of capacitively loaded gyrator to realize a grounded inductor; (b) differential g_m-C implementation; and (c) a floating inductor requires two gyrators.

$$C_{eff} = C + \frac{1}{2}\left(\frac{C_s}{2} + C_i + C_o\right) \tag{16.37}$$

where C_i is the parasitic capacitance at the input terminals of the OTA, see Figure 16.15, and C is assumed to be split as in Figure 16.19. Note also that a parasitic capacitor of value $C_{in} = C_i + C_o$ is measured across the inductor of Figure 16.21b so that the inductor

$$L_{eff} = \frac{C_{eff}}{g_m^2} \tag{16.38}$$

has a self-resonance frequency*

$$\omega_0 = \frac{1}{\sqrt{L_{eff}C_{in}}} \tag{16.39}$$

Finally, to complete the inductor model, recall that the OTAs have finite differential output conductances,[†] g_o (see Figure 16.15), which appear across the input and load terminals of the gyrator in Figure 16.21b. Consequently, the full inductive admittance realized by Figure 16.21b equals

$$Y_L(s) = g_o + sC_{in} + \frac{g_m^2}{sC_{eff} + g_o} \tag{16.40}$$

to yield the equivalent circuit in Figure 16.22 for the inductor L. The designer should keep this circuit in mind when using this method to develop a filter.

Evidently, the realized quality factor of the inductor equals $Q_L = \omega C_{eff}/g_o$. This means OTAs with large output resistance $r_o = 1/g_o$ are needed for high-quality inductor simulations.

---

* With modern submicron CMOS processes, the self-resonance frequency of electronic inductors can be as high as a few gigahertz with quality factors in the range of a few 100. See Ref. [9].
† As was mentioned, the input conductances can normally be neglected.

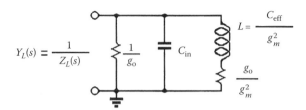

FIGURE 16.22 Passive equivalent circuit for the grounded inductor simulated in Figure 16.21b.

Next we discuss first- and second-order g_m-C sections used as building blocks for the cascade approach to high-order filter design. As was the case for the MOSFET-C method, we will see that g_m-C sections are constructed by interconnecting integrators.

16.3.1.3 First- and Second-Order Sections

Consider the integrator in Figure 16.18a; we make it lossy by loading it with the resistor in Figure 16.17b. Let the input signals also be fed through capacitors into the integrator output nodes as shown in Figure 16.23. The circuit is readily analyzed by writing Kirchhoff's current law at the output node to yield

$$\frac{V_o}{V_i} = -\frac{sC_1 + g_{m1}}{s(C_1 + C) + g_{m2}} \tag{16.41}$$

The circuit may realize any desired first-order function by choosing the appropriate values for the transconductances and the capacitors. For example, a low-pass can be obtained by setting $C_1 = 0$; a high-pass results from $g_{m1} = 0$ and possibly $C = 0$; and $C = 0$ and $g_{m1} = -g_{m2}$ results in an all-pass function.

A second-order block can be designed by interconnecting two integrators in a variety of feedback configurations. To keep the method more transparent, we show one such possibility in Figure 16.24 with single-ended outputs. A differential structure is obtained by mirroring the circuit at ground and duplicating the appropriate components as was demonstrated in connection with Figure 16.4. Let us disregard for now the dashed OTA with inputs V_a and V_b and apply an input signal V_i; writing the node equations for Figure 16.24 we obtain

$$g_{m3}(V_i - V_2) + g_{m4}(V_1 - V_3) = 0; \quad sC_1 V_1 = -g_{m1} V_3; \quad sC_2 V_2 = -g_{m2} V_1 \tag{16.42}$$

By eliminating two of the three voltages V_1, V_2, or V_3 from these equations, we obtain the bandpass, low-pass, and high-pass functions, respectively,

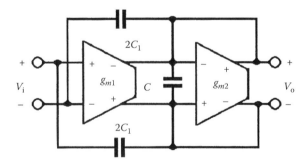

FIGURE 16.23 First-order g_m-C section.

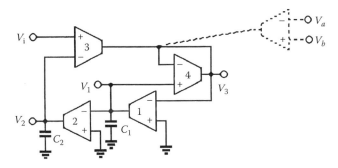

FIGURE 16.24   General biquadratic g_m-C section.

$$\frac{V_1}{V_i} = H_{BP}(s) = \frac{-sC_2 g_{m1} g_{m3}}{s^2 C_1 C_2 g_{m4} + sC_2 g_{m1} g_{m4} + g_{m1} g_{m2} g_{m3}} \tag{16.43a}$$

$$\frac{V_2}{V_i} = H_{LP}(s) = \frac{g_{m1} g_{m2} g_{m3}}{s^2 C_1 C_2 g_{m4} + sC_2 g_{m1} g_{m4} + g_{m1} g_{m2} g_{m3}} \tag{16.43b}$$

$$\frac{V_3}{V_i} = H_{HP}(s) = \frac{s^2 C_1 C_2 g_{m3}}{s^2 C_1 C_2 g_{m4} + sC_2 g_{m1} g_{m4} + g_{m1} g_{m2} g_{m3}} \tag{16.43c}$$

In *any* electrical network, one can generate different numerator polynomials, i.e., different transmission zeros, without disturbing the poles, i.e., the system polynomial, by applying an input *voltage* to any node that is lifted off *ground*, or by sending an input *current* into any *floating* node. The second of these possibilities is illustrated in Figure 16.24 in dashed form where a current $g_m(V_b - V_a)$ is sent into Node 3. We leave the analysis to the reader. We demonstrate the first possibility by applying V_i to the noninverting terminals of OTA$_1$ and OTA$_2$, which are grounded in Figure 16.24. The reader may show by routine analysis that lifting these two nodes off ground and then applying V_i to both of them (in addition to the original input) results in the complete biquadratic transfer function

$$\frac{V_3}{V_i} = \frac{s^2 C_1 C_2 g_{m3} + sg_{m1}(C_2 g_{m4} - C_1 g_{m3}) + g_{m1} g_{m2} g_{m3}}{s^2 C_1 C_2 g_{m4} + sC_2 g_{m1} g_{m4} + g_{m1} g_{m2} g_{m3}} \tag{16.44}$$

with which, for example, a notch filter may be realized by setting $C_2 g_{m4} = C_1 g_{m3}$.

A great variety of second-order sections can be found in the literature, e.g., [3, Chapter 16.3]. Many are designed for specific transfer functions rather than the general circuit in Figure 16.24, and are often simpler and contain fewer OTAs. Readers are well advised to scan the available literature for the best circuit for their applications.

16.3.2 Cascade Realizations

Apart from modularity and simple design methods, the main advantage of the cascade approach is its generality: a cascade structure can realize a transfer function with arbitrary zero locations, whereas simulations of lossless *LC ladders* discussed below are restricted to $j\omega$-axis transmission zeros. Implementing a prescribed transfer function by the cascade method with g_m-C integrated filters follows the same principles that were discussed in Chapter 15 for discrete active *RC* filters and leads to the filter structure in Figure 16.9. The difference lies only in the final realization of the sections in monolithic form. We demonstrate the principle with the example of a high-frequency filter for the read/write channel of a

TABLE 16.2   Filter Parameters for Equation 16.45

	Pole Frequency (Normalized to 10^7 s^{-1})	Pole Quality Factor (Normalized to 10^7 s^{-1})	Zero
Biquad 1	$\omega_1 = 1.14762$	$Q_1 = 0.68110$	$\sigma_z = 0.95$
Biquad 2	$\omega_2 = 1.71796$	$Q_2 = 1.11409$	
Biquad 3	$\omega_3 = 2.31740$	$Q_3 = 2.02290$	
Section 4	$\sigma = 0.86133$	—	

magnetic disk storage system,[*] where the most critical specification is constant delay, i.e., linear phase. To this end, let us discuss the design of a seventh-order cascade low-pass with a constant delay approximated in the Chebyshev (equiripple) sense. The specifications call for a delay variation of maximally 1 ns over the passband and a bandwidth $f_c = 10$ MHz. The transfer function to be implemented is

$$H_7(s) = \frac{K_0\left(s^2 - \sigma_z^2\right)}{(s + \sigma)(s^2 + s\omega_1/Q_1 + \omega_1^2)(s^2 + s\omega_2/Q_2 + \omega_2^2)(s^2 + s\omega_3/Q_3 + \omega_3^2)} \tag{16.45}$$

with the required parameters given in Table 16.2. The purpose of the symmetrical pair of zeros at $\pm\sigma_z$ is magnitude equalization to effect a gain boost for pulse slimming. Note that these zeros do not affect the phase or the delay because the factor $(\omega^2 + \sigma_z^2)$ is real on the $j\omega$-axis. The fully differential second-order circuit chosen for the low-pass sections is shown in Figure 16.25a. Simple analysis shows that it realizes the function

$$H_{LP}(s) = \frac{V_{LP}}{V_i} = \frac{g_{m1}g_{m2}}{s^2C_1C_2 + sC_1g_{m2} + g_{m1}g_{m2}} = \frac{\omega_0^2}{s^2 + s\omega_0/Q_0 + \omega_0^2} \tag{16.46}$$

where

$$\omega_0 = \sqrt{\frac{g_{m1}g_{m2}}{C_1C_2}} \tag{16.47a}$$

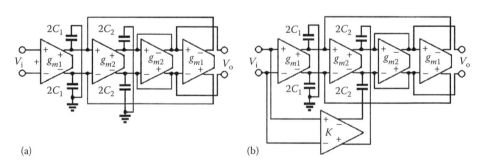

(a)   (b)

FIGURE 16.25   (a) Second-order low-pass g_m-C section; (b) an equalizer section.

---

[*] A similar commercially successful design in bipolar technology is discussed in Ref. [12].

and

$$Q_0 = \sqrt{\frac{g_{m1} C_2}{g_{m2} C_1}} \tag{16.47b}$$

are the pole frequency and pole quality factor, respectively. There remains the question of how to obtain the two real transmission zeros at $\pm\sigma_z$. For this purpose we recall that the zeros of a transfer function can be changed without destroying the poles by feeding the input signal or a fraction thereof into any of the ground nodes lifted off ground. For the situation at hand this can be accomplished by lifting the capacitors $2C_2$ in the low-pass off ground and feeding KV_i with the appropriate polarity into the terminals so generated. Figure 16.25b shows the resulting circuit, which may be analyzed to yield the transfer function

$$\frac{V_o}{V_i} = \frac{-Ks^2 C_1 C_2 + g_{m1} g_{m2}}{s^2 C_1 C_2 + s C_1 g_{m2} + g_{m1} g_{m2}} \tag{16.48}$$

with the zero at $\sigma_z = \pm\omega_0 K^{-1/2}$. Observe that the two transconductors labeled g_{m1} and g_{m2}, respectively, have common output terminals; they can, therefore, be merged into one double-input transconductor each as discussed in connection with Figure 16.16c. Further, we need the circuit in Figure 16.23 with $C_1 = 0$ to realize a first-order section, a lossy integrator with the function

$$\frac{V_o}{V_i} = -\frac{g_{m1}}{sC + g_{m2}} \tag{16.49}$$

Finally, we must determine the component values. To this end we compute from Equations 16.47 and 16.49 the relationships

$$C_1 = \frac{g_m}{\omega_i Q_i} = \frac{12.5}{\omega_i Q_i} \text{ pF}; \quad C_2 = g_m \frac{Q_i}{\omega_i} = 12.5 \frac{Q_i}{\omega_i} \text{ pF};$$

$$C = \frac{g_m}{\sigma} = \frac{12.5}{\sigma} \text{ pF}; \quad K = \left(\frac{\omega_1}{\sigma_z}\right)^2 \tag{16.50}$$

where we used that ω_i in Table 16.2 was normalized by a factor 10^7 s^{-1} and we assumed for simplicity that all OTAs have the same transconductance values,[*] $g_{m1} = g_{m2} = g_m = 125 \ \mu\text{S}$. Lastly, to illustrate the need to account for parasitics, we observe that in Figure 16.25 the capacitor C_1 is paralleled by $2C_o$ and $1C_i$, and the capacitor C_2 by $2C_o$ and $3C_i$. The third input capacitor in parallel with C_2 is arrived at by the fact that each biquad must drive the next biquad in the cascade connection and there sees an OTA input. The capacitor C in Figure 16.23 is in parallel with $2C_o$ and $1C_i$. To arrive at numerical values we will assume $C_o = 0.09$ pF and $C_i = 0.04$ pF. Consequently, the capacitor values are obtained from Equation 16.50 as

$$C_1 = \frac{12.5}{\omega_i Q_i} \text{ pF} - 0.22 \text{ pF}; \quad C_2 = 12.5 \frac{Q_i}{\omega_i} \text{ pF} - 0.3 \text{ pF}; \quad C = \frac{12.5}{\sigma} \text{ pF} - 0.22 \text{ pF} \tag{16.51}$$

Table 16.3 contains the computed capacitor values and Figure 16.26 shows the final cascade block diagram with the equalizer section leading and the first-order low-pass at the end. The two control

---

[*] This assumption is for convenience of design and layout because it permits a given transconductance cell to be used throughout the circuit. However, using different transconductance values in different sections has the advantage that capacitor values may be equalized, since by Equation 17.50 the capacitors are proportional to g_m.

TABLE 16.3   Component Values of the Cascade Filter for Equation 16.45

Section	$i = 1$	$i = 2$	$i = 3$	$i = 4$
C_{1i}	15.77 pF	6.31 pF	2.45 pF	—
C_{2i}	7.12 pF	7.81 pF	10.61 pF	—
C	—	—	—	14.29 pF
K	1.459	—	—	—

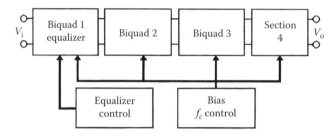

FIGURE 16.26   Structure of the seventh-order low-pass filter including control circuitry.

blocks are necessary to be able to tune the filter electronically: the gain K is varied via the *Equalizer control* block to set the position of the zeros and thereby the amount of gain boost, i.e., pulse slimming. Electronic tuning of the frequency parameter g_m/C is accomplished via adjusting the bias currents of the OTAs by the block labeled *Bias f_c control*. Thereby uncontrollable changes in the value g_m/C due to process variations or temperature can be accounted for. Details of such a control scheme are discussed in Section 16.4.

16.3.3 Ladder Simulations

As mentioned earlier, the main reason for using the popular *LC* ladder simulation method for filter design is the generally lower passband sensitivity of this topology to component tolerances [1, Chapter 3], [3, Chapter 13], see also Chapter 7. As before, the procedures are, in principle, identical to those discussed in connection with discrete circuits and are best illustrated with the help of a generic example. Let us consider again the classical ladder structure in Figure 16.11, which realizes the fifth-order elliptic low-pass characteristic Equation 16.21 and is described by Equation 16.22. Two methods are available to simulate the ladder. The first and most intuitive method replaces the inductors L_2 and L_4 by capacitively loaded gyrators (Figure 16.21). The second method recognizes that the inductors and the grounded capacitors in Figure 16.11 perform the function of integration. This signal-flow graph method is completely analogous to the process discussed earlier in Chapter 15 and for MOSFET-C filters, and will be presented in Section 16.3.3.2. We point out here that the element replacement and the signal-flow graph methods lead to the same circuitry [3, Chapter 16.4] so that either method can be used at the designer's convenience.

16.3.3.1 Element Replacement Methods

Replacing the inductors L_2 and L_4 in Figure 16.11 by capacitively loaded gyrators and using differential balanced circuitry leads to the circuit in Figure 16.27a. It is obtained by first converting the voltage source to a current source (Norton transformation), which also converts the series source resistor R into a shunt resistor. The first OTA in Figure 16.27a performs the source transformation, the second OTA is the grounded resistor. Since the two inductors are floating, the implementation of each requires two gyrators, i.e., four OTAs (see Figure 16.21c). Note that all OTAs are identical, and that all

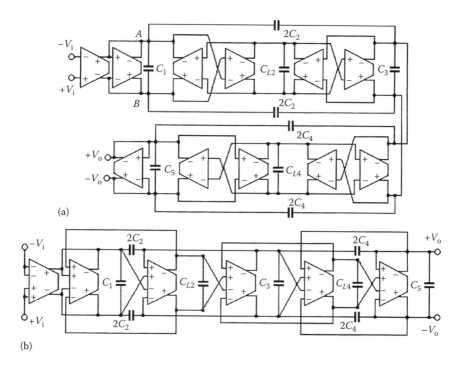

(a)

(b)

FIGURE 16.27 (a) Transconductor-C simulation by the element replacement method of the fifth-order elliptic low-pass ladder in Figure 16.11, including source and load resistors. All g_m-cells are identical. Note that the floating inductors require two gyrators for implementation. (b) The circuit with dual-input OTAs.

capacitors except C_2 and C_4 could be grounded; for example, instead of connecting C_1 *between* nodes A and B in Figure 16.27a, capacitors of value $2C_1$ could be connected from *both* nodes A and B to ground. Comparing the active circuit with the LC prototype readily identifies both structure* and components. The element values are obtained directly from the prototype LC ladder, e.g., from published tables, such as Ref. [2], or from appropriate synthesis software. Labeling for the moment the normalized components in the prototype LC ladder by the subscript n, i.e., R_n, $C_{i,n}$, and $L_{i,n}$, the transformation into the real component values with units of [F] and [H] is achieved by the following equations:

$$R = \frac{R_n}{g_m}; \quad C_i = C_{i,n}\frac{g_m}{\omega_c}; \quad L_i = L_{i,n}\frac{1}{g_m\omega_c} = \frac{C_{Li}}{g_m^2} \rightarrow C_{Li} = L_{i,n}\frac{g_m}{\omega_c} \tag{16.52}$$

where

g_m is the transconductance value chosen by the designer
ω_c is the normalizing frequency (usually the specified passband corner frequency)
R_n is in most prototype designs normalized to $R_n = 1$

Naturally, as discussed earlier, all capacitor values must be predistorted to account for the parasitic capacitors appearing at the capacitor nodes. For example, note that C_{L2} is paralleled by two OTA input and two OTA output capacitors, and for symmetrical layout (Figure 16.19), by $0.25C_s$. We see that the

---

* The number of floating capacitors is of course doubled because of the balanced differential structure of the active implementation.

element replacement method is a very straightforward design process; it has been found to work very well in practice.

Figure 16.27b shows the same circuit but realized with dual-input OTAs (Figure 16.16c). As was pointed out earlier, this merging of OTAs can always be used when two (or more) OTAs share a common output node. It results in simplified circuitry, and possibly reduced power consumption and chip area. In such cases, only the linearized input stages of the OTA must be duplicated, but bias, output, and common-mode rejection circuitry can be shared. Observe also that the input voltage is applied to both inputs of the first dual-input OTA in Figure 16.27b, thereby doubling its value of g_m. This multiplies the transfer function by a factor of two and eliminates the 6 dB loss inherent in the *LC* ladder (see Figure 16.12).

A small example will show the design process. For an antialiasing application we need to realize a third-order elliptic low-pass filter with $f_c = 10$ MHz bandwidth, 0.5 dB passband ripple, 17.5 MHz stopband edge, and 23 dB stopband attenuation. It leads to the transfer function [2]

$$H(s) = \frac{0.28163(s^2 + 3.2236)}{(s + 0.7732)(s^2 + 0.5016s + 1.1742)} \quad (16.53)$$

and the normalized element values $R_n = 1$, $C_{1n} = C_{3n} = 1.293$, $C_{2n} = 0.3705$, and $L_{2n} = 0.8373$. The topology is as in Figure 16.11 with $C_{4n} = L_{4n} = C_{5n} = 0$. Figure 16.28 shows the active circuit. Observe that we have realized each of the floating capacitors C_j as $0.5C_j + 0.5C_j$, $j = 1, 2, 3, L2$, with inverted bottom-plate connections. This design choice preserves symmetry in the balanced differential layout by placing the unavoidable substrate capacitors of value $\approx 0.1 \times (0.5C_{L2})$ *at each* of the upper and lower nodes of C_{L2} and $\approx 0.1 \times (0.5C_k + 0.5C_2)$ at the upper and lower nodes of C_k, $k = 1, 3$. Choosing the value of transconductance as $g_m \approx 180$ μS, using Equation 16.51, $\omega_c = 2\pi \times 10 \times 10^6$ s^{-1}, and observing the necessary predistortion for the *differential* parasitic OTA input and output capacitors, $C_i = 0.03$ pF and $C_o = 0.08$ pF, respectively, and for the bottom-plate capacitors C_s assumed to be 10% of the corresponding circuit capacitor value, results in

$$C_1 = C_{1n}\frac{g_m}{\omega_c} - 3C_o - 2C_i - 0.1\frac{1}{2}\left(C_2 + \frac{C_1}{2}\right) = 3.191 \text{ pF}$$

$$C_2 = C_{2n}\frac{g_m}{\omega_c} = 1.061 \text{ pF}$$

$$C_3 = C_{3n}\frac{g_m}{\omega_c} - 2(C_o + C_i) - 0.1\left(\frac{C_2}{2} + \frac{C_3}{4}\right) = 3.268 \text{ pF} \quad (16.54)$$

$$C_{L2} = L_{2n}\frac{g_m}{\omega_c} - 2(C_o + C_i) - 0.1\frac{C_L}{4} = 2.120 \text{ pF}$$

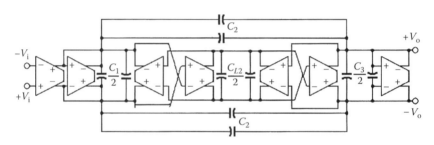

FIGURE 16.28 Realization of the third-order elliptic low-pass ladder of Equation 16.53. The capacitor values indicated refer to *each* of the pair of capacitors.

Notice that in a ladder structure all parasitic capacitors can be absorbed in the circuit capacitors. Consequently, no new parasitic poles or zeros are created that might destroy the transfer function shape. This is a further important advantage of the ladder-simulation method for g_m-C filters.

16.3.3.2 Signal-Flow Graph Methods

As we have seen earlier, the signal-flow graph (SFG) or operational-simulation method takes the circuit equations describing the ladder (Kirchhoff's laws and the $I-V$ relationships for the elements) and realizes them directly via summers (for Kirchhoff's laws) and integrators (for inductors and grounded capacitors). The procedure was detailed in connection with Equations 16.22 through 16.24. Recall that Equation 16.23 represents *integrations* of voltages into currents and currents into voltages. As was the case for the op-amp–based active *RC* filters in Chapter 15 and the MOSFET-C design, all signals in the SFG g_m-C circuit are voltages. They are summed at the OTA inputs, then are multiplied by g_m to produce an output current that is integrated by a capacitor to produce a voltage as input for the OTA of the next stage. To reflect these facts in the relevant equations, we scale Equation 16.23 analogously to Equation 16.24 to obtain

$$\frac{I_{L2}}{g_m} = \frac{g_m(V_1 - V_3)}{s(L_{2n}g_m)g_m}; \quad \frac{I_{L4}}{g_m} = \frac{g_m(V_3 - V_5)}{s(L_{4n}g_m)g_m} \tag{16.55a}$$

$$V_1 = \frac{g_m V_i - (I_{l2}/g_m)g_m + sC_{2n}V_3}{s(C_{1n} + C_{2n}) + g_m} \tag{16.55b}$$

$$V_3 = \frac{(I_{L2}/g_m)g_m - (I_{L4}/g_m)g_m + sC_{2n}V_1 + sC_{4n}V_5}{s(C_{2n} + C_{3n} + C_{4n})} \tag{16.55c}$$

$$V_5 = V_o = \frac{(I_{L4}/g_m)g_m + sC_{4n}V_3}{s(C_{4n} + C_{5n}) + g_m} \tag{16.55d}$$

The scaling factor in this case is the design transconductance g_m. Note that $L_i\, g_m^2$ in Equation 16.55a has units of capacitance and that source and load resistors in Equations 16.55b and d have the value $1/g_m$. Implementing these equations with lossless or lossy, as appropriate, integrators in fully differential form results in the circuit in Figure 16.29, where we used a signal notation similar to that in Figure 16.14 and chose all integrating capacitors grounded. Starting from a normalized *LC* prototype, the actual component values are obtained again via Equation 16.52. Note that the OTA at the input performs the voltage-to-current conversion (V_i to I_1) and that the last OTA both here and in Figure 16.27 implements the load resistor. The second OTA in Figure 16.27, realizing the source resistor, is saved in Figure 16.29 by sending the current I_1 directly into the integrating node (the capacitor C_1), as suggested by Equation 16.55b.* Also observe that circuit complexity in Figure 16.29 was kept low by resorting to OTAs with dual inputs. We note again that all transconductors in Figures 16.27 through 16.29 are identical[†] so that a single optimized g_m-cell (an *analog gate*) can be used throughout the filter chip for an especially simple IC design process, analogous to that of a gate array. The inherent 6 dB loss of the *LC* ladder can also be eliminated, if desired, by lifting the two grounded inputs of the second OTA in Figure 16.29 off ground and connecting them to the input voltage $\pm V_i/2$ as indicated by the dashed connections.

---

* Had we used this "resistor," the current into the integration node would have been realized as $g_m V_i = g_m V_i \times (1/g_m) \times g_m$, clearly a redundant method.

[†] This can generally be achieved in g_m-C ladder simulations; the only exception occurs for those *LC* ladders that require unequal terminating resistors, such as even-order Chebyshev filters; in that case, one of the transconductors will also be different.

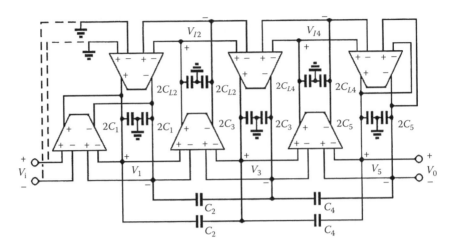

FIGURE 16.29 Signal-flow graph g_m-C realization of a fifth-order elliptic low-pass function. The circuit is an active simulation of the LC-ladder in Figure 16.11, including source and load resistors. All g_m-cells are identical.

As a simple design example, consider again the realization of the third-order elliptic low-pass filter described by Equation 16.53. The nominal LC elements were given earlier; the circuit is described by Equation 16.55 with $C_{4n} = C_{5n} = L_{4n} = 0$, i.e.,

$$\frac{I_{L2}}{g_m} = \frac{g_m(V_1 - V_3)}{s(L_{2n}g_m)g_m} \tag{16.56a}$$

$$V_1 = \frac{2g_mV_i - (I_{L2}/g_m)g_m + sC_{2n}V_3}{s(C_{1n} + C_{2n}) + g_m} \tag{16.56b}$$

$$V_3 = V_o = \frac{(I_{L2}/g_m)g_m + sC_{2n}V_1}{s(C_{2n} + C_{3n}) + g_m} \tag{16.56c}$$

which leads to the circuit in Figure 16.30a, as the reader may readily verify. Note the connection of the input, which realizes the term $2g_mV_i$ in Equation 16.56b and eliminates the 6 dB loss as mentioned

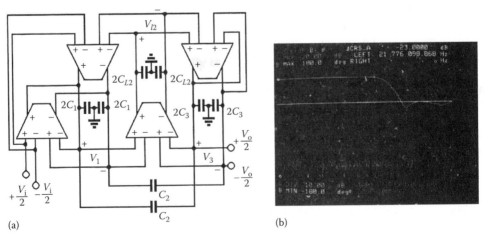

(a)   (b)

FIGURE 16.30 (a) SFG g_m-C third-order elliptic low-pass filter; (b) experimental performance in 2-μm CMOS technology.

earlier. We observe again that the capacitors must be predistorted to account for parasitics as was discussed in connection with Equation 16.54; specifically, labeling as before the *differential* OTA input and output capacitors as C_i and C_o, respectively, we find

$$C_1 \approx C_{1,\,\text{nominal}} - 2C_o - 2C_i; \quad C_{L2} \approx C_{L2,\,\text{nominal}} - C_o - 2C_i; \quad C_3 \approx C_{3,\,\text{nominal}} - C_o - 2C_i$$

Figure 16.30b shows the experimental performance of the circuit fabricated in 2-μm CMOS technology with design-automation software that uses the OTAs from a design library as "analog gates," and *automatically* lays out the chip and predistorts the capacitors according to the process file. The transconductance value used is $g_m \approx 180$ μS, and the capacitors are approximately the same as the ones for the example in Figure 16.28. We will see in Section 16.5 that apart from differences due to layout parasitics, it is no coincidence that the element values are essentially the same as before. Notice that the filter meets all design specifications. The lower trace is the thermal noise at ≈ 70 dB below the signal; the measured total harmonic distortion was THD $< 1\%$ for a 2-$V_{p\text{-}p}$ input signal at 3 MHz.

16.4 Tuning

To obtain accurate filter performance with frequency-parameters set by RC products or C/g_m ratios, accurate *absolute* values of components are required. These must be realized by the IC process *and maintained during operation*. Although IC processing is very reliable in realizing accurate *ratios* of like components on a chip, the processing tolerances of *absolute values* must be expected to be of the order of 20%–50% or more. Component tolerances of this magnitude are generally far too large for an untuned filter to perform within specifications. Consequently, filters *must be tuned* to their desired performance by adjusting element values. Clearly, in fully integrated filters, where all components are on a silicon chip, tuning must be performed electronically by some suitably designed automatic control circuitry that is part of the total continuous-time filter system. Tuning implies *measuring* filter performance, *comparing* it with a known standard, *calculating* the errors, and *applying a correction* to the system to reduce the errors. An accurate *reference frequency*, e.g., a system clock (V_{REF} in Figure 16.31), is usually used as a convenient standard. From the filter's response to the signal at this known frequency, errors are detected and the appropriate correction signals, typically dc bias voltages, are applied via the control circuitry [1, Chapter 7], [3, Chapter 16.5]. Reference [7, pt. 6] contains many papers showing practical approaches to specific tuning problems.

Figure 16.31 shows a block diagram of this so-called Master–Slave architecture that is followed by most currently proposed designs. The Master–Slave system is used because the reference signal, V_{REF}, cannot be applied to the Main Filter simultaneously with the main signal, V_{in}, because of undesirable interactions (intermodulation) between V_{in} and V_{REF}. The approach is based on the premise that the filter must operate continuously, i.e., that V_{in} cannot be switched off occasionally to permit tuning in a signal-free

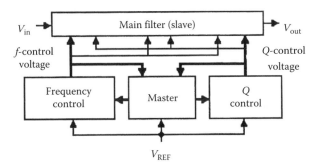

FIGURE 16.31 Block diagram of Master–Slave control system for integrated *c-t* filters.

environment. Therefore, V_{REF} is applied to a circuit, the "Master" filter, that is carefully designed to model the relevant behavior of the "Slave." Master–Slave design and layout require great care to avoid serious matching, noise, and crosstalk problems.

Figure 16.31 should help the reader understand the principle of the operation without having to consider the actual circuitry that is very implementation-specific, see Ref. [7, pt. 6] for design examples. The Main Filter (the "Slave") performs the required signal processing. Since the Master is designed to model the Slave, the behaviors of Master and Slave are assumed to match and track. Tuning is then accomplished by applying the generated correction or control signal simultaneously to both Master and Slave. The system contains a *Frequency-Control* block that detects frequency-parameter errors in the Master's response to the reference signal and generates a frequency-control voltage that is applied to the Master in a closed-loop control scheme such that any detected errors are minimized.* Since the Master is designed to be an accurate model of the Slave, their errors, too, can be assumed to match and track. Consequently, when the frequency-control voltage is applied at the appropriate locations to the Slave, it can be expected to correct any frequency errors in the main filter.

The purpose of the additional block in Figure 16.31, labeled *Q-Control*, is to tune the *shape* of the transfer characteristic. Once the frequency parameters are correct as tuned by the *f*-control loop, the transfer function shape is determined only by the quality factors Q_i of the poles and zeros. *Q*, as a ratio of two frequencies, is a *dimensionless* number and as such is determined by a *ratio* of like components (resistor, capacitor, and/or g_m-ratio; see e.g., Equation 16.47b). *At fairly low frequencies and moderate values of Q*, the quality factor is realizable quite accurately in an IC design. For high-frequency, high-*Q* designs, however, *Q* is found to be a very sensitive function of parasitic elements and phase shifts so that it is unreasonable to expect *Q* to turn out correctly without tuning. Therefore, including for generality a *Q*-Control block[†] permits automatic tuning of the transfer function shape by a scheme that is completely analogous to the *f*-control method, as illustrated in Figure 16.31.

With few exceptions, all currently proposed automatic tuning schemes follow the Master–Slave approach. A different concept proposed in Ref. [11] uses adaptive techniques to tune all poles *and* zeros of a filter function for improved tuning accuracy. Although the Master–Slave and the adaptive techniques work well, generally, the necessary circuitry has been found to be too large, noisy, and power hungry for many practical applications. Thus, alternative choices use a simple post-fabrication trim step[‡] to eliminate the large fabrication tolerances, possibly together with careful design to make the electronic devices (OTAs) independent of temperature variations [12,13].

We also point out that at the highest frequencies, higher than, say, 100 MHz, circuit capacitors to realize the prescribed frequency parameters, g_m/C, become very small ($C \approx 0.15$ pF at 100 MHz and $g_m = 100\ \mu S$). Filter behavior is then essentially determined by parasitics, and matching required between Master and Slave circuitry is unreliable. In that case, a form of direct tuning [7, paper 6.4] must be used, where the main filter is tuned during time periods when the signal is absent. The system is essentially as depicted in Figure 16.31, except that the Master is absent and V_{REF} is applied to the main filter. When tuning is complete (taking a few milliseconds), V_{REF} is turned off, V_{in} is again applied to the filter, and the generated dc control signals are held on a capacitor until the next tuning update.

---

* In the designs reported to date, frequency-control blocks built around some type of phase-locked loop scheme, using a multiplier or an EXOR gate as phase detector, have been found most successful, but magnitude-locking approaches similar to those employed for *Q* control have also been used. A largely digital approach to frequency and *Q* tuning is discussed in Ref. [10].

† Because magnitude errors can be shown in many cases to be proportional to errors in quality factor, the QControl block is normally implemented around an amplitude-locking scheme, where the filter's magnitude response at a given frequency is locked to a known reference level.

‡ For example, an on-chip laser-trimmed resistor or an external resistor set to determine the bias for OTAs. It works for applications with very low quality factors or where no component drifts are expected during operation.

16.5 Discussion and Conclusions

In this chapter we discussed one of the fastest growing research and development areas in the topic of continuous-time filters, the field of fully integrated filters. Growing pressures to reduce costs and size, and improve reliability lead to increasingly larger parts of electronic systems being placed onto integrated circuits, a trend that *c-t* filters need to follow. As might have been expected, in most respects the methods for IC *c-t* filter design are identical to well-known active *RC* techniques, and follow directly the well-understood and proven standard active *RC* biquad-cascade or ladder-simulation methodologies. The difference lies in the final implementation, which the designer may adapt to any IC process and power supply level appropriate for the implementation of the system. Signal-to-noise ratios of the order of 65–80 dB and better, and distortion levels of less than 0.5% at signal levels of 1 V are obtainable. The two most prominent design approaches are the MOSFET-*C* and the g_m-*C* methods, both of which lead to easy and systematic designs and result in filters that have proven themselves in practice. At the time of this writing, g_m-*C* filters appear to have the edge in high-frequency performance: with OTA bandwidths reaching gigahertz frequencies even for CMOS technology, *c-t* filters can be designed readily for applications in the range of several hundred megahertz, i.e., to about 40%–60% of the active-device bandwidth. A second important difference that requires the designer's attention is that the IC filter must be automatically tunable. This implies electronically variable components, OTAs or MOSFET "resistors," in the filter and some automatic control scheme for detecting errors and providing adjustments. Only area- and power-efficient designs, and simple low-noise circuitry for such tuning schemes will be acceptable and guarantee the ultimate commercial success of integrated filters. At the present time, many approaches have been proposed that the reader may modify or adapt to his/her system [7].

We point out again that the cascade design is the more general procedure because it permits the realization of arbitrary transmission zeros anywhere in the complex *s*-plane, as required, for example, in gain or phase equalizers. Lossless ladders, on the other hand, are more restrictive because their transmission zeros are constrained to lie on the $j\omega$ axis, but they have lower passband sensitivities to component tolerances than cascade filters. Since good gyrators are readily available for g_m-*C* filters—in contrast to MOSFET-*C* designs—two competing implementation methods appear to suggest themselves in the g_m*C* approach: the signal-flow graph and the element-substitution methods. As mentioned, there is in fact no discernible difference between the two methods; indeed, they lead to the same structures. For instance, the reader may readily verify that apart from a minor wiring change at the inputs, the circuits in Figures 16.27b and 16.29 are identical, as are the ones in Figures 16.28 (after redrawing it for dual-input OTAs) and 16.30. The wiring change is illustrated in Figure 16.32. Figure 16.32a shows

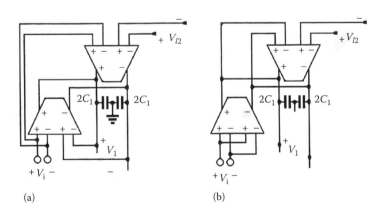

(a)　　　　　　　　　　　　　(b)

FIGURE 16.32　Excerpt from Figure 16.28 (a) to illustrate alternative wiring and (b) at the filter input 0.

the input section of Figure 16.29 (wired for 0-dB dc gain) and the wiring in Figure 16.32b is that of Figure 16.27b. Notice that both circuits realize

$$(sC_1 + g_m)V_1 = 2g_m V_i - g_m V_{I2} \tag{16.57}$$

The designer, therefore, may choose a ladder method based on his/her familiarity with the procedure, available tables, or prototype designs; the final circuits are the same.

A further item merits reemphasizing at this point: As the reader may verify from the examples presented, g_m-C ladder structures generally have at all circuit nodes a design capacitor that can be used to absorb parasitics by predistortion as was discussed earlier. It may be preferable, therefore, to avoid the g_m-C–op-amp integrator of Figure 16.20 with its increased noise level, power consumption, and with its associated parasitic poles (and zeros) and use parasitics-absorption as we have done in our examples. This method will *not* introduce any new parasitic critical frequencies into the filter and result in lower distortion of the transfer function shape and easier tuning. As we mentioned earlier, this feature is a substantial advantage of g_m-C ladder filters.

References

1. R. Schaumann, M. S. Ghausi, and K. R. Laker, *Design of Analog Filters: Passive, Active RC and Switched Capacitor*, Englewood Cliffs, NJ: Prentice-Hall, 1990.
2. A. I. Zverev, *Handbook of Filter Synthesis,* New York: Wiley, 1967.
3. R. Schaumann and M. Van Valkenburg, *Design of Analog Filters*, New York: Oxford University Press, 2001.
4. D. Li and Y. Tsividis, Active *LC* filters on silicon, *IEE Proc. Circuits Devices Syst.*, 147(1), 49–56, Feb. 2000.
5. E. Sanchez-Sinencio and J. Silva-Martinez, CMOS transconductance amplifiers, architectures and active filters: a tutorial, *IEEE Proc. Circuits, Devices Syst.*, 147(1), 3–12, Feb. 2000.
6. Y. Tsividis, M. Banu, and J. Khoury, Continuous-time MOSFET-C filters in VLSI, *IEEE Trans. Circuits Syst.*, CAS-33, Special Issue on VLSI Analog and Digital Signal Processing, 125–140, 1986; see also *IEEE J. Solid-State Circuits*, SC-21, 15–30, 1986.
7. Y. Tsividis and J. A. Voorman, Eds., *Integrated Continuous-Time Filters: Principles, Design and Implementations*, New York: IEEE, 1993.
8. C. Toumazou, F. J. Lidgey, and D. G. Haigh, Eds., *Analogue IC Design: The Current-Mode Approach,* London: IEE—Peter Peregrinus, 1990.
9. A. Thanachayanont and A. Payne, CMOS floating active inductor and its applications to bandpass filter and oscillator designs, *IEE Proc. Circuits Devices Syst.*, 147(1), 41–48, Feb. 2000.
10. A. I. Karsilayan and R. Schaumann, Mixed-mode automatic tuning scheme for high-Q continuous-time filters, *IEE Proc. Circuits Devices Syst.*, 147(1), 57–64, Feb. 2000.
11. K. A. Kozma, D. A. Johns, and A. S. Sedra, Automatic tuning of continuous-time filters using an adaptive tuning technique, *IEEE Trans. Circuits Syst.*, CAS-38, 1241–1248, 1991.
12. G. A. De Veirman and R. G. Yamasaki, Design of a bipolar 10-MHz continuous-time 0.05° equiripple linear phase filter, *IEEE J. Solid State Circuits*, SC-27, 324–331, March 1992.
13. C. A. Laber and P. R. Gray, A 20-MHz BiCMOS parasitic-insensitive continuous-time filter and second-order equalizer optimized for disk-drive read channels, *IEEE J. Solid State Circuits*, SC-28, 462–470, April 1993.

17

Switched-Capacitor
Filters

Jose Silva-Martinez
Texas A&M University

Edgar Sánchez-Sinencio
Texas A&M University

17.1 Introduction

The need to have monolithic high-performance analog filters motivated circuit designers in the late 1970s to investigate alternatives to conventional active-RC filters. A practical alternative appeared in the form of switched-capacitor (SC) filters [1–3]. The original idea was to replace a resistor by an SC simulating the resistor. Thus, this equivalent resistor could be implemented with a capacitor and two switches operating with two-clock phases. SC filters consist of switches, capacitors, and op-amps. They are characterized by difference equations in contrast to differential equations for continuous-time filters. Simultaneously, the mathematical operator to handle sample-data systems such as SC circuits is the z-transform, while the Laplace transform is used for continuous-time circuits. Several key properties of SC circuits have made them very popular in industrial environments:

1. Time constants (RC products) from active-RC filters become capacitor ratios multiplied by the clock period T. That is

$$\text{RC} \Rightarrow \frac{C}{C_R} T = \frac{C}{C_R f_c}$$

where f_c is the clock frequency used to drive the SC equivalent resistor.
2. Reduced silicon area, since the equivalent of large resistors can be simulated using small-size capacitors. Furthermore, positive and/or negative equivalent resistors can be easily implemented with SC techniques.
3. Above expression can be realized in real applications with a good accuracy of nearly 0.1%.
4. Typically, the load of an SC circuit is mainly capacitive; therefore the required op-amps do not require a low-impedance output stage. This allows the use of unbuffered cascode operational transconductance amplifiers, which is especially useful in high-speed applications.
5. SC filters can be implemented in digital circuit process technologies; metal–metal capacitors are quite often used in deep submicron technologies. Thus, useful mixed-mode signal circuits can be economically realized.
6. SC design technique has matured. In the audio range, SC design techniques have become the dominant design approach. Many circuits in communication applications and data converters use SC implementations.

In what follows, we will discuss basic building blocks involved in low- and high-order filters. Limitations and practical design considerations will be presented. Furthermore, due to the industrial push for lower power supply voltages, a brief discussion on low-voltage circuit design is included.

17.2 Basic Building Blocks

The basic building blocks involved in SC circuits are voltage gain amplifiers, integrators, second-order filters, and nonoverlapping clock generators. A key building block is the integrator. By means of a two-integrator loop a second-order (biquadratic) filter can be realized. Furthermore, a cascade connection of biquadratic filters yields higher order filters.

17.2.1 Voltage Gain Amplifiers

The gain amplifier is a basic building block in SC circuits. Both the peak gain of a second-order filter and the link between the resonators in a ladder filter are controlled by voltage amplifier stages rather than by integrators. Many other applications such as pipeline data converters require voltage gain stages as well. A voltage amplifier can be implemented by using two capacitors and an operational amplifier, as shown in Figure 17.1a. Ideally, the gain of this amplifier is given by $-C_S/C_i$. This topology is compact, versatile, and time continuous. Although this gain amplifier is quite simple, several second-order effects present in the op-amp make its design more complex. A major drawback of this topology is the lack of dc feedback.

For dc, the capacitors behave as open circuits; hence the operating point of the operational amplifier is not stabilized by the integrating capacitor C_i. In addition, the leakage current present at the input of the op-amp is integrated by the integrating capacitor, whose voltage eventually saturates the circuit. The leakage current I_{leak} in SCs circuits is a result of the diodes associated with the bottom plate of the capacitors and the switches (drain and source junctions). The leakage current is typically about 1 nA/cm^2. Analysis of Figure 17.1a considering the leakage current I_{leak} present at the inverting terminal of the amplifier yields

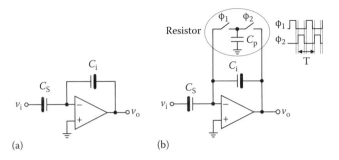

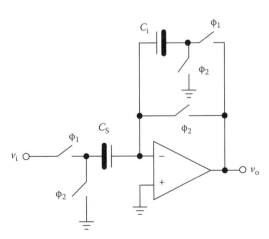

FIGURE 17.1 Voltage amplifiers. (a) Without dc feedback and (b) with dc feedback.

$$v_o(t) = v_o(t_o) - \frac{C_S}{C_i} V_i(t) - \frac{I_{leak}}{C_i}(t - t_o) \qquad (17.1)$$

Note that the dc output voltage is defined by the input signal and the initial conditions at $t - t_o$. The leakage current present at the input of the op-amp is integrated by C_i and eventually saturates the circuit. To overcome this drawback, an SC resistor can be added as shown in Figure 17.1b. ϕ_1 and ϕ_2 are two nonoverlapping clock phases with nearly 50% duty cycle. The SC resistor gives a dc path for the leakage current but reduces further the low-frequency gain. A detailed analysis of this topology shows that the dc output voltage is equal to $-I_{leak}T/C_p$, with C_p the parasitic capacitor associated with the feedback path. Employing charge conservation analysis method for SC networks, it can also be shown that the z-domain transfer function of this topology is

$$H(z) = \frac{V_o(z)}{V_i(z)} = -\frac{C_S}{C_i} \frac{1 - z^{-1}}{1 - \left(1 - \frac{C_p}{C_i}\right)z^{-1}} \qquad (17.2)$$

with $z = e^{j2\pi fT}$. For low frequencies, $z \approx 1$, the magnitude of the transfer function is very small, and only for high frequencies, $z \approx -1$, the circuit behaves as an inverting voltage amplifier.

An offset free voltage amplifier is shown in Figure 17.2. During the clock phase ϕ_2, the output voltage is equal to the op-amp offset voltage and it is sampled by both the integrating and sampling capacitors. Because of the sampling of the offset voltage during the previous clock phase, during the integrating clock phase the charge injected by the sampling capacitor is equal to $C_S V_i$, and the charge extracted from the integrating capacitor becomes equal to $C_i V_o$ if the offset voltage of the amplifier is the same in both clock phases. In this case, the offset voltage does not affect the charge recombination, and the z-domain transfer function during the clock phase ϕ_1 becomes

$$H(z) = -\frac{C_S}{C_i} \qquad (17.3)$$

FIGURE 17.2 Voltage amplifier available during the ϕ_1 clock phase.

Equation 17.3 shows that ideally this topology is insensitive to the op-amp offset voltage, and does not have any low-frequency limitation. The topology behaves as an inverting amplifier if the clock phases shown in Figure 17.2 are used. A noninverting amplifier is obtained if the clock phases associated with the sampling capacitor are interchanged.

Because during ϕ_2 the op-amp output is short circuited with the inverting input, this topology presents two drawbacks. First, the amplifier output is only available during the clock phase ϕ_1. This limitation could be important in complex applications wherein the output of the amplifier is required during both clock phases. The second disadvantage of this topology is the large excursion of the op-amp output voltage. During the first clock phase, the op-amp output voltage is equal to the offset voltage and in the next clock phase this voltage is equal to $-(C_S/C_i)V_i$. Hence an op-amp with large slew rate may be required; this may demand significant amount of power since the load capacitor is driving during this phase too.

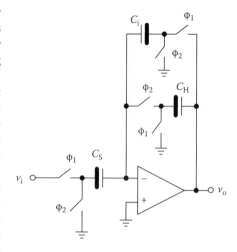

FIGURE 17.3 Voltage amplifier available during both clock phases.

Another interesting topology is shown in Figure 17.3. During ϕ_2, the op-amp output voltage is equal to the previous voltage plus the op-amp offset voltage plus V_o/A_V, where A_V is the open-loop dc gain of the op-amp. In this clock phase, both capacitors C_i and C_S are charged to the voltage at the inverting terminal of the op-amp. This voltage is approximately equal to the op-amp offset voltage plus V_o/A_V. During the next clock phase, the sampling capacitor is charged to $C_S(V_i - V_-)$, but because it was precharged to $-C_S V_-$, the injected charge to C_i is equal to $C_S V_i$. As a result of this charge cancellation, the op-amp output voltage is equal to $-(C_S/C_i)V_i$. Therefore, this topology has low sensitivity to the op-amp offset voltage and to the op-amp dc gain. A minor drawback of this topology is that the op-amp stays in open loop during the nonoverlapping clock phase transition times. This fact produces spikes during these time intervals. A solution for this is to connect a small capacitor between the op-amp output and the left-hand plate of C_S.

17.2.2 First-Order Blocks

The standard stray-insensitive integrators are shown in Figure 17.4. Note that in sampled data systems both input and output signals can be sampled at different clock periods. This yields different transfer functions; this property increases the flexibility of the architecture when used in complex systems. We will assume a two-phase nonoverlapping clock: an odd clock phase ϕ_1 and an even clock phase ϕ_2.

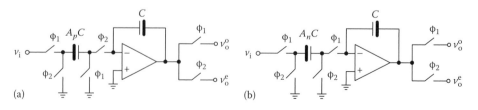

FIGURE 17.4 Conventional stray-insensitive SC integrators. (a) Noninverting and (b) inverting.

For the noninverting integrator the following transfer functions are often used

$$H^{oo}(z) = \frac{V_o^o(z)}{V_{in}^o(z)} = \frac{A_p z^{-1}}{1 - z^{-1}} = \frac{A_p}{z - 1} \tag{17.4a}$$

$$H^{oe}(z) = \frac{V_o^e(z)}{V_{in}^o(z)} = \frac{A_p z^{-1/2}}{1 - z^{-1}} = \frac{A_p z^{1/2}}{z - 1} \tag{17.4b}$$

where H^{xy} stands for the output defined at phase x while the input is sampled during phase y. For the inverting integrator

$$H^{oe}(z) = \frac{V_o^o(z)}{V_{in}^o(z)} = -\frac{A_n}{1 - z^{-1}} = -\frac{A_n z}{z - 1} \tag{17.5a}$$

$$H^{oe}(z) = \frac{V_o^e(z)}{V_{in}^o(z)} = -\frac{A_n z^{-1/2}}{1 - z^{-1}} = -\frac{A_n z^{1/2}}{z - 1} \tag{17.5b}$$

where z^{-1} represents a unit delay. A crude demonstration showing the integration nature of these SC integrators [1,8] in the *s*-domain is to consider high-sampling rate, that is, a clock frequency ($f_c = 1/T$) much higher than the operating signal frequencies. Let us consider Equation 17.4a, and assuming high-sampling rate ($\omega T \ll 1$) we can write a mapping from the *z*- to the *s*-domain:

$$z = e^{j2\pi fT} = e^{sT} \cong 1 + sT \tag{17.6}$$

$$H(s) = \frac{A_p}{z - 1}\bigg|_{z \cong 1 + sT} \cong \frac{1}{(T/A_p)s} \tag{17.7}$$

where $s\ (=j2\pi f)$ is the complex frequency variable. This last expression corresponds to a continuous-time noninverting integrator with a time constant of $T/A_p = 1/f_c A_p$, that is, a capacitance ratio times the clock period.

In many applications the capacitor ratios associated with the integrators are very large, thus the total capacitance becomes excessive. This is particularly critical for biquadratic filters with high Q, where the ratio between the largest and smallest capacitance is proportional to the quality factor Q. A suitable inverting SC integrator for high-Q applications [11,13] is shown in Figure 17.5; the corresponding transfer function during ϕ_2 is

$$H^{oe}(z) = \frac{V_o^e(z)}{V_{in}^o(z)}$$

$$= -\frac{C_1 C_3}{C_2(C_2 + C_3)}\left(\frac{z^{-1/2}}{1 - z^{-1}}\right) \tag{17.8}$$

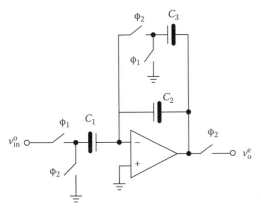

FIGURE 17.5 A SC integrator with reduced capacitance spread.

This integrator is comparable in performance to the conventional of Figure 17.3, in terms of stray sensitivity and finite gain error. Note from Equation 17.8 that the transfer function is only defined during ϕ_2. During ϕ_1, the circuit behaves as a voltage amplifier plus the initial conditions of the previous clock phase, and in the following clock phase the same amount of charge is extracted but the integrating capacitor is reduced due to the switches connected to C_1, thus relatively high-slew-rate op-amps could be required. A serious drawback could be the increased offset in comparison with the standard SC integrators. However, in typical two integrator loop filters, the other integrator can be chosen to be offset and low dc gain compensated as shown in Figure 17.6.

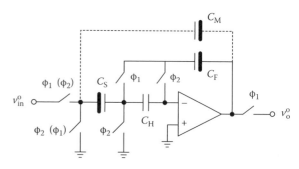

FIGURE 17.6   Offset and gain compensated integrator.

The SC integrator performs the integration during ϕ_1 by means of C_S and C_F and the hold capacitor C_H stores the offset voltage. The voltage across C_H compensates the offset voltage and the dc gain error of the op-amp. Note that the SC integrator of Figure 17.6 can operate as a noninverting integrator if the clocking in parenthesis is employed. C_M provides a time-continuous feedback around the op-amp. The transfer function for infinite op-amp gain is

$$H^{oo}(z) = \frac{V_o^o(z)}{V_{in}^o(z)} = -\frac{C_S}{C_F}\left(\frac{1}{1-z^{-1}}\right) \tag{17.9}$$

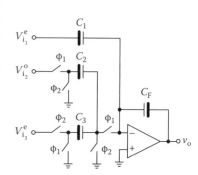

FIGURE 17.7   General form of a first-order building block.

Next we discuss a general form of a first-order building block (see Figure 17.7). Observe that some switches can be shared. The output voltage during ϕ_1 can be expressed as

$$V_o^o = -\frac{C_1}{C_F}V_{i_1}^e - \frac{C_2}{C_F}\left(\frac{1}{1-z^{-1}}\right)V_{i_2}^o + \frac{C_3}{C_F}\left(\frac{z^{-1/2}}{1-z^{-1}}\right)V_{i_3}^e \tag{17.10}$$

Observe that the capacitor C_3 and switches can be considered as the implementation of a negative resistor leading to a noninverting amplifier. Also note that $V_{i_2}^o$ could be V_o^e, this connection would make the integrator a lossy one. In that case Equation 17.10 can be written as

$$V_o^o\frac{\left(1+\frac{C_2}{C_F}\right)z-1}{z-1} = -\frac{C_1}{C_F}V_{i_1}^e + \frac{C_3}{C_F}\frac{z^{1/2}}{z-1}V_{i_3}^e, \quad \text{for } V_{i_2}^o = V_o^o \tag{17.11}$$

The building block of Figure 17.7 is the basis of higher order filters.

17.2.3 Switched-Capacitor Biquadratic Sections

The circuit shown in Figure 17.8 can implement any pair of poles and zeros in the z-domain. For $C_A = C_B = 1$ we can write

$$H^{ee}(z) = \frac{V_o^e(z)}{V_{in}^e(z)} = -\frac{(C_5 + C_6)z^2 + (C_1 C_2 - C_5 - 2C_6)z + C_6}{z^2 + (C_2 C_3 + C_2 C_4 - 2)z + (1 - C_2 C_4)} \tag{17.12}$$

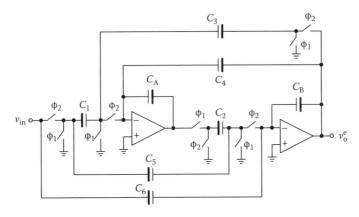

FIGURE 17.8 A SC biquadratic section [Martin-Sedra].

Simple design equations for a particular type of filter follows:

$$\text{Low-pass case: } C_5 = C_6 = 0 \tag{17.13a}$$

$$\text{High-pass case: } C_1 = C_5 = 0 \tag{17.13b}$$

$$\text{Bandpass case: } C_1 = C_6 = 0 \tag{17.13c}$$

Comparing the coefficients of the denominator of Equation 17.12 with the general z-domain expression $z^2 - (2r \cos \theta)z + r^2$ of a second-order system, we can obtain the following expressions:

$$C_2 C_4 = 1 - r^2 \tag{17.14a}$$

$$C_2 C_3 = 1 - 2r \cos \theta + r^2 \tag{17.14b}$$

Note that in this expression it has been assumed that the filter poles are complex conjugate located at $z = re^{\pm j\omega_{od}T}$. ω_{od} and Q are the poles frequency and the filter's quality factor, respectively. For equal voltage levels at the two integrator outputs, and assuming Q greater than 3, and high-sampling rate $(\theta = \omega_{od}T \ll 1)$ we can write

$$C_2 = C_3 = \sqrt{1 + r^2 - 2r \cos \theta} \cong \sqrt{(1 - r)^2 + r\theta^2} \cong \theta = \omega_{od}T \tag{17.14c}$$

$$C_4 = \frac{1 - r^2}{C_2} \cong \frac{1}{Q} \tag{17.14d}$$

where $\cos(\theta) \cong 1 - \theta^2/2$ and $r \cong 1$. The capacitance spread for high-sampling rate and high-Q unity DC gain low-pass filter can be expressed as

$$\frac{C_{max}}{C_{min}} = \max\left\{\frac{1}{C_2}, \frac{1}{C_4}\right\} = \max\left\{\frac{1}{\omega_{od}T}, Q\right\} \tag{17.15}$$

In some particular cases this capacitance spread becomes prohibited large. For such cases the SC integrators shown in Figures 17.5 and 17.6 can be used to replace the conventional building blocks.

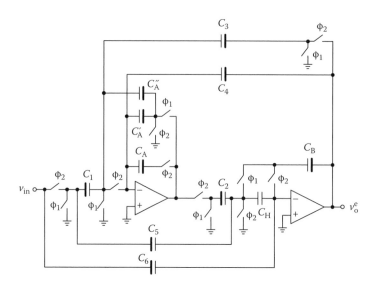

FIGURE 17.9   An improved capacitance area SC biquadratic section.

This combination yields the SC biquadratic section shown in Figure 17.9. This structure, besides offering a reduction of the total capacitance, also yields a reduction of the offset voltage of the op-amps. Note that the capacitors C_H does not play an important role in the design, and can be chosen to have a small value. The design equations for the zeros are similar to Equation 17.13a through c. For the poles, comparing with $z^2 - (2r \cos \theta)z + r^2$ and the analysis of Figure 17.9, we can write

$$\frac{C_2 C_3}{C_A + C_B} = 1 + r^2 - 2r \cos \theta \tag{17.16a}$$

$$\frac{C'_A C_2 C_4}{C_{A_1} C_A C_B} = 1 - r^2 \tag{17.16b}$$

where $C_{A1} = C'_A + C''_A$. Simple design equations can be obtained assuming high-sampling rate, large Q, and $C_2 = C_3 = C_4 = C'_A = C_h = 1$; then

$$C_A + C_B \cong \frac{1}{\omega_{od} T} \tag{17.17a}$$

$$C''_A \cong Q \omega_{od} T - 1 \tag{17.17b}$$

Another common situation is the use of SC filters at high frequencies; in such cases a structure with minimum gain-bandwidth product ($GB = \omega_u$) requirements is desirable. This structure is shown in Figure 17.10 and is often referred to as a decoupled structure. It is worthwhile to mention that two SC architectures can have ideally the same transfer function; however, with real op-amps their frequency (and time) response differs significantly. A rule of thumb to reduce the GB effects in SC filters is to minimize the direct connections between the output of one op-amp to the input of another op-amp. It is desirable to transfer the output of an op-amp to a grounded capacitor, and in the next clock phase, transfer the capacitor charge into the op-amp input. If required, the connection to the next op-amp has to be done during the hold phase of the previous op-amp. If the connection is done during the charge

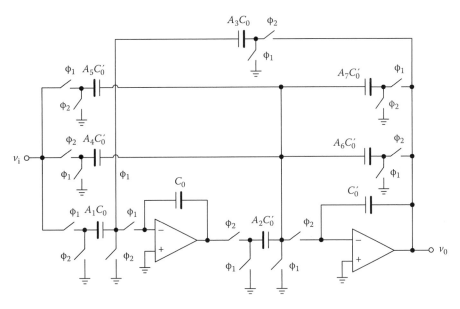

FIGURE 17.10   A decoupled SC biquadratic section.

injection phase, the output of the first integrator is a ramp (step response of the integrator), and the settling time for the next integrator increases further. More discussion on ω_u effects is given in Section 17.3.

If the input signal is sampled during ϕ_2 and held during ϕ_1, the ideal transfer function is given by

$$H^{ee}(z) = \frac{V_o^e(z)}{V_i^e(z)} = -\left(\frac{A_4}{1+A_6}\right)\left(\frac{z^2 - z\left(\frac{A_4+A_5-A_1A_2}{A_4}\right) + \frac{A_5}{A_4}}{z^2 - z\left(\frac{2+A_6+A_7-A_2A_3}{1+A_6}\right) + \frac{1+A_7}{1+A_6}}\right)$$ (17.18)

Analysis of this equation yields the following expressions:

$$r^2 = \frac{1+A_7}{1+A_6}$$ (17.19a)

$$2r\cos\theta = \frac{2+A_6+A_7-A_2A_3}{1+A_6}$$ (17.19b)

The capacitor A_7C_0' can be used as a design trade-off parameter to optimize the biquad performance. A simple set of design equations follows:

$$A_6 = \frac{(1+A_7)-r^2}{r^2}$$ (17.20a)

$$A_2 = A_3 = (1+A_7)\sqrt{\frac{1+r^2-2r\cos\theta}{r^2}}$$ (17.20b)

Under high-sampling rate and high-Q conditions, the following expressions can be obtained:

$$\omega_o \cong f_c\sqrt{\frac{A_2A_3}{1+A_6}}$$ (17.21a)

and

$$Q \cong \sqrt{\frac{A_2 A_3 (1 + A_6)}{A_6 - A_7}} \qquad (17.21b)$$

A trade-off between Q-sensitivity and total capacitance is given by A_6 and A_7.

17.3 Effects of the Op-Amp Finite Parameters

17.3.1 Finite Op-Amp DC Gain Effects

The effect of finite op-amp dc voltage gain A_o in a lossless SC integrator is to transform a lossless integrator into a lossy one. This brings degradation in the transfer function both in magnitude and phase. Typically, the magnitude deviation due to the integrator amplitude variation is not critical. By contrast, the phase deviation from the ideal integrator has a very important influence on the overall performance. When the real SC integrators are used to build a two-integrator biquadratic filter, it can be proved that the actual quality factor becomes

$$Q_A = \frac{1}{\frac{1}{Q} + \frac{2}{A_o}} \cong \left(1 - \frac{2Q}{A_o}\right) Q \qquad (17.22)$$

where Q and A_o are the desired filter's quality factor and op-amp dc gain, respectively. The actual frequency of the poles suffers small deviations as well

$$\omega_{o_A} = \frac{A_o}{1 + A_o} \omega_o \qquad (17.23)$$

From this equation we can conclude that the ω_o deviations are negligible if $A_o > 100$. However, the Q deviations can be significant depending on the Q and A_o values, e.g., Q_{error} is given by $-2Q/A_o$, which is more critical for high-Q applications.

17.3.2 Finite Op-Amp Gain-Bandwidth Product Effects

The op-amp bandwidth is very critical for high-frequency applications. The analysis is carried out when the op-amp voltage gain is modeled with one dominant pole, i.e.,

$$A_v(s) = \frac{A_o}{1 + s/\omega_p} = \frac{A_o \omega_p}{s + \omega_p} = \frac{\omega_u}{s + \omega_p} \cong \frac{\omega_u}{s} \qquad (17.24)$$

where
    A_o is the dc gain
    ω_u is approximately the unity-gain bandwidth (GB)
    ω_p is the op-amp bandwidth

The analysis taking into account $A_v(s)$ is rather cumbersome since the op-amp input–output characterization is a continuous-time system modeled by a first-order differential equation, and the rest of the SC circuit is characterized by discrete-time systems modeled by difference equations. It can be shown that the step response of a single SC circuit with an op-amp gain modeled by Equation 17.24 due to step inputs applied at $t = t_1$ is

$$V_o(t - t_1) = V_o(t_1) e^{-(t - t_1)\alpha\omega_u} + V_{od}\{1 - e^{-(t - t_1)\alpha\omega_u}\} \qquad (17.25)$$

where

$V_o(t_1)$ and V_{od} are the initial and the final output voltages, respectively
α is a topology dependent voltage divider, $0 < \alpha \leq 1$, given by the following expression:

$$\alpha = \frac{\sum C_f}{\sum_i C_i} \tag{17.26}$$

where

C_f sum consists of all feedback capacitors connected directly between the op-amp output and the inverting input terminal
C_i sum comprises the capacitors connected to the negative op-amp terminal

Note that $\alpha \omega_u$ should be maximized for fast settling time. For the common case the SC circuit is driven by 50% duty cycle clock phases, the integrating time is close to $T/2$, hence the figure of merit to be maximized is $\alpha T \omega_u / 2$. The output voltage will settle to its final value within an error of 0.67% after five equivalent time constants, hence a rule of thumb yielding reduced GB product effects requires

$$\frac{\alpha T \omega_u}{2} > 5 \tag{17.27a}$$

or

$$f_u > \frac{1.6}{\alpha} f_s \tag{17.27b}$$

For the multiple op-amp case the basic concept prevails. For a two-clock phase SC filters the system can be described by an $\mathbf{M} \times \mathbf{M}$ matrix $\mathbf{A}$ for each clock-phase, where $\mathbf{M}$ is the number of op-amps in the filter architecture. For a fixed clock frequency, $f_c = 1/T$, there are two components to be optimized: α and ω_u. effects biquadratic SC filter presented in the previous section (Figure 17.10) has the property that matrix $\mathbf{A}$ has been optimized. For illustration on how to determine the $\mathbf{A}$ matrix of an SC filter, let us consider the filter depicted in Figure 17.10; for each clock phases matrix $\mathbf{A}$ is given by

$$\mathbf{A}\big|_{\varphi_1} = \begin{bmatrix} \frac{1}{1+A_1+A_3} & 0 \\ 0 & 1 \end{bmatrix} \tag{17.28}$$

$$\mathbf{A}\big|_{\varphi_2} = \begin{bmatrix} 1 & \frac{A_2}{1+A_2+A_4+A_5+A_6+A_7} \\ 0 & \frac{1+A_8}{1+A_2+A_4+A_5+A_6+A_7} \end{bmatrix} \tag{17.29}$$

Thus, the designer should use the extra degrees of freedom to maximize the diagonal entries of $\mathbf{A}$ during all phases. Thus, the worst case for any entry of $\mathbf{A}$ determines the minimum value of ω. Another design consideration could be to maximize ω_u. This last consideration should be carefully addressed since very large ω_u values besides extra power consumption can cause excessive phase margin degradation or excessive noise if the capacitors are reduced. Therefore, a judiciously trade-off must be used in choosing the ω_u of each op-amp in the filter topology.

17.4 Noise and Clock Feedthrough

The lower range of signals that can be processed by the electronic devices is limited by several unwanted signals that appear at the output of the circuit. The RMS values of these electrical signals determine the system noise level, and it represents the lowest limit for the incoming signals to be processed. In most of the cases the circuit cannot properly detect input signals smaller than the noise level.

The most critical noise sources are those due to (1) the employed elements (transistors, diodes, resistors, etc); (2) the noise induced by the clocks; (3) the harmonic distortion components generated due to the intrinsic nonlinear characteristics of the devices; and (4) the noise induced by the surrounding circuitry. In this section, types (1), (2), and (3) are considered. Using fully differential structures can further reduce the noise generated by the surrounding circuitry and coupled to the output of the SC circuit. These structures are partially treated in Section 17.5 but excellent references can be found in the literature [2,4,6,9,13].

17.4.1 Noise due to the Op-Amp

In an MOS transistor, the noise is generated by different mechanisms but there are two dominant noise sources: channel thermal noise and $1/f$ or flicker noise. A discussion of the nature of these noise sources follows.

17.4.2 Thermal Noise

The flow of the carriers due to the drain-source voltage takes places on the source-drain channel, mostly like in a typical resistor. Therefore, due to the random flow of the carriers, thermal noise is generated. For an MOS transistor biased in the linear region the spectral density of the thermal noise is approximately given by [1–4,12]

$$V_{eqth}^2 = 4kTR_{on} \tag{17.30}$$

where R_{on}, k, and T are the drain-source resistance of the transistor, the Boltzmann constant, and the temperature (in Kelvin degrees), respectively. In saturation region, the gate-referred spectral noise density can be calculated by the same expression but with R_{on} equal to δ/g_m, being g_m the small-signal transconductance of the transistor and δ a fitting parameter usually in the range of 0.7–1.

17.4.3 $1/f$ Noise

This type of noise is mainly due to the imperfections in the silicon–silicon oxide interface. The surface states and the traps in this interface randomly interfere with the charges flowing through the channel; hence the generated noise is strongly dependent of the technology. The $1/f$ noise (or flicker noise) is inversely proportional to the gate area because at larger areas more traps and surface states are present and some averaging occurs. The spectral density of the gate referred $1/f$ noise is commonly characterized as

$$V_{eq1/f}^2 = \frac{k_F}{WLf} \tag{17.31}$$

where the product WL, f, and k_F are the gate area of the transistor, the frequency in hertz, and the flicker constant, respectively. The spectral noise density of the MOS transistor is composed by both components, therefore the input referred spectral noise density of a transistor operated in its saturation region becomes

$$V_{eq}^2 = \frac{4\delta kT}{g_m} + \frac{k_F}{WLf} \tag{17.32}$$

17.4.4 Op-Amp Noise Contributions

In an op-amp, the output referred noise density is composed by the noise contribution of all transistors; hence the noise level is function of the op-amp architecture. A typical unbuffered folded cascode op-amp (folded cascode OTA) is shown in Figure 17.11. For the computation of the noise level, the contribution of each transistor has to be evaluated. Obtaining the OTA output current generated by the gate-referred noise of all the transistors can do this. For instance, the spectral density of the output-referred noise current due to M_1 is straightforwardly determined because the gate-referred noise is at the input of the OTA, leading to

$$i_{o1}^2 = G_m^2 V_{eq1}^2 \qquad (17.33)$$

where

    G_m (equal to g_{m1} at low frequencies) is the OTA transconductance
    V_{eq1}^2 is the input (gate) referred noise density of M_1

Similarly, the contributions of M_2 and M_5 to the spectral density of the output referred noise current are given by

$$\begin{aligned} i_{o2}^2 &= g_{m2}^2 v_{eq2}^2 \\ i_{o5}^2 &= g_{m5}^2 v_{eq5}^2 \end{aligned} \qquad (17.34)$$

Usually the noise contributions of transistors M_3 and M_4 at medium frequencies are very small in comparison to the other components; this is a nice property of the cascode transistors. This is because their noise drain currents, due to the source degeneration implicit in the cascode transistors, are determined by the equivalent conductance associated with their sources instead of those by their transconductance. Since in a saturated MOS transistor the equivalent conductance is much smaller than the transistor transconductance, this noise drain current contribution can be neglected at low and medium frequencies; this is however not the case at very high frequencies where the parasitic capacitors play an important role. The noise contribution of M_6 is mainly common-mode noise evenly split in the

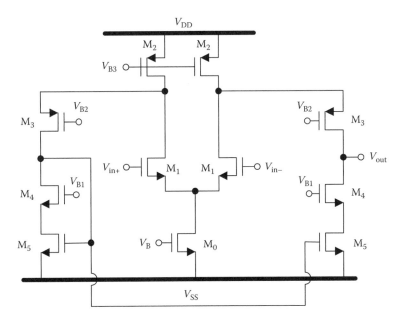

FIGURE 17.11 A folded cascode operational transconductance amplifier.

two arms by transistor M_1, therefore it is ideally canceled at the OTA output due to the current subtraction. At low frequencies, the spectral density of the total output referred noise current can be approximately calculated as

$$i_o^2 = 2\left(g_{m1}^2 V_{eq1}^2 + g_{m2}^2 v_{eq2}^2 + g_{m5}^2 v_{eq5}^2\right) \tag{17.35}$$

The factor 2 is the result of the pair of transistors M_1, M_2, and M_5. From this equation, the OTA input referred noise density is obtained by dividing the output noise current by the square of $G_m = g_{m1}$, yielding

$$v_{OTAin}^2 = 2v_{eq1}^2\left(1 + \frac{g_{m2}^2 v_{eq2}^2 + g_{m5}^2 v_{eq5}^2}{g_{m1}^2 v_{eq1}^2}\right) \tag{17.36}$$

According to this result, if g_{m1} is larger than g_{m2} and g_{m5}, the OTA input referred noise density is mainly determined by the OTA input stage. In that case and using Equation 17.32 and 17.36 yields

$$v_{OTAin}^2 \cong 2v_{eq1}^2 = \frac{2k_F}{W_1 L_1 f} + \frac{8\delta kT}{g_{m1}} = v_{eq1/f1}^2 + 4kTR_{eqth} \tag{17.37}$$

where the factor 2 has been included in $v_{eq1/f}$ and R_{eqth}. In Equation 17.37, $v_{eq1/f}^2$ is the equivalent $1/f$ noise density and R_{eqth} is the equivalent resistance for noise, equal to 2 δ/g_{m1}.

17.4.5 Noise in a Switched-Capacitor Integrator

In an SC lossless integrator, the output referred noise density component due to the OTA is frequency limited by the GB product of the OTA. In order to avoid misunderstandings, in this section f_u (the unity-gain frequency of the OTA in hertz) is used instead of ω_u (in radians per second). Since f_u must be higher than the clock frequency, high-frequency noise is folded back into the integrator baseband. In the case of the SC integrator and assuming that the flicker noise is not folded back, the output referred spectral noise density becomes

$$V_{oeq1}^2 = \left(v_{eq1/f}^2 + 4KTR_{eqth}\left(1 + \frac{2f_u}{f_c}\right)\right)|1 + H(z)|^2 \tag{17.38}$$

where the folding factor is equal to f_u/f_c, f_c is the clock frequency and $H(z)$ is the z-domain transfer function of the integrator. The factor $2f_u/f_c$ is the result of both positive and negative folding. Typically, the frequency range of the signal to be processed is around and below the unity-gain frequency of the integrator, therefore $|H(z)| > 1$ and Equation 17.38 can be approximated as

$$V_{oeq1}^2 = \left(v_{eq1/f}^2 + 4kTR_{eqth}\left(1 + \frac{2f_u}{f_c}\right)\right)|H(z)|^2 \tag{17.39}$$

17.4.6 Noise due to the Switches

In SC networks, the switches are implemented with single or complementary MOS transistors. These transistors are biased in the cutoff and ohmic region for open and close operations, respectively. In cutoff region, the drain-source resistance of the MOS transistor is very high, then the noise contribution of the

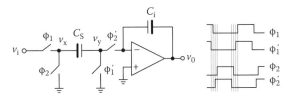

FIGURE 17.12 Typical switched-capacitor lossless integrator.

switch is confined to very low frequencies and it can be considered as dc offset. When the switch is in the on state, the transistor is biased in linear region and its spectral noise distribution is characterized by Equation 17.30. This noise contribution is the most fundamental limit for the signal-to-noise ratio of SC networks. The effects of these noise sources are better appreciated if an SC integrator is considered.

Let us consider the SC integrator of Figure 17.12 and assume that $\phi_1 = \phi_1'$ and $\phi_2 = \phi_2'$. The spectral noise density of the ϕ_1-driven switches are low-pass filtered by the on resistance of the switches R_{on} and the sampling capacitor C_s. The cutoff frequency of this continuous-time filter is given by $f_{on} = 1/(2\pi R_{on} C_s)$. Typically, f_{on} is higher than the clock frequency, therefore, the high-frequency noise is folded back into the baseband of the integrator when the SC integrator samples it. Taking into account the folding effects, the spectral noise density component of the switch-capacitor combination becomes

$$v_{eq2}^2 = 4kTR_{on}\left(1 + \frac{2f_{on}}{f_c}\right) \tag{17.40}$$

with R_{on} the switch resistance. Considering that the noise bandwidth is given by $2\pi f_{on} = 1/R_{on}C_s$, then the integrated noise level yields:

$$v_{eq-integrated}^2 = \frac{kT}{C_s}\left(1 + \frac{2f_{on}}{f_c}\right) \tag{17.41}$$

For the noise induced by the ϕ_2 driven switches, the situation is slightly different, but for practical purposes the same equation is used. According to Equation 17.41, the integrated noise power is inversely proportional to C_s. Therefore, for low-noise applications it is desirable to design the integrators with large capacitors. However, the costs for the noise reduction are larger silicon area and higher power consumption. This last result is because the slew rate of the OTA is inversely proportional to the load capacitance and in order to maintain the specifications it is necessary to increase the current drive capability of the OTA. Clearly, there is a trade-off between noise level, silicon area, and power consumption.

17.4.7 Clock Feedthrough

Another factor that limits the accuracy of SC networks is the charge induced by the clocking of the switches. These charges are induced by the gate-source capacitance, the gate-drain capacitance, and the charge stored in the channel. Furthermore, some of these charges are input signal dependent and introduce distortion in the circuit. Although these errors cannot be canceled, there are some techniques that reduce these effects.

The analysis of the clock feedthrough is not straightforward because it depends on the order of the clock phases, the relative delay of the clock phases, as well as the speed of the clock transition [13]. For instance, let us consider in Figure 17.12 the case when ϕ_1 goes down before ϕ_1'. This situation is shown in Figure 17.13a.

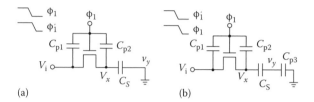

FIGURE 17.13 Charge induced due to the clocks. (a) If ϕ_1 goes down before ϕ_1' and (b) If ϕ_1' goes down before ϕ_1.

While C_{p1} is connected between two low-impedance nodes, C_{p2} is connected between ϕ_1, a low-impedance node, and v_x. For $\phi_1 > v_i + V_T$ the transistor is on and the current injected by C_{p2} is absorbed by the drain-source resistance; then v_x remains at a voltage equal to v_i. When the transistor is turned off, $\phi_1 < v_i + V_T$, charge conservation at node v_x leads to

$$v_x = v_i + \frac{C_{p2}}{C_S + C_{p2}}(\text{VSS} - v_i - V_T) \tag{17.42}$$

where VSS is the low level of ϕ_1 and ϕ_2. During the next clock phase, the charge of C_S is injected to C_i. Thus, C_{p2} induces a charge error proportional to $v_i C_{p2}/(C_S + C_{p2})$. In addition, an offset voltage proportional to VSS $- V_T$ is also generated. Because the threshold voltage V_T is a nonlinear function of v_i, an additional error in the transfer function and harmonic distortion components appear at the output of the integrator. The same effect occurs when the clock phases ϕ_2 and ϕ_2' have a similar sequence.

Let us consider the case when ϕ_1' is opened before ϕ_1; the situation is shown in the Figure 17.13b. Before the transistors turn off, $v_x = v_i$ and $v_y = 0$. When the transistor is turned off, $\phi_1 < v_i + V_T$, the charge is recombined between C_S, C_{p2}, and C_{p3}. After the charge redistribution, the charge injected into C_{p3} is approximately given by

$$\Delta Q_{C_{p3}} = \frac{C_{p2} C_{p3}}{C_{p2} + C_{p3}}(\text{VSS} - v_i - V_T) \tag{17.43}$$

Notice that the current feedthrough generated by C_{p2} flows through both capacitors C_S and C_{p3}, hence $\Delta Q_{C_{p3}} = -\Delta Q_{C_S}$. During the next clock phase, both capacitors C_S and C_{p3} transfer the ideal charge $C_S v_i + \Delta Q_{C_S}$ and $\Delta Q_{C_{p3}}$, making the clock feedthrough induced error close to zero. The conclusion is that if the clock phase ϕ_1' is a bit delayed than ϕ_1 the clock-induced error is further reduced. This is also true for the clock phases ϕ_2 and ϕ_2'.

In Figure 17.12, the right-hand switch also introduces clock feedthrough, but unlike the clock feedthrough previously analyzed, this is input signal independent. When the clock phase ϕ_2 goes down, the gate-source overlap capacitor extracts from the summing node the following charge:

$$\Delta Q = C_{GS}(\text{VSS} - V_T) \tag{17.44}$$

In this case, V_T does not introduce distortion because v_y is almost at zero voltage for both clock phases. The main effect of this charge is to introduce an offset voltage. The same analysis reveals that the bottom right-hand switch introduces similar offset voltage. From the previous analysis it can be seen that using minimum transistors dimension can reduce the clock feedthrough. This implies minimum parasitic capacitors and minimum induced charge from the channel. If possible, the clock phases should be arranged for minimum clock feedthrough. The use of extra dummy switches driven by the complementary clock phase may alleviate the clock feedthrough issue, but its effectiveness must be carefully evaluated by postlayout simulations.

17.4.8 Channel Mobile Charge

Charge injection also occurs due to the mobile charge in the channel. If the transistor is biased in linear region, the channel mobile charge can be approximated by

$$Q_{ch} = C_{GS}(V_{GS} - V_T) \tag{17.45}$$

where C_{GC} is the gate-channel capacitor; see Figure 17.14.

When the switch is turned off this charge is released and part of it goes to the sampling capacitor. Fortunately, the previously discussed early clock phase technique for the reduction of clock feedthrough also reduces the effects of the channel mobile charge injection. The mobile charge injected to the sampling capacitor is a function of several parameters, e.g., input signal, falling rate of the clock, overlap capacitors, gate capacitor, threshold voltage, integrating capacitor, and the supply voltages.

The effects of the channel mobile charges are severe when the ϕ_1-driven transistor is opened before the ϕ_1'-driven transistor. In this case, the situation is similar to that shown in Figure 17.13a, in which one terminal of C_S is still grounded. While ϕ_1 is higher than $v_i + V_T$ the channel resistance is small and v_i absorbs most of the channel-released charge. For $\phi_1 < v_i + V_T$, the channel resistance increases further and a substantial amount of charge released by the channel will flow through C_S, introducing a charge error. If the clock phases are arranged as shown in Figure 17.13b, most of the charge released by the channel returns back to v_i. The main reason is because the equivalent capacitor seen at the right-hand side of the transistor is nearly equal to C_{p3}, if C_{p2} is neglected, making this a high-impedance node. Because this parasitic capacitor is smaller than the sampling capacitor, a small amount of extracted (or injected) charge will produce a huge variation on v_y, see Figure 17.13b, pushing back most of the mobile charges. Equation 17.45 shows that the mobile charge is a function of V_{GS} ($=V_{clock} - V_{in}$ for the sampling switch), hence the part of the charge injected into the sampling capacitor is signal dependent. This issue can be alleviated if the clock is correlated with the incoming signal such that VGS is maintained constant using charge pumps circuits; the technique proposed in Ref. [14] implements this concept.

17.4.9 Dynamic Range

The dynamic range is defined as the ratio of the maximum signal that the circuit can drive without significant distortion to the noise level. The maximum distortion tolerated by the circuit depends on the application, but −60 dB is commonly used. Since the linearity of the capacitor is good enough and if the harmonic distortion components introduced by the OTA input stage are small, the major limitation for

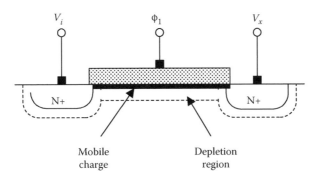

FIGURE 17.14 Cross section of the MOS transistor showing the mobile charge.

the linearity is determined by the output stage of the OTA. For the folded cascode OTA of Figure 17.11 this limit is

$$v_{omax} \cong V_{R2} + V_{TP3} \tag{17.46}$$

If the reference voltage V_{R2} is maximized, Equation 17.46 yields

$$V_{omax} \cong V_{DD} - 2V_{DSATP} \tag{17.47}$$

A similar expression can be obtained for the lowest limit. Assuming a symmetrical single-side output stage, from Equation 17.47, the maximum rms value of the OTA output voltage is given by

$$v_{oRMS} \cong \frac{V_{DD} - 2V_{DSATP}}{\sqrt{2}} \tag{17.48}$$

If the in-band noise, integrated up to $\omega = 1/R_{int}C_I$, is considered and if the most important term of Equation 17.41 is retained, the dynamic range of the single-ended SC integrator becomes

$$DR \cong \frac{(V_{DD} - 2V_{DSATP})}{\sqrt{2}\sqrt{2kT/C_S}} = \frac{(V_{DD} - 2V_{DSATP})}{2\sqrt{kT}}\sqrt{C_S} \tag{17.49a}$$

For this equation, the alias effects have been neglected and only the noise contribution of the switches driven by ϕ_1 and ϕ_2 are considered. At room temperature, this equation is reduced to the following expression:

$$DR \cong 7.9 \times 10^9 \sqrt{C_S}(V_{DD} - 2V_{DSATP}) \tag{17.49b}$$

According to this result, the dynamic range of the SC integrator is reduced when the power supplies are scaled down and minimum capacitors are employed. Clearly, there is a trade-off between power consumption, silicon area, and dynamic range. As an example, for the case of $C_S = 1.0$ pF and supply voltages of ± 1.5 V and $V_{DSATP} = 0.25$ V, the dynamic range of a single integrator is around 78 dB. For low-frequency applications, however, the dynamic range is lower due to the low-frequency flicker noise component. This is a very optimistic result since neither the op-amp noise not aliasing effects nor other noise sources were considered.

17.5 Design Considerations for Low-Voltage Switched-Capacitor Circuits

For the typical digital supply voltages, 0−5 V, SCs achieve dynamic ranges of the order of 80–100 dB. As long as the power supplies are reduced, the swing of the signal decreases and the switch resistance increases further. Both effects reduce the dynamic range of the SC networks. A discussion of these topics follows.

17.5.1 Low-Voltage Operational Amplifiers

The design techniques for low-voltage low-power amplifiers for SC circuits have been addressed by several authors [4,7–9,11–13]. The implementation of op-amps for low-voltage applications does not seem to be a fundamental limitation as long as the transistor threshold voltage is smaller than $(V_{DD} - V_{SS})/2$. This limitation will become clear in the design example presented in this section.

The design of the operational amplifier is strongly dependent on the application. For high-frequency circuits, the folded cascode op-amp is suitable, but the cascode transistors limit the swing of the signals at the output stage. If large output voltage swing is needed, complementary stages are desirable. To illustrate the design trade-offs involved in the design of a low-voltage OTA, let us consider the folded cascode OTA of Figure 17.11. For low-voltage applications and small signals, the transistors have to be biased with very low V_{DSAT} ($=V_{GS} - V_T$).

For the case of supply voltages limited to ± 0.75 V and $V_T = 0.5$ V, $V_{DSAT1} + V_{DSAT6}$ must be lower than 0.25 V, otherwise the transistor M_6 goes to the triode region. For large signals, however, the variations of the input signal produce variations at the source voltage of M_1. It is well known that linear range of the differential pair is of the order of $\pm 1.4\, V_{DSAT1}$. Hence, if the noninverting input of the OTA is connected to the common-mode level V_{CM}, for proper operation of the OTA input stage (see Figure 17.11) it is desirable to satisfy

$$V_{CM} - V_T - VSS > 1.4 V_{DSAT1} + V_{DSAT6} \tag{17.50}$$

It has to be taken into account that the threshold voltage of M_1 increases if an N-well process is used, due to the body effects. In critical applications, PMOS transistors fabricated in a different well with their source tied to their own well can be used. Equation 17.50 shows that it is beneficial to select a proper common-mode level V_{CM} to extend the linear range of the amplifier input stage, especially for low-voltage applications.

The dimensioning of the transistors and the bias conditions are directly related to the application. For instance, if the SC integrator must slew 1 V into 4 µs and the load capacitor is of the order of 10 pF, the OTA output current must be equal to or higher than 2.5 µA. Typically, for the folded cascode OTA the dc current of both output and input stages are the same. Therefore, the bias current for M_1, M_3, M_4, and M_5 can be equal to 2.5 µA. The bias current for M_2 and M_6 is 5 µA. If $V_{GS1} - V_{T1}$ is fixed at 100 mV the dimensions of M_1 can be computed according to the required small signal transconductance. Similarly, the dimension of the other transistors can be calculated, most of them designed to maximize the output swing and dc gain. A very important issue in the design of low-voltage amplifiers is the reference voltage. In the folded cascode of Figure 17.11, the values of the reference voltages V_{R1} and V_{R2} must be optimized for maximum output swing. OTA design techniques are fully covered in Refs. [11] and [13].

17.5.2 Analog Switches

For low-voltage applications, the highest voltage that can be processed is limited by the analog switches rather than by the op-amps. For a single NMOS transistor, the switch resistance is approximately given by

$$R_{DS} = \frac{1}{\mu_n C_{OX} \frac{W}{L} (V_{GS} - V_T)} \tag{17.51}$$

where μ_n and C_{OX} are technological parameters. According to Equation 17.51, the switch resistance increases further when V_{GS} approaches to V_T. This effect is shown in Figure 17.15 for the case $V_{DD} = -V_{SS} = 0.75$ V and $V_T = 0.5$ V. From this figure, the switch resistance is higher than 300 kΩ for input signals of 0.2 V. However, for a drain-source voltage higher than $V_{GS} - V_T$ the transistor saturates and does not behave as a switch anymore. This limitation clearly reduces further the dynamic range of the SC circuits.

A solution for this drawback is to generate the clocks from higher voltage supplies. A simplified diagram of a voltage doubler is depicted in Figure 17.16a. During the clock phase ϕ_1, the capacitor C_1 is charged to V_{DD} and during the next clock phase its negative plate is connected to V_{DD}. Hence, at the

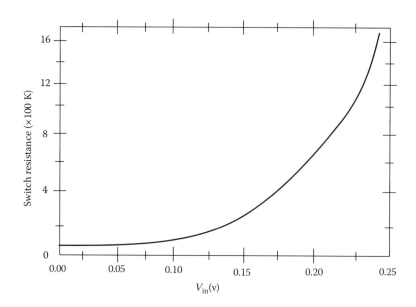

FIGURE 17.15 Typical switch resistance for an NMOS transistor.

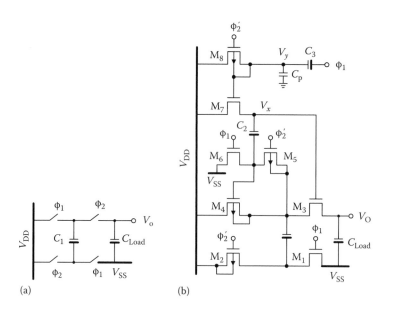

FIGURE 17.16 Voltage doubler. (a) Simplified diagram and (b) transistor level diagram.

beginning of ϕ_2, the voltage at the top plate of C_1 is equal to $2V_{DD} - V_{SS}$. After several clock cycles, if C_{LOAD} is not further discharged, the charge is recombined leading to an output voltage equal to $2V_{DD} - V_{SS}$. A practical implementation for an N-well process is shown in Figure 17.16b.

In this circuit, the transistors M_1, M_2, M_3, and M_4 behave as the switches S_1, S_2, S_3, and S_4 respectively, of Figure 17.16a. While for M_1 and M_2 the normal clocks are used, special clock phases are generated for M_3 and M_4 because they drive higher voltages. The circuit operates as follows.

During ϕ_1, M_8 is opened because ϕ_2' is high. The voltage at node v_y is higher than V_{DD} because the capacitors C_3 and C_p were charged to V_{DD} during the previous clock phase ϕ_2 and, when ϕ_1 goes up, charge is injected to the node through the capacitor C_3. Since v_y is higher than V_{DD}, M_7 is turned on. The bottom plate of C_2 is connected to V_{SS} by M_6 and C_2 is charged to $V_{DD}-V_{SS}$.

During ϕ_2, the refresh clock phase, the bottom plate of C_1 is connected to V_{DD} by the PMOS transistor M_2. Note that if an NMOS transistor is employed, the voltage at the bottom plate of C_1 is equal to $V_{DD}-V_T$, resulting in lower output voltage. During ϕ_2 if C_1 is not discharged, the voltage at its top plate is $2V_{DD} - V_{SS}$. The voltage at node v_x becomes close to $3V_{DD}-2V_{SS}$ volts, turning M_3 on and enabling the charge recombination of C_1 and C_{LOAD}. As a result, after several clock periods, the output voltage v_o becomes equal to $2V_{DD}-V_{SS}$. To avoid discharges the gate of M_4 is connected to the bottom plate of C_2. Thus, M_4 is turned off during the refresh phase. An efficient solution using boosted clock switches is described in Ref. [14]. Another technique using switched op-amps, described in Ref. [15], have been also successfully used.

17.6 Fully Differential Filters

An important issue when designing a high-frequency filter is the proper selection of the clock phases for the switches. It is desirable that the input to each integrator is as close as possible to a step function, and to minimize the loading capacitors during the integration clock phase. Making the OTAs to perform the hold and integrate functions at alternate phases ensure this. Unfortunately this approach cannot be implemented in the conventional single-ended topologies unless an additional inverter is added to the loop. Two backward integrators with a half-delay in each and by crossing the outputs of the fully differential integrators in order to have an odd number of inversions in the loops is one of the best configurations for high-frequency applications as depicted in Figure 17.17. The first op-amp may process the information

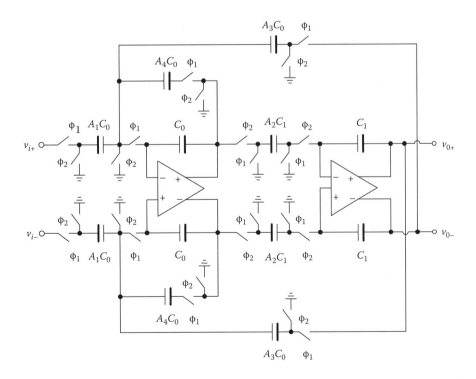

FIGURE 17.17 Fully decoupled, fully differential biquadratic filter.

during ϕ_1 and hold it during ϕ_2 while the second op-amp do the same functions during the complementary clock phases. In general, having the differential outputs available makes the system more flexible and gives several advantages while designing high-performance analog signal processors.

Fully differential amplifiers are widely used technique to reduce the effect of charge injection, clock feedthrough, and better rejection to common-mode signals such as substrate and power supply noise and an improved dynamic range when compared to single-ended circuits. As long as the noise present at the output of the amplifier is present in both outputs of the amplifier with the same amplitude and same phase, it will be rejected by the differential nature of the following stage. In addition, an additional loop that operates on the common-mode signals presents low output impedance for these signals, it fix the operating point of the differential outputs at the desired level and further minimizes the common-mode output fluctuations up to the unity-gain frequency of the loop. Their disadvantage comes from the fact that they require a common-mode feedback (CMFB) circuit.

The fully differential version of the folded cascode amplifier is depicted in Figure 17.18a. It requires the same amount of power as the single-ended version shown in Figure 17.11. Since the current mirror present in the single-ended amplifier for the conversion of the fully differential current of the differential pair into single-ended output is eliminated, the fully differential amplifier has a single internal pole located at the source terminal of M_3. Since several parasitic poles and phantom zeros are eliminated, the fully differential amplifier has inherently better phase margin than its single-ended counterpart. The circuit is fully symmetric and any noise injected through M_0 is evenly split by the differential pair and will appear at the two outputs as a pure common-mode signal. The common-mode noise present at the transistors M_1 is further attenuated by the small sensitivity of the differential pair to common-mode

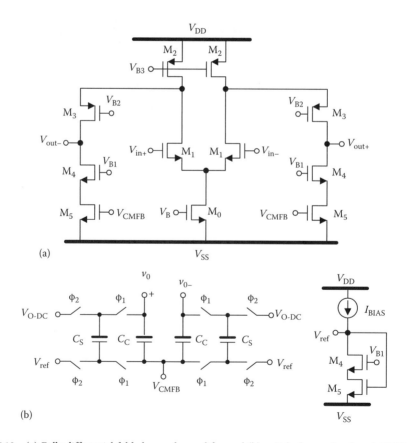

FIGURE 17.18 (a) Fully differential folded cascode amplifier and (b) switched-capacitor based CMFB circuit.

signals; the small common-mode current resulting at the drain of M_1 will be reflected at the output as a common-mode signal. Similarly, any V_{DD} noise generates a current noise through M_2 that appears as a common-mode noise at both outputs of the amplifier; notice that to minimize these components V_{B3} must be further attached to V_{DD}. V_{SS} noise generates current that appears at the output as common-mode noise, as well thanks to the differential nature of the topology. The gate voltage of transistors M_5 is used to accommodate the control of the CMFB loop; V_{B3} can also be used for that purpose.

The CMFB is a block that extracts the common-mode signal present at both outputs and compares its dc value with a reference voltage in a loop with enough gain for the common-mode signals. If the loop has enough gain, the error between the detected common-mode signal and the reference voltage is minimized forcing the common-mode output signals (DC operating point of each amplifier output) to be at the reference level. Several CMFB circuits have been proposed in the literature [4,11,12] but for SC applications it has been preferred to use SC-based CMFB (SC–CMFB) circuits [4] since they do not consume significant power and have better linearity when compared to their continuous-time counterpart. The CMFB circuits consist of a common-mode voltage detector (output is proportional to $\frac{v_{out+} + v_{out-}}{2}$ and a circuit that generates the common-mode error after comparing the common-mode output voltage and a proper and stable reference voltage. The efficient SC detector, shown in Figure 17.18b, uses two SC resistors (C_C and switches) that detect the common-mode output voltage and compare it with V_{O-DC} [4,9,13]. Using conventional charge redistribution analysis and ignoring the effect of C_C, the voltage at node V_{CMFB} during ϕ_1 is given by

$$V_{CMFB} = \left(\frac{v_{out+} + v_{out-}}{2} - V_{O-DC}\right) + V_{ref} \tag{17.52}$$

The circuit makes the comparison of the common-mode output voltage and the desired output DC level V_{O-DC}. A remarkable advantage of this circuit is that in addition to the desired comparison, it translates the ideal voltage V_{ref} needed at the gate of M_5 to generate the DC current in transistors M_5. In steady state, V_{CMFB} will be very close to V_{ref}. Equation 17.52 is also valid for the common-mode noise; in the absence of noise in V_{ref} and V_{O-DC}, the common-mode output noise is compared with zero and for the frequencies where the CMFB loop is high, it is suppressed, thanks to the feedback. Hence it is desirable not only to have large low-frequency gain but wide-band CMFB loop to reject high-frequency noise as well.

Since the SC resistors and the parasitic capacitors generate a relatively low-frequency pole at V_{CMFB} node, the capacitors C_S are added to introduce a compensating zero that helps with the stabilization of the CMFB loop. One of the drawbacks of this topology is that the common detector loads significantly the output of the amplifier, reducing its unity-gain frequency and slew rate. The value of C_S is typically one-quarter to one-tenth of C_C. To prevent significant offset voltages due to charge injection from the switches, C_S is commonly chosen to be greater than 200 fF. Therefore, the capacitive loading on the amplifier due to the CMFB can go from 1 pF to 2 pF in most designs. As demonstrated in Ref. [16] this SC–CMFB suffers of poor rejection to the supply noise used as reference for the common-mode loop. This can be partially alleviated by combining passive common-mode detector and active signal comparator at the expenses of additional power consumption or using continuous-time CMFB systems.

17.7 Design Examples

17.7.1 Biquadratic Filter

In this section, a second-order bandpass filter (BPF) is designed. The specifications for this biquad are

Center frequency 1.6 kHz
Quality factor 16
Peak gain 10 dB
Clock frequency 8 KHz

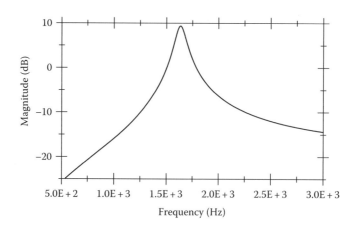

FIGURE 17.19   Frequency response of the second-order bandpass filter.

A transfer function in the z-domain that realizes this filter is given by the following expression:

$$H(z) = \frac{0.1953(z-1)z}{z^2 + 0.5455z + 0.9229} \qquad (17.53)$$

The equivalent $H(s)$ is obtained and then mapped into the z-domain using the LDI transformation. This transfer function can be implemented by using the biquads presented in Section 17.2. For the biquad of Figure 17.10, and employing $A_1 = A_4 = A_7 = 0$, the circuit behaves as a BPF. Equating the terms of Equation 17.53 with the terms of Equation 17.19, the following equations are obtained:

$$A_6 = \frac{1}{0.9229} - 1$$

$$A_5 = \frac{0.1953}{0.9229} \qquad (17.54)$$

$$A_2 A_3 = 2 + A_6 - \frac{0.5455}{0.9229}$$

Solving these equations, the following values are obtained: $A_6 = 0.0835$, $A_5 = 0.2116$, and $A_2 A_3 = 1.4924$. A typical design procedure employs $A_2 = 1$. For this case, after node scaling the total capacitance is of the order of 32 unit capacitances. The frequency response of the filters is shown in Figure 17.19.

17.7.2 Sixth-Order Bandpass Ladder Filter

In this section, the design procedure for a sixth-order BPF based on an RLC prototype is considered. The ladder filters are very attractive because of their low passband sensitivity to tolerances in the filter components. Let us consider the following specifications:

Center frequency 100 kHz
Bandwidth 2.5 kHz
Passband ripple <0.5 dB
Clock frequency 2 MHz

The design starts with an RLC low-pass prototype, see Figure 17.20.

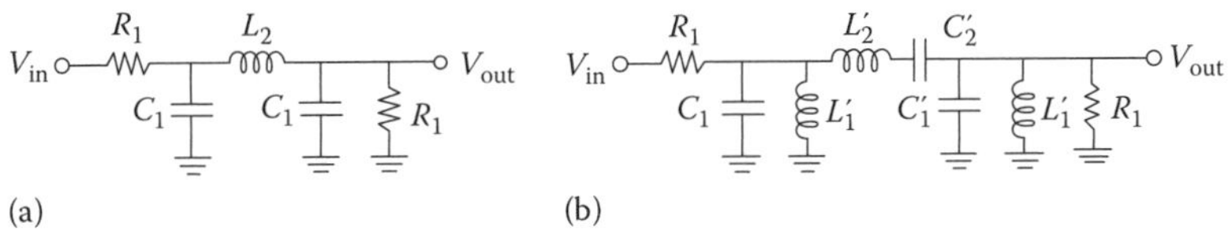

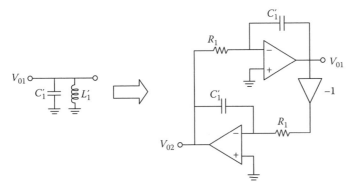

FIGURE 17.20   RLC prototypes. (a) Third-order low-pass filter and (b) sixth-order bandpass filter.

The values of the components for a prototype with 1 rad/s passband frequency can be obtained from tables or computed from well-known expressions. For this example, the passive components are $R_1 = 1\ \Omega$, $C_1 = 1.5963$ F, and $L_2 = 1.0967$ H. Using the low-pass to bandpass transformation, the capacitor C_1 is transformed into a parallel of a capacitor and an inductor. The values of the resulting components are

$$C_1' = \frac{C_1}{BW}$$
$$L_1' = \frac{1}{\omega_0^2 C'}$$

(17.55)

where ω_0 and BW are the center frequency in radians per second and the filter bandwidth in radians per second, respectively. The inductor L_2 is transformed in a series of an inductor and a capacitor whose values become

$$L_2' = \frac{L_2}{BW}$$
$$C_2' = \frac{1}{\omega_0^2 L_2'}$$

(17.56)

The bandpass prototype is shown in Figure 17.20b. Before making the denormalizations it is desirable to transform the passive prototype to an active implementation. The grounded LC tank circuit can be simulated by the RC active implementation shown in Figure 17.21.

The value of the simulated inductance is

$$L_1' = R_1^2 C_1'$$

(17.57)

FIGURE 17.21   RC active implementation of a grounded LC tank circuit.

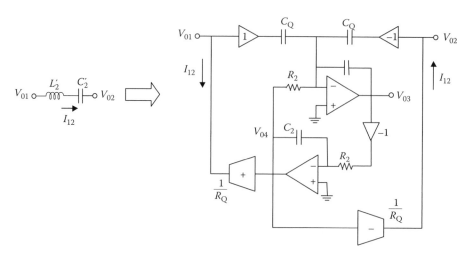

FIGURE 17.22   RC active implementation of a floating LC tank circuit.

It is important to note that the output of the active-RC circuit is a low-impedance node. Therefore, the inverting input of the op-amps must be used for current injection. Similarly, the circuit shown in Figure 17.22 can simulate a floating resonator. Obviously, this is an expensive implementation but typically several active elements can be shared in the final design. Using typical circuit analysis techniques it can be shown that i_{12} for the passive LC tank is related to the components by the following expression:

$$i_{12} = \frac{sC_2'}{1 + s^2 L_2' C_2'} (v_{o1} - v_{o2}) \tag{17.58}$$

and for the active implementation

$$i_{12} = \frac{\left(\frac{R_2}{R_Q} C_Q\right) s}{1 + s^2 R_2^2 C_2^2} (v_{o1} - v_{o2}) \tag{17.59}$$

Comparing Equations 17.57 and 17.58, the active and passive implementations are equivalents if the following constraints are satisfied:

$$C_2' = \frac{R_2}{R_Q} C_Q \tag{17.60}$$

$$L_2' C_2' = R_2^2 C_2^2 \tag{17.61}$$

Obviously, the resistors R_2 can be replaced by SC equivalents. Using these building blocks, the implementation of the sixth-order BPF is straightforward. The sixth-order SC BPF is shown in Figure 17.23. The implementation of the $L_1' C_1'$ tank circuit is straightforward from Figure 17.21. The resonant frequency of the SC resonator is determined by the capacitors θC_1. The node voltage v_{o1} is taken at the output of the first op-amp. The input resistor is implemented by the capacitor C_R and its associated switches. In Figure 17.23, the resonator associated with v_{o3} and v_{o4} and the capacitors C_Q implement the $L_2' C_2'$ floating tank circuit. The final step of the design is to compute the values of the components. A simplified design follows, but more detailed procedures are presented in Refs. [1,2,6].

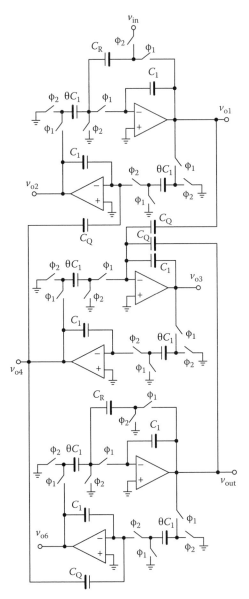

FIGURE 17.23   Sixth-order bandpass ladder filter.

The loops of the SC filter implementation are of the lossless discrete integrator (LDI) type and the resistors are of the serial type. Hence, if an LDI prewarping scheme is used the analog center frequency of the filter should be mapped to the desired value, mainly for high-Q filters wherein the interaction between resonators is very small. Since the resistors are not LDI, a distortion in the quality factor of the filter sections occurs, thus increasing the passband ripple. The LDI transformation relates the analog and discrete frequencies by the following expression [1,6]:

$$\omega_{analog} = 2f_c \sin\left(\frac{\omega_{discrete}}{2f_c}\right) \tag{17.62}$$

Applying the LDI transformation to the center frequency of the filter, the predistorted center frequency is equal to 99,056 kHz. Using this prewarped frequency and Equations 17.54 through 17.61, the following component values are obtained:

$$R_1 = 1\ \Omega$$
$$C_1' = 1.01623 \times 10^{-4}\ \text{F}$$
$$L_1' = 2.51316 \times 10^{-8}\ \text{H}$$
$$C_2' = 3.65803 \times 10^{-8}\ \text{F}$$
$$L_2' = 6.98181 \times 10^{-5}\ \text{H}$$

In addition, the continuous-time and SC resistors are approximately related by the following expression:

$$C_{eq} \cong \frac{1}{f_c R_{cont}} \tag{17.63}$$

In the SC filter, the transconductor $1/R_Q$ is implemented by $C_Q/C_1 R$. Hence, the final values of the capacitors are

$$C_R = 0.0738\ \text{pF}$$
$$C_1 = 15.000\ \text{pF}$$
$$\theta C_1 = 4.6900\ \text{pF}$$
$$C_Q = 0.2833\ \text{pF}$$

While the filter center frequency of this design is accurate, the passband ripple is increased to around 0.8 dB instead of 0.5 dB. This is because the resistors used in the terminals of the filter are not implemented as LDI resistors. However, this effect can be partially corrected adjusting the resonant frequency of the second loop. If θC_1 is equal to 4.685 pF for the second resonator, the ripple is decreased to around 0.55 dB. The results for this case are shown in Figure 17.24.

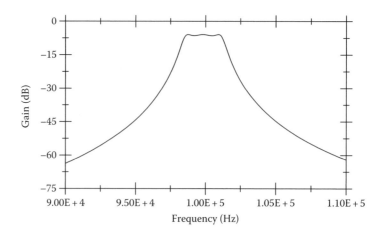

FIGURE 17.24   Frequency response for the sixth-order bandpass filter.

17.7.3 Programmable Switched-Capacitor Filter

In most of the programmable filters, the important parameters are the resonant frequencies, the pole-quality factor, and sometimes, the peak gain. Thus, programmable low-pass, bandpass, high-pass, and bump equalizers are typically designed. Nevertheless, in some applications it is more important to control the gain at the frequency bands instead of that at the resonant frequency, the quality factor, or the peak gain. A typical approach for the implementation of these systems is to employ a parallel of a low-pass, bandpass, and a high-pass filter with programmable peak gain. For a second-order system, the implementation of this approach needs at least six operational amplifiers, three capacitor banks, and the implementation of six poles. Therefore, the number of switches, the power dissipation, and the silicon area needed for these structures are considerable. Another approach follows in this section.

In order to independently control the low-, medium-, and high-frequency bands it is required to realize the following transfer function:

$$H(s) = \frac{K_1 s^2 + K_2 \mathrm{BW} s + K_3 \omega_o^2}{s^2 + \mathrm{BW} s + \omega_o^2} \tag{17.64}$$

where BW and ω_o are the bandwidth and the frequency of the poles, respectively. The control of the frequency bands is carried out by the parameters K_1, K_2, and K_3. From Equation 17.64 it is clear that the dc gain and the high-frequency gain are equal to K_3 and K_1, respectively. A disadvantage of this expression is the lack of good control in the medium frequencies. If K_1 is equal to K_3, the filter gain at the resonant frequency depends on K_2. However, for the general case it is affected by the parameters K_1 and K_3. In addition, the shape of the transfer function is not well behaved for moderate- and high-Q applications. The behavior of the transfer function is improved if the following equation is employed:

$$H(s) = K_1 \left(\frac{s^2 + K_2 K_3 \mathrm{BW} s + K_3 \omega_o^2}{s^2 + K_1 \mathrm{BW} s + K_1 \omega_o^2} \right) \tag{17.65}$$

For low and high frequencies the behavior of this equation is similar to that of Equation 17.64. At medium frequencies the behavior of the transfer function is better controlled by the parameter K_2 than for the case of Equation 17.64. For the medium-frequency band, the gain is related to the product K_2K_3 instead of the absolute value of K_2. Therefore, the effect of parameter K_2 on the transfer function is related to K_3.

A block diagram representation of Equation 17.65 is presented in Figure 17.25. In this figure, the filter bandwidth BW is equal to ω_o/Q, with Q equal to the quality factor of the filter. The control parameters are the gain factors K_1, K_2, and K_3. The implementation of this block diagram can be carried out by using various techniques, e.g., OTA-C, MOSFET-C, or SC. An SC implementation is shown in Figure 17.26. For high-sampling rate $2\pi f_c \gg \omega_o$, the capacitors θC_1 are related with the pole frequency by the equation $\theta = \omega_o T$. The capacitor banks control the gain of the three frequency bands. For the computation of the total capacitance, the capacitors can be associated in the following groups: the first group (K_3C_1, C_1), second group (θC_2, K_2C_2/Q, C_2), and third group ($\theta C_3/Q$, θC_3, C_3/K_1, C_3).

The versatility of the topology is shown in Figure 17.27. For these results, the following design parameters have been employed: clock frequency $= 128$ kHz, $\omega_o = 6.2832 * 350$ rad/s, and $Q = 0.5$. For Figure 17.26, the parameter K_3 is equal to 10, fixing the low-frequency gain at 20 dB. The medium-frequency band is controlled by the parameter K_2, which corresponds to the values 0.5 (attenuation of 6 dB), 1 (0 dB attenuation), and 2 (gain of 6 dB). The high-frequency gain is controlled by the parameter K_1, which corresponds to the values 1, 2.5, 5, and 10, giving a high-frequency gain of 0, 8, 14, and 20 dB, respectively.

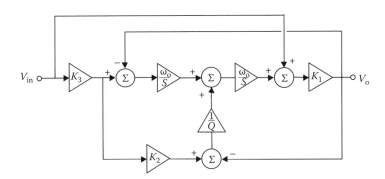

FIGURE 17.25 A flow diagram representation of Equation 17.65.

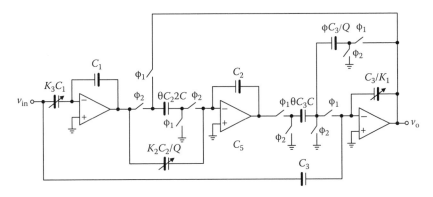

FIGURE 17.26 Switched-capacitor realization for the programmable filter.

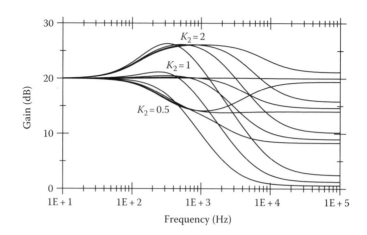

FIGURE 17.27   Effect of the parameter K_2 on the filter transfer function. For this plot $K_3 = 10$ and K_1 has been varied (1, 2.5, 5 and 10).

17.7.4 Switched-Capacitor Filters with Reduced Area

Most of the high-performance filters, sigma–delta modulators, and data converters are based on SC techniques [11,13]. Several architectures for narrow-band applications have been reported; many of them use several paths to relax the OTA specifications. In this approach two biquadratic filters connected in parallel running at $f_s/2$ each but acting in complementary clock phases lead to filters working at twice the clock speed. Gain-compensated single-stage OTAs and double sampling techniques are also available for the design of efficient narrow-band (highly selective) filters.

For high-Q filters, the scenario is even more complex because large capacitive spreads (proportional to Q) are required. Area efficient high-Q filters require special design strategies such as judiciously use of partial positive feedback and use of slower clocks in critical filter building blocks. The use of slower clocks allows us to reduce the capacitive spread as well. The resistance of an SC resistor driven by periodic clocks is approximately given by T/C ($=1/f_sC$). If the signal is sampled once every N-periods, the equivalent resistance increases by a factor of N [17]. The basic idea of this approach is shown in Figure 17.28 for the case $N=4$; the frequency of the clock phases ϕ_{11} and ϕ_{22} is $f_s/4$. Each time A_2C_0 is activated, small amount of charge is taken by the capacitor controlling the losses of the lossy integrator.

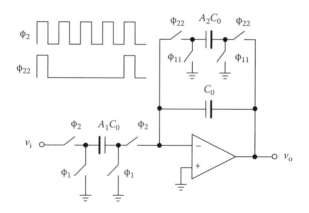

FIGURE 17.28   SC lossy integrator using a slower clock to reduce capacitor area.

If this technique is used for high-Q applications ($A_2 \ll 1$), those losses generate some spurious tones as well, but usually they are very small [17–18]. This approach results in lower capacitance spread without sacrificing sensitivity of critical filter parameters to component variations. The combination of several design techniques results in significant reduction in the total capacitance savings both silicon area and power consumption [18]. As a consequence of the use of slower clocks, additional alias components appear at integer multiples of f_s/N that may limit the filter's performance and increase the requirements on the antialias filter that precedes the SC network. It is critical to estimate the amount of alias components when using these techniques.

17.7.5 Spectrum Analyzer Using Switched-Capacitor Techniques

The growth in complexity and number of functions that can be integrated on a single chip have increased rapidly in the last years, making the testing a difficult task. Even though the majority of these systems are mainly digital, the analog section is always an important part of such mixed-mode architectures. Testing in analog circuits faces many problems; e.g., sweeping frequency and amplitude, measurement of magnitudes, and phase for a number of frequencies. Spectrum and network analyzers are routinely used for the characterization of mixed-mode systems. A cheap on-chip spectrum analyzer has been developed in Ref. [19]. Due to the flexibility of SC techniques, a gracious synchronization is ensured between the frequency of the input stimuli f_{in} and the center frequency of the circuits that perform the measurements, controlled by clocks ϕ_1 and ϕ_2.

A conceptual schematic diagram of the base-band SC network analyzer is shown in Figure 17.29. It consists of a digital frequency synthesizer, an SC sinewave generator, two VGAs, an SC BPF, and an analog-to-digital converter. The frequency synthesizer generates the master clock used as the sinusoidal generator sampling frequency as well as the nonoverlapping clock phases for the SC blocks. The sinewave generator, based on SC techniques, delivers a sinusoidal signal with a frequency of 1/16 of the master clock frequency. The amplitude of the signal coming from the sinusoidal generator is adjustable to provide the proper level to the stimuli. The output of the device under test (DUT) is bandpass filtered and conditioned by the second VGA to accommodate it to the proper input range of the ADC. The narrowband SC filter is a key building block of the spectrum-vector analyzer, its function is twofold: one, it can be centered at the center frequency f_0 to obtain the DUT transfer function; second, to select the proper frequency component ($f_0, 2f_0, 3f_0,$) for harmonic distortion characterization. The main advantage of this system is its inherent synchronization between the sinewave input signal and center frequency of the BPF.

The sinewave generator is based on SC circuits, with an oscillating frequency equal to $f_s/16$, where f_s is the clock frequency generated by the digital frequency synthesizer. Figure 17.30 shows the schematic diagram of a flexible sinewave generator using an SC circuit. It consists of a programmable gain amplifier whose preset gain stages correspond to the values of an ideal sampled and held sinewave. The SC circuit has four different gain stages, which generate a sinusoidal output with 16 steps per period generated from four capacitors and two reference voltages. The switch *PZ* sets the zero of the sinusoidal waveform.

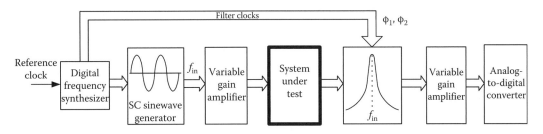

FIGURE 17.29 On-chip spectrum/vector analyzer block diagram using switched-capacitor techniques.

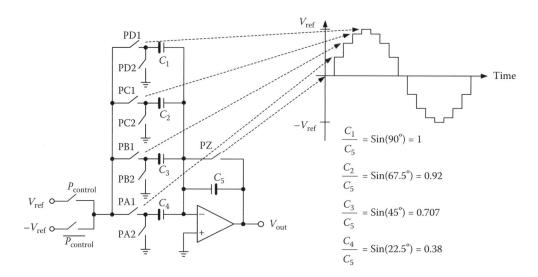

FIGURE 17.30 SC amplifier used to generate the discrete sinewave with low distortion.

The switches PA1 through PD1 are closed sequentially with each clock cycle to generate the first quarter-period of the sinusoidal waveform. Once the maximum value is obtained, the switches close in the opposite direction (from PD1 to PA1) in order to generate the second quarter-period. In the second zero crossing, $P_{control}$ switches from V_{ref} to $-V_{ref}$ so the lower half of the signal is generated. The capacitors are weighted such that the ideal values of a discrete sinewave signal are generated. The advantage of this implementation against direct digital synthesizers lies in the simplicity of the digital logic required to generate the required clock phases, resulting in a very compact implementation. The reset switch PZ also eliminates the accumulation of offset voltages. Since the peak value of the discrete sinewave signal is determined by V_{ref}, making it programmable allows obtaining an amplitude programmable sinewave generator. A modification of the structure allows generating the sinewave signal using a single reference voltage. Figure 17.31 shows the sinewave generator output for three different reference voltages: ± 200 mV, ± 100 mV, and ± 50 mV. The oscillating frequency is 1 kHz, and the clock frequency is 16 MHz. The measured HD3 of this block is in the order of -51 dB for an input signal of 200 mV$_{peak}$.

17.7.6 Programmable Switched-Capacitor Oscillator

Sinewave oscillators are essential parts in many electronic systems and in a host of applications. Integrating the oscillator with the other circuit blocks on a single chip makes it easy and reliable to implement several applications including built-in testing. Among the various types of oscillator, a BPF-based oscillator is an attractive and practical implementation due to its many advantages such as the programmability of the oscillation frequency by means of changing the center frequency (f_0) of the BPF, and the fact that the oscillation amplitude can be controlled with the help of a comparator. In base-band applications, BPF can be implemented with conventional SC design techniques that are preferred because of its accuracy, simple implementation, and reduced sensitivity to process and temperature variations. A block diagram of a BPF-based SC oscillator is shown in Figure 17.32 [20].

For highly linear oscillator, conventional approaches to minimize the frequency harmonics of the SC BPF-based oscillator requires a high-quality-factor (Q-factor) BPF, which involves high capacitor spread and, hence, leads to large silicon area. Also, improving the linearity by increasing Q-factor is not that efficient considering the fact that the nth-order harmonic distortion (HD$_n$) can be approximated as

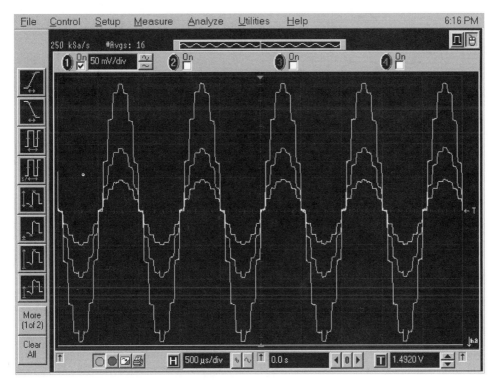

FIGURE 17.31 Measured output signal of the sinewave generator.

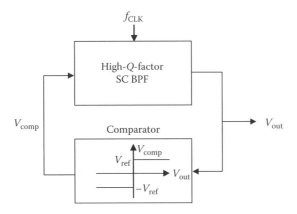

FIGURE 17.32 Block diagram of conventional BPF-based SC oscillator.

$1/(n^2 Q)$, where Q is a Q-factor of BPF in second-order BPF-based oscillator. If HD3 of -60 dB is needed, then the required Q-factor becomes more than 100 that may not be practical for an IC realization.

The linearity can be significantly improved without requiring a high-Q-factor BPF if a technique based on nonlinear shaping of the frequency spectrum is used. For this purpose, a comparator must exhibit multilevel outputs. The four-level square wave with a certain condition (as depicted for $f(t)$ in Figure 17.33) rejects the third-order harmonic component. Implementation of a multilevel comparator

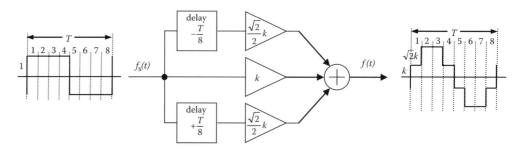

FIGURE 17.33 Conceptual diagram of a four-level quasi-sinusoidal wave generator.

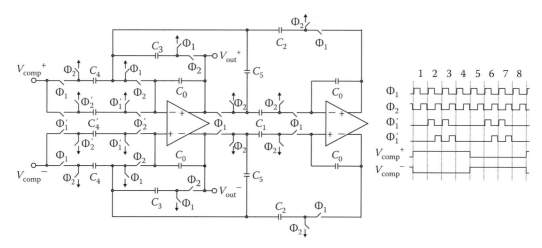

FIGURE 17.34 SC BPF implementation with an embedded FIR filter.

can be explained as adding finite-impulse-response (FIR) filter to a conventional comparator as shown in Figure 17.33. In this figure, $f_s(t)$ is an output of a conventional comparator and $f(t)$ is an output of FIR filter. A transfer function of an FIR filter in discrete-time domain, assuming that $T/8$ is one sampling period, can be expressed

$$H(z) = \frac{F(z)}{F_s(z)} = \frac{k}{\sqrt{2}} z(1 + \sqrt{2}z^{-1} + z^{-2}) \tag{17.66}$$

Equation 17.66 shows zeros at $z = 0$, $e^{\pm j3\pi/4}$. Note that $z = e^{\pm j3\pi/4}$ corresponds to the third-order harmonic frequency since $8/T$ is used for a sampling frequency. FIR filter can be easily embedded into SC BPF with the minimum cost, because BPF readily has an adder and a multiplier at its input. Figure 17.34 shows the BPF with the embedded FIR filter. In Figure 17.34, C_4' should be $\sqrt{2}\,C_4$ for the cancellation of third-order harmonic, and all other capacitors should be determined by the requirements as of a conventional BPF.

The SC BPF-based oscillator was fabricated using CMOS 0.35 μm technology. BPF has a center frequency of $f_0 = 10$ MHz, a Q-factor of 10, and a master clock frequency of $f_C = 80$ MHz. The measured frequency spectrums for both oscillators are shown in Figure 17.35. Over -54 dB of HD3 for the SC BPF-based with FIR filter is achieved, which is 20 dB smaller than the conventional oscillator. HD3 improvement of the proposed oscillator is mainly determined by the accuracy of the multiplying factor

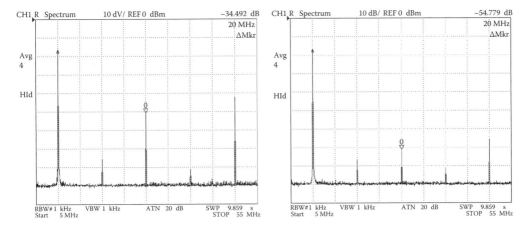

FIGURE 17.35 Measured frequency spectrum of the SC oscillators.

($\sqrt{2}$ in this case). In practical implementation, the ratio of 4:3 (=1.33) can be used for approximation of $\sqrt{2}$ (=1.4142), which limits the amount of harmonic cancellation. The implementation of the FIR filter requires minimum additional area (10% of the total area).

References

1. P. E. Allen and E. Sánchez-Sinencio, *Switched-Capacitor Circuits.* New York: Van Nostrand, 1984.
2. R. Unbehauen and A. Cichocki, *MOS Switched-Capacitor and Continuous-Time Integrated Circuits and Systems.* Berlin: Springer-Verlag, 1989.
3. R. W. Brodersen, P. R. Gray, and D. A. Hodges, MOS switched-capacitor filters, *Proceedings on IEEE,* 657, 61–75, January 1979.
4. R. Castello and P. R. Gray, A high-performance micropower switched-capacitor filter, *IEEE Journal on Solid-State Circuits,* SC-20, 1122–1132, December 1985.
5. E. Sánchez-Sinencio, J. Silva-Martinez, and R. L. Geiger, Biquadratic SC filters with small GB effects, *IEEE Transactions on Circuits Systems,* 31, 876–884, October 1984.
6. R. Gregorian and G. Temes, *Analog MOS Integrated Circuits,* New York: Wiley, 1986.
7. E. Vittoz, Very low power circuit design: Fundamentals and limits, in *Proc. IEEE/ISCAS'93,* Chicago, IL, May 1993, 1451–1453.
8. R. L. Geiger, P. E. Allen, and N. R. Strader, *VLSI Design Techniques for Analog and Digital Circuits,* New York: McGraw-Hill, 1990.
9. R. Castello and P. R. Gray, Performance limitations in switched-capacitor filters, *IEEE Transactions Circuits Systems,* CAS-32, 865–876, September 1985.
10. K. Nagaraj, A parasitic insensitive area-efficient approach for realizing very large time constants in switched-capacitor filters, *IEEE Transactions on Circuits Systems,* 36, 1210–1216, September 1989.
11. E. Sánchez-Sinencio and A. Andreou, editors, *Low-Voltage/Low-Power Integrated Circuits and Systems,* New York: IEEE Press, 1999.
12. P. R. Gray and R. Meyer, MOS operational amplifier design—a tutorial overview, *IEEE Journal on Solid-State Circuits,* 17, 6, 969–982, December 1982.
13. D. A. Johns and K. W. Martin, *Analog Integrated Circuit Design,* John Wiley & Sons, New York, 1997.
14. A. M. Abo and P. R. Gray, A 1.5 V, 10-bit, 14.3-MS/s CMOS pipeline analog-to-digital converter, *IEEE Journal on Solid-State Circuits,* 34, 5, 599–606, May 1999.
15. J. Crols and M. Steyaert, Switched-opamp: An approach to realize full CMOS switched-capacitor circuits at very low power supply voltages, *IEEE Journal on Solid-State Circuits,* 29, 8, 936–942, August 1994.

16. D. Hernandez-Garduno and J. Silva-Martinez, Continuous-time common-mode feedback for high-speed switched-capacitor networks, *IEEE Journal of Solid-State Circuits*, 40, 1610–1617, August 2005.
17. J. L. Ausin et al., Switched-capacitor circuits with periodical nonuniform individual sampling, *IEEE Transactions on Circuits and Systems-II: Analog and Digital Signal Processing*, 50, 8, 404–414. August 2003.
18. J. Adut, M. Rocha-Perez, and J. Silva-Martinez, A 6th order broadband 10.7 MHz SC ladder filter, *IEEE Transactions on Circuits and Systems, part I.* 53, 1625–1635, August 2006.
19. M. Mendez-Rivera, J. Silva-Martinez, E. Sanchez-Sinencio, and A. Valdes-Garcia, An on-chip spectrum analyzer for analog built-in testing, *Journal of Electronic Testing: Theory and Applications*, 21, 205–219, February 2005.
20. S. W. Park, José L. Ausín, F. Bahmani, and E. Sánchez-Sinencio, Nonlinear shaping SC oscillator with enhanced linearity, *IEEE Journal of Solid-State Circuits*, 42, 11, 2421–2431, November 2007.

Digital Filters

Rashid Ansari
University of Illinois at Chicago

A. Enis Cetin
Bilkent University

18

FIR Filters

M. H. Er
Nanyang Technological University

Andreas Antoniou
University of Victoria

Yong Ching Lim
National University of Singapore

Tapio Saramäki
Tampere University of Technology

18.1 Properties of FIR Filters

M.H. Er

18.1.1 Linear Phase Property

The finite-impulse response (FIR) filter is characterized by a unit-sample response that has a finite duration. One of the advantages of FIR filters compared to their infinite-impulse response (IIR) counterparts is that FIR filters can be designed with exactly linear phase. Linear phase response is important for applications where phase distortion due to n onlinear phase can degrade performance, such as in data transmission and television applications.

A FIR causal filter can be characterized by the transfer function [1]

$$H(z) = \sum_{n=0}^{N-1} h(nT)z^{-n} \tag{18.1}$$

where
$h(nT)$ is the impulse response of the filter
N is the filter length
T is the sampling interval

Using the relationship that

$$H(z) = \frac{Y(z)}{X(z)} \tag{18.2}$$

the difference equation of a FIR filter can be obtained by taking the inverse Z-transform of Equation 18.1, that is

$$y(iT) = \sum_{n=0}^{N-1} h(nT)X(iT - nT) \tag{18.3}$$

which says that the current output of a FIR causal filter is the weighted sum of the current and past inputs. The weighting coefficients are given by the impulse response of the filter.

From Equation 18.1, the frequency response can be obtained by replacing $z = e^{j\omega T}$ as

$$H(e^{j\omega T}) = \sum_{n=0}^{N-1} h(nT)e^{-j\omega n T} = M(\omega)e^{j\phi(\omega)} \tag{18.4}$$

where $M(\omega)$ and $\phi(\omega)$ are the magnitude and phase responses, respectively, defined as

$$M(\omega) = \left| H(e^{j\omega T}) \right| \tag{18.5a}$$

$$\phi(\omega) = \arg\, H(e^{j\omega T}) \tag{18.5b}$$

The phase delay and group (time) delay functions of a filter are defined as

$$\tau_p = -\frac{\phi(\omega)}{\omega} \tag{18.6}$$

and

$$\tau_g = -\frac{d\phi(\omega)}{d\omega} \tag{18.7}$$

respectively. Filters for which τ_p and τ_g are independent of frequency are referred to as constant time delay or linear phase filters. Hence, the phase response of a linear phase filter is given by

$$\phi(\omega) = -\tau\omega, \quad -\pi < \omega < \pi \tag{18.8}$$

where τ is a constant phase delay in samples.

From Equations 18.4, 18.5b and 18.8, the phase response can be expressed as

$$\phi(\omega) = -\tau\omega = \tan^{-1} \frac{-\sum_{n=0}^{N-1} h(nT) \sin(\omega n T)}{\sum_{n=0}^{N-1} h(nT) \cos(\omega n T)} \tag{18.9}$$

Consequently,

$$\tan(\omega\tau) = \frac{\sum_{n=0}^{N-1} h(nT)\sin(\omega nT)}{\sum_{n=0}^{N-1} h(nT)\cos(\omega nT)} \tag{18.10}$$

Using the definition $\tan(\omega\tau) = \sin(\omega\tau)/\cos(\omega\tau)$, Equation 18.10 can be reexpressed as

$$\sum_{n=0}^{N-1} h(nT)\sin(\omega\tau - \omega nT) = 0 \tag{18.11}$$

It can be shown [1] that a solution to Equation 18.11 is given by

$$\tau = \frac{(N-1)T}{2} \tag{18.12}$$

and

$$h(nT) = h[(N-1-n)T], \quad 0 \le n \le N-1 \tag{18.13}$$

Hence, FIR filters can be designed to have constant phase and group delays if the conditions of Equations 18.12 and 18.13 are satisfied. The symmetry property of Equation 18.13 can also lead to efficient filter realizations.

In applications where only constant group delay is needed, the phase response can have the form

$$\phi(\omega) = \phi_0 - \tau\omega \tag{18.14}$$

where ϕ_0 is a constant. With $\phi_0 = \pm\pi/2$, it can be shown [1] that the impulse response is of the form

$$h(nT) = -h[(N-1-n)T], \quad 0 \le n \le N-1 \tag{18.15}$$

In this case, the impulse response exhibits antisymmetrical property.

18.1.2 Frequency Response of Linear Phase FIR Filters

The frequency response of a causal linear phase FIR filter can be simplified to some simple forms using Equations 18.13 and 18.15 and the values of N as follows:

(1) *Symmetric impulse response and $N = odd$*. In this case,

$$H(e^{j\omega T}) = \sum_{n=0}^{(N-3)/2} h(nT)e^{-j\omega nT} + h\left[\frac{(N-1)T}{2}\right]e^{-j\omega(N-1)T/2} + \sum_{n=(N+1)/2}^{N-1} h(nT)e^{-j\omega nT} \tag{18.16}$$

Using Equation 18.13, letting $m = N-1-n$ and changing the limits of summation, and finally letting $m = n$, the last summation in Equation 18.16 can be reexpressed as

$$\sum_{n=(N+1)/2}^{N-1} h(nT)e^{-j\omega nT} = \sum_{n=0}^{(N-3)/2} h(nT)e^{-j\omega(N-1-n)T} \tag{18.17}$$

Substituting Equation 18.17 into Equation 18.16, one obtains

$$H(e^{j\omega T}) = \sum_{n=0}^{(N-3)/2} h(nT)\left[e^{-j\omega nT} + e^{-j\omega(N-1-n)T}\right] + h\left[\frac{(N-1)T}{2}\right]e^{-j\omega(N-1)T/2} \tag{18.18}$$

Factoring $e^{-j\omega(N-1)T/2}$ in Equation 18.18 and letting $k=(N-1)/2-n$, Equation 18.18 can be reexpressed as

$$H(e^{j\omega T}) = e^{-j\omega(N-1)T/2}\left\{\sum_{k=1}^{(N-1)/2} h\left[\left(\frac{N-1}{2}-k\right)T\right](e^{j\omega kT} + e^{-j\omega kT}) + h\left[\frac{(N-1)T}{2}\right]\right\} \tag{18.19}$$

Using the property that $e^{j\theta} + e^{-j\theta} = 2\cos\theta$, Equation 18.19 can be simplified to the form

$$H(e^{j\omega T}) = e^{-j\omega(N-1)T/2}\left\{\sum_{k=1}^{(N-1)2} 2h\left[\left(\frac{N-1}{2}-k\right)T\right]\cos(\omega kT) + h\left[\frac{(N-1)T}{2}\right]\right\} \tag{18.20}$$

Letting $a(o)=h[(N-1)T/2]$ and $a(k)=2h[((N-1)/2-k)T]$, Equation 18.20 can be simplified further to

$$H(e^{j\omega T}) = e^{-j\omega(N-1)T/2}\left[\sum_{k=0}^{(N-1)/2} a(k)\cos(\omega kT)\right] \tag{18.21}$$

(2) *Symmetric impulse response and N= even.* For this case, the frequency response takes the form

$$H(e^{j\omega T}) = e^{-j\omega(N-1)T/2}\left\{\sum_{k=0}^{N/2-1} 2h(kT)\cos\left[\omega\left(\frac{N}{2}-k-\frac{1}{2}\right)T\right]\right\} \tag{18.22}$$

Letting $b(k)=2h[N/2-k)T]$, $k=1, 2, \ldots, N/2$, Equation 18.22 can be expressed as

$$H(e^{j\omega T}) = e^{-j\omega(N-1)T/2}\left\{\sum_{k=1}^{N/2} b(k)\cos\left[\omega\left(k-\frac{1}{2}\right)T\right]\right\} \tag{18.23}$$

An interesting feature of this frequency response is that $H(e^{j\omega T})$ is always equal to zero for $\omega=\pi$, independent of $b(k)$. This implies that high-pass filter characteristics cannot be realized with this type of filter.

(3) *Antisymmetric impulse response and N= odd.* For this case, the derivation of the frequency response is the same as that in (1) except that the cosine summations are replaced by the sine summations multiplied by j because of Equation 18.15. Hence, the frequency response is given by

$$H(e^{j\omega T}) = je^{-j\omega(N-1)T/2}\left\{\sum_{k=1}^{(N-1)/2} 2h\left[\left(\frac{N-1}{2}-k\right)T\right]\sin(\omega kT) + h\left[\frac{(N-1)T}{2}\right]\right\} \tag{18.24}$$

It should be noted that for odd values of N, Equation 18.15 requires that $h[(N-1)T/2]=0$. Letting $c(k)=2h[((N-1)/2-k)T]$, $k=1, 2,\ldots, (N-1)/2$, Equation 18.24 becomes

TABLE 18.1 Frequency Response of Linear Phase FIR Filters

$h(nT)$	N	$H(e^{i\omega T})$
Symmetrical	Odd	$e^{-j\omega(N-1)T/2} \displaystyle\sum_{k=0}^{(N-1)/2} a(k) \cos(\omega kT)$
	Even	$e^{-j\omega(N-1)T/2} \displaystyle\sum_{k=1}^{N/2} b(k) \cos\left[\omega\left(k - \tfrac{1}{2}\right)T\right]$
Antisymmetrical	Odd	$je^{-j\omega(N-1)T/2} \displaystyle\sum_{k=1}^{(N-1)/2} c(k) \sin(\omega kT)$
	Even	$je^{-j\omega(N-1)T/2} \displaystyle\sum_{k=1}^{N/2} d(k) \sin\left[\omega\left(k - \tfrac{1}{2}\right)T\right]$

Where $a(o) = h\left[\frac{(N-1)T}{2}\right]$, $a(k) = c(k) = 2h\left[\left(\frac{N-1}{2} - k\right)T\right]$, $b(k) = d(k) = 2h\left[\left(\frac{N}{2} - k\right)T\right]$.

$$H(e^{j\omega T}) = je^{-j\omega(N-1)T/2}\left\{\sum_{k=1}^{(N-1)/2} c(k) \sin(\omega kT)\right\} \qquad (18.25)$$

A notable feature of this frequency response is that at frequencies $\omega = 0$ and $\omega = \pi$, the frequency response is always zero, independent of $c(k)$.

(4) *Antisymmetric impulse response and N = even.* For this case, the frequency response is the same as that in (2) except the cosine summations become sine summations multiplied by j as follows:

$$H(e^{j\omega T}) = je^{-j\omega(N-1)T/2}\left\{\sum_{k=0}^{N/2-1} 2h(kT) \sin\left[\omega\left(\frac{N}{2} - k - \frac{1}{2}\right)T\right]\right\} \qquad (18.26)$$

Letting $d(k) = 2h[N/2 - k)T]$, $k = 1, 2, \ldots, N/2$, Equation 18.26 becomes

$$H(e^{j\omega T}) = je^{-j\omega(N-1)T/2}\left\{\sum_{k=1}^{N/2} d(k) \sin\left[\omega\left(k - \frac{1}{2}\right)T\right]\right\} \qquad (18.27)$$

In this case, the frequency response is zero at $\omega = 0$, independent of $d(k)$.

In summary, the frequency responses of the four possible types of FIR filters with linear phase are given in Table 18.1.

18.1.3 Locations of Zeros of Linear Phase FIR Filters

The symmetric and antisymmetric conditions of the impulse response given by Equations 18.13 and 18.15 impose certain constraints on the zeros of the transfer function $H(z)$ [2]. For the case where N is an odd value, $H(z)$ can be written as

$$H(z) = z^{-(N-1)/2} \sum_{k=0}^{(N-1)/2} \frac{a(k)}{2}(z^k \pm z^{-k}) \qquad (18.28)$$

where the $\pm$ sign corresponds to symmetry and antisymmetry in the impulse response respectively, and $a(o)$ and $a(k)$ are defined in Table 18.1.

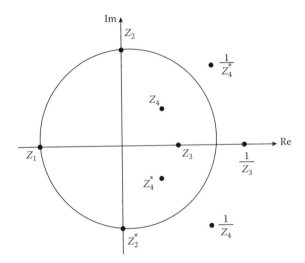

FIGURE 18.1 Typical zero positions of linear phase filters.

Substituting z^{-1} for z in Equation 18.28, one obtains

$$H(z^{-1}) = z^{(N-1)/2} \sum_{k=0}^{(N-1)/2} \frac{a(k)}{2}(z^{-k} \pm z^k)$$ (18.29)

It follows from Equations 18.28 and 18.29 that

$$H(z^{-1}) = \pm z^{(N-1)} H(z)$$ (18.30)

Equation 18.30 shows that $H(z)$ and $H(z^{-1})$ are identical to within a delay of $(N-1)$ samples and a multiplier of ± 1. Thus, the zeros of $H(z^{-1})$ are identical to the zeros of $H(z)$. Therefore, if $z_i = r_i e^{j\phi i}$ is a zero of $H(z)$, then $z_i^{-1} = (1/r_i)e^{-j\phi i}$ must also be a zero of $H(z)$. This has the following implications on the zero locations:

1. If $r_i = 1$ and $\phi_i = 0$ or π, then the zeros lie at either $z = +1$ or $z = -1$. In these cases, the zero is its own complex conjugate.
2. If $r_i = 1$ and $\phi_i \neq 0$ or π, then the zeros of $H(z)$ that are on the unit circle are also zeros of $H(z^{-1})$ that are on the unit circle. Hence, the zeros occur in complex conjugate pairs on the unit circle.
3. If $r_i \neq 1$ and $\phi_i = 0$ or π, then the zeros are real and occur in reciprocal pairs on the unit circle.
4. If $r_i \neq 1$ and $\phi_i \neq 0$ or π, then the zeros occur in quadruplets with complex conjugate reciprocal pairs off the unit circle.

Figure 18.1 shows the possible types of zeros for linear phase FIR filters.

18.2 Windowing Techniques

M.H. Er

Windowing is one of the earliest techniques for designing FIR filters [3,4]. The technique is simple because the filter coefficients can be obtained in closed form without the need for solving complex optimization problems as in some other sophisticated FIR design techniques. Hence, the design time is very short and the technique remains an attractive tool for FIR filter design.

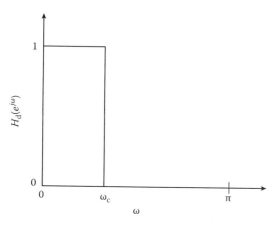

FIGURE 18.2   An ideal low-pass filter specification.

To understand the windowing technique, first consider the process of obtaining a finite-length impulse response by truncating an infinite-duration impulse response sequence. Suppose $H_d(e^{j\omega T})$ is an ideal desired low-pass response with cutoff frequency ω_c. As the frequency response of an FIR filter is a periodic function, it can be expressed as a Fourier series as

$$H_d(e^{j\omega T}) = \sum_{n=-\infty}^{\infty} h_d(nT)e^{-j\omega nT} \qquad (18.31)$$

where

$$h_d(nT) = \frac{T}{2\pi} \int_{-\pi/T}^{\pi/T} H_d(e^{j\omega T})e^{j\omega nT}\,d\omega \qquad (18.32)$$

In general, $H_d(e^{j\omega T})$ is piecewise constant with a certain passband and stopband and with discontinuities at the boundaries between bands. Hence, the impulse sequence $h_d(nT)$ is of infinite duration. For example, for the ideal low-pass response shown in Figure 18.2, the corresponding impulse response sequence is given by

$$h_d(nT) = \frac{\omega_c T}{\pi}\left(\frac{\sin \omega_c nT}{\omega_c nT}\right), \qquad -\infty \le n \le \infty \qquad (18.33)$$

It is clear that Equation 18.33 is a noncausal IIR filter. Also it is unstable and therefore unrealizable.

The rectangular window. One way to obtain a finite-duration causal impulse response is to simply truncate $h_d(nT)$ and introduce sufficient delay to obtain a causal impulse response, i.e., define

$$h(nT) = \begin{cases} h_d'(nT) & 0 \le n \le N-1 \\ 0 & \text{elsewhere} \end{cases} \qquad (18.34)$$

where $h_d'(nT)$ is a delay version of $h_d(nT)$.

This can be represented as the product of the desired impulse response and a finite-duration window $w_r(nT)$, i.e.,

$$h(nT) = h_d'(nT)w_r(nT) \qquad (18.35)$$

where $w_r(nT)$ is the rectangular window function defined as

$$w_r(nT) = \begin{cases} 1 & 0 \le n \le N-1 \\ 0 & \text{elsewhere} \end{cases} \qquad (18.36)$$

Let $\theta = \omega T$ and using the fact that multiplication of two discrete-time sequences corresponds to a convolution of their Fourier transforms. Hence,

$$H(e^{j\omega}) = \frac{1}{2\pi} \int\limits_{-\pi}^{\pi} H'_d(e^{j\theta}) W_r(e^{j(\omega-\theta)}) d\theta \qquad (18.37)$$

where $W_r(e^{j\theta})$ is the spectrum of the rectangular window.

Since the two functions in the integral are periodic, a circular convolution results and the limits of integration are taken over one period. Thus the frequency response $H(e^{j\omega})$ will be a "smeared" version of the desired response $H'_d(e^{j\omega})$ and the discontinuities in the desired frequency response become transition bands of $H(e^{j\omega})$. To understand this, it is instructive to examine the frequency response for the causal rectangular window, that is,

$$W_r(e^{j\omega T}) = \sum_{n=0}^{N-1} e^{-j\omega nT}$$

$$= e^{-j\omega(N-1)T/2} \frac{\sin(\omega NT/2)}{\sin(\omega T/2)} \qquad (18.38)$$

The spectrum $W_r(e^{j\omega T})$ for $N = 31$ is shown in Figure 18.3. The spectrum $W_r(e^{j\omega T})$ has two features that are worth noting, the mainlobe width and the sidelobe amplitude. The mainlobe width is defined as the distance between the two points closest to $\omega = 0$, where $W_r(e^{j\omega T})$ is zero. For a rectangular window, the mainlobe width is equal to $4\pi/N$. The maximum sidelobe amplitude for $W_r(e^{j\omega T})$ is equal to approximately -13 dB relative to the maximum value at $\omega = 0$.

Figure 18.4 shows the log–magnitude response of applying a 31-point rectangular window to approximate an ideal low-pass filter with a cutoff frequency equal to $\pi/4$. It can be seen that the sharp transition in the ideal response at $\omega = \omega_c$ has been converted into a gradual transition. Also, in the passband a series of overshoots and undershoots occur, and in the stopband, where the desired response is zero, the FIR filter has a nonzero response. These are the results of the convolution between $W_r(e^{j\omega T})$ and $H_d(e^{j\omega T})$. The mainlobe of $W_r(e^{j\omega T})$ causes the smearing of the desired response and the sidelobes of $W_r(e^{j\omega T})$ appear as overshoots and undershoots to the desired response. It is interesting to note that there will

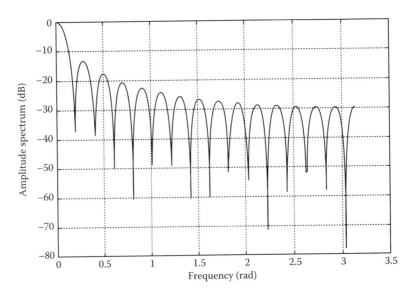

FIGURE 18.3   Fourier transform of the rectangular window.

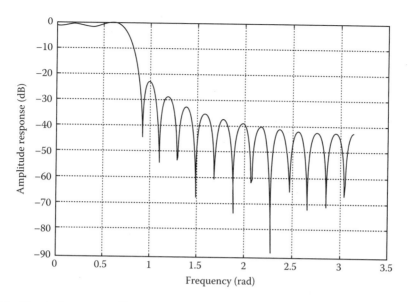

FIGURE 18.4 Magnitude response of low-pass FIR filter design using a 31-point rectangular window.

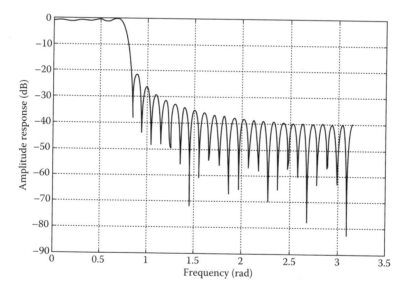

FIGURE 18.5 Magnitude response of low-pass FIR filter design using a 61-point rectangular window.

always be oscillations in the function $H(e^{j\omega T})$ in the vicinity of the steep transitions in $H'_d(e^{j\omega T})$, no matter how large the value of N as shown in Figure 18.5 for $N = 61$. This result is known as the Gibbs phenomenon [2] in the theory of Fourier series.

To reduce the oscillations in $H(e^{j\omega T})$, other window functions having spectra exhibiting smaller sidelobes must be used. To understand how the form of windows should be selected, it is observed that the sidelobes of the rectangular window represent the high-frequency components and are due to the sharp transitions from one to zero at the edges of the window. Therefore, the amplitudes of these sidelobes can be reduced by replacing the sharp transitions by more gradual ones. Some of the most frequently used window functions are described below.

The Bartlett window. The Bartlett window, also known as the triangular window, is defined as

$$w_t(nT) = \begin{cases} \frac{2n}{N-1} & 0 \leq n \leq \frac{N-1}{2} \\ 2 - \frac{2n}{N-1} & \frac{N-1}{2} \leq n \leq N-1 \\ 0 & \text{elsewhere} \end{cases} \tag{18.39}$$

The spectrum $W_t(e^{j\omega T})$ is shown in Figure 18.6. As expected, the sidelobe level is smaller than that of the rectangular window, being reduced from -13 to -25 dB relative to the maximum. However, the mainlobe width is now $8\pi/N$, twice that of the rectangular window. Hence, there is a trade-off between mainlobe width and sidelobe level.

Figure 18.7 illustrates the FIR low-pass magnitude response obtained by using the Bartlett window. Comparing Figure 18.7 to Figure 18.4, it is observed that the Bartlett window produces a smoother magnitude response.

The Hanning window. The Hanning window, also known as the raised-cosine window, is given by

$$w_c(nT) = \begin{cases} \frac{1}{2}\left[1 - \cos\left(\frac{2\pi n}{N-1}\right)\right] & 0 \leq n \leq N-1 \\ 0 & \text{elsewhere} \end{cases} \tag{18.40}$$

The amplitude spectrum of this window is shown in Figure 18.8 for $N = 31$. The magnitude of the first sidelobe level is -31 dB, down with respect to the peak value at $\omega = 0$. Comparing to the triangular window, there is an improvement of 6 dB. Since the mainlobe widths of both windows are the same, the Hanning window is preferred over the triangular one.

The amplitude response of the FIR low-pass filter with $\omega_c = \pi/4$ produced by applying the Hanning window with $N = 31$ is shown in Figure 18.9. The largest peak in the stopband is now reduced to -44 dB relative to the passband level.

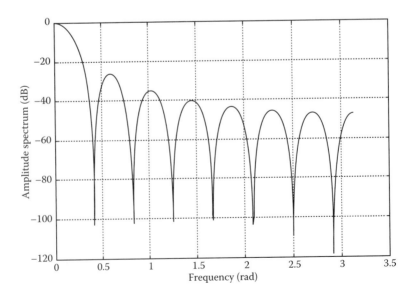

FIGURE 18.6 Fourier transform of the Bartlett window.

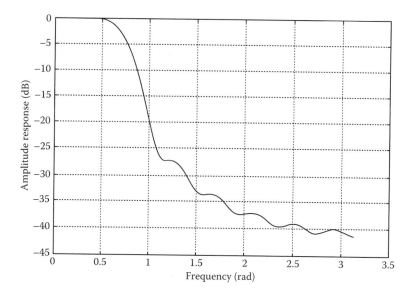

FIGURE 18.7 Magnitude response of low-pass FIR filter design using a 31-point Bartlett window.

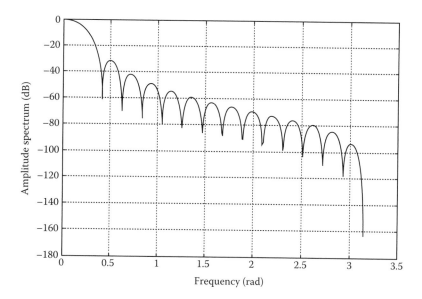

FIGURE 18.8 Fourier transform of the Hanning window.

The Hamming window. The Hamming window is given by

$$w_h(nT) = \begin{cases} 0.54 - 0.46\cos\left(\frac{2\pi n}{N-1}\right) & 0 \le n \le N - 1 \\ 0 & \text{elsewhere} \end{cases} \tag{18.41}$$

Figure 18.10 shows the amplitude spectrum of the Hamming window, The magnitude of the highest sidelobe is about −41 dB, a reduction of 10 dB relative to the Hanning window. This reduction is achieved at the expense of higher sidelobes at the higher frequencies.

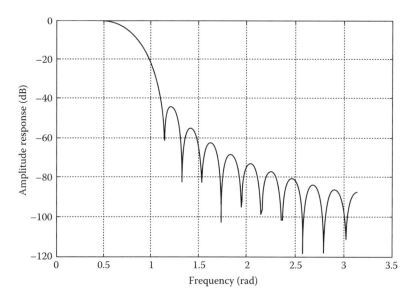

FIGURE 18.9   Magnitude response of low-pass FIR filter design using a 31-point Hanning window.

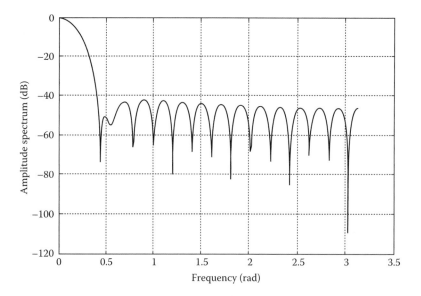

FIGURE 18.10   Fourier transform of the Hamming window.

Figure 18.11 illustrates the amplitude response of the FIR low-pass filter with $\omega_c = \pi/4$ obtained by applying the Hamming window for $N = 31$. The first sidelobe peak is -51 dB, a -7 dB improvement with respect to that using the Hanning window. However, it is noted that as frequency increases, the stopband attenuation does not increase as much as that produced by using the Hanning window.

The Blackman window. The Blackman window is given by

$$w_b(nT) = \begin{cases} 0.42 - 0.5\cos\left(\frac{2\pi n}{N-1}\right) + 0.08\cos\left(\frac{4\pi n}{N-1}\right) & 0 \le n \le N-1 \\ 0 & \text{elsewhere} \end{cases} \tag{18.42}$$

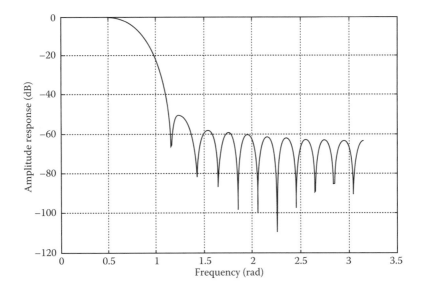

FIGURE 18.11 Magnitude response of low-pass FIR filter design using a 31-point Hamming window.

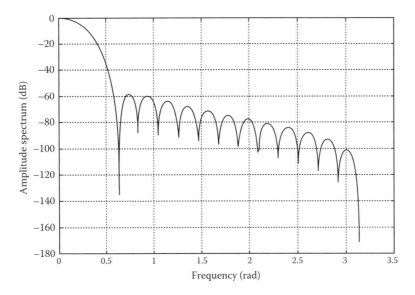

FIGURE 18.12 Fourier transform of the Blackman window.

The amplitude response of the Blackman window is shown in Figure 18.12. It can be seen that it has the highest sidelobe level, down -57 dB from the mainlobe peak. However, the mainlobe width has increased to $12\pi/N$.

The amplitude response of the FIR low-pass filter obtained when applying the Blackman window is shown in Figure 18.13. The minimum attenuation in the stopband is about -74 dB, but it occurs for $\omega > \pi/2$.

The Kaiser window. For the foregoing windows, the width of the mainlobe is inversely proportional to N. However, the minimum stopband attenuation is independent of the window length and is a function of

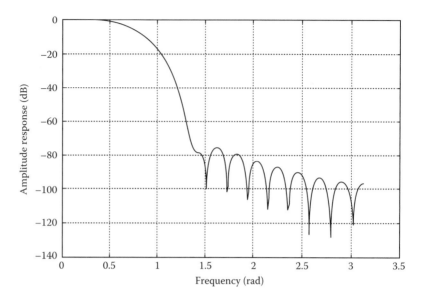

FIGURE 18.13   Magnitude response of low-pass FIR filter design using a 31-point Blackman window.

the selected window. Hence, to meet a desired stopband attenuation, the designer is forced to select a window that meets the design specifications. It is worth noting that windows with low sidelobe levels have broader mainlobe widths, hence requiring an increase in the order of the filter N to achieve the desired transition width.

In 1974, Kaiser [4] introduced a new window, now known as the Kaiser window, based on discrete-time approximations of the prolate spheroidal wave functions. This window has a variable parameter β, which can be varied to control the sidelobe level with respect to the mainlobe peak. As in other windows,

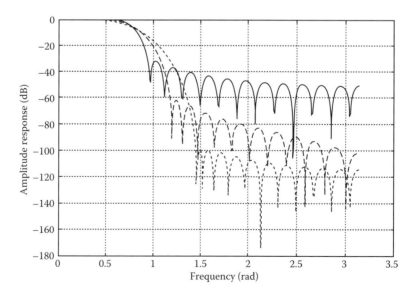

FIGURE 18.14   Magnitude responses of low-pass FIR filter design using a 31-point Kaiser window with $\beta = 1$ (solid line), $\beta = 6$ (dash line), and $\beta = 10$ (dotted line).

TABLE 18.2 Spectral Properties of *N*-Point Windows

Window	Mainlobe Width	Peak Amplitude of Sidelobe (dB)
Rectangular	$4\pi/N$	-13
Bartlett	$8\pi/N$	-25
Hanning	$8\pi/N$	-31
Hamming	$8\pi/N$	-41
Blackman	$12\pi/N$	-57

the mainlobe width can be adjusted by changing the length of the window, which in turn adjusts the transition width of the filter. Therefore, FIR filters can be efficiently designed using the Kaiser window.

The Kaiser windows are defined by

$$w_k(nT) = \begin{cases} \dfrac{I_0\left(2\beta\sqrt{\frac{n}{N-1}-\left(\frac{n}{N-1}\right)^2}\right)}{I_0(\beta)} & 0 \le n \le N-1 \\ 0 & \text{elsewhere} \end{cases} \tag{18.43}$$

where $I_0(x)$ is the modified zeroth-order Bessel function of the first kind. Kaiser has shown that these windows are nearly optimum in the sense of having the largest energy in the mainlobe for a given peak sidelobe amplitude.

To give an impression of the results that can be obtained by using Kaiser windows, Figure 18.14 shows the amplitude responses of an FIR low-pass filter design with $N = 31$ and $\omega_c = \pi/4$ and with $\beta = 1$, $\beta = 6$, and $\beta = 10$. As the value of β increases, the stopband attenuation of the low-pass filter increases and the transition band widens. Proper choice of N then leads to the final design.

The windows discussed above are compared in terms of their mainlobe width and the maximum sidelobe level in Table 18.2.

References

1. L. R. Rabiner and B. Gold, *Theory and Application of Digital Signal Processing*, Englewood Cliffs, NJ: Prentice-Hall, 1975.
2. A. V. Oppenheim and R. W. Schafer, *Digital Signal Processing*, Englewood Cliffs, NJ: Prentice-Hall, 1975.
3. L. R. Rabiner, Techniques for designing finite duration impulse response digital filters, *IEEE Trans. Commun. Technol.*, COM-19, 188–195, April 1971.
4. J. F. Kaiser, Nonrecursive digital filter design using the I_O-sinh window function, in *Selected Papers in Digital Signal Processing, II*, New York: IEEE, 1976.

18.3 Design of FIR Filters by Optimization

Andreas Antoniou

18.3.1 Equiripple FIR Filters

The design of FIR filters can be accomplished either through noniterative or iterative methods. Non-iterative methods entail the use of a small set of closed-form formulas and are, as a consequence, simple to apply. A frequently used method of this class is through the use of the Fourier series in conjunction with window functions. Iterative methods are based on the application of optimization techniques. These are characterized by a substantial increase in the computational complexity, but often lead to designs that are optimal in some respect.

This section deals with an iterative method for the design of FIR filters known as the *weighted-Chebyshev* method. In this approach, an error function is formulated for the desired filter in terms of a linear combination of cosine functions and is then minimized by using a very efficient multivariable optimization algorithm known as the *Remez exchange algorithm*. When convergence is achieved, the error function becomes equiripple, as in other Chebyshev solutions. The amplitude of the error in different frequency bands of interest is controlled by applying weighting to the error function.

The weighted-Chebyshev method is very flexible and can be used to obtain optimal solutions for most types of FIR filters, e.g., digital differentiators, Hilbert transformers, and low-pass, high-pass, bandpass, bandstop, and multiband filters with piecewise-constant amplitude responses. Furthermore, it can be used to design filters with arbitrary amplitude responses. Consequently, it is widely used. In common with other optimization methods, the weighted-Chebyshev method requires a large amount of computation; however, as the cost of computation is becoming progressively cheaper with time, this disadvantage is not a serious one.

The underlying principles of the weighted-Chebyshev method were proposed during the early 1970s [1–3] and a series of developments soon after [4–8] led to the well-known computer program of McClellan et al. [9]. Some more recent enhancements to the method are reported in Refs. [10,11]. A detailed treatment of the subject can be found in Ref. [12].

18.3.2 Problem Formulation

An FIR filter with a symmetrical impulse response and odd length N can be represented by the transfer function

$$H(z) = \sum_{n=0}^{N-1} h(nT)z^{-n}$$

If we assume a sampling rate $\omega_s = 2\pi$, we have $T = 2\pi/\omega_s = 1$ s, and hence the frequency response of the filter can be expressed as

$$H(e^{j\omega T}) = e^{-jc\omega} P_c(\omega)$$

where

$$P_c(\omega) = \sum_{k=0}^{c} a_k \, \cos k\omega \tag{18.44}$$

with

$$a_0 = h(c)$$
$$a_k = 2h(c - k) \quad \text{for } k = 1, 2, \ldots, c$$
$$c = (N - 1)/2$$

For a desired frequency response $e^{-jc\omega}D(\omega)$ and a specified weighting function $W(\omega)$, an error function $E(\omega)$ can be constructed as

$$E(\omega) = W(\omega)[D(\omega) - P_c(\omega)] \tag{18.45}$$

If it were possible to minimize the magnitude of the above error such that

$$|E(\omega)| \leq \delta_p$$

with respect to some compact subset of the frequency interval $[0, \pi]$, say Ω, a filter would be obtained in which

$$|E_0(\omega)| = |D(\omega) - P_c(\omega)| \leq \frac{\delta_p}{|W(\omega)|} \quad \text{for } \omega \in \Omega \tag{18.46}$$

In an equiripple filter, the magnitude of the error oscillates uniformly between zero and some maximum in each passband and stopband. In a low-pass equiripple filter, the amplitude response assumes the form depicted in Figure 18.15, where δ_p and δ_a are the amplitudes of the passband and stopband ripples, and ω_p and ω_a are the passband and stopband edges, respectively. Hence, we require

$$D(\omega) = \begin{cases} 1 & \text{for } 0 \leq \omega \leq \omega_p \\ 0 & \text{for } \omega_a \leq \omega \leq \pi \end{cases}$$

with

$$|E_0(\omega)| \leq \begin{cases} \delta_p & \text{for } 0 \leq \omega \leq \omega_p \\ \delta_a & \text{for } \omega_a \leq \omega \leq \pi \end{cases} \tag{18.47}$$

Therefore, from Equations 18.46 and 18.47 we deduce

$$W(\omega) = \begin{cases} 1 & \text{for } 0 \leq \omega \leq \omega_p \\ \delta_p/\delta_a & \text{for } \omega_a \leq \omega \leq \pi \end{cases}$$

Similarly, for high-pass filters, we obtain

$$D(\omega) = \begin{cases} 0 & \text{for } 0 \leq \omega \leq \omega_a \\ 1 & \text{for } \omega_p \leq \omega \leq \pi \end{cases}$$

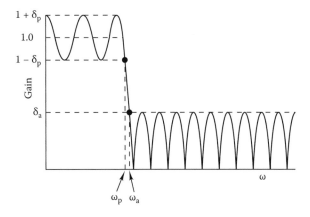

FIGURE 18.15 Amplitude response of equiripple low-pass filter. (Reproduced from Antoniou, A. *Digital Filters: Analysis, Design, and Applications*, McGraw-Hill, New York, 1993. With permission.)

and

$$W(\omega) = \begin{cases} \delta_p/\delta_a & \text{for } 0 \leq \omega \leq \omega_a \\ 1 & \text{for } \omega_p \leq \omega \leq \pi \end{cases}$$

Bandpass and bandstop filters. The above formulation can be easily extended to other types of filters. For bandpass filters, we have

$$D(\omega) = \begin{cases} 0 & \text{for } 0 \leq \omega \leq \omega_{a1} \\ 1 & \text{for } \omega_{p1} \leq \omega \leq \omega_{p2} \\ 0 & \text{for } \omega_{a2} \leq \omega \leq \pi \end{cases}$$

and

$$W(\omega) = \begin{cases} \delta_p/\delta_a & \text{for } 0 \leq \omega \leq \omega_{a1} \\ 1 & \text{for } \omega_{p1} \leq \omega \leq \omega_{p2} \\ \delta_p/\delta_a & \text{for } \omega_{a2} \leq \omega \leq \pi \end{cases}$$

where δ_p and δ_a are the amplitudes of the passband and stopband ripples, respectively, ω_{p1} and ω_{p2} are the passband edges, and ω_{a1} and ω_{a2} are the stopband edges, as depicted in Figure 18.16. On the other hand, for bandstop filters

$$D(\omega) = \begin{cases} 1 & \text{for } 0 \leq \omega \leq \omega_{p1} \\ 0 & \text{for } \omega_{a1} \leq \omega \leq \omega_{a2} \\ 1 & \text{for } \omega_{p2} \leq \omega \leq \pi \end{cases}$$

and

$$W(\omega) = \begin{cases} 1 & \text{for } 0 \leq \omega \leq \omega_{p1} \\ \delta_p/\delta_a & \text{for } \omega_{a1} \leq \omega \leq \omega_{a2} \\ 1 & \text{for } \omega_{p2} \leq \omega \leq \pi \end{cases}$$

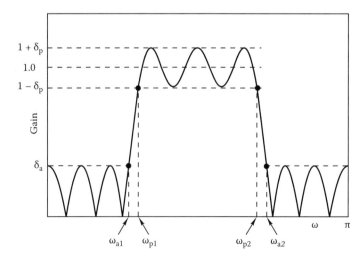

FIGURE 18.16 Amplitude response of equiripple bandpass filter. (Reproduced from Antoniou, A. *Digital Filters: Analysis, Design, and Applications*, McGraw-Hill, New York, 1993. With permission.)

The alternation theorem. An effective approach for the design of equiripple filters is to solve the minimax problem

$$\underset{x}{\text{minimize}}\left\{\max_{\omega}|E(\omega)|\right\} \tag{18.48}$$

where

$$x = \begin{bmatrix} a_0 & a_1 & \cdots & a_c \end{bmatrix}^{\text{T}}$$

is a column vector whose elements are the coefficients of the transfer function of the filter, which happen to be the values of the impulse response. The solution of this problem exists by virtue of the so-called alternation theorem [13], which is as follows.

THEOREM 18.1

If $P_c(\omega)$ is a linear combination of $r = c + 1$ cosine functions of the form

$$P_c(\omega) = \sum_{k=0}^{c} a_k \cos k\omega$$

then a necessary and sufficient condition that $P_c(\omega)$ be the unique, best, weighted-Chebyshev approximation to a continuous function $D(\omega)$ on Ω, where Ω is a compact subset of the frequency interval $[0, \pi]$, is that the weighted error function $E(\omega)$ exhibit at least $r + 1$ extremal frequencies in Ω, i.e., there must exist at least $r + 1$ points $\hat{\omega}$ in Ω such that

$$\hat{\omega}_0 < \hat{\omega}_1 < \cdots < \hat{\omega}_r$$
$$E(\hat{\omega}_i) = -E(\hat{\omega}_{i+1}) \quad \text{for } i = 0, 1, \ldots, r - 1$$

and

$$|E(\hat{\omega}_i)| = \max_{\omega \in \Omega} |E(\omega)| \quad \text{for } i = 0, 1, \ldots, r$$

From the alternation theorem and Equation 18.45 we can write

$$E(\hat{\omega}_i) = W(\hat{\omega}_i)[D(\hat{\omega}_i) - P_c(\hat{\omega}_i)] = (-1)^i \delta \tag{18.49}$$

for $i = 0, 1, \ldots, r$, where δ is a constant. This system of equations can be put in matrix form as

$$\begin{bmatrix} 1 & \cos \hat{\omega}_0 & \cos 2\hat{\omega}_0 & \cdots & \cos c\hat{\omega}_0 & \frac{1}{W(\hat{\omega}_0)} \\ 1 & \cos \hat{\omega}_1 & \cos 2\hat{\omega}_1 & \cdots & \cos c\hat{\omega}_1 & \frac{-1}{W(\hat{\omega}_1)} \\ \vdots & \vdots & \vdots & \vdots & \vdots & \vdots \\ 1 & \cos \hat{\omega}_r & \cos 2\hat{\omega}_r & \cdots & \cos c\hat{\omega}_r & \frac{(-1)^r}{W(\hat{\omega}_r)} \end{bmatrix} \begin{bmatrix} a_0 \\ a_1 \\ \vdots \\ a_c \\ \delta \end{bmatrix} = \begin{bmatrix} D(\hat{\omega}_0) \\ D(\hat{\omega}_1) \\ \vdots \\ D(\hat{\omega}_r) \end{bmatrix} \tag{18.50}$$

If the extremal frequencies (or extremals for short) were known, coefficients a_k and, in turn, the frequency response of the filter could be computed using Equation 18.44. The solution of this system exists since the above $(r + 1) \times (r + 1)$ matrix is nonsingular [13].

18.3.2.1 Remez Exchange Algorithm

The Remez exchange algorithm is an *iterative multivariable* algorithm, which is naturally suited for the solution of the minimax problem stated in Equation 18.48. It is based on the second optimization method of Remez [14] and involves the following basic steps.

ALGORITHM 18.1:   Basic Remez Exchange Algorithm

1. Initialize extremals $\hat{\omega}_0$, $\hat{\omega}_1$, ..., $\hat{\omega}_r$ ensuring that an extremal is assigned at each band edge.
2. Locate the frequencies $\hat{\omega}_0$, $\hat{\omega}_1$, ..., $\hat{\omega}_\rho$ at which $|E(\omega)|$ is maximum and $|E(\hat{\omega}_i)| \geq \delta$. These frequencies are *potential* extremals for the next iteration.
3. Compute the convergence parameter

$$Q = \frac{\max\left|E(\hat{\omega}_i)\right| - \min\left|E(\hat{\omega}_i)\right|}{\max\left|E(\hat{\omega}_i)\right|}$$

   where $i = 0, 1, \ldots, \rho$.
4. Reject $\rho - r$ *superfluous* potential extremals $\hat{\omega}_i$ according to an appropriate rejection criterion and renumber the remaining $\hat{\omega}_i$ sequentially; then set $\hat{\omega}_i = \hat{\omega}_i$ for $i = 0, 1, \ldots, r$.
5. If $Q > \varepsilon$, where ε is a convergence tolerance (say $\varepsilon = 0.01$), repeat from step 2; otherwise continue to step 6.
6. Compute $P_c(\omega)$ using the last set of extremals; then deduce $h(n)$, the impulse response of the required filter, and stop.

The amount of computation required by the algorithm tends to depend quite heavily on the initialization scheme used in step 1, on the search method used for the location of the maxima of the error function in step 2, and on the criterion used to reject superfluous frequencies $\hat{\omega}_i$ in step 4.

Initialization of extremals. The simplest scheme for the initialization of extremals $\hat{\omega}_i$ for $i = 0, 1, \ldots, r$ is to assume that they are uniformly spaced in the frequency bands of interest. If there are J distinct bands in the required filter of widths $B_1, B_2, \ldots, B_J$ and extremals are to be located at the left-hand and right-hand band edges of each band, the sum of these bandwidths should be divided into $r + 1 - J$ intervals. Thus the average interval between adjacent extremals is

$$W_0 = \frac{1}{r + 1 - J} \sum_{j=1}^{J} B_j$$

Since the quantities B_j/W_0 need not be integers, the use of W_0 for the generation of the extremals will almost always result in a fractional interval in each band. This problem can be avoided by rounding the number of intervals B_j/W_0 to the nearest integer and then readjusting the frequency interval for the corresponding band accordingly. This can be achieved by letting the number of intervals in bands j and J be

$$m_j = \text{Int}\left(\frac{B_j}{W_0} + 0.5\right) \quad \text{for } j = 1, 2, \ldots, J - 1$$

and

$$m_J = r - \sum_{j=1}^{J-1} (m_j + 1)$$

respectively, and then recalculating the frequency intervals for the various bands as

$$W_j = \frac{B_j}{m_j} \quad \text{for } j = 1, 2, \ldots, J$$

A more sophisticated initialization scheme, which was found to give good results, is described in Ref. [15].

Location of maxima of the error function. The frequencies, $\hat{\omega}_i$, which *must include maxima at band edges if* $|E(\hat{\omega}_i)| \geq |\delta|$, can be located by simply evaluating $|E(\omega)|$ over a dense set of frequencies. A reasonable number of frequency points that yield sufficient accuracy in the determination of the frequencies $\hat{\omega}_i$ are $8(N+1)$. This corresponds to about 16 frequency points per ripple of $|E(\omega)|$. A suitable frequency interval for the jth band is $w_j = W_j/S$ with $S = 16$.

The above *exhaustive* search can be implemented in terms of Algorithm 18.2 below, where ω_{Lj} and ω_{Rj} are the left-hand and right-hand edges in band j; W_j is the interval between adjacent extremals and m_j is the number of intervals W_j in band j; w_j is the interval between successive samples of $|E(\omega)|$ in interval W_j and S is the number of intervals w_j in each interval W_j; N_j is the total number of intervals w_j in band j; and J is the number of bands.

ALGORITHM 18.2: Exhaustive Step-by-Step Search

1. Set $N_j = m_j S$, $w_j = B_j/N_j$, and $e = 0$.
2. For each of bands $1, 2, \ldots, j, \ldots, J$ do the following. For each of frequencies $\omega_{1j} = \omega_{Lj}$, $\omega_{2j} = \omega_{Lj} + \omega_{j}, \ldots, \omega_{ij} = \omega_{Lj} + (i-1)\omega_{j}, \ldots, \omega_{(N_j+1)j} = \omega_{Rj}$, set $\hat{\omega}_e = \omega_{ij}$ and $e = e + 1$ provided that $|E(\omega_{ij})| \geq |\delta|$ and one of the following conditions holds:
   (a) Case $\omega_{ij} = \omega_{Lj}$: if $|E(\omega_{ij})|$ is maximum at $\omega_{ij} = \omega_{Lj}$ (i.e., $|E(\omega_{Lj})| > |E(\omega_{Lj} + \varepsilon)|$)
   (b) Case $\omega_{Lj} < \omega_{ij} < \omega_{Rj}$: if $|E(\omega)|$ is maximum at $\omega = \omega_{ij}$ (i.e., $|E(\omega_{ij} - \omega_j)| < |E(\omega_{ij})| > |E(\omega_{ij} + \omega_j)|$)
   (c) Case $\omega_{ij} = \omega_{Rj}$: if $|E(\omega_{ij})|$ is maximum at $\omega_{ij} = \omega_{Rj}$ (i.e., $|E(\omega_{Rj})| > |E(\omega_{Rj} - \varepsilon)|$)

The parameter ε in steps 2(a) and 2(c) is a small positive constant and a value $10^{-2}w_j$ yields satisfactory results.

In practice, $|E(\omega)|$ is maximum at an interior left-hand band edge* if its first derivative at the band edge is negative, and a mirror-image situation applies at an interior right-hand band edge. In such cases, $|E(\omega)|$ has a zero immediately to the right or left of the band edge and the inequality in step 2(a) or 2(c) may sometimes fail to identify a maximum. However, the problem can be avoided by using the inequality $|E(\omega_{Lj} - \varepsilon)| > |E(\omega_{Lj})|$ in step 2(a) and $|E(\omega_{Rj})| < |E(\omega_{Rj} + \varepsilon)|$ in step 2(c) for interior band edges.

In rare circumstances, a maximum of $|E(\omega)|$ may occur between a band edge and the first sample point. Such a maximum may be missed by Algorithm 18.2, but the problem can be easily identified since the number of potential extremals will then be less than the minimum. The remedy is to check the number of potential extremals at the end of each iteration and if it is found to be less than $r + 1$, the density of sample points, i.e., S is doubled and the iteration is repeated. If the problem persists, the process is repeated until the required number of potential extremals is obtained. If a value of S equal to or less than 256 does not resolve the problem, the loss of potential extremals is most likely due to some other reason.

An important precaution in the implementation of the preceding search method is to ensure that extremals belong to the dense set of frequency points to avoid numerical ill-conditioning in the computation of $E(\omega)$ (see Equations 18.49 and 18.51). In addition, the condition $|E(\omega_{ij})| \leq |\delta|$ should

---

* An interior band edge is one in the range $0 < \omega < \pi$, i.e., not at $\omega = 0$ or π.

be replaced by $|E(\omega_{ij})| > |\delta| - \varepsilon_1$, where ε_1 is a small positive constant, say 10^{-6}, to ensure that no maxima are missed owing to roundoff errors.

The search method is very reliable and its use in Algorithm 18.1 leads to a *robust* algorithm since the entire frequency axis is searched using a dense set of frequency points. Its disadvantage is that it requires a considerable amount of computation and is, therefore, inefficient.

A more efficient version of Algorithm 18.2 is obtained by maintaining all the interior band edges as extremals throughout the optimization independently of the behavior of the error function at the band edges. However, the algorithm obtained tends to be somewhat less robust, i.e., it tends to fail more frequently than Algorithm 18.2.

Computation of $|E(\omega)|$ and $P_c(\omega)$. In steps 2 and 6 of the basic Remez algorithm (Algorithm 18.1), $|E(\omega)|$ and $P_c(\omega)$ need to be evaluated. This can be done by determining coefficients a_k by inverting the matrix in Equation 18.50. However, this approach is inefficient and may be subject to numerical ill-conditioning, in particular, if δ is small and N is large. An alternative and more efficient approach is to deduce δ analytically and then interpolate $P_c(\omega)$ on the r frequency points using the *barycentric* form of the *Lagrange interpolation* formula. The necessary formulation is as follows.

Parameter δ can be deduced as

$$\delta = \frac{\sum_{k=0}^{r} \alpha_k D(\hat{\omega}_k)}{\sum_{k=0}^{r} \frac{(-1)^k \alpha_k}{W(\hat{\omega}_k)}}$$

and $P_c(\omega)$ is given by

$$P_c(\omega) = \begin{cases} C_k & \text{for } \omega = \hat{\omega}_0, \hat{\omega}_1, \ldots, \hat{\omega}_{r-1} \\ \dfrac{\sum_{k=0}^{r-1} \frac{\beta_k C_k}{x - x_k}}{\sum_{k=0}^{r-1} \frac{\beta_k}{x - x_k}} & \text{otherwise} \end{cases} \qquad (18.51)$$

where

$$\alpha_k = \prod_{i=0, i \neq k}^{r} \frac{1}{x_k - x_i}$$

$$C_k = D(\hat{\omega}_k) - (-1)^k \frac{\delta}{W(\hat{\omega}_k)}$$

$$\beta_k = \prod_{i=0, i \neq k}^{r-1} \frac{1}{x_k - x_i}$$

with

$$x = \cos \omega \quad \text{and} \quad x_i = \cos \hat{\omega}_i \quad \text{for } i = 0, 1, 2, \ldots, r$$

In step 2 of the Remez algorithm, $|E(\omega)|$ often needs to be evaluated at a frequency that was an extremal during the previous iteration. For these cases, the magnitude of the error function is simply $|\delta|$ according to Equation 18.49, and need not be evaluated. This would reduce the amount of computation to some extent.

An alternative formulation that simplifies the implementation of the Remez exchange algorithm can be found in Ref. [12].

Rejection of superfluous potential extremals. The solution of Equation 18.50 can be obtained only if precisely $r+1$ extremals are available. By differentiating $E(\omega)$, one can show that in a filter with one frequency band of interest (e.g., a digital differentiator) the number of maxima in $|E(\omega)|$ (potential extremals in step 2 of Algorithm 18.1) can be as high as $r+1$. In the weighted-Chebyshev method, band edges at which $|E(\omega)|$ is maximum or $|E(\omega)| \geq |\delta|$ are treated as potential extremals (see Algorithm 18.2). Therefore, whenever the number of frequency bands is increased by one, the number of potential extremals is increased by 2, i.e., for a filter with J bands there can be as many as $r+2J-1$ frequencies $\hat{\omega}_i$ and a maximum of $2J-2$ superfluous $\hat{\omega}_i$ may occur. This problem is overcome by rejecting $\rho - r$ of the potential extremals $\hat{\omega}_i$, if $\rho > r$, in step 4 of the algorithm.

A simple rejection scheme is to reject the $\rho - r$ frequencies $\hat{\omega}_i$ that yield the lowest $|E(\hat{\omega}_i)|$ and then renumber the remaining $\hat{\omega}_i$ from 0 to r [8]. This strategy is based on the well-known fact that the magnitude of the error in a given band is inversely related to the density of extremals in that band, i.e., a low density of extremals results in a large error and a high density results in a small error. Conversely, a low band error is indicative of a high density of extremals, and rejecting superfluous $\hat{\omega}_i$ in such a band is the appropriate course of action.

A problem with the scheme just described is that whenever a frequency remains an extremal in two successive iterations, $|E(\omega)|$ assumes the value of $|\delta|$ in the second iteration by virtue of Equation 18.49. In practice, there are almost always several frequencies that remain extremals from one iteration to the next, and the value of $|E(\omega)|$ at these frequencies will be the same. Consequently, the rejection of potential extremals on the basis of the magnitude of the error can become arbitrary and may lead to the rejection of potential extremals in bands where the density of extremals is low. This tends to increase the number of iterations, and it may even prevent the algorithm from converging on occasion. This problem can to some extent be alleviated by rejecting only potential extremals that are not band edges.

An alternative rejection scheme based on the aforementioned strategy, which gives excellent results for two-band and three-band filters, involves ranking the frequency bands in the order of lowest average band error, dropping the band with the highest average error from the list, and then rejecting potential extremals, one per band, in a cyclic manner starting with the band with the lowest average error [11]. The steps involved are as follows.

ALGORITHM 18.3: Rejection of Superfluous Potential Externals

1. Compute the average band errors

$$E_j = \frac{1}{\nu_j} \sum_{\hat{\omega}_i \in \Omega_j} \left| E(\hat{\omega}_i) \right| \quad \text{for } j = 1, 2, \ldots, J$$

   where Ω_j is the set of potential externals in band j given by

$$\Omega_j = \left\{ \hat{\omega}_j : \omega_{Lj} \leq \hat{\omega}_j \leq \omega_{Rj} \right\}$$

   ν_j is the number of potential externals in band j, and J is the number of bands.
2. Rank the J bands in the order of lowest average error and let $l_1, l_2, \ldots, l_J$ be the ranked list obtained, i.e., l_1 and l_J are the bands with the lowest and highest average errors, respectively.
3. Reject one $\hat{\omega}_i$ in each of bands $l_1, l_2, \ldots, l_{J-1}, l_1, l_2, \ldots, l_1$ until $\rho - r$ superfluous $\hat{\omega}_i$ are rejected. In each case, reject the $\hat{\omega}_i$, other than a band edge, that yields the lowest $|E(\hat{\omega}_i)|$ in the band.

For example, if $J = 3$, $\rho - r = 3$, and the average errors for bands 1, 2, and 3 are, respectively, 0.05, 0.08, and 0.02, then $\hat{\omega}_i$ are rejected in bands 3, 1, and 3. Note that potential extremals are not rejected in band 2, which is the band of highest average error.

Computation of impulse response. The impulse response in step 6 of Algorithm 18.1 can be determined by noting that function $P_c(\omega)$ is the frequency response of a noncausal version of the required filter. The impulse response of this filter, represented by $h_0(n)$ for $-c \leq n \leq c$, can be determined by computing $P_c(k\Omega)$ for $k = 0, 1, 2, \ldots, c$, where $\Omega = 2\pi/N$, and then using the *inverse discrete Fourier transform*. It can be shown that

$$h_0(n) = h_0(-n) = \frac{1}{N} \left[P_c(0) + \sum_{k=1}^{c} 2P_c(k\Omega) \cos\left(\frac{2\pi kn}{N}\right) \right]$$

for $n = 0, 1, 2, \ldots, c$. Therefore, the impulse response of the required causal filter is given by

$$h(n) = h_0(n - c)$$

for $n = 0, 1, 2, \ldots, N - 1$.

18.3.2.2 Improved Search Methods

For a filter of length N, with the number of intervals w_j in each internal W_j equal to S, the exhaustive step-by-step search described (Algorithm 18.2) requires about $S \times (N + 1)/2$ function evaluations, where each function evaluation entails $N - 1$ additions, $(N + 1)/2$ multiplications, and $(N + 1)/2$ divisions (see Equation 18.51).

A Remez optimization usually requires 4 to 8 iterations for low-pass or high-pass filters, 4 to 10 iterations for bandpass filters, and 4 to 12 iterations for bandstop filters. Further, if prescribed specifications are to be achieved and the appropriate value of N is unknown, typically two to four Remez optimizations have to be performed.* Thus, if $N = 101$, $S = 16$, number of Remez optimizations $= 4$, iterations per optimization $= 6$, the design would entail 24 iterations, 19,200 function evaluations, 1.92×10^6 additions, 0.979×10^6 multiplications, and 0.979×10^6 divisions. This is in addition to the computation required for the evaluation of δ and coefficients α_k, C_k, and β_k once per iteration. In effect, the amount of computation required to complete a design is quite substantial.

The bulk of the computation in Algorithm 18.2 is carried out to locate the maxima of $|E(\omega)|$ and the large amount of computation is a consequence of the exhaustive character of the search. Therefore, any attempt to reduce the computational complexity of the Remez exchange algorithm must of necessity involve a more efficient search for the maxima of $|E(\omega)|$.

The error function in the weighted-Chebyshev method is well behaved in practice, and is normally unimodal between successive zeros, as can be seen in Figures 18.15 and 18.16. Hence, the maxima of $|E(\omega)|$ can be located through more sophisticated search methods that utilize gradient information. Two such methods are the so-called *selective step-by-step search* and *cubic-interpolation search* reported in Refs. [10,11]. Collectively, the two search methods can reduce the amount of computation to about one-fifth the amount required by the exhaustive search.

Selective step-by-step search. The underlying principle in the development of the selective step-by-step search is that normally there is strict alternation between the maxima and the zeros of $|E(\omega)|$. In a given iteration, the maxima of $|E(\omega)|$ are either old maxima from the previous iteration that have moved or new maxima introduced at band edges. New interior maxima may also arise, in theory, but such occurrences are quite rare in practice.

---

* See Section 18.3.2.3.

The selective step-by-step search involves three distinct parts as follows:

(1) Maxima that correspond to previous maxima are located by searching in the neighborhoods of the most recent set of extremals in a step-by-step fashion using the first derivative of $|E(\omega)|$. If the first derivative is positive at an extremal, the search is carried out to the right of the extremal; otherwise, the search is carried out to the left of the extremal.

(2) New maxima at band edges can be located by noting the circumstances under which new maxima can arise. These are as follows:

   a. To the right of $\omega = 0$ (first band), if there is an extremal and $|E(\omega)|$ has a minimum at $\omega = 0$

   b. To the left of $\omega = \pi$ (last band), if there is an extremal and $|E(\omega)|$ has a minimum at $\omega = \pi$

   c. At $\omega = 0$, if there is no extremal at $\omega = 0$

   d. At $\omega = \pi$, if there is no extremal at $\omega = \pi$

   e. To the right of an interior left-hand edge

   f. To the left of an interior right-hand edge

   g. At $\omega = \omega_{Lj}$, if there is no extremal at $\omega = \omega_{Lj}$

   h. At $\omega = \omega_{Rj}$, if there is no extremal at $\omega = \omega_{Rj}$

(3) New interior maxima, which cannot be located by the checks in (1) and (2) can be found by noting the presence of large gaps in the set of potential extremals identified in (1) and (2). If the difference between two consecutive potential extremals exceeds 1.5 to 2 times the initial interval between extremals (i.e., W_j), then the interval is checked for additional maxima.

If a selective step-by-step search based on the above principles is used in Algorithm 18.1, then at the start of the optimization the distance between a typical extremal $\hat{\omega}_i$ and the nearby maximum point $\bar{\omega}_i$ will be less than half the period of the corresponding ripple of $|E(\omega)|$, owing to the relative symmetry of the ripples of the error function. In effect, during the first iteration less than half of the combined width of the different bands needs to be searched. Thus the number of function evaluations required would be reduced from about 16 to less than 8 per extremal in practice. This will reduce the number of function evaluations by more than 50% relative to that required by the exhaustive search of Algorithm 18.1 without degrading the accuracy of the optimization in any way. As the optimization progresses and the solution is approached, extremal $\hat{\omega}_i$ and maximum point $\bar{\omega}_i$ tend to coincide and, therefore, the cumulative length of the frequency range that has to be searched is progressively reduced, thereby resulting in further economies in the number of function evaluations. In the last iteration, only two or three function evaluations are needed (including derivatives) per ripple. As a result, the total number of function evaluations can be reduced by 65%–70% relative to that required by the exhaustive search [10,11].

Cubic-interpolation search. The maxima in item (1) of the above method can also be found through the use of one stage of polynomial interpolation. Either quadratic or cubic interpolation can be used. In these methods, a polynomial approximation is obtained for the magnitude of the error function in the neighborhood of a given extremal and the location of the maximum is determined by finding the point at which the first derivative is zero. Although cubic interpolation entails a more complicated formulation than quadratic interpolation, it leads to improved accuracy, which tends to translate into improved efficiency.

Several choices are possible in setting up a cubic-interpolation search for the problem at hand. One that was found to work well in practice entails evaluating $|E(\omega)|$ at three frequency points and its derivative at one point. Choosing the extremal itself as one of the points reduces the computation further since the value of $|E(\omega)|$ is known to be $|\delta|$ from the previous iteration. Thus this scheme entails three function evaluations per extremal.

The computational complexity of the cubic-interpolation search described remains constant from iteration to iteration since the number of function evaluations required to perform an interpolation is constant. At the start of the optimization, the cubic-interpolation search is more efficient than the selective step-by-step search. However, as the solution is approached the number of function evaluations required by the selective search is progressively reduced, as was stated earlier, and at some point the selective search becomes more efficient. A prudent strategy under the circumstances is to use the cubic-interpolation search at the start of the optimization and switch over to the selective step-by-step search when some suitable criterion is satisfied. Extensive experimentation has shown that computational advantage can be gained by using the cubic-interpolation search if parameter Q (see Algorithm 18.1) is greater than about 0.65 and the selective search otherwise. The use of the cubic-interpolation search along with the selective step-by-step search of the preceding section can reduce the number of function evaluations by 70%–85% relative to that required by the exhaustive search [10,11].

More information, including the necessary formulation as well as a practical and efficient implementation of the Remez exchange algorithm in terms of the above search methods, can be found in Ref. [12].

Example 18.1

The Remez algorithm was used with (1) the exhaustive search, (2) the selective step-by-step search, and (3) the selective search in conjunction with the cubic-interpolation search to design an FIR equiripple high-pass filter satisfying the following specifications:

Filter length N: 23
Passband edge ω_p: 2.0 rad/s
Stopband edge ω_a: 1.0 rad/s
Ratio δ_p/δ_a: 15.0
Sampling frequency ω_s: 2π rad/s

The progress of the design is illustrated in Table 18.3. As can be seen, the exhaustive and selective search methods required six iterations each, whereas the selective search in conjunction with cubic interpolation required seven iterations. However, the number of function evaluations (evaluations of $P_c(\omega)$ using Equation 18.51 plus evaluations of its first or second derivative) decreased from 1013 in the first method to 350 in the second method to 259 in the third method. In the Remez algorithm, approximately 80%–90% of the computational effort involves function evaluations. In effect, relative to that required by the exhaustive search, the use of the selective step-by-step search reduced the amount of computation by about 65.4%, and the use of the selective step-by-step search in conjunction with the cubic-interpolation search reduced the amount of computation by about 74.4%.

The three methods resulted in approximately the same impulse responses, as can be seen in Table 18.4, and the passband ripple and minimum stopband attenuation obtained in each case were 0.043 and 75.7 dB, respectively. The amplitude response of the filter is illustrated in Figure 18.17.

TABLE 18.3   Progress in Design of High-Pass Filter (Example 18.1)

Iteration Number	Exhaustive Search		Selective Search		Selective Search with Cubic Interpolation	
	Q	FEs	Q	FEs	Q	FEs
1	0.9912	169	0.9912	93	0.9912	66
2	0.9207	168	0.9207	86	0.9406	44
3	0.9480	169	0.9480	55	0.8830	42
			$\hat{\omega}_{32}$ rejected			
4	0.7249	169	0.7249	62	0.6952	31
5	0.0923	169	0.0923	31	0.1417	31
6	0.0017	169	0.0017	23	0.0102	23
7	—	—	—	—	0.0000	22
Total FE's		1013		350		259

TABLE 18.4 Impulse Response of High-Pass Filter (Example 18.1)

| | $h_0(n) = h_0(-n)$ | |
| | Exhaustive or | Selective Search with |
n	Selective Search	Cubic Interpolation
0	5.034954×10^{-1}	5.035077×10^{-1}
1	-3.123538×10^{-1}	-3.123535×10^{-1}
2	-3.085731×10^{-3}	-3.097829×10^{-3}
3	8.932914×10^{-2}	8.932911×10^{-2}
4	2.053235×10^{-3}	2.063564×10^{-3}
5	-3.898118×10^{-2}	-3.898177×10^{-2}
6	-8.467375×10^{-4}	-8.540079×10^{-4}
7	1.660800×10^{-2}	1.660858×10^{-2}
8	5.401585×10^{-5}	5.892008×10^{-5}
9	-6.100465×10^{-3}	-6.101979×10^{-3}
10	7.298411×10^{-4}	7.281192×10^{-4}
11	9.275654×10^{-4}	9.285482×10^{-4}

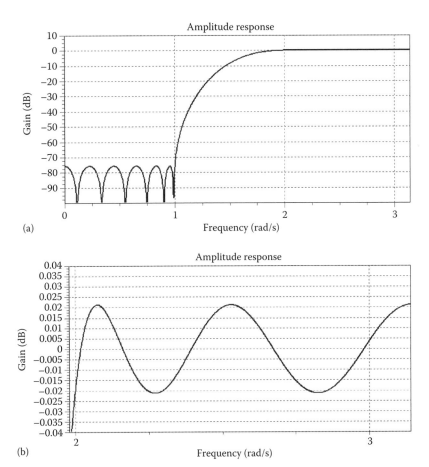

FIGURE 18.17 Amplitude response of equiripple high-pass filter (Example 18.1): (a) baseband and (b) passband.

Example 18.2

In this example, the Remez algorithm was used with (1) the exhaustive search, (2) the selective step-by-step search, and (3) the selective step-by-step search in conjunction with the cubic-interpolation search to design an FIR equiripple bandstop filter satisfying the following specifications:

  Filter length N: 29
  Lower passband edge ω_{p1}: 0.8 rad/s
  Upper passband edge ω_{p2}: 2.1 rad/s
  Lower stopband edge ω_{a1}: 1.1 rad/s
  Upper stopband edge ω_{a2}: 1.8 rad/s
  Ratio δ_{p1}/δ_a: 5.0
  Ratio δ_{p1}/δ_{p2}: 2.0
  Sampling frequency ω_s: 2π rad/s

The progress of the design is illustrated in Table 18.5. In this example, each of the three methods required four iterations. The number of function evaluations decreased from 804 in the first method to 190 in the second method to 131 in the third method. In effect, the use of the selective step-by-step search reduced the amount of computation by about 76.4%, and the use of the

TABLE 18.5   Progress in Design of Bandstop Filter (Example 18.2)

Iteration Number	Exhaustive Search		Selective Search		Selective Search with Cubic Interpolation	
	Q	FEs	Q	FEs	Q	FEs
1	0.6836	201	0.6836	79	0.6940	36
2	0.3138	201	0.3138	51	0.2378	36
3	0.0804	201	0.0804	34	0.0675	32
4	0.0000	201	0.0000	26	0.0007	27
Total FE's		804		190		131

TABLE 18.6   Impulse Response of Bandstop Filter (Example 18.2)

	$h_0(n) = h_0(-n)$	
n	Exhaustive or Selective Search	Selective Search with Cubic Interpolation
0	6.656629×10^{-1}	6.656478×10^{-1}
1	-4.187327×10^{-2}	-4.186510×10^{-2}
2	2.635370×10^{-1}	2.635297×10^{-1}
3	8.005521×10^{-2}	8.005307×10^{-2}
4	-1.131284×10^{-1}	-1.131056×10^{-1}
5	-3.691645×10^{-2}	-3.691932×10^{-2}
6	-9.914085×10^{-4}	-1.013621×10^{-3}
7	-3.018917×10^{-2}	-3.017017×10^{-2}
8	2.931776×10^{-2}	2.930006×10^{-2}
9	5.022490×10^{-2}	5.022450×10^{-2}
10	-9.715988×10^{-3}	-9.687345×10^{-3}
11	-2.550790×10^{-2}	-2.553543×10^{-2}
12	-4.023265×10^{-4}	-3.827029×10^{-4}
13	-3.410741×10^{-2}	-3.412007×10^{-2}
14	-1.421939×10^{-2}	-1.424189×10^{-2}

selective step-by-step search in conjunction with the cubic-interpolation search reduced the amount of computation by about 83.7%, relative to that required by the exhaustive search.

The three methods resulted in approximately the same impulse responses, as can be seen in Table 18.6. The amplitude response of the filter is illustrated in Figure 18.18; the passband ripples obtained for the two passbands were 1.78 and 0.89 dB, respectively, and the minimum stopband attenuation was 33.79 dB.

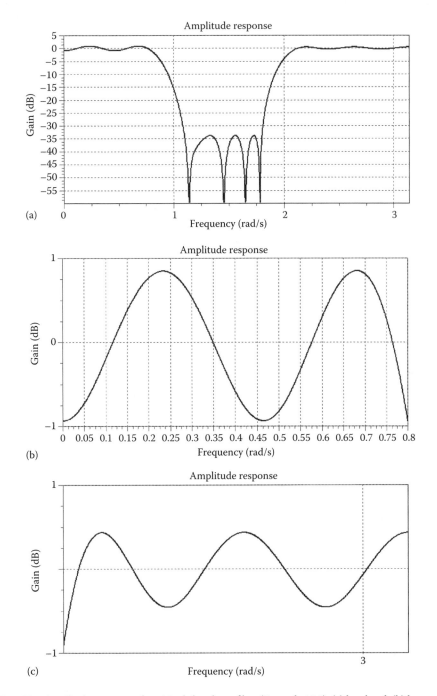

FIGURE 18.18 Amplitude response of equiripple bandstop filter (Example 18.2): (a) baseband, (b) lower passband, and (c) upper passband.

18.3.2.3 Prescribed Specifications

Given a filter length N, a set of passband and stopband edges, and a ratio δ_p/δ_a, an FIR filter with approximately piecewise-constant amplitude-response specifications can be readily designed. While the filter obtained will have passband and stopband edges at the correct locations and the ratio δ_p/δ_a will be as required, the amplitudes of the passband and stopband ripples are highly unlikely to be precisely as specified. An acceptable design can be obtained by predicting the value of N on the basis of the required specifications and then designing filters for increasing or decreasing values of N until the lowest value of N that satisfies the specifications is found.

A reasonably accurate *empirical* formula for the prediction of N for the case of low-pass and high-pass filters, due to Hermann et al. [16], is

$$N = \text{Int}\left[\frac{(D - FB^2)}{B} + 1.5\right] \tag{18.52}$$

where

$$B = |\omega_a - \omega_p|/2\pi$$
$$D = \left[0.005309(\log\delta_p)^2 + 0.07114\log\delta_p - 0.4761\right]\log\delta_a$$
$$\quad - \left[0.00266(\log\delta_p)^2 + 0.5941\log\delta_p + 0.4278\right]$$
$$F = 0.51244(\log\delta_p - \log\delta_a) + 11.012$$

Interestingly this formula can also be used to predict the filter length in the design of bandpass, bandstop, and multiband filters in general. In these filters, a value of N is computed for each transition band between a passband and stopband or a stopband and passband using Equation 18.52 and the largest value of N so obtained is taken to be the predicted filter length. *Prescribed specifications* can be achieved by using the following design algorithm.

ALGORITHM 18.4: Design of Filters Satisfying Prescribed Specifications

1. Compute N using Equation 18.52; if N is even, set $N = N + 1$.
2. Design a filter of length N and determine the minimum value of δ, say $\breve{\delta}$.
   A. If $\breve{\delta} > \delta_p$, then do the following:
      i. Set $N = N + 2$, design a filter of length N, and find $\breve{\delta}$.
      ii. If $\breve{\delta} \leq \delta_p$, then go to step 3; else, go to step 2(A)(i).
   B. If $\breve{\delta} < \delta_p$, then do the following:
      i. Set $N = N - 2$, design a filter of length N, and find $\breve{\delta}$.
      ii. If $\breve{\delta} > \delta_p$ then go to step 4; else, go to step 2(B)(i).
3. Use the last set of extremals and the corresponding value of N to obtain the impulse response of the required filter and stop.
4. Use the last but one set of extremals and the corresponding value of N to obtain the impulse response of the required filter and stop.

Example 18.3

Algorithm 18.4 was used to design an FIR equiripple bandpass filter that would satisfy the following specifications:

Odd filter length
Maximum passband ripple A_p: 0.5 dB
Minimum stopband attenuation A_{a1}: 50.0 dB
Minimum stopband attenuation A_{a2}: 30.0 dB
Lower passband edge ω_{p1}: 1.2 rad/s
Upper passband edge ω_{p2}: 1.8 rad/s
Lower stopband edge ω_{a1}: 0.9 rad/s
Upper stopband edge ω_{a2}: 2.1 rad/s
Sampling frequency ω_s: 2π rad/s

The progress of the design is illustrated in Table 18.7. As can be seen, a filter of length 41 was initially predicted, which was found to have a passband ripple of 0.47 dB, a minimum stopband attenuation of 50.4 in the lower stopband, and 30.4 dB in the upper stopband, i.e., the required specifications were satisfied. Then a filter length of 39 was tried and found to violate the specifications. Hence the first design is the required filter. The impulse response is given in Table 18.8. The corresponding amplitude response is depicted in Figure 18.19.

18.3.2.4 Generalization

There are four types of constant-delay FIR filters. The impulse response can be *symmetrical* or *anti-symmetrical*, and the filter length can be *odd* or *even*. In the preceding sections, we considered the design of filters with symmetrical impulse response and odd length. In this section, we show that the Remez algorithm can also be applied for the design of other types of filters.

Antisymmetrical impulse response and odd filter length. Assuming that $\omega_s = 2\pi$, the frequency response of an FIR filter with *antisymmetrical* impulse and *odd* length can be expressed as

$$H(e^{j\omega T}) = e^{-jc\omega}jP_c'(\omega)$$

TABLE 18.7 Progress in Design of Bandpass Filter (Example 18.3)

n	Iterations	FEs	A_p, dB	A_{a1}, dB	A_{a2}, dB
41	8	550	0.47	50.4	30.4
39	7	527	0.67	47.5	27.5

TABLE 18.8 Impulse Response of Bandpass Filter (Example 18.3)

n	$h_0(n) = h_0(-n)$	n	$h_0(n) = h_0(-n)$
0	2.761666×10^{-1}	11	2.726816×10^{-2}
1	1.660224×10^{-2}	12	-2.663859×10^{-2}
2	-2.389235×10^{-1}	13	-1.318252×10^{-2}
3	-3.689501×10^{-2}	14	6.312944×10^{-3}
4	1.473038×10^{-1}	15	-5.820976×10^{-3}
5	2.928852×10^{-2}	16	5.827957×10^{-3}
6	-4.770552×10^{-2}	17	1.528658×10^{-2}
7	-2.008131×10^{-3}	18	-8.288708×10^{-3}
8	-1.875082×10^{-2}	19	-1.616904×10^{-2}
9	-2.262965×10^{-2}	20	1.092728×10^{-2}
10	3.860990×10^{-2}	—	—

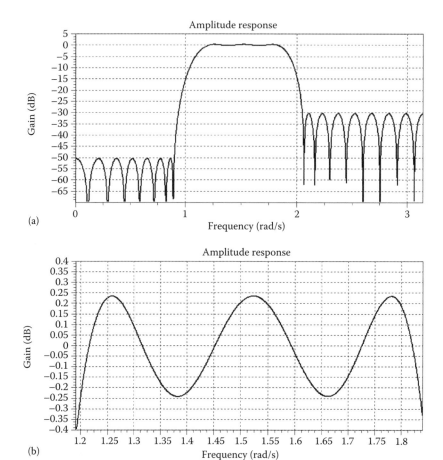

FIGURE 18.19 Amplitude response of equiripple bandpass filter (Example 18.3): (a) baseband and (b) passband.

where

$$P'_c(\omega) = \sum_{k=1}^{c} a_k \sin k\omega$$
$$a_k = 2h(c - k) \quad \text{for } k = 1, 2, \ldots, c$$
$$c = (N - 1)/2$$

(18.53)

A filter with a desired frequency response $e^{-jc\omega} jD(\omega)$ can be designed by constructing the error function

$$E(\omega) = W(\omega)[D(\omega) - P'_c(\omega)]$$

(18.54)

and then minimizing $|E(\omega)|$ with respect to some compact subset of the frequency interval $[0, \pi]$. From Equation 18.53, $P'_c(\omega)$ can be expressed as [6]

$$P'_c(\omega) = (\sin \omega) \, P_{c-1}(\omega)$$

(18.55)

where

$$P_{c-1}(\omega) = \sum_{k=0}^{c-1} \tilde{c}_k \cos k\omega$$

and

$$a_1 = \tilde{c}_0 - \frac{1}{2}\tilde{c}_2$$

$$a_k = \frac{1}{2}(\tilde{c}_{k-1} - \tilde{c}_{k+1}) \quad \text{for } k = 2, 3, \ldots, c-2$$

$$a_{c-1} = \frac{1}{2}\tilde{c}_{c-2}$$

$$a_c = \frac{1}{2}\tilde{c}_{c-1}$$

Hence Equation 18.54 can be put in the form

$$E(\omega) = \tilde{W}(\omega)[\tilde{D}(\omega) - \tilde{P}(\omega)] \tag{18.56}$$

where

$$\tilde{W}(\omega) = Q(\omega)W(\omega)$$
$$\tilde{D}(\omega) = D(\omega)/Q(\omega)$$
$$\tilde{P}(\omega) = P_{c-1}(\omega)$$
$$Q(\omega) = \sin\omega$$

Evidently, Equation 18.56 is of the same form as Equation 18.45, and upon proceeding as in Section 18.3.2, one can obtain the system of equations

$$
\begin{bmatrix}
1 & \cos\hat{\omega}_0 & \cos 2\hat{\omega}_0 & \cdots & \cos(c-1)\hat{\omega}_0 & \frac{1}{\tilde{W}(\hat{\omega}_0)} \\
1 & \cos\hat{\omega}_1 & \cos 2\hat{\omega}_1 & \cdots & \cos(c-1)\hat{\omega}_1 & \frac{-1}{\tilde{W}(\hat{\omega}_1)} \\
\vdots & \vdots & \vdots & & \vdots & \vdots \\
1 & \cos\hat{\omega}_r & \cos 2\hat{\omega}_r & \cdots & \cos(c-1)\hat{\omega}_r & \frac{(-1)^r}{\tilde{W}(\hat{\omega}_r)}
\end{bmatrix}
\begin{bmatrix}
a_0 \\
a_1 \\
\vdots \\
a_{c-1} \\
\delta
\end{bmatrix}
=
\begin{bmatrix}
\tilde{D}(\hat{\omega}_0) \\
\tilde{D}(\hat{\omega}_1) \\
\vdots \\
\tilde{D}(\hat{\omega}_r)
\end{bmatrix}
$$

where $r = c$ is the number of cosine functions in $P_{c-1}(\omega)$. The above system is the same as that in Equation 18.50 except that the number of extremals has been reduced from $c + 2$ to $c + 1$; therefore, the application of the Remez algorithm follows the methodology detailed in Sections 18.3.2 and 18.3.2.1 formulation and the Remez exchange algorithm.

The use of Algorithm 18.1 yields the optimum $P_{c-1}(\omega)$ and from Equation 18.55, the cosine function $P_c'(\omega)$ can be formed. Now $jP_c'(\omega)$ is the frequency response of a noncausal version of the required filter. The impulse response of this filter can be obtained as

$$h_0(n) = -h_0(-n) = -\frac{1}{N}\left[\sum_{k=1}^{c} 2P_c'(k\Omega)\,\sin\left(\frac{2\pi kn}{N}\right)\right]$$

for $n = 0, 1, 2, \ldots, c$, where $\Omega = 2\pi/N$, by using the inverse discrete Fourier transform. The impulse response of the corresponding causal filter is given by

$$h(n) = h_0(n - c)$$

for $n = 0, 1, 2, \ldots, N-1$.

The Remez algorithm can also be applied for the design of filters with symmetrical or antisymmetrical impulse response and even N. However, these filters are used less frequently. The reader is referred to Refs. [6,12] for more details.

18.3.2.5 Digital Differentiators

The Remez algorithm can be easily applied for the design of *equiripple digital differentiators*. The ideal frequency response of a causal differentiator is of the form $e^{-jc\omega} jD(\omega)$ where

$$D(\omega) = \omega \quad \text{for } 0 < |\omega| < \pi \tag{18.57}$$

and

$$c = (N - 1)/2$$

Since the frequency response is antisymmetrical, differentiators can be designed in terms of filters with antisymmetrical impulse response of either odd or even length.

Problem formulation. Assuming odd filter length, Equations 18.54 and 18.57 give the error function

$$E(\omega) = W(\omega)\big[\omega - P'_c(\omega)\big] \quad \text{for } 0 < \omega \le \omega_p$$

where ω_p is the required bandwidth. Equiripple absolute or relative error may be required, depending on the application at hand. Hence, $W(\omega)$ can be chosen to be either unity or $1/\omega$. In the latter case, which is the more meaningful of the two in practice, $E(\omega)$ can be expressed as

$$E(\omega) = 1 - \frac{1}{\omega} P'_c(\omega) \quad \text{for } 0 < \omega \le \omega_p$$

and from Equation 18.55

$$E(\omega) = 1 - \frac{\sin \omega}{\omega} P_{c-1}(\omega) \quad \text{for } 0 < \omega \le \omega_p \tag{18.58}$$

Therefore, the error function can be expressed as in Equation 18.56 with

$$\tilde{W}(\omega) = \frac{1}{\tilde{D}(\omega)} = \frac{\sin \omega}{\omega}$$
$$\tilde{P}(\omega) = P_{c-1}(\omega)$$

Prescribed specifications. A digital differentiator is fully specified by the constraint

$$|E(\omega)| \le \delta_p \quad \text{for } 0 < \omega \le \omega_p$$

where
   δ_p is the maximum passband error
   ω_p is the bandwidth of the differentiator

The differentiator length N that will just satisfy the required specifications is not normally known *a priori* and, although it may be determined on a hit-or-miss basis, a large number of designs may need to be carried out. In filters with approximately piecewise-constant amplitude responses, N can be predicted using the empirical formula of Equation 18.52. In the case of differentiators, N can be predicted by noting a useful property of digital differentiators. If δ and δ_1 are the maximum passband errors in differentiators of lengths N and N_1, respectively, then the quantity $\ln(\delta/\delta_1)$ is *approximately linear* with respect to $N - N_1$ for a wide range of values of N_1 and ω_p. Assuming linearity, we can show that [17]

$$N = N_1 + \frac{\ln(\delta/\delta_1)}{\ln(\delta_2/\delta_1)}(N_2 - N_1) \qquad (18.59)$$

where δ_2 is the maximum passband error in a differentiator of length N_2.

By designing two low-order differentiators, a fairly accurate prediction of the required value of N can be obtained by using Equation 18.59. Once a filter order is predicted a series of differentiators can be designed with increasing or decreasing N until a design that just satisfies the specifications is obtained.

Example 18.4

The selective step-by-step search with cubic interpolation was used in Algorithm 18.4 to design a digital differentiator that should satisfy the following specifications:

Odd differentiator length
Bandwidth ω_p: 2.75 rad/s
Maximum passband ripple δ_p: 1.0×10^{-5}
Sampling frequency ω_s: 2π rad/s

The progress of the design is illustrated in Table 18.9. First, differentiators of lengths 21 and 23 were designed and the required N to satisfy the specifications was predicted to be 55 using Equation 18.59. This differentiator length was found to satisfy the specifications, and a design for length 53 was then carried out. The second design was found to violate the specifications and hence the first design is the required differentiator. The impulse response of this differentiator is given in Table 18.10. The amplitude response and passband relative error of the differentiator are plotted in Figure 18.20a and b.

18.3.2.6 Arbitrary Amplitude Responses

Very frequently FIR filters are required whose amplitude responses cannot be described by analytical functions. For example, in the design of two-dimensional filters through the singular-value decomposition [18,19], the required two-dimensional filter is obtained by designing a set of one-dimensional digital filters whose amplitude responses turn out to have arbitrary shapes. In these applications, the desired amplitude response $D(\omega)$ is specified in terms of a table that lists a prescribed set of frequencies and the corresponding values of the required filter gain. Filters of this class can be readily designed by employing some interpolation scheme that can be used to evaluate $D(\omega)$ and its first derivative with respect to ω at any ω. A suitable scheme is to fit a set of third-order polynomials to the prescribed amplitude response.

18.3.2.7 Multiband Filters

The algorithms presented in the previous sections can also be used to design *multiband* filters. While there is no theoretical upper limit on the number of bands, in practice, the design tends to become more and more difficult as the number of bands is increased. The reason is that the difference between the number of possible maxima in the error function and the number of extremals increases linearly

TABLE 18.9 Progress in Design of Digital Differentiator (Example 18.4)

N	Iterations	FE's	δ_p
21	4	145	1.075×10^{-2}
23	4	162	6.950×10^{-3}
55	7	815	8.309×10^{-6}
53	7	757	1.250×10^{-5}

TABLE 18.10   Impulse Response of Digital Differentiator (Example 18.4)

n	$h(n) = -h_0(-n)$	n	$h_0(n) = -h_0(-n)$
0	0.0	14	1.762268×10^{-2}
1	-9.933416×10^{-1}	15	-1.313097×10^{-2}
2	4.868036×10^{-1}	16	9.615295×10^{-3}
3	-3.138353×10^{-1}	17	-6.902518×10^{-3}
4	2.245441×10^{-1}	18	4.844090×10^{-3}
5	-1.690252×10^{-1}	19	-3.312235×10^{-3}
6	1.306918×10^{-1}	20	2.197502×10^{-3}
7	-1.024631×10^{-1}	21	-1.407064×10^{-3}
8	8.081083×10^{-2}	22	8.632670×10^{-4}
9	-6.377426×10^{-2}	23	-5.023168×10^{-4}
10	5.016708×10^{-2}	24	2.729367×10^{-4}
11	-3.921782×10^{-2}	25	-1.349790×10^{-4}
12	3.039137×10^{-2}	26	5.859128×10^{-5}
13	-2.329439×10^{-2}	27	-1.634535×10^{-5}

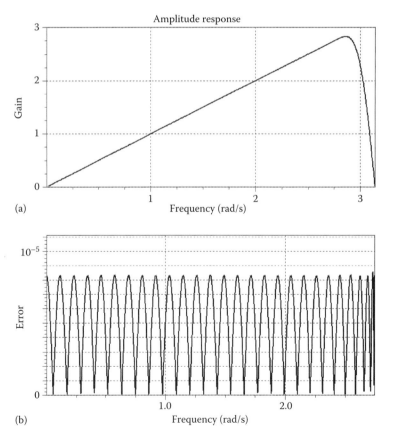

FIGURE 18.20   Design of digital differentiator (Example 18.4): (a) amplitude response and (b) passband relative error.

with the number of bands, e.g., if the number of bands is 8, then the difference is 14. As a consequence, the number of potential extremals that need to be rejected is large and the available rejection techniques become inefficient. The end result is that the number of iterations is increased quite significantly, and convergence is slow and sometimes impossible.

In mathematical terms, the above difficulty is attributed to the fact that, in the weighted-Chebyshev methods considered here, the approximating polynomial becomes seriously *underdetermined* if the number of bands exceeds three. The problem can be overcome by using the generalized Remez method described in Ref. [15]. This approach was found to yield better results for filters with more than four or five bands.

References

1. O. Hermann, Design of nonrecursive digital filters with linear phase, *Electron. Lett.*, 6, 182–184, May 1970.
2. E. Hofstetter, A. Oppenheim, and J. Siegel, A new technique for the design of non-recursive digital filters, *5th Annu. Princeton Conf. Informat. Sci. Syst.*, 64–72, March 1971.
3. T. W. Parks and J. H. McClellan, Chebyshev approximation for nonrecursive digital filters with linear phase, *IEEE Trans. Circuit Theory*, CT-19, 189–194, March 1972.
4. T. W. Parks and J. H. McClellan, A program for the design of linear phase finite impulse response digital filters, *IEEE Trans. Audio Electroacoust.*, AU-20, 195–199, August 1972.
5. L. R. Rabiner and O. Herrmann, On the design of optimum FIR low-pass filters with even impulse response duration, *IEEE Trans. Audio Electroacoust.*, AU-21, 329–336, August 1973.
6. J. H. McClellan and T. W. Parks, A unified approach to the design of optimum FIR linear-phase digital filters, *IEEE Trans. Circuit Theory*, CT-20, 697–701, November 1973.
7. J. H. McClellan, T. W. Parks, and L. R. Rabiner, A computer program for designing optimum FIR linear phase digital filters, *IEEE Trans. Audio Electroacoust*, AU-21, 506–526, December 1973.
8. L. R. Rabiner, J. H. McClellan, and T. W. Parks, FIR digital filter design techniques using weighted Chebyshev approximation, *Proc. IEEE*, 63, 595–610, April 1975.
9. J. H. McClellan, T. W. Parks, and L. R. Rabiner, FIR linear phase filter design program, in *Programs for Digital Signal Processing*, New York: IEEE, 1979, pp. 5.1-1–5.1-13.
10. A. Antoniou, Accelerated procedure for the design of equiripple nonrecursive digital filters, *IEE Proc., Pt. G*, 129, 1–10, February 1982 (see *IEE Proc., Pt. G*, vol.129, p.107, June 1982 for errata).
11. A. Antoniou, New improved method for the design of weighted-Chebyshev, nonrecursive, digital filters, *IEEE Trans. Circuits Syst.*, CAS-30, 740–750, October 1983.
12. A. Antoniou, *Digital Signal Processing: Signals, Systems, and Filters*, New York: McGraw-Hill, 2005.
13. E. W. Cheney, *Introduction to Approximation Theory*, New York: McGraw-Hill, 1966, 72–100.
14. E. Ya. Remez, *General Computational Methods for Tchebycheff Approximation*, Kiev: Atomic Energy Comm., 1957, Translation 4491, 1–85, 1957.
15. D. J. Shpak and A. Antoniou, A generalized Reméz method for the design of FIR digital filters, *IEEE Trans. Circuits Syst.*, 37, 161–174, February 1990.
16. O. Herrmann, L. R. Rabiner, and D. S. K. Chan, Practical design rules for optimum finite impulse response low-pass digital filters, *Bell Syst. Tech. J.*, 52, 769–799, July–August 1973.
17. A. Antoniou and C. Charalambous, Improved design method for Kaiser differentiators and comparison with equiripple method, *IEE Proc., Pt. E*, 128, 190–196, September 1981.
18. A. Antoniou and W. -S. Lu, Design of two-dimensional digital filters by using the singular value decomposition, *IEEE Trans. Circuits Syst.*, CAS-34, 1191–1198, October 1987.
19. W. -S. Lu, H. -P. Wang, and A. Antoniou, Design of two-dimensional FIR digital filters using the singular-value decomposition, *IEEE Trans. Circuits Syst.*, CAS-37, 35–46, January 1990.

18.4 Design of Computationally Efficient FIR Filters Using Periodic Subfilters as Building Blocks

Tapio Saramäki

For many digital signal processing applications, FIR filters are preferred over their IIR counterparts as the former can be designed with exactly linear phase and they are free of stability problems and limit cycle oscillations. The major drawback of FIR filters is that they require, especially in applications demanding narrow transition bands, considerably more arithmetic operations and hardware components than do comparable IIR filters. Ignoring the correction term for very low-order filters, the minimum order of an optimum linear-phase low-pass FIR filter can be approximated [1] by

$$N \approx \Phi(\delta_{\mathrm{p}}, \delta_{\mathrm{s}})/(\omega_{\mathrm{s}} - \omega_{\mathrm{p}}) \tag{18.60a}$$

where

$$\Phi(\delta_{\mathrm{p}}, \delta_{\mathrm{s}}) = 2\pi \left[0.005309(\log_{10} \delta_{\mathrm{p}})^2 + 0.07114 \log_{10} \delta_{\mathrm{p}} - 0.4761 \right] \log_{10} \delta_{\mathrm{s}}$$
$$- 2\pi \left[0.00266(\log_{10} \delta_{\mathrm{p}})^2 + 0.5941 \log_{10} \delta_{\mathrm{p}} + 0.4278 \right] \tag{18.60b}$$

Here, ω_{p} and ω_{s} are the passband and stopband edge angles, whereas δ_{p} and δ_{s} are the passband and stopband ripple magnitudes. From the above estimate, it is seen that as the transition bandwidth $\omega_{\mathrm{s}} - \omega_{\mathrm{p}}$ is made smaller, the required filter order increases inversely proportionally to it. Since the direct-form implementation exploiting the coefficient symmetry requires approximately $N/2$ multipliers, this kind of implementation becomes very costly if the transition bandwidth is small.

The cost of implementation of a narrow transition-band FIR filter can be significantly reduced by using multiplier-efficient realizations, fast convolution algorithms, or multirate filtering. This section considers those multiplier-efficient realizations that use as basic building blocks the transfer functions obtained by replacing each unit delay in a conventional transfer function by multiple delays. We concentrate on the synthesis techniques described in Refs. [2–4,6,8–10].

18.4.1 Frequency-Response Masking Approach

A very elegant approach to significantly reducing the implementation cost of an FIR filter has been proposed by Lim [3]. In this approach, the overall transfer function is constructed as

$$H(z) = F(z^L)G_1(z) + \left[z^{-LN_F/2} - F(z^L) \right] G_2(z) \tag{18.61a}$$

where

$$F(z^L) = \sum_{n=0}^{N_F} f(n)z^{-nL}, \quad f(N_F - n) = f(n) \tag{18.61b}$$

$$G_1(z) = z^{-M_1} \sum_{n=0}^{N_1} g_1(n)z^{-n}, \quad g_1(N_1 - n) = g_1(n) \tag{18.61c}$$

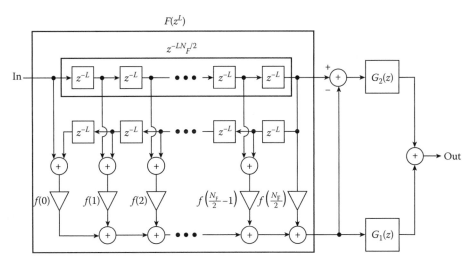

FIGURE 18.21 Efficient implementation for a filter synthesized using the frequency-response masking approach.

and

$$G_2(z) = z^{-M_2} \sum_{n=0}^{N_2} g_2(n)z^{-n}, \quad g_2(N_2 - n) = g_2(n) \tag{18.61d}$$

Here, N_F is even, whereas both N_1 and N_2 are either even or odd. For $N_1 \le N_2$, $M_1 = 0$ and $M_2 = (N_1 - N_2)/2$, whereas for $N_1 < N_2$, $M_1 = (N_2 - N_1)/2$ and $M_2 = 0$. These selections guarantee that the delays of both of the terms of $H(z)$ are equal. An efficient implementation for the overall filter is depicted in Figure 18.21, where the delay term $z^{-LN_F/2}$ is shared with $F(z^L)$. Also, $G_1(z)$ and $G_2(z)$ can share their delays if a transposed direct-form implementation (exploiting the coefficient symmetry) is used.

The frequency response of the overall filter can be written as

$$H(e^{j\omega}) = H(\omega)e^{-j(LN_F + \max[N_1, N_2])\omega/2} \tag{18.62}$$

where $H(\omega)$ denotes the *zero-phase frequency response* of $H(z)$ and can be expressed as

$$H(\omega) = H_1(\omega) + H_2(\omega) \tag{18.63a}$$

where

$$H_1(\omega) = F(L\omega)G_1(\omega) \tag{18.63b}$$

and

$$H_2(\omega) = [1 - F(L\omega)]G_2(\omega) \tag{18.63c}$$

with

$$F(\omega) = f(N_F/2) + 2\sum_{n=1}^{N_F/2} f(N_F/2 - n)\cos n\omega \tag{18.63d}$$

and

$$
G_k(\omega) = \begin{cases} g_k(N_k/2) + 2 \displaystyle\sum_{n=1}^{N_k/2} g_k(N_k/2 - n) \cos n\omega & N_k \text{ even} \\[2ex] 2 \displaystyle\sum_{n=0}^{(N_k-1)/2} g_k[(N_k - 1)/2 - n] \cos[(n + 1/2)\omega] & N_k \text{ odd} \end{cases}
\tag{18.63e}
$$

for $k = 1, 2$.

The efficiency as well as the synthesis of $H(z)$ are based on the properties of the pair of transfer functions $F(z^L)$ and $z^{-LN_F/2} - F(z^L)$, which can be generated from the pair of *prototype* transfer functions

$$
F(z) = \sum_{n=0}^{N_F} f(n)z^{-n}
\tag{18.64}
$$

and $z^{-N_F/2} - F(z)$ by replacing z^{-1} by z^{-L}, that is, by substituting for each unit delay L unit delays. The order of the resulting filters is increased to LN_F, but since only every Lth impulse response value is nonzero, the filter complexity (number of adders and multipliers) remains the same. The above prototype pair forms a *complementary* filter pair since their zero-phase frequency responses, $F(\omega)$ and $1 - F(\omega)$ with $F(\omega)$ given by Equation 18.63d, add up to unity. Figure 18.22a illustrates the relations between these responses in the case where $F(z)$ and $z^{-N_F/2} - F(z)$ is a low-pass–high-pass filter pair with edges at θ and ϕ.

The substitution $z^{-L} \to z^{-1}$ preserves the complementary property resulting in the *periodic* responses $F(L\omega)$ and $1 - F(L\omega)$, which are frequency-axis compressed versions of the prototype responses such that the interval $[0, L\pi]$ is shrunk onto $[0, \pi]$ (see Figure 18.22b). Since the periodicity of the prototype responses is 2π, the periodicity of the resulting responses is $2\pi/L$ and they contain several passband and stopband regions in the interval $[0, \pi]$.

For a low-pass filter $H(z)$, one of the transition bands provided by $F(z^L)$ or $z^{-LN_F/2} - F(z^L)$ is used as that of the overall filter. In the first case, denoted by Case A, the edges are given by (Figure 18.23)

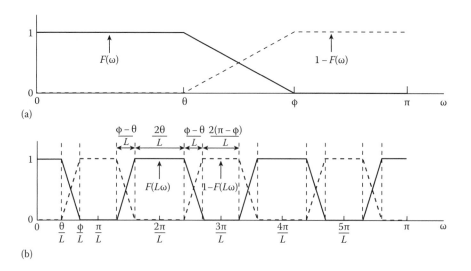

FIGURE 18.22 Generation of a complementary periodic filter pair by starting with a low-pass–high-pass complementary pair. (a) Prototype filter responses $F(\omega)$ and $1 - F(\omega)$. (b) Periodic responses $F(L\omega)$ and $1 - F(L\omega)$ for $L = 6$.

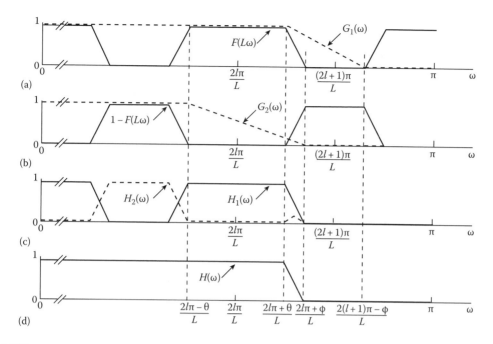

FIGURE 18.23 Case A design of a low-pass filter using the frequency-response masking technique.

$$\omega_p = (2l\pi + \theta)/L \quad \text{and} \quad \omega_s = (2l\pi + \phi)/L \tag{18.65}$$

where l is a fixed integer, and in the second case, referred to as Case B, by (Figure 18.24)

$$\omega_p = (2l\pi - \phi)/L \quad \text{and} \quad \omega_s = (2l\pi - \theta)/L \tag{18.66}$$

The widths of these transition bands are $(\phi - \theta)/L$, which is only $1/L$th of that of the prototype filters. Since the filter order is roughly inversely proportional to the transition bandwidth, this means that the arithmetic complexity of the periodic transfer functions to provide one of the transition bands is only $1/L$th of that of a conventional nonperiodic filter. Note that the orders of both the periodic filters and the corresponding nonperiodic filters are approximately the same, but the conventional filter does not contain zero-valued impulse response samples.

In order to exploit the attractive properties of the periodic transfer functions, the two low-order masking filters $G_1(z)$ and $G_2(z)$ are designed such that the subresponses $H_1(\omega)$ and $H_2(\omega)$ as given by Equations 18.63b and c approximate in the passband $F(L\omega)$ and $1 - F(L\omega)$, respectively, so that their sum approximates unity, as is desired. In the filter stopband, the role of the masking filters is to attenuate the extra unwanted passbands and transition bands of the periodic responses. In Case A, this is achieved by selecting the edges of $G_1(z)$ and $G_2(z)$ as (see Figure 18.23)

$$\omega_p^{(G_1)} = \omega_p = [2l\pi + \theta]/L \quad \text{and} \quad \omega_s^{(G_1)} = [2(l+1)\pi - \phi]/L \tag{18.67a}$$

$$\omega_p^{(G_2)} = [2l\pi - \theta]/L \quad \text{and} \quad \omega_s^{(G_2)} = \omega_s = [2l\pi + \phi]/L \tag{18.67b}$$

Since $F(L\omega) \approx 0$ on $[\omega_s, \omega_s^{(G1)}]$, the stopband region of $G_1(z)$ can start at $\omega = \omega_s^{(G1)}$, instead of $\omega = \omega_s$. Similarly, since $H_1(\omega) \approx F(L\omega) \approx 1$ and $[1 - F(L\omega)] \approx 0$ on $[\omega_p^{(G2)}, \omega_p]$, the passband region of $G_2(z)$ can start at $\omega = \omega_p^{(G2)}$, instead of $\omega = \omega_p$.

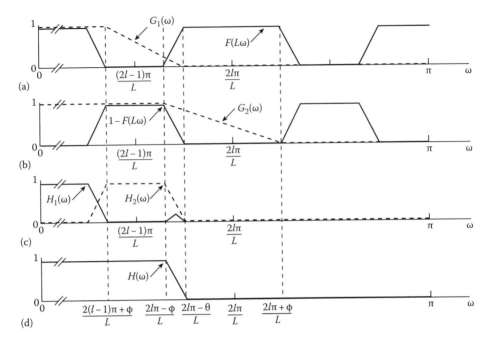

FIGURE 18.24 Case B design of a low-pass filter using the frequency-response masking technique.

For Case B designs, the required edges of the two masking filters, $G_1(z)$ and $G_2(z)$ are (see Figure 18.24)

$$\omega_p^{(G_1)} = [2(l-1)\pi + \phi]/L \quad \text{and} \quad \omega_s^{(G_1)} = \omega_s = [2l\pi - \theta]/L \tag{18.68a}$$

$$\omega_p^{(G_2)} = \omega_p = [2l\pi - \phi]/L \quad \text{and} \quad \omega_s^{(G_2)} = [2l\pi + \theta]/L \tag{18.68b}$$

The effects of the ripples of the subresponses on the ripples of the overall response $H(\omega)$ have been studied carefully in Ref. [3]. Based on these observations, the design of $H(z)$ with passband and stopband ripples of δ_p and δ_s can be accomplished for both Case A and Case B in the following two steps:

1. Design $G_k(z)$ for $k = 1, 2$ using either the Remez algorithm or linear programming such that $G_k(\omega)$ approximates unity on $[0, \omega_p^{(Gk)}]$ with tolerance $0.85\delta_p \cdots 0.9\delta_p$ and zero on $[\omega_s^{(Gk)}, \pi]$ with tolerance $0.85\delta_s \cdots 0.9\delta_s$.
2. Design $F(L\omega)$ such that the overall response $H(\omega)$ approximates unity on

$$\Omega_p^{(F)} = \begin{cases} \left[\omega_p^{(G_2)}, \omega_p\right] = [[2l\pi - \theta]/L, [2l\pi + \theta]/L] & \text{for Case A} \\ \left[\omega_p^{(G_1)}, \omega_p\right] = [[2(l-1)\pi + \phi]/L, [2l\pi - \phi]/L] & \text{for Case B} \end{cases} \tag{18.69a}$$

with tolerance δ_p and approximates zero on

$$\Omega_s^{(F)} = \begin{cases} \left[\omega_s, \omega_s^{(G_1)}\right] = [[2l\pi + \phi]/L, [2(l+1)\pi - \phi]/L] & \text{for Case A} \\ \left[\omega_s, \omega_s^{(G_2)}\right] = [[2l\pi - \theta]/L, [2l\pi + \theta]/L] & \text{for Case B} \end{cases} \tag{18.69b}$$

with tolerance δ_s.

TABLE 18.11 Error Function for Designing $F(\omega)$ Using the Remez Algorithm

---

$E_F(\omega) = W_F(\omega)[F(\omega) - D_F(\omega)]$,

where

$D_F(\omega) = [u(\omega) + l(\omega)]/2$, $W_F(\omega) = 2/[u(\omega) - l(\omega)]$

with

$u(\omega) = \min(\Psi_1(\omega) + \psi_1(\omega), \Psi_2(\omega) + \psi_2(\omega))$

$l(\omega) = \max(\Psi_1(\omega) - \psi_1(\omega), \Psi_2(\omega) - \psi_2(\omega))$

$\Psi_k(\omega) = \dfrac{D_H[h_k(\omega)] - G_2[h_k(\omega)]}{G_1[h_k(\omega)] - G_2[h_k(\omega)]}$, $k = 1, 2$

$\psi_k(\omega) = \dfrac{1/W_H[h_k(\omega)]}{|G_1[h_k(\omega)] - G_2[h_k(\omega)]|}$, $k = 1, 2$

and

$h_1(\omega) = (2l\pi + \omega)/L$, $\quad h_2(\omega) = \begin{cases} (2l\pi - \omega)/L & \text{for } \omega \in [0, \theta] \\ [2(l+1)\pi - \omega]/L & \text{for } \omega \in [\phi, \pi] \end{cases}$

for Case A and

$h_1(\omega) = (2l\pi - \omega)/L$, $\quad h_2(\omega) = \begin{cases} (2l\pi + \omega)/L & \text{for } \omega \in [0, \theta] \\ [2(l-1)\pi + \omega]/L & \text{for } \omega \in [\phi, \pi] \end{cases}$

for Case B

The design of $F(L\omega)$ can be performed conveniently using linear programming [3]. Another, computationally more efficient, alternative is to use the Remez algorithm [10]. Its use is based on the fact that

$$|E_H(\omega)| \leq 1 \text{ for } \omega \in \Omega_p^{(F)} \cup \Omega_s^{(F)} \tag{18.70a}$$

where

$$E_H(\omega) = W_H(\omega)[H(\omega) - D_H(\omega)] \tag{18.70b}$$

is satisfied when $F(\omega)$ is designed such that the maximum absolute value of the error function given in Table 18.11 becomes less than or equal to unity on $[0, \theta] \cup [\phi, \pi]$.

For step 2 of the above algorithm, $D_H(\omega) = 1$ and $W_H(\omega) = 1/\delta_p$ on $\Omega_p^{(F)}$, whereas $D_H(\omega) = 0$ and $W_H(\omega) = 1/\delta_s$ on $\Omega_s^{(F)}$, giving for $k = 1, 2$

$$D_H[h_k(\omega)] = \begin{cases} 1 & \text{for } \omega \in [0, \theta] \\ 0 & \text{for } \omega \in [\phi, \pi] \end{cases} \quad \text{and} \quad W_H[h_k(\omega)] = \begin{cases} 1/\delta_p & \text{for } \omega \in [0, \theta] \\ 1/\delta_s & \text{for } \omega \in [\phi, \pi] \end{cases} \tag{18.71a}$$

for Case A and

$$D_H[h_k(\omega)] = \begin{cases} 1 & \text{for } \omega \in [0, \theta] \\ 0 & \text{for } \omega \in [\phi, \pi] \end{cases} \quad \text{and} \quad W_H[h_k(\omega)] = \begin{cases} 1/\delta_s & \text{for } \omega \in [0, \theta] \\ 1/\delta_p & \text{for } \omega \in [\phi, \pi] \end{cases} \tag{18.71b}$$

for Case B. Even though the resulting error function looks very complicated, it is straightforward to use the subroutines EFF and WATE in the Remez algorithm described in Ref. [5] for optimally designing $F(z)$.

The order of $G_1(z)$ can be considerably reduced by allowing larger ripples on those regions of $G_1(z)$ where $F(L\omega)$ has one of its stopbands. As a rule of thumb, the ripples on these regions can be selected to be 10 times larger [3]. Similarly, the order $G_2(z)$ can be decreased by allowing (ten times) larger ripples on those regions where $F(L\omega)$ has one of its passbands.

In practical filter synthesis problems, ω_p and ω_s are given and l, L, θ, and ϕ must be determined. To ensure that Equation 18.65 yields a desired solution with $0 \le \theta < \phi \le \pi$, it is required that (see Figure 18.23)

$$\frac{2l\pi}{L} \le \omega_p \quad \text{and} \quad \omega_s \le \frac{(2l+1)\pi}{L} \tag{18.72a}$$

for some positive integer l, giving

$$l = \lfloor L\omega_p/(2\pi) \rfloor, \quad \theta = L\omega_p - 2l\pi, \quad \text{and} \quad \phi = L\omega_s - 2l\pi \tag{18.72b}$$

where $\lfloor x \rfloor$ stands for the largest integer that is smaller than or equal to x. Similarly, to ensure that Equation 18.66 yields a desired solution with $0 \le \theta < \phi \le \pi$, it is required that (see Figure 18.23)

$$\frac{(2l-1)\pi}{L} \le \omega_p \quad \text{and} \quad \omega_s \le \frac{2l\pi}{L} \tag{18.73a}$$

for some positive integer l, giving

$$l = \lceil L\omega_s/(2\pi) \rceil, \quad \theta = 2l\pi - L\omega_s, \quad \text{and} \quad \phi = 2l\pi - L\omega_p \tag{18.73b}$$

where $\lceil x \rceil$ stands for the smallest integer that is larger than or equal to x. For any set of ω_p, ω_s, and L, either Equation 18.72b or Equation 18.73b (not both) will yield the desired θ and ϕ, provided that L is not too large. If $\theta = 0$ or $\phi = \pi$, then the resulting specifications for $F(\omega)$ are meaningless and the corresponding value of L cannot be used.

The remaining problem is to determine L to minimize the number of multipliers, which is $N_P/2 + 1 + 1\lfloor (N_1 + 2)/2 \rfloor + \lfloor (N_2 + 2)/2 \rfloor$ or $N_F + N_1 + N_2 + 3$ depending on whether the symmetries in the filter coefficients are exploited or not. Hence, in both cases, a good measure of the filter complexity is the sum of the orders of the subfilters. Instead of determining the actual minimum filter orders for various values of L, the computational workload can be significantly reduced based on the use of the estimation formula given by Equations 18.60a and b. Since the widths of transition bands of $F(z)$, $G_1(z)$, and $G_2(z)$ are $\phi - \theta$, $(2\pi - \phi - \theta)/L$ and $(\phi + \theta)/L$, respectively, good estimates for the corresponding filter orders are

$$N_F \approx \frac{\Phi(\delta_p, \delta_s)}{\phi - \theta}, \quad N_1 \approx \frac{L\Phi(\delta_p, \delta_s)}{2\pi - \phi - \theta}, \quad \text{and} \quad N_2 \approx \frac{L\Phi(\delta_p, \delta_s)}{\phi + \theta} \tag{18.74}$$

For the optimum nonperiodic direct-form design, the transition bandwidth is $\omega_s - \omega_p = (\phi - \theta)/L$, giving

$$N_{opt} \approx \frac{L\Phi(\delta_p, \delta_s)}{\phi - \theta} \tag{18.75}$$

The sum of the subfilter orders can be expressed in terms of N_{opt} as follows:

$$N_{ove} = N_{opt} \left[\frac{1}{L} + \frac{\phi - \theta}{2\pi - \phi - \theta} + \frac{\phi - \theta}{\phi + \theta} \right] \tag{18.76}$$

The smallest values of N_{ove} are typically obtained at those values of L for which $\theta + \phi \approx \pi$ and, correspondingly, $2\pi - \theta - \phi \approx \pi$. In this case, $N_1 \approx N_2$ and Equation 18.76 reduces, after substituting $\phi - \theta = L(\omega_s - \omega_p)$, to

$$N_{ove} = N_{opt}\left[\frac{1}{L} + 2L(\omega_s - \omega_p)/\pi\right] \tag{18.77}$$

At these values of L, N_F decreases and $N_1 \approx N_2$ increases inversely proportionally to L with the minimum of N_{ove} given by

$$N_{ove} = 2N_{opt}\sqrt{\frac{2(\omega_s - \omega_p)}{\pi}} \tag{18.78}$$

taking place at

$$L_{opt} = 1\Big/\sqrt{\frac{2(\omega_s - \omega_p)}{\pi}} \tag{18.79}$$

If for $L = L_{opt}$, $\theta + \phi$ is not approximately equal to π, then L minimizing the filter complexity can be found in the near vicinity of L_{opt}. The following example illustrates the use of the above estimation formulas.

Example 18.1

Consider the specifications: $\omega_p = 0.4\pi$, $\omega_s = 0.402\pi$, $\delta_p = 0.01$, and $\delta_s = 0.001$. For the optimum conventional direct-form design, $N_{opt} = 2541$, requiring 1271 multipliers when the coefficient symmetry is exploited. Equation 18.79 gives $L_{opt} = 16$. Table 18.12 shows, for the admissible values of L in the vicinity of this value, l, θ, ϕ, the estimated orders for the subfilters, and the sum of the subfilter orders as well as whether the overall filter is a Case A or Case B design. For N_F, the minimum even order larger than or equal to the estimated order is used, whereas N_2 is forced to be even (odd) if N_1 is even (odd).

Also with the estimated filter orders of Table 18.12, $L = 16$ gives the best result. The actual filter orders are $N_F = 162$, $N_1 = 70$, $N_2 = 98$. The responses of the subfilters as well as that of the overall design are depicted in Figure 18.25. The overall number of multipliers and adders for this design are 168 and 330, respectively, which are 13 percent of those required by an equivalent conventional direct-form design (1271 and 2541). The overall filter order is 2690, which is only 6 percent higher than that of the direct-form design (2541).

TABLE 18.12 Estimated Filter Orders for the Admissible Values of L in the Vicinity of $L_{opt} = 16$

L	Case	l	θ	ϕ	N_F	N_1	N_2	$N_F + N_1 + N_2$
8	B	2	0.784π	0.8π	318	98	26	442
9	B	2	0.382π	0.4π	282	38	58	378
11	A	2	0.4π	0.422π	232	47	69	348
12	A	2	0.8π	0.824π	212	162	38	412
13	B	3	0.774π	0.8π	196	155	43	394
14	B	3	0.372π	0.4π	182	58	92	332
16	A	3	0.4π	0.432π	160	70	98	328
17	A	3	0.8π	0.834π	150	236	54	440
18	B	4	0.764π	0.8π	142	210	58	410
19	B	4	0.362π	0.4π	134	78	128	340
21	A	4	0.4π	0.442π	122	92	128	342
22	A	4	0.8π	0.844π	116	314	68	498

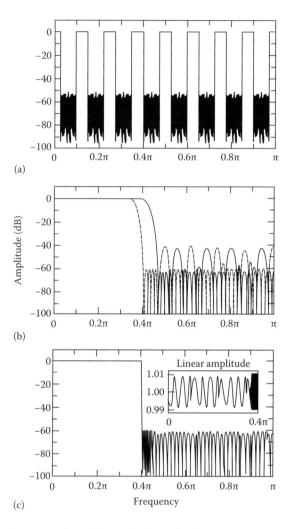

FIGURE 18.25 Amplitude responses for a filter synthesized using the frequency-response masking approach. (a) Periodic response $F(L\omega)$. (b) Responses $G_1(\omega)$ (solid line) and $G_2(\omega)$ (dashed line). (c) Overall response.

18.4.2 Multistage Frequency-Response Masking Approach

If the order of $F(z)$ is too high, its complexity can be reduced by implementing it using the frequency-response masking technique. Extending this to an arbitrary number of stages results in the multistage frequency-response masking approach [3,4], where $H(z)$ is generated iteratively as

$$H(z) \equiv F^{(0)}(z) = F^{(1)}(z^{L_1})G_1^{(1)}(z) + \left[z^{-L_1 N_F^{(1)}/2} - F^{(1)}(z^{L_1})\right]G_2^{(1)}(z) \tag{18.80a}$$

$$F^{(1)}(z) = F^{(2)}(z^{L_2})G_1^{(2)}(z) + \left[z^{-L_2 N_F^{(2)}/2} - F^{(2)}(z^{L_2})\right]G_2^{(2)}(z) \tag{18.80b}$$

$$\vdots$$

$$F^{(R-1)}(z) = F^{(R)}\left(z^{L_R}\right)G_1^{(R)}(z) + \left[z^{-L_R N_F^{(R)}/2} - F^{(R)}\left(z^{L_R}\right)\right]G_2^{(R)}(z) \tag{18.80c}$$

TABLE 18.13 Implementation Form for the Transfer Function in the Multistage Frequency-Response Masking Approach

$$H(z) \equiv F^{(0)}(z^{\hat{L}_0}) = F^{(1)}(z^{\hat{L}_1})G_1^{(1)}(z^{\hat{L}_0}) + [z^{-M_1} - F^{(1)}(z^{\hat{L}_1})]G_2^{(1)}(z^{\hat{L}_0})$$

$$F^{(1)}(z^{\hat{L}_1}) = F^{(2)}(z^{\hat{L}_2})G_1^{(2)}(z^{\hat{L}_1}) + [z^{-M_2} - F^{(2)}(z^{\hat{L}_2})]G_2^{(2)}(z^{\hat{L}_1})$$

$$\vdots$$

$$F^{(R-1)}(z^{\hat{L}_{R-1}}) = F^{(R)}(z^{\hat{L}_R})G_1^{(R)}(z^{\hat{L}_{R-1}}) + [z^{-M_R} - F^{(R)}(z^{\hat{L}_R})]G_2^{(R)}(z^{\hat{L}_{R-1}}),$$

where

$$\hat{L}_0 = 1, \hat{L}_r = \prod_{k=1}^{r} L_k, \quad r = 1, 2, \cdots, R$$

$$M_R = \hat{L}_R N_F^{(R)}/2, \quad M_{R-r} = M_{R-r+1} + m_{R-r}, \quad r = 1, 2, \cdots, R-1$$

$$m_{R-r} = \hat{L}_{R-r} \max\left\{N_1^{(R-r+1)}, N_2^{(R-r+1)}\right\}/2, \quad r = 1, 2, \cdots, R-1$$

$N_F^{(R)}$ is the order of $F^{(R)}(z)$

$N_1^{(r)}$ and $N_2^{(r)}$ are the orders of $G_1^{(r)}(z)$ and $G_2^{(r)}(z)$, respectively

Here, the $G_1^{(r)}(z)$'s and $G_2^{(r)}(z)$'s for $r = 1, 2, \ldots, R$ as well as $F^{(R)}(z)$ are the filters to be designed. For implementation purposes, $H(z)$ can be expressed in the form shown in Table 18.13. Figure 18.26 shows an efficient implementation for a three-stage filter, where the delay terms z^{-M3}, z^{-m2}, and z^{-m1}, can be shared with $F^{(3)}(\hat{z}^{L_3})$. In order to obtain a desired overall solution, the orders of the $G_1^{(r)}(z)$'s and $G_2^{(r)}(z)$'s for $r = 2, 3, \ldots, R$, denoted by $N_1^{(r)}$ and $N_2^{(r)}$ in Table 18.13, have to be even.

Given the filter specifications and the L_r's for $r = 1, 2, \ldots, R$, the $G_1^{(r)}(z)$'s and $G_2^{(r)}(z)$'s as well as $F^{(R)}(z)$ can be synthesized in the following steps:

1. Set $r = 1, L = L_1$, and

$$D_H(\omega) = \begin{cases} 1 & \text{for } \omega \in [0, \omega_p] \\ 0 & \text{for } \omega \in [\omega_s, \pi] \end{cases}, \quad W_H(\omega) = \begin{cases} 1/\delta_p & \text{for } \omega \in [0, \omega_p] \\ 1/\delta_s & \text{for } \omega \in [\omega_s, \pi] \end{cases} \qquad (18.81)$$

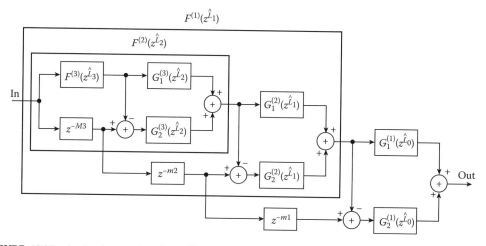

FIGURE 18.26 An implementation for a filter synthesized using the three-stage frequency-response masking approach.

2. Determine whether $F^{(r-1)}(z)$ is a Case A or Case B design as well as θ, ϕ, and l for $F^{(r)}(z)$ according to Equation 18.72b or Equation 18.73b. Also, determine $\omega_p^{(Gk)}$ and $\omega_s^{(Gk)}$ for $k = 1, 2$ from Equations 18.67a and b or Equations 18.68a and b.
3. Design $G_k^{(r)}(z)$ for $k = 1, 2$, using either the Remez algorithm or linear programming, in such a way that

$$\max_{\omega \in \left[0, \omega_p^{(Gk)}\right] \cup \left[\omega_s^{(Gk)}, \pi\right]} \left| W_H(\omega) \left[G_k^{(r)}(\omega) - D_H(\omega) \right] \right| \leq 0.9 \tag{18.82}$$

4. Determine $W_F(\omega)$ and $D_F(\omega)$ from Table 18.11.
5. If $r = R$, then go to the next step. Otherwise, set $r = r + 1$, $L = L_r$, $W_H(\omega) = W_F(\omega)$, $D_H(\omega) = D_F(\omega)$, $\omega_p = \theta$, $\omega_s = \phi$, and go to step 2.
6. Design $F^{(R)}(z)$, using either the Remez algorithm or linear programming, in such a way that

$$\max_{\omega \in [0, \theta] \cup [\phi, \pi]} \left| W_F(\omega) \left[F^{(R)}(\omega) - D_F(\omega) \right] \right| \leq 1 \tag{18.83}$$

In the above algorithm, $G_1^{(1)}(z)$ and $G_2^{(1)}(z)$ are determined like in the one-stage frequency-response masking technique. The remaining filter part as given by Equation 18.80b has then to be designed such that the maximum absolute value of the error function given in Table 18.11 becomes less than or equal to unity on $[0, \theta] \cup [\phi, \pi]$. Using the substitutions $\omega_p = \theta$ and $\omega_s = \phi$, the synthesis problem for $F^{(1)}(z)$ becomes the same as for the overall filter with the only exception that the desired function $D_F(\omega)$ and the weighting function $W_F(\omega)$ are not constants in the passband and stopband regions. Therefore, the following $G_1^{(r)}(z)$'s and $G_2^{(r)}(z)$'s can be designed in the same manner. Finally, $F^{(R)}(z)$ is determined at step 6 like $F(z)$ in one-stage designs.

Given the filter specifications, the remaining problem is to select R as well as the L_r's to minimize the filter complexity. This problem has been considered in Ref. [4]. Assuming that for all the selected L_r's, $\theta + \phi \approx \pi$, the sum of the estimated orders of $F^{(R)}(z)$ and the $G_1^{(r)}(z)$'s and $G_2^{(r)}(z)$'s becomes

$$N_{\text{ove}}(R) = \left[1 \bigg/ \prod_{r=1}^{R} L_r + [2(\omega_s - \omega_p)/\pi] \sum_{r=1}^{R} L_r \right] N_{\text{opt}} \tag{18.84}$$

The minimum of $N_{\text{ove}}(R)$ taking place at

$$L_1 = L_2 = \cdots = L_R = L_{\text{opt}}(R) = \left[\frac{2(\omega_s - \omega_p)}{\pi} \right]^{-1/(R+1)} \tag{18.85}$$

is

$$N_{\text{ove}}(R) = (R + 1) \left[\frac{(2\omega_s - \omega_p)}{\pi} \right]^{R/(R+1)} N_{\text{opt}} \tag{18.86}$$

The derivation of the above formula is based on the assumption that the orders of all the $G_1^{(r)}(z)$'s and $G_2^{(r)}(z)$'s for $r = 1, 2, \ldots, R$ are equal, which is seldom true. Therefore, in order to minimize the overall filter complexity, the values of the L_r's should be varied in the vicinity of $L_{\text{opt}}(R)$. Given ω_p, ω_s, and R, good values for the L_r's can be obtained by the following procedure:

1. Set $r = 1$.
2. Determine $L = L_{\text{opt}}(R + 1 - r)$ from Equation 18.85.
3. For values of $\tilde{L}_r$ in the vicinity of L determine $\theta(\tilde{L}_r)$ and $\phi(\tilde{L}_r)$.

4. If $r = R$, then go to step 7. Otherwise, go to the next step.
5. Determine $L_r = \tilde{L}_r$ minimizing

$$(R + 1 - r)\left[\frac{2[\phi(\tilde{L}_r) - \theta(\tilde{L}_r)]}{\pi}\right]^{(R-r)/(R+1-r)} + \frac{\phi(\tilde{L}_r) - \theta(\tilde{L}_r)}{\theta(\tilde{L}_r) + \phi(\tilde{L}_r)} + \frac{\phi(\tilde{L}_r) - \theta(\tilde{L}_r)}{2\pi - \theta(\tilde{L}_r) - \phi(\tilde{L}_r)} \qquad (18.87)$$

6. Set $r = r + 1$, $\omega_p = \theta(L_r)$, $\omega_s = \phi(L_r)$, and go to step 2.
7. Determine $L_R = \tilde{L}_r$ minimizing

$$\frac{1}{\tilde{L}_R} + \frac{\phi(\tilde{L}_R) - \theta(\tilde{L}_R)}{\theta(\tilde{L}_R) + \phi(\tilde{L}_R)} + \frac{\phi(\tilde{L}_R) - \theta(\tilde{L}_R)}{2\pi - \theta(\tilde{L}_R) - \phi(\tilde{L}_R)} \qquad (18.88)$$

At the first step in this procedure, L_1 is determined to minimize the estimated overall complexity of $G_1^{(1)}(z)$, $G_2^{(1)}(z)$, and the remaining $F_1^{(1)}(z)$, which is given by Equation 18.87 as a fraction of N_{opt}. Compared to Equation 18.76 for the one-stage design, $1/\tilde{L}_r$ is replaced in Equation 18.87 by the first term. This term estimates the complexity of $F_1^{(1)}(z)$ based on the use of Equation 18.86 with $\omega_p = \theta(\tilde{L}_r)$ and $\omega_s = \phi(\tilde{L}_r)$ and the fact that it is an $R - 1$ stage design. Also, L_2 is redetermined based on the same assumptions and the process is continued in the same manner. Finally, L_R is determined to minimize the sum of the estimated orders of $G_1^{(R)}(z)$, $G_2^{(R)}(z)$, and $F^{(R)}(z)$ like in the one-stage design (cf. Equation 18.76).

Example 18.2

Consider the specifications of Example 18.1. For a two-stage design, the above procedure gives $L_1 = L_2 = 6$. For these values, $F^{(0)}(z) + H(z)$ and $F^{(1)}(z)$ are Case A designs ($l = 1$) with $\theta = 0.4\pi$ and $\phi = 0.412\pi$; and $\theta = 0.4\pi$ and $\phi = 0.472\pi$, respectively. The minimum orders of $G_1^{(1)}(z)$, $G_2^{(1)}(z)$, $G_1^{(2)}(z)$, $G_2^{(2)}(z)$, and $F^{(2)}(z)$ are 26, 40, 28, 36, and 74, respectively. Compared with the conventional direct-form FIR filter of order 2541, the number of multipliers and adders required by this design (107 and 204) are only 8 percent at the expense of a 15 percent increase in the overall filter order (to 2920). For a three-stage design, we get $L_1 = L_2 = L_3 = 4$. In this case, $F^{(0)}(z)$, $F^{(1)}(z)$, and $F^{(2)}(z)$ are Case B designs ($l = 1$) with $\theta = 0.392\pi$ and $\phi = 0.4\pi$; $\theta = 0.4\pi$ and $\phi = 0.432\pi$; and $\theta = 0.272\pi$ and $\phi = 0.4\pi$, respectively. The minimum orders of $G_1^{(1)}(z)$, $G_2^{(1)}(z)$, $G_1^{(2)}(z)$, $G_2^{(2)}(z)$, $G_1^{(3)}(z)$, $G_2^{(3)}(z)$, and $F^{(3)}(z)$ are 16, 28, 18, 24, 16, 32, and 40, respectively. The number of multipliers and adders (94 and 174) are only 7 percent of those required by the direct-form equivalent at the expense of a 26 percent increase in the overall filter order (to 3196). The amplitude responses of the resulting two-stage and three-stage designs are depicted in Figure 18.27.

18.4.3 Design of Narrowband Filters

Another general approach for designing multiplier-efficient FIR filters has been proposed by Jing and Fam [2]. This design technique is based on iteratively using the fact that there exist efficient implementation forms for filters with $\omega_s < \pi/2$ and for filters with $\omega_p > \pi/2$. A filter with $\omega_s < \pi/2$ is called a *narrowband* filter while that with $\omega_p > \pi/2$ is called a *wideband* filter. This section considers the design of narrowband filters, whereas Section 18.4.4 is devoted to the design of wideband filters. Finally, these techniques are combined, resulting in the Jing–Fam approach.

When the stopband edge of $H(z)$ is less than $\pi/2$, the first transition band of $F(z^L)$ can be used as that of $H(z)$ (Figure 18.28), that is,

$$\omega_p = \theta/L \quad \text{and} \quad \omega_s = \phi/L \qquad (18.89)$$

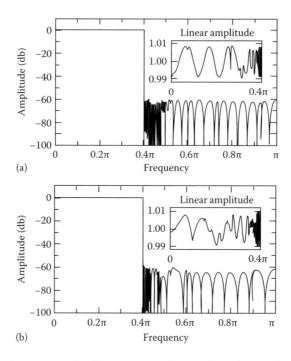

(a)

(b)

FIGURE 18.27 Amplitude responses for filters synthesized using the multistage frequency-response masking approach. (a) Two-stage filter. (b) Three-stage filter.

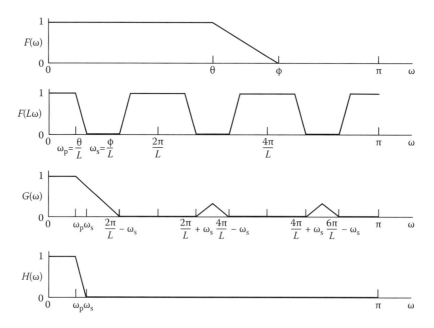

FIGURE 18.28 Synthesis of a narrowband filter as a cascade of a periodic and a nonperiodic filter.

In this case, the overall transfer function can be written in the following simplified form [6,8]:

$$H(z) = F(z^L)G(z) \tag{18.90}$$

where the orders of both $F(z)$ and $G(z)$ can be freely selected to be either even or odd. As shown in Figure 18.28, the role of $G(z)$ is to provide the desired attenuation on those regions where $F(z^L)$ has extra unwanted passband and transition band regions, that is, on

$$\Omega_s(L, \omega_s) = \bigcup_{k=1}^{\lfloor L/2 \rfloor} \left[k\frac{2\pi}{L} - \omega_s, \ \min\left(k\frac{2\pi}{L} + \omega_s, \pi\right) \right] \tag{18.91}$$

There exist two ways of designing the subfilters $F(z^L)$ and $G(z)$. In the first case, they are determined, by means of the Remez algorithm, to satisfy

$$1 - \delta_p^{(F)} \le F(\omega) \le 1 + \delta_p^{(F)} \quad \text{for } \omega \in [0, L\omega_p] \tag{18.92a}$$

$$-\delta_s \le F(\omega) \le \delta_s \quad \text{for } \omega \in [L\omega_s, \pi] \tag{18.92b}$$

$$1 - \delta_p^{(G)} \le G(\omega) \le 1 + \delta_p^{(G)} \quad \text{for } \omega \in [0, \omega_p] \tag{18.92c}$$

$$-\delta_s \le G(\omega) \le \delta_s \quad \text{for } \omega \in \Omega_s(L, \omega_s) \tag{18.92d}$$

where

$$\delta_p^{(G)} + \delta_p^{(F)} = \delta_p \tag{18.92e}$$

The ripples $\delta_p^{(F)}$ and $\delta_p^{(G)}$ can be selected, e.g., to be half the overall ripple δ_p. In the above specifications, both $F(z^L)$ and $G(z)$ have $[0, \omega_p]$ as a passband region.

Another approach, leading to a considerable reduction in the order of $G(z)$, is to design simultaneously $F(\omega)$ to satisfy

$$1 - \delta_p \le F(\omega)G(\omega/L) \le 1 + \delta_p \quad \text{for } \omega \in [0, L\omega_p] \tag{18.93a}$$

$$-\delta_s \le F(\omega)G(\omega/L) \le \delta_s \quad \text{for } \omega \in [L\omega_s, \pi] \tag{18.93b}$$

and $G(\omega)$ to satisfy

$$G(0) = 1 \tag{18.94a}$$

$$-\delta_s | leF(L\omega)G(\omega) \le \delta_s \quad \text{for } \omega \in \Omega_s(L, \omega_s) \tag{18.94b}$$

The desired overall solution can be obtained by iteratively determining, by means of the Remez algorithm, $F(z)$ to meet the criteria of Equations 18.93a and b and $G(z)$ to meet the criteria of Equations 18.94a and b. Typically, only three to five designs of both of the subfilters are required to arrive at a solution that does not change if further iterations are used. For more details, see Ref. [8] or [10]. Figure 18.29 shows typical responses for $G(z)$ and $F(z^L)$ and for the overall optimized design. As seen in this figure, $G(z)$ has all its zeros on the unit circle concentrating on providing the desired attenuation for the overall response on $\Omega_s(L, \omega_s)$, whereas $F(z^L)$ makes the overall response equiripple in the passband and in the stopband portion $[\omega_s, \pi/L]$.

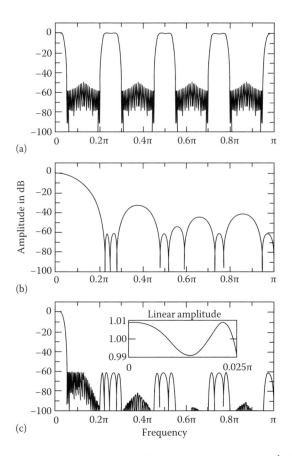

FIGURE 18.29 Typical amplitude responses for a filter of the form $H(z) = F(z^L)G(z)$. $L = 8$, $\omega_p = 0.025\pi$, $\omega_s = 0.05\pi$, $\delta_p = 0.01$, and $\delta_s = 0.001$. (a) $F(z^L)$ of order 26 in z^L. (b) $G(z)$ of order 19. (c) Overall filter.

For the order of $F(z)$, a good estimate is

$$N_F \approx \frac{\Phi(\delta_p, \delta_s)/L}{\omega_s - \omega_p} \tag{18.95}$$

so that it is $1/L$th of that of an optimum conventional nonperiodic filter meeting the given overall criteria. The order of $G(z)$, in turn, can be estimated accurately by [10]

$$N_G = \cos h^{-1}(1/\delta_s) \left[\frac{1}{X\left(\omega_p, \frac{2\pi}{L} - \frac{\omega_p + 2\omega_s}{3}\right)} + \frac{L/2}{X\left(\frac{L\omega_p}{2}, \pi - \frac{L(\omega_p + 2\omega_s)}{6}\right)} \right] \tag{18.96a}$$

where

$$X(\omega_1, \omega_2) = \cos h^{-1}[(2\cos\omega_1 - \cos\omega_2 + 1)/(1 + \cos\omega_2)] \tag{18.96b}$$

The minimization of the number of multipliers, $[(N_F + 2)/2] + [(N_G + 2)/2]$, with respect to L can be performed conveniently by evaluating the sum of the above estimated orders for the admissible values of L,

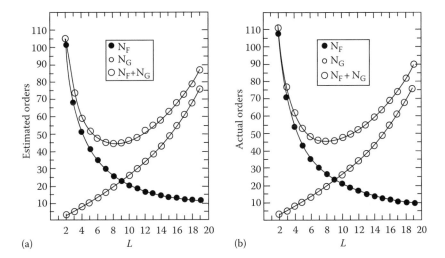

FIGURE 18.30 Estimated and actual subfilter orders as well as the sum of the subfilter orders vs. L in a typical narrowband case.

$2 \leq L < \pi/\omega_s$. The upper limit is a consequence of the fact that the stopband edge angle of $F(z)$, $\phi = L\omega_s$, must be less than π. The following example illustrates the minimization of the filter complexity.

Example 18.3

The narrowband specifications are $\omega_p = 0.025\pi$, $\omega_s = 0.05\pi$, $\delta_p = 0.01$, and $\delta_s = 0.001$. Figure 18.30a shows the estimated N_F, N_G, and $N_F + N_G$ as functions of L, whereas Figure 18.30b shows the corresponding actual minimum orders. It is seen that the estimated orders are so close to the actual ones that the minimization of the filter complexity can be accomplished based on the use of the above estimation formulas. It is also observed that $N_F + N_G$ is a unimodal function of L. With the estimates, $L = 8$ gives the best result. The estimated orders are $N_F = 25$ and $N_G = 19$, whereas the actual orders are $N_F = 26$ and $N_G = 19$. The amplitude responses for the subfilters and for the overall filter are depicted in Figure 18.29. This design requires 24 multipliers and 45 adders. The minimum order of a conventional direct-form design is 216, requiring 109 multipliers and 216 adders. The price paid for these 80% reductions in the filter complexity is a 5 percent increase in the overall filter order (from 216 to 227).

In the cases where L can be factored into the product

$$L = \prod_{r=1}^{R} L_r \qquad (18.97)$$

where the L_r's are integers, further savings in the filter complexity can be achieved by designing $G(z)$ in the following multistage form [8]:

$$G(z) = G_1(z)G_2(z^{L_1})G_3(z^{L_1 L_2}) \cdots G_R(z^{L_1 L_2 \cdots L_{R-1}}) \qquad (18.98)$$

Another alternative to reduce the number of adders and multipliers is to use special structures for implementing $G(z)$ [8–10].

18.4.4 Design of Wideband Filters

The synthesis of a wideband filter $H(z)$ can be converted into the design of a narrowband filter based on the following fact. If $\hat{H}(z)$ of even order $2M$ is a low-pass design with the following edges and ripples:

$$\hat{\omega}_p = \pi - \omega_s, \quad \hat{\omega}_s = \pi - \omega_p, \quad \hat{\delta}_p = \delta_s, \quad \hat{\delta}_s = \delta_p \tag{18.99}$$

then

$$H(z) = z^{-M} - (-1)^M \hat{H}(-z) \tag{18.100}$$

is a low-pass filter having the passband and stopband edge angles at ω_p and ω_s and the passband and stopband ripples of δ_p and δ_s. Hence, if ω_p and ω_s of $H(z)$ are larger than $\pi/2$, then $\hat{\omega}_p$ and $\hat{\omega}_s$ of $\hat{H}(z)$ are smaller than $\pi/2$ [10]. This enables us to design $\hat{H}(z)$ in the form

$$\hat{H}(z) = F(z^L)G(z) \tag{18.101}$$

using the techniques of Section 18.4.3, yielding

$$H(z) = z^{-M} - (-1)^M F[(-z)^L] G(-z) \tag{18.102a}$$

where

$$M = (LN_F + N_G)/2 \tag{18.102b}$$

is half the order of $F(z^L)G(z)$. For implementation purposes, $H(z)$ is expressed as

$$H(z) = z^{-M} - \hat{F}(z^L)\hat{G}(z), \quad \hat{F}(z^L) = (-1)^M F[(-z)^L], \quad \hat{G}(z) = G(-z) \tag{18.103}$$

An implementation of this transfer function is shown in Figure 18.31, where the delay term z^{-M} can be shared with $\hat{F}(z^L)$. To avoid half-sample delays, the order of $\hat{F}(z^L)\hat{G}(z)$ has to be even.

Example 18.4

The wideband specifications are $\omega_p = 0.95\pi$, $\omega_s = 0.975\pi$, $\delta_p = 0.001$, and $\delta_s = 0.01$. From Equation 18.99, the specifications of $\hat{H}(z)$ become $\omega_p = 0.025\pi$, $\hat{\omega}_s = 0.05\pi$, $\hat{\delta}_p = 0.01$, and $\hat{\delta}_s = 0.001$. These are the narrowband specifications of Example 18.3. The desired wideband design is thus obtained by using the subfilters $F(z^L)$ and $G(z)$ of Figure 18.29 ($L = 8$, $N_F = 26$, and $N_G = 19$). However, the overall order is odd (227). A solution with even order is achieved by increasing the order of $G(z)$ by one ($N_G = 20$). Figure 18.32 shows the amplitude response of the resulting filter, requiring 25 multipliers, 46 adders, and 228 delay elements. The corresponding numbers for a conventional direct-form equivalent of order 216 are 109, 216, and 216, respectively.

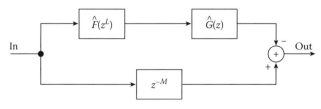

FIGURE 18.31 Implementation for a wideband filter in the form $H(z) = z^{-M} - (-1)^M F[(-z^L)G(-z)$. $\hat{F}(z^L) = (-1)^M \cdot F[(-z)^L]$ and $\hat{G}(z) = G(-z)$.

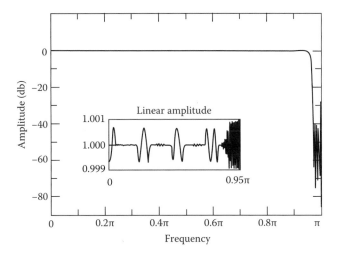

FIGURE 18.32 Amplitude response for a wideband filter implemented as shown in Figure 18.38.

18.4.5 Generalized Designs

The Jing–Fam approach [2] is based on iteratively using the facts that a narrowband filter can be implemented effectively as $H(z) = F(z^L)G(z)$ and a wideband filter in the form of Equations 18.102a and b. In this approach, a narrowband filter is generated [9] as

$$H(z) \equiv \hat{H}_1(z) = G_1(z)F_1(z^{L_1}) \qquad (18.104a)$$

where

$$F_1(z) = z^{-M_1} - (-1)^{M_1}\hat{H}_2(-z) \quad \hat{H}_2(z) = G_2(z)F_2(z^{L_2}) \qquad (18.104b)$$

$$F_2(z) = z^{-M_2} - (-1)^{M_2}\hat{H}_3(-z) \quad \hat{H}_3(z) = G_3(z)F_3(z^{L_3}) \qquad (18.104c)$$

$$\vdots$$

$$F_{R-2}(z) = z^{-M_{R-2}} - (-1)^{M_{R-2}}\hat{H}_{R-1}(-z) \quad \hat{H}_{R-1}(z) = G_{R-1}(z)F_{R-1}(z^{L_{R-1}}) \qquad 18.104d)$$

$$F_{R-1}(z) = z^{-M_{R-1}} - (-1)^{M_{R-1}}\hat{H}_R(-z) \quad \hat{H}_R(z) = G_R(z) \qquad (18.104e)$$

with M_r for $r = 1, 2, \ldots, R-1$ being half the order of $\hat{H}_{r+1}(z)$. Here, the basic idea is to convert iteratively the design of the narrowband overall filter into the designs of narrowband transfer function $\hat{H}_r(z)$ for $r = 2, 3, \ldots, R$ until the transition bandwidth of the remaining $\tilde{H}_r(z) = G_R(z)$ becomes large enough and, correspondingly, its complexity (the number of multipliers) is low enough. The desired conversion is performed by properly selecting the L_r's and designing the lower-order filters $G_r(z)$ for $r = 1, 2, \ldots, R-1$.

In order to determine the conditions for the L_r's as well as the design criteria for the $G_r(z)$'s, we consider the rth iteration, where

$$\hat{H}_r(z) = G_r(z)F_r(z^{L_r}) \qquad (18.105a)$$

with

$$F_r(z) = z^{-M_r} - (-1)^{M_r}\hat{H}_{r+1}(-z) \qquad (18.105b)$$

Let the ripples of $\hat{H}_r(z)$ be $\hat{\delta}_p^{(r)}$ and $\hat{\delta}_s^{(r)}$ and the edges be located at $\omega_p^{(r)} < \pi/2$ and $\omega_s^{(r)} < \pi/2$. Since $F_r(z)$ is implemented in the form of Equation 18.105b, it cannot alone take care of shaping the passband response of $\hat{H}_r(z)$. Therefore, the simultaneous criteria for $G_r(z)$ and $F_r(z)$ are stated according to Equations 18.92a through e so that the passband and stopband regions of $G_r(z)$ are, respectively, $[0,\omega_p^{(r)}]$ and $\Omega_s(L_r\omega_s^{(r)})$ with $\Omega_s(L, \omega_s)$ given by Equation 18.91. L_r has to be determined such that the edges of $F_r(z)$, $L_{r\omega p}^{(r)}$ and $L_{r\omega s}^{(r)}$, become larger than $\pi/2$ and, correspondingly, the edges of $\hat{H}_{r+1}(z)$, $\omega_p^{(r+1)} = \pi - L_{r\omega s}^{(r)}$ and $\omega_s^{(r+1)} = \pi - L_{r\omega p}^{(r)}$, become less than $\pi/2$.

In the case of the specifications of Equations 18.92a through e, the stopband ripple of $G_r(z)$, denoted for later use by $\delta_s^{(r)}$, and that of $F_r(z)$ are equal to $\hat{\delta}_s^{(r)}$, whereas the sum of the passband ripples is equal to $\hat{\delta}_p^{(r)}$. Denoting by $\hat{\delta}_p^{(r)}$ the passband ripple selected for $G_r(z)$, the corresponding ripple of $F_r(z)$ is $\hat{\delta}_p^{(r)} - \delta_p^{(r)}$. Since $F_r(z)$ and $\hat{H}_{(r+1)}(z)$ interchange the ripples, the ripple requirements for $\hat{H}_{r+1}(z)$ are $\hat{\delta}_p^{(r+1)} = \hat{\delta}_s^{(r)}$ and $\hat{\delta}_s^{(r+1)} = \hat{\delta}_p^{(r)} - \delta_p^{(r)}$.

The criteria for the $G_r(z)$'s for $r = 1, 2, \ldots, R$ can thus be stated as

$$1 - \delta_p^{(r)} \le G_r(\omega) \le 1 + \delta_p^{(r)} \quad \text{for } \omega \in [0, \omega_p^{(r)}] \tag{18.106a}$$

$$-\delta_s^{(r)} \le G_r(\omega) \le \delta_s^{(r)} \quad \text{for } \omega \in \Omega_s^{(r)} \tag{18.106b}$$

where

$$\Omega_s^{(r)} = \begin{cases} \bigcup_{k=1}^{\lfloor L_r/2 \rfloor} \left[k\frac{2\pi}{L_r} - \omega_s^{(r)}, \ \min\left(k\frac{2\pi}{L_r} + \omega_s^{(r)}, \pi \right) \right] & \text{for } r < R \\ \left[\omega_s^{(R)}, \pi \right] & \text{for } r = R \end{cases} \tag{18.106c}$$

Here, the $\omega_p^{(r)}$'s and $\omega_s^{(r)}$'s for $R = 2, 3, \ldots, R$ are determined iteratively as

$$\omega_p^{(r)} = \pi - L_{r-1}\omega_s^{(r-1)}, \quad \omega_s^{(r)} = \pi - L_{r-1}\omega_p^{(r-1)} \tag{18.106d}$$

where $\omega_p^{(1)} = \omega_p$ and $\omega_s^{(1)} = \omega_s$ are the edges of the overall design, and the $\delta_s^{(r)}$'s as

$$\delta_s^{(r)} = \begin{cases} \delta_p - \sum_{\substack{k=1 \\ k \text{ odd}}}^{r-1} \delta_p^{(k)} & \text{for } r \text{ even} \\ \delta_s - \sum_{\substack{k=2 \\ k \text{ even}}}^{r-1} \delta_p^{(k)} & \text{for } r \text{ odd} \end{cases} \tag{18.106e}$$

where δ_p and δ_s are the ripple values of the overall filter and $\delta_p^{(r)}$ is the passband ripple selected for $G_r(z)$. In order for the overall filter to meet the given ripple requirements, $\delta_s^{(R)}$ and the $\delta_p^{(r)}$'s have to satisfy for R even

$$\sum_{\substack{k=2 \\ k \text{ even}}}^{R} \delta_p^{(k)} = \delta_s, \quad \delta_s^{(R)} + \sum_{\substack{k=1 \\ k \text{ odd}}}^{R-1} \delta_p^{(k)} = \delta_p \tag{18.107a}$$

or for R odd

$$\sum_{\substack{k=1 \\ k \text{ odd}}}^{R} \delta_p^{(k)} = \delta_p, \quad \delta_s^{(R)} + \sum_{\substack{k=2 \\ k \text{ even}}}^{R-1} \delta_p^{(k)} = \delta_s \tag{18.107b}$$

In the above, the L_r's have to be determined such that the $\omega_s^{(r)}$'s for $r < R$ become smaller than $\pi/2$. It is also desired that for the last filter stage $G_R(z)$, $\omega_s^{(R)}$ is smaller than $\pi/2$.

If $2\pi/L_r - \omega_s^{(r)} < \pi/2$ for $r < R$ or $\omega_s^{(R)} < \pi/2$, then the arithmetic complexity of $G_r(z)$ can be reduced by designing it, using the techniques of previous sections, in the form

$$G_r(z) = G_r^{(1)}(z^{K_r})G_r^{(2)}(z) \tag{18.108}$$

It is preferred to design the subfilters of $G_r(z)$ in such a way that the passband shaping is done entirely by $G_r^{(1)}(z^{K_r})$. The number of multipliers in the $G_r(z)$'s for $r = 1, 2, \ldots, R-1$ can be reduced by the experimentally observed fact that the overall filter still meets the given criteria when the stopband regions of these filters are decreased by using in Equation 18.106c the substitution

$$\left(2\omega_s^{(r)} + \omega_p^{(r)}\right)/3 \mapsto \omega_s^{(r)} \tag{18.109}$$

After some manipulations, $H(z)$ as given by Equation 18.104a through e and Equation 18.106 can be rewritten in the explicit form shown in Table 18.14. If $G_r(z)$ is a single-stage design, then $G_r^{(1)}(z^{K_r}) + 1$. In order to obtain the desired overall solution, the overall order of $G_r(z)$ for $r \geq 2$, denoted by N_r in Table 18.14, has to be even. Realizations for the overall transfer function are given in Figure 18.33, where

$$m_r = \hat{M}_r - \hat{M}_{r+1} = \frac{1}{2}\hat{L}_rN_r, \quad r = 2, 3, \ldots, R-1, \quad m_R = \hat{M}_R \tag{18.110}$$

The structure of Figure 18.33b is preferred since the delay terms z^{-m_r} can be shared with $H_R^{(1)}(z^{K_R\hat{L}_R})$ or, if this filter stage is not present, with $H_R^{(2)}(z^{K_R\hat{L}_R})$. This is because the overall order of this filter stage is usually larger than the sum of the m_r's.

If the edges ω_p and ω_s of the overall filter are larger than $\pi/2$, then we set $H(z) + F_1(z)$ in Equation 18.104a. In this case, $\delta_p^{(1)} + 0$, $L_1 + 1$, and $G_1(z)$,, $\omega_p^{(1)}$ and $\omega_s^{(1)}$ are absent. Furthermore, $\omega_p^{(2)} = \pi - \omega_s$, and $\omega_s^{(2)} = \pi - \omega_p$, and $H_1(z)$ is absent in Figure 18.33 and in Table 18.14.

TABLE 18.14   Explicit Form for the Transfer Function in the Jing–Fam Approach

$$H(z) = H_1(z^{\hat{L}_1})\left\{I_2 z^{-\hat{M}_2} + H_2(z^{\hat{L}_2})\left[I_3 z^{-\hat{M}_3} + H_3(z^{\hat{L}_3})(\cdots)\right.\right.$$
$$\left.\left.\left\{I_{R-1}z^{-\hat{M}_{R-1}} + H_{R-1}(z^{\hat{L}_{R-1}})\left[I_R z^{-\hat{M}_R} + H_R(z^{\hat{L}_R})\right]\right\}\cdots\right)\right]\right\},$$

where

$$H_r(z^{\hat{L}_r}) = H_r^{(1)}(z^{K_r\hat{L}_r})H_r^{(2)}(z^{\hat{L}_r})$$
$$H_r^{(1)}(z) = G_r^{(1)}(J_r^{(1)}z), \quad H_r^{(2)}(z) = S_rG_r^{(2)}(J_r^{(2)}z)$$
$$S_1 = 1, \quad S_r = -(-1)^{\hat{M}_r/\hat{L}_r}, \quad r = 2, 3, \ldots, R$$
$$J_1^{(2)} = 1, \quad J_2^{(2)} = -1, \quad J_r^{(2)} = -\left[J_{r-1}^{(2)}\right]^{\hat{L}_{r-1}}, \quad r = 3, 4, \ldots, R$$
$$J_r^{(1)} = \left[J_r^{(2)}\right]^{K_r}$$
$$\hat{L}_1 = 1, \quad \hat{L}_r = \prod_{k=1}^{r-1} L_k, \quad r = 2, 3, \ldots, R$$
$$\hat{M}_R = \frac{1}{2}\hat{L}_RN_R, \quad \hat{M}_{R-r} = \hat{M}_{R-r+1} + \frac{1}{2}\hat{L}_{R-r}N_{R-r}, \quad r = 1, 2, \ldots, R-2$$
$$I_2 = 1, \quad I_r = \left[J_{r-1}^{(2)}\right]^{\hat{M}_r/\hat{L}_{r-1}}, \quad r = 3, 4, \ldots, R$$
$$N_r = K_rN_r^{(1)} + N_r^{(2)}$$
$$N_r^{(1)} \text{ and } N_r^{(2)} \text{ are the orders of } G_r^{(1)}(z) \text{ and } G_r^{(2)}(z), \text{ respectively}$$

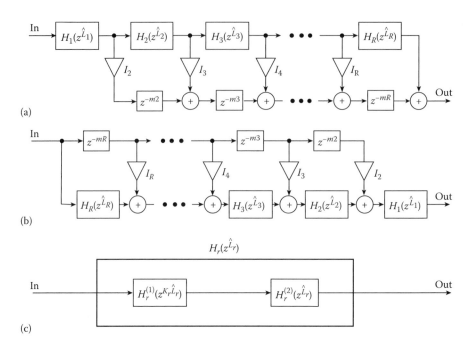

(a)

(b)

(c)

FIGURE 18.33 Implementations for a filter synthesized using the Jing–Fam approach. (a) Basic structure. (b) Transposed structure. (c) Structure for the subfilter $H_r(z^{\circ L_r})$.

The remaining problem is to select R, the L_r's, the K_r's, and the ripple values such that the filter complexity is minimized. The following example illustrates this.

Example 18.5

Consider the specifications of Example 18.1, that is, $\omega_p = 0.4\pi$, $\omega_s = 0.402\pi$, $\delta_p = 0.01$, and $\delta_s = 0.001$. In this case, the only alternative is to select $L_1 = 2$. The resulting passband and stopband regions for $G_1(z)$ are (the substitution of Equation 18.109 is used)

$$\Omega_p^{(1)} = [0, 0.4\pi] \quad \text{and} \quad \Omega_s^{(1)} = [0.5987\pi, \pi]$$

For $\hat{H}_2(z)$, the edges become $\omega_p^{(2)} = \pi - L_{1\omega s} = 0.196\pi$ and $\omega_s^{(2)} = \pi - L_{1\omega p} = 0.2\pi$. For L_2, there are two alternatives to make the edges of $\hat{H}_3(z)$, $\omega_p^{(3)} = \pi - L_{2\omega s}^{(2)}$ and $\omega_s^{(3)} = \pi - L_{2\omega p}^{(2)}$, less than $\pi/2$. These are $L_2 = 3$ and $L_2 = 4$. For $R = 5$ stages, there are the following four alternatives to make all the $\omega_s^{(r)}$'s smaller than $\pi/2$:

$$L_1 = 2, \quad L_2 = 4, \quad L_3 = 3, \quad L_4 = 2$$
$$L_1 = 2, \quad L_2 = 4, \quad L_3 = 4, \quad L_4 = 4$$
$$L_1 = 2, \quad L_2 = 3, \quad L_3 = 2, \quad L_4 = 4$$
$$L_1 = 2, \quad L_2 = 3, \quad L_3 = 2, \quad L_4 = 3$$

Among these alternatives, the first one results in an overall filter with the minimum complexity. In this case, the edges of $\hat{H}_3(z)$, $\hat{H}_4(z)$, and $\hat{H}_5(z) + G_5(z)$ become as shown in Table 18.15. The corresponding passband and stopband regions for $G_2(z)$, $G_3(z)$, $G_4(z)$, and $G_5(z)$ are

TABLE 18.15 Data for a Filter Designed Using the Jing–Fam Approach

	$r=1$	$r=2$	$r=3$	$r=4$	$r=5$
$\omega p^{(r)}$	0.4π	0.196π	0.2π	0.352π	0.2π
$\omega s^{(r)}$	0.402π	0.2π	0.216π	0.4π	0.296π
$\delta p^{(r)}$	$\frac{1}{3}\times10^{-2}$	$\frac{1}{3}\times10^{-3}$	$\frac{1}{3}\times10^{-2}$	$\frac{1}{3}\times10^{-3}$	$\frac{1}{3}\times10^{-2}$
$\delta s^{(r)}$	10^{-3}	$\frac{2}{3}\times10^{-2}$	$\frac{2}{3}\times10^{-3}$	$\frac{1}{3}\times10^{-2}$	$\frac{1}{3}\times10^{-3}$
L_r	2	4	3	2	—
K_r	—	3	2	—	3
$N_r^{(1)}$	—	20	11	—	22
$N_r^{(2)}$	31	10	8	26	14
N_r	31	70	30	26	80
$\hat{L}_r$	1	2	8	24	48
$J_r^{(1)}$	—	-1	1	—	-1
$J_r^{(2)}$	1	-1	-1	1	-1
$\hat{M}_r$	—	2422	2352	2232	1920
I_r	—	1	1	-1	1
S_r	1	1	-1	1	-1
m_r	—	70	120	312	1920

$$\Omega_p^{(2)} = [0, 0.196\pi], \quad \Omega_s^{(2)} = [0.3013\pi, 0.6987\pi] \cup [0.8013\pi, \pi]$$
$$\Omega_p^{(3)} = [0, 0.2\pi], \quad \Omega_s^{(3)} = [0.4560\pi, 0.8773\pi]$$
$$\Omega_p^{(4)} = [0, 0.352\pi], \quad \Omega_s^{(4)} = [0.616\pi, \pi]$$
$$\Omega_p^{(5)} = [0, 0.2\pi], \quad \Omega_s^{(5)} = [0.296\pi, \pi]$$

What remains is to determine the ripple requirements. From Equation 18.107b, it follows for $R=5$, $\delta_p^{(1)} + \delta_p^{(3)} + \delta_p^{(5)} = \delta_p$ and $\delta_p^{(2)} + \delta_p^{(4)} + \delta_p^{(6)} = \delta_s$. By simply selecting the ripple values in these summations to be equal, the required ripples for the $G_r(z)$'s become as shown in Table 18.15.

The first and fourth subfilter are single-stage filters since their stopband edges are larger than $\pi/2$, whereas the remaining three filters are two-stage designs. The parameters describing the overall filter are shown in Table 18.15, whereas Figure 18.34a depicts the response of this filter. The number of multipliers and the order of this design are 78 and 4875, whereas the corresponding numbers for the direct-form equivalent are 1271 and 2541. The number of multipliers required by the proposed design is thus only 6 percent of that of the direct-form filter. Since the complexity of $H_5(z^{\hat{L}_5})$ is similar to those of the earlier filter stages, $R=5$ is a good selection in this example.

The overall filter order as well as the number of multipliers can be decreased by selecting smaller ripple values for the first stages, thereby allowing larger ripples for the last stages. Proper selections for the ripple requirements and filter orders are shown in Table 18.16. The first four filters have been optimized such that their passband variations are minimized. The first criteria are met by a half-band filter of order 34, having the passband and stopband edges at 0.4013π and 0.5987π. Since every second impulse response coefficient of this filter is zero-valued except for the central coefficient with an easily

TABLE 18.16 Data for Another Filter Designed Using the Jing–Fam Approach

$\delta p^{(r)}$	7.3×10^{-4}	7.1×10^{-5}	3.5×10^{-4}	12.1×10^{-5}	89.2×10^{-4}
$\delta s^{(r)}$	10^{-3}	92.7×10^{-4}	92.9×10^{-5}	89.2×10^{-4}	80.8×10^{-5}
K_r	—	3	2	—	2
$N_r^{(1)}$	—	22	13	—	27
$N_r^{(2)}$	34	10	8	24	6

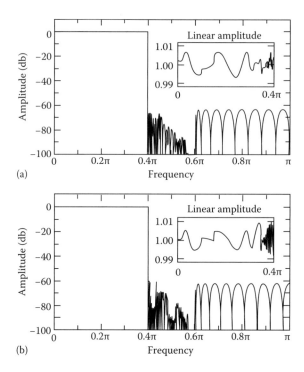

FIGURE 18.34   Amplitude responses for filters synthesized using the Jin–Fam approach.

implementable value of 1/2, this filter requires only nine multipliers. For the last stage, K_5 is reduced to 2 to decrease the overall filter order. The order of the resulting overall filter [see Figure 18.34b] is 3914, which is 54 percent higher than that of the direct-form equivalent. The number of multipliers is reduced to 70.

The Jing–Fam approach cannot be applied directly for synthesizing filters whose edges are very close to $\pi/2$. This problem can, however, be overcome by slightly changing the sampling rate or, if this is not possible, by shifting the edges by a factor 3/2 by using decimation by this factor at the filter input and interpolation by the same factor at the filter output [2]. One attractive feature of the Jing–Fam approach is that it can be combined with multirate filtering to reduce the filter complexity even further [7].

When comparing the above designs with the filters synthesized using the multistage frequency-response masking technique (Example 18.2), it is observed that the above designs require slightly fewer multipliers at the expense of an increased overall filter order. Both of these general approaches are applicable to those specifications that are not very narrowband or very wideband. For most very narrowband and wideband cases, filters synthesized in the simplified forms $H(z) = F(z^L)G(z)$ and $H(z) = z^{-M} - (-1)^M F[(-z)^L]G(-z)$, respectively, give the best results (see Examples 18.3 and 18.4).

References

1. O. Herrmann, L.R. Rabiner, and D.S.K. Chan, Practical design rules for optimum finite impulse response lowpass digital filters, *Bell System Tech. J.*, 52, 769–799, July–August 1973.
2. Z. Jing and A. T. Fam, A new structure for narrow transition band, lowpass digital filter design, *IEEE Trans. Acoust., Speech, Signal Processing*, ASSP-32, 362–370, April 1984.
3. Y. C. Lim, Frequency-response masking approach for the synthesis of sharp linear phase digital filters, *IEEE Trans. Circuits Syst.*, CAS-33, 357–364, April 1986.

4. Y. C. Lim and Y. Lian The optimum design of one- and two-dimensional FIR filters using the frequency response masking technique, *IEEE Trans. Circuits Syst.—II: Analog and Digital Signal Processing,* 40, 88–95, February 1993.

5. J. H. McClellan, T. W. Parks, and L. R. Rabiner, A computer program for designing optimum FIR linear phase digital filters, *IEEE Trans. Audio Electroacoust.,* AU-21, 506–526, December 1973; also reprinted in *Selected Papers in Digital Signal Processing, II,* IEEE Digital Signal Processing Comm. and IEEE-ASSP, Eds., New York: IEEE Press, 1976, 97–117.

6. Y. Neuvo, C.-Y. Dong, and S. K. Mitra, Interpolated finite impulse response filters, *IEEE Trans. Acoust., Speech, Signal Processing,* ASSP-32, 563–570, June 1984.

7. T. Ramstad and T. Saramäki, Multistage, multirate FIR filter structures for narrow transition-band filters, in *Proc. 1990 IEEE Int. Symp. Circuits Syst.,* New Orleans, LA, 1990, 2017–2021.

8. T. Saramäki, Y. Neuvo, and S. K. Mitra, Design of computationally efficient interpolated FIR filters, *IEEE Trans. Circuits Syst.,* CAS-35, 70–88, January 1988.

9. T. Saramäki and A. T. Fam, Subfilter approach for designing efficient FIR filters, in *Proc. 1988 IEEE Int. Symp. Circuits Syst.,* Espoo, Finland, 2903–2915, 1988.

10. T. Saramäki, Finite impulse response filter design, in *Handbook for Digital Signal Processing,* S. K. Mitra and J. F. Kaiser, Eds., New York: Wiley, 1993, Ch. 4, 155–277.

19

IIR Filters

Sawasd Tantaratana
Thammasat University

Chalie Charoenlarpnopparut
Thammasat University

Phakphoom Boonyanant
*National Electronics and Computer
Technology Center*

Yong Ching Lim
Nanyang Technological University

19.1 Properties of IIR Filters

Sawasd Tantaratana

19.1.1 System Function and Impulse Response

A digital filter with impulse response having infinite length (i.e., its values outside a finite interval cannot all be zero) is termed infinite impulse response (IIR) filter. The most important class of IIR filters can be described by the difference equation

$$y(n) = b_0 x(n) + b_1 x(n-1) + \cdots + b_M x(n-M)$$
$$- a_1 y(n-1) - a_2 y(n-2) - \cdots - a_N y(n-N) \tag{19.1}$$

where

    $x(n)$ is the input
    $y(n)$ is the output of the filter
    $\{a_1, a_2, \ldots, a_N\}$ and $\{b_0, b_1, \ldots, b_M\}$ are constant coefficients

We assume that $a_N \neq 0$. The impulse response is the output of the system when it is driven by a unit impulse at $n = 0$, with the system being initially at rest, i.e., the output being zero prior to applying the input. We denote the impulse response by $h(n)$. With $x(0) = 1, x(n) = 0$ for $n \neq 0$, and $y(n) = 0$ for

$n < 0$, we can compute $h(n)$, $n \geq 0$, from Equation 19.1 in a recursive manner. Taking the z-transform of Equation 19.1, we obtain the system function

$$H(z) = \frac{Y(z)}{X(z)} = \frac{b_0 + b_1 z^{-1} + \cdots + b_M z^{-M}}{1 + a_1 z^{-1} + \cdots + a_N z^{-N}} \tag{19.2}$$

where N is the order of the filter. The system function and the impulse response are related through the z-transform and its inverse, i.e.,

$$H(z) = \sum_{n=-\infty}^{\infty} h(n) z^{-1} \quad h(n) = \frac{1}{2\pi j} \oint_C H(z) z^{n-1} dz \tag{19.3}$$

where C is a closed counterclockwise contour in the region of convergence. See Chapter 5 of *Fundamentals of Circuits and Filters* for a discussion of z-transform. We assume that $M \leq N$. Otherwise, the system function can be written as

$$H(z) = \left[c_0 + c_1 z^{-1} + \cdots + c_{M-N} z^{-(M-N)} \right] + \left[\frac{b_0' + b_1' z^{-1} + \cdots + b_M' z^{-M}}{1 + a_1 z^{-1} + \cdots + a_N z^{-N}} \right] \tag{19.4}$$

which is a finite impulse response (FIR) filter in parallel with an IIR filter, or as

$$H(z) = \left[c_0' + c_1' z^{-1} + \cdots + c_{M-N}' z^{-(M-N)} \right] \left[\frac{b_0'' + b_1'' z^{-1} + \cdots + b_M'' z^{-M}}{1 + a_1 z^{-1} + \cdots + a_N z^{-N}} \right] \tag{19.5}$$

which is an FIR filter in cascade with an IIR filter. FIR filters are covered in Chapter 18. This chapter covers IIR filters.

For ease of implementation, it is desirable that the coefficients $\{a_1, a_2, \ldots, a_N\}$ and $\{b_0, b_1, \ldots, b_M\}$ be real numbers (as opposed to complex numbers), which is another assumption that we make, unless specified otherwise.

19.1.2 Causality (Physical Realizability)

A causal (physically realizable) filter is one whose output value does not depend on the future input values. A noncausal filter cannot be realized in real time since some future inputs are needed in computing the current output value. The difference equation in Equation 19.1 can represent a causal system or a noncausal system. If the output $y(n)$ is calculated, for an increasing value of n, from $x(n), x(n-1), \ldots, x(n-M)$, $y(n-1), \ldots, y(n-N)$, according to the right-hand side of Equation 19.1 then the difference equation represents a causal system. On the other hand, we can rewrite Equation 19.1 as

$$y(n-N) = \frac{1}{a_N} [b_0 x(n) + b_1 x(n-1) + \cdots + b_M x(n-M)$$
$$- y(n) - a_1 y(n-1) - \cdots - a_{N-1} y(n-N+1)] \tag{19.6}$$

If the system calculates $y(n-N)$, for a decreasing due of n, using the right-hand side of Equation 19.6, then the system is noncausal since $y(n-N)$ depends on $x(n), \ldots, x(n-M)$, which are future input values. We shall assume that the IIR filter represented by Equation 19.1 is causal. It follows from this assumption that $h(n) = 0$ for $n < 0$ and that the convergence region of $H(z)$ is of the form: $|z| > r_0$, where r_0, is a nonnegative constant.

Noncausal filters are useful in practical application where the output need not be calculated in real time or where the variable n does not represent time, such as in image processing where n is a spatial variable. Generally, a noncausal filter can be modified to be causal by adding sufficient delay at the output.

19.1.3 Poles and Zeros

Rewriting Equation 19.2 we have

$$H(z) = z^{N-M} \frac{b_0 z^M + b_1 z^{M-1} + \cdots + b_{M-1} z + b_M}{z^N + a_1 z^{N-1} + \cdots + a_{N-1} z^{-N}} \tag{19.7}$$

Assuming $b_0, b_M \neq 0$, then there are N poles given by the roots of the denominator polynomial and M zeros given by the roots of the numerator polynomial. In addition, there are $N - M$ zeros at the origin on the complex plane. The locations of the poles and zeros can be plotted on the complex z plane. Denoting the poles by $p_1, p_2, \ldots, p_N$, and the nonzero zeros by $q_1, q_2, \ldots, q_M$, we can write

$$H(z) = b_0 z^{N-M} \frac{(z - q_1)(z - q_2) \cdots (z - q_M)}{(z - p_1)(z - p_2) \cdots (z - p_N)} \tag{19.8}$$

Since we assume that the coefficients $\{a_1, a_2, \ldots, a_N\}$ and $\{b_0, b_1, \ldots, b_M\}$ are each complex-valued pole (i.e., pole off the real axis on the z plane), there must be another pole that is the complex conjugate of the first. Similarly, complex-valued zeros must exist in complex–conjugate pairs. The combination of a complex–conjugate pole pair (or zero pair) yields a second-order polynomial with real coefficients. Real-valued pole (or zero) can appear single in Equation 19.8.

It is clear from Equation 19.8 that knowing all the pole and zero locations, we can write the system function to within a constant factor. Since the constant factor is only a gain, which can be adjusted as desired, specifying the locations of the poles and zeros essentially specifies the system function of the IIR filter.

19.1.4 Stability

A causal IIR filter is stable (in the sense that a bounded input gives rise to a bounded output) if all the poles lie inside the unit circle. If there are one or more simple poles on the unit circle (and all the others lie inside the unit circle), then the filter is marginally stable, giving a sustained oscillation. If there are multiple poles (more than one pole at the same location) on the unit circle or if there is at least one pole outside the unit circle, a slight input will give rise to an output with increasing magnitude. For most practical filters, all the poles are designed to lie inside the unit circle. In some special systems (such as oscillators), poles are placed on the unit circle to obtain the desired result.

Given the system function in the form of Equation 19.2 or Equation 19.7, the stability can be verified by finding all the poles of the filters and checking to see if all of them are inside the unit circle. Equivalently, stability can be verified directly from the coefficients $\{a_i\}$, using the Schür-Cohn algorithm [1]. For a second-order system, if the coefficient a_1, and a_2, lie inside the triangle in Figure 19.1, then the system is stable.

19.1.5 Frequency Response

The frequency response of the IIR filter is the due of the system function evaluated on the unit circle on the complex plane, i.e., with $z = e^{j2\pi f}$, where f varies from 0 to 1, or from $-1/2$ to $1/2$. The variable f represents the digital frequency. For simplicity, we write $H(f)$ for $H(z)|_{z=\exp(j2\pi f)}$. Therefore,

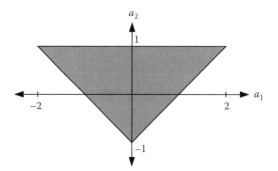

FIGURE 19.1   Region for the coefficients a_1 and a_2 that yield a stable second-order IIR filter.

$$H(f) = b_0 e^{j2\pi(M-N)f} \frac{(e^{j2\pi f} - q_1)(e^{j2\pi f} - q_2) \cdots (e^{j2\pi f} - q_M)}{(e^{j2\pi f} - p_1)(e^{j2\pi f} - p_2) \cdots (e^{j2\pi f} - p_N)} \tag{19.9}$$

$H(f)$ is generally a complex function of f, consisting of the real part $H_R(f)$ and the imaginary part $H_I(f)$. It can also be expressed in terms of the magnitude $|H(f)|$ and the phase $\theta(f)$

$$H(f) = H_R(f) + jH_I(f) = |H(f)|e^{j\theta(f)} \tag{19.10}$$

From Equation 19.9 we see that the magnitude response $|H(f)|$ equals the product of the magnitudes of the individual factors in the numerator, divided by the product of the magnitudes of the individual factors in the denominator. The magnitude square can be written as

$$|H(f)|^2 = H(f)H^\star(f) = [H_R(f)]^2 + [H_I(f)]^2 \tag{19.11}$$

Since $H^\star(f) = H^\star(1/z^\star)|_{z=\exp(j2\pi f)}$ and $H^\star(1/z^\star) = H(z^{-1})$ when all the coefficients of $H(z)$ are real, we have

$$|H(f)|^2 = H(z) \cdot H(z^{-1})|_{z=\exp(j2\pi f)} \tag{19.12}$$

Using Equation 19.12, the magnitude square can be put in the form

$$|H(f)|^2 = \frac{\sum_{k=0}^{M} \tilde{b}_k \cos(2\pi kf)}{\sum_{k=0}^{N} \tilde{a}_k \cos(2\pi kf)} \tag{19.13}$$

where the coefficients are given by

$$\tilde{b}_0 = \sum_{j=0}^{M} b_j^2 \quad \tilde{b}_k = 2\sum_{j=k}^{M} b_j b_{j-k} \quad k = 1, \ldots, M$$
$$\tilde{a}_0 = \sum_{j=0}^{N} a_j^2 \quad \tilde{a}_k = 2\sum_{j=k}^{N} a_j a_{j-k} \quad k = 1, \ldots, N \tag{19.14}$$

with the understanding that $a_0 = 1$. Given $\{\tilde{b}_0, \tilde{b}_1, \ldots, \tilde{b}_M\}$ we can find $\{b_0, b_1, \ldots, b_M\}$ and vice versa. Similarly, $\{\tilde{a}_1, \tilde{a}_2, \ldots, \tilde{a}_N\}$ and $\{a_1, a_2, \ldots, a_N\}$ can be computed from each other. The form in Equation 19.13 is useful in computer-aided design of IIR filters using linear programming [2].

We see from Equation 19.9 that the phase response $\theta(f)$ equals the sum of the phases of the individual factors in the numerator minus the sum of the phases of the individual factors in the denominator. The phase can be written in terms of the real and imaginary parts of $H(f)$ as

$$\theta(f) = \arctan\left[\frac{H_I(f)}{H_R(f)}\right] \qquad (19.15)$$

A filter having linear phase in a frequency band (e.g., in the passband) means that there is no phase distortion in that band.

The group delay is defined as

$$\tau(f) = -\frac{1}{2\pi}\frac{d}{df}\theta(f) \qquad (19.16)$$

The group delay corresponds to the delay, from the input to the output, of the envelope of a narrowband signal [3]. A linear phase gives rise to a constant group delay. Nonlinearity in the phase appears as deviation of the group delay from a constant value.

The magnitude response of IIR filter does not change, except for a constant factor, if a zero is replaced by the reciprocal of its complex conjugate, i.e., if $(z - q)$ is replaced with $(z - 1/q^*)$. This can be seen as follows. Letting $\hat{H}(z)$ be the system function without the factor $(z - q)$, we have

$$|H(f)|^2 = H(z)H^*(1/z^*)|_{z=\exp(j2\pi f)}$$
$$= \hat{H}(z)\hat{H}^*(1/z^*)\frac{(z - q)(z^{-1} - q^*)}{(z - 1/q^*)(z^{-1} - 1/q)}\Big|_{z=\exp(2j\pi f)}$$
$$= |q|^2$$

Similarly, replacing the pole at p with a pole at $1/p^*$ will not alter the magnitude of the response except for a constant factor. This property is useful in changing an unstable IIR filter to a stable one without altering the magnitude response.

Compared to an FIR filter, an IIR filter requires a much lower order to achieve the same requirement of the magnitude response. However, the phase of a stable casual IIR filter cannot be made linear. This is the major reason not to use an IIR filter in applications where linear phase is essential. Nevertheless, using phase compensation such as allpass filters (see Section 19.1.8), the phase of an IIR filter can be adjusted close to linear. This process increases the order of the overall system, however. Note that if causality is not required, then a linear-phase IIR filter can be obtained using a time-reversal filter [2].

19.1.6 Realizations

A realization of an IIR filter according to Equation 19.1 is shown in Figure 19.2a, which is called Direct Form I. By rearranging the structure, we can obtain Direct Form II, as shown in Figure 19.2b. Through transposition, we can obtain Transposed Direct Form I and Transposed Direct Form II as shown in Figure 19.2c and d.

The system function can be put in the form

$$H(z) = \prod_{i=1}^{K}\frac{b_{i0} + b_{i1}z^{-1} + b_{i2}z^{-2}}{1 + a_{i1}z^{-1} + a_{i2}z^{-2}} \qquad (19.17)$$

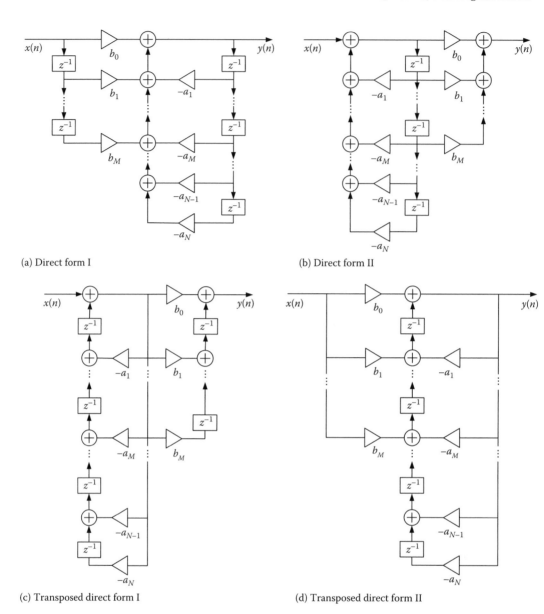

(a) Direct form I

(b) Direct form II

(c) Transposed direct form I

(d) Transposed direct form II

FIGURE 19.2   Direct form realizations of IIR filters.

by factoring the numerators and denominators into second-order factors, or in the form

$$H(z) = \frac{b_N}{a_N} + \sum_{i=1}^{K} \frac{b_{i0} + b_{i1}z^{-1}}{1 + a_{i1}z^{-1} + a_{i2}z^{-2}} \tag{19.18}$$

by partial fraction expansion. The value of K is $N/2$ when N is even and is $(N+1)/2$ when N is odd. When N is odd, one of a_{i2} must be zero, as well as one of b_{i2}, in Equation 19.17 and one of b_{i1} in Equation 19.18. All the coefficients in Equations 19.17 and 19.18 are real numbers. According to Equation 19.17, the IIR filter can be realized by K second-order IIR filters in cascade, as shown in Figure 19.3a. According

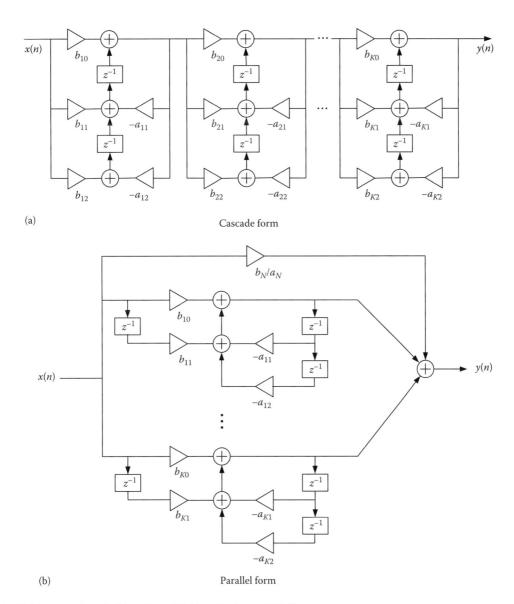

(a)   Cascade form

(b)   Parallel form

FIGURE 19.3 Cascade (a) and parallel (b) realizations of IIR filters.

to Equation 19.18, the IIR filter is realized by K second-order IIR filters and one scaler (i.e., b_N/a_N) in parallel, as depicted in Figure 19.3b. Each second-order subsystem can use any of the structures given in Figure 19.2.

There are many other realizations for IIR filters, such as state-space structures [4], wave structures, and lattice structures (Section 19.3).

Actual implementation of IIR filters requires that the signals and the coefficients be represented in a finite number of bits (or digits). Quantization of the coefficients to a finite number of bits essentially changes the filter coefficients, hence the frequency response changes. Coefficient quantization of a stable IIR filter may yield an unstable filter. For example, consider a second-order IIR filter with $a_1 = 1.26$ and $a_2 = 0.3$, which correspond to pole locations of -0.9413 and -0.3187, respectively. Suppose that we quantize these coefficients to two bits after the decimal point, yielding a quantized a_1 of 1.01 in binary or

1.25 and a quantized a_2 of 0.01 in binary or 0.25. This pair corresponds to pole locations at -1.0 and -0.25, respectively. Since one pole is on the unit circle, the IIR filter with quantized coefficients produces an oscillation. In this example, the quantization is equivalent to moving a point inside the triangle in Figure 19.1 to a point on the edge of the triangle. Different realizations are affected differently by coefficient quantization. Chapter 20 investigates coefficient quantization and roundoff noise in detail.

19.1.7 Minimum Phase

An IIR filter is a minimum-phase filter if all the zeros and poles are inside the unit circle. A minimum-phase filter introduces the smallest group delay among all filters that have the same magnitude response. A minimum-phase IIR filter can be constructed from a nonminimum-phase filter by replacing each zero (or pole) outside the unit circle with a zero (or pole) that is the reciprocal of its complex conjugate, as illustrated in Figure 19.4. This process moves all zeros and poles outside the unit circle to the inside. The magnitude response does not change, except for a constant factor, which is easily adjusted. Given an IIR filter $H(z)$ with input $x(n)$ and output $y(n)$, the inverse filter $1/H(z)$ can reconstruct $x(n)$ from $y(n)$ by feeding $y(n)$ to the input of $1/H(z)$. Assuming that both the filter and the inverse filter are causal, both of them can be stable only if $H(z)$ is a minimum-phase filter.

19.1.8 Allpass Filters

An allpass filter has a magnitude response of unity (or constant). An Nth-order IIR allpass filter with real coefficients has a system function given by

$$H(z) = z^{-N} \frac{D(z)}{D(z^{-1})} = z^{-N} \frac{a_N z^N + \cdots + a_2 z^2 + a_1 z + 1}{1 + a_1 z^{-1} + a_2 z^{-2} + \cdots + a_N z^{-N}} \tag{19.19}$$

$$= z^{-N} \frac{(1 - p_1 z)(1 - p_2 z) \cdots (1 - p_N z)}{(1 - p_1 z^{-1})(1 - p_2 z^{-1}) \cdots (1 - p_N z^{-1})} \tag{19.20}$$

Since $H(z)(z^{-1}) = 1$, it follows that $|H(f)|^2 = 1$. The factor z^{-N} is included so that the filter is causal. Equation 19.20 implies that zeros and poles come in reciprocal pairs: if there is a pole at $z = p$, then there is a zero at $z = 1/p$, as illustrated in Figure 19.5.

Since the coefficients are real, poles and zeros off the real axis must exist in quadruplets: poles at p and p^* and zeros at $1/p$ and $1/p^*$, where $|p| < 1$ for stability. For poles and zeros on the real axis, they exist in reciprocal pairs: pole at p and zero at $1/p$, where p is real and $|p| < 1$ for stability. Since the numerator and the denominator in (19.19) share the same set of coefficients, we need only N multiplications in

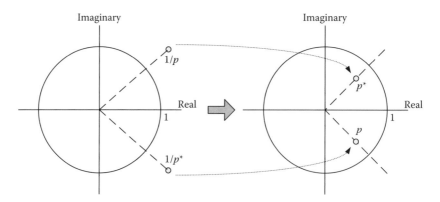

FIGURE 19.4   Changing a zero location to obtain a minimum-phase filter.

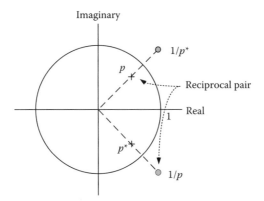

FIGURE 19.5   Pole-zero reciprocal pair in an allpass IIR filter.

realizing an Nth-order allpass filter. The system function in Equation 19.19 can be written as the product (or sum) of first- and second-order allpass filters. The system function and the phase response of a first-order allpass filter is given by

$$H(z) = \frac{a_1 z + 1}{z + a_1} \tag{19.21}$$

$$\theta(f) = \arctan\left[\frac{(a_1^2 - 1)\sin(\omega)}{2a_1 + (a_1^2 + 1)\cos(\omega)}\right] \tag{19.22}$$

where $\omega = 2\pi f$. For a second-order allpass filter, these are

$$H(z) = \frac{a_2 z^2 + a_1 z + 1}{z^2 + a_1 z + a_2} \tag{19.23}$$

$$\theta(f) = \arctan\left[\frac{2a_1(a_2 - 1)\sin(\omega) + (a_2^2 - 1)\sin(2\omega)}{2a_2 + a_1^2 + 2a_1(a_2 + 1)\cos(\omega) + (a_2^2 + 1)\cos(2\omega)}\right] \tag{19.24}$$

The group delay $\tau(f)$ of an allpass filter is always ≥ 0. The output signal energy of an allpass filter is the same as the input signal energy, i.e., $\sum_{n=-\infty}^{\infty} |y(n)|^2 = \sum_{n=-\infty}^{\infty} |x(n)|^2$, which means that the allpass filter is a lossless system. Note that if we attempt to find a minimum-phase filter from a stable allpass filter, by moving all the zeros inside the unit circle, all poles and zeros would cancel out, yielding the trivial filter with a system function of unity.

A more general form of Equation 19.19, allowing the coefficients to be complex, is Nth-order allpass filer with system function

$$H(z) = z^{-N}\frac{D^*(z^*)}{D(z^{-1})} = z^{-N}\frac{a_N^* + \cdots + a_2^* z^2 + a_1^* z + 1}{1 + a_1 z^{-1} + a_2 z^{-2} + \cdots + a_N z^{-N}} \tag{19.25}$$

$$= z^{-N}\frac{(1 - p_1^* z)(1 - p_2^* z)\cdots(1 - p_N^* z)}{(1 - p_1 z^{-1})(1 - p_2 z^{-1})\cdots(1 - p_N z^{-1})} \tag{19.26}$$

Therefore, for a pole at $z = p$ there is a zero at $z = 1/p^*$, i.e., poles and zeros exist in reciprocal–conjugate pairs.

Allpass filters have been used as building blocks for various applications [5]. Particularly, an allpass filter can be designed to approximate a desired phase response. Therefore, an allpass filter in cascade with an IIR filter can be used to compensate the nonlinear phase of the IIR filter. Such a cascade filter has

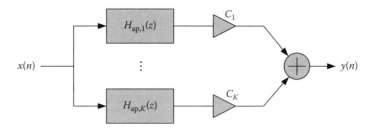

FIGURE 19.6 Block diagram of an IIR filter, using allpass filters.

system function of the form $H(z) = H_{\text{IIR}}(z)H_{\text{ap}}(z)$, where $H_{\text{IIR}}(z)$ is an IIR filter satisfying some magnitude response and $H_{\text{ap}}(z)$ is an allpass filter that compensates for the nonlinearity of the phase response of $H_{\text{IIR}}(z)$. Allpass filters in parallel connection can be used to approximate a desired magnitude response. For this, the system function is in the form $H(z) = \sum_{i=1}^{K} c_i H_{\text{ap},i}(z)$, where $H_{\text{ap},i}(z)$ is an allpass filter and c_i is a coefficient. A block diagram is shown in Figure 19.6.

References

1. J. G. Proakis and D. G. Manolakis, *Digital Signal Processing Principles, Algorithms, and Applications*, 2nd ed. New York: Macmillan, 1992.
2. L. R. Rabiner and B. Gold, *Theory and Application of Digital Signal Processing*, Englewood Cliffs, NJ: Prentice-Hall, 1975.
3. A. V. Oppenheim and R. W. Schafer, *Discrete-Time Signal Processing*, Englewood Cliffs, NT: Prentice-Hall, 1989.
4. R. A. Roberts and C. T. Mullis, *Digital Signal Processing*, Reading, MA: Addison-Wesley, 1987.
5. P. A. Regalia, S. K. Mitra, and P. P. Vaidyanathan, The digital all-pass filter: A versatile signal processing building block, *Proc. IEEE*, 76, 19–37, Jan. 1988.

19.2 Design of IIR Filters

Sawasd Tantaratana, Chalie Charoenlarpnopparut, and Phakphoom Boonyanant

19.2.1 Introduction

A filter is generally designed to satisfy a frequency response specification. IIR filter design normally focuses on satisfying a magnitude response specification. If the phase response is essential, it is usually satisfied by a phase compensation filter, such as an allpass filter (see Section 19.1.8). We will adopt a magnitude specification that is normalized so that the maximum magnitude is 1. The magnitude square in the passband must be at least $1/(1 + \varepsilon^2)$ and at most 1; while it must be no larger than δ^2 in the stopband, where ε and δ are normally small. The passband edge is denoted by f_p and the stopband edge by f_s. Figure 19.7a shows such a specification for a low-pass filter (LPF). The region between the passband and the stopband is the transition band. There is no constraint on the response in the transition band. Another specification that is often used is shown in Figure 19.7b using δ_1, and δ_2, to specify the acceptable magnitude. Given δ_1, and δ_2, they can be converted to ε and δ using $\varepsilon = 2\delta_1^{0.5}/(1 - \delta_1)$ and $\delta = \delta_2/(1 + \delta_1)$. The magnitude is often specified in decibels, which is $20 \log_{10} |H(f)|$. Specifications for other types of filters (high-pass, bandpass, and bandstop) are similar.

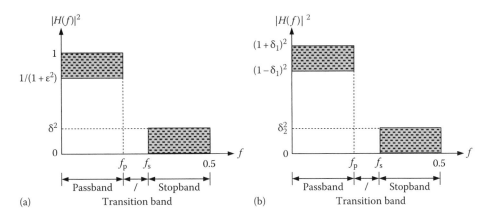

FIGURE 19.7 Specifications for a digital LPF. (a) Specification using ε and δ; (b) specification using δ_1 and δ_2.

We can classify various IIR filter design methods into three categories: the design using analog prototype filter, the design using digital frequency transformation, and computer-aided design. In the first category, an analog filter is designed to the (analog) specification and the analog filter transfer function is transformed to digital system function using some kind of transformation. The second category assumes that a digital LPF can be designed. The desired digital filter is obtained from the digital LPF by a digital frequency transformation. The last category uses some algorithm to choose the coefficients so that the response is as close (in some sense) as possible to the desired filter. Design methods in the first two categories are simple to do, requiring only a handheld calculator. Computer-aided design requires some computer programming, but it can be used to design nonstandard filters.

19.2.2 Analog Filters

Here, we describe four basic types of analog LPFs that can be used as prototype for designing IIR filters. For each type, we give the transfer function, its magnitude response, and the order N needed to satisfy the (analog) specification. We will use $H_a(s)$ for the transfer function of an analog filter, where s is the variable in the Laplace transform. Each of these filters have all its poles on the left-half s plane, so that it is stable. We will use the variable λ to represent the analog frequency in radians/second. The frequency response $H_a(\lambda)$ is the transfer function evaluated at $s = j\lambda$. The analog LPF specification is given by

$$(1 + \varepsilon^2)^{-1} \le |H_a(\lambda)|^2 \le 1 \quad \text{for } 0 \le (\lambda/2\pi) \le (\lambda_p/2\pi)$$
$$0 \le |H_a(\lambda)|^2 \le \delta^2 \quad \text{for } (\lambda_s/2\pi) \le (\lambda/2\pi) \le \infty \tag{19.27}$$

where λ_p and λ_s are the passband edge and stopband edge, respectively. The specification is sketched in Figure 19.8.

19.2.2.1 Butterworth Filters

The transfer function of an Nth-order Butterworth filter is given by

$$H_a(s) = \begin{cases} \displaystyle\prod_{i=1}^{N/2} \frac{1}{(s/\lambda_c)^2 - 2\text{Re}(s_i)(s/\lambda_c) + 1} & N = \text{even} \\[4ex] \displaystyle\frac{1}{(s/\lambda_c) + 1} \prod_{i=1}^{(N-1)/2} \frac{1}{(s/\lambda_c)^2 - \text{Re}(s_i)(s/\lambda_c) + 1} & N = \text{odd} \end{cases} \tag{19.28}$$

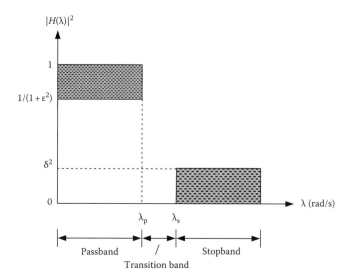

FIGURE 19.8   Specification for an analog LPF.

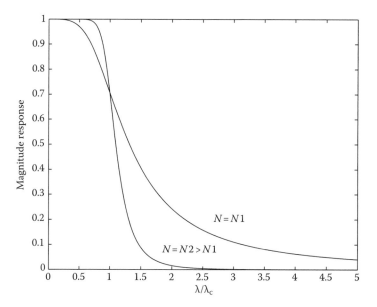

FIGURE 19.9   Magnitude responses of Butterworth filters.

where $\lambda_p \leq \lambda_c \leq \lambda_s$ and $s_i = \exp\{j[1 + (2i - 1)/N]\pi/2\}$. The magnitude response square is

$$|H_a(\lambda)|^2 = \frac{1}{1 + (\lambda/\lambda_c)^{2N}} \tag{19.29}$$

Figure 19.9 shows the magnitude response $|H_a(\lambda)|$, with $\lambda_c = 1$. Note that a Butterworth filter is an all-pole (no zero) filter, with the poles being at $s = \lambda_c s_i$ and $s = \lambda_c s_i^*, i = 1, \ldots, N/2$ if N is even or $i = 1, \ldots, (N - 1)/2$ if N is odd, where x^* denotes the complex conjugate of x. When N is odd, there

TABLE 19.1 Factorized, Normalized, Butterworth Polynomial $B_N(s)$

N	$B_N(s)$
1	$(s + 1)$
2	$(s^2 + \sqrt{2}s + 1)$
3	$(s + 1)(s^2 + s + 1)$
4	$(s^2 + \sqrt{2 - \sqrt{2}}s + 1)(s^2 + \sqrt{2 + \sqrt{2}}s + 1)$
5	$(s + 1)\left(s^2 + \left(\frac{\sqrt{5}-1}{2}\right)s + 1\right)\left(s^2 + \left(\frac{\sqrt{5}+1}{2}\right)s + 1\right)$
6	$\left(s^2 + \left(\frac{\sqrt{3}-1}{\sqrt{2}}\right)s + 1\right)(s^2 + \sqrt{2}s + 1)\left(s^2 + \left(\frac{\sqrt{3}+1}{\sqrt{2}}\right)s + 1\right)$
7	$(s + 1)(s^2 + 0.4450s + 1)(s^2 + 1.2470s + 1)(s^2 + 1.8019s + 1)$
8	$(s^2 + 0.3902s + 1)(s^2 + 1.1111s + 1)(s^2 + 1.6629s + 1)(s^2 + 1.9616s + 1)$
9	$(s + 1)^2(s^2 + 0.3473s + 1)(s^2 + 1.5321s + 1)(s^2 + 1.8794s + 1)$
10	$(s^2 + 0.3219s + 1)(s^2 + 0.9080s + 1)(s^2 + 1.4142s + 1)(s^2 + 1.7820s + 1)(s^2 + 1.9754s + 1)$

is another pole at $s = -\lambda_c$. All N poles are on the left-half s plane, located on the circle with radius λ_c. Therefore, the filter in Equation 19.28 is stable.

To satisfy the specification in Equation 19.27, the filter order can be calculated from

$$N = \text{integer} \geq \frac{\log[\varepsilon/(\delta^{-2} - 1)^{1/2}]}{\log[\lambda_p/\lambda_s]} \tag{19.30}$$

The value of λ_c can be chosen as any value in the following range:

$$\lambda_p \varepsilon^{-1/N} \leq \lambda_c \leq \lambda_s(\delta^{-2} - 1)^{-1/(2N)} \tag{19.31}$$

If we choose $\lambda_c = \lambda_p \varepsilon^{-1/N}$, then the magnitude response square passes through $1/(1 + \varepsilon^2)$ at $\lambda = \lambda_p$. If we choose $\lambda_c = \lambda_s(\delta^{-2} - 1)^{-1/(2N)}$, then the magnitude response square passes through δ^2 at $\lambda = \lambda_s$. If λ_c is between these two values, then the magnitude square will be $\geq 1/(1 + \varepsilon^2)$ at $\lambda = \lambda_p$ and $\leq \delta^2$ at $\lambda = \lambda_s$.

For the sake of convenience, the transfer function of the Nth-order Butterworth filter with $\lambda_c = 1$ can be found by

$$H_a(s) = \frac{1}{B_N(s)}$$

where $B_N(s)$ is the normalized Butterworth polynomial of order N shown in Table 19.1. For other values of λ_c', the transfer function can be computed similarly by analog–analog frequency transformation, i.e., replacing s in $H_a(s)$ with s/λ_c'.

19.2.2.2 Chebyshev Filters (Type-I Chebyshev Filters)

A Chebyshev filter is also an all-pole filter. The Nth-order Chebyshev filter has a transfer function given by

$$H_a(s) = C \prod_{i=1}^{N} \frac{1}{(s - p_i)} \tag{19.32}$$

where

$$
C = \begin{cases} -\displaystyle\prod_{i=1}^{N} p_i & N \text{ is odd} \\[2ex] (1+\varepsilon^2)^{-1/2} \displaystyle\prod_{i=1}^{N} p_i & N \text{ is even} \end{cases} \tag{19.33}
$$

$$
p_i = -\lambda_p \sinh(\phi) \sin\left(\frac{2i-1}{2N}\pi\right) + j\lambda_p \cosh(\phi) \cos\left(\frac{2i-1}{2N}\pi\right) \tag{19.34}
$$

$$
\phi = \frac{1}{N} \ln\left[\frac{1+(1+\varepsilon^2)^{1/2}}{\varepsilon}\right] \tag{19.35}
$$

The value of C normalizes the magnitude so that the maximum magnitude is 1. Note that C is always a positive constant. The poles are on the left-half s plane, lying on an ellipse centered at the origin with a minor radius of $\lambda_p \sinh(\phi)$ and major radius of $\lambda_p \cosh(\phi)$. Except for one pole when N is odd, all the poles have a complex–conjugate pair. Specifically, $p_i = p_{N-i+1}^*, i = 1, 2, \cdots, N/2$ or $(N-1)/2$. Combining each complex–conjugate pair in Equation 19.32 yields a second-order factor with real coefficients. The magnitude response can be computed from Equations 19.33 through 19.35 with $s = j\lambda$. Its square can also be written as

$$
|H_a(\lambda)|^2 = \frac{1}{1+\varepsilon^2 T_N^2(\lambda/\lambda_p)} \tag{19.36}
$$

where $T_N(x)$ is the Nth degree Chebyshev polynomial of the first kind, which is shown in Table 19.2 and also given recursively by

$$
\begin{aligned} T_0(x) &= 1 \quad T_1(x) = x \\ T_{n+1}(x) &= 2xT^n(x) - T_{n-1}(x) \quad n \geq 1 \end{aligned} \tag{19.37}
$$

Notice that $T_N^2(\pm 1) = 1$. Therefore, we have from Equation 19.36 that the magnitude square passes through $1/(1+\varepsilon^2)$ at $\lambda = \lambda_p$, i.e., $|H_a(\lambda_p)|^2 = 1/(1+\varepsilon^2)$. Note also that $T_N(0) = (-1)^{N/2}$ for even N and it is 0 for odd N. Therefore, $|H_a(0)|^2$ equals $1/(1+\varepsilon^2)$ for even N and it equals 1 for odd N. Figure 19.10 shows some examples of magnitude response square.

TABLE 19.2   Coefficients of Chebyshev Polynomials $T_n(x)$ of the First Kind, of Order n, in Ascending Powers of Variable x

n	Coefficients of $T_n(x)$
1	0, 1
2	−1, 0, 2
3	0, −3, 0, 4
4	1, 0, −8, 0, 8
5	0, 5, 0, −20, 0, 16
6	−1, 0, 18, 0, −48, 0, 32
7	0, −7, 0, 56, 0, −112, 0, 64
8	1, 0, −32, 0, 160, 0, −256, 0, 128
9	0, 9, 0, −120, 0, 432, 0, −576, 0, 256
10	−1, 0, 50, 0, −400, 0, 1120, 0, −1280, 0, 512

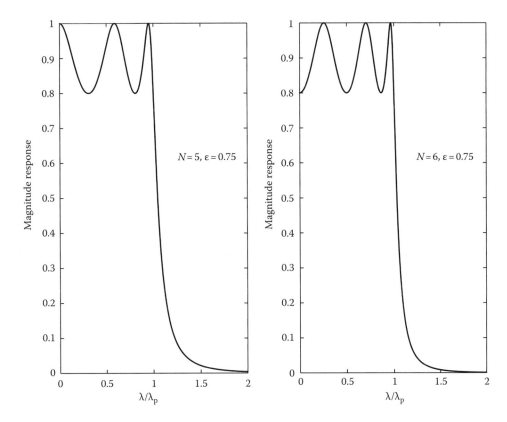

FIGURE 19.10 Magnitude responses of Chebyshev filters.

The filter order required to satisfy the specification in Equation 19.27 is

$$N \geq \frac{\cosh^{-1}[(\delta^{-2} - 1)^{1/2}/\varepsilon]}{\cosh^{-1}(\lambda_s/\lambda_p)}$$
$$= \frac{\log\{[(\delta^{-2} - 1)^{1/2}/\varepsilon] + [(\delta^{-2} - 1)/\varepsilon^2 - 1]^{1/2}\}}{\log\{(\lambda_s/\lambda_p) + [(\lambda_s/\lambda_p)^2 - 1]^{1/2}\}} \qquad (19.38)$$

which can be computed knowing ε, δ, λ_p, and λ_s.

Table 19.3 shows the normalized Chebyshev polynomials $C_N(s)$ for different values of N and passband ripples. The transfer function of a unity dc gain, unity cutoff frequency Chebyshev Type-I filter is given by

$$H_a(s) = \frac{1}{C_N(s)}$$

19.2.2.3 Inverse Chebyshev Filters (Type-II Chebyshev Filters)

Notice from Figure 19.10 that the Chebyshev filter has magnitude response containing equiripples in the passband. The equiripples can be arranged to go inside the stopband, for which case we obtain inverse Chebyshev filters. The magnitude response square of the inverse Chebyshev filter is

TABLE 19.3 Normalized Chebyshev Polynomials $C_N(s)$ for Passband Ripple = 0.5, 1, and 3 dB

N	$C_N(s)$ for Passband Ripple 0.5 dB
1	$0.3493s + 1$
2	$0.6595s^2 + 0.9403s + 1$
3	$1.3972s^3 + 1.7506s^2 + 2.1446s + 1$
4	$2.6382s^4 + 3.1589s^3 + 4.5294s^2 + 2.7053s + 1$
5	$5.589s^5 + 6.5530s^4 + 10.8279s^3 + 7.3192s^2 + 4.2058s + 1$
6	$10.5527s^6 + 12.2324s^5 + 22.9188s^4 + 16.7763s^3 + 12.3663s^2 + 4.5626s + 1$

N	$C_N(s)$ for Passband Ripple 1 dB
1	$0.5088s + 1$
2	$0.9070s^2 + 0.9957s + 1$
3	$2.0354s^3 + 2.0117s^2 + 2.5206s + 1$
4	$3.6281s^4 + 3.4569s^3 + 5.2750s^2 + 2.6943s + 1$
5	$8.1416s^5 + 7.6272s^4 + 13.7496s^3 + 7.9331s^2 + 4.7265s + 1$
6	$14.5123s^6 + 13.4711s^5 + 28.0208s^4 + 17.4459s^3 + 13.6321s^2 + 4.4565s + 1$

N	$C_N(s)$ for Passband Ripple 3 dB
1	$0.9976s + 1$
2	$1.4125s^2 + 0.9109s + 1$
3	$3.9905s^3 + 2.3833s^2 + 3.7046s + 1$
4	$5.6501s^4 + 3.2860s^3 + 6.6057s^2 + 2.287s + 1$
5	$15.9621s^5 + 9.1702s^4 + 22.5867s^3 + 8.7622s^2 + 6.5120s + 1$
6	$22.6005s^6 + 12.8981s^5 + 37.5813s^4 + 15.6082s^3 + 15.8000s^2 + 3.6936s + 1$

$$|H_a(\lambda)|^2 = \frac{1}{1 + \frac{(\delta^{-2}-1)}{T_N^2(\lambda_s/\lambda)}} \qquad (19.39)$$

Since $T_N^2(\pm 1) = 1$, Equation 19.39 gives $|H_a(\lambda_s)|^2 = \delta^2$. Figure 19.11 depicts some examples of Equation 19.38. Note that $|H_a(\infty)|$ equals 0 if N is odd and it equals δ if N is even.

The transfer function giving rise to Equation 19.39 is given by

$$H_a(s) = \begin{cases} C \prod\limits_{i=1}^{N} \frac{(s-q_i)}{(s-p_i)} & N = \text{even} \\[2mm] \frac{C}{(s-p_{(N+1)/2})} \prod\limits_{i=1,i\neq(N+1)/2}^{N} \frac{(s-q_i)}{(s-p_i)} & N = \text{odd} \end{cases} \qquad (19.40)$$

where

$$C = \begin{cases} \prod\limits_{i=1}^{N} \frac{p_i}{q_i} & N \text{ is even} \\[2mm] -p_{(N+1)/2} \prod\limits_{i=1,i\neq(N+1)/2}^{N} \frac{p_i}{q_i} & N \text{ is odd} \end{cases} \qquad (19.41)$$

$$p_i = \frac{\lambda_s}{\alpha_i^2 + \beta_i^2}(\alpha_i - j\beta_i) \qquad q_i = j\frac{\lambda_s}{\cos\left(\frac{2i-1}{2N}\pi\right)} \qquad (19.42)$$

$$\phi = \frac{1}{N}\cosh^{-1}(\delta^{-1}) = \frac{1}{N}\ln\left[\delta^{-1} + (\delta^{-2}-1)^{1/2}\right] \qquad (19.43)$$

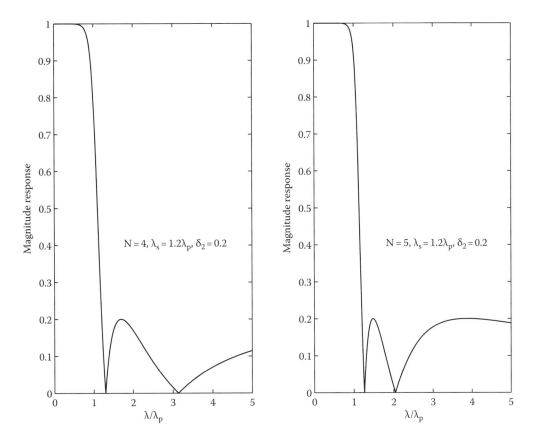

FIGURE 19.11   Magnitude responses of inverse Chebyshev filters.

Note that the zeros are on the imaginary axis on the s plane. The filter order N required to satisfy the specification in Equation 19.27 is the same as the order for the Chebyshev filter, given by Equation 19.38.

Another form for the inverse Chebyshev filter has magnitude response square given by

$$|H_a(\lambda)|^2 = \frac{1}{1 + \varepsilon^2 \frac{T_N^2(\lambda_s/\lambda_p)}{T_N^2(\lambda_s/\lambda)}} \tag{19.44}$$

which passes through $1/(1 + \varepsilon^2)$ at $\lambda = \lambda_p$. For further details of this form see Ref. [1].

19.2.2.4 Elliptic Filters (Cauer Filters)

Elliptic filters have equiripples in both the passband and the stopband. We summarize the magnitude response and the transfer function of an elliptic filter as follows. Detail of derivation can be found in Refs. [2,3].

The magnitude response square is given by

$$|H_a(\lambda)|^2 = \frac{1}{1 + \varepsilon^2 R_N^2(\lambda)} \tag{19.45}$$

where $R_N(\lambda)$ is the Chebyshev rational function given by

$$R_N(\lambda) = \begin{cases} \dfrac{(\delta^{-2}-1)^{1/4}}{\varepsilon^{1/2}} \lambda \displaystyle\prod_{i=1}^{(N-1)/2} \dfrac{\lambda^2 - \lambda_r \operatorname{sn}^2\left[\frac{2iK(\lambda_r)}{N}, \lambda_r\right]}{\lambda^2 \lambda_r \operatorname{sn}^2\left[\frac{2iK(\lambda_r)}{N}, \lambda_r\right] - 1} & N = \text{odd} \\[3em] \dfrac{(\delta^{-2}-1)^{1/4}}{\varepsilon^{1/2}} \lambda \displaystyle\prod_{i=1}^{(N-1)/2} \dfrac{\lambda^2 - \lambda_r \operatorname{sn}^2\left[\frac{(2i-1)K(\lambda_r)}{N}, \lambda_r\right]}{\lambda^2 \lambda_r \operatorname{sn}^2\left[\frac{(2i-1)(\lambda_r)}{N}, \lambda_r\right] - 1} & N = \text{even} \end{cases} \tag{19.46}$$

Here, $\lambda_r = \lambda_p/\lambda_s$, $K(t)$ is the complete elliptic integral of the first kind given by

$$K(t) = \int_0^{\pi/2} \frac{d\theta}{(1 - t^2 \sin^2 \theta)^{1/2}} = \int_0^1 \frac{dx}{[(1 - x^2)(1 - t^2 x^2)]^{1/2}} \tag{19.47}$$

The Jacobian elliptic sine function $\operatorname{sn}[u, t]$ is defined as

$$\operatorname{sn}[u, t] = \sin \phi \quad \text{if } u = \int_0^\phi \frac{d\theta}{(1 - t^2 \sin^2 \theta)^{1/2}} \tag{19.48}$$

The integral

$$F(\phi, t) = \int_0^\phi \frac{d\theta}{(1 - t^2 \sin^2 \theta)^{1/2}} = \int_0^{\sin \phi} \frac{dx}{[(1 - x^2)(1 - t^2 x^2)]^{1/2}} \tag{19.49}$$

is called the elliptic integral of the first kind. Note that $K(t) = F(\pi/2, t)$.

The transfer function corresponding to the magnitude response in Equation 19.45 is

$$H_a(s) = \begin{cases} \dfrac{C}{(s+p_0)} \displaystyle\prod_{i=1}^{(N-1)/2} \dfrac{(s^2 + B_i)}{(s^2 + A_{i1}s + A_{i2})} & N \text{ odd} \\[3em] C \displaystyle\prod_{i=1}^{N/2} \dfrac{(s^2 + B_i)}{(s^2 + A_{i1}s + A_{i2})} & N \text{ even} \end{cases} \tag{19.50}$$

$$C = \begin{cases} p_0 \displaystyle\prod_{i=1}^{(N-1)/2} \dfrac{A_{i2}}{B_i} & N \text{ odd} \\[3em] \dfrac{1}{(1+\varepsilon^2)^{1/2}} \displaystyle\prod_{i=1}^{N/2} \dfrac{A_{i2}}{B_i} & N \text{ even} \end{cases} \tag{19.51}$$

The pole p_0 and the coefficients B_i, A_{i1} are calculated as follows:

$$\lambda_r = \frac{\lambda_p}{\lambda_s} \quad \lambda_c = \sqrt{\lambda_p \lambda_s} \quad \alpha = 0.5 \frac{1 - (1 - \lambda_r^2)^{1/4}}{1 + (1 - \lambda_r^2)^{1/4}} \tag{19.52}$$

$$\beta = e^{-\pi K[(1-\lambda_r^2)^{1/2}]/K(\lambda_r)} \approx \alpha + 2\alpha^5 + 15\alpha^9 + 150\alpha^{13} \tag{19.53}$$

$$\gamma = \frac{1}{2N} \ln \left[\frac{(1+\varepsilon^2)^{1/2} + 1}{(1+\varepsilon^2)^{1/2} - 1} \right] \tag{19.54}$$

$$\sigma = \left| \frac{2\beta^{1/4} \prod_{k=0}^{\infty} (-1)^k \beta^{k(k+1)} \sinh [(2k+1)\gamma]}{1 + 2 \prod_{k=1}^{\infty} (-1)^k \beta^{k^2} \cosh [2k\gamma]} \right| \tag{19.55}$$

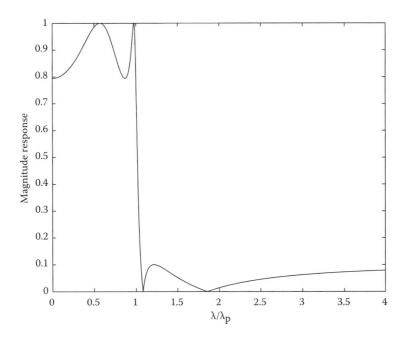

FIGURE 19.12 Magnitude response of elliptic filter.

$$\zeta = (1 + \lambda_r \sigma^2)\left(1 + \frac{\sigma^2}{\lambda_r}\right) \quad \eta = \begin{cases} i & N \text{ odd} \\ i - 0.5 & N \text{ even} \end{cases} \tag{19.56}$$

$$\psi_i = \frac{2\beta^{1/4} \sum_{k=0}^{\infty} (-1)^k \beta^{k(k+1)} \sin\left[(2k+1)\pi\eta/N\right]}{1 + 2\sum_{k=1}^{\infty} (-1)^k \beta^{k^2} \cos\left[2k\pi\eta/N\right]} \tag{19.57}$$

$$\mu_i = \left[(1 - \lambda_r \psi_i^2)\left(1 - \frac{\psi_i^2}{\lambda_r}\right)\right]^{1/2} \tag{19.58}$$

$$p_0 = \lambda_c \sigma \quad B_i = \frac{\lambda_c^2}{\psi_i^2} \quad A_{i1} = \frac{2\lambda_c \sigma \mu_i}{1 + \sigma^2 \psi_i^2} \quad A_{i2} = \lambda_c^2 \frac{\sigma^2 \mu_i^2 + \zeta \psi_i^2}{\left[1 + \sigma^2 \psi_i^2\right]^2} \tag{19.59}$$

The infinite summations above converge very quickly, so that only a few terms are needed in actual calculation. A simple program can be written to compute the values in Equations 19.52 through 19.59. The filter order required to satisfy Equation 19.27 is calculated from

$$N \geq \frac{1}{\log(\beta)} \log\left[\frac{\varepsilon^2}{16(\sigma^{-2} - 1)}\right] \tag{19.60}$$

where β is given by Equation 19.53. An example of the magnitude response is plotted in Figure 19.12. We see that there are ripples in both the passband and the stopband.

19.2.2.5 Comparison

In comparing the filters given above, the Butterworth filter requires the highest order and the elliptic filter requires the smallest order to satisfy the same passband and stopband specifications. The Butterworth filter and the inverse Chebyshev filter have nicer (closer to linear) phase characteristics in the passband than Chebyshev and elliptic filters. The magnitude responses of the Butterworth and Chebyshev filters decrease monotonically in the stopband to zero, which reduces the aliasing caused by some analog-to-digital transformation.

19.2.3 Design Using Analog Prototype Filters

In this subsection, we consider designing IIR filters using analog prototype filters. This method is suitable for designing the standard types of filters: low-pass filter (LPF), high-pass filter (HPF), bandpass filter (BPF), and bandstop filter (BSF). The basic idea is to transform the digital specification to analog specification, design an analog flter, and then transform the analog filter transfer function to digital filter system function. Several types of transformation have been studied.

The design steps are outlined in Figure 19.13. Given the desired magnitude response $|H^x(f)|$ of digital LPF, HPF, BPF, or BSF, it is transformed to analog magnitude specification (of the corresponding type: LPF, HPF, BPF, or BSF) $|H_a^x(\lambda)|$. The analog magnitude specification is then transformed to analog LPF magnitude specification $|H_a(\lambda)|$. We then design an analog prototype filter as discussed in Section 19.2.2, obtaining analog LPF transfer function $H_a(s)$. Next, the analog LPF transfer function is transformed to analog transfer function $H_a^x(s)$ of the desired type (LPF, HPF, BPF, or BSF), followed by a transformation to digital filter system function $H^x(z)$. By combining the appropriate steps, we can obtain transformations to go directly from $|H^x(f)|$ to $|H_a(\lambda)|$ and directly from $H_a(s)$ to $H(z)$, as indicated by the dotted lines in Figure 19.13. Note that for designing digital LPF, the middle steps involving $|H_a^x(\lambda)|$ and $H_a^x(s)$ are not applicable.

19.2.3.1 Transformations

There are several types of transformations. They arise from approximating continuous-time signals and systems by discrete-time signals and system. Table 19.4 shows several transformations, with their advantages and disadvantages. The constant T is the sampling interval. The resulting mapping is used for transforming $H_a(s)$ to $H(z)$. For example, in the backward difference approximation we obtain $H(z)$ by replacing the variable s with $(1 - z^{-1})/T$ in $H_a(s)$, i.e., $H(z) = H_a(s)|_{s=1-z^{-1}/T}$. The bilinear

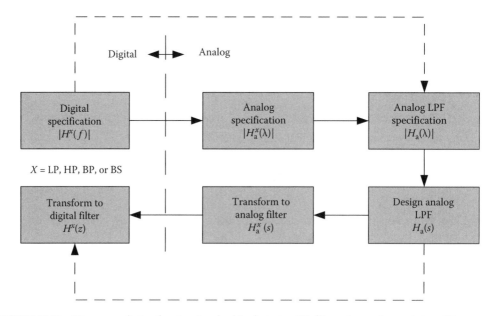

FIGURE 19.13   Diagram outlining the steps involved in designing IIR filter using analog prototype filter.

TABLE 19.4 Various Types of Analog-to-Digital Transformation

Type of Transformation	Principle	Resulting Mapping	Advantages	Disadvantages
Backward difference approximation	$\dfrac{dy}{dt} \approx \dfrac{y(n) - y(n-1)}{T}$	$s = \dfrac{1 - z^{-1}}{T},\quad z = \dfrac{1}{1 - sT}$ s plane / z plane / Unit circle	Stable analog filter yields stable digital filter	Left-half s plane is not mapped onto unit circle on z plane. Pole locations will be in the circle centered at 0.5 with radius 0.5 on the z plane
Forward difference approximation	$\dfrac{dy}{dt} \approx \dfrac{y(n+1) - y(n)}{T}$	$s = \dfrac{z - 1}{T},\quad z = 1 + sT$ s plane / z plane / Unit circle		Stable analog filler does not yield stable digital filter. Left-half s plane is not mapped onto unit circle on z plane
Impulse invariant method	Sample the analog impulse response: $h(n) = h_a(nT)$	Transform $$H_a(s) = \sum_{i=1}^{N} \frac{b_i}{s - p_i}$$ to $$H(z) = \sum_{i=1}^{N} \frac{b_i}{[1 - z^{-1}\exp(p_i T)]}$$	Preserve shape of impulse response. Stable analog filter yields stable digital filter. The analog frequency and digital frequency are linearly related, $\lambda T = 2\pi f,\ -0.5 \le f \le 0.5$	Aliasing in the frequency domain (due to many-to-one mapping from s to z plane) $$T \cdot H(f) = \sum_{k=-\infty}^{\infty} H_a\left(2\pi \frac{f + k}{T}\right)$$

(continued)

TABLE 19.4 (continued)   Various Types of Analog-to-Digital Transformation

Type of Transformation	Principle	Resulting Mapping	Advantages	Disadvantages
Bilinear transformation	Approximation $y(t) = \int_{t-T}^{t} y'(\tau)d\tau + y(t-T)$ by $y(n) = \frac{T}{2}[y'(n) + y'(n-1) + y(n-1)]$	 $s = \frac{2}{T}\left(\frac{1-z^{-1}}{1+z^{-1}}\right)$, $z = \frac{(2/T)+s}{(2/T)-s}$ Mapping of pole locations	Stable analog filter yields stable digital filter Left-half s plane is mapped onto unit circle z plane, on a one-to-one mapping	Frequency warping nonlinear relation between analog frequency and digital frequency
Matched z transformation	Map zero on s plane directly to pole and zero on z plane	Transform $H_a(s) = \sum_{i=1}^{N}\dfrac{s-q_i}{s-p_i}$ to $H(z) = \sum_{i=1}^{N}\dfrac{[1-z^{-1}\exp(q_iT)]}{[1-z^{-1}\exp(p_iT)]}$ 	Stable analog filter yields stable digital filter	Aliasing in the frequency domain (due to many-to-one mapping from s to z plane)

transformation is the best all-around method, followed by the impulse invariant method. Therefore, we describe these two transformations in more detail.

19.2.3.1.1 Bilinear Transformations

Using this transformation, the analog filter is converted to digital filter by replacing s in the analog filter transfer function with $(2/T)(1 - z^{-1})/(1 + z^{-1})$, i.e.,

$$H(z) = H_a(s)|_{s=(2/T)(1-z^{-1})/(1+z^{-1})} \tag{19.61}$$

From the mapping, we can show as follows that the imaginary axis on the s plane is mapped to the unit circle on the z plane. Letting $s = j\lambda$, we have

$$
\begin{aligned}
z &= \frac{(2/T) + s}{(2/T) - s} = \frac{(2/T) + j\lambda}{(2/T) - j\lambda} \\
&= \frac{\sqrt{(2/T)^2 + (\lambda)^2} e^{j \arctan(\lambda/(2/T))}}{\sqrt{(2/T)^2 + (\lambda)^2} e^{j \arctan(-\lambda/(2/T))}}
\end{aligned}
\tag{19.62}
$$

which is the unit circle on the z plane as λ goes from $-\infty$ to ∞. Writing $z = e^{j2\pi f}$ in Equation 19.62, we obtain the relation between the analog frequency λ and the digital frequency f:

$$f = \frac{1}{\pi} \arctan\left(\frac{\lambda T}{2}\right) \qquad \lambda = \frac{2}{T} \tan(\pi f) \tag{19.63}$$

which is plotted in Figure 19.14. Equation 19.63 is used for converting digital specification to analog specification, i.e., $\lambda_s = (2/T) \tan(\pi f_s)$ and $\lambda_p = (2/T) \tan(\pi f_p)$. In a complete design process, starting from the digital specification and ending at the digital filter system function, as outlined in Figure 19.13, the sampling interval T is canceled out in the process. Hence, it has no effect and any convenient value (such as 1 or 2) can be used.

19.2.3.1.2 Impulse Invariance Method

This method approximates the analog filter impulse response $h_a(t)$ by its samples separated by T seconds. The result is the impulse response $h(n)$ of the digital filter, i.e., $h(n) = h_a(nT)$. From this relation, it can be shown that

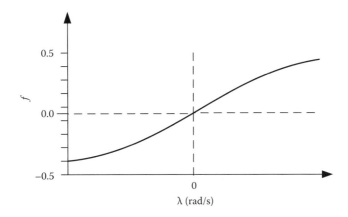

FIGURE 19.14   Relation between λ and f for bilinear transformation.

$$H(f) = \frac{1}{T} \sum_{k=-\infty}^{\infty} H_a(\lambda) \Bigg|_{\lambda = 2\pi(f+k)/T} = \frac{1}{T} \sum_{k=-\infty}^{\infty} H_a\left(2\pi \frac{f+k}{T}\right) \tag{19.64}$$

The analog and digital frequencies are related by

$$f = \frac{\lambda T}{\pi}, \quad |f| \le 0.5 \tag{19.65}$$

From Equation 19.64, the digital filter frequency response is the sum of shifted versions of the analog filter frequency response. There is aliasing if $H_a(\lambda)$ is not zero for $|\lambda/2\pi| > 1/(2T)$. Therefore, the analog filter used in this method should have a frequency response that goes to zero quickly as λ goes to ∞. Because of the aliasing, this method cannot be used for designing a HPF. Writing the analog filter transfer function in the form

$$H_a(s) = \sum_{i=1}^{N} \frac{b_i}{(s - p_i)} \tag{19.66}$$

it follows that the analog impulse response is given by $h_a(t) = \sum_{i=1}^{N} b_i e^{p_i t}$ and the digital filter can be obtained as

$$H(z) = \sum_{n=0}^{\infty} h(n) z^{-n} = \sum_{n=0}^{\infty} h(nT) z^{-n}$$

$$= \sum_{i=1}^{N} b_i \sum_{n=0}^{\infty} (e^{p_i T} z^{-1})^n = \sum_{i=1}^{N} \frac{b_i}{1 - e^{p_i T} z^{-1}} \tag{19.67}$$

Therefore, an analog filter transfer function $H_a(s) = \sum_{i=1}^{N} b_i/(s - p_i)$ gets transformed to a digital filter system function $H(z) = \sum_{i=1}^{N} b_i/(1 - e^{p_i T} z^{-1})$, as shown in Table 19.4. Similar to the bilinear transformation, in a complete design process the choice of T has no effect (except for the final magnitude scaling factor).

19.2.3.2 Low-Pass Filters

We give one example in designing an LPF using the impulse invariant method and one example using the bilinear transformation. In this example, suppose that we wish to design a digital filter using an analog Butterworth prototype filter. The digital filter specification is

$$20 \log |H(f)| \ge -2\,\text{dB} \quad \text{for } 0 \le f \le 0.11$$
$$20 \log |H(f)| \le -10\,\text{dB} \quad \text{for } 0.2 \le f \le 0.5$$

where the log is of base 10. Therefore, we have $\varepsilon = 0.7648$, $\delta = 0.3162$, $f_p = 0.11$, and $f_s = 0.2$. Let us use the impulse invariant method. Therefore, the analog passband edge and stopband edge are $\lambda_p = 0.22\pi/T$ and $\lambda_s = 0.4\pi/T$, respectively. We use the same ripple requirements: $\varepsilon = 0.7648$ and $\delta = 0.3162$. Using these values, a Butterworth filter order is calculated from Equation 19.30, yielding $N \ge 2.3$. So, we choose $N = 3$. With $\lambda_c = \lambda_p \varepsilon^{-1/N} = 0.2406\pi/T$, we find the analog filter transfer function to be

$$H_a(s) = \frac{\lambda_c^3}{(s + \lambda_c)(s^2 + \lambda_c s + \lambda_c^2)}$$

$$= \lambda_c \left[\frac{1}{s + \lambda_c} + \frac{-0.5 - j0.5/\sqrt{3}}{s + 0.5(1 - j\sqrt{3})\lambda_c} + \frac{-0.5 + j0.5/\sqrt{3}}{s + 0.5(1 + j\sqrt{3})\lambda_c} \right]$$

$$= \frac{0.7559}{T} \left[\frac{1}{s + 0.7559/T} + \frac{-0.5 - j0.5/\sqrt{3}}{s + 0.3779(1 - j\sqrt{3})/T} + \frac{-0.5 + j0.5/\sqrt{3}}{s + 0.3779(1 + j\sqrt{3})/T} \right]$$

Using Equations 19.66 and 19.67 we obtain the digital filter system function:

$$H(z) = \frac{0.7559}{T}\left[\frac{1}{1 - e^{-0.7559}z^{-1}} + \frac{-0.5 - j0.5/\sqrt{3}}{1 - e^{-0.3779(1-j\sqrt{3})}z^{-1}} + \frac{-0.5 + j0.5/\sqrt{3}}{1 - e^{-0.3779(1-j\sqrt{3})}z^{-1}}\right]$$

$$= \frac{0.7559}{T}\left[\frac{1}{1 - 0.4696z^{-1}} - \frac{1 - 0.7846z^{-1}}{1 - 1.0873z^{-1} + 0.4696z^{-2}}\right]$$

Due to aliasing, the maximum value of the resulting magnitude response (which is at $f = 0$ or $z = 1$) is no longer equal to 1, although the analog filter has maximum magnitude (at $\lambda = 0$ or $s = 0$) of 1. Note that the choice of T affects only the scaling factor, which is only a constant gain factor. If we adjust the system function so that the maximum magnitude is 1, that is, $|H(f)| = 1$, we have

$$H(z) = \frac{0.7565}{T}\left[\frac{1}{1 - 0.4696z^{-1}} - \frac{1 - 0.7846z^{-1}}{1 - 1.0873z^{-1} + 0.4696z^{-2}}\right]$$

The magnitude response in decibels and the phase response in degrees are plotted in Figure 19.15. From the result, $|H(f)| = -1.97\,\text{dB}$ at $f = 0.11$ and $|H(f)| = -13.42\,\text{dB}$ at $f = 0.2$; both satisfy the desired specification. The aliasing in this example is small enough that the resulting response still meets the specification. It is possible that the aliasing is large enough that the designed filter does not meet the specification. To compensate for the unknown aliasing, we may want to use smaller ε and δ in designing the analog prototype filter.

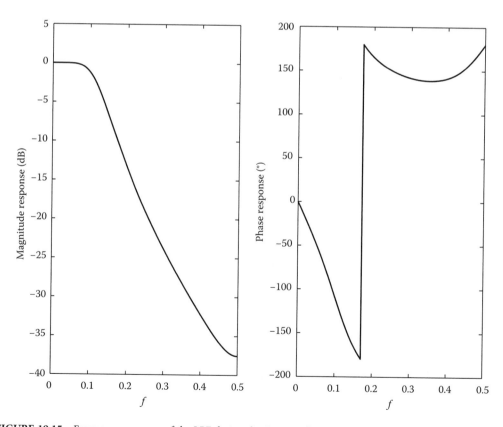

FIGURE 19.15 Frequency response of the LPF designed using impulse invariant method.

In this next example, we demonstrate the design method using bilinear transformation, with an analog elliptic prototype filter. Let the desired filter specification be

$$|H(f)|^2 \geq 0.8 \ (\text{or } -0.97\,\text{dB}) \quad \text{for } 0 \leq f \leq 0.1125$$

$$|H(f)|^2 \leq 2.5 \times 10^{-5} \ (\text{or } -46.02\,\text{dB}) \quad \text{for } 0.15 \leq f \leq 0.5$$

which means $\varepsilon = 0.5$, $\delta = 0.005$, $f_p = 0.1125$, and $f_s = 0.15$. For bilinear transformation, we calculate the analog passband and stopband edges as $\lambda_p = (2/T)\tan(\pi f_p) = 0.7378/T$ and $\lambda_s = (2/T)\tan(\pi f_s) = 1.0190/T$, respectively. Therefore, $\lambda_p/\lambda_s = 0.7240$. From Equation 19.60, we obtain the order $N \geq 4.8$. So, we use $N = 5$. The analog elliptic filter transfer function is calculated from Equations 19.50 through 19.59 to be

$$H_a(s) = \frac{7.8726 \times 10^{-3}\left[\left(\frac{sT}{2}\right)^2 + 0.6006\right]\left[\left(\frac{sT}{2}\right)^2 + 0.2782\right]}{\left[\left(\frac{sT}{2}\right) + 0.1311\right]\left[\left(\frac{sT}{2}\right)^2 + 0.1689\left(\frac{sT}{2}\right) + 0.0739\right]\left[\left(\frac{sT}{2}\right)^2 - 0.0457\left(\frac{sT}{2}\right) + 0.1358\right]}$$

To convert to digital filter system function, we replace s with $(2/T)(1 - z^{-1})/(1 + z^{-1})$. Equivalently, we replace $sT/2$ with $(1 - z^{-1})/(1 + z^{-1})$, yielding

$$H(z) = \frac{1.0511 \times 10^{-2}(1 + z^{-1})(1 - 0.4991Z^{-1} + z^{-2})(1 - 1.1294z^{-1} + z^{-2})}{(1 - 0.7682z^{-1})(1 - 1.4903z^{-1} + 0.7282z^{-2})(1 - 1.5855z^{-1} + 1.0838z^{-2})}$$

Note that the choice of T has no effect on the resulting system function. The magnitude response in decibels and the phase response are plotted in Figure 19.16, which satisfies the desired magnitude specification. Note the equiripples in both the passband and the stopband.

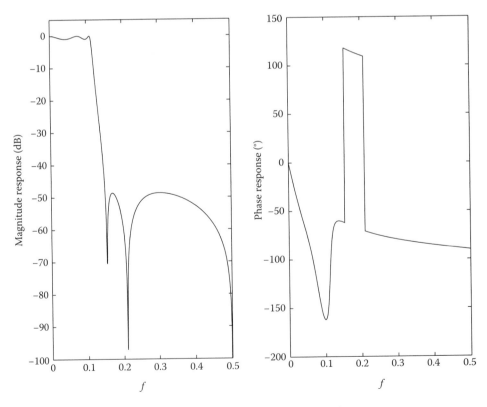

FIGURE 19.16 Frequency response of the LPF designed using bilinear transformation.

19.2.3.3 High-Pass Filters

As mentioned above, the impulse invariant method is not suitable for HPFs due to aliasing. Therefore, we only discuss the bilinear transformation. In addition to the procedure used with designing an LPF, we need to transform the analog high-pass specification to analog low-pass specification and transform the resulting analog LPF to analog HPF. There is a simple transformation for this job: replacing s in the analog LPF transfer function with $1/s$. In terms of the frequency, $j\lambda$ becomes $1/j\lambda = j(-1/\lambda)$, i.e., a low frequency is changed to a (negative) high frequency. Therefore, an analog LPF becomes an analog HPF. When combined with the bilinear transformation, this process gives the transformation

$$s = \frac{T}{2}\frac{(1+z^{-1})}{(1-z^{-1})} \quad \text{or} \quad z = \frac{s+(T/2)}{s-(T/2)} \tag{19.68}$$

Writing $s = j\lambda$, we can show that $z = \exp\{j[2\arctan(2\lambda/T) - \pi]\}$. With $z = \exp(j2\pi f)$, we have

$$\lambda = \frac{T}{2}\tan[\pi(f + 0.5)] \tag{19.69}$$

To write f in terms of λ, we can show that, after adjusting the range of f to $[-1/2, 1/2]$,

$$f = \begin{cases} -\frac{1}{2} + \frac{1}{\pi}\arctan(2\lambda/T) & \lambda > 0 \\ +\frac{1}{2} + \frac{1}{\pi}\arctan(2\lambda/T) & \lambda < 0 \end{cases} \tag{19.70}$$

Equations 19.69 and 19.70 give the relation between the digital frequency and the analog frequency, corresponding to the transformation in Equation 19.68. This relation is plotted in Figure 19.17, from which we see that a low digital frequency corresponds to a high analog frequency and vice versa.

We can summarize the design steps as follows. Given a digital HPF specification as in Figure 19.18, it is converted to an analog LPF specification using Equation 19.70 to obtain the passband and stopband edges λ_p and λ_s, from f_p and f_s, respectively. With λ_p, λ_s, ε, and δ, we design the low-pass analog prototype filter. Let the transfer function be $H_a(s)$. This transfer function is then converted to digital HPF system function by replacing s with $(T/2)(1 + z^{-1})/(1 - z^{-1})$.

Note that this corresponds to the procedure in Figure 19.13, with the bypass of the "analog specification" block and the "transform to analog filter" block, as indicated by the dotted lines in Figure 19.13.

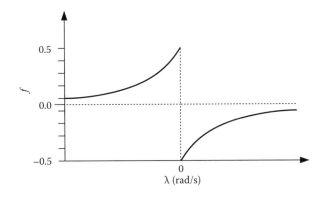

FIGURE 19.17   The relation for designing an HPF.

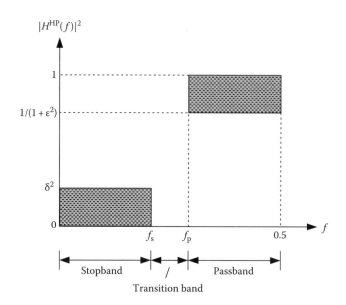

FIGURE 19.18   Digital HPF specification.

As an example, consider designing a digital HPF with the following specification:

$$|H^{HP}(f)|^2 \geq 0.8 \ (\text{or} \ -0.97 \, \text{dB}) \quad \text{for} \ 0.4 \leq f \leq 0.5$$
$$|H^{HP}(f)|^2 \leq 2.5 \times 10^{-5} \ (\text{or} \ -46.02 \, \text{dB}) \quad \text{for} \ 0 \leq f \leq 0.3$$

Since T does not affect the result, we let $T = 2$ for convenience. We calculate the analog LPF passband and stopband edges as $\lambda_p = \tan[\pi(0.5 + f_p)] = -0.3249 \, \text{rad/s}$ and $\lambda_s = \tan[\pi(0.5 + f_s)] = -0.7265 \, \text{rad/s}$. Since the magnitude response is in symmetry with respect to $\lambda = 0$, we use $\lambda_p = 0.3249 \, \text{rad/s}$ and $\lambda_s = 0.7265 \, \text{rad/s}$. Therefore, $\lambda_s/\lambda_p = 2.2361$. Suppose that we choose the inverse Chebyshev filter as the analog prototype filter. From Equation 19.37, we obtain the order $N \geq 4.6$. So, we use $N = 5$. From Equation 19.40, the low-pass analog inverse Chebyshev filter transfer function is

$$H_a(s) = \frac{1.8160 \times 10^{-2}(s^2 + 0.5835)(s^2 + 1.5276)}{(s + 0.4822)(s^2 + 0.6772s + 0.2018)(s^2 - 0.2131s + 0.1663)}$$

To convert to digital filter system function, we replace s with $(1 + z^{-1})/(1 - z^{-1})$, yielding

$$H^{HP}(s) = \frac{1.8160 \times 10^{-2}(1 - z^{-1})(1 + 0.5261z^{-1} + z^{-2})(1 - 0.4175z^{-1} + z^{-2})}{(1 + 0.3493z^{-1})(1 + 0.8498z^{-1} + 0.2792z^{-2})(1 + 1.2088z^{-1} + 0.6910z^{-2})}$$

The magnitude response and the phase response are plotted in Figure 19.19.

19.2.3.4 Bandpass Filters

A magnitude response specification for a digital BPF is depicted in Figure 19.20a. Note that there are two passband edges (f_{p1} and f_{p2}) and two stopband edges (f_{s1} and f_{s2}). For the bilinear transformation $s = (2/T)(1 - z^{-1})/(1 - z^{-1})$ we can transform the digital BPF specification to an analog BPF specification by letting

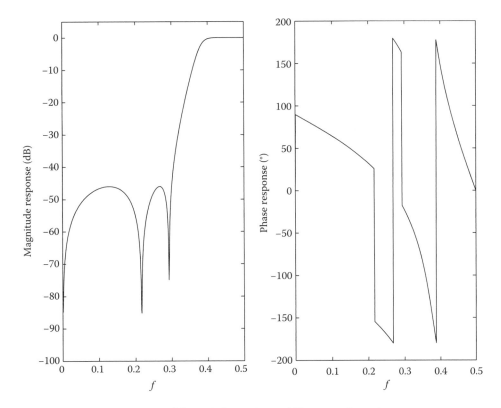

FIGURE 19.19 Frequency response of the HPF designed using bilinear transformation.

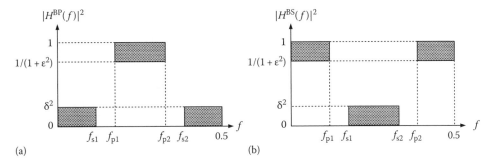

FIGURE 19.20 Magnitude specifications for digital BPF and BSF: (a) digital BPF specification and (b) digital BSF specification.

$$\lambda_{p1} = 2T \tan{(\pi f_{p1})} \quad \lambda_{p2} = 2T \tan{(\pi f_{p2})}$$
$$\lambda_{s1} = 2T \tan{(\pi f_{s1})} \quad \lambda_{s2} = 2T \tan{(\pi f_{s2})} \tag{19.71}$$

and keeping the same ε and δ.

Now, we need a transformation between an analog BPF and an analog LPF. To distinguish between the variable s and λ for the two filters, let us use s' and λ' for the analog LPF and s and λ for the analog BPF, respectively. A transformation for converting an analog LPF to an analog BPF is given by

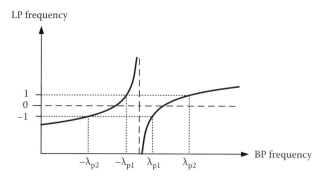

FIGURE 19.21   Relation between λ and λ′ for bandpass-to-low-pass conversion.

$$s' = \frac{s^2 + \lambda_0^2}{Ws} \quad \text{or} \quad \lambda' = \frac{\lambda^2 - \lambda_0^2}{W\lambda} \tag{19.72}$$

where

$$W = \lambda_{p2} - \lambda_{p1} \quad \text{and} \quad \lambda_0^2 = \lambda_{p1}\lambda_{p2} \tag{19.73}$$

Figure 19.21 depicts an example of the relation between λ and λ′. Note that λ_{p1} and λ_{p2} get mapped to $\lambda' = -1$ and $+1$, respectively. Therefore, the analog LPF has a passband edge of 1. The values of λ_{s1} and λ_{s2} get mapped to $\lambda'_{s1} = -|(\lambda_{s1}^2 - \lambda_0^2)/(W\lambda_{s1})|$ and $\lambda'_{s2} = |(\lambda_{s2}^2 - \lambda_0^2)/(W\lambda_{s2})|$. However, these two values may not be negative of each other. Since the analog LPF must have a symmetric magnitude response, we must use the more stringent of the two stopband edges, i.e., the smaller of $|\lambda'_{s1}|$ and $|\lambda'_{s2}|$. Letting

$$\lambda'_s = \min\{|\lambda'_{s1}|, |\lambda'_{s2}|\} = \min\left\{\left|\frac{\lambda_{s1}^2 - \lambda_0^2}{W\lambda_{s1}}\right|, \left|\frac{\lambda_{s2}^2 - \lambda_0^2}{W\lambda_{s2}}\right|\right\} \tag{19.74}$$

we now have the analog LPF specification. Therefore, a prototype analog LPF can be designed.

The design process can be summarized as follows. First, the desired digital BPF magnitude specification is converted to an analog BPF magnitude specification using Equation 19.71. Then the analog BPF specification is converted to an analog LPF specification using λ'_s calculated from Equation 19.74 and $\lambda'_p = 1$. Next, a prototype analog LPF is designed with the values of $\varepsilon, \delta, \lambda'_p = 1$, yielding an analog LPF transfer function $H_a(s')$. The LPF transfer function is converted to an analog BPF transfer function $H_a^{HP}(s)$, using the transformation (from s' to s) given in Equation 19.72. Finally, the analog BPF transfer function is converted to a digital BPF transfer function $H^{BP}(z)$ using the bilinear transformation $s = (2/T)(1 - z^{-1})/(1 + z^{-1})$. As before, the value of T does not affect the result.

For example, let the desired digital BPF have the following specification:

$$|H^{BP}(f)|^2 \begin{cases} \geq 0.8 \ (-0.97 \ \text{dB}) & \text{for } 0.25 \leq f \leq 0.3 \\ \leq 2.5 \times 10^{-5} \ (-46.02 \ \text{dB}) & \text{for } 0 \leq f \leq 0.2 \text{ and } 0.35 \leq f \leq 0.5 \end{cases}$$

which means $\varepsilon = 0.5, \delta = 0.005, f_{p1} = 0.25, f_{p2} = 0.3, f_{s1} = 0.2$, and $f_{s2} = 0.35$. Let $T = 2$ for convenience. Using $\lambda = \tan(\pi f)$, we obtain the analog BPF passband and stopband edges as $\lambda_{p1} = \tan(\pi f_{p1}) = 1.0 \ \text{rad/s}$, $\lambda_{p2} = \tan(\pi f_{p2}) = 1.3764 \ \text{rad/s}$, $\lambda_{s1} = \tan(\pi f_{s1}) = 0.7265 \ \text{rad/s}$, and $\lambda_{s2} = \tan(\pi f_{s2}) = 1.9626 \ \text{rad/s}$. Therefore, $\lambda_0^2 = 1.3764$ and $W = 0.3764$. So, we have $\lambda'_s = \min\{3.1030, 3.3509\} = 3.1030 \ \text{rad/s}$. Suppose that we use the elliptic LPF as an analog prototype filter. With $\varepsilon = 0.5$, $\delta = 0.005$, $\lambda_p' = 1$, and $\lambda_s' = 3.1030\,\text{rad/s}$, we need an elliptic filter of order $N = 3$. The low-pass analog elliptic filter transfer function is

$$H_a(s') = \frac{4.1129 \times 10^{-2}(s'^2 + 12.6640)}{(s' + 0.5174)(s'^2 + 0.763s' + 1.0067)}$$

Replacing s' with $(s^2 + 1.3764)/(0.3764s)$ yields the analog BPF transfer function

$$H_a^{BP}(s) = \frac{1.5480 \times 10^{-2}s(s^4 + 4.5467s^2 + 1.8944)}{(s^2 + 0.1947s + 1.3764)(s^4 + 0.1793s^3 + 2.8953s^2 + 0.2467s + 1.8944)}$$

Note that an Nth-order LPF becomes a $2N$th-order BPF. To convert to digital filter system function, we replace s with $(1 - z^{-1})/(1 + z^{-1})$, yielding

$$H^{BP}(z) = \frac{7.2077 \times 10^{-3}(1 - z^{-2}) \times (1 + 0.4807z^{-1} + 1.1117z^{-2} + 0.4807z^{-3} + z^{-4})}{(1 + 0.2928z^{-1} + 0.8485z^{-2}) \times (1 + 0.5973z^{-1} + 1.8623z^{-2} + 0.5539z^{-3} + 0.8629z^{-4})}$$

The magnitude and phase responses are plotted in Figure 19.22.

Note that for the transformation in Equation 19.72, we can also let $W = \lambda_{s2} - \lambda_{s1}$ and $\lambda_0^2 = \lambda_{s1}\lambda_{s2}$, instead of Equation 19.73. Such a choice will give $\lambda_s' = 1$. The passband edge for the prototype LPF is now calculated from $\lambda_p' = \min\{|(\lambda_{p1}^2 - \lambda_0^2)/(W\lambda_{p1})|\ |\lambda_{p2}^2 - \lambda_0^2)/(W\lambda_{p2})|\}$.

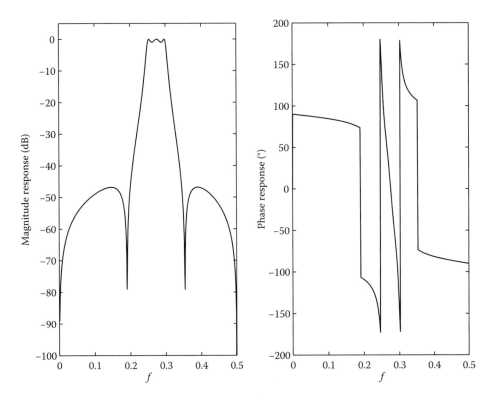

FIGURE 19.22 Frequency response of the designed digital BPF.

19.2.3.5 Bandstop Filters

A digital BSF specification is depicted in Figure 19.20b. As in the case of the BPF there are two passband edges (f_{p1} and f_{p2}) and two stopband edges (f_{s1} and f_{s2}). A transformation from analog BSF to analog LPF is given by

$$s' = \frac{Ws}{s^2 + \lambda_0^2} \quad \text{or} \quad \lambda' = \frac{W\lambda}{\lambda_0^2 - \lambda^2} \tag{19.75}$$

where W and λ_0^2 are given by Equation 19.73. Note that the expression for s in Equation 19.75 is the reciprocal of that in Equation 19.72. The relation between the LPF frequency λ' and the BSF frequency λ is depicted in Figure 19.23. The passband edges λ_{p1} and λ_{p2} get mapped to $\lambda' = 1$ and -1, respectively. The values of λ_{s1} and λ_{s2} get mapped to $\lambda'_{s1} = W\lambda_{s1}/(\lambda_0^2 - \lambda_{s1}^2)$ and $\lambda'_{s2} = W\lambda_{s2}/(\lambda_0^2 - \lambda_{s2}^2)$. Therefore, the passband edge and stopband edge of the prototype analog LPF are 1 and λ'_s, respectively, where

$$\lambda'_s = \min\{|\lambda'_{s1}|, |\lambda'_{s2}|\} = \min\left\{\left|\frac{W_{s1}\lambda}{\lambda_0^2 - \lambda_{s1}^2}\right|, \left|\frac{W\lambda_{s2}}{\lambda_0^2 - \lambda_{s2}^2}\right|\right\} \tag{19.76}$$

The design process for the BSF can follow the same process as the design for the BPF, except that we use Equations 19.75 and 19.76 instead of Equations 19.62 and 19.64.

Similar to the case of the BPF, we can also let $W = \lambda_{s2} - \lambda_{s1}$ and $\lambda_0^2 = \lambda_{s1}\lambda_{s2}$ instead of Equation 19.73, for the transformation in Equation 19.75. The stopband edge and the passband edge for the prototype LPF are now $\lambda'_s = 1$ and $\lambda'_p = \min\{|W\lambda_{p1}/(\lambda_0^2 - \lambda_{p1}^2)|, |W\lambda_{p2}/(\lambda_0^2 - \lambda_{p2}^2)|\}$.

19.2.4 Design Using Digital Frequency Transformations

This method assumes that we can design a digital LPF. The desired filter is then obtained from the digital LPF by transforming the digital LPF in the z domain. Let us denote the z variable for the digital LPF by z' and that for the desired digital filter by z. Similarly, we use f' for the digital frequency of the digital LPF and f for the frequency of the desired digital filter. Suppose that the digital LPF has system function $H(z')$ and the desired digital filter has system function $H^x(z)$, where x stands for LP, HP, BP, or BS. The system function $H^x(z)$ is obtained from $H(z')$ by replacing z' with an appropriate function of z. The LPF $H(z')$ can be designed using the method discussed in the Section 19.2.3, or by some other means. The specification for the digital LPF is obtained from the specification of the desired digital filter through the relation between f' and f. The relation depends on the specific transformation. Note that the

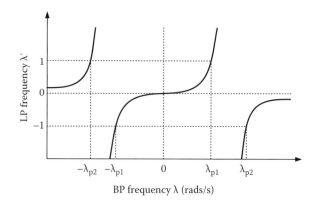

FIGURE 19.23   Relation between λ and λ' for bandstop-to-lowpass conversion.

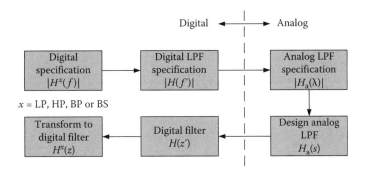

FIGURE 19.24 Design process using digital frequency transformation.

difference between the method in this section and the method described above is that the transformation between the desired type of filter and the LPF is in the digital domain (the z domain) for the current method whereas it is in the analog domain (the s domain) in the previous method. Figure 19.24 shows the design process using digital frequency transformation. The advantage of the current method is that in designing a desired digital HPF, BPF, or BSF, we design a digital LPF, which can make use of the impulse invariant method, in addition to the bilinear transformation. This is not the case for the method discussed previously, due to excessive aliasing.

19.2.4.1 Low-Pass Filters

We can transform a digital LPF to a digital LPF using the transformation

$$z' = \frac{z + \alpha}{1 + \alpha z} \quad |\alpha| < 1 \tag{19.77}$$

With $z = \exp(j2\pi f)$ and $z' = \exp(j2\pi f')$, we can show that the digital frequencies are related by

$$f' = \frac{1}{2\pi} \arctan \left[\frac{(1 - \alpha^2) \sin 2\pi f}{2\alpha + (1 + \alpha^2) \cos 2\pi f} \right] \tag{19.78}$$

The relation given by Equation 19.78 is plotted in Figure 19.25a. If $\alpha = 0$, then $z' = z$ and $f' = f$, which is the trivial case. When $\sigma \neq 0$, there is frequency warping introduced by the transformation. After choosing a, we can transform the desired digital LPF specification to another digital LPF specification, i.e., calculate f'_p and f'_s from f_p and f_s. With f_p, f_s, ε, and δ, a digital LPF can then be designed to satisfy the specification. The resulting system function is then transformed to the desired LPF using Equation 19.77. This method may yield a filter of lower or higher order (due to the frequency warping) compared to the case that there is no digital frequency transformation ($\alpha = 0$).

As an alternative to specifying α, we can specify f'_p, which, together with f_p, specifies the value of α, according to Equation 19.68. With the value of α, we can calculate f'_s from f_s. We can also exchange the role of the passband edge with the stopband edge, i.e., we specify f'_s and compute α from the values of f'_s and f_s. With α, f'_p can be determined from f_p.

19.2.4.2 High-Pass Filters

We can transform a digital LPF to a digital HPF using the transformation

$$z' = -\frac{z + \alpha}{1 + \alpha z} \quad |\alpha| < 1 \tag{19.79}$$

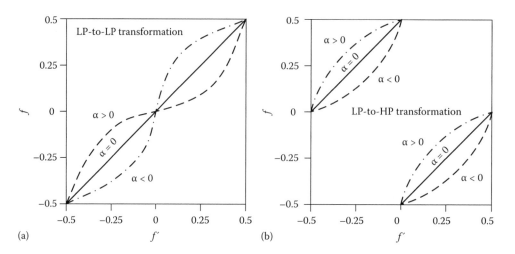

FIGURE 19.25 Frequency relation for digital frequency transformations: (a) lowpass to lowpass and (b) lowpass to highpass.

With $z = \exp(j2\pi f)$ and $z' = \exp(j2\pi f')$, we obtain the relation between the two digital frequencies:

$$f' = \frac{1}{2\pi} \arctan\left[\frac{-(1-\alpha^2)\sin 2\pi f}{-2\alpha - (1+\alpha^2)\cos 2\pi f}\right] \tag{19.80}$$

The relation is plotted in Figure 19.25b. If $\alpha = 0$, then $z' = -z$ and $f' = f + 0.5$. The design process proceeds as follows. After choosing the value of α, the desired digital HPF specification is transformed to the digital LPF specification, using the relation in Equation 19.80. Using the resulting values of f'_p and f'_s, together with the ripple specifications (ε and δ), a digital LPF is designed to satisfy the specification. The resulting LPF system function $H(z')$ is then transformed to the desired HPF by substituting z' with $-(z + \alpha)/(1 + \alpha z)$, given by Equation 19.79. Similar to the case of the LPF, we can specify f'_p (instead of α) and calculate the required value of α from f'_p and f_p.

19.2.4.3 Bandpass Filters

To transform a digital LPF to a digital BPF using the transformation

$$z' = -\frac{1 + \frac{2\alpha k}{k+1} z + \frac{k-1}{k+1} z^2}{\frac{k-1}{k+1} + \frac{2\alpha k}{k+1} z + z^2} \quad |\alpha| < 1, k > 0 \tag{19.81}$$

This implies the following relation between the two digital frequencies:

$$f' = \frac{1}{2\pi} \arctan\left[\frac{(1-b)\{2a\sin 2\pi f + (1+b)\sin 4\pi f\}}{-a^2 - 2b - 2a(1+b)\cos 2\pi f - (b^2+1)\cos 4\pi f}\right] \tag{19.82}$$

where $a = 2\alpha k/(k+1)$ and $b = (k-1)/(k+1)$. An example of Equation 19.82 is plotted in Figure 19.26. If $\alpha = 0$, the curve would be odd symmetric with respect to $f = 0.25$. The design process is similar to the case of the HPF, except that there are now two passband edges, f'_{pi} and $|f'_{p2}|$, and two stopband

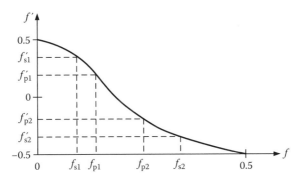

FIGURE 19.26 Frequency relation for BP to LP digital frequency transformation.

edges, f'_{s1} and $|f'_{p2}|$, for the digital LPF to satisfy (see Figure 19.25). To satisfy both sets, we let $f'_p = \max\{f'_{p1}, |f'_{p2}|\}$ and $f'_s = \min\{f_{s1}, |f'_{s2}|\}$ be the passband and stopband edges for the digital LPF filter. If we specify f'_p, then together with f_{p1} and f_{p2}, they determine the values of α and k:

$$\alpha = \frac{\cos[\pi(f_{p2} + f_{p1})]}{\cos[\pi(f_{p2} - f_{p1})]} \quad k = \cot[\pi(f_{p2} - f_{p1})]\tan(\pi f'_p) \tag{19.83}$$

With the values of α and k, we calculate the values of f'_{s1} and f'_{s2}, from Equation 19.82 and let $f'_s = \min\{f'_{s1}, |f'_{s2}|\}$. Thus we have f'_p, f'_s, ε, and δ as the digital LPF specification. After a digital LPF is designed to satisfy this specification, it is converted to digital BPF by the transformation in Equation 19.81.

19.2.4.4 Bandstop Filters

To transform a digital LPF to a digital BSF, we can use

$$z' = \frac{1 + \frac{2\alpha k}{k+1}z + \frac{k-1}{k+1}z^2}{\frac{k-1}{k+1} + \frac{2\alpha k}{k+1}z + z^2} \quad |\alpha| < 1, k > 0 \tag{19.84}$$

The corresponding relation between the two digital frequencies is

$$f' = \frac{1}{2\pi}\arctan\left[-\frac{(1-b)\{2a\sin 2\pi f + (1+b)\sin 4\pi f\}}{a^2 + 2b + 2a(1+b)\cos 2\pi f + (b^2 + 1)\cos 4\pi f}\right] \tag{19.85}$$

where $a = 2\alpha k/(k+1)$ and $b = (k-1)/(k+1)$. An example is plotted in Figure 19.27. The design process is the same as described in Section 19.2.4.3.

When f'_{pi} is specified, together with f_{p1} and f_{p2}, the values of α and k can be calculated from

$$\alpha = \frac{\cos[\pi(f_{p2} + f_{p1})]}{\cos[\pi(f_{p2} - f_{p1})]} \quad k = \tan[\pi(f_{p2} - f_{p1})]\tan(\pi f'_p) \tag{19.86}$$

With these values, we calculate the values of f'_{s1} and f'_{s2} from Equation 19.85. Letting $f'_s = \min\{f'_{s1}, |f'_{s2}|\}$, we have f'_p, f'_s, ε, and δ, which constitute the digital LPF specification. A digital LPF is then designed and converted to digital BSF by the transformation in Equation 19.84.

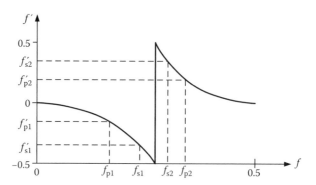

FIGURE 19.27 Frequency relation for BS to LP digital frequency transformation.

19.2.5 Computer-Aided Design

The general idea is to use an algorithm to search for coefficients such that the resulting response (magnitude and/or phase) is "close" to the desired response. The "closeness" is in some well-defined sense. The advantage of such method is that it can be used to design nonstandard filters such as multiband filters, phase equalizers, differentiators, etc. However, it requires a computer program to execute the algorithm. In addition, it usually cannot directly determine the filter order such that the passband and stopband ripples are within the desired ranges. The order is usually determined through several trials.

A large number of procedures have been studied for designing IIR filter in literatures. They can be classified into two categories. One is the indirect approach [2] which can be divided into two steps. First, an FIR filter that meets the required specifications is designed. Then, a lower order IIR filter is obtained, which maintains the original magnitude and phase specifications by applying model reduction techniques. The other is the direct approach in which IIR filters are directly designed to meet the frequency response specification using the least square criterion or Chebyshev (minimax) criterion. Several techniques are employed to achieve this propose. We review some of them in this section.

19.2.5.1 Least Squares ($\mathcal{L}_2$-Norm) Criterion

Consider the transfer function of an IIR filter given as

$$H(z) = \frac{Y(z)}{X(z)} = \frac{b_0 + b_1 z^{-1} + \cdots + b_M z^{-M}}{1 + a_1 z^{-1} + \cdots + a_N z^{-N}} \qquad (19.87)$$

where $X(z)$ and $Y(z)$ are the denominator and numerator polynomials, respectively. We wish to design the frequency response $H(e^{j\omega})$ such that it optimally approximates the desired frequency response $D(e^{j\omega})$. The filter coefficients can be found by minimizing the weighted square error

$$E(\omega) = \int W^2(\omega)|D(e^{j\omega}) - H(e^{j\omega})|^2 d\omega \qquad (19.88)$$

where $W(\omega)$ denotes a nonnegative weighting function, which is zero in the transition band. The weighted least squares (WLS) method has been successfully applied to design an FIR filter [1] where its coefficients are obtained from a well-known least squares solution. However, for IIR filter design, the WLS criterion in Equation 19.88 is no longer a quadratic form of the filter coefficients. Therefore, nonlinear optimization techniques with some stability constraints are required to solve the cost function (Equation 19.88). However, the global minimum of $E(\omega)$ may not be guaranteed. We will leave the stability issue for the last part of this section.

To avoid the complexity of using nonlinear optimization techniques, the iterative procedure replaces the cost function as

$$E_k(\omega) = \int |\varepsilon_k(\omega)|^2 d\omega$$

$$= \int \frac{W^2(\omega)}{|X_{k-1}(e^{j\omega})|^2} |D(e^{j\omega})X_k(e^{j\omega}) - Y_k(e^{j\omega})|^2 d\omega \qquad (19.89)$$

where $X_k(e^{j\omega})$ and $Y_k(e^{j\omega})$ are polynomials to be determined in the kth iteration. Several methods have been proposed to solve this problem in the literature and we present a few of them here.

19.2.5.1.1 Adaptive Weighted Least Squares

The adaptive WLS method for IIR filter design can be formulated from the error function $E_k(\omega)$ in Equation 19.89

$$E_k(\omega) = \frac{W(\omega)}{X_{k-1}(e^{j\omega})} [D(e^{j\omega})X_k(e^{j\omega}) - Y_k(e^{j\omega})] \qquad (19.90)$$

By defining the following vectors

$$\mathbf{c}_0 = \left[e^{-j\omega} \cdots e^{-j\omega N} \right]^{\mathrm{T}}$$

$$\mathbf{c}_1 = \left[1 \ e^{-j\omega} \cdots e^{-j\omega M} \right]^{\mathrm{T}}$$

$$\mathbf{a}_k = \left[a_1^{(k)} \cdots a_N^{(k)} \right]^{\mathrm{T}}$$

$$\mathbf{b}_k = \left[b_0^{(k)} \cdots b_M^{(k)} \right]^{\mathrm{T}}$$

the error function can be represented in matrix form as

$$E_k(\omega) = \frac{W(\omega)}{1 + \mathbf{c}_0^{\mathrm{T}}(\omega)\mathbf{a}_{k-1}} [D(e^{j\omega}) + D(e^{j\omega})\mathbf{c}_0^{\mathrm{T}}(\omega)\mathbf{a}_k - \mathbf{c}_1^{\mathrm{T}}(\omega)\mathbf{b}_k] \qquad (19.91)$$

where $X_k(e^{j\omega}) = 1 + \mathbf{c}_0^{\mathrm{T}}(\omega)\mathbf{a}_k$ and $Y_k(e^{j\omega}) = \mathbf{c}_1^{\mathrm{T}}(\omega)\mathbf{b}_k$. Note that the superscript T denotes the transposition operation. Furthermore, if the above error function is evaluated on a dense frequency grid, the following vector equation can be formed

$$\mathbf{e}_k = \mathbf{W}_k(\mathbf{d} - \mathbf{C}\mathbf{x}_k) \qquad (19.92)$$

where

$$\mathbf{e}_k = \left[\varepsilon_k(e^{j\omega_1}) \ \varepsilon_k(e^{j\omega_2}) \cdots \right]^{\mathrm{T}}$$

$$\mathbf{W}_k = \mathrm{diag}[W_k(\omega_1) W_k(\omega_2) \cdots]$$

$$\mathbf{d} = \left[D(e^{j\omega_1}) \ D(e^{j\omega_2}) \cdots \right]$$

$$\mathbf{C} = \begin{bmatrix} -e^{-j\omega_1}D(e^{j\omega_1}) & \cdots & -e^{-jN\omega_1}D(e^{j\omega_1}) & 1 & e^{-j\omega_1} & \cdots & e^{-jM\omega_1} \\ -e^{-j\omega_2}D(e^{j\omega_2}) & \cdots & -e^{-jN\omega_2}D(e^{j\omega_2}) & 1 & e^{-j\omega_2} & \cdots & e^{-jM\omega_2} \\ \vdots & \vdots & \vdots & \vdots & \vdots & \vdots & \vdots \end{bmatrix}$$

$$\mathbf{x}_k = \begin{bmatrix} \mathbf{a}_k \\ \mathbf{b}_k \end{bmatrix}$$

Here, the weighting function $\hat{W}_k(\omega)$ is defined as

$$\hat{W}_k(\omega) = \frac{W(\omega)}{|X_{k-1}(e^{j\omega})|} \tag{19.93}$$

The cost function in Equation 19.89 can be approximated as

$$E_k(\omega) = \|\varepsilon(\omega)\|_2^2 \approx \mathbf{e}_k^H \mathbf{e}_k \tag{19.94}$$

where superscript $(\cdot)^H$ denotes the conjugate transpose operation. Minimizing the cost function $E_k(\omega)$ leads to the well-known least squares solution,

$$\mathbf{x}_k = [\text{Re}(\mathbf{C}^H)\mathbf{W}_k\text{Re}(\mathbf{C}) + \text{Im}(\mathbf{C}^H)\mathbf{W}_k\text{Im}(\mathbf{C})]^{-1}[\text{Re}(\mathbf{C}^H)\mathbf{W}_k\text{Re}(\mathbf{d}) + \text{Im}(\mathbf{C}^H)\mathbf{W}_k\text{Im}(\mathbf{d})] \tag{19.95}$$

Once the filter coefficients are obtained, some weight updating procedures [1] are then applied to achieve the equiripple filter response.

19.2.5.1.2 Quadratic Programming

Using the same cost function in Equation 19.89, the problem can be formulated in a standard quadratic programming form as

$$E = \mathbf{x}_k^T \mathbf{H}_k \mathbf{x}_k + \mathbf{x}_k^T \mathbf{p}_k + \text{constant} \tag{19.96}$$

where

$$\mathbf{x}_k = \begin{bmatrix} \mathbf{a}_k \\ \mathbf{b}_k \end{bmatrix}, \quad \mathbf{H}_k = \begin{bmatrix} \mathbf{H}_{11} & \mathbf{H}_{12} \\ \mathbf{H}_{12}^T & \mathbf{H}_{22} \end{bmatrix}, \quad \text{and} \quad \mathbf{p}_k = \begin{bmatrix} \mathbf{p}_1 \\ \mathbf{p}_2 \end{bmatrix}$$

with

$$\mathbf{H}_{11} = \int \hat{W}_k(\omega)|D(e^{j\omega})|^2 (\mathbf{c}_0 \mathbf{c}_0^H) d\omega$$

$$\mathbf{H}_{12} = \int \hat{W}_k(\omega)\text{Re}(D(e^{j\omega})(\mathbf{c}_0 \mathbf{c}_1^H)) d\omega$$

$$\mathbf{H}_{22} = \int \hat{W}_k(\omega)(\mathbf{c}_1 \mathbf{c}_1^H) d\omega$$

$$\mathbf{p}_1 = \int \hat{W}_k(\omega)|D(e^{j\omega})|^2 \text{ Re}(\mathbf{c}_0) d\omega$$

$$\mathbf{p}_2 = -\int \hat{W}_k(\omega)\text{Re}(D(e^{j\omega})\mathbf{c}_1) d\omega$$

$$\hat{W}_k(\omega) = \frac{W^2(\omega)}{|X_{k-1}(e^{j\omega})|^2}$$

The stability and amplitude requirements are taken care of by imposing linear inequality constraints [4]. Filter coefficients can be found by minimizing the cost function (Equation 19.96) using standard quadratic programming tools.

19.2.5.1.3 Eigenfilter

The eigenfilter approach was proposed in Ref. [5] for designing linear-phase FIR filters. The goal is to express the cost function (Equation 19.89) in form of

$$E = \mathbf{h}^T \mathbf{P} \mathbf{h} \tag{19.97}$$

where
    $\mathbf{h}$ is a vector containing the unknown filter coefficients matrix
    $\mathbf{P}$ is symmetric, real, and positive-definite

If $\mathbf{h}$ has unit norm, i.e., $\mathbf{h}^T\mathbf{h} = 1$, to avoid trivial solution, then the optimal $\mathbf{h}$ which minimizes the cost function E is simply the eigenvector corresponding to the minimum eigenvalue of $\mathbf{P}$. The eigenfilter method can also be applied to design IIR filter. The problem may be formulated in two ways.

1. Time-domain approach
   In this method, the objective error function is formulated in time domain as

$$e(n) = \sum_{k=0}^{N} a(k)h_1(n-k) - \sum_{k=0}^{M} b(k)\delta(n-k) \tag{19.98}$$

   where $h_1(n)$ is the impulse response of a target transfer function $H(z) = H_1(z) + H_1(z^{-1})$. The filter $H_1(z)$ is stable and causal so that a noncausal implementation of the system is necessary. With some mathematical manipulation, the error function (Equation 19.98) can be expressed in the standard form of Equation 19.97. Therefore, the IIR filter coefficients $a(n)$ and $b(n)$ can be found. This method can also be extended to design 2-D IIR filters. However, to achieve good approximation in filter design, computation must be performed on matrices of very large size.

2. Frequency-domain approach
   In this approach, the error function in Equation 19.90 is used to formulate the problem in another way as

$$\varepsilon(\omega) = W(\omega)\mathbf{h}^T\mathbf{c} \tag{19.99}$$

   where

$$\begin{aligned}\mathbf{c} &= [D(e^{j\omega})\ D(e^{j\omega})e^{-j\omega}\ \cdots\ D(e^{j\omega})e^{-j\omega N} - 1 - e^{-j\omega}\ \cdots\ e^{-j\omega M}]^T \\ \mathbf{h} &= [a_0\ a_1\ \cdots\ a_N\ b_0\ b_1\ \cdots\ b_M]^T \end{aligned} \tag{19.100}$$

   Consequently, the cost function in Equation 19.89 can be expressed as

$$E = \int \mathbf{h}^T\mathbf{c}^*\mathbf{c}^T\mathbf{h}W(\omega)d\omega = \mathbf{h}^T\mathbf{P}\mathbf{h}, \quad \omega \in \Omega \tag{19.101}$$

   where
       superscript * denotes the conjugation operation
       Ω is the frequency region of interest

   the matrix

$$\mathbf{P} = \int \mathbf{c}^*\mathbf{c}^T W(\omega)d\omega \tag{19.102}$$

   is a symmetric, real, positive-definite matrix.

By subjecting $\mathbf{h}$ to the usual unit norm condition $\mathbf{h}^T\mathbf{h} = 1$, the optimum filter coefficients that minimize the cost function are the elements of the eigenvector of the matrix $\mathbf{P}$ corresponding to the minimum eigenvalue. The eigenfilter method can solve constrained filter design problem. However, to obtain equiripple filters, a weight adaptive procedure is needed.

19.2.5.2 Weighted Chebyshev ($\mathcal{L}_\infty$-Norm) Criterion

An IIR filter can also be formulated on weighted Chebyshev criterion. Filter coefficients are chosen such that its weighted Chebyshev (minimax) error between desired and actual frequency response is minimized. The iterative cost function in Equation 19.89 is now evaluated on Chebyshev criterion as

$$E = \max_{\omega \in \Omega} \frac{W(\omega)}{|X_{k-1}(e^{j\omega})|} |D(e^{j\omega})X_k(e^{j\omega}) - Y_k(e^{j\omega})|$$
$$= \max_{\omega \in \Omega} W_k(\omega)|D(e^{j\omega}) + \mathbf{x}_k^T\mathbf{s}| \qquad (19.103)$$

where

$$\mathbf{s} = [D(e^{j\omega})\mathbf{c}_0 - \mathbf{c}_1]^T$$

$\mathbf{x}_k$ is the coefficient vector previously defined in Equation 19.92.

The solution of the minimax problem in Equation 19.103 can be found by solving the following equivalent linear programming problem

$$\min_{\mathbf{x}_k} \varepsilon_k \qquad (19.104)$$

subject to

$$W_k(\omega)|D(e^{j\omega}) + \mathbf{x}_k^T\mathbf{s}| \leq \varepsilon_k \qquad (19.105)$$

With a stability constraint, the above linear programming problem can be arranged in standard form of linear programming technique and solved using off-the-shelf linear programming software.

19.2.6 Stability Issues

An IIR filter can be unstable if there are some poles outside the unit circle. However, in case that the phase response is not important in the design, an unstable IIR filter obtained from a design algorithm can be stabilized by conjugate reciprocal substitution of the unstable factor without changing its amplitude. If, however, phase response is a part of design specification, some other techniques must be used. There are four major approaches that can be applied for filter stabilization when optimization techniques previously described are used. The following is a summary. More details and references can be found in Ref. [6].

1. First approach is proposed by Deczky [3]. In this method, a standard gradient-based optimization is modified so that the searching trajectory is only inside the border of stability. However, the algorithm has high computation complexity. A standard optimization tool cannot be used.
2. In the second approach, the target frequency response is chosen such that the desired filter is stable as in Ref. [7]. The target filter is restricted and may be too difficult to obtain filter stabilization.

3. Third approach is to impose constraints on the denominator $X(z)$ of the filter transfer function. A stable IIR filter is achieved by imposing the real part of $X(z)$ to be nonnegative. However, using these methods, some optimal solution that is excluded from the constraints may not be found.
4. Fourth approach is to transform a transfer function such that its optimum solution always lies in the stable area [6]. However, this method leads to a very complicated cost function that requires nonlinear optimization methods.

References

1. Y. C. Lim, J. H. Lee, C. K. Chen, and R. H. Yang, A weighted least squares algorithm for quasiequiripple FIR and IIR digital filter design, *IEEE Trans. Signal Process.*, 40, 551–558, Mar. 1992.
2. B. Beliczynki, J. Kale, and G. D. Cain, Approximation of FIR by IIR digital filter: An algorithm based on balanced model reduction, *IEEE Trans. Signal Process.*, 40, 532–542, Mar. 1992.
3. A. Deczky, Synthesis of recursive filters using the minimum p-error criterion, *IEEE Trans. Audio Electroacoust*, AU-20, 257–263, 1972.
4. W. S. Lu, Design of stable IIR digital filter with equirripple passbands and peak-constrained least-squares stopbands, *IEEE Trans. Circuits Syst. II*, 46, 1421–1426, Nov. 1999.
5. P. P. Vaidyanathan and T. Q. Nguyen, Eigenfilter: A new approach to least-squares FIR filter design and applications including Nyquist filters, *IEEE Trans. Circuits Syst.*, CAS-34, 11–23, Jan. 1987.
6. A. Tarczynski, G. D. Cain, E. Hermanowicz, and M. Rojewski, A WISE method for designing IIR filters, *IEEE Trans. Signal Process.*, 49, 1421–1432, July 2001.
7. T. Kobayashi and S. Imai, Design of IIR digital filters with arbitrary log magnitude function by WLS techniques, *IEEE Trans. Acoust., Speech, Signal Process.*, ASSP-38, 247–252, Feb. 1990.

19.3 Lattice Digital Filters

Yong Ching Lim

19.3.1 Lattice Filters

There are several families of lattice structures for the implementation of IIR filters. Two of the most commonly encountered families are the tapped numerator structure shown in Figure 19.28 [3] and the injected numerator structure shown in Figure 19.29 [3]. It should be noted that not all the taps and injectors of the filters are nontrivial. For example, if $\lambda_i = 0$ for all i, the structure of Figure 19.28 simplifies to that of Figure 19.30 [1]. If $\phi_i = 0$ for $i > 0$, the structure of Figure 19.29 reduces to that of Figure 19.31. For both families, the denominator of the filter's transfer function is synthesized using a lattice network. The transfer function's numerator of the tapped numerator structure is realized by a

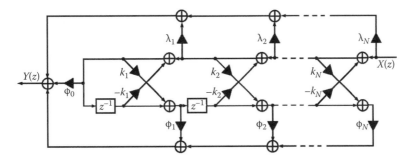

FIGURE 19.28 General structure of a tapped numerator lattice filter.

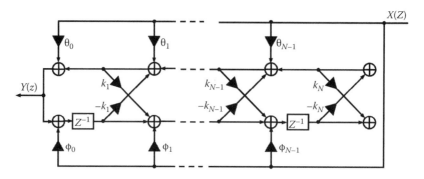

FIGURE 19.29   General structure of an injected numerator lattice filter.

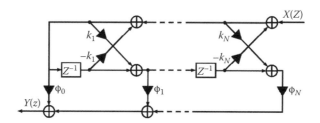

FIGURE 19.30   Structure of a tapped numerator lattice filter with $\lambda_i = 0$ for all i.

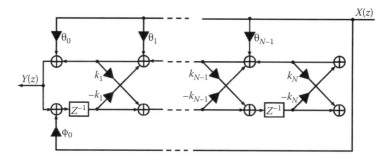

FIGURE 19.31   Structure of an injected lattice filter with $\phi_i = 0$ for $i > 0$.

weighted sum of the signals tapped from $N + 1$ appropriate points of the lattice. For the injected numerator structure, the transfer function's numerator is realized by weighting and injecting the input into $N + 1$ appropriate points on the lattice. The lattice itself may appear in several forms as shown in Figure 19.32 [1]. Figure 19.33 shows the structure of a third-order injected numerator filter synthesized using the one-multiplier lattice of Figure 19.32b.

19.3.2 Evaluation of the Reflection Coefficients k_n [2]

The nth reflection coefficient k_n for both families of filters may be evaluated as follows: Let the transfer function of the filter $H(z)$ be given by

$$H(z) = \frac{B(z)}{A(z)} \tag{19.106}$$

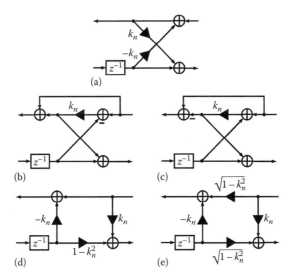

FIGURE 19.32 (a) Two-multiplier lattice; (b) and (c) one-multiplier lattice; (d) three-multiplier lattice; and (e) four-multiplier lattice.

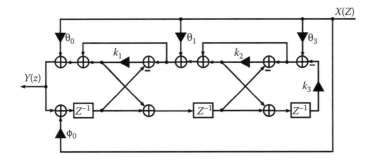

FIGURE 19.33 Third-order one-multiplier injected numerator lattice filter.

where

$$B(z) = \sum_{n=0}^{N} b_n z^{-n} \tag{19.107}$$

$$A(z) = 1 + \sum_{n=1}^{N} a_n z^{-n} \tag{19.108}$$

Define

$$D_N(z) = A(z) \tag{19.109}$$

$$D_{n-1}(z) = \frac{D_n(z) - k_n z^{-n} D_n(z^{-1})}{1 - k_n^2} \tag{19.110}$$

$$= 1 + \sum_{r=1}^{n-1} d_{n-1}(r) z^{-r} \tag{19.111}$$

The algorithm for computing k_n for all n runs as follows:

1. Set $n = N$.
2. Compute $D_n(z)$.
3. $k_n = d_n(n)$.
4. Decrement n.
5. If $n = 0$, stop; otherwise, go to 2.

19.3.3 Evaluation of the Tap Gains φ_n and λ_n [3]

For the tapped numerator filters, φ_n and λ_n may be computed as follows:
Define

$$\Gamma_N(z) = B(z) \tag{19.112}$$

$$\Gamma_{n-1}(z) = \Gamma_n(z) - \lambda_n D_n(z) - \psi_n z^{-n} D_n\left(z^{-1}\right) \tag{19.113}$$

$$= \sum_{r=0}^{n-1} \gamma_{n-1}(r) z^{-r} \tag{19.114}$$

The algorithm for computing φ_n and λ_n for all n runs as follows:

1. Set $n = N$.
2. Compute $\Gamma_n(z)$.
   Set either $\psi_n = 0$, $\lambda_n = 0$, or $\lambda_n = 0$.
   If $\psi_n = 0$, $\lambda_n = \gamma_n(n)/k_n$.
   If $\lambda_n = 0$, $\psi_n = \gamma_n(n)$.
3. Decrement n.
4. If $n = -1$, stop; otherwise, go to 2.

19.3.4 Evaluation of the Injector Gains θ_n and ϕ_n [3]

For the injected numerator filters, θ_n and ϕ_n may be computed as follows:
Define

$$L_0^0(z) = \begin{bmatrix} 1 & 0 \\ 0 & 1 \end{bmatrix} \tag{19.115}$$

$$L_m^n(z) = \begin{bmatrix} 1 & k_n z^{-1} \\ k_n & z^{-1} \end{bmatrix} L_m^{n-1}(z), \quad n > m \tag{19.116}$$

$$= \begin{bmatrix} P_m^n(z) & Q_m^n(z) \\ R_m^n(z) & S_m^n(z) \end{bmatrix} \tag{19.117}$$

$$P_m^n(z) = 1 + \sum_{r=1}^{n-m-1} p_m^n(r) z^{-r} \tag{19.118}$$

$$Q_m^n(z) = \sum_{r=1}^{n-m} q_m^n(r) z^{-r} \tag{19.119}$$

$$\Xi_{N-1} = B(z) + \phi_0 Q_0^N(z) \tag{19.120}$$

$$\Xi_{n-1}(z) = \Xi_n(z) + \phi_{N-n}Q^N_{N-n}(z) - \theta_{N-n-1}P^N_{N-n-1}(z) \qquad (19.121)$$

$$= \sum_{r=0}^{n-1} \xi_{n-1}(r)z^{-r} \qquad (19.122)$$

The algorithm for computing θ_n and ϕ_n for all n runs as follows:

1. $\phi_0 = -b_N/q^N_0(N)$.
   Set $n = 0$.
2. Increment n.
   Compute $\Xi_{N-n}(z)$.
   Set either $\phi_n = 0$ or $\theta_{n-1} = 0$.
   If $\phi_n = 0$, $\theta_{n-1} = \xi_{N-n}(N-n)/p^N_{n-1}(N-n)$
   If $\theta_{n-1} = 0$, $\phi_n = -\xi_{N-n}(N-n)/q^N_n(N-n)$
3. If $n = N - 1$ go to 4; otherwise, go to 2.
4. $\theta_{N-1} = \xi_0(0)$. Stop.

References

1. A. H. Gray, Jr. and J. D. Markel, Digital lattice and ladder filter synthesis, *IEEE Trans. Audio Electroacoust.*, AU-21, 491–500, Dec. 1973.
2. A. H. Gray, Jr. and J. D. Markel, A normalized digital filter structure, *IEEE Trans. Acoustics, Speech, Signal Process.*, ASSP-23, 268–277, June 1975.
3. Y. C. Lim, On the synthesis of the IIR digital filters derived from single channel AR lattice network, *IEEE Trans. Acoustics, Speech, Signal Process.*, ASSP-32, 741–749, Aug. 1984.

20

Finite Wordlength Effects

Bruce W. Bomar
University of Tennessee
Space Institute

Practical digital filters must be implemented with finite precision numbers and arithmetic. As a result, both the filter coefficients and the filter input and output signals are in discrete form. This leads to four types of finite wordlength effects.

Discretization (quantization) of the filter coefficients has the effect of perturbing the location of the filter poles and zeros. As a result, the actual filter response differs slightly from the ideal response. This *deterministic* frequency response error is referred to as coefficient quantization error.

The use of finite precision arithmetic makes it necessary to quantize filter calculations by rounding or truncation. Roundoff noise is that error in the filter output that results from rounding or truncation calculations within the filter. As the name implies, this error looks like low-level noise at the filter output.

Quantization of the filter calculations also renders the filter slightly nonlinear. For large signals this nonlinearity is negligible and roundoff noise is the major concern. However, for recursive filters with a zero or constant input, this nonlinearity can cause spurious oscillations called limit cycles.

With fixed-point arithmetic it is possible for filter calculations to overflow. The term overflow oscillation, sometimes also called adder overflow limit cycle, refers to a high-level oscillation that can exist in an otherwise stable filter due to the nonlinearity associated with the overflow of internal filter calculations.

In this chapter, we examine each of these finite wordlength effects. Both fixed-point and floating-point number representations are considered.

20.1 Number Representation

In digital signal processing, $(B + 1)$-bit fixed-point numbers are usually represented as two's-complement signed fractions in the format

$$b_0 \cdot b_{-1} b_{-2} \cdots b_{-B}$$

The number represented is then

$$X = -b_0 + b_{-1} 2^{-1} + b_{-2} 2^{-2} + \cdots + b_{-B} 2^{-B} \qquad (20.1)$$

where b_0 is the sign bit and the number range is $-1 \leq X < 1$. The advantage of this representation is that the product of two numbers in the range from -1 to 1 is another number in the same range.

Floating-point numbers are represented as

$$X = (-1)^s m 2^c \qquad (20.2)$$

where

s is the sign bit
m is the mantissa
c is the characteristic *or* exponent

To make the representation of a number unique, the mantissa is *normalized* so that $0.5 \leq m < 1$.

Although floating-point numbers are always represented in the form of Equation 20.2, the way in which this representation is actually *stored* in a machine may differ. Since $m \geq 0.5$, it is not necessary to store the 2^{-1}-weight bit of m, which is always set. Therefore, in practice numbers are usually stored as

$$X = (-1)^s (0.5 + f) 2^c \qquad (20.3)$$

where f is an unsigned fraction, $0 \leq f < 0.5$.

Most floating-point processors now use the IEEE Standard 754 32-bit floating-point format for storing numbers. According to this standard, the exponent is stored as an unsigned integer p where

$$p = c + 126 \qquad (20.4)$$

Therefore, a number is stored as

$$X = (-1)^s (0.5 + f) 2^{p-126} \qquad (20.5)$$

where s is the sign bit, f is a 23-b unsigned fraction in the range $0 \leq f < 0.5$, and p is an 8-b unsigned integer in the range $0 \leq p \leq 255$. The total number of bits is $1 + 23 + 8 = 32$. For example, in IEEE format $3/4$ is written $(-1)^0 (0.5 + 0.25) 2^0$ so $s = 0$, $p = 126$, and $f = 0.25$. The value $X = 0$ is a unique case and is represented by all bits zero (i.e., $s = 0$, $f = 0$, and $p = 0$). Although the 2^{-1}-weight mantissa bit is not actually stored, it does exist so the mantissa has 24 b plus a sign bit.

20.2 Fixed-Point Quantization Errors

In fixed-point arithmetic, a multiply doubles the number of significant bits. For example, the product of the two 5-b numbers 0.0011 and 0.1001 is the 10-b number 00.00011011. The extra bit to the left of the decimal point can be discarded without introducing any error. However, the least significant four of the remaining bits must ultimately be discarded by some form of quantization so that the result can be stored to 5 b for use in other calculations. In the example above this results in 0.0010 (quantization by rounding) or 0.0001 (quantization by truncating). When a sum of products calculation is performed, the quantization can be performed either after each multiply or after all products have been summed with double-length precision.

We will examine three types of fixed-point quantization—rounding, truncation, and magnitude truncation. If X is an exact value then the rounded value will be denoted $Q_r(X)$, the truncated value $Q_t(X)$, and the magnitude truncated value $Q_{mt}(X)$. If the quantized value has B bits to the right of the decimal point, the quantization step size is

$$\Delta = 2^{-B} \tag{20.6}$$

Since rounding selects the quantized value nearest the unquantized value, it gives a value which is never more than $\pm\Delta/2$ away from the exact value. If we denote the rounding error by

$$\varepsilon_r = Q_r(X) - X \tag{20.7}$$

then

$$-\frac{\Delta}{2} \le \varepsilon_r \le \frac{\Delta}{2} \tag{20.8}$$

Truncation simply discards the low-order bits giving a quantized value that is always less than or equal to the exact value so

$$-\Delta < \varepsilon_t \le 0 \tag{20.9}$$

Magnitude truncation chooses the nearest quantized value that has a magnitude less than or equal to the exact value so

$$-\Delta < \varepsilon_{mt} < \Delta \tag{20.10}$$

The error resulting from quantization can be modeled as a random variable uniformly distributed over the appropriate error range. Therefore, calculations with roundoff error can be considered error-free calculations that have been corrupted by additive white noise. The mean of this noise for rounding is

$$m_{\varepsilon_r} = E\{\varepsilon_r\} = \frac{1}{\Delta} \int_{-\Delta/2}^{\Delta/2} \varepsilon_r \, d\varepsilon_r = 0 \tag{20.11}$$

where $E\{\ \}$ represents the operation of taking the expected value of a random variable. Similarly, the variance of the noise for rounding is

$$\sigma_{\varepsilon_r}^2 = E\{(\varepsilon_r - m_{\varepsilon_r})^2\} = \frac{1}{\Delta} \int_{-\Delta/2}^{\Delta/2} (\varepsilon_r - m_{\varepsilon_r})^2 \, d\varepsilon_r = \frac{\Delta^2}{12} \tag{20.12}$$

Likewise, for truncation,

$$m_{\varepsilon_t} = E\{\varepsilon_t\} = -\frac{\Delta}{2}$$

$$\sigma_{\varepsilon_t}^2 = E\{(\varepsilon_t - m_{\varepsilon_t})^2\} = \frac{\Delta^2}{12} \tag{20.13}$$

and, for magnitude truncation

$$m_{\varepsilon_{mt}} = E\{\varepsilon_{mt}\} = 0$$

$$\sigma_{\varepsilon_{mt}}^2 = E\{(\varepsilon_{mt} - m_{\varepsilon_{mt}})^2\} = \frac{\Delta^2}{3} \qquad (20.14)$$

20.3 Floating-Point Quantization Errors

With floating-point arithmetic it is necessary to quantize after both multiplications and additions. The addition quantization arises because, prior to addition, the mantissa of the smaller number in the sum is shifted right until the exponent of both numbers is the same. In general, this gives a sum mantissa that is too long and so must be quantized.

We will assume that quantization in floating-point arithmetic is performed by rounding. Because of the exponent in floating-point arithmetic, it is the relative error that is important. The relative error is defined as

$$\varepsilon_r = \frac{Q_r(X) - X}{X} = \frac{\varepsilon_r}{X} \qquad (20.15)$$

Since $X = (-1)^s m 2^c$, $Q_r(X) = (-1)^s Q_r(m) 2^c$ and

$$\varepsilon_r = \frac{Q_r(m) - m}{m} = \frac{\varepsilon}{m} \qquad (20.16)$$

If the quantized mantissa has B bits to the right of the decimal point, $|\varepsilon_r| < \Delta/2$ where, as before, $\Delta = 2^{-B}$. Therefore, since $0.5 \le m < 1$,

$$|\varepsilon_r| < \Delta \qquad (20.17)$$

If we assume that ε is uniformly distributed over the range from $-\Delta/2$ to $\Delta/2$ and m is uniformly distributed over 0.5 to 1,

$$m_{\varepsilon_r} = E\left\{\frac{\varepsilon}{m}\right\} = 0$$

$$\sigma_{\varepsilon_r}^2 = E\left\{\left(\frac{\varepsilon}{m}\right)^2\right\} = \frac{2}{\Delta} \int_{1/2}^{1} \int_{-\Delta/2}^{\Delta/2} \frac{\varepsilon^2}{m^2} \, d\varepsilon \, dm$$

$$= \frac{\Delta^2}{6} = (0.167)2^{-2B} \qquad (20.18)$$

In practice, the distribution of m is not exactly uniform. Actual measurements of roundoff noise in Ref. [1] suggested that

$$\sigma_{\varepsilon_r}^2 \approx 0.23\Delta^2 \qquad (20.19)$$

while a detailed theoretical and experimental analysis in Ref. [2] determined

$$\sigma_{\varepsilon_r}^2 \approx 0.18\Delta^2 \qquad (20.20)$$

From Equation 20.15 we can represent a quantized floating-point value in terms of the unquantized value and the random variable ε_r using

$$Q_r(X) = X(1 + \varepsilon_r) \tag{20.21}$$

Therefore, the finite-precision product $X_1 X_2$ and the sum $X_1 + X_2$ can be written

$$fl(X_1 X_2) = X_1 X_2 (1 + \varepsilon_r) \tag{20.22}$$

and

$$fl(X_1 + X_2) = (X_1 + X_2)(1 + \varepsilon_r) \tag{20.23}$$

where ε_r is zero-mean with the variance of Equation 20.20.

20.4 Roundoff Noise

To determine the roundoff noise at the output of a digital filter we will assume that the noise due to a quantization is stationary, white, and uncorrelated with the filter input, output, and internal variables. This assumption is good if the filter input changes from sample to sample in a sufficiently complex manner. It is not valid for zero or constant inputs for which the effects of rounding are analyzed from a limit cycle perspective.

To satisfy the assumption of a sufficiently complex input, roundoff noise in digital filters is often calculated for the case of a zero-mean white noise filter input signal $x(n)$ of variance σ_x^2. This simplifies calculation of the output roundoff noise because expected values of the form $E\{x(n)x(n-k)\}$ are zero for $k \neq 0$ and give σ_x^2 when $k = 0$. This approach to analysis has been found to give estimates of the output roundoff noise that are close to the noise actually observed for other input signals.

Another assumption that will be made in calculating roundoff noise is that the product of two quantization errors is zero. To justify this assumption, consider the case of a 16-b fixed-point processor. In this case a quantization error is of the order 2^{-15}, while the product of two quantization errors is of the order 2^{-30}, which is negligible by comparison.

If a linear system with impulse response $g(n)$ is excited by white noise with mean m_x and variance σ_x^2, the output is noise of mean [3, pp. 788–790]

$$m_y = m_x \sum_{n=-\infty}^{\infty} g(n) \tag{20.24}$$

and variance

$$\sigma_y^2 = \sigma_x^2 \sum_{n=-\infty}^{\infty} g^2(n) \tag{20.25}$$

Therefore, if $g(n)$ is the impulse response from the point where a roundoff takes place to the filter output, the contribution of that roundoff to the variance (mean-square value) of the output roundoff noise is given by Equation 20.25 with σ_x^2 replaced with the variance of the roundoff. If there is more than one source of roundoff error in the filter, it is assumed that the errors are uncorrelated so the output noise variance is simply the sum of the contributions from each source.

20.4.1 Roundoff Noise in FIR Filters

The simplest case to analyze is a finite impulse response (FIR) filter realized via the convolution summation

$$y(n) = \sum_{k=0}^{N-1} h(k)x(n-k) \tag{20.26}$$

When fixed-point arithmetic is used and quantization is performed after each multiply, the result of the N multiplies is N-times the quantization noise of a single multiply. For example, rounding after each multiply gives, from Equations 20.6 and 20.12, an output noise variance of

$$\sigma_o^2 = N\frac{2^{-2B}}{12} \tag{20.27}$$

Virtually all digital signal processor integrated circuits contain one or more double-length accumulator registers which permit the sum-of-products in Equation 20.26 to be accumulated without quantization. In this case only a single quantization is necessary following the summation and

$$\sigma_o^2 = \frac{2^{-2B}}{12} \tag{20.28}$$

For the floating-point roundoff noise case we will consider Equation 20.26 for $N=4$ and then generalize the result to other values of N. The finite-precision output can be written as the exact output plus an error term $e(n)$. Thus,

$$
\begin{aligned}
y(n) + e(n) = (\{[&h(0)x(n)[1 + \varepsilon_1(n)]] \\
&+ h(1)x(n-1)[1 + \varepsilon_2(n)]\,[1 + \varepsilon_3(n)] \\
&+ h(2)x(n-2)[1 + \varepsilon_4(n)]\}\{1 + \varepsilon_5(n)\} \\
&+ h(3)x(n-3)[1 + \varepsilon_6(n)])[1 + \varepsilon_7(n)]
\end{aligned}
\tag{20.29}
$$

In Equation 20.29, $\varepsilon_1(n)$ represents the error in the first product, $\varepsilon_2(n)$ the error in the second product, $\varepsilon_3(n)$ the error in the first addition, etc. Notice that it has been assumed that the products are summed in the order implied by the summation of Equation 20.26.

Expanding Equation 20.29, ignoring products of error terms, and recognizing $y(n)$ gives

$$
\begin{aligned}
e(n) = h(0)&x(n)[\varepsilon_1(n) + \varepsilon_3(n) + \varepsilon_5(n) + \varepsilon_7(n)] \\
&+ h(1)x(n-1)[\varepsilon_2(n) + \varepsilon_3(n) + \varepsilon_5(n) + \varepsilon_7(n)] \\
&+ h(2)x(n-2)[\varepsilon_4(n) + \varepsilon_5(n) + \varepsilon_7(n)] \\
&+ h(3)x(n-3)[\varepsilon_6(n) + \varepsilon_7(n)]
\end{aligned}
\tag{20.30}
$$

Assuming that the input is white noise of variance σ_x^2 so that $E\{x(n)x(n-k)\}$ is zero for $k \neq 0$, and assuming that the errors are uncorrelated,

$$E\{e^2(n)\} = [4h^2(0) + 4h^2(1) + 3h^2(2) + 2h^2(3)]\sigma_x^2\sigma_{\varepsilon_r}^2 \tag{20.31}$$

In general, for any N,

$$\sigma_o^2 = E\{e^2(n)\} = \left[Nh^2(0) + \sum_{k=1}^{N-1} (N+1-k)h^2(k)\right]\sigma_x^2\sigma_{\varepsilon_r}^2 \qquad (20.32)$$

Notice that if the order of summation of the product terms in the convolution summation is changed, then the order in which the $h(k)$'s appear in Equation 20.32 changes. If the order is changed so that the $h(k)$ with smallest magnitude is first, followed by the next smallest, etc., then the roundoff noise variance is minimized. However, performing the convolution summation in nonsequential order greatly complicates data indexing and so may not be worth the reduction obtained in roundoff noise.

20.4.2 Roundoff Noise in Fixed-Point IIR Filters

To determine the roundoff noise of a fixed-point infinite impulse response (IIR) filter realization, consider a causal first-order filter with impulse response

$$h(n) = a^n u(n) \qquad (20.33)$$

realized by the difference equation

$$y(n) = ay(n-1) + x(n) \qquad (20.34)$$

Due to roundoff error, the output actually obtained is

$$\hat{y}(n) = Q\{ay(n-1) + x(n)\} = ay(n-1) + x(n) + e(n) \qquad (20.35)$$

where $e(n)$ is a random roundoff noise sequence. Since $e(n)$ is injected at the same point as the input, it propagates through a system with impulse response $h(n)$. Therefore, for fixed-point arithmetic with rounding, the output roundoff noise variance from Equations 20.6, 20.12, 20.25, and 20.33 is

$$\sigma_o^2 = \frac{\Delta^2}{12} \sum_{n=-\infty}^{\infty} h^2(n) = \frac{\Delta^2}{12} \sum_{n=0}^{\infty} a^{2n} = \frac{2^{-2B}}{12} \frac{1}{1-a^2} \qquad (20.36)$$

With fixed-point arithmetic there is the possibility of overflow following addition. To avoid overflow it is necessary to restrict the input signal amplitude. This can be accomplished by either placing a *scaling* multiplier at the filter input or by simply limiting the maximum input signal amplitude. Consider the case of the first-order filter of Equation 20.34. The transfer function of this filter is

$$H(e^{j\omega}) = \frac{Y(e^{j\omega})}{X(e^{j\omega})} = \frac{1}{e^{j\omega} - a} \qquad (20.37)$$

so

$$\left|H(e^{j\omega})\right|^2 = \frac{1}{1 + a^2 - 2a\cos(\omega)} \qquad (20.38)$$

and

$$\left|H(e^{j\omega})\right|_{\max} = \frac{1}{1 - |a|} \qquad (20.39)$$

The peak gain of the filter is $1/(1 - |a|)$ so limiting input signal amplitudes to $|x(n)| \leq 1 - |a|$ will make overflows unlikely.

An expression for the output roundoff noise-to-signal ratio can easily be obtained for the case where the filter input is white noise, uniformly distributed over the interval from $-(1 - |a|)$ to $(1 - |a|)$ [4,5]. In this case

$$\sigma_x^2 = \frac{1}{2(1 - |a|)} \int_{-(1-|a|)}^{1-|a|} x^2 \, dx = \frac{1}{3}(1 - |a|)^2 \tag{20.40}$$

so, from Equation 20.25,

$$\sigma_y^2 = \frac{1}{3} \frac{(1 - |a|)^2}{1 - a^2} \tag{20.41}$$

Combining Equations 20.36 and 20.41 then gives

$$\frac{\sigma_o^2}{\sigma_y^2} = \left(\frac{2^{-2B}}{12} \frac{1}{1 - a^2}\right) \left(3 \frac{1 - a^2}{(1 - |a|)^2}\right) = \frac{2^{-2B}}{12} \frac{3}{(1 - |a|)^2} \tag{20.42}$$

Notice that the noise-to-signal ratio increases without bound as $|a| \to 1$.

Similar results can be obtained for the case of the causal second-order filter realized by the difference equation

$$y(n) = 2r \cos(\theta)y(n - 1) - r^2 y(n - 2) + x(n) \tag{20.43}$$

This filter has complex–conjugate poles at $re^{\pm j\theta}$ and impulse response

$$h(n) = \frac{1}{\sin(\theta)} r^n \sin[(n + 1)\theta] \, u(n) \tag{20.44}$$

Due to roundoff error, the output actually obtained is

$$\hat{y}(n) = 2r \cos(\theta)y(n - 1) - r^2 y(n - 2) + x(n) + e(n) \tag{20.45}$$

There are two noise sources contributing to $e(n)$ if quantization is performed after each multiply, and there is one noise source if quantization is performed after summation. Since

$$\sum_{n=-\infty}^{\infty} h^2(n) = \frac{1 + r^2}{1 - r^2} \frac{1}{(1 + r^2)^2 - 4r^2 \cos^2(\theta)} \tag{20.46}$$

the output roundoff noise is

$$\sigma_o^2 = \nu \frac{2^{-2B}}{12} \frac{1 + r^2}{1 - r^2} \frac{1}{(1 + r^2)^2 - 4r^2 \cos^2(\theta)} \tag{20.47}$$

where $\nu = 1$ for quantization after summation, and $\nu = 2$ for quantization after each multiply.

To obtain an output noise-to-signal ratio we note that

$$H(e^{j\omega}) = \frac{1}{1 - 2r\cos(\theta)e^{-j\omega} + r^2 e^{-j2\omega}} \tag{20.48}$$

and, using the approach of [6],

$$\left|H(e^{j\omega})\right|^2_{max} = \frac{1}{4r^2\left\{\left[sat\left(\frac{1+r^2}{2r}\cos(\theta)\right) - \frac{1+r^2}{2r}\cos(\theta)\right]^2 + \left[\frac{1-r^2}{2r}\sin(\theta)\right]^2\right\}} \tag{20.49}$$

where

$$sat(\mu) = \begin{cases} 1 & \mu > 1 \\ \mu & -1 \le \mu \le 1 \\ -1 & \mu < -1 \end{cases} \tag{20.50}$$

Following the same approach as for the first-order case then gives

$$\frac{\sigma_o^2}{\sigma_y^2} = \nu \frac{2^{-2B}}{12} \frac{1+r^2}{1-r^2} \frac{3}{(1+r^2)^2 - 4r^2\cos^2(\theta)}$$

$$\times \frac{1}{4r^2\left\{\left[sat\left(\frac{1+r^2}{2r}\cos(\theta)\right) - \frac{1+r^2}{2r}\cos(\theta)\right]^2 + \left[\frac{1-r^2}{2r}\sin(\theta)\right]^2\right\}} \tag{20.51}$$

Figure 20.1 is a contour plot showing the noise-to-signal ratio of Equation 20.51 for $\nu = 1$ in units of the noise variance of a single quantization $2^{-2B}/12$. The plot is symmetrical about $\theta = 90°$, so only the range from $0°$ to $90°$ is shown. Notice that as $r \to 1$, the roundoff noise increases without bound. Also notice that the noise increases as $\theta \to 0°$.

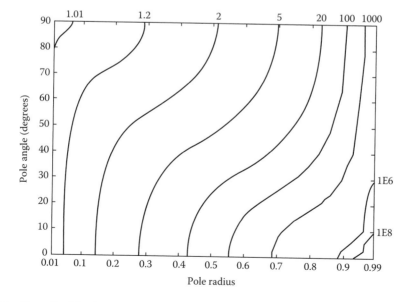

FIGURE 20.1   Normalized fixed-point roundoff noise variance.

It is possible to design state-space filter realizations that minimize fixed-point roundoff noise [7–10]. Depending on the transfer function being realized, these structures may provide a roundoff noise level that is orders-of-magnitude lower than for a nonoptimal realization. The price paid for this reduction in roundoff noise is an increase in the number of computations required to implement the filter. For an Nth-order filter the increase is from roughly $2N$ multiplies for a direct form realization to roughly $(N+1)^2$ for an optimal realization. However, if the filter is realized by the parallel or cascade connection of first- and second-order optimal subfilters, the increase is only to about $4N$ multiplies. Furthermore, near-optimal realizations exist that increase the number of multiplies to only about $3N$ [10].

20.4.3 Roundoff Noise in Floating-Point IIR Filters

For floating-point arithmetic it is first necessary to determine the injected noise variance of each quantization. For the first-order filter this is done by writing the computed output as

$$y(n) + e(n) = [ay(n-1)(1 + \varepsilon_1(n)) + x(n)](1 + \varepsilon_2(n)) \tag{20.52}$$

where
  $\varepsilon_1(n)$ represents the error due to the multiplication
  $\varepsilon_2(n)$ represents the error due to the addition

Neglecting the product of errors, Equation 20.52 becomes

$$\begin{aligned} y(n) + e(n) &\approx ay(n-1) + x(n) + ay(n-1)\varepsilon_1(n) \\ &\quad + ay(n-1)\varepsilon_2(n) + x(n)\varepsilon_2(n) \end{aligned} \tag{20.53}$$

Comparing Equations 20.34 and 20.53, it is clear that

$$e(n) = ay(n-1)\varepsilon_1(n) + ay(n-1)\varepsilon_2(n) + x(n)\varepsilon_2(n) \tag{20.54}$$

Taking the expected value of $e^2(n)$ to obtain the injected noise variance then gives

$$\begin{aligned} E\{e^2(n)\} &= a^2 E\{y^2(n-1)\}E\{\varepsilon_1^2(n)\} + a^2 E\{y^2(n-1)\}E\{\varepsilon_2^2(n)\} \\ &\quad + E\{x^2(n)\}E\{\varepsilon_2^2(n)\} + E\{x(n)y(n-1)\}E\{\varepsilon_2^2(n)\} \end{aligned} \tag{20.55}$$

To carry this further it is necessary to know something about the input. If we assume the input is zero-mean white noise with variance σ_x^2, then $E\{x^2(n)\} = \sigma_x^2$ and the input is uncorrelated with past values of the output so $E\{x(n)y(n-1)\} = 0$ giving

$$E\{e^2(n)\} = 2a^2\sigma_y^2\sigma_{\varepsilon_r}^2 + \sigma_x^2\sigma_{\varepsilon_r}^2 \tag{20.56}$$

and

$$\begin{aligned} \sigma_o^2 &= \left(2a^2\sigma_y^2\sigma_{\varepsilon_r}^2 + \sigma_x^2\sigma_{\varepsilon_r}^2\right)\sum_{n=-\infty}^{\infty} h^2(n) \\ &= \frac{2a^2\sigma_y^2 + \sigma_x^2}{1 - a^2}\sigma_{\varepsilon_r}^2 \end{aligned} \tag{20.57}$$

However,

$$\sigma_y^2 = \sigma_x^2 \sum_{n=-\infty}^{\infty} h^2(n) = \frac{\sigma_x^2}{1 - a^2} \tag{20.58}$$

$$\sigma_o^2 = \frac{1 + a^2}{(1 - a^2)^2} \sigma_{\varepsilon_r}^2 \sigma_x^2 = \frac{1 + a^2}{1 - a^2} \sigma_{\varepsilon_r}^2 \sigma_y^2 \tag{20.59}$$

and the output roundoff noise-to-signal ratio is

$$\frac{\sigma_o^2}{\sigma_y^2} = \frac{1 + a^2}{1 - a^2} \sigma_{\varepsilon_r}^2 \tag{20.60}$$

Similar results can be obtained for the second-order filter of Equation 20.43 by writing

$$y(n) + e(n) = \big([2r \cos (\theta)y(n - 1)(1 + \varepsilon_1(n)) - r^2 y(n - 2)(1 + \varepsilon_2(n))] \\ \times [1 + \varepsilon_3(n)] + x(n))(1 + \varepsilon_4(n)) \tag{20.61}$$

Expanding with the same assumptions as before gives

$$e(n) \approx 2r \cos (\theta)y(n - 1)[\varepsilon_1(n) + \varepsilon_3(n) + \varepsilon_4(n)] \\ - r^2 y(n - 2)[\varepsilon_2(n) + \varepsilon_3(n) + \varepsilon_4(n)] + x(n)\varepsilon_4(n) \tag{20.62}$$

and

$$E\{e^2(n)\} = 4r^2 \cos^2 (\theta)\sigma_y^2 \sigma_{\varepsilon_r}^2 + r^2 \sigma_y^2 3\sigma_{\varepsilon_r}^2 \\ + \sigma_x^2 \sigma_{\varepsilon_r}^2 - 8r^3 \cos (\theta)\sigma_{\varepsilon_r}^2 E\{y(n - 1)y(n - 2)\} \tag{20.63}$$

However,

$$E\{y(n - 1)y(n - 2)\} \\ = E\{2r \cos (\theta)y(n - 2) - r^2 y(n - 3) + x(n - 1)]y(n - 2)\} \\ = 2r \cos (\theta)E\{y^2(n - 2)\} - r^2 E\{y(n - 2)y(n - 3)\} \\ = 2r \cos (\theta)E\{y^2(n - 2)\} - r^2 E\{y(n - 1)y(n - 2)\} \\ = \frac{2r \cos (\theta)}{1 + r^2} \sigma_y^2 \tag{20.64}$$

so

$$E\{e^2(n)\} = \sigma_{\varepsilon_r}^2 \sigma_x^2 + \left[3r^4 + 12r^2 \cos^2(\theta) - \frac{16r^4 \cos^2(\theta)}{1 + r^2} \right] \sigma_{\varepsilon_r}^2 \sigma_y^2 \tag{20.65}$$

and

$$\sigma_o^2 = E(n) \sum_{n=-\infty}^{\infty} h^2(n) \\ = \xi \left[\sigma_{\varepsilon_r}^2 \sigma_x^2 + \left[3r^4 + 12r^2 \cos^2(\theta) - \frac{16r^4 \cos^2(\theta)}{1 + r^2} \right] \sigma_{\varepsilon_r}^2 \sigma_y^2 \right] \tag{20.66}$$

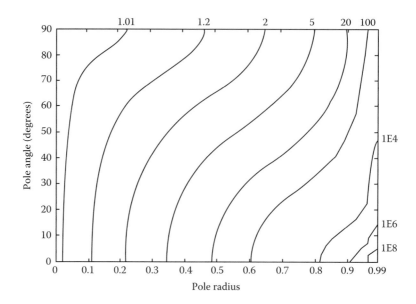

FIGURE 20.2   Normalized floating-point roundoff noise variance.

where from Equation 20.46,

$$\xi = \sum_{n=-\infty}^{\infty} h^2(n) = \frac{1+r^2}{1-r^2} \frac{1}{(1+r^2)^2 - 4r^2 \cos^2(\theta)} \tag{20.67}$$

Since $\sigma_y^2 = \xi \sigma_x^2$, the output roundoff noise-to-signal ratio is then

$$\frac{\sigma_o^2}{\sigma_y^2} = \xi \left[1 + \xi \left[3r^4 + 12r^2 \cos^2(\theta) - \frac{16r^4 \cos^2(\theta)}{1+r^2} \right] \right] \sigma_{\varepsilon_r}^2 \tag{20.68}$$

Figure 20.2 is a contour plot showing the noise-to-signal ratio of Equation 20.68 in units of the noise variance of a single quantization $\sigma_{\varepsilon_r}^2$. The plot is symmetrical about $\theta = 90°$, so only the range from $0°$ to $90°$ is shown. Notice the similarity of this plot to that of Figure 20.1 for the fixed-point case. It has been observed that filter structures generally have very similar fixed-point and floating-point roundoff characteristics [2]. Therefore, the techniques of [7–10], which were developed for the fixed-point case, can also be used to design low-noise floating-point filter realizations. Furthermore, since it is not necessary to scale the floating-point realization, the low-noise realizations need not require significantly more computation than the direct form realization.

20.5 Limit Cycles

A limit cycle, sometimes referred to as a multiplier roundoff limit cycle, is a low-level oscillation that can exist in an otherwise stable filter as a result of the nonlinearity associated with rounding (or truncating) internal filter calculations [11]. Limit cycles require recursion to exist and do not occur in nonrecursive FIR filters.

As an example of a limit cycle, consider the second-order filter realized by

$$y(n) = Q_r \left\{ \frac{7}{8} y(n-1) - \frac{5}{8} y(n-2) + x(n) \right\} \tag{20.69}$$

where $Q_r\{\ \}$ represents quantization by rounding. This is a stable filter with poles at $0.4375 \pm j0.6585$. Consider the implementation of this filter with 4-b (3 b and a sign bit) two's complement fixed-point arithmetic, zero initial conditions $(y(-1) = y(-2) = 0)$, and an input sequence $x(n) = \frac{3}{8}\delta(n)$, where $\delta(n)$ is the unit impulse or unit sample. The following sequence is obtained:

$$y(0) = Q_r\left\{\frac{3}{8}\right\} = \frac{3}{8}$$

$$y(1) = Q_r\left\{\frac{21}{64}\right\} = \frac{3}{8}$$

$$y(2) = Q_r\left\{\frac{3}{32}\right\} = \frac{1}{8}$$

$$y(3) = Q_r\left\{-\frac{1}{8}\right\} = -\frac{1}{8}$$

$$y(4) = Q_r\left\{-\frac{3}{16}\right\} = -\frac{1}{8}$$

$$y(5) = Q_r\left\{-\frac{1}{32}\right\} = 0$$

$$y(6) = Q_r\left\{\frac{5}{64}\right\} = \frac{1}{8}$$

$$y(7) = Q_r\left\{\frac{7}{64}\right\} = \frac{1}{8}$$

$$y(8) = Q_r\left\{\frac{1}{32}\right\} = 0$$

$$y(9) = Q_r\left\{-\frac{5}{64}\right\} = -\frac{1}{8}$$

$$y(10) = Q_r\left\{-\frac{7}{64}\right\} = -\frac{1}{8}$$

$$y(10) = Q_r\left\{-\frac{1}{32}\right\} = 0$$

$$y(12) = Q_r\left\{\frac{5}{64}\right\} = \frac{1}{8}$$

$$\vdots$$

(20.70)

Notice that while the input is zero except for the first sample, the output oscillates with amplitude $1/8$ and period 6.

Limit cycles are primarily of concern in fixed-point recursive filters. As long as floating-point filters are realized as the parallel or cascade connection of first- and second-order subfilters, limit cycles will generally not be a problem since limit cycles are practically not observable in first- and second-order systems implemented with 32-b floating-point arithmetic [12]. It has been shown that such systems must have an extremely small margin of stability for limit cycles to exist at anything other than underflow levels, which are at an amplitude of less than 10^{-38} [12].

There are at least three ways of dealing with limit cycles when fixed-point arithmetic is used. One is to determine a bound on the maximum limit cycle amplitude, expressed as an integral number of quantization steps [13]. It is then possible to choose a word length that makes the limit cycle amplitude acceptably low. Alternately, limit cycles can be prevented by randomly rounding calculations up or down [14]. However, this approach is complicated to implement. The third approach is to properly choose the filter realization structure and then quantize the filter calculations using magnitude truncation [15,16]. This approach has the disadvantage of producing more roundoff noise than truncation or rounding [see Equations 20.12 through 20.14].

20.6 Overflow Oscillations

With fixed-point arithmetic it is possible for filter calculations to overflow. This happens when two numbers of the same sign add to give a value having magnitude greater than one. Since numbers with magnitude greater than one are not representable, the result overflows. For example, the two's complement numbers 0.101 (5/8) and 0.100 (4/8) add to give 1.001 which is the two's complement representation of $-7/8$.

The overflow characteristic of two's complement arithmetic can be represented as $R\{\}$ where

$$R\{X\} = \begin{cases} X - 2 & X \geq 1 \\ X & -1 \leq X < -1 \\ X + 2 & X < -1 \end{cases} \tag{20.71}$$

For the example just considered, $R\{9/8\} = -7/8$.

An overflow oscillation, sometimes also referred to as an *adder overflow limit cycle*, is a high-level oscillation that can exist in an otherwise stable fixed-point filter due to the gross nonlinearity associated with the overflow of internal filter calculations [17]. Like limit cycles, overflow oscillations require recursion to exist and do not occur in nonrecursive FIR filters. Overflow oscillations also do not occur with floating-point arithmetic due to the virtual impossibility of overflow.

As an example of an overflow oscillation, once again consider the filter of Equation 20.69 with 4-b fixed-point two's complement arithmetic and with the two's complement overflow characteristic of Equation 20.71:

$$y(n) = Q_r \left\{ R \left[\frac{7}{8} y(n-1) - \frac{5}{8} y(n-2) + x(n) \right] \right\} \tag{20.72}$$

In this case we apply the input

$$x(n) = -\frac{3}{4}\delta(n) - \frac{5}{8}\delta(n-1)$$

$$= \left\{ -\frac{3}{4}, -\frac{5}{8}, 0, 0, \ldots \right\} \tag{20.73}$$

giving the output sequence

$$y(0) = Q_r \left\{ R \left[-\frac{3}{4} \right] \right\} = Q_r \left\{ -\frac{3}{4} \right\} = -\frac{3}{4}$$

$$y(1) = Q_r \left\{ R \left[-\frac{41}{32} \right] \right\} = Q_r \left\{ \frac{23}{32} \right\} = \frac{3}{4}$$

$$y(2) = Q_r \left\{ R \left[\frac{9}{8} \right] \right\} = Q_r \left\{ -\frac{7}{8} \right\} = -\frac{7}{8}$$

$$y(3) = Q_r \left\{ R \left[-\frac{79}{64} \right] \right\} = Q_r \left\{ \frac{49}{64} \right\} = \frac{3}{4}$$

$$y(4) = Q_r\left\{R\left[\frac{77}{64}\right]\right\} = Q_r\left\{-\frac{51}{64}\right\} = -\frac{3}{4}$$

$$y(5) = Q_r\left\{R\left[-\frac{9}{8}\right]\right\} = Q_r\left\{\frac{7}{8}\right\} = \frac{7}{8}$$

$$y(6) = Q_r\left\{R\left[\frac{79}{64}\right]\right\} = Q_r\left\{\frac{-49}{64}\right\} = -\frac{3}{4}$$

$$y(7) = Q_r\left\{R\left[-\frac{77}{64}\right]\right\} = Q_r\left\{\frac{51}{64}\right\} = \frac{3}{4}$$

$$y(8) = Q_r\left\{R\left[\frac{9}{8}\right]\right\} = Q_r\left\{-\frac{7}{8}\right\} = -\frac{7}{8}$$

$$\vdots$$

(20.74)

This is a large-scale oscillation with nearly full-scale amplitude.

There are several ways to prevent overflow oscillations in fixed-point filter realizations. The most obvious is to scale the filter calculations so as to render overflow impossible. However, this may unacceptably restrict the filter dynamic range. Another method is to force completed sums-of-products to saturate at ±1, rather than overflowing [18,19]. It is important to saturate only the completed sum, since intermediate overflows in two's complement arithmetic do not affect the accuracy of the final result. Most fixed-point digital signal processors provide for automatic saturation of completed sums if their *saturation arithmetic* feature is enabled. Yet another way to avoid overflow oscillations is to use a filter structure for which any internal filter transient is guaranteed to decay to zero [20]. Such structures are desirable anyway, since they tend to have low roundoff noise and be insensitive to coefficient quantization [21].

20.7 Coefficient Quantization Error

Each filter structure has its own finite, generally nonuniform grids of realizable pole and zero locations when the filter coefficients are quantized to a finite word length. In general the pole and zero locations desired in a filter do not correspond exactly to the realizable locations. The error in filter performance (usually measured in terms of a frequency response error) resulting from the placement of the poles and zeros at the nonideal but realizable locations is referred to as coefficient quantization error. Consider the second-order filter with complex–conjugate poles

$$\begin{aligned} \lambda &= re^{\pm j\theta} \\ &= \lambda_r \pm j\lambda_i \\ &= r\cos(\theta) \pm jr\sin(\theta) \end{aligned}$$

(20.75)

and transfer function

$$H(z) = \frac{1}{1 - 2r\cos(\theta)z^{-1} + r^2 z^{-2}}$$

(20.76)

realized by the difference equation

$$y(n) = 2r\cos(\theta)y(n-1) - r^2 y(n-2) + x(n)$$

(20.77)

Figure 20.3 from Ref. [5] shows that quantizing the difference equation coefficients results in a nonuniform grid of realizable pole locations in the *z* plane. The grid is defined by the intersection of vertical lines

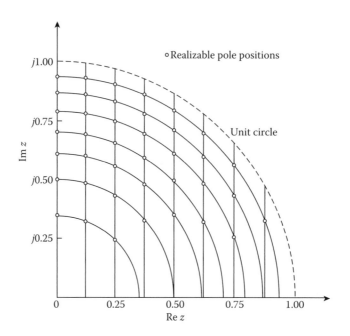

FIGURE 20.3   Realizable pole locations for the difference Equation of 20.76.

corresponding to quantization of $2\lambda_r$ and concentric circles corresponding to quantization of $-r^2$. The sparseness of realizable pole locations near $z = \pm 1$ will result in a large coefficient quantization error for poles in this region.

Figure 20.4 gives an alternative structure to Equation 20.77 for realizing the transfer function of Equation 20.76. Notice that quantizing the coefficients of this structure corresponds to quantizing λ_r and λ_i. As shown in Figure 20.5 from Ref. [5], this results in a uniform grid of realizable pole locations. Therefore, large coefficient quantization errors are avoided for all pole locations.

It is well established that filter structures with low roundoff noise tend to be robust to coefficient quantization, and vice versa [22–24]. For this reason, the uniform grid structure of Figure 20.4 is also popular because of its low roundoff noise. Likewise, the low-noise realizations of Ref. [7–10] can be expected to be relatively insensitive to coefficient quantization, and digital wave filters

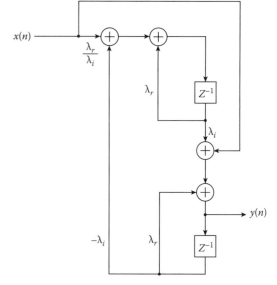

FIGURE 20.4   Alternate realization structure.

and lattice filters that are derived from low-sensitivity analog structures tend to have not only low coefficient sensitivity, but also low roundoff noise [25,26].

It is well known that in a high-order polynomial with clustered roots, the root location is a very sensitive function of the polynomial coefficients. Therefore, filter poles and zeros can be much more accurately controlled if higher order filters are realized by breaking them up into the parallel or cascade connection of first- and second-order subfilters. One exception to this rule is the case of linear-phase

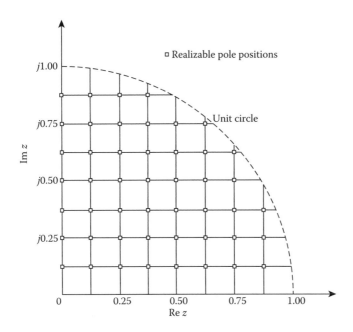

FIGURE 20.5   Realizable pole locations for the alternate realization structure.

FIR filters in which the symmetry of the polynomial coefficients and the spacing of the filter zeros around the unit circle usually permits an acceptable direct realization using the convolution summation.

Given a filter structure it is necessary to assign the ideal pole and zero locations to the realizable locations. This is generally done by simply rounding or truncating the filter coefficients to the available number of bits, or by assigning the ideal pole and zero locations to the nearest realizable locations. A more complicated alternative is to consider the original filter design problem as a problem in discrete optimization, and choose the realizable pole and zero locations that give the best approximation to the desired filter response [27–30].

20.8 Realization Considerations

Linear-phase FIR digital filters can generally be implemented with acceptable coefficient quantization sensitivity using the direct convolution sum method. When implemented in this way on a digital signal processor, fixed-point arithmetic is not only acceptable but may actually be preferable to floating-point arithmetic. Virtually all fixed-point digital signal processors accumulate a sum of products in a double-length accumulator. This means that only a single quantization is necessary to compute an output. Floating-point arithmetic, on the other hand, requires a quantization after every multiply and after every add in the convolution summation. With 32-b floating-point arithmetic these quantizations introduce a small enough error to be insignificant for many applications.

When realizing IIR filters, either a parallel or cascade connection of first- and second-order subfilters is almost always preferable to a high-order direct-form realization. With the availability of very low-cost floating-point digital signal processors, like the Texas Instruments TMS320C32, it is highly recommended that floating-point arithmetic be used for IIR filters. Floating-point arithmetic simultaneously eliminates most concerns regarding scaling, limit cycles, and overflow oscillations. Regardless of the arithmetic employed, a low roundoff noise structure should be used for the second-order sections. Good choices are given in Refs. [2,10]. Recall that realizations with low fixed-point roundoff noise also have low floating-point roundoff noise. The use of a low roundoff noise structure for the second-order sections also tends to give a realization with low coefficient quantization sensitivity. First-order sections are not as critical in determining the roundoff noise and coefficient sensitivity of a realization, and so can generally be implemented with a simple direct form structure.

References

1. C. Weinstein and A. V. Oppenheim, A comparison of roundoff noise in floating-point and fixed-point digital filter realizations, *Proc. IEEE,* 57, 1181–1183, June 1969.
2. L. M. Smith, B. W. Bomar, R. D. Joseph, and G. C. Yang, Floating-point roundoff noise analysis of second-order state-space digital filter structures, *IEEE Trans. Circuits Syst. II,* 39, 90–98, February 1992.
3. J. G. Proakis and D. G. Manolakis, *Introduction to Digital Signal Processing,* 1st edn., New York: Macmillan, 1988.
4. A. V. Oppenheim and R.W. Schafer, *Digital Signal Processing,* Englewood Cliffs, NJ: Prentice-Hall, 1975.
5. A. V. Oppenheim and C. J. Weinstein, Effects of finite register length in digital filtering and the fast Fourier transform, *Proc. IEEE,* 60, 957–976, August 1972.
6. B. W. Bomar and R. D. Joseph, Calculation of L_∞ norms for scaling second-order state-space digital filter sections, *IEEE Trans. Circuits Syst.,* CAS-34, 983–984, August 1987.
7. C. T. Mullis and R. A. Roberts, Synthesis of minimum roundoff noise fixed-point digital filters, *IEEE Trans. Circuits Syst.,* CAS-23, 551–562, September 1976.
8. L. B. Jackson, A. G. Lindgren, and Y. Kim, Optimal synthesis of second-order state-space structures for digital filters, *IEEE Trans. Circuits Syst.,* CAS-26, 149–153, March 1979.
9. C. W. Barnes, On the design of optimal state-space realizations of second-order digital filters, *IEEE Trans. Circuits Syst.,* CAS-31, 602–608, July 1984.
10. B. W. Bomar, New second-order state-space structures for realizing low roundoff noise digital filters, *IEEE Trans. Acoust., Speech, Signal Process.,* ASSP-33, 106–110, February 1985.
11. S. R. Parker and S. F. Hess, Limit-cycle oscillations in digital filters, *IEEE Trans. Circuit Theory,* CT-18, 687–697, November 1971.
12. P. H. Bauer, Limit cycle bounds for floating-point implementations of second-order recursive digital filters, *IEEE Trans. Circuits Syst. II,* 40, 493–501, August 1993.
13. B. D. Green and L. E. Turner, New limit cycle bounds for digital filters, *IEEE Trans. Circuits Syst.,* 35, 365–374, April 1988.
14. M. Buttner, A novel approach to eliminate limit cycles in digital filters with a minimum increase in the quantization noise, in *Proc. 1976 IEEE Int. Symp. Circuits Syst.,* April 1976, 291–294.
15. P. S. R. Diniz and A. Antoniou, More economical state-space digital filter structures which are free of constant-input limit cycles, *IEEE Trans. Acoust., Speech, Signal Process.,* ASSP-34, 807–815, August 1986.
16. B. W. Bomar, Low-rounoff-noise limit-cycle-free implementation of recursive transfer functions on a fixed-point digital signal processor, *IEEE Trans. Ind. Electron.,* 41, 70–78, February 1994.
17. P. M. Ebert, J. E. Mazo, and M. G. Taylor, Overflow oscillations in digital filters, *Bell Syst., Tech. J.,* 48. 2999–3020, November 1969.
18. A. N. Willson, Jr., Limit cycles due to adder overflow in digital filters, *IEEE Trans. Circuit Theory,* CT-19, 342–346, July 1972.
19. J. H. F. Ritzerfield, A condition for the overflow stability of second-order digital filters that is satisfied by all scaled state-space structures using saturation, *IEEE Trans. Circuits Syst.,* 36, 1049–1057, August 1989.
20. W. T. Mills, C. T. Mullis, and R. A. Roberts, Digital filter realizations without overflow oscillations, *IEEE Trans. Acoust., Speech, Signal Process.,* ASSP-26, 334–338, August 1978.

21. B. W. Bomar, On the design of second-order state-space digital filter sections, *IEEE Trans. Circuits Syst.*, 36, 542–552, April 1989.
22. L. B. Jackson, Roundoff noise bounds derived from coefficient sensitivities for digital filters, *IEEE Trans. Circuits Syst.*, CAS-23, 481–485, August 1976.
23. D. B. V. Rao, Analysis of coefficient quantization errors in state-space digital filters, *IEEE Trans. Acoust., Speech, Signal Process.*, ASSP-34, 131–139, February 1986.
24. L. Thiele, On the sensitivity of linear state-space systems, *IEEE Trans. Circuits Syst.*, CAS-33, 502–510, May 1986.
25. A. Antoniou, *Digital Filters: Analysis and Design*, New York: McGraw-Hill, 1979.
26. Y. C. Lim, On the synthesis of IIR digital filters derived from single channel AR lattice network, *IEEE Trans. Acoust., Speech, Signal Process.*, ASSP-32, 741–749, August 1984.
27. E. Avenhaus, On the design of digital filters with coefficients of limited wordlength, *IEEE Trans. Audio Electroacoust.*, AU-20, 206–212, August 1972.
28. M. Suk and S. K. Mitra, Computer-aided design of digital filters with finite wordlengths, *IEEE Trans. Audio Electroacoust.*, AU-20, 356–363, December 1972.
29. C. Charalambous and M. J. Best, Optimization of recursive digital filters with finite wordlengths, *IEEE Trans. Acoust., Speech, Signal Process.*, ASSP-22, 424–431, December 1979.
30. Y. C. Lim, Design of discrete-coefficient-value linear-phase FIR filters with optimum normalized peak ripple magnitude, *IEEE Trans. Circuits Syst.*, 37, 1480–1486, December 1990.

21

VLSI Implementation of Digital Filters

Joseph B. Evans
University of Kansas

Timothy R. Newman
University of Kansas

21.1 Introduction

Digital implementations of filters are preferred over analog realizations for many reasons. Improvements in VLSI technology have enabled digital filters to be used in an increasing number of application domains.

There are a variety of methods that can be used to implement digital filters. In this chapter we focus on the use of traditional VLSI digital logic families such as CMOS, rather than more exotic approaches. The vast majority of implementations encountered in practice make use of traditional technologies because the performance and cost characteristics of these approaches are so favorable.

Digital filter implementations can be classified into several categories based on the architectural approach used: general purpose, special purpose, and programmable logic implementations. The choice of a particular approach should be based upon the flexibility and performance required by a particular application. General-purpose architectures possess a great deal of flexibility, but are somewhat limited in performance, being best suited for relatively low sampling frequencies, usually under 10 MHz. Special-purpose architectures are capable of much higher performance, with sampling frequencies as high as 1 GHz, but are often only configurable for one application domain. Programmable logic implementations lie somewhere between these extremes, providing both flexibility and reasonably high performance, with sampling rates as high as 200 MHz.

Digital filtering implementations have been strongly influenced by evolution of VLSI technology. The regular computational structures encountered in filters are well suited for VLSI implementation. This regularity often translates into efficient parallelism and pipelining. Further, the small set of computational structures required in digital filtering makes automatic synthesis of special-purpose and programmable logic designs feasible. The design automation of digital filter implementation is relatively simple compared to the general design synthesis problem. For this reason, digital filters are often the test case for evaluating new device and computer-aided design technologies.

21.2 General-Purpose Processors

General-purpose digital signal processors are by far the most commonly used method for digital filter implementation, particularly at audio bandwidths. These systems possess architectures well suited to digital filtering, as well as other digital signal processing (DSP) algorithms.

21.2.1 Historical Perspective

General-purpose digital signal processors trace their lineage back to the microprocessors of the early 1980s. The generic microprocessors of that period were ill suited for the implementation of DSP algorithms, due to the lack of hardware support for numerical algorithms of significant complexity in those architectures. The primary requirement for DSP implementation was identified to be hardware support for multiplication, due to the large number of multiply-accumulate (MAC) operations in DSP algorithms and their large contribution to computational delays. The earliest widely available single chip general-purpose DSP implementation was from AT&T, which evolved into the AT&T DSP20 family. Products such as the Texas Instruments TMS32010 and NEC 7720 soon followed this. The early DSP chips exhibited several shortcomings, such as difficult programming paradigms, awkward architectures for many applications, and limited numerical precision. Many of these difficulties were imposed by the limits of the VLSI technology of the time, and some by inexperience in this particular application area. Despite these shortcomings, however, the early processors were well suited to the implementation of digital filter algorithms, because digital filtering was identified as one of the target areas for these architectures. This match between architecture and algorithms continues to be exhibited in current general-purpose DSP chips.

21.2.2 Current Processors

There are a variety of general-purpose digital signal processors currently commercially available. We will look at several of the most common architectural families in detail, although this discussion will not be comprehensive by any means. The processors are best classified into two categories, fixed-point processors and floating-point processors. In both cases, these architectures are commonly (although not exclusively) based on a single arithmetic unit shared among all computations, which leads to constraints on the sampling rates that may be attained.

Fixed-point processors exhibit extremely high performance in terms of maximum throughput as compared to their floating-point counterparts. In addition, fixed-point processors are typically inexpensive as compared to floating-point options, due to the smaller integrated circuit die area occupied by fixed-point processing blocks. A major difficulty encountered in implementing filters on fixed-point processors is that overflow and underflow need to be prevented by careful attention to scaling, and round-off effects may be significant.

Floating-point processors, on the other hand, are significantly easier to program, particularly in the case of complex algorithms, at the cost of lower performance and larger die area. Given the regular structure of most digital filtering algorithms and computer-aided design support for filters based on limited precision arithmetic, fixed-point implementations may be the more cost effective option for this type of algorithm. Because of the prevalence of both types of general-purpose processor, examples of each will be examined in detail.

Two widely used floating-point processor families will be studied, although there are many contenders in this field. These families are the Texas Instruments family of floating-point DSPs, in particular the TI TMS320C3x (TI, 1992) family, and the Analog Devices ADSP-21020 family (Schweber, 1993). More

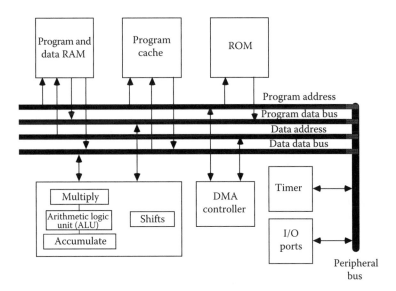

FIGURE 21.1 Texas Instruments TMS320C30 architecture.

recent examples of the TI family are the TMS320C67x DSP chips (TI, 2006a), and more recent Analog Devices parts are the ADSP-2116x SHARC chips (Analog Devices, 2006).

The architecture of the TI TMS320C30 is illustrated in Figure 21.1. The floating-point word size used by this processor was 32 bits. The most prominent feature of this chip was the floating-point arithmetic unit, which contains a floating-point multiplier and adder. This unit was highly pipelined to support high throughput, at the cost of latency; when data is input to the multiplier, for example, the results will not appear on the output from that unit until several clock cycles later. Other features included a separate integer unit for control calculations, and significant amounts (2k words) of SRAM for data and on-chip instruction memory. On-chip ROM (4k words) was also optionally provided in order to eliminate the need for an external boot ROM in some applications. This chip also included a 64-word instruction cache to allow its use with lower speed memories. The modified Harvard architecture, that is, the separate data and instruction buses, provided for concurrent instruction and data word transfers within one cycle time. The TMS320C30 offered instruction cycle times as low as 60 ns. A code segment which implements portions of an finite impulse response (FIR) filter on this device was

```
     RPTS    RC
     MPYF3   *AR0++(1),*AR1++(1)%,R0
||   ADDF3   R0,R2,R2
     ADDF    R0,R2,R0
```

where the MPYF3 instruction performs a pipelined multiply operation in parallel with data and coefficient pointer increments. The ADDF3 instruction is performed in parallel with the MPYF3 instruction, as denoted by the "||" symbol. Because these operations are in parallel, only one instruction cycle per tap is required. An FIR filter tap was benchmarked at 60 ns on this chip. Similarly, a typical biquad infinite impulse response (IIR) filter code segment was

```
     MPYF3   *AR0,*AR1,R0
     MPYF3   *++AR0(1),*AR1--(1)%,R1
     MPYF3   *++AR0(1),*AR1,R0
```

```
||  ADDF3   R0,R2,R2
    MPYF3   *++AR0(1),*AR1--(1)%,R0
||  ADDF3   R0,R2,R2
    MPYF3   *++AR0(1),R2,R2
||  STF     R2,*AR1++(1)%
    ADDF    R0,R2
    ADDF    R1,R2,R0
```

where the `MPYF3` and `ADDF3` instructions implement the primary filter arithmetic and memory pointer modification operations in parallel, as in the previous example. The biquad IIR benchmark on this processor was 300 ns. More recent members of the TI floating-point family such as the TMS320C6727 support two parallel FIR filter taps at 2.86 ns and two IIR biquad sections at 106 ns.

Another floating-point chip worthy of note is the Analog Devices ADSP-21020 series. The architecture of the ADSP-21020 chip is shown in Figure 21.2. This chip can be seen to share a number of features with the TMS320C3x family, that is, a 32 bit by 32 bit floating-point MAC unit (not pipelined), modified Harvard architecture, and 16 words of on-chip memory. In this case, the scratchpad memory was organized into register files, much like a general-purpose RISC architecture register set. The memory capacity of this device was significantly smaller than that of its competitors. As in the case of the TMS320C3x, on the other hand, an instruction cache (32 words) was also provided. The cycle time for the ADSP-21020 was 40 ns. An N tap FIR filter code segment illustrates the operation of this device,

```
            i0 = coef;
            f9 = 0.0;
            f1 = 0; f4 = dm(i0,m0); f5 = pm(i8,m8);
            lcntr = N, DO bottom until lce;
bottom:         f1 = f1+f9; f9 = f4*f5; f4 = dm(i0,m0); f5 = pm(i8,m8);
            f1 = f1+f9;
```

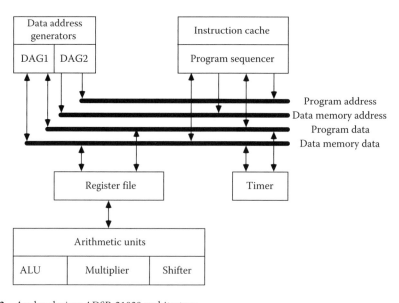

FIGURE 21.2 Analog devices ADSP-21020 architecture.

where the "*" and "+" instructions perform the MAC operations, and the dm() and pm() instructions perform the memory address update operations in parallel. An FIR filter tap thus executed in one instruction per tap on the ADSP-21020, or in 40 ns. An IIR filter biquad section required 200 ns on this chip. More recent members of this family such as the ADSP-21367 support FIR filter taps at 3 ns and IIR biquad sections at 60 ns. Note that while the assembly language for the Analog Devices chip was significantly different from that of the Texas Instruments chip, the architectural similarities are striking.

Two families of fixed-point digital signal processors will also be examined and compared. These are the Texas Instruments TMS320C5x family (TI, 1993) and the Motorola DSP56000 series of devices (Motorola, 1989). More recent examples of the TI family are the TMS320C62x DSP chips (TI, 2006b) and more recent Motorola parts are the DSP56300 series of chips (Motorola, 2007).

The Texas Instruments TMS320C5x series devices were high performance digital signal processors derived from the original TI DSP chip, the TMS32010, and its successor, the TMS320C2x. The architecture of the TMS320C50 is shown in Figure 21.3. This chip was based on the Harvard architecture, that is, separate data and instruction buses. This additional bandwidth between processing elements supported rapid concurrent transfers of data and instructions. This chip used a 16 bit by 16 bit fixed-point multiplier and a 32-bit accumulator, and up to 10k words on-chip scratchpad RAM. This architecture supported instruction rates of 50 ns. An FIR filter code segment is shown below, where the primary filter tap operations were performed by the MACD instruction,

```
RPTK  N
MACD  *-,COEFFP
```

This exhibits a general similarity with that for the TI floating-point chips, in particular a single instruction cycle per tap, although in this case a single instruction was executed as opposed to two parallel instructions on the TMS320C3x. The memory addressing scheme was also significantly different. An FIR filter on the TMS320C5x could thus be implemented in 25 ns per tap. An Nth-order IIR filter code segment is show below, where the MACD and AC instructions performed the primary multiplication operations,

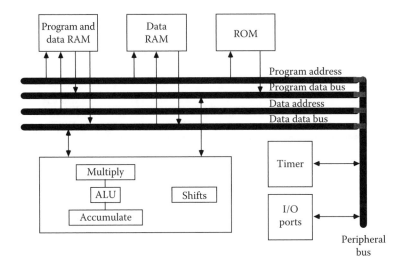

FIGURE 21.3   Texas Instruments TMS320C50 architecture.

```
ZPR
LACC   *,15,AR1
RPT    #(N-2)
AC     COEFFB,*-
APAC
SACH   *,1
ADRK   N-1
RPTZ   #(N-1)
MACD   COEFFA,*-
LTA    *,AR2
SACH   *,1
```

A single IIR biquad section could be performed in 250 ns on this chip. More recent members of the TI family such as the TMS320C6203C support two parallel FIR filter taps at 3.33 ns speeds and IIR biquad sections at 66.6 ns.

The Motorola 56001 series was a fixed-point architecture with 24-bit word size, as opposed to the smaller word sizes in most fixed-point DSP chips. The architecture of the 56001 is depicted in Figure 21.4. This chip shared many of the same features as other DSP chips, that is, Harvard architecture, on-chip scratchpad memory (512 words), and hardware MAC support, in this case 24 bit by 24 bit operators which form a 56 bit result. The instruction cycle time of the Motorola 56001 was 97.5 ns. An FIR filter implemented on the 56001 might use the code segment shown below,

```
MOVE   #AADDR, R0
MOVE   #BADDR+n, R4
NOP
```

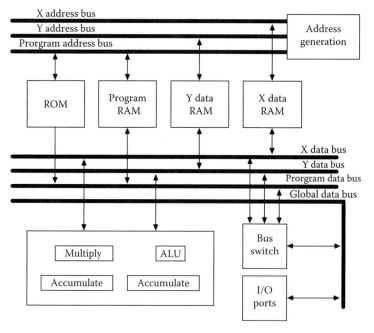

FIGURE 21.4   Motorola 56001 architecture.

```
CLR  A           X:(R0)+,X0  Y:(R4)-,Y0
REP  #N
MAC  X0,Y0,A  X:(R0)+,X0  Y:(R4)-,Y0
RND  A
```

where the MAC instruction retrieves data from the appropriate registers, loads it into the multiplier, and leaves the result in the accumulator. The 56001 could perform FIR filtering at a rate of one instruction per tap, or 97.5 ns per tap. An IIR filter code segment used the MAC instruction, as well as several others to set up the registers for the arithmetic unit, as shown below.

```
OR    #$08,MR
RND   A           X:(R0)-,X0   Y:(R4)+,Y0
MAC   -Y0,X0,A   X:(R0)-,X1   Y:(R4)+,Y0
MAC   -Y0,X1,A   X1,X:(R0)+   Y:(R4)+,Y0
MAC   Y0,X0,A    A,X:(R0)     Y:(R4),Y0
MAC   Y0,X1,A                 1
MOVE          A,X:OUTPUT
```

The 56001 could compute a second-order IIR biquad in seven instruction cycles, or 682.5 ns. More recent members of this family such as the DSP56L307 support FIR filter taps at an asymptotic speed of 6.25 ns and IIR biquad sections at 56.25 ns.

From these examples, it can be seen that general-purpose DSP processors possess many common features which make them well suited for digital filtering. The hardware MAC unit, Harvard architecture, and on-chip memory are consistent characteristics of these devices. The major shortcoming of such architectures for digital filtering is the necessity to multiplex a single arithmetic unit (or very small number of ALUs), which implies that sampling rates above $1/NT$ are not possible, where N is the number of atomic operations (e.g., FIR filter taps) and T is the time to complete those operations.

21.2.3 Future Directions

Several trends have become apparent as VLSI technology has improved. One trend of note is the increasing use of parallelism, both on-chip and between chips. The support for multiprocessor communications in the TI TMS320C80 provides an avenue for direct parallel implementation of algorithms. Architectures based upon multiple fixed-point DSP processors on a single chip have also been fielded.

Another trend has been the development of better programming interfaces for the general-purpose chips. In particular, high level language compilers have improved to the point where they provide for reasonably good performance for complex algorithms, although still not superior to that obtained by manual assembly language programming.

This trend is being accelerated by an initiative called GNURadio (GNU Radio, 2007), which is making it easier to implement all types of digital filters purely in software in order to enable creation of software radios. GNURadio provides signal processing blocks implemented in C++ that are connected together to form a complete communications system. The advantage of this software implementation is that it is generally hardware independent and can run on any general-purpose processor, opening up the world of digital implementations to anyone with a PC. The added flexibility comes at a loss of efficiency. Due to the generalized nature of the hardware independent system, the computations will not be implemented as efficiently as they would be if written in assembly for a specific processor. An IIR filter code segment from the GNURadio project is shown below:

```
acc = d_fftaps[0] * input;
for (i = 1; i < n; i ++)
  acc += (d_fftaps[i] * d_prev_input[latest + i]
        + d_fbtaps[i] * d_prev_output[latest + i]);
```

This C++ code segment from the GNURadio library implement a direct form 1 IIR filter where fftaps are the feed-forward taps and fbtaps are the feed-back taps. To use the GNURadio software for practical signal analysis, some sort of analog to digital apparatus is required. This can be as simple as a standard PC sound card or as complex as the universal software radio peripheral (USRP), which is sold by the maintainers of the GNURadio project as an RF front end that contains an onboard field programmable gate arrays (FPGA) and several optional daughter boards for the frequency band of interest (USRP, 2007).

Another trend that is worthy of note is the development of low-power DSP implementations. These devices are targeted at the wireless personal communications system (PCS) marketplace, where minimum power usage is critical. The developments in this area have been particularly striking, given the strong dependence of power consumption on clock frequency, which is usually high in DSP implementations. Through a combination of careful circuit design, power supply voltage reductions, and architectural innovations, extremely low power implementations have been realized.

A final trend is related to the progress of general-purpose processors relative to DSP chips. The evolution of general-purpose DSP implementations may have come full circle, as general-purpose processors such as the Intel Pentium family and digital equipment company (DEC) Alpha family possess on-chip floating-point multiplication units, as well as memory bandwidths equaling or exceeding that of the DSP chips. These features are reflected in the performance of these chips on standard benchmarks (Stewart et al., 1992), in which the DEC Alpha outperforms the fastest DSP engines. Similar results were obtained from the Pentium upon the implementation of the MMX capabilities; even older Pentium chipsets outperform most floating-point and fixed-point DSP chips (BTDI, 2000).

These trends, particularly multiple processor parallelism and general-purpose ease of programming, are illustrated by the Cell Broadband Engine Architecture developed via a joint effort between Sony, Toshiba, and IBM (IBM, 2007). The Cell combines a standard Power architecture core with streamlined coprocessing components, with the first major use of the Cell being in Sony's Playstation 3 console. It was designed to bridge the gap between general-purpose processors and more specialized high-performance processors. The architecture consists of a main processor called the power processing element (PPE), eight coprocessors called the synergistic processing elements (SPEs), and a high-bandwidth bus connecting all of these elements. The PPE is able to run a conventional operating system because of its Power Architecture core, and has control over the eight SPEs allowing high performance and mathematically intensive tasks to be achieved.

21.3 Special-Purpose Implementations

The tremendous growth in the capabilities of VLSI technology and the corresponding decrease in the fabrication costs have lead to the wide availability advent of application-specific integrated circuits (ASICs). These devices are tailored to a particular application or domain of applications in order to provide the highest possible performance at low per-unit costs.

Although it is difficult to generalize, special-purpose implementations share some common features. The first is the high degree of parallelism in these designs. For example, a typical special-purpose FIR filter implementation will contain tens or hundreds of MAC units, each of which executes a filter tap operation at the same time. This is in contrast to most general-purpose architectures, in which a single MAC unit is shared. Another common feature is extensive pipelining between arithmetic operators; this leads to high sampling rates and high throughput, at some cost in latency. Finally, these implementations are often lacking in flexibility, being designed for specific application domains. The number of filter taps may be fixed, or the filter coefficients themselves may be fixed. In almost all instances, these implementations are based on fixed-point arithmetic.

Because the implementation cost of multiplication operations is so large compared to other operations, significant research effort has been expended on developing fast and efficient multiplier architectures, as well as digital filter design techniques that can be used to reduce the number of multiplications. A large number of multiplier architectures have been developed, ranging from bit-serial structures to bit and word level pipelined array designs (Ma and Taylor, 1990). The most appropriate architecture is a function of the application requirements, as various area versus speed options are available. The other major research direction is the minimization of multiplication operations. In this case, multiplications are eliminated by conscientious structuring of the realization, as in linear phase filters, circumvented by use of alternate number systems such as the residue number system (RNS), or simplified to a limited number of shift-and-add operations. The later option has been used successfully in a large number of both FIR and IIR realizations, some of which will be discussed below.

Historically, bit-serial implementations of digital filters have been of some interest to researchers and practitioners in the early days of VLSI because of the relatively high cost of silicon area devoted to both devices and routing (Denyer and Renshaw, 1985). Even in primitive technologies, bit-serial implementations could exploit the natural parallelism in digital filtering algorithms.

As clock rates have risen, and silicon area has become more economical, parallel implementations have become the most effective way of implementing high-performance digital filters. The concept of the systolic array has strongly influenced the implementation of both FIR and IIR filters (Kung, 1988). Systolic arrays are characterized by spatial and temporal locality, that is, algorithms and processing elements should be structured to minimize interconnection distances between nodes and to provide at least a single delay element between nodes. Interconnection distances need to be kept to a minimum to reduce delays associated with signal routing, which is becoming the dominant limiting factor in VLSI systems. Imposing pipeline delays between nodes minimizes computational delay paths and leads to high throughput.

These characteristic features of special-purpose digital filter designs will be illustrated by examples of FIR and IIR filter implementations. It should be noted that it is increasingly difficult to identify ASICs that only perform digital filtering; as VLSI capabilities increase, this functionality is more typically embedded with other functions in very application-focused devices.

21.3.1 FIR Filter Examples

FIR filters may be implemented in a number of ways, depending on application requirements. The primary factors that must be considered are the filter length, sampling rate, and area, which determine the amount of parallelism that can be applied. Once the degree of parallelism and pipelining are determined, the appropriate general filter structure can be determined.

A typical high-performance FIR filter implementation (Khoo et al., 1993) provided sampling rates of 180 MHz for 32 linear phase taps. This chip used canonical signed digit (CSD) coefficients. This representation is based on a number system in which the digits take the values $(-1, 0, 1)$. A filter tap can be implemented with a small number of these digits, and hence that tap requires a small number of shift-and-add operations. Each coefficient was implemented based on two bit shift-and-add units, as depicted in Figure 21.5. Delay elements are bypassed during configuration to allow realization of coefficients with additional bits. This chip also made use of extensive pipelining, carry-save addition, and advanced single-phase clocking techniques to provide high throughput.

In part due to the highly structured nature of FIR filtering algorithms, automatic design tools have been used to successfully implement high-performance FIR filters similar to that just presented. These methods often integrate the filter and architectural design into a unified process which can effectively utilize silicon area to provide the desired performance.

At the other extreme of performance was the Motorola 56200 FIR filter chip (Motorola, 1988) (Figure 21.6). This chip, although quite old, represents an approach to the custom implementation of

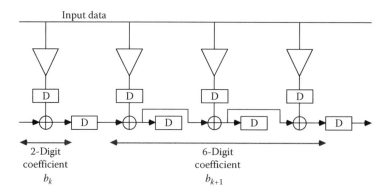

FIGURE 21.5   Custom FIR filter architecture for 180 MHz sampling rates.

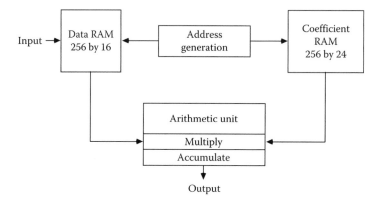

FIGURE 21.6   Motorola 56200 architecture.

long (several hundred taps) FIR filters. In this case, a single processing element was multiplexed among all of the filter taps, similar in concept to the approach used in general-purpose DSP processors. Due to the regularity of the filter structure, extensive pipelining in the arithmetic unit could be used to support a large number of taps at audio rates. This chip could be used to realize a 256 tap FIR filter at sampling rates up to 19 kHz, with higher performance for shorter filters. Longer filters could be implemented using cascaded processors.

A comparison of implementations (Hartley et al., 1989; Hatamian and Rao, 1990; Khoo et al., 1993; Laskowski and Samueli 1992; Ruetz, 1989; Yassa et al., 1987; Yoshino et al., 1990) illustrates the range of design and performance options. This is illustrated in Table 21.1, where the "score" is calculated

TABLE 21.1   FIR Filter ASIC Comparison

Design	Taps	Area (mm^2)	Rate (MHz)	Technology (μm)	Score
Khoo 93	32	20.1	180.0	1.2	495.2
Laskowski 92	43	40.95	150.0	1.2	139.3
Yoshino 90	64	48.65	100.0	0.8	33.68
Ruetz 89	64	225.0	22.0	1.5	21.12
Yassa 87	16	17.65	30.0 (est.)	1.25	53.12
Hartley 89	4	25.8 (est.)	37.0	1.25	11.20
Hatamian 90	40	22.0	100.0	0.9	132.5

according to the sampling rate multiplied by the number of taps per unit area, with normalization for the particular technology used. This simplistic comparison does not consider differences in word length or coefficient codings, but it does provide some insight into the results of the various design approaches. A significant number of other digital FIR filtering chips exist, both research prototypes and commercial products; this exposition only outlines some of the architectural options.

21.3.2 IIR Filter Examples

Custom IIR filter implementations are also most commonly based on parallel architectures, although there are somewhat fewer custom realizations of IIR filters than FIR filters. A significant difficulty in the implementation of high performance IIR filters in the need for feedback in the computation of an IIR filter section. This limits the throughput that can be attained to at least one MAC cycle in a straightforward realization. Another difficulty is the numerical stability of IIR filters with short coefficients, which makes aggressive quantization of coefficients less promising.

In order to address the difficulties with throughput limitation due to feedback, structures based on systolic concepts have been developed. Although the feedback problem imposes a severe constraint on the implementation, use of bit and word level systolic structures which pipeline data most significant digit first can minimize the impact of this restriction (Woods et al., 1990). Using these techniques, and a Signed Binary Number Representation (SBNR) similar to a CSD code, first-order sections with sampling rates of 15 MHz were demonstrated in a 1.5 μm standard cell process in an area of 21.8 mm^2. This particular design used fairly large coefficient and data words, however, at 12 bits and 11 bits, respectively.

The numerical stability problem has been addressed through a variety of techniques. One of these is based on minimizing limited precision effects by manipulation of traditional canonical filter structures and clever partitioning of arithmetic operations. A more recent and general approach is based on modeling the digital implementations of filters after their analog counterparts; these classes of filters are known as wave digital filters (WDFs) (Fettweis, 1986). WDFs exhibit good passband ripple and stopband attenuation, with high tolerance to limited wordlength effects. Because of the latter property, efficient implementations based on short word sizes are feasible. A WDF design for a second-order section in custom 1.5 μm CMOS based on a restricted coefficient set akin to CSD supported 10 MHz sampling rates in an area of 12.9 mm^2 (Wicks and Summerfield, 1993).

21.3.3 Future Trends

The future trends in digital filter implementation appear to be a fairly straightforward function of the increasing capability of VLSI devices. In particular, more taps and filter sections per chip and higher sampling rates are becoming achievable. Related to these trends are higher degrees of on-chip parallelism. Further, programmability is more reasonable as density and speed margins increase, although there is still a high cost in area and performance. Finally, special-purpose implementations show extraordinary promise in the area of low power systems, where custom circuit design techniques and application-specific architectural features can be combined to best advantage.

21.4 Programmable Logic Implementations

The rapid evolution of VLSI technology has enabled the development of several high-density programmable logic architectures. There are several novel features that make these devices of interest beyond their traditional field-of-state machine implementation. In particular, the density of the largest of these devices is over 12,000,000 gates (Xilinx, 2007), which encompasses the level of complexity found in the majority of ASICs (although some ASICs are significantly more complex). This level of complexity is sufficient to support many designs that would traditionally need to be implemented as ASICs. The speed of programmable logic devices (PLDs) and FPGAs is quite reasonable, with toggle rates on the order of 550 MHz (Xilinx, 2007). While this is not as great as custom implementations, it does allow many applications to be realized in this new technology.

One of the most significant features of FPGA implementations is the capability for in-system reprogrammability in many FPGA families. Unlike traditional field programmable parts based on anti-fuse technology and which can only be programmed once, many of the new architectures are based on memory technology. This means that entirely new computational architectures can be implemented simply by reprogramming the logic functions and interconnection routing on the chip. Ongoing research efforts have been directed toward using FPGAs as generalized coprocessors for supercomputing and signal processing applications.

The implications of programmable device technology for filter implementation are significant. These devices provide an enormous amount of flexibility, which can be used in the implementation of a variety of novel architectures on a single chip. This is particularly useful for rapid prototyping of digital filtering algorithms, where several high-performance designs can be evaluated in a target environment on the same hardware platform. Further, complex adaptive systems based on this technology and which use a variety of signal processing and digital filter techniques are becoming increasingly popular in a variety of applications.

Because many of the programmable logic architectures are based on SRAM technology, the density of these devices can be expected to grow in parallel with the RAM growth curve, that is, at approximately 60% per year. Further, since these devices may be used for a large variety of applications, they have become high-volume commodity parts, and hence prices are relatively low compared to more specialized and low-volume DSP chips. This implies that new DSP systems that were not previously technically and economically feasible to implement in this technology are now feasible.

One of the extra costs of this approach, as opposed to the full custom strategy, is the need for support chips. Several chips are typically needed, including memory to store the programmable device configuration, as well as logic to control the downloading of the program. These issues are generally outweighed by the flexibility provided by programmable solutions.

We will next examine the implementation of several FIR and IIR digital filtering architectures based on FPGAs.

21.4.1 FIR Filter Implementations

Several approaches to the FPGA implementation of FIR filters can be taken. Due to the flexibility of these parts, switching from one architecture to the next only requires reprogramming the device, subject to constraints on I/O pin locations. Two fundamental strategies for realizing FIR filters will be illustrated here, one which is suited to relatively short filters (or longer filters cascaded across several chips) operating at high rates, and another which is suited for longer filters at lower rates.

A high-performance FIR filter example (Evans, 1993), illustrated in Figure 21.7, was based on the observation that since the entire device is reprogrammable, architectures in which filter coefficient multiplications are implemented as "hardwired" shifts can be easily reconfigured depending on the desired filter response. In this example, each of the coefficients was represented in a CSD code with a limited number of nontrivial (e.g., nonzero) bits, which allowed each tap to be implemented as a small number of shift-and-add operations. A filter tap could be implemented in two columns of logic blocks on a Xilinx 3100-series FPGA (Xilinx, 1993), where the two columns of full adders and associated delays implement a tap based on CSD coefficients with two nontrivial bits. With this architecture, up to 11 taps could be implemented on a single Xilinx XC3195 FPGA at sampling rates of above 40 MHz. Longer filters could be implemented by a cascade of FPGA devices.

An FIR filter architecture for longer filters was based upon implementation of several traditional MAC units on one chip, as shown in Figure 21.8. Each of these MAC units could then be shared among a large number of filter tap computations, much as the single MAC unit in the Motorola 56200 was multiplexed.

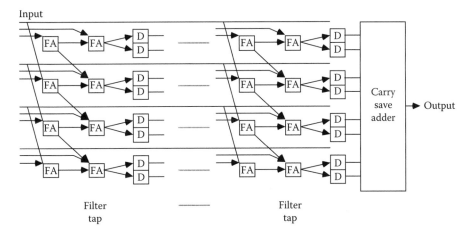

FIGURE 21.7   High-performance FIR architecture on FPGAs.

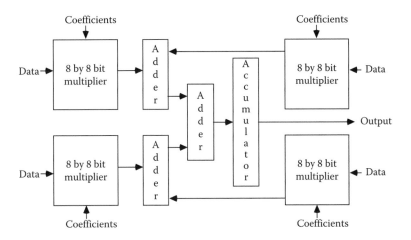

FIGURE 21.8   FIR architecture on FPGAs for large number of taps.

Since four multipliers could be implemented in the Xilinx 4000-series, the inherent parallelism of FIR filters can be exploited to support sampling rates of up to 1.25 MHz for 32 taps in that technology.

21.4.2 IIR Filter Implementations

As in the case of FIR filters, IIR filters can be implemented using a "hardwired" architecture suited to high performance, or a more traditional approach based on general MAC units. In the case of IIR filters, however, the hardwired implementation is significantly more desirable than the alternate approach due to the difficulty in rescheduling multiplexed processing elements in a system with feedback.

An architecture which is reconfigured to implement different filters will generally provide both high-performance and good area efficiency. An example of such a system is shown in Figure 21.9, in which two IIR biquad sections were implemented on a single FPGA using a traditional canonical filter structure (Chou et al., 1993). Each of the columns realized a shift-and-add for one nontrivial bit of a coefficient, where the shaded blocks also contained delay registers. This implementation yielded sampling rates of better than 10 MHz for typical coefficients.

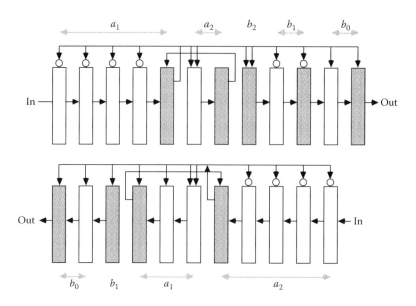

FIGURE 21.9   High-performance IIR architecture on FPGAs.

A more traditional approach to the realization of IIR filters using MAC units is also possible, but may be less efficient. The general architecture is similar to that of the FIR filter in Figure 21.8, with slight modifications to the routing between arithmetic units and support for scaling necessary in an IIR biquad section.

21.4.3 Future Trends

Because of the rapid advances in FPGA technology, higher performance digital filtering may in fact be possible with programmable logic than with typical custom ASIC approaches (Moeller, 1999). In addition, there are a wide variety of DSP core functions being offered by FPGA manufacturers (Xilinx, 2000) which will further accelerate this revolution in DSP implementation.

References

Analog Devices, 2006. *ADSP-21367 SHARC DSP Hardware Reference*, Rev. A., Analog Devices, Norwood, MA.

BTDI, 2000. *BDTImark Scores*, Berkeley Design Technologies, Inc., Berkeley, CA.

Chou, C. -J., S. Mohanakrishnan, and J. Evans, 1993. FPGA implementation of digital filters, *International Conference on Signal Processing Applications and Technology*, pp. 80–88, Santa Clara, CA.

Denyer, P. and D. Renshaw, 1985. *VLSI Signal Processing: A Bit-Serial Approach*, Addison-Wesley, Reading, MA.

Evans, J., 1993. An efficient FIR filter architecture, *IEEE International Symposium on Circuits and Systems*, pp. 627–630, Chicago, IL.

Fettweis, A., 1986. Wave digital filters: Theory and practice, *Proceedings of the IEEE*, 74(2), 270–327.

GNU Radio, 2007. GNU Radio 3.0.3 Release, http://www.gnu.org/software/gnuradio/index.html

Hartley, R., P. Corbett, et al., 1989. A high speed FIR filter designed by Compiler, *IEEE Cust. IC Conf.*, pp. 20.2.1–20.2.4, San Diego, CA.

Hatamian, M. and S. Rao, 1990. A 100 MHz 40-tap programmable FIR filter chip, *IEEE International Symposium on Circuits and Systems*, pp. 3053–3056, New Orleans, LA.

IBM, 2007, *The CELL project at IBM Research*, http://www.research.ibm.com/cell/

Khoo, K.-Y., A. Kwentus, and A. Willson, Jr., 1993. An efficient 175 MHz programmable FIR digital filter, *IEEE International Symposium on Circuits and Systems*, pp. 72–75, Chicago, IL.

Kung, S. Y., 1988. *VLSI Array Processors*, Prentice-Hall, Englewood Cliffs, NJ.

Laskowski, J. and Samueli, H., 1992. A 150-mhz 43-tap half-band fir digital filter in 1.2-μm CMOS generated by silicon compiler, *IEEE Custom Integrated Circuits Conference*, pp. 11.4.1–11.4.4.

Ma, G.-K. and F. Taylor, 1990. Multiplier policies for digital signal processing, *IEEE ASSP Mag.*, January, pp. 6–20.

Moeller, T. J., 1999. Field programmable gate arrays for radar front-end digital signal processing, MS thesis, MIT, Cambridge, MA.

Motorola, 1988. *DSP56200 Cascadable Adaptive Finite Impulse Response Digital Filter Chip*, Motorola, Phoenix, AZ.

Motorola, 1989. *DSP56001 Digital Signal Processor User's Manual*, Motorola, Phoenix, AZ.

Motorola, 2007. *DSP56374 24-Bit Digital Signal Processor User's Manual*, Motorola, Austin, TX.

Ruetz, P., 1989. The architectures and design of a 20-MHz real-time DSP chip set, *IEEE J. Solid State Circuits*, 24(2), 338–348.

Schweber, W., 1993. Floating-point DSP for high-speed signal processing, *Analog Dialogue*, 25(4), 3–5.

Stewart, L., A. Payne, and T. Levergood, 1992. Are DSP chips obsolete?, *International Conference on Signal Processing Applications and Technology*, Nov. 1992, pp. 178–187, Cambridge, MA.

Texas Instruments (TI), 1992. *TMS320C3x User's Manual*, Texas Instruments, Dallas, TX.

Texas Instruments (TI), 1993. *TMS320C5x User's Manual*, Texas Instruments, Dallas, TX.

Texas Instruments (TI), 2006a. *TMS320C6727 Floating-Point Digital Signal Processor*, Texas Instruments, Dallas, TX.

Texas Instruments (TI), 2006b. *TMS320C6205 Fixed-Point Digital Signal Processors*, Texas Instruments, Dallas, TX.

USRP, 2007, *Universal Software Radio Peripheral User's Manual*, Ettus Research LLC, Mountain View, CA.

Wicks, A. and S. Summerfield, 1993. VLSI implementation of high speed wave digital filters based on a restricted coefficient set, *IEEE International Symposium on Circuits and Systems*, pp. 603–606, Chicago, IL.

Woods, R., J. McCanny, S. Knowles, and O. McNally, 1990. A high performance IIR digital filter chip, *IEEE International Symposium on Circuits and Systems*, pp. 1410–1413, New Orleans, LA.

Xilinx, 1993. *The Programmable Logic Data Book*, Xilinx, San Jose, CA.

Xilinx, 2000. *XtremeDSP Technical Backgrounder*, Xilinx, San Jose, CA.

Xilinx, 2007. *Virtex-5 Platform FPGA Handbook*, Xilinx, San Jose, CA.

Yassa, F., J. Jasica, et al., 1987. A silicon compiler for digital signal processing: methodology, implementation, and applications, *Proc. IEEE*, 75(9), 1272–1282.

Yoshino, T., R. Jain, et al., 1990. A 100-MHz 64-Tap FIR digital filter in 0.8 μm BiCMOS gate array, *IEEE J. Solid State Circuits*, 25(6), 1494–1501.

Further Information

The publication IEEE Transactions on Circuits and Systems—II: Analog and Digital Signal Processing frequently contains articles on the VLSI implementation of digital filters as well as design methods for efficient implementation. The *IEEE Transactions on Signal Processing* often includes articles in these areas as well. Articles in the *IEEE Journal on Solid State Circuits*, the *IEEE Transactions on VLSI Systems*, and the *IEE Electronics Letters* regularly cover particular implementations of digital filters.

The conference proceedings for the IEEE International Symposium on Circuits and Systems and the IEEE International Conference on Acoustics, Speech, and Signal Processing also contain a wealth of information on digital filter implementation.

The textbook *VLSI Array Processors* by S. Y. Kung discusses the concept of systolic arrays at length. The textbook *Software Radio: A Modern Approach to Radio Engineering* by Jeffrey Reed covers issues engineers must understand in order to utilize DSP in software radio subsystems. The textbook *Digital Signal Processing with Field Programmable Gate Arrays* by Uwe Meyer-Baese discusses how FPGA implementations are revolutionizing digital signal processing and covers implementations of FIR and IIR filters along with several other DSP processing systems on FPGAs.

22

Two-Dimensional
FIR Filters

Rashid Ansari
University of Illinois at Chicago

A. Enis Cetin
Bilkent University

22.1 Introduction

In this chapter, methods of designing two-dimensional (2-D) finite-extent impulse response (FIR) discrete-time filters are described. 2-D FIR filters offer the advantages of phase linearity and guaranteed stability, which makes them attractive in applications. Over the years an extensive array of techniques for designing 2-D FIR filters has been accumulated [14,23,30]. These techniques can be conveniently classified into the two categories of general and specialized designs. Techniques in the category of general design are intended for approximation of *arbitrary* desired frequency responses usually with no structural constraints on the filter. These techniques include approaches such as windowing of the ideal impulse response (IIR) [22] or the use of suitable optimality criteria possibly implemented with iterative algorithms. On the other hand, techniques in the category of special design are applicable to restricted classes of filters, either due to the nature of the response being approximated or due to imposition of structural constraints on the filter used in the design. The specialized designs are a consequence of the observation that commonly used filters have characteristic underlying features that can be exploited to simplify the problem of design and implementation. The stopbands and passbands of filters encountered in practice are often defined by straight-line, circular, or elliptical boundaries. Specialized design methodologies have been developed for handling these cases and they are typically based on techniques such as the transformation of one-dimensional (1-D) filters or the rotation and translation of separable filter responses. If the desired response possesses symmetries, then the symmetries imply relationships among the filter coefficients, which are exploited in both the design and the implementation of the filters. In some design problems it may be advantageous to impose structural constraints in the form of parallel and cascade connections.

The material in this chapter is organized as follows. A preliminary discussion of characteristics of 2-D FIR filters and issues relevant to the design methods appears in Section 22.2. Following this, methods of general and special FIR filter design are described in Sections 22.3 and 22.4, respectively. Several examples of design illustrating the procedure are also presented. Issues in 2-D FIR filter implementation are briefly discussed in Section 22.5. Finally, two-dimensional filter banks are briefly discussed in Section 22.6, and a list of sources for further information is provided.

22.2 Preliminary Design Considerations

In any 2-D filter design there is a choice between FIR and IIR filters, and their relative merits are briefly examined next. 2-D FIR filters offer certain advantages over 2-D IIR filters as a result of which FIR filters have found widespread use in applications such as image and video processing. One key attribute of an FIR filter is that it can be designed with strictly linear passband phase, and it can be implemented with small delays without the need to reverse the signal array during processing. A 2-D FIR filter impulse response has only a finite number of nonzero samples which guarantee stability. On the other hand, stability is difficult to test in the case of 2-D IIR filters due to the absence of a 2-D counterpart of the fundamental theorem of algebra, and a 2-D polynomial is almost never factorizable. If a 2-D FIR filter is implemented nonrecursively with finite precision, then it does not exhibit limit cycle oscillations. Arithmetic quantization noise and coefficient quantization effects in FIR filter implementation are usually very low. A key disadvantage of FIR filters is that they typically have higher computational complexity than IIR filters for meeting the same magnitude specifications, especially in cases where the specifications are stringent.

The term 2-D FIR filter refers to a linear shift-invariant system whose input–output relation is represented by a convolution [14]

$$y(n_1, n_2) = \sum_{(k_1, k_2) \in I} \sum h(k_1, k_2) x(n_1 - k_1, n_2 - k_2), \tag{22.1}$$

where
$x(n_1, n_2)$ and $y(n_1, n_2)$ are the input and the output sequences, respectively
$h(n_1, n_2)$ is the impulse response sequence
I is the support of the impulse response sequence

FIR filters have compact support, meaning that only a finite number of coefficients are nonzero. This makes the impulse response sequence of FIR filters absolutely summable, thereby ensuring filter stability. Usually the filter support, I, is chosen to be a rectangular region centered at the origin, e.g., $I = \{(n_1, n_2): -N_1 \leq n_1 \leq N_1, -N_2 \leq n_2 \leq N_2\}$. However, there are some important cases where it is more advantageous to select a nonrectangular region as the filter support [32].

Once the extent of the impulse response support is determined, the sequence $h(n_1, n_2)$ should be chosen in order to meet given filter specifications under suitable approximation criteria. These aspects are elaborated in Section 22.2.1. This is followed by a discussion of phase linearity and filter response symmetry considerations and then some guidelines on using the design methods are provided (Sections 22.2.2 and 22.2.3).

22.2.1 Filter Specifications and Approximation Criteria

The problem of designing a 2-D FIR filter consists of determining the impulse response sequence, $h(n_1, n_2)$, or its system function, $H(z_1, z_2)$, in order to satisfy given requirements on the filter response. The filter requirements are usually specified in the frequency domain, and only this case is considered here. The frequency response,* $H(\omega_1, \omega_2)$, corresponding to the impulse response $h(n_1, n_2)$, with a support, I, is expressed as

$$H(\omega_1, \omega_2) = \sum_{(n_1, n_2) \in I} h(n_1, n_2) e^{-j(n_1 \omega_1 + n_2 \omega_2)}. \tag{22.2}$$

Note that $H(\omega_1, \omega_2) = H(\omega_1 + 2\pi, \omega_2) = H(\omega_1, \omega_2 + 2\pi)$ for all (ω_1, ω_2). In other words, $H(\omega_1, \omega_2)$ is a periodic function with a period 2π in both ω_1 and ω_2. This implies that by defining $H(\omega_1, \omega_2)$ in the region $\{-\pi < \omega_1 \leq \pi, -\pi < \omega_2 \leq \pi\}$, the frequency response of the filter for all (ω_1, ω_2) is determined.

For 2-D FIR filters the specifications are usually given in terms of the magnitude response, $|H(\omega_1, \omega_2)|$. Attention in this chapter is confined to the case of a two-level magnitude design, where the desired magnitude levels are either 1.0 (in the passband) or 0.0 (in the stopband). Some of the procedures can be easily modified to accommodate multilevel magnitude specifications, as, for instance, in a case that requires the magnitude to increase linearly with distance from the origin in the frequency domain.

Consider the design of a 2-D FIR low-pass filter whose specifications are shown in Figure 22.1. The magnitude of the low-pass filter ideally takes the value 1.0 in the passband region, F_p, which is centered around the origin, $(\omega_1, \omega_2) = (0, 0)$, and 0.0 in the stopband region, F_s. As a magnitude discontinuity is not possible with a finite filter support, I, it is necessary to interpose a transition region, F_t, between F_p and F_s. Also, magnitude bounds $|H(\omega_1, \omega_2) - 1| \leq \delta_p$ in the passband and $|H(\omega_1, \omega_2)| \leq \delta_s$ in the stopband are specified, where the parameters δ_p and δ_s are positive real numbers, typically much less than 1.0. The frequency response $H(\omega_1, \omega_2)$ is assumed to be real. Consequently, the low-pass filter is specified in the frequency domain by the regions, F_p, F_s, and the tolerance parameters, δ_p and δ_s.

A variety of stopband and passband shapes can be specified in a similar manner.

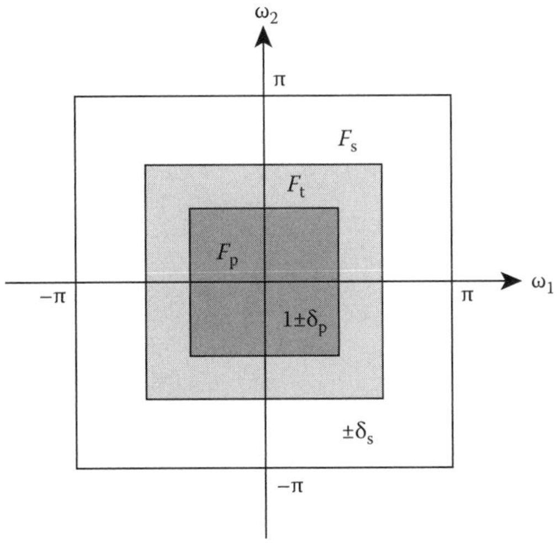

In order to meet given specifications, an adequate filter order (defined here to be the number of nonzero impulse response samples) needs to be determined. If the specifications are stringent, with tight tolerance parameters and small transition regions, then the filter support region, I, must be large. In other words, there is a trade-off between the filter support region, I, and the frequency domain specifications. In the general case the filter order is not known *a priori*, and may be determined either through an iterative process or using estimation rules if available. If the filter order is given, then in order to determine an optimum solution to the design problem, an appropriate optimality criterion is needed. Commonly used criteria in 2-D filter design are minimization of the L_p norm, p finite, of the approximation error, or the L_∞ norm. If desired, a maximal flatness requirement at desired frequencies can be imposed

FIGURE 22.1 Frequency response specifications for a 2D low-pass filter ($|H(\omega_1, \omega_2) - 1| \leq \delta_p$ for $(\omega_1, \omega_2) \in F_p$ and $|H(\omega_1, \omega_2)| \leq \delta_s$ for $(\omega_1, \omega_2) \in F_s$).

---

* Here $\omega_1 = 2\pi f_1$ and $\omega_2 = 2\pi f_2$ are the horizontal and vertical angular frequencies, respectively.

[24]. It should be noted that if the specifications are given in terms of the tolerance bounds on magnitude, as described above, then the use of L_∞ criterion is appropriate. However, the use of other criteria such as a weighted L_2 norm can serve to arrive at an almost minimax solution [2,10].

22.2.2 Zero-Phase FIR Filters and Symmetry Considerations

Phase linearity is important in many filtering applications. As in the 1-D case, a number of conditions for phase linearity can be obtained depending on the nature of symmetry. But the discussion here is limited to the case of "zero phase" design, with a purely real frequency response. A salient feature of 2-D FIR filters is that realizable FIR filters, which have purely real frequency responses, are easily designed. The term "zero phase" is somewhat misleading in the sense that the frequency response may be negative at some frequencies. The term should be understood in the sense of "zero phase in passband" because the passband frequency response is within a small deviation of the value 1.0. The frequency response may assume negative values in the stopband region where phase linearity is immaterial. In frequency domain, the zero-phase or real frequency response condition corresponds to

$$H(\omega_1, \omega_2) = H^*(\omega_1, \omega_2), \tag{22.3}$$

where $H^*(\omega_1, \omega_2)$ denotes the complex conjugate of $H(\omega_1, \omega_2)$. The condition Equation 22.3 is equivalent to

$$h(n_1, n_2) = h^*(-n_1, -n_2) \tag{22.4}$$

in the spatial-domain. Making a common practical assumption that $h(n_1, n_2)$ is real, the above condition reduces to

$$h(n_1, n_2) = h(-n_1, -n_2), \tag{22.5}$$

implying a region of support with the above symmetry about the origin.

Henceforth, only the design of zero-phase FIR filters is considered. With $h(n_1, n_2)$ real, and satisfying Equation 22.5, the frequency response, $H(\omega_1, \omega_2)$, is expressed as

$$\begin{aligned} H(\omega_1, \omega_2) &= h(0,0) + \sum_{(n_1, n_2) \in I_1} h(n_1, n_2) e^{-j(\omega_1 n_1 + \omega_2 n_2)} + \sum_{(n_1, n_2) \in I_2} h(n_1, n_2) e^{-j(\omega_1 n_1 + \omega_2 n_2)} \\ &= h(0,0) + \sum_{(n_1, n_2) \in I_1} 2h(n_1, n_2) \cos(\omega_1 n_1 + \omega_2 n_2), \end{aligned} \tag{22.6}$$

where I_1 and I_2 are disjoint regions such that $I_1 \cup I_2 \cup \{(0,0)\} = I$, and if $(n_1, n_2) \in I_1$, then $(-n_1, -n_2) \in I_2$.

In order to understand the importance of phase linearity in image processing, consider an example that illustrates the effect of nonlinear-phase filters on images. In Figure 22.2a, an image that is corrupted by white Gaussian noise is shown. This image is filtered with a nonlinear-phase low-pass filter and the resultant image is shown in Figure 22.2b. It is observed that edges and textured regions are severely distorted in Figure 22.2b. This is due to the fact that the spatial alignment of frequency components that define an edge in the original is altered by the phase nonlinearity. The same image is also filtered with a zero-phase low-pass filter, $H(\omega_1, \omega_2)$, which has the same magnitude characteristics as the nonlinear-phase filter. The resulting image is shown in Figure 22.2c. It is seen that the edges are perceptually preserved in Figure 22.2c, although blurred due to the low-pass nature of the filter. In this example, a separable zero-phase low-pass filter, $H(\omega_1, \omega_2) = H_1(\omega_1) H_1(\omega_2)$, is used, where $H_1(\omega)$ is a 1-D Lagrange filter with a cutoff $\pi/2$. In spatial domain $h(n_1, n_2) = h_1(n_1) h_1(n_2)$ where $h_1(n) = \{\ldots, 0, -1/32, 0, 9/32,$

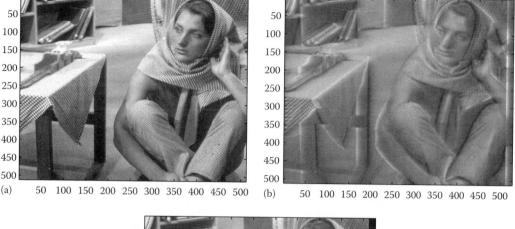

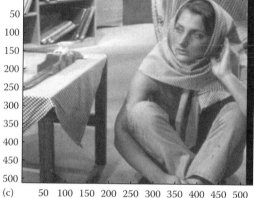

FIGURE 22.2 (a) Original image of 696 × 576 pixels; (b) nonlinear-phase low-pass filtered image; and (c) zero-phase low-pass filtered image.

1/2, 9/32, 0, −1/32, 0,...} is the impulse response of the seventh-order symmetric (zero-phase) 1-D Lagrange filter. The nonlinear-phase filter is a cascade of the above zero-phase filter with an allpass filter.

In some filter design problems, symmetries in frequency domain specifications can be exploited by imposing restrictions on the filter coefficients and the shape of the support region for the impulse response. A variety of symmetries that can be exploited is extensively studied in Refs. [3,32,44,45]. For example, a condition often encountered in practice is the symmetry with respect to each of the two frequency axes. In this case, the frequency response of a zero-phase filter satisfies

$$H(\omega_1, \omega_2) = H(-\omega_1, \omega_2) = H(\omega_1, -\omega_2). \tag{22.7}$$

This yields an impulse response that is symmetric with respect to the n_1 and n_2 axes, i.e.,

$$h(n_1, n_2) = h(-n_1, n_2) = h(n_1, -n_2). \tag{22.8}$$

By imposing symmetry conditions, one reduces the number of independently varying filter coefficients that must be determined in the design. This can be exploited in reducing both the computational complexity of the filter design and the number of arithmetic operations required in the implementation.

22.2.3 Guidelines on the Use of the Design Techniques

The design techniques described in this chapter are classified into the two categories of general and specialized designs. The user should use the techniques of general design in cases requiring approximation of arbitrary desired frequency responses, usually with no structural constraints on the filter. The specialized designs are recommended in cases where filters exhibit certain underlying features that can be exploited to simplify the problem of design and implementation.

In the category of general design, four methods are described. Of these, the windowing procedure is quick and simple. It is useful in situations where implementation efficiency is not critical, especially in single-use applications. The second procedure is based on linear programming, and is suitable for design problems where equiripple solutions are desired to meet frequency domain specifications. The remaining two procedures may also be used for meeting frequency domain specifications, and lead to nearly equiripple solution. The third procedure provides solutions for L_p approximations. The fourth procedure is an iterative procedure that is easy to implement, and is convenient in situations where additional constraints are to be placed on the filter.

In the category of specialized design described here, the solutions are derived from 1-D filters. These often lead to computationally efficient implementation, and are recommended in situations where low implementation complexity is critical, and the filter characteristics possess features that can be exploited in the design. An important practical class of filters is one where specifications can be decomposed into a set of separable filter designs requiring essentially the design of suitable 1-D filters. Here the separable design procedure should be used. Another class of filters is one where the passbands and stopbands are characterized by circular, elliptical, or special straight-line boundaries. In this case a frequency transformation method called the McClellan transformation procedure proves effective. The desired 2-D filter constant-magnitude contours are defined by a proper choice of parameters in a transformation of variables applied to a 1-D zero-phase filter. Finally, in some cases filter specifications are characterized by ideal frequency responses in which passbands and stopbands are separated by straight-line boundaries that are not suitable for applying the McClellan transformation procedure. In this case the design may be carried out by nonrectangular transformations and sampling grid conversions. The importance of this design method stems from the implementation efficiency that results from a generalized notion of separable processing.

22.3 General Design Methods for Arbitrary Specifications

Some general methods of meeting arbitrary specifications are now described. These are typically based on extending techniques of 1-D design. However, there are important differences. The Parks–McClellan procedure for minimax approximation based on the alternation theorem does not find a direct extension. This is because the set of cosine functions used in the 2-D approximation does not satisfy the Haar condition on the domain of interest [25], and the Chebyshev approximation does not have a unique solution. However, techniques that employ exchange algorithms have been developed for the 2-D case [20,25,36].

Here we consider four procedures in some detail. The first technique is based on windowing. It is simple, but not optimum for Chebyshev approximation. The second technique is based on frequency sampling, and this can be used to arrive at equiripple solutions using linear programming. Finally, two techniques for arriving iteratively at a nearly equiripple solution are described. The first of these is based on L_p approximations using nonlinear optimization. The second is based on the use of alternating projections in the sample and the frequency domains.

22.3.1 Design of 2-D FIR Filters by Windowing

This design method is basically an extension of the window-based 1-D FIR filter design to the case of 2-D filters. An IIR sequence, which is usually an infinite-extent sequence, is suitably windowed to make the support finite. 1-D FIR filter design by windowing and classes of 1-D windows are described in detail in Section 22.2.

Let $h_{id}(n_1, n_2)$ and $H_{id}(\omega_1, \omega_2)$ be the impulse and frequency responses of the ideal filter, respectively. The impulse response of the required 2-D filter, $h(n_1, n_2)$, is obtained as a product of the IIR sequence and a suitable 2-D window sequence which has a finite-extent support, I, that is,

$$h(n_1, n_2) = \begin{cases} h_{id}(n_1, n_2)w(n_1, n_2), & (n_1, n_2) \in I, \\ 0, & \text{otherwise,} \end{cases} \tag{22.9}$$

where $w(n_1, n_2)$ is the window sequence. The resultant frequency response, $H(\omega_1, \omega_2)$, is a smoothed version of the ideal frequency response as $H(\omega_1, \omega_2)$ is related to the $H_{id}(\omega_1, \omega_2)$ via the periodic convolution, that is,

$$H(\omega_1, \omega_2) = \frac{1}{4\pi^2} \int_{-\pi}^{\pi} \int_{-\pi}^{\pi} H_{id}(\Omega_1, \Omega_2) W(\omega_1 - \Omega_1, \omega_2 - \Omega_2) d\Omega_1 d\Omega_2, \tag{22.10}$$

where $W(\omega_1, \omega_2)$ is the frequency response of the window sequence, $w(n_1, n_2)$.

As in the 1-D case, a 2-D window sequence, $w(n_1, n_2)$, should satisfy three requirements:

1. It must have a finite-extent support, I.
2. Its discrete-space Fourier transform should in some sense approximate the 2-D impulse function, $\delta(\omega_1, \omega_2)$.
3. It should be real, with a zero-phase discrete-space Fourier transform.

Usually 2-D windows are derived from 1-D windows. Three methods of constructing windows are briefly examined. One method consists of obtaining a separable window from two 1-D windows, that is,

$$w_r(n_1, n_2) = w_1(n_1)w_2(n_2), \tag{22.11}$$

where $w_1(n)$ and $w_2(n)$ are the 1-D windows. Thus, the support of the resultant 2-D window, $w_r(n_1, n_2)$, is a rectangular region. The frequency response of the 2-D window is also separable, i.e., $W_r(\omega_1, \omega_2) = W_1(\omega_1) W_2(\omega_2)$.

The second method of constructing a window, due to Huang [22], consists of sampling the surface generated by rotating a 1-D continuous-time window, $w(t)$, as follows:

$$w_c(n_1, n_2) = w\left(\sqrt{n_1^2 + n_2^2}\right), \tag{22.12}$$

where $w(t) = 0$, $t \geq N$. The impulse response support is $I = \left\{n_1, n_2: \sqrt{n_1^2 + n_2^2} < N\right\}$. Note that the 2-D Fourier transform of the $w_c(n_1, n_2)$ is not equal to the circularly rotated version of the Fourier transform of $w(t)$.

Finally, in the third method, proposed by Yu and Mitra [53], the window is constructed by using a 1-D to 2-D transformation belonging to a class called the McClellan transformations [33]. These transformations are discussed in greater detail in Section 22.4. Here we consider a special case of the transform that produces approximately circular contours in the 2-D frequency domain. Briefly, the discrete-space frequency transform of the 2-D window sequence obtained with a McClellan transformation applied to a 1-D window is given by

$$T(\omega_1, \omega_2) = \left. \sum_{n=-N}^{N} w(n)e^{-j\omega n} \right|_{\cos(\omega)=0.5\cos(\omega_1)+0.5\cos(\omega_2)+0.5\cos(\omega_1)\cos(\omega_2)-0.5}$$

$$= \left. w(0) + 2\sum_{n=1}^{N} w(n)\cos(n\omega) \right|_{\cos(\omega)=0.5\cos(\omega_1)+0.5\cos(\omega_2)+0.5\cos(\omega_1)\cos(\omega_2)-0.5}$$

$$= \left. \sum_{n=0}^{N} b(n)\cos^n(\omega) \right|_{\cos(\omega)=0.5\cos(\omega_1)+0.5\cos(\omega_2)+0.5\cos(\omega_1)\cos(\omega_2)-0.5} \tag{22.13}$$

where $w(n)$ is an arbitrary symmetric 1-D window of duration $2N+1$ centered at the origin, and the coefficients, $b(n)$, are obtained from $w(n)$ via Chebyshev polynomials [33]. After some algebraic manipulations it can be shown that

$$T(\omega_1, \omega_2) = \sum_{n_1=-N}^{N} \sum_{n_2=-N}^{N} w_t(n_1, n_2)e^{-j(n_1\omega_1+n_2\omega_2)}, \tag{22.14}$$

where $w_t(n_1, n_2)$ is a zero-phase 2-D window of size $(2N+1) \times (2N+1)$ obtained by using the McClellan transformation.

The construction of 2-D windows using the above three methods is now examined. In the case of windows obtained by the separable and the McClellan transformation approaches, the 1-D prototype is a Hamming window,

$$w_h(n) = \begin{cases} 0.54 + 0.46\cos(\pi n/N), & |n| < N, \\ 0, & \text{otherwise.} \end{cases} \tag{22.15}$$

In the second case $w_c(n_1, n_2) = 0.54 + 0.46\cos(\pi\sqrt{n_1^2 + n_2^2}/N)$. By selecting $w_1(n) = w_2(n) = w_h(n)$ in Equation 22.11 we get a 2-D window, $w_r(n_1, n_2)$, of support $I = \{|n_1| < N, |n_2| < N\}$ which is a square-shaped symmetric region centered at the origin. For $N = 6$ the region of support, I, contains $11 \times 11 = 121$ points. Figure 22.3a shows the frequency response of this window. A second window is designed by using Equation 22.12, i.e., $w_c(n_1, n_2) = w_h(\sqrt{n_1^2 + n_2^2})$. For $N = 6$ the frequency response of this filter is shown in Figure 22.3b. The region of support is almost circular and it contains 113 points. From these examples, it is seen that the 2-D windows may not behave as well as 1-D windows. Speake and Mersereau [46] compared these two methods and observed that the main-lobe width and the highest attenuation level of the side-lobes of the 2-D windows differ from their 1-D prototypes.

Let us construct a 2-D window by the McClellan transformation with a 1-D Hamming window of order 13 ($N = 6$) as the prototype. The frequency response of the 2-D window, $w_t(n_1, n_2)$, is shown in Figure 22.3c. The frequency response of this window is almost circularly symmetric and it preserves the features of its 1-D prototype.

Consider the design of a circularly symmetric low-pass filter. The ideal frequency response for $(\omega_1, \omega_2) \in [-\pi, \pi] \times [-\pi, \pi]$ is given by

$$H_{id}(\omega_1, \omega_2) = \begin{cases} 1, & \sqrt{\omega_1^2 + \omega_2^2} \leq \omega_c, \\ 0, & \text{otherwise,} \end{cases} \tag{22.16}$$

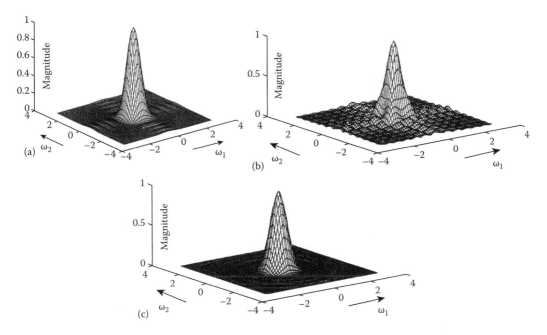

FIGURE 22.3 Frequency responses of the (a) separable, (b) Huang, and (c) McClellan 2D windows generated from a Hamming window of order 13 ($N=6$).

whose impulse response is given by

$$h_{\text{id}}(n_1, n_2) = \frac{\omega_c J_1 \left(\omega_c \sqrt{n_1^2 + n_2^2} \right)}{2\pi \sqrt{n_1^2 + n_2^2}}, \tag{22.17}$$

where

$J_1(\cdot)$ is the first-order Bessel function of the first kind

ω_c is the cutoff frequency

The frequency response of the 2-D FIR filter obtained with a rectangular window of size $2 \times 5 + 1$ by $2 \times 5 + 1$ is shown in Figure 22.4a. Note the Gibbs-phenomenon type ripples at the passband edges.

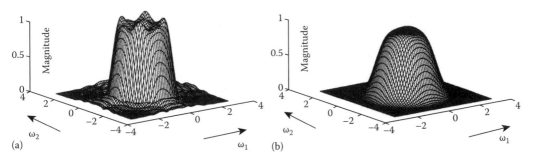

FIGURE 22.4 Frequency responses of the 2D filters designed with (a) a rectangular window and (b) a separable window of Figure 22.3a.

In Figure 22.4b the separable window of Figure 22.3a, derived from a Hamming window, is used to design the 2-D filter. Note that this 2-D filter has smaller ripples at the passband edges.

In windowing methods, it is often assumed that $H_{id}(\omega_1, \omega_2)$ is given. However, if the specifications are given as described in Section 22.2, then a proper $H_{id}(\omega_1, \omega_2)$ should be constructed. The ideal magnitudes are either 1.0 (in passband) or 0.0 (in stopband). However, there is a need to define a *cutoff* boundary, which lies within the transition band. This can be accomplished by using a suitable notion of "midway" cutoff between the transition boundaries. In practical cases where transition boundaries are given in terms of straight-line segments or smooth curves such as circles and ellipses, the construction of "midway" cutoff boundary is relatively straightforward. The IIR, $h_{id}(n_1, n_2)$, is computed from the desired frequency response, $H_{id}(\omega_1, \omega_2)$, either analytically (if possible), or by using the discrete Fourier transform (DFT). In the latter case the desired response, $H_{id}(\omega_1, \omega_2)$, is first sampled on a rectangular grid in the Fourier domain, then an inverse DFT computation is carried out via a 2-D fast Fourier transform (FFT) algorithm to obtain an approximation to the sequence $h_{id}(n_1, n_2)$. The resulting sequence is an aliased version of the IIR. Therefore, a sufficiently dense grid should be used in order to reduce the effects of aliasing.

In practice, several trials may be needed to design the final filter satisfying bounds both in the passbands and stopbands. The filter support is adjusted to obtain the smallest order to meet given requirements.

Filter design with windowing is a simple approach that is suitable for applications where a quick and nonoptimal design is needed. Additional information on windowing can be found in Refs. [26,46].

22.3.2 Frequency Sampling and Linear Programming-Based Method

This method is based on the application of the sampling theorem in the frequency domain. Consider the design of a 2-D filter with impulse response support of $N_1 \times N_2$ samples. The frequency response of the filter can be obtained from a conveniently chosen set of its samples on an $N_1 \times N_2$ grid. For example, the DFT of the impulse response can be used to interpolate the response for the entire region $[0, 2\pi] \times [0, 2\pi]$. The filter design then becomes a problem of choosing an appropriate set of DFT coefficients [21].

One choice of DFT coefficients consists of the ideal frequency response values, assuming a suitable cutoff. However, the resultant filters usually exhibit large magnitude deviations away from the DFT sample locations in the filter passbands and stopbands. The approximation error can be reduced by allowing the DFT values in the transition band to vary, and choosing them to minimize the deviation of the magnitude from the desired values. Another option is to allow all the DFT values to vary, and pick the optimal set of values for minimum error. The use of DFT-based interpolation allows for a computationally efficient implementation. The implementation cost of the method basically consists of a 2-D array product and inverse discrete Fourier transform (IDFT) computation, with appropriate addition.

Let us consider the set $S \subset Z^2$ that defines the equispaced frequency locations $\left(\frac{2k_1 \pi}{N_1}, \frac{2k_2 \pi}{N_2} \right)$:

$$S = \{(k_1, k_2): k_1 = 0, 1, \ldots, N_1, k_2 = 0, 1, \ldots, N_2\}. \tag{22.18}$$

The DFT values can be expressed as

$$H_{DFT}[k_1, k_2] = |H(\omega_1, \omega_2)| \Big|_{(\omega_1, \omega_2) = \left(\frac{2k_1 \pi}{N_1}, \frac{2k_2 \pi}{N_2} \right)}, \quad (k_1, k_2) \in S. \tag{22.19}$$

The filter coefficients, $h(n_1, n_2)$, are found by using an IDFT computation

$$h(n_1, n_2) = \frac{1}{N_1 N_2} \sum_{k_1=0}^{N_1-1} \sum_{k_2=0}^{N_2-1} H_{DFT}[k_1, k_2] e^{j \left(\frac{2\pi}{N_1} k_1 n_1 + \frac{2\pi}{N_2} k_2 n_2 \right)}, \quad (n_1, n_2) \in S. \tag{22.20}$$

If Equation 22.20 is substituted in the expression for frequency response

$$H(\omega_1, \omega_2) = \sum_{n_1=0}^{N_1-1} \sum_{n_2=0}^{N_2-1} h(n_1, n_2) e^{-j(\omega_1 n_1 + \omega_2 n_2)}, \quad (22.21)$$

we arrive at the interpolation formula

$$H(\omega_1, \omega_2) = \sum_{k_1=0}^{N_1-1} \sum_{k_2=0}^{N_2-1} H_{\text{DFT}}[k_1, k_2] A_{k_1 k_2}(\omega_1, \omega_2), \quad (22.22)$$

where

$$A_{k_1 k_2}(\omega_1, \omega_2) = \frac{1}{N_1 N_2} \left(\frac{1 - e^{-jN_1 \omega_1}}{1 - e^{-j(\omega_1 - 2\pi k_1/N_1)}} \right) \left(\frac{1 - e^{-jN_2 \omega_2}}{1 - e^{-j(\omega_2 - 2\pi k_2/N_2)}} \right). \quad (22.23)$$

Equation 22.22 serves as the basis of the frequency sampling design. As mentioned before, if the H_{DFT} are chosen directly according to the ideal response, then the magnitude deviations are usually large. To reduce the ripples, one option is to express the set S as the disjoint union of two sets S_t and S_c, where S_t contains indices corresponding to the transition band F_t, and S_c contains indices corresponding to the care-bands, i.e., the union of the passbands and stopbands, $F_p \cup F_s$. The expression for frequency response in Equation 22.22 can be split into two summations, one over S_t and the other over S_c

$$H(\omega_1, \omega_2) = \sum_{S_t} H_{\text{DFT}}[k_1, k_2] A_{k_1 k_2}(\omega_1, \omega_2) + \sum_{S_c} H_{\text{DFT}}[k_1, k_2] A_{k_1 k_2}(\omega_1, \omega_2), \quad (22.24)$$

where the first term on the right-hand side is optimized. The design equations can be put in the form

$$1 - \alpha\delta \le H(\omega_1, \omega_2) \le 1 + \alpha\delta, \quad (\omega_1, \omega_2) \in F_p \quad (22.25)$$

and

$$-\delta \le H(\omega_1, \omega_2) \le \delta, \quad (\omega_1, \omega_2) \in F_s, \quad (22.26)$$

where
  δ is the peak approximation error in the stopband
  $\alpha\delta$ is the peak approximation error in the passband, where α is any positive constant defining the
    relative weights of the deviations

The problem is readily cast as a linear programming problem with a sufficiently dense grid of points.
  For equiripple design, all the DFT values H_{DFT} over S_t and S_c are allowed to vary. Following is an example of this design.

Example 22.1

The magnitude response for the approximation of a circularly symmetric response is shown in Figure 22.5. Here the passband is the interior of the circle $R_1 = \pi/3$ and the stopband is the exterior of the circle $R_2 = 2\pi/3$. With $N_1 = N_2 = 9$, the passband ripple is 0.08 dB and the minimum stopband attenuation is 32.5 dB.

22.3.3 FIR Filters Optimal in L_p Norm

A criterion different from the minimax criterion is briefly examined. Let us define the error at the frequency pair (ω_1, ω_2) as follows:

$$E(\omega_1, \omega_2) = H(\omega_1, \omega_2) - H_{\text{id}}(\omega_1, \omega_2). \quad (22.27)$$

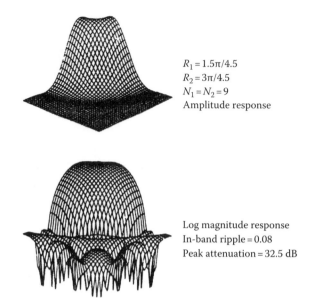

$R_1 = 1.5\pi/4.5$
$R_2 = 3\pi/4.5$
$N_1 = N_2 = 9$
Amplitude response

Log magnitude response
In-band ripple = 0.08
Peak attenuation = 32.5 dB

FIGURE 22.5 Frequency response of the circularly symmetric filter obtained by using the frequency sampling method. (From Hu, J.V. and Rabiner, L.R., *IEEE Trans. Audio Electroacoust.*, 20, 249, 1972. With permission. © 1972 IEEE.)

One design approach is to minimize the L_p norm of the error

$$\varepsilon_p = \left(\frac{1}{4\pi^2} \int_{-\pi}^{\pi} \int_{-\pi}^{\pi} |E(\omega_1, \omega_2)|^p \, d\omega_1 d\omega_2 \right)^{\frac{1}{p}}. \tag{22.28}$$

Filter coefficients are selected by a suitable algorithm. For $p = 2$ Parseval's relation implies that

$$\varepsilon_2^2 = \sum_{n_1=-\infty}^{\infty} \sum_{n_2=-\infty}^{\infty} [h(n_1, n_2) - h_{\mathrm{id}}(n_1, n_2)]^2. \tag{22.29}$$

By minimizing Equation 22.29 with respect to the filter coefficients, $h(n_1, n_2)$, which are nonzero only in a finite-extent region, I, one gets

$$h(n_1, n_2) = \begin{cases} h_{\mathrm{id}}(n_1, n_2), & (n_1, n_2) \in I, \\ 0, & \text{otherwise,} \end{cases} \tag{22.30}$$

which is the filter designed by using a straightforward rectangular window. Due to the Gibbs phenomenon it may have large variations at the edges of passband and stopband regions. A suitable weighting function can be used to reduce the ripple [2], and an approximately equiripple solution can be obtained.

For the general case of $p \neq 2$ [32], the minimization of Equation 22.28 is a nonlinear optimization problem. The integral in Equation 22.28 is discretized and minimized by using an iterative nonlinear optimization technique. The solution for $p = 2$ is easy to obtain using linear equations. This serves as an excellent initial estimate for the coefficients in the case of larger values of p. As p increases, the solution becomes approximately equiripple. The error term, $E(\omega_1, \omega_2)$, in Equation 22.28 is nonuniformly weighted in passbands and stopbands, with larger weight given close to band-edges where deviations are typically larger.

22.3.4 Iterative Method for Approximate Minimax Design

We now consider a simple procedure based on alternating projections in the sample and frequency domains, which leads to an approximately equiripple response. In this method, the zero-phase FIR filter design problem is formulated to alternately satisfy the frequency domain constraints on the magnitude response bounds and spatial domain constraints on the impulse response support [1,11,12]. The algorithm is iterative and each iteration requires two 2-D FFT computations.

As pointed out in Section 22.2, 2-D FIR filter specifications are given as requirements on the magnitude response of the filter. It is desirable that the frequency response, $H(\omega_1, \omega_2)$, of the zero-phase FIR filter be within prescribed upper and lower bounds in its passbands and stopbands. Let us specify bounds on the frequency response $H(\omega_1, \omega_2)$ of the minimax FIR filter, $h(n_1, n_2)$, as follows:

$$H_{id}(\omega_1, \omega_2) - E_d(\omega_1, \omega_2) \le H(\omega_1, \omega_2) \le H_{id}(\omega_1, \omega_2) + E_d(\omega_1, \omega_2) \quad \omega_1, \omega_2 \in R, \tag{22.31}$$

where

$H_{id}(\omega_1, \omega_2)$ is the ideal filter response

$E_d(\omega_1, \omega_2)$ is a positive function of (ω_1, ω_2) which may take different values in different passbands and stopbands

R is a region defined in Equation 22.28 consisting of passbands and stopbands of the filter (note that $H(\omega_1, \omega_2)$ is real for a zero-phase filter)

Usually, $E_d(\omega_1, \omega_2)$ is chosen constant in a passband or a stopband. Inequality equation (Equation 22.31) is the frequency domain constraint of the iterative filter design method.

In spatial domain the filter must have a finite-extent support, I, which is symmetric region around the origin. The spatial domain constraint requires that the filter coefficients must be equal to zero outside the region, I.

The iterative method begins with an arbitrary finite-extent, real sequence $h_0(n_1, n_2)$ that is symmetric ($h_0(n_1, n_2) = h_0(-n_1, n_2)$). Each iteration consists of making successive imposition of spatial and frequency domain constraints onto the current iterate. The kth iteration consists of the following steps:

- Compute the Fourier transform of the kth iterate $h_k(n_1, n_2)$ on a suitable grid of frequencies by using a 2-D FFT algorithm.
- Impose the frequency domain constraint as follows:

$$G_k(\omega_1, \omega_2) = \begin{cases} H_{id}(\omega_1, \omega_2) + E_d(\omega_1, \omega_2) & \text{if } H_k(\omega_1, \omega_2) > H_{id}(\omega_1, \omega_2) + E_d(\omega_1, \omega_2), \\ H_{id}(\omega_1, \omega_2) - E_d(\omega_1, \omega_2) & \text{if } H_k(\omega_1, \omega_2) < H_{id}(\omega_1, \omega_2) - E_d(\omega_1, \omega_2), \\ H_k(\omega_1, \omega_2) & \text{otherwise.} \end{cases} \tag{22.32}$$

- Compute the inverse Fourier transform of $G_k(\omega_1, \omega_2)$.
- Zero out $g_k(n_1, n_2)$ outside the region I to obtain h_{k+1}.

The flow diagram of this method is shown in Figure 22.6. It can be proven that the algorithm converges for all symmetric input sequences. This method requires the specification of the bounds or equivalently, $E_d(\omega_1, \omega_2)$, and the filter support, I. In 2-D filter design, filter order estimates for prescribed frequency domain specifications are not available. Therefore, successive reduction of bounds is used. If the specifications are too tight, then the algorithm does not converge. In such cases one can either progressively enlarge the filter support region, or relax the bounds on the ideal frequency response.

The size of the 2-D FFT must be chosen sufficiently large. The passband and stopband edges are very important for the convergence of the algorithm. These edges must be represented accurately on the frequency grid of the FFT algorithm.

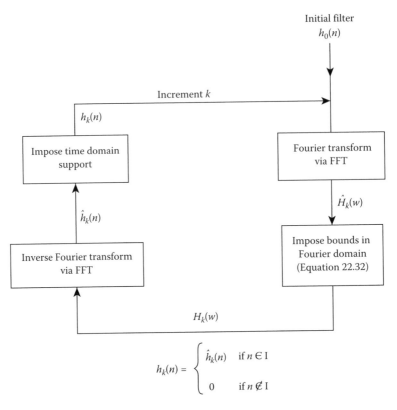

FIGURE 22.6   Flow diagram of the iterative filter design algorithm.

The shape of the filter support is very important in any 2-D filter design method. The support should be chosen to exploit the symmetries in the desired frequency response. For example, diamond-shaped supports show a clear advantage over the commonly assumed rectangular regions in designing diamond filters or 90° fan filters [4,17].

Since there are efficient FFT routines, 2-D FIR filters with large orders can be designed by using this procedure.

Example 22.2

Let us consider the design of a circularly symmetric low-pass filter. Maximum allowable deviation is $\delta_p = \delta_s = 0.05$ in both passband and the stopband. The passband and stopband cutoff boundaries have radii of $0.43\,\pi$ and $0.63\,\pi$, respectively. This means that the functions $E_d(\omega_1, \omega_2) = 0.05$ in the passband and the stopband. In the transition band the frequency response is conveniently bounded by the lower bound of the stopband and the upper bound of the passband. The filter support is a square-shaped 17×17 region. The frequency response of this filter is shown in Figure 22.7.

Example 22.3

Let us now consider an example in which we observe the importance of filter support. We design a fan filter whose specifications are shown in Figure 22.8. Maximum allowable deviation is $\delta_p = \delta_s = 0.1$ in both passband and the stopband. If one uses a 7×7 square-shaped support which has 49 points, then it cannot meet the design specifications. However, a diamond-shaped support,

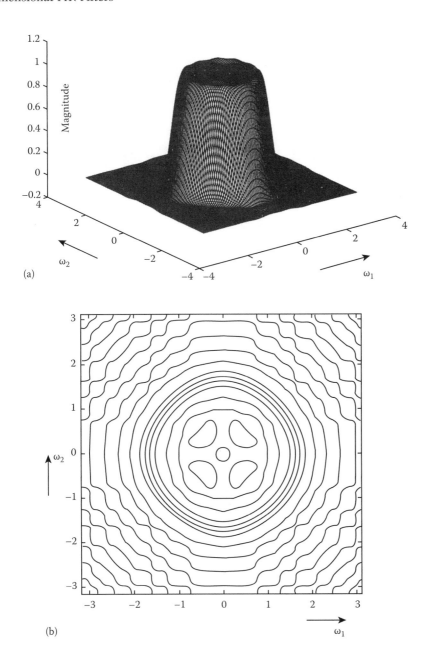

FIGURE 22.7 (a) Frequency response and (b) contour plot of the low-pass filter of Example 22.1.

$$I_d = \{-5 \leq n_1 + n_2 \leq 5\} \cap \{-5 \leq n_1 - n_2 \leq 5\}, \tag{22.33}$$

together with the restriction that ("de" for diamond-support with some eliminated samples)

$$I_{de} = I_d \cap \{n_1 + n_2 = \text{odd or } n_1 = n_2 = 0\} \tag{22.34}$$

produces a filter satisfying the bounds. The filter support region, I_{de}, contains 37 points. The resultant frequency response is shown in Figure 22.8.

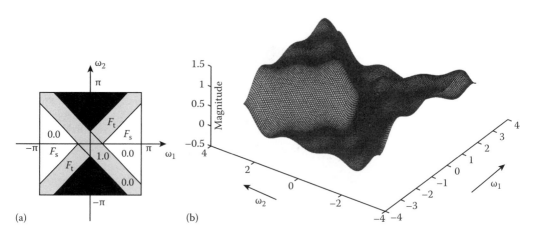

FIGURE 22.8   (a) Specifications and (b) perspective frequency response of the fan filter designed in Example 22.2.

22.4 Special Design Procedure for Restricted Classes

Many cases of practical importance typically require filters belonging to restricted classes. The stopbands and passbands of these filters are often defined by straight-line, circular, or elliptical boundaries. In these cases, specialized procedures lead to efficient design and low-cost implementation. The filters in these cases are derived from 1-D prototypes.

22.4.1 Separable 2-D FIR Filter Design

The design of 2-D FIR filters composed of 1-D building blocks is briefly discussed. In cases where the specifications are given in terms of multiple passbands in the shapes of rectangles with sides parallel to the frequency axes, the design problem can be decomposed into multiple designs. The resulting filter is a parallel connection of component filters that are themselves separable filters. The separable structure was encountered earlier in the construction of 2-D windows from 1-D windows in Section 22.3. The design approach is essentially the same. We will confine the discussion to cascade structures, which is a simple and very important practical case.

The frequency response of the 2-D separable FIR filter is expressed as

$$H(\omega_1, \omega_2) = H_1(\omega_1)H_2(\omega_2), \tag{22.35}$$

where $H_1(\omega)$ and $H_2(\omega)$ are frequency responses of two 1-D zero-phase FIR filters of durations N_1 and N_2. The corresponding 2-D filter is also a zero-phase FIR filter with $N \times M$ coefficients, and its impulse response is given by

$$h(n_1, n_2) = h_1(n_1)h_2(n_2), \tag{22.36}$$

where $h_1(n)$ and the $h_2(n)$ are the impulse responses of the 1-D FIR filters.

If the ideal frequency response can be expressed in a separable cascade form as in Equation 22.35, then the design problem is reduced to the case of appropriate 1-D filter designs. A simple but important example is the design of a 2-D low-pass filter with a symmetric square-shaped passband, PB = {(ω_1, ω_2): $|\omega_1| < \omega_c$, $|\omega_2| < \omega_c$}. Such a low-pass filter can be designed from a single 1-D FIR filter with a cutoff frequency of ω_c by using Equation 22.36. A low-pass filter constructed this way is used in Figure 22.2c.

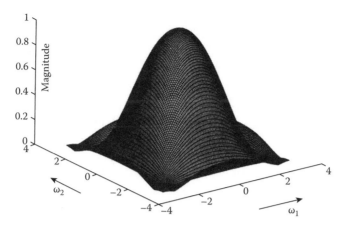

FIGURE 22.9 Frequency response of the separable low-pass filter $H(\omega_1, \omega_2) = H_1(\omega_1)H_1(\omega_2)$ where $H_1(\omega)$ is a seventh-order Lagrange filter.

The frequency response of this 2-D filter whose 1-D prototypes are seventh-order Lagrange filters is shown in Figure 22.9.

This design method is also used in designing 2-D filter banks which are utilized in subband coding of images and video signals [49,51,52]. The design of 2-D filter banks is discussed in Section 22.6.

22.4.2 Frequency Transformation Method

In this method a 2-D zero-phase FIR filter is designed from a 1-D zero-phase filter by a clever substitution of variables. The design procedure was first proposed by McClellan [33] and the frequency transformation is usually called the McClellan transformation [14,35,37,38].

Let $H_1(\omega)$ be the frequency response of a 1-D zero-phase filter with $2N+1$ coefficients. The key idea of this method is to find a suitable transformation $\omega = G(\omega_1, \omega_2)$ such that the 2-D frequency response, $H(\omega_1, \omega_2)$, which is given by

$$H(\omega_1, \omega_2) = |H_1(\omega)|_{\omega=G(\omega_1, \omega_2)} \tag{22.37}$$

approximates the desired frequency response, $H_{id}(\omega_1, \omega_2)$.

Since the 1-D filter is a zero-phase filter, its frequency response is real, and it can be written as follows:

$$H_1(\omega) = h_1(0) + \sum_{n=1}^{N} 2h_1(n) \cos(\omega n), \tag{22.38}$$

where the term $\cos(\omega n)$ can be expressed as a function of $\cos(\omega)$ by using the nth-order Chebyshev polynomial, T_n,[*] i.e.,

$$\cos(\omega n) = T_n[\cos(\omega)]. \tag{22.39}$$

---

[*] Chebyshev polynomials are recursively defined as follows: $T_0(x) = 1$, $T_1(x) = x$, and $T_n(x) = 2xT_{n-1}(x) - T_{n-2}(x)$.

Using Equation 22.39, the 1-D frequency response can be written as

$$H_1(\omega) = \sum_{n=0}^{N} 2b(n)[\cos(\omega)]^n, \tag{22.40}$$

where the coefficients, $b(n)$, are related to the filter coefficients, $h(n)$.

In this design method the key step is to substitute a transformation function, $F(\omega_1, \omega_2)$, for $\cos(\omega)$ in Equation 22.40. In other words, the 2-D frequency response, $H(\omega_1, \omega_2)$, is obtained as follows:

$$H(\omega_1, \omega_2) = H_1(\omega)\big|_{\cos(\omega)=F(\omega_1, \omega_2)}$$

$$= \sum_{n=0}^{N} 2b(n)[F(\omega_1, \omega_2)]^n. \tag{22.41}$$

The function, $F(\omega_1, \omega_2)$, is called the McClellan transformation.

The frequency response, $H(\omega_1, \omega_2)$, of the 2-D FIR filter is determined by two free functions, the 1-D prototype frequency response, $H_1(\omega)$, and the transformation, $F(\omega_1, \omega_2)$. In order to have $H(\omega_1, \omega_2)$ be the frequency response of an FIR filter, the transformation, $F(\omega_1, \omega_2)$, must itself be the frequency response of a 2-D FIR filter. McClellan proposed $F(\omega_1, \omega_2)$ to be the frequency response of a 3×3 zero-phase filter in Ref. [33]. In this case the transformation, $F(\omega_1, \omega_2)$, can be written as follows:

$$F(\omega_1, \omega_2) = A + B \, \cos(\omega_1) + C \, \cos(\omega_2) + D \, \cos(\omega_1 - \omega_2) + E \, \cos(\omega_1 + \omega_2), \tag{22.42}$$

where the real parameters, A, B, C, D, and E, are related to the coefficients of the 3×3 zero-phase FIR filter. For $A = -(1/2)$, $B = C = (1/2)$, $D = E = (1/4)$, the contour plot of the transformation, $F(\omega_1, \omega_2)$, is shown in Figure 22.10. Note that in this case the contours are approximately circularly symmetric around

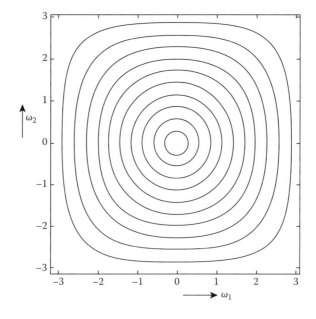

FIGURE 22.10   Contour plot of the McClellan transformation, $F(\omega_1, \omega_2) = 0.5 \, \cos(\omega_1) + 0.5 \, \cos(\omega_2) + 0.5 \, \cos(\omega_1) \cos(\omega_2) - 0.5$.

the origin. It can be seen that the deviation from the circularity, expressed as a fraction of the radius, decreases with the radius. In other words, the distortion from a circular response is larger for large radii. It is observed from Figure 22.10 that, with the above choice of parameters, A, B, C, D, and E, the transformation is bounded ($|F(\omega_1, \omega_2)| \leq 1$), which implies that $H(\omega_1, \omega_2)$ can take only the values that are taken by the 1-D prototype filter, $H_1(\omega)$. Since $|\cos(\omega)| \leq 1$, the transformation, $F(\omega_1, \omega_2)$, which replaces $\cos(\omega)$ in Equation 22.41 must also take values between 1 and -1. If a particular transformation does not obey these bounds, then it can be scaled such that the scaled transformation satisfies the bounds.

If the transformation, $F(\omega_1, \omega_2)$, is real (it is real in Equation 22.42) then the 2-D filter, $H(\omega_1, \omega_2)$, will also be real or, in other words, it will be a zero-phase filter. Furthermore, it can be shown that the 2-D filter, $H(\omega_1, \omega_2)$, is an FIR filter with a support containing $(2M_1N + 1) \times (2M_2N + 1)$ coefficients, if the transformation, $F(\omega_1, \omega_2)$, is an FIR filter with $(2M_1 + 1) \times (2M_2 + 1)$ coefficients, and the order of the 1-D prototype filter is $2N + 1$. In (19.42) $M_1 = M_2 = 1$. As it can be intuitively guessed, one can design a 2-D approximately circularly symmetric low-pass (high-pass) (bandpass) filter with the above McClellan transformation by choosing the 1-D prototype filter, $H_1(\omega)$, a low-pass (high-pass) (bandpass) filter.

We will present some examples to demonstrate the effectiveness of the McClellan transformation.

Example 22.4

2-D window design by transformations [53]: In this example we design 2-D windows by using the McClellan transformation. Actually, we briefly mentioned this technique in Section 22.3. The 1-D prototype filter is chosen as an arbitrary 1-D symmetric window centered at the origin. Let $w(n)$ be the 1-D window of size $2N + 1$, and $W(\omega) = \sum_{n=-N}^{N} w(n) \exp(-j\omega n)$ be its frequency response. The transformation, $F(\omega_1, \omega_2)$, is chosen as in Equation 22.42 with the parameters $A = -(1/2)$, $B = C = (1/2)$, $D = E = (1/4)$, of Figure 22.10. This transformation, $F(\omega_1, \omega_2)$, can be shown to be equal to

$$F(\omega_1, \omega_2) = 0.5 \cos(\omega_1) + 0.5 \cos(\omega_2) + 0.5 \cos(\omega_1) \cos(\omega_2) - 0.5. \tag{22.43}$$

The frequency response of the McClellan window, $H_t(\omega_1, \omega_2)$, is given by

$$H_t(\omega_1, \omega_2) = W(\omega)|_{\cos(\omega) = F(\omega_1, \omega_2)}. \tag{22.44}$$

The resultant 2-D zero-phase window, $w_t(n_1, n_2)$, is centered at the origin and of size $(2N + 1) \times (2N + 1)$ because $M_1 = M_2 = 1$. The window coefficients can be computed either by using the inverse Chebyshev relation,* or by using the inverse Fourier transform of Equation 22.44. The frequency response of a 2-D window constructed from a 1-D Hamming window of order 13 is shown in Figure 22.3c. The size of the window is 13×13.

Example 22.5

Let us consider the design of a circularly symmetric low-pass filter and a bandpass filter by using the transformation of Equation 22.43. In this case, if one starts with a 1-D low-pass (bandpass) filter as the prototype filter, then the resulting 2-D filter will be a 2-D circularly symmetric low-pass (bandpass) filter due to the almost circularly symmetric nature of the transformation. In this example, the Lagrange filter of order 7 considered in Section 22.2 is used as the prototype. The prototype 1-D bandpass filter of order 15 is designed by using the Parks–McClellan algorithm [41]. Frequency response and contour plots of the low-pass and bandpass filters are shown in Figures 22.11 and 22.12, respectively.

---

* $1 = T_0(x)$, $x = T_1(x) - T_0(x)$, $x^2 = (t_0(x) + T_2(x))$, $x^3 = (3T_1(x) + T_3(x))$ etc.

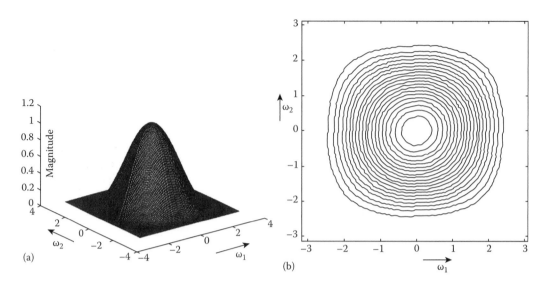

(a)

(b)

FIGURE 22.11   Frequency response and contour plots of the low-pass filter of Example 22.5.

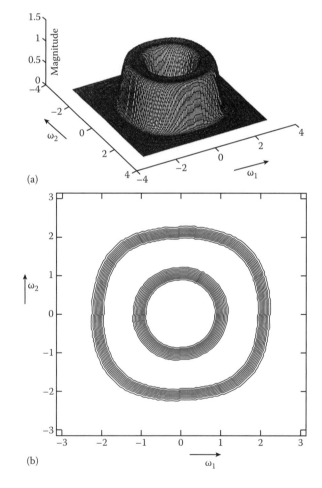

(a)

(b)

FIGURE 22.12   Frequency response and contour plots of the bandpass filter of Example 22.5.

It is seen from the above examples that filters designed by the transformation method appear to have better frequency responses than those designed by the windowing or frequency sampling methods. In other words, one can control the 2-D frequency response by controlling the frequency response of the 1-D prototype filter and choosing a suitable 2-D transformation. Furthermore, in some special cases it was shown that minimax optimal filters can be designed by the transformation method [20].

We have considered specific cases of the special transformations given by Equation 22.42. By varying the parameters in Equation 22.42 or expanding the transformation to include additional terms, a wider class of contours can be approximated. Ideally, the frequency transformation approach requires the simultaneous optimal selection of the transformation, $F(\omega_1, \omega_2)$, and the 1-D prototype filter $H_1(\omega)$ to approximate a desired 2-D frequency response. This can be posed as a nonlinear optimization problem. However, a suboptimal two-stage design by separately choosing $F(\omega_1, \omega_2)$ and $H_1(\omega)$ works well in practice. The transformation $F(\omega_1, \omega_2)$ should approximate $1(-1)$ in the passband (stopband) of the desired filter. The contour produced by the transformation corresponding to the 1-D passband (stop-band) edge frequency, $\omega_p(\omega_s)$, should ideally map to the given passband (stopband) boundary in the 2-D specifications. However, this cannot be achieved in general given the small number of variable parameters in the transformation. The parameters are therefore selected to minimize a suitable norm of the error between actual and ideal (constant) values of the transformation over the boundaries.

Various transformations and design considerations are described in Refs. [37,38,40,42,43]. The use of this transformation in exact reconstruction filter bank design was proposed in Ref. [7].

Filters designed by the transformation method can be implemented in a computationally efficient manner [14,30]. The key idea is to implement Equation 22.41 instead of implementing the filter by using the direct convolution sum. By implementing the transformation, $F(\omega_1, \omega_2)$, which is an FIR filter of low-order, in a modular structure realizing Equation 22.41 is more advantageous than ordinary convolution sum [14,34].

In the case of circular passband design, it was observed that for low-order transformation, the transformation contours exhibit large deviations from circularity. A simple artifice to overcome this problem in approximating wideband responses is to use decimation of a 2-D narrowband filter impulse response [18]. The solution consists of transforming the specifications to an appropriate narrowband design, where the deviation from circularity is smaller. The narrow passband can be expanded by decimation while essentially preserving the circularity of the passband.

22.4.3 Design Using Nonrectangular Transformations and Sampling Rate Conversions

In some filter specifications the desired responses are characterized by ideal frequency responses in which passbands and stopbands are separated by straight-line boundaries that are not necessarily parallel to the frequency axes. Examples of these are the various kinds of fan filters [4,15,17,27] and diamond-shaped filters [6,48]. Other shapes with straight-line boundaries are also approximated [8,9,13,28,29,50]. Several design methods applicable in such cases have been developed and these methods are usually based on transformations related to concepts of sampling rate conversions. Often alternate frequency domain interpretations are used to explain the design manipulations. A detailed treatment of these methods is beyond the scope of this chapter. However some key ideas are described and a specific case of a diamond filter is used to illustrate the methods. The importance of these design methods stems from the implementation efficiency that results from a generalized notion of separable processing.

In the family of methods considered here, manipulations of a separable 2-D response using a combination of several steps are carried out. In the general case of designing filters with straight-line boundaries, it is difficult to describe a systematic procedure. However, in a given design problem, an appropriate set of steps in the design is suggested by the nature of the desired response.

Some underlying ideas can be understood by examining the problem of obtaining a filter with a parallelogram-shaped passband region. The sides of the parallelogram are assumed to be tilted with respect to the frequency axes. One approach to solving this problem is to perform the following series of manipulations on a separable prototype filter with a rectangular passband. The prototype filter impulse response is upsampled on a *nonrectangular* grid. The upsampling is done by an integer factor greater than one and it is defined by a nondiagonal nonsingular integer matrix [39]. The upsampling produces a parallelogram by a rotation and compression of the frequency response of the prototype filter together with a change in the periodicity. The matrix elements are chosen to produce the desired orientation in the resulting response. Depending on the desired response, cascading to eliminate unwanted portions of the passband in the frequency response, along with possible shifts and additions, may be used. The nonrectangular upsampling is then followed by a rectangular decimation of the sequence to expand the passband out to the desired size. In some cases, the operations of the upsampling transformation and decimation can be combined by the use of nonrectangular decimation of impulse response samples. Results using such procedures produce efficient filter structures that are implemented with essentially 1-D techniques but where the orientations of processing are not parallel to the sample coordinates.

Consider the case of a diamond filter design shown in Figure 22.13. Note that the filter in Figure 22.13 can be obtained from the filter in Figure 22.14a by a transformation of variables. If $F_a(z_1, z_2)$ is the transfer function of the filter approximating the response in Figure 22.14a, then the diamond filter transfer function $D(z_1, z_2)$ given by

$$D(z_1, z_2) = F_a\left(z_1^{\frac{1}{2}} z_2^{\frac{1}{2}}, z_1^{-\frac{1}{2}} z_2^{\frac{1}{2}}\right) \tag{22.45}$$

will approximate the response in Figure 22.1a. The response in Figure 22.2a can be expressed as the sum of the two responses shown in Figure 22.2b and c. We observe that if $F_b(z_1, z_2)$ is the transfer function of the filter approximating the response in Figure 22.2b then

$$F_c(z_1, z_2) = F_b(-z_1, -z_2) \tag{22.46}$$

will approximate the response in Figure 22.14c. This is due to the fact that negating the arguments shifts the (periodic) frequency response of F_b by (π, π). The response in Figure 22.14b can be expressed as the product of two ideal 1-D low-pass filters, one horizontal and one vertical, which have the response shown in Figure 22.14d. This 1-D frequency response can be approximated by a half-band filter. Such an approximation will produce a response in which the transition band straddles both sides of the cutoff frequency boundaries in Figure 22.14a. If we wish to constrain the transition band to lie within the boundaries of the diamond-shaped region in Figure 22.13a, then we should choose a 1-D filter whose stopband interval is $(\pi/2, \pi)$. Let $H(z)$ be the transfer function of the prototype 1-D low-pass filter approximating the response in Figure 22.14d with a suitably chosen transition boundary. The transfer function $H(z)$ can be expressed as

$$H(z) = T_1(z^2) + zT_2(z^2). \tag{22.47}$$

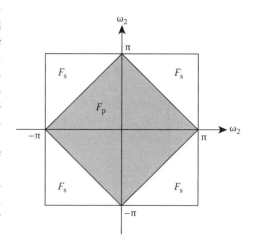

FIGURE 22.13 Ideal frequency response of a diamond filter.

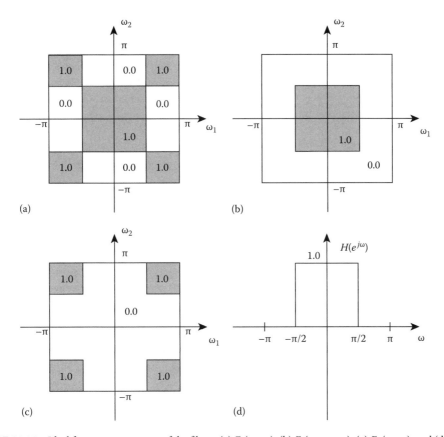

FIGURE 22.14   Ideal frequency responses of the filters: (a) $F_a(z_1, z_2)$, (b) $F_b(-z_1, -z_2)$, (c) $F_c(z_1, z_2)$, and (d) $H(z)$ in obtaining a diamond filter.

The transfer function F_a is given by

$$F_a(z_1, z_2) = H(z_1)H(z_2) + H(-z_1)H(-z_2). \qquad (22.48)$$

Combining Equations 22.45, 22.47, and 22.48 we get

$$D(z_1, z_2) = 2T_1(z_1, z_2)T_1\left(z_1^{-1}z_2\right) + 2z_2 T_2(z_1, z_2)T_2\left(z_1^{-1}z_2\right). \qquad (22.49)$$

As mentioned before, $H(z)$ can be chosen to be a half-band filter with

$$T_1(z^2) = 0.5. \qquad (22.50)$$

The filter T_2 can be either FIR or IIR. It should be noted that the result can also be obtained as a nonrectangular downsampling, by a factor of 2, of the impulse response of the filter $F_b(-z_1, -z_2)$.

Another approach that utilizes multirate concepts is based on a novel idea of applying frequency masking in the 2-D case [31].

22.5 2-D FIR Filter Implementation

The straightforward way to implement 2-D FIR filters is to evaluate the convolution sum given in Equation 22.1. Let us assume that the FIR filter has L nonzero coefficients in its region of support I. In order to get an output sample, L multiplications and L additions need to be performed. The number of arithmetic operations can be reduced by taking advantage of the symmetry of the filter coefficients, that is, $h(n_1, n_2) = h(-n_1, -n_2)$. For example, let the filter support be a rectangular region, $I = \{n_1 = -N_1, \ldots, 0, 1, \ldots, N_1, n_2 = -N_2, \ldots, 0, 1, \ldots, N_2\}$. In this case,

$$y(n_1, n_2) = \sum_{k_1=-N_1}^{N_1} \sum_{k_2=1}^{N_2} [h(k_1, k_2)x(n_1 - k_1, n_2 - k_2) + x(n_1 + k_1, n_2 + k_2)]$$

$$+ h(0,0)x(n_1, n_2) + \sum_{k_1=1}^{N_1} h_1(k_1, 0)[x(n_1 - k_1, n_2) + x(n_1 + k_1, n_2)], \qquad (22.51)$$

which requires approximately half of the multiplications required in the direct implementation equation 22.1.

Any 2-D FIR filter can also be implemented by using an FFT algorithm. This is the direct generalization of 1-D FFT-based implementation [14,30]. The number of arithmetic operations may be less than the space domain implementation in some cases.

Some 2-D filters have special structures that can be exploited during implementation. As we pointed out in Section 22.4, 2-D filters designed by McClellan-type transformations can be implemented in an efficient manner [14,34,35] by building a network whose basic module is the transformation function which is usually a low order 2-D FIR filter.

2-D FIR filters that have separable system responses can be implemented in a cascade structure. In general, an arbitrary 2-D polynomial cannot be factored into subpolynomials due to the absence of a counterpart of fundamental theorem of algebra in two or higher dimensions (whereas in 1-D any polynomial can be factored into polynomials of lower orders). Since separable 2-D filters are constructed from 1-D polynomials, they can be factored and implemented in a cascade form. Let us consider Equation 22.35 where $H(x_1, x_2) = H_1(x_2)H_2(x_2)$ which corresponds to $h(n_1, n_2) = h_1(n_1) h_2(n_2)$ in space domain. Let us assume that orders of the 1-D filters $h_1(n)$ and $h_2(n)$ are $2N_1 + 1$ and $2N_2 + 1$, respectively. In this case the 2-D filter, $h(n_1, n_2)$, has the same rectangular support, I, as in Equation 22.51. Therefore,

$$y(n_1, n_2) = \sum_{k_2=-N_2}^{N_2} h_2(k_2) \sum_{k_1=-N_1}^{N_1} h(k_1)x(n_1 - k_1, n_2 - k_2). \qquad (22.52)$$

The 2-D filtering operation in Equation 22.52 is equivalent to a two-stage 1-D filtering in which the input image, $x(n_1, n_2)$, is first filtered horizontally line by line by $h_1(n)$, then the resulting output is filtered vertically column by column by $h_2(n)$. In order to produce an output sample, the direct implementation requires $(2N_1 + 1) \times (2N_2 + 1)$ multiplications, whereas the separable implementation requires $(2N_1 + 1) + (2N_2 + 1)$ multiplications, which is computationally much more efficient than the direct form realization. This is achieved at the expense of memory space (separable implementation needs a buffer to store the results of first stage during the implementation). By taking advantage of the symmetric nature of h_1 and h_2, the number of multiplications can be further reduced.

Filter design methods by imposing structural constraints like cascade, parallel, and other forms are proposed by several researchers including Refs. [16,47]. These filters can be efficiently implemented because of their special structures. Unfortunately, the design procedure requires nonlinear optimization techniques which may be very complicated.

With advances in VLSI technology, the implementation of 2-D FIR filters using high-speed digital signal processors is becoming increasingly common in complex image processing systems.

22.6 Two-Dimensional Filter Banks

2-D subband decomposition of signals using filter banks (that implement a 2-D wavelet transform) find applications in a wide range of tasks including image and video coding, restoration, denoising, and signal analysis. For example, in recently finalized JPEG-2000 image coding standard an image is first processed by a 2-D filter bank. Data compression is then carried out in the subband domain. In this section we briefly discuss the case of four-channel separable filter banks. Chapter 24 provides a detailed description of 1-D filter banks.

In most cases, 2-D filter banks are constructed in a separable form with the use of the filters of 1-D filter banks, i.e., as a product of two 1-D filters [49,52]. We confine our attention to a 2-D four-channel filter bank obtained from a 1-D two-channel filter bank. Let h_0 and h_1 denote the analysis filters of a 1-D two-channel filter bank. The four analysis filters of the separable 2-D filter bank are given by

$$h_{i,j}(n_1, n_2) = h_i(n_1)h_j(n_2), \quad i,j = 0, 1. \tag{22.53}$$

The filters h_0 and h_1 can be either FIR or IIR. If they are FIR (IIR), then the 2-D filters, $h_{i,j}$, are also FIR (IIR). Frequency responses of these four filters, $H_{i,j}(\omega_1, \omega_2)$, $i,j=0,1$, are given by

$$H_{i,j}(\omega_1, \omega_2) = H_i(\omega_1)H_j(\omega_2), \quad i,j = 0, 1, \tag{22.54}$$

where $H_0(\omega_1)$ and $H_1(\omega_2)$ are the frequency responses of the 1-D low-pass (approximating an ideal cutoff frequency at $\pi/2$) and high-pass filters of a 1-D subband filter bank, respectively [51]. Any 1-D filter bank described in Chapter 24 can be used in Equation 22.53 to design 2-D filter banks. Feature-rich structures for 1-D filter banks are described in Ref. [6].

The 2-D signal is decomposed by partitioning its frequency domain support into four rectangular regions of equal areas. The ideal passband regions of the filters, $H_{i,j}(\omega_1, \omega_2)$, are shown in Figure 22.15. For example, the ideal passband of $H_{0,0}(\omega_1, \omega_2)$ is the square region $[-\pi/2, \pi/2] \times [-\pi/2, \pi/2]$. The 2-D filter bank is shown in Figure 22.16.

Corresponding 2-D synthesis filters are also constructed in a separable manner from the synthesis filters of the 1-D filter bank. If the 1-D filter bank has the perfect reconstruction (PR) property, then the 2-D filter bank also has the PR property. Subband decomposition filter banks (or filter banks implementing the 2-D wavelet transform) consist of analysis and synthesis filters, upsamplers, and downsamplers as discussed in Chapter 24. In the separable 2-D filter bank, downsampling is carried out both horizontally and vertically as follows:

$$x_0(n_1, n_2) = x_a(2n_1, 2n_2). \tag{22.55}$$

Here we consider the input 2-D signal x_a to be an image. The downsampled image x_0 is a quarter-size version of x_a. Only one sample out of four is retained in the downsampling operation described in Equation 22.55. The upsampling operation is the dual of the downsampling

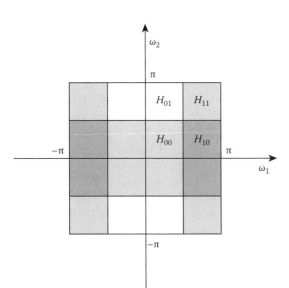

FIGURE 22.15 Ideal passband regions of the separable filters of a rectangular filter bank.

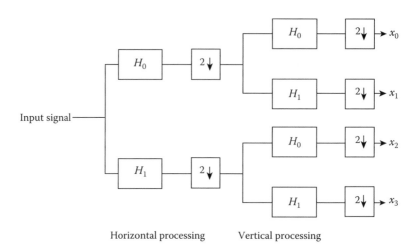

FIGURE 22.16   Block diagram of separable processing in 2D filter bank.

operation. In other words, a zero valued sample is inserted in upsampling corresponding to the location of each dropped sample during downsampling.

The implementation of the above filter bank can be carried out separably in a computationally efficient manner as described in Chapter 24 and Refs. [49,52]. The input image is first processed horizontally row by row by a 1-D filter bank with filters, h_0 and h_1. After the input signal is horizontally filtered with the 1-D two-channel filter bank, the signal in each channel is downsampled row-wise to yield two images. Each image is then filtered and downsampled vertically by the filter bank. As a result, four quarter-size subimages, x_i, $i = 0, 1, 2, 3$, are obtained. These images are the same as the images obtained by direct implementation of the analysis filter bank shown in Figure 22.16. The synthesis filter bank is also implemented in a separable manner.

Nonseparable 2-D filter banks [14] are not as computationally efficient as separable filter banks as discussed in Section 22.5.

In two or higher dimensions, downsampling and upsampling are not restricted to the rectangular grid, but can be carried out in a variety of ways. One example of this is quincunx downsampling where the downsampled image x_q is related to the input image x_a as follows:

$$x_q(n_1, n_2) = x_a(n_1 + n_2, n_2 - n_1). \tag{22.56}$$

In this case only the samples for which $n_1 + n_2$ is even are retained in the output. Filter banks employing quincunx and other downsampling strategies are described in Refs. [6,8,9,19,50,52] and in Chapter 25. Filter banks that employ quincunx downsampling have only two channels and the frequency support is partitioned in a diamond-shaped manner as shown in Figure 22.13. Filters of a quincunx filter bank which have diamond-shaped passbands and stopbands can be designed from a 1-D subband decomposition filter bank using the transformation method described in Equations 22.45 through 22.48. Chapter 25 provides a detailed discussion of the theory and applications of directional filter banks.

Acknowledgments

The authors would like to express their sincere thanks to Omer Nezih Gerek, Imad Abdel Hafez, and Ahmet Murat Bagci for the help they provided in preparing the figures for the chapter.

References

1. A. Abo-Taleb and M. M. Fahmy, Design of FIR two-dimensional digital filters by successive projections, *IEEE Transactions on Circuits and Systems*, CAS-31, 801–805, 1984.
2. V. R. Algazi, M. Suk, C.-S. Rim, Design of almost minimax FIR filters in one and two dimensional by WLS techniques, *IEEE Transactions on Circuits and Systems*, 33(6), 590–596, June 1986.
3. S. A. H. Aly and M. M. Fahmy, Symmetry in two-dimensional rectangularly sampled digital filters, *IEEE Transactions on Acoustics, Speech, and Signal Processing*, ASSP-29, 794–805, 1981.
4. R. Ansari, Efficient IIR and FIR fan filters, *IEEE Transactions on Circuits System*, CAS-34, 941–945, August 1987.
5. R. Ansari, A. E. Cetin, and S. H. Lee, Subband coding of images using nonrectangular filter banks, *Proceedings of the SPIE 32nd Annual International Technical Symposium: Applications of Digital Signal Processing*, San Diego, CA, Vol. 974, August 1988.
6. R. Ansari, C. W. Kim, and M. Dedovic, Structure and design of two-channel filter banks derived from a triplet of halfband filters, *IEEE Transactions on Circuits and Systems II*, 46, 1487–1496, December 1999.
7. R. Ansari and C. Guillemot, Exact reconstruction filter banks using diamond FIR filters, *Proceedings of 1990 Bilkent International Conference on New Trends in Communication, Control, and Signal Processing*, E. Arikan (Ed.), pp. 1412–1421, Elsevier, Holland, 1990.
8. R. H. Bamberger and M. J. T. Smith, Efficient 2-D analysis/synthesis filter banks for directional image component representation, *Proceedings of IEEE International Symposium on Circuits and Systems*, New Orleans, LA, pp. 2009–2012, May 1990.
9. R. H. Bamberger and M. J. T. Smith, A filter bank for the directional decomposition of images: Theory and design, *IEEE Transactions on Acoustics, Speech, Signal Processing*, 40(4), 882–892, April 1992.
10. C. Charalambous, The performance of an algorithm for minimax design of two-dimensional linear-phase FIR digital filters, *IEEE Transactions on Circuits and Systems*, CAS-32, 1016–1028, 1985.
11. A. E. Cetin and R. Ansari, An iterative procedure for designing two dimensional FIR filters, *Proceedings of IEEE International Symposium on Circuits and Systems (ISCAS)*, Philadelphia, PA, pp. 1044–1047, 1987.
12. A. E. Cetin and R. Ansari, Iterative procedure for designing two dimensional FIR filters, *Electronics Letters, IEE*, 23, 131–133, January 1987.
13. T. Chen and P. P. Vaidyanathan, Multidimensional multirate filters and filter banks derived from one-dimensional filters, *IEEE Transactions on Signal Processing*, 41, 1035–1047, March 1993.
14. D. Dudgeon and R.M. Mersereau, *Multidimensional Digital Signal Processing*, Prentice-Hall, Englewood Cliffs, NJ, 1984.
15. P. Embree, J. P. Burg, and M. M. Backus, Wideband velocity filtering—The pie slice process, *Geophysics*, 28, 948–974, 1963.
16. O. D. Faugeras and J. F. Abramatic, 2-D FIR filter design from independent 'small' generating kernels using a mean square and Tchebyshev error criterion, *Proceedings of IEEE International Conference on Acoustics, Speech, and Signal Processing*, Washington, DC, pp. 1–4, 1979.
17. A. P. Gerheim, Synthesis procedure for 90° fan filters, *IEEE Transactions on Circuits and Systems*, CAS30, 858–864, December 1983.
18. C. Guillemot and R. Ansari, Two-dimensional filters with wideband circularly symmetric frequency response, *IEEE Transactions on Circuits and Systems II: Analog and Digital Signal Processing*, 41(10), 703–707, October 1994.
19. C. Guillemot, A. E. Cetin, and R. Ansari, Nonrectangular wavelet representation of 2-D signals: Application to image coding, *Wavelets and Application to Image Coding*, M. Barlaud (Ed.), p. 2764, Elsevier Publications, Amsterdam, Holland, 1994.
20. D. B. Harris and R. M. Mersereau, A comparison of algorithms for minimax design of two-dimensional linear phase FIR digital filters, *IEEE Transactions on Acoustics, Speech, and Signal Processing*, ASSP-25, 492–500, 1977.

21. J. V. Hu and L. R. Rabiner, Design techniques for two-dimensional digital filters, *IEEE Transactions on Audio Electroacoustics,* 20, 249–257, 1972.
22. T. S. Huang, Two-dimensional windows, *IEEE Transactions on Audio and Electroacoustics,* AU-20(1), 88–90, 1972.
23. T. S. Huang (Ed)., *Two-Dimensional Digital Signal Processing I: Linear Filters,* Springer-Verlag, New York, 1981.
24. Y. Kamp and J. P. Thiran, Maximally flat nonrecursive two-dimensional digital filters, *IEEE Transactions on Circuits and Systems,* CAS-21, 437–449, May 1974.
25. Y. Kamp and J. P. Thiran, Chebyshev approximation for two-dimensional nonrecursive digital filters, *IEEE Transactions on Circuits and Systems,* CAS-22, 208–218, 1975.
26. H. Kato and T. Furukawa, Two-dimensional type-preserving circular windows, *IEEE Transactions on Acoustics, Speech, and Signal Processing,* ASSP-29, 926–928, 1981.
27. A. H. Kayran and R. A. King, Design of recursive and nonrecursive fan filters with complex transformations, *IEEE Transactions on Circuits and Systems,* CAS-30, 849–857, December 1983.
28. C. -L. Lau and R. Ansari, Two dimensional digital filter and implementation based on generalized decimation, *Princeton Conference,* Princeton, NJ, March 1986.
29. C. -L. Lau and R. Ansari, Design of two-dimensional filters using sampling rate alteration, *IEEE International Symposium on Circuits and Systems,* Montreal, Canada, pp. 474–477, 1984.
30. J. S. Lim, *Two-Dimensional Signal and Image Processing,* Prentice-Hall, Englewood Cliffs, NJ, 1990.
31. Y. C. Lim and Y. Lian, The optimum design of one- and two-dimensional FIR filters using the frequency response masking technique, *IEEE Transactions on Circuits and Systems II: Analog and Digital Signal Processing,* 40, 88–95, 1993.
32. J. H. Lodge and M. M. Fahmy, An efficient l_p optimization technique for the design of two-dimensional linear-phase FIR digital filters, *IEEE Transactions on Acoustics, Speech, and Signal Processing,* ASSP-28, 308–313, 1980.
33. J. H. McClellan, The design of two-dimensional filters by transformations, *Proceedings of the 7th Annual Princeton Conf. Information Sciences and Systems,* Princetone, NJ, 247–251, 1973.
34. J. H. McClellan and D. S. K. Chan, A 2-D FIR filter structure derived from the Chebyshev recursion, *IEEE Transactions on Circuits and Systems,* 24, 372–378, 1977.
35. W. F. G. Mecklenbrauker and R. M. Mersereau, McClellan transformation for 2-D digital filtering: II—Implementation, *IEEE Transactions on Circuits and Systems,* CAS-23, 414–422, 1976.
36. R. M. Mersereau, D. B. Harris, and H. S. Hersey, An efficient algorithm for the design of two-dimensional digital filters, *Proceedings of the 1974 International Symposium Circuits and Systems,* Newton, MA, pp. 443–446, 1975.
37. R. M. Mersereau, W. F. G. Mecklenbrauker, and T. F. Quatieri, Jr., McClellan transformation for 2-D digital filtering: I—Design, *IEEE Transactions on Circuits and Systems,* CAS-23, 405–414, 1976.
38. R. M. Mersereau, The design of arbitrary 2-D zero-phase FIR filters using transformations, *IEEE Transactions on Circuits and Systems,* 27, 372–378, 1980.
39. R. M. Mersereau and T. C. Speake, The processing of periodically sampled multidimensional signals, *IEEE Transactions on Acoustics, Speech, and Signal Processing,* ASSP-31, 188–194, February 1983.
40. D. T. Nguyen and M. N. S. Swamy, Scaling free McClellan transformation for 2-D digital filters, *IEEE Transactions on Circuits and Systems,* CAS-33, 108–109, January 1986.
41. T. W. Parks and J. H. McClellan, Chebyshev approximation for nonrecursive digital filters with linear phase, *IEEE Transactions on Circuits Theory,* 19, 189–194, 1972.
42. S. -C. Pei and J. -J. Shyu, Design of 2-D FIR digital filters by McClellan transformation and least squares eigencontour mapping, *IEEE Transactions on Circuits and Systems II: Analog and Digital Signal Processing,* 40, 546–555, 1993.
43. E. Z. Psarakis and G. V. Moustakides, Design of two-dimensional zero-phase fir filters via the generalized McClellan transform, *IEEE Transactions on Circuits and Systems,* CAS-38, 1355–1363, November 1991.

44. P. K. Rajan, H. C. Reddy, and M. N. S. Swamy, Fourfold rational symmetry in two-dimensional FIR digital filters employing transformations with variable parameters, *IEEE Transactions on Acoustics, Speech, and Signal Processing*, ASSP-31, 488–499, 1982.

45. V. Rajaravivarma, P. K. Rajan, and H. C. Reddy, Design of multidimensional FIR digital filters using the symmetrical decomposition technique, *IEEE Transactions on Signal Processing*, 42, 164–174, January 1994.

46. T. Speake and R. M. Mersereau, A comparison of different window formulas for two-dimensional FIR filter design, *Proceedings on IEEE International Conference on Acoustics, Speech, and Signal Processing*, Washington, DC, pp. 5–8, 1979.

47. S. Treitel and J. L. Shanks, The design of multistage separable planar filters, *IEEE Transactions on Geoscience Electronics*, GE-9, 10–27, 1971.

48. G. J. Tonge, The sampling of television images, Independent Broadcasting Authority, Experimental and Development Report 112/81.

49. M. Vetterli, A theory of multirate filter banks, *Signal Processing*, 6, 97–112, February 1984.

50. E. Viscito and J. P. Allebach, The analysis and design of multidimensional FIR perfect reconstruction filter banks for arbitrary sampling lattices, *IEEE Transactions on Circuits and Systems*, CAS-38, 29–41, January 1991.

51. J. W. Woods and S. D. O'Neill, Subband coding of images, *IEEE Transactions on Acoustics, Speech, and Signal Processing*, ASSP-34, 1278–1288, 1986.

52. J. W. Wood (Ed.), *Subband Image Coding*, Kluwer Academic Publishers, Norwell, MA, 1990.

53. T.-H. Yu and S. K. Mitra, A new two-dimensional window, *IEEE Transactions on Acoustics, Speech, and Signal Processing*, ASSP-33, 1058–1061, 1985.

Further Information

Most of the research articles describing the advances in 2-D FIR filter design methods appear in *IEEE Transactions on Signal Processing*, *IEEE Transactions on Image Processing*, *IEEE Transactions on Circuits and Systems*, and *Electronics Letters*, and *International Conference on Acoustics, Speech, and Signal Processing (ICASSP)* and *International Symposium on Circuits and Systems*.

23

Two-Dimensional IIR Filters

A. G. Constantinides
Imperial College of Science,
Technology, and Medicine

Xiaojian Xu
Beihang University

23.1 Introduction

A linear 2-D IIR digital filter can be characterized by its transfer function

$$H(z_1, z_2) = \frac{N(z_1, z_2)}{D(z_1, z_2)} = \frac{\sum_{i=0}^{N_2} \sum_{j=0}^{M_2} a_{ij} z_1^{-i} z_2^{-j}}{\sum_{i=0}^{N_1} \sum_{j=0}^{M_1} b_{ij} z_1^{-i} z_2^{-j}} \tag{23.1}$$

where the sampling period $T_i = 2\pi/\omega_{si}$ for $i = 1, 2$ with ω_{si} and the sampling frequencies a_{ij} and b_{ij} are real numbers known as the coefficients of the filter. Without loss of generality we can assume $M_1 = M_2 = N_1 = N_2 = M$ and $T_1 = T_2 = T$. Designing a 2-D filter is to calculate the filter coefficients a_{ij} and b_{ij} in such a way that the amplitude response and/or the phase response (group delay) of the designed filter approximates to some ideal responses while maintaining the stability of the designed filter. The latter requires that

$$D(z_1, z_2) \neq 0 \quad \text{for } |z_i| \geq 1, \, i = 1, 2 \tag{23.2}$$

The amplitude response of the 2-D filter is expressed as

$$M(\omega_1, \omega_2) = \left| H(e^{j\omega_1 T}, \, e^{j\omega_2 T}) \right| \tag{23.3}$$

the phase response as

$$\phi(\omega_1, \omega_2) = \arg H(e^{j\omega_1 T}, e^{j\omega_2 T}) \tag{23.4}$$

and the two group delay functions as

$$\tau_i(\omega_1, \omega_2) = \frac{d\phi(\omega_1, \omega_2)}{d\omega_i}, \ i = 1, 2 \tag{23.5}$$

Equation 23.1 is the general form of transfer functions of the nonseparable numerator and denominator 2-D IIR filters. It can involve two subclasses, namely, the separable product transfer function

$$H(z_1, z_2) = H_1(z_1)H_2(z_2)$$
$$= \frac{\sum_{i=0}^{N_2} a_{1i} z_1^{-i} \sum_{j=0}^{M_2} a_{2i} z_2^{-j}}{\sum_{i=0}^{N_1} b_{1i} z_1^{-i} \sum_{j=0}^{M_1} b_{2i} z_2^{-j}} \tag{23.6}$$

and the separable denominator, nonseparable numerator transfer function given by

$$H(z_1, z_2) = \frac{\sum_{i=0}^{N_2} \sum_{j=0}^{M_2} a_{ij} z_1^{-i} z_2^{-j}}{\sum_{i=0}^{N_1} b_{1i} z_1^{-i} \sum_{j=0}^{M_1} b_{2j} z_2^{-j}} \tag{23.7}$$

The stability constraints for the above two transfer functions are the same as those for the individual two 1-D cases. These are easy to check and correspondingly the transfer function is easy to stabilize if the designed filter is found to be unstable. Therefore, in the design of the above two classes, in order to reduce the stability problem to that of the 1-D case, the denominator of the 2-D transfer function is chosen to have two 1-D polynomials in z_1 and z_2 variables in cascade. However, in the general formulation of nonseparable numerator and denominator filters, this oversimplification is removed. The filters of this type are generally designed either through transformation of 1-D filters, or through optimization approaches, as is discussed in the following.

23.2 Transformation Techniques

23.2.1 Analog Filter Transformations

In the design of 1-D analog filters, a group of analog filter transformations of the form $s = g(s')$ is usually applied to normalized continuous low-pass transfer functions like those obtained by using the Bessel, Butterworth, Chebyshev, and elliptic approximations. These transformations can be used to design low-pass, high-pass, bandpass, or bandstop filters satisfying piecewise-constant amplitude response specifications. Through the application of the bilinear transformation, corresponding 1-D digital filters can be designed, and since 2-D digital filters can be designed in terms of 1-D filters, these transformations are of considerable importance in the design of 2-D digital filters as well. In the 2-D cases, the transformations have a form of

$$s = g(s_1, s_2) \tag{23.8}$$

As a preamble, in this section two groups of transformations of interest in the design of 2-D digital filters are introduced, which essentially produce 2-D continuous transfer functions from 1-D ones.

23.2.1.1 Rotated Filter

The first group of transformations, suggested by Shanks, Treitel, and Justice [1], are of the form

$$g_1(s_1, s_2) = -s_1 \sin \beta + s_2 \cos \beta \tag{23.9a}$$

$$g_2(s_1, s_2) = s_1 \cos \beta + s_2 \sin \beta \tag{23.9b}$$

They map 1-D into 2-D filters with arbitrary directionality in a 2-D frequency response plane. These filters are called rotated filters because they are obtained by rotating 1-D filters.

If $H(s)$ is a 1-D continuous transfer function, then a corresponding 2-D continuous transfer function can be generated as

$$H_{D1}(s_1, s_2) = H(s)\Big|_{s=g_1(s_1,s_2)} \tag{23.10a}$$

$$H_{D2}(s_1, s_2) = H(s)\Big|_{s=g_2(s_1,s_2)} \tag{23.10b}$$

by replacing the s in $H(s)$ with $g_1(s_1, s_2)$ and $g_2(s_1, s_2)$, respectively.

It is easy to show [2] that a transformation of $g_1(s_1, s_2)$ or $g_2(s_1, s_2)$ will give rise to a contour in the amplitude response of the 2-D analog filter that is a straight line rotated by an angle β with respect to the s_1 or s_2 axis, respectively. Figure 23.1 illustrates an example of 1-D to 2-D analog transformation by (23.10a) for $\beta = 0°$ and $\beta = 45°$.

The rotated filters are of special importance in the design of circularly symmetric filters, as will be discussed in Section 23.4.

23.2.1.2 Transformation Using a Two-Variable Reactance Function

The second group of transformations is based on the use of a two-variable reactance function. One of the transformations was suggested by Ahmadi, Constantinides, and King [3,4]. This is given by

$$g_3(s_1, s_2) = \frac{a_1 s_1 + a_2 s_2}{1 + b s_1 s_2} \tag{23.11}$$

where a_1, a_2, and b are all positive constants.

Let us consider a 2-D filter designed by using a 1-D analog low-pass filter with cutoff frequency Ω_c. Equation 23.11 results in

$$\Omega_2 = \frac{\Omega_c - a_1 \Omega_1}{a_2 - b \Omega_c \Omega_1} \tag{23.12}$$

The mapping of $\Omega = \Omega_c$ onto the (Ω_1, Ω_2) plane for various values of b is depicted in Figure 23.2 [5]. The cutoff frequencies along the Ω_1 and Ω_2 axes can be adjusted by simply varying a_1 and a_2. On the other hand, the convexity of the boundary can be adjusted by varying b. We note that b must be greater than zero to preserve stability. Also, it should be noted that $g_3(s_1, s_2)$ becomes a low-pass to bandpass transformation along $s_1 = s_2$, and therefore the designed filter will behave like a bandstop filter along $\Omega_1 = \Omega_2$. This problem can be overcome by using a guard filter of any order.

King and Kayran [6] have extended the above technique by using a higher order reactance function of the form

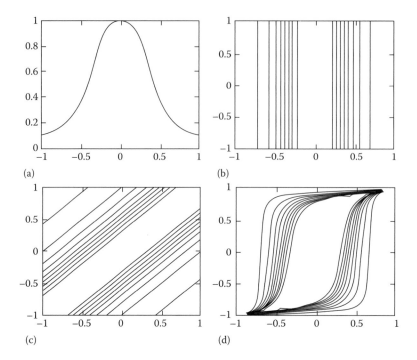

FIGURE 23.1 1D to 2D analog filter transformation. (a) Amplitude response of 1D analog filter. (b) Contour plot of 2D filter, $\beta = 0°$. (c) Contour plot of 2D filter, $\beta = 45°$. (d) Contour plot of 2D filter, $\beta = 45°$, after applying double bilinear transformation.

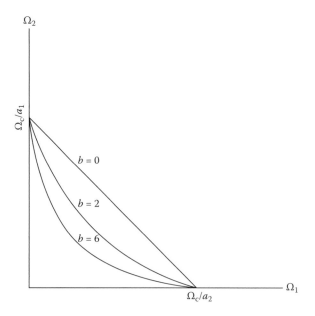

FIGURE 23.2 Plot of Equation 23.12 for different values of b.

$$g_4(s_1, s_2) = \frac{a_1 s_1 + a_2 s_2}{1 + b_1(s_1^2 + s_2^2) + b_2 s_1 s_2} \tag{23.13}$$

and they proved that the stability of $g_4(s_1, s_2)$ is ensured if

$$b_1 > 0 \tag{23.14}$$

and

$$b_1 > \frac{b_2^2}{4} - b_1^2 > 0 \tag{23.15}$$

However, it is necessary as earlier to include a guard filter, which may have the simple form of

$$G(z_1, z_2) = \frac{(1 + z_1)(1 + z_2)}{(d_1 + z_1)(d_2 + z_2)} \tag{23.16}$$

in order to remove the high-pass regions along all radii except the coordinate axes.

Then, through an optimization procedure, the coefficients of $g_4(s_1, s_2)$ and $G(z_1, z_2)$ are calculated subject to the constraints of Equations 23.14 and 23.15, so that the cutoff frequency of the 1-D filter is mapped into a desired cutoff boundary in the (Ω_1, Ω_2) plane.

23.2.2 Spectral Transformations

Spectral transformation is another kind of important transformation in the design of both 1-D and 2-D IIR filters. In this section, three groups of spectral transformations are discussed. Among them, the linear transformations map frequency axes onto frequency axes in the (Ω_1, Ω_2) plane, the complex transformation is of wide applications to the design of fan filters, and the Constantinides transformations transform a discrete function into another discrete function and through which any transformation of a low-pass filter to another low-pass, high-pass, bandpass, or bandstop filter becomes possible.

23.2.2.1 Linear Transformations

Consider a group of linear transformations that map frequency axes onto themselves in the (Ω_1, Ω_2) plane. There are eight possible such transformations [7,8] and they have the algebraic structure of a finite group under the operation of multiplication [2]; each transformation can be expressed as

$$\begin{bmatrix} \omega_1 \\ \omega_2 \end{bmatrix} := D(T) \begin{bmatrix} \omega_1 \\ \omega_2 \end{bmatrix} \tag{23.17}$$

where $D(T)$ is a 2×2 unitary matrix representing transformation $T\delta$. The eight transformations and their effect on the frequency response of the digital filter are as follows with a multiplication table being illustrated in Table 23.1 [2].

1. Identity (I):

$$D(I) = \begin{bmatrix} 1 & 0 \\ 0 & 1 \end{bmatrix}$$

2. Reflection about the ω_1 axis ($\rho_{\omega1}$):

$$D(\rho_{\omega1}) = \begin{bmatrix} 1 & 0 \\ 0 & -1 \end{bmatrix}$$

TABLE 23.1   Multiplication Table of Group

	I	$\rho_\omega 1$	$\rho_\omega 2$	$\rho_\psi 1$	$\rho_\psi 2$	R_4	R_4^2	R_4^3
I	I	$\rho_\omega 1$	$\rho_\omega 2$	$\rho_\psi 1$	$\rho_\psi 2$	R_4	R_4^2	R_4^3
$\rho_\omega 1$	$\rho_\omega 1$	I	R_4^2	R_4^3	R_4	$\rho_\psi 2$	$\rho_\omega 2$	$\rho_\psi 1$
$\rho_\omega 2$	$\rho_\omega 2$	R_4^2	I	R_4	R_4^3	$\rho_\psi 1$	$\rho_\omega 1$	$\rho_\psi 2$
$\rho_\psi 1$	$\rho_\psi 1$	R_4	R_4^3	I	R_4^2	$\rho_\omega 1$	$\rho_\psi 2$	$\rho_\omega 2$
$\rho_\psi 2$	$\rho_\psi 2$	R_4^3	R_4	R_4^2	I	$\rho_\omega 2$	$\rho_\psi 1$	$\rho_\omega 1$
R_4	R_4	$\rho_\psi 1$	$\rho_\psi 2$	$\rho_\omega 2$	$\rho_\omega 1$	R_4^2	R_4^3	I
R_4^2	R_4^2	$\rho_\omega 2$	$\rho_\omega 1$	$\rho_\psi 2$	$\rho_\psi 1$	R_4^3	I	R_4
R_4^3	R_4^3	$\rho_\psi 2$	$\rho_\psi 1$	$\rho_\omega 1$	$\rho_\omega 2$	I	R_4	R_4^2

3. Reflection about the ω_2 axis ($\rho_{\omega 2}$):

$$D(\rho_{\omega 2}) = \begin{bmatrix} -1 & 0 \\ 0 & 1 \end{bmatrix}$$

4. Reflection about the ($\rho_{\psi 1}$):

$$D(\rho_{\psi 1}) = \begin{bmatrix} 0 & 1 \\ 1 & 0 \end{bmatrix}$$

5. Reflection about the ($\rho_{\psi 2}$):

$$D(\rho_{\psi 2}) = \begin{bmatrix} 0 & -1 \\ -1 & 0 \end{bmatrix}$$

6. Counterclockwise rotation by 90° (R_4):

$$D(R_4) = \begin{bmatrix} 0 & -1 \\ 1 & 0 \end{bmatrix}$$

7. Counterclockwise rotation by 180° (R_4^2):

$$D(R_4^2) = \begin{bmatrix} -1 & 0 \\ 0 & -1 \end{bmatrix}$$

8. Counterclockwise rotation by 270° (R_4^3):

$$D(R_4^3) = \begin{bmatrix} 0 & 1 \\ -1 & 0 \end{bmatrix}$$

In the above symbolic representation, ψ_1 and ψ_2 represent axes that are rotated by 45° in the counter-clockwise sense relative to the ω_1 and ω_2 axes, respectively, and R_k denotes rotation by $360°/k$ in the counterclockwise sense. These transformations could equivalently be defined in the (z_1, z_2) domain by complex conjugating and/or interchanging the complex variables z_1 and z_2 in the filter transfer function.

An important property of the group is that each transformation distributes over a product of functions of ω_1 and ω_2, that is,

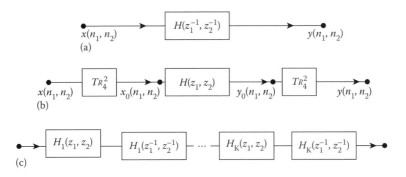

FIGURE 23.3 Design of zero-phase filters: (a) noncausal filter; (b) equivalent causal implementation; (c) cascade zero-phase filter.

$$T_\delta \left[\prod_{i=1}^{K} F_i(\omega_1, \omega_2) \right] = \prod_{i=1}^{K} T_\delta[F_i(\omega_1, \omega_2)] \tag{23.18}$$

where $T\delta$ represents any of the eight transformation operators. The validity of this property follows the definition of the transformations [8].

In the (z_1, z_2) domain, if $H(z_1, z_2)$ represents a causal filter with impulse response $h(n_1, n_2)$, then the filter represented by $H(z_1^{-1}, z_2^{-1})$ will have an impulse response $h(-n_1, -n_2)$, and is therefore, noncausal, since $h(-n_1, -n_2) \neq 0$ for $n_1 < 0$, $n_2 < 0$. Such a filter can be implemented in terms of causal transfer function $H(z_1, z_2)$, i.e., by rotating the n_1 and n_2 axes of the input signal by 180°, processing the rotated signal by the causal filter, and then rotating the axes of the output signal by 180°, as illustrated in Figure 23.3b.

Noncausal filters can be used for the realization of zero-phase filters by cascading K pairs of filters whose transfer functions are $H_i(z_1, z_2)$ and $H(z_1^{-1}, z_2^{-1})$ for $i = 1, 2, \ldots, K$, as depicted in Figure 23.3c.

23.2.2.2 Complex Transformation and 2-D Fan Filters

Complex Transformation. A complex transformation is of the form [9]

$$z = e^\phi z_1^{\alpha_1/\beta_1} z_2^{\alpha_2/\beta_1} \tag{23.19}$$

by which a 2-D filter $H(z_1, z_2)$ can be derived from 1-D filter $H_1(z)$. The corresponding frequency transformation of Equation 23.19 is

$$\exp(j\omega) \rightarrow \exp\left\{ j\left(\phi + \omega_1 \frac{\alpha_1}{\beta_1} + \omega_2 \frac{\alpha_2}{\beta_2} \right) \right\} \tag{23.20}$$

or

$$\omega \rightarrow \phi + \omega_1 \frac{\alpha_1}{\beta_1} + \omega_2 \frac{\alpha_2}{\beta_2} \tag{23.21}$$

There are three major effects of transformation Equation 23.20 on the resulting filter:

1. Frequency shifting along the ω_1 axis. The frequency response of the resulting filter will be shifted by ϕ along the ω_1 axis.
2. Rotation of the frequency response. The angle of rotation is

$$\theta = \arctan\left(\frac{\alpha_2}{\beta_2}\right) \tag{23.22}$$

Since the original filter is 1-D and a function of z_1, the angle of rotation will be defined by the fractional power of z_2.

3. Scaling the frequency response along the ω_1 axis. The fractional power of z_1 will scale the frequency response by a factor β_1/α_1. However, the periodicity of the frequency response will be $(\alpha_1/\beta_1)2\pi$ instead of 2π. Other effects may also be specified [9].

The complex transformation is of importance in the design of fan filters. By using a prototype lowpass filter with a cutoff frequency at $\omega_c = \pi/2$, and the transformation Equation 23.19, one obtains the shifted, scaled, and rotated characteristics in the frequency domain. We denote the transformed filter by

$$H\left(z_1, z_2; \frac{\alpha_1}{\beta_1}, \frac{\alpha_2}{\beta_2}\right) = H_1(z)\Big|_{z=e^{\phi} z_1^{\alpha_1/\beta_1} z_2^{\alpha_2/\beta_2}} \tag{23.23}$$

In general, the filter coefficients in function H will be complex and the variables z_1 and z_2 will have rational noninteger powers. However, appropriate combinations of transformed filters will remove both of these difficulties, as will be shown in the following.

Symmetric Fan Filters. An ideal symmetric fan filter has the specification of

$$H_{f1}(e^{j\omega_1 T}, e^{j\omega_2 T}) = \begin{cases} 1 & \text{for } |\omega_1| \geq |\omega_2| \\ 0 & \text{otherwise} \end{cases} \tag{23.24}$$

We introduce transfer function $\hat{H}_1(z_1, z_2)$, $\hat{H}_2(z_1, z_2)$, $\hat{H}_3(z_1, z_2)$, and $\hat{H}_4(z_1, z_2)$, of four filters generated by Equation 23.23 with $(\alpha_1/\beta_1, \alpha_1/\beta_2) = (1/2, 1/2)$, $(-1/2, 1/2)$, $(1/2, -1/2)$, and $(-1/2, -1/2)$, respectively, and $\phi = \pi/2$. The responses of the transformed filters, $\hat{H}_i(z_1, z_2)$, $i = 1, 2, 3, 4$, can be found in Figure 23.4 together with the prototype filter.

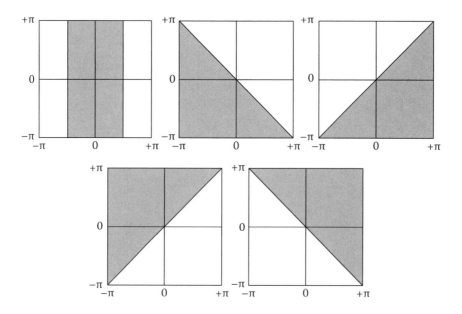

FIGURE 23.4 Basic building blocks of symmetric fan filters.

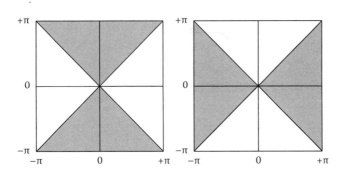

FIGURE 23.5 Amplitude characteristics of $G_{11}(z_1, z_2)$ and $G_{22}(z_1, z_2)$.

In this design procedure, the filters in Figure 23.4 will be used as the basic building blocks for a fan filter specified by Equation 23.24. One can construct the following filter characteristics:

$$G_{11}(z_1, z_2) = \hat{H}_1(z_1, z_2)\hat{H}_1^*(z_1^{-1}, z_2^{-1})\hat{H}_2(z_1, z_2)\hat{H}_2^*(z_1^{-1}, z_2^{-1})$$
$$+ \hat{H}_3(z_1, z_2)\hat{H}_3^*(z_1^{-1}, z_2^{-1})\hat{H}_4(z_1, z_2)\hat{H}_4^*(z_1^{-1}, z_2^{-1}) \tag{23.25a}$$

$$G_{22}(z_1, z_2) = \hat{H}_1(z_1, z_2)\hat{H}_1^*(z_1^{-1}, z_2^{-1})\hat{H}_3(z_1, z_2)\hat{H}_3^*(z_1^{-1}, z_2^{-1})$$
$$+ \hat{H}_2(z_1, z_2)\hat{H}_2^*(z_1^{-1}, z_2^{-1})\hat{H}_4(z_1, z_2)\hat{H}_4^*(z_1^{-1}, z_2^{-1}) \tag{23.25b}$$

which are shown in Figure 23.5.

Quadrant Fan Filters. The frequency characteristic of a quadrant fan filter is specified as

$$H_{f2}(e^{j\omega_1 T}, e^{j\omega_2 T}) = \begin{cases} 1 & \text{for } \omega_1\omega_2 \geq 0 \\ 0 & \text{otherwise} \end{cases} \tag{23.26}$$

We consider the same ideal prototype filter. Then the transformed filters $\hat{H}_{14}(z_1, z_2)$, $\hat{H}_{12}(z_1, z_2)$, $\hat{H}_{23}(z_1, z_2)$, and $\hat{H}_{34}(z_1, z_2)$, are obtained via Equation 23.23 with $(\alpha_1/\beta_1, \alpha_2/\beta_2)$ equal to $(1, 0)$, $(0, 1)$, $(-1, 0)$, and $(0, -1)$, respectively, and $\phi = \pi/2$. The subscripts on $\hat{H}$ refer to quadrants to which the low-pass filter characteristics have been shifted. Figure 23.6 illustrates the amplitude responses of these transformed filters together with the prototype.

The filters in Figure 23.6 will be used as the basic building blocks for fan filters specified by Equation 23.26. In a similar manner to Equations 23.25a and b, the filter characteristics $G_{13}(z_1, z_2)$ and $G_{23}(z_1, z_2)$ can be constructed as follows:

$$G_{13}(z_1, z_2) = \hat{H}_{12}(z_1, z_2)\hat{H}_{12}^*(z_1^{-1}, z_2^{-1})\hat{H}_{14}(z_1, z_2)\hat{H}_{14}^*(z_1^{-1}, z_2^{-1})$$
$$+ \hat{H}_{23}(z_1, z_2)\hat{H}_{23}^*(z_1^{-1}, z_2^{-1})\hat{H}_{34}(z_1, z_2)\hat{H}_{34}^*(z_1^{-1}, z_2^{-1}) \tag{23.27a}$$

$$G_{24}(z_1, z_2) = \hat{H}_{12}(z_1, z_2)\hat{H}_{12}^*(z_1^{-1}, z_2^{-1})\hat{H}_{23}(z_1, z_2)\hat{H}_{23}^*(z_1^{-1}, z_2^{-1})$$
$$+ \hat{H}_{14}(z_1, z_2)\hat{H}_{14}^*(z_1^{-1}, z_2^{-1})\hat{H}_{34}(z_1, z_2)\hat{H}_{34}^*(z_1^{-1}, z_2^{-1}) \tag{23.27b}$$

whose amplitude responses are depicted in Figure 23.7.

23.2.2.3 Constantinides Transformations

The so-called Constantinides [10,19] are of importance in the design of 1-D digital filters, and are of the form

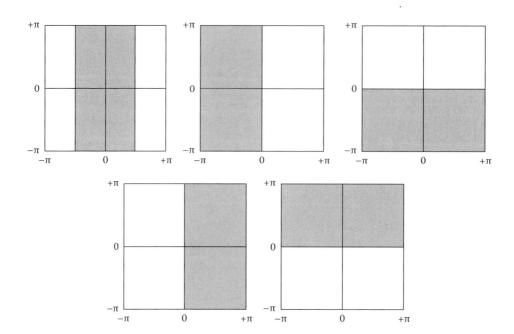

FIGURE 23.6   Basic building blocks for quadrant fan filter design.

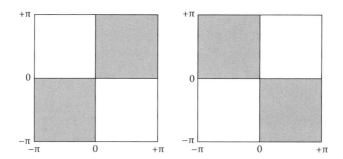

FIGURE 23.7   Two quadrant pass filters.

$$z = f(\bar{z}) = e^{jl\pi} \prod_{t=1}^{m} \frac{\bar{z} - a_t^*}{1 - a_t \bar{z}} \qquad (23.28)$$

where
   l and m are integers
   a_t^* is the complex conjugate of a_t

Pendergrass, Mitra, and Jury [11] showed that, in the decision of 2-D IIR filters, this group of transformations is as useful as in 1-D cases. By choosing the parameters l, m, and a_t in Equation 23.28 properly, a set of four specific transformations can be obtained that can be used to transform a low-pass transfer function into a corresponding low-pass, high-pass, bandpass, or a bandstop transfer function. These transformations are summarized in Table 23.2, where subscript i is included to facilitate the application of the transformation to 2-D discrete transfer functions.

Let Ω_i and ω_i for $i = 1, 2$ be the frequency variables in the original and transformed transfer function, respectively. Suppose $H_L(z_1, z_2)$ is a low-pass transfer function with respect to z_i, if each of z_1 and z_2 is

TABLE 23.2 Constantinides Transformations

Type	Transformation	Parameters
LP to LP	$z_i = \dfrac{\bar{z}_i - \alpha_i}{1 - \alpha_i \bar{z}_i}$	$\alpha_i = \dfrac{\sin\left[(\Omega_{pi} - \omega_{pi})T_i/2\right]}{\sin\left[(\Omega_{pi} + \omega_{pi})T_i/2\right]}$
LP to HP	$z_i = \dfrac{\bar{z}_i - \alpha_i}{1 - \alpha_i \bar{z}_i}$	$\alpha_i = \dfrac{\cos\left[(\Omega_{pi} - \omega_{pi})T_i/2\right]}{\cos\left[(\Omega_{pi} + \omega_{pi})T_i/2\right]}$
LP to BP	$z_i = -\dfrac{\bar{z}_i^2 - \dfrac{2\alpha_i k_i}{k_i + 1}\bar{z}_i + \dfrac{k_i - 1}{k_i + 1}}{1 - \dfrac{2\alpha_i k_i}{k_i + 1}\bar{z}_i + \dfrac{k_i - 1}{k_i + 1}\bar{z}_i^2}$	$\alpha_i = \dfrac{\cos\left[(\omega_{p2i} + \omega_{p1i})T_i/2\right]}{\cos\left[(\omega_{p2i} + \omega_{p1i})T_i/2\right]}$ $k_i = \tan\dfrac{\Omega_{pi}T_i}{2}\cot\dfrac{(\omega_{p2i} - \omega_{p1i})T_i}{2}$
LP to BS	$z_i = -\dfrac{\bar{z}_i^2 - \dfrac{2\alpha_i}{1 + k_i}\bar{z}_i + \dfrac{1 - k_i}{1 + k_i}}{1 - \dfrac{2\alpha_i}{1 + k_i}\bar{z}_i + \dfrac{1 - k_i}{1 + k_i}\bar{z}_i^2}$	$\alpha_i = \dfrac{\cos\left[(\omega_{p2i} - \omega_{p1i})T_i/2\right]}{\cos\left[(\omega_{p2i} + \omega_{p1i})T_i/2\right]}$ $k_i = \tan\dfrac{\Omega_{pi}T_i}{2}\tan\dfrac{(\omega_{p2i} - \omega_{p1i})T_i}{2}$

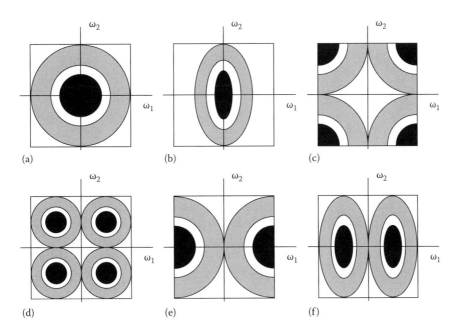

(a)   (b)   (c)

(d)   (e)   (f)

FIGURE 23.8 Application of Constantinides transformations to 2D IIR filters. (a) Circularly symmetric low-pass filter. (b) LP to LP for z_1 and z_2. (c) LP to HP for z_1 and z_2. (d) LP to BP for z_1 and z_2. (e) LP to HP for z_1 and LP to LP for z_2. (f) LP to BP for z_1 and LP to LP for z_2.

transformed to be low-pass, bandpass, or high-pass, then a number of different 2-D filter combinations can be achieved. As an example, some of the possible amplitude responses are illustrated in Figure 23.8a through f [2].

23.3 Design of Separable Product Filters

A 2-D IIR filter is characterized by a separable product transfer function of the form

$$H(z_1, z_2) = H_1(z_1)H_2(z_2) \tag{23.29}$$

if its passband or stopband is of the shape of a rectangular domain. The design of such a class of filters can be accomplished by using the method proposed by Hirano and Aggarwal [12]. The method can be used to design filters with quadrantal or half-plane symmetry.

23.3.1 Design of Quadrantally Symmetric Filters

A 2-D filter is said to be quadrantally symmetric if its amplitude response satisfies the equality

$$|H(z_1, z_2)| = |H(z_1^*, z_1^*)| = |H(z_1^*, z_2)| = |H(z_1, z_1^*)| \tag{23.30}$$

Consider two 1-D bandpass filters specified by two transfer functions $H_1(z)$ and $H_2(z)$, respectively, and let $z = z_1$ in the first one and $z = z_2$ in the second. If their frequency responses can be expressed by

$$\left|H_1(e^{j\omega_1 T_1})\right| = \begin{cases} 1 & \omega_{12} \leq \omega_1 < \omega_{13} \\ 0 & 0 \leq \omega_1 < \omega_{11} \text{ or } \omega_{14} \leq \omega_1 < \infty \end{cases} \tag{23.31a}$$

and

$$\left|H_2(e^{j\omega_2 T_2})\right| = \begin{cases} 1 & \omega_{22} \leq \omega_2 < \omega_{23} \\ 0 & 0 \leq \omega_2 < \omega_{21} \text{ or } \omega_{24} \leq \omega_2 \leq \infty \end{cases} \tag{23.31b}$$

respectively, and since

$$\left|H(e^{j\omega_1 T_1}, e^{j\omega_2 T_2})\right| = \left|H_1(e^{j\omega_1 T_1})\right|\left|H_2(e^{j\omega_2 T_2})\right|$$

Equations 23.31a and b give

$$\left|H(e^{j\omega_1 T_1}, e^{j\omega_2 T_2})\right| = \begin{cases} 1 & \omega_{12} \leq \omega_1 < \omega_{13} \text{ and } \omega_{22} \leq \omega_2 < \omega_{23} \\ 0 & \text{otherwise} \end{cases}$$

Evidently, the 2-D filter obtained will pass frequency components that are in both passband of $H_1(z_1)$ and $H_2(z_2)$; that is, the passband $H_1(z_1, z_2)$ will be a rectangle with sides $\omega_{13} - \omega_{12}$ and $\omega_{23} - \omega_{22}$. On the other hand, frequency components that are in the stopband of either the first filter or the second filter will be rejected. Hence, the stopband of $H(z_1, z_2)$ consists of the domain obtained by combining the stopbands of the two filters.

By a similar method, if each of the two filters is allowed to be a low-pass, bandpass, or high-pass 1-D filter, then nine different rectangular passbands can be achieved, as illustrated in Figure 23.9. The cascade arrangement of any two of those filters may be referred to as a generalized bandpass filter [12].

Another typical simple characteristic is specified by giving the rectangular stopband region, which is referred to as the rectangular stop filter. Also, there are nine possible types of such a filter which are complementary to that of the generalized bandpass filter shown in Figure 23.9 (i.e., considering the shadowed region as the stopband). This kind of bandstop filter can be realized in the form of

$$H(z_1, z_2) = H_A(z_1)H_A(z_2) - e^{jk\pi}[H_1(z_1)H_2(z_2)]^2 \tag{23.32}$$

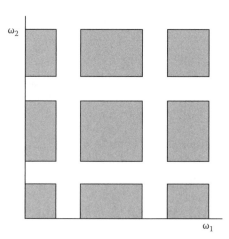

FIGURE 23.9   Idealized amplitude response of generalized bandpass filter.

where $H_1(z_1)H_2(z_2)$ is a generalized bandpass filter described above, and $H_A(z_1)$ and $H_A(z_2)$ are allpass 1-D filters [13] whose poles are poles of $H_1(z_1)$ and $H_2(z_2)$, respectively. Equation 23.32 can be referred to as a generalized bandstop filter.

Extending the principles discussed above, if a 2-D filter is constructed by cascading K 2-D filters with passbands P_i, stopbands S_i, and transfer function $H_i(z_1, z_2)$, the overall transfer function is obtained as

$$H(z_1, z_2) = \prod_{i=1}^{k} H_i(z_i, z_2)$$

and the passband P and stopband S of the cascaded 2-D filter are defined by

$$P = \bigcap P_i \quad \text{and} \quad S = \bigcup_{i=1}^{K} S_i$$

that is, the only frequency components not to be rejected will be those that will be passed by each and every filter in the cascade arrangement.

On the other hand, if a 2-D filter is constructed by connecting K 2-D filters in parallel, then

$$H(z_1, z_2) = \sum_{i=1}^{K} H_i(z_1, z_2)$$

Assuming that all the parallel filters have the same phase shift, and the passband $P_i = 1$ of the various filters are not overlapping, then the passband and stopband of the parallel arrangement is given by

$$P = \bigcup_{i=1}^{K} P_i \quad \text{and} \quad S = \bigcap S_i$$

Parallel IIR filters are more difficult to design than cascade ones, due to the requirement that the phase shifts of the various parallel filters be equal. However, if all the data to be filtered are available at the start of the processing, zero-phase filters can be used.

By combining parallel and cascade subfilters, 2-D IIR filters whose passbands or stopbands are combinations of rectangular regions can be designed, as illustrated by the following example.

Example 23.1

Design a 2-D IIR filter whose passband is the area between two overlapping rectangles, as depicted in Figure 23.10a.

Solution

1. Construct a first 2-D low-pass filter with rectangular passband $(\omega_{12}, \omega_{22})$ by using 1-D low-pass filters, as shown in Figure 23.10a.
2. Construct a second 2-D low-pass filter with rectangular passband $(\omega_{11}, \omega_{21})$ by 1-D low-pass filters. Then using Equation 23.32 to construct a 2-D high-pass filter with rectangular stopband $(\omega_{11}, \omega_{21})$, as shown in Figure 23.10c.
3. Cascade the first 2-D low-pass filter with the 2-D high-pass filter to obtain the required filter. The amplitude response of a practically designed 2-D filter is in Figure 23.10d.

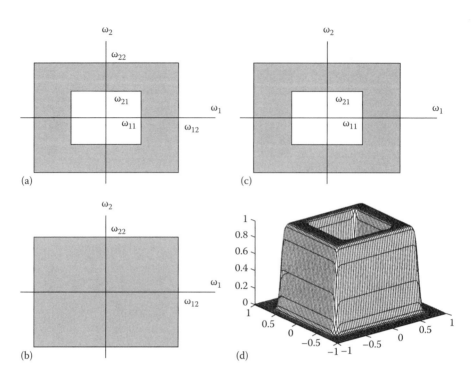

FIGURE 23.10   Amplitude response for the filter of Example 23.1: (a) specified response; (b) response of the 2D low-pass filter; (c) response of the 2D high-pass filter; (d) 3D plot of the amplitude response of a real 2D IIR filter.

23.3.2 Design of Half-Plane Symmetric Filters

A 2-D filter is said to be half-plane symmetric if its amplitude response satisfies

$$\left| H(z_1, z_2) \right| = \left| H(z_1^*, z_2^*) \right| \tag{23.33}$$

but

$$\left| H(z_1, z_2) \right| \neq \left| H(z_1^*, z_2) \right| \neq \left| H(z_1, z_2^*) \right| \tag{23.34}$$

that is, half-plane symmetry does not imply quadrantal symmetry.

The design of those filters can be accomplished by cascading two quadrant pass filters derived previously in Section 23.2.2.2 with quadrantally symmetric filters, as is demonstrated in Example 23.2.

Example 23.2

Design a 2-D half-plane symmetric filter whose passband is defined in Figure 23.11a.

Solution

1. Follow the steps 1 to 3 in Example 23.1 to construct a 2-D bandpass filter with the passband being the area between two overlapping rectangles, as shown in Figure 23.10a.

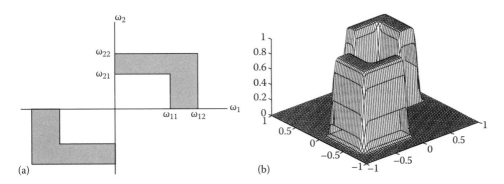

FIGURE 23.11 Amplitude response for the filter of Example 23.2: (a) specified response; (b) 3D plot of the amplitude response of a real 2D IIR filter.

    2. Construct a two quadrant (first and third quadrants) pass filter as depicted in Figure 23.7a.
    3. Cascade the 2-D bandpass filter with the two quadrant pass filters to obtain the required filter.

The amplitude response of a practically designed filter is shown in Figure 23.11b.

23.3.3 Design of Filters Satisfying Prescribed Specifications

The problem of designing filters that satisfy prescribed specifications to a large extent has been solved for the case of 1-D filters, and by extending the available methods, 2-D IIR filters satisfying prescribed specifications can also be designed.

    Assume that we are to design a 2-D low-pass filter with the following specification:

$$|H(\omega_1, \omega_2)| = \begin{cases} 1 \pm \Delta p & \text{for } 0 \leq |\omega_i| \leq \omega_{pi} \\ \Delta a & \text{for } \omega_{ai} \leq |\omega_i| \leq \omega_{si}/2, \quad i = 1, 2 \end{cases} \tag{23.35}$$

where ω_{pi} and ω_{ai}, $i = 1, 2$ are passband and stopband edges along the ω_1 and ω_2 axes, respectively. The two 1-D filters that are cascaded to form a 2-D one are specified by ω_{pi}, ω_{ai}, δ_{pi}, and δ_{ai} ($i = 1, 2$) as the passband edge, stopband edge, passband ripple, and stopband loss, respectively. From Equation 23.29 we have

$$\max \{M(\omega_1, \omega_2)\} = \max \{M_1(\omega_1)\} \max \{M_2(\omega_2)\}$$

and

$$\min \{M(\omega_1, \omega_2)\} = \min \{M_1(\omega_1)\} \min \{M_2(\omega_2)\}$$

Hence the derived 2-D filter will satisfy the specifications of Equation 23.35 if the following constraints are satisfied

$$(1 + \delta_{p1})(1 + \delta_{p2}) \leq 1 + \Delta p \tag{23.36}$$

$$(1 - \delta_{p1})(1 - \delta_{p2}) \geq 1 - \Delta p \tag{23.37}$$

$$(1 + \delta_{p1})\delta_{a2} \leq \Delta a \tag{23.38}$$

$$(1 + \delta_{p2})\delta_{a1} \leq \Delta a \tag{23.39}$$

$$\delta_{a1}\delta_{a2} \leq \Delta a \tag{23.40}$$

Constraints 23.36 and 23.37 can be expressed respectively in the alternative form

$$\delta_{p1} + \delta_{p2} + \delta_{p1}\delta_{p2} \le \Delta p \tag{23.41}$$

and

$$\delta_{p1} + \delta_{p2} - \delta_{p1}\delta_{p2} \le \Delta p \tag{23.42}$$

Hence if Equation 23.41 is satisfied, Equation 23.42 is also satisfied. Similarly, Constraints 23.38 through 23.40 will be satisfied if

$$\max\left\{(1 + \delta_{p1})\delta_{a2}, (1 + \delta_{p2})\delta_{a1}\right\} \le \Delta a \tag{23.43}$$

since $(1 + \delta_{p1}) \gg \delta_{a1}$ and $(1 + \delta_{p2}) \gg \delta_{a2}$. Now if we assume that $\delta_{p1} = \delta_{p2} = \delta_p$ and $\delta_{a1} = \delta_{a2} = \delta_a$, then we can assign

$$\delta_p = (1 + \Delta p)^{1/2} - 1 \tag{23.44}$$

and

$$\delta_a = \frac{\Delta a}{(1 + \Delta p)^{1/2}} \tag{23.45}$$

so as to satisfy Constraints 23.36 through 23.40. And since $\Delta p \ll 1$, we have

$$\delta_p \approx \frac{\Delta p}{2} \tag{23.46}$$

$$\delta_a \approx \Delta a \tag{23.47}$$

Consequently, if the maximum allowable passband and stopband errors Δp and Δa are specified, the maximum passband ripple A_p and the maximum stopband attenuation A_a and dB for the two 1-D filters can be determined as

$$A_p = 20 \log \frac{1}{1 - \delta_p} = 20 \log \frac{2}{2 - \Delta p} \tag{23.48}$$

and

$$A_a = 20 \log \frac{1}{\delta_p} = 20 \log \frac{1}{\Delta a} \tag{23.49}$$

Finally, if the passband and stopband edges ω_{pi} and ω_{ai} are also specified, the minimum order and the transfer function of each of the two 1-D filters can readily be obtained using any of the approaches of the previous sections. Similar treatments for bandpass, bandstop, and high-pass filters can be carried out.

Example 23.3 [12]

Design a zero-phase filter whose amplitude response is specified in Figure 23.12a, with $A_p = 3$ dB, $A_a = 20$ dB.

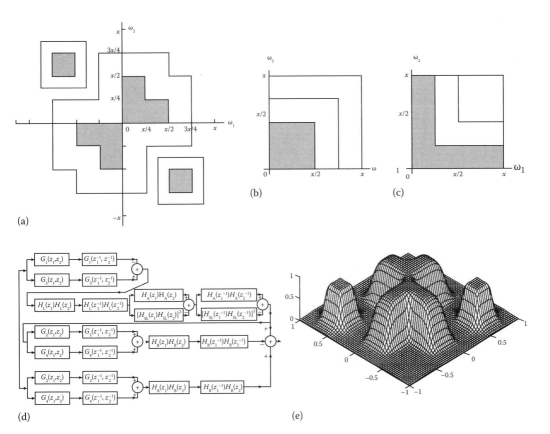

FIGURE 23.12 Amplitude responses of the filters in Example 23.3: (a) the given characteristics; (b) rectangular pass subfilter; (c) rectangular step subfilter; (d) final configuration of the 2D filter; (e) 3D plot of the amplitude response of the resulting 2D filter.

Solution

1. Decomposition: Because there are no contiguous passbands between the characteristics of the first and third quadrants and that of the second and fourth quadrants, in the first step of design, the required 2-D filter can be decomposed into two subfilters $H_{13}(z_1, z_2)$ and $H_{24}(z_1, z_2)$, which represent the characteristics of the first and third quadrants and that of the second and fourth quadrants, respectively. By connecting $H_{13}(z_1, z_2)$ and $H_{24}(z_1, z_2)$, in parallel, the required characteristics of the 2-D filter specified in Figure 23.12a can then be realized.

$$H(z_1, z_2) = H_{13}(z_1, z_2) + H_{24}(z_1, z_2)$$

(a) Decomposition of $H_{13}(z_1, z_2)$: To accomplish the design of $H_{13}(z_1, z_2)$, further decompositions should be made. The characteristics of the first and third quadrants can be realized by cascading three subfilters, i.e., a two quadrant filter $G_{13}(z_1, z_2)$ [as shown in Figure 23.7a], a low-pass filter $H_L(z_1, z_2)$ [as shown in Figure 23.12b], and a rectangular stop filter $H_S(z_1, z_2)$ [as shown in Figure 23.12c],

$$H_{13}(z_1, z_2) = G_{13}(z_1, z_2)H_L(z_1, z_2)H_S(z_1, z_2)$$

furthermore, the rectangular stop filter $H_S(z_1, z_2)$ can be designed by allpass filter $H_A(z_1, z_2)$ and low-pass filter $H_{SL}(z_1, z_2)$, using Equation 23.32

$$H_S(z_1, z_2) = H_A(z_1, z_2) - [H_{SL}(z_1, z_2)]^2$$

(b) Decomposition of $H_{24}(z_1, z_2)$: Similarly, $H_{24}(z_1, z_2)$ can be decomposed into the cascade of a two quadrant filter $G_{24}(z_1, z_2)$ [as is shown in Figure 23.7b], and a bandpass filter $H_B(z_1, z_2)$ which can be realized by using two 1-D bandpass filters for both directions.

$$H_{24}(z_1, z_2) = G_{24}(z_1, z_2)H_B(z_1, z_2)$$

The final configuration of the desired filter $H(z_1, z_2)$ is illustrated in Figure 23.12d, with the final transfer function being of the form

$$H(z_1, z_2) = H_{13}(z_1, z_2) + H_{24}(z_1, z_2) - H_{13}(z_1, z_2)H_{24}(z_1, z_2)$$

where the purpose of the term $H_{13}(z_1, z_2)H_{24}(z_1, z_2)$, is to remove the overlap that may be created by adding $H_{13}(z_1, z_2)$ and $H_{24}(z_1, z_2)$.

(2) Design of all the subfilters: At this point, the problem is to derive the two quadrant subfilters $G_{13}(z_1, z_2)$ and $G_{24}(z_1, z_2)$, the low-pass subfilters $H_L(z_1, z_2)$ and $H_{SL}(z_1, z_2)$, the allpass subfilter $H_A(z_1, z_2)$, and the bandstop filter $H_B(z_1, z_2)$.

Note the symmetry of the given characteristics, the identical 1-D sections can be used to develop all the above 2-D subfilters, and the given specifications can easily be combined into the designs of all the 1-D sections.

(3) By connecting all the 2-D subfilters in cascade or parallel as specified in Figure 23.12d, the required 2-D filter is obtained. The 3-D plot of the amplitude response of the final resulting 2-D filter is depicted in Figure 23.12e.

23.4 Design of Circularly Symmetric Filters

23.4.1 Design of LP Filters

As mentioned in Section 23.2.1, rotated filters can be used to design circularly symmetric filters. Costa and Venetsanopoulos [14] and Goodman [15] proposed two methods of this class, based on transforming an analog transfer function or a discrete one by rotated filter transformation, respectively. The two methods lead to filters that are, theoretically, unstable but by using an alternative transformation suggested by Mendonca et al. [16], this problem can be eliminated.

23.4.1.1 Design Based on 1-D Analog Transfer Function

Costa and Venetsanopoulos [14] proposed a method to design circularly symmetric filters. In their method, a set of 2-D analog transfer functions is first obtained by applying the rotated filter transformation in Equation 23.9a for several different values of the rotation angle β to a 1-D analog low-pass transfer function. A set of 2-D discrete low-pass functions are then deduced through the application of the bilinear transformation. The design is completed by cascading the set of 2-D digital filters obtained. The steps involved are as follows.

Step 1. Obtain a stable 1-D analog low-pass transfer function

$$H_{A1}(s) = \frac{Ns}{Ds} = K_0 \frac{\prod_{i=1}^{M} (s - z_{ai})}{\prod_{i=1}^{N} (s - p_{ai})} \tag{23.50}$$

where
   z_{ai} and p_{ai} for $i = 1, 2, \ldots$, are the zeros and poles of $H_{A1}(s)$, respectively
   K_0 is a multiplier constant

Step 2. Let β_k for $k = 1, 2, \ldots, K$ be a set of rotation angles defined by

$$\beta_k = \begin{cases} \left(\dfrac{2k-1}{2K} + 1\right)\pi & \text{for even } K \\[2ex] \left(\dfrac{k-1}{K} + 1\right)\pi & \text{for even } K \end{cases} \tag{23.51}$$

Step 3. Apply the transformation of Equation 23.9a to obtain a 2-D analog transfer function as

$$|H_{A2k}(s_1, s_2) = H_{A1}(s)|_{s=-s_1 \sin\beta_k + s_2 \cos\beta_k} \tag{23.52}$$

for each rotation angle β_k identified in Step 2.

Step 4. Apply the double bilinear transformation to $H_{A2k}(s_1, s_2)$ to obtain

$$H_{D2k}(z_1, z_2) = |H_{A2k}(s_1, s_2)|_{s_i = 2(z_i - 1)/T_i(z_i + 1)}, \quad i = 1, 2 \tag{23.53}$$

Assuming that $T_1 = T_2 = T$, Equations 23.50 and 23.53 yield

$$H_{D2k}(z_1, z_2) = K_1 \prod_{i=1}^{M_0} H_{2i}(z_1, z_2) \tag{23.54}$$

where

$$H_{2i}(z_1, z_2) = \frac{a_{11i} + a_{21i}z_1 + a_{12i}z_2 + a_{22i}z_1 z_2}{b_{11i} + b_{21i}z_1 + b_{12i}z_2 + b_{22i}z_1 z_2} \tag{23.55}$$

$$K_1 = K_0 \left(\frac{T}{2}\right)^{N-M} \tag{23.56}$$

$$\begin{aligned}
a_{11i} &= -\cos\beta_k + \sin\beta_k - \frac{Tz_{ai}}{2} \\
a_{21i} &= -\cos\beta_k - \sin\beta_k - \frac{Tz_{ai}}{2} \quad \text{for } 1 \leq i \leq M \\
a_{12i} &= \cos\beta_k + \sin\beta_k - \frac{Tz_{ai}}{2} \\
a_{22i} &= \cos\beta_k - \sin\beta_k - \frac{Tz_{ai}}{2} \\
a_{11i} &= a_{12i} = a_{21i} = a_{22i} = 1 \quad \text{for } M \leq i \leq M_0
\end{aligned} \tag{23.57a}$$

$$\begin{aligned}
b_{11i} &= -\cos\beta_k + \sin\beta_k - \frac{Tp_{ai}}{2} \\
b_{21i} &= -\cos\beta_k - \sin\beta_k - \frac{Tp_{ai}}{2} \quad \text{for } 1 \leq i \leq N \\
b_{12i} &= \cos\beta_k + \sin\beta_k - \frac{Tp_{ai}}{2} \\
b_{22i} &= \cos\beta_k - \sin\beta_k - \frac{Tp_{ai}}{2} \\
b_{11i} &= b_{12i} = b_{21i} = b_{22i} = 1 \quad \text{for } N \leq i \leq M_0
\end{aligned} \tag{23.57b}$$

and

$$M_0 = \max(M, N)$$

Step 5. Cascade the filters obtained in Step 4 to yield an overall transfer function

$$H(z_1, z_2) = \prod_{k=1}^{K} H_{D2k}(z_1, z_2)$$

It is easy to find that, at point $(z_1, z_2) = (-1, -1)$, both the numerator and denominator polynomials of $H_{2i}(z_1, z_2)$ assume the value of zero. And thus each $H_{2i}(z_1, z_2)$ has nonessential singularity of the second kind on the unit bicircle

$$U^2 = \{(z_1, z_2): |z_1| = 1, |z_2| = 1\}$$

The nonessential singularity of each $H_{2i}(z_1, z_2)$ can be eliminated and, furthermore, each subfilter can be stabilized by letting

$$b'_{12i} = b_{12i} + \varepsilon b_{11i} \tag{23.58a}$$

$$b'_{22i} = b_{22i} + \varepsilon b_{21i} \tag{23.58b}$$

where ε is a small positive constant. With this modification, the denominator polynomial of each $H_{2i}(z_1, z_2)$ is no longer zero and, furthermore, the stability of the subfilter can be guaranteed if

$$\mathrm{Re}(p_{ai}) < 0 \tag{23.59}$$

and

$$270° < \beta_k < 360° \tag{23.60}$$

As can be seen in Equation 23.51, half of the rotation angles are in the range $180° < \beta_k < 270°$ and according to the preceding stable conditions they yield unstable subfilters. However, the problem can easily be overcome by using rotation angles in the range given by Equation 23.60 and the rotating the transfer function of the subfilter by $-90°$ using linear transformations described in Section 23.2.1. For example, an effective rotation angle $\beta_k = 225°$ is achieved by rotating the input data by $90°$, filtering using a subfilter rotated by $315°$, and then rotating the output data by $-90°$, as shown in Figure 23.13.

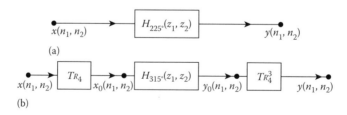

FIGURE 23.13   Realization of subfilter for rotation angle in the range $180° < \beta_k < 270°$. (a) An effective rotation angle $\beta_k = 225°$ achieved by rotating the input by $90°$. (b) Filtering using a subfilter rotated by $315°$, and then rotating the output data by $-90°$.

In addition, a 2-D zero-phase filter can be designed by cascading subfilters for rotation angles $\pi + \beta_k$ for $k = 1, 2, \ldots, N$. The resulting transfer function is given by

$$H(z_1, z_2) = \prod_{k=1}^{K} H_{D2k}(z_1, z_2) H_{D2k}(z_1, z_2)$$

where the noncausal sections can be realized as illustrated in Figure 23.3.

23.4.1.2 Design Based on 1-D Discrete Transfer Function

The method proposed by Goodman [15] is based on the 1-D discrete transfer function transformation. In the method, a 1-D discrete transfer is first obtained by applying the bilinear transformation to a 1-D analog transfer function. Then, through the application of an allpass transformation that rotates the contours of the amplitude of the 1-D discrete transfer function, a corresponding 2-D transfer function is obtained. The steps involved are as follows.

Step 1. Obtain a stable 1-D analog low-pass transfer function $H_{A1}(s)$ of the form given by Equation 23.50.

Step 2. Apply the bilinear transformation to $H_{A1}(s)$ to obtain

$$H_{D1}(z) = H_{A1}(s)\big|_{s=2(z-1)/T(z+1)} \tag{23.61}$$

Step 3. Let β_k for $k = 1, 2, \ldots, K$ be a set of rotation angles given by Equation 23.51.

Step 4. Apply the allpass transformation defined by

$$z = f_k(z_1, z_2) = \frac{1 + c_k z_1 + d_k z_2 + e_k z_1 z_2}{e_k + d_k z_1 + c_k z_2 + z_1 z_2} \tag{23.62}$$

where

$$c_k = \frac{1 + \sin\beta_k + \cos\beta_k}{1 - \sin\beta_k + \cos\beta_k} \tag{23.63a}$$

$$d_k = \frac{1 - \sin\beta_k - \cos\beta_k}{1 - \sin\beta_k + \cos\beta_k} \tag{23.63b}$$

$$e_k = \frac{1 + \sin\beta_k - \cos\beta_k}{1 - \sin\beta_k + \cos\beta_k} \tag{23.63c}$$

to obtain the 2-D discrete transfer function

$$H_{D2k}(z_1, z_2) = H_{D1}(z)\big|_{z=f_k(z_1, z_2)} \tag{23.64}$$

for $k = 1, 2, \ldots, K$. The procedure yields the 2-D transfer function of Equation 23.64, as can be easily demonstrated, and by cascading the rotated subfilters $H_{D2k}(z_1, z_2)$ the design can be completed.

The method of Goodman is equivalent to that of Costa and Venetsanopoulos and consequently, the resulting filter is subject to the same stability problem due to the nonessential singularity of the second kind at point $(z_1, z_2) = (-1, -1)$. To achieve a stable design, Goodman suggested that the transfer function $H_{D2k}(z_1, z_2)$ for $k = 1, 2, \dots, K$ be obtained directly by minimizing an appropriate objective function subject to the constraints

$$c_k + d_k - e_k \leq 1 - \varepsilon$$
$$c_k - d_k + e_k \leq 1 - \varepsilon$$
$$-c_k + d_k + e_k \leq 1 - \varepsilon$$
$$-c_k - d_k - e_k \leq 1 - \varepsilon$$

through an optimization procedure. If ε is a small positive constant, the preceding constraints constitute necessary and sufficient conditions for stability and, therefore, such an approach will yield a stable filter.

23.4.1.3 Elimination of Nonessential Singularities

To eliminate the nonessential singularities in the preceding two methods, Mendonca et al. [16] suggested a new transformation of the form

$$s = g_5(s_1, s_2) = \frac{\cos \beta_k s_1 + \sin \beta_k s_2}{1 + c s_1 s_2} \tag{23.65}$$

by combining the transformations in Equations 23.9a and 23.11 to replace the transformation Equation 23.9a. If we ensure that

$$\cos \beta_k > 0, \quad \sin \beta_k > 0, \quad \text{and} \quad c > 0$$

then the application of this transformation followed by the application of the double bilinear transformation yields stable 2-D digital filters that are free of nonessential singularities of the second kind. If, in addition

$$c = \frac{1}{\omega_{\max}^2}$$

then local-type preservation can be achieved on the set Ω_2 given by

$$\Omega_2 = \{(\omega_1, \omega_2): \omega_1 \geq 0, \omega_2 \geq 0, \omega_1 \omega_2 \leq \omega_{\max}\}$$

and if $\omega_{\max} \to \infty$, then a global-type preservation can be approached as closely as desired.

By using the transformation of Equation 23.65 instead of that in Equation 23.9a in the method of Costa and Venetsanopoulos, the transfer function of Equation 23.54 becomes

$$H_{D2k}(z_1, z_2) = K_1 P_{D2}(z_1, z_2) \times \prod_{i=1}^{M_0} \frac{a_{11i} + a_{21i} z_1 + a_{12i} z_2 + a_{22i} z_1 z_2}{b_{11i} + b_{21i} z_1 + b_{12i} z_2 + b_{22i} z_1 z_2} \tag{23.66}$$

where

$$K_1 = K_0 \left(\frac{T}{2}\right)^{N-M} \tag{23.67a}$$

$$P_{D2}(z_1, z_2) = \left[1 + \frac{4c}{T^2} + \left(1 - \frac{4c}{T^2}\right) z_1 + \left(1 - \frac{4c}{T^2}\right) z_2 + \left(1 + \frac{4c}{T^2}\right) z_1 z_2\right]^{N-M} \tag{23.67b}$$

and

$$a_{11i} = -\cos \beta_k - \sin \beta_k - \left(\frac{T}{2} + \frac{2c}{T}\right) z_{ai}$$

$$a_{21i} = \cos \beta_k - \sin \beta_k - \left(\frac{T}{2} + \frac{2c}{T}\right) z_{ai} \quad \text{for } 1 \le i \le M$$

$$a_{12i} = -\cos \beta_k + \sin \beta_k - \left(\frac{T}{2} - \frac{2c}{T}\right) z_{ai}$$

$$a_{22i} = \cos \beta_k + \sin \beta_k - \left(\frac{T}{2} + \frac{2c}{T}\right) z_{ai}$$

$$a_{11i} = a_{21i} = a_{12i} = a_{22i} = 1 \quad \text{for } M \le i \le M_0$$

$$b_{11i} = -\cos \beta_k - \sin \beta_k - \left(\frac{T}{2} + \frac{2c}{T}\right) p_{ai}$$

$$b_{21i} = \cos \beta_k - \sin \beta_k - \left(\frac{T}{2} - \frac{2c}{T}\right) p_{ai} \quad \text{for } 1 \le i \le N$$

$$b_{12i} = -\cos \beta_k + \sin \beta_k - \left(\frac{T}{2} - \frac{2c}{T}\right) p_{ai}$$

$$b_{22i} = \cos \beta_k + \sin \beta_k - \left(\frac{T}{2} - \frac{2c}{T}\right) p_{ai}$$

$$b_{11i} = b_{21i} = b_{12i} = b_{22i} = 1 \quad \text{for } N \le i \le M_0$$

$$M_0 = \max(M, N)$$

An equivalent design can be obtained by applying the allpass transformation of Equation 23.62 in Goodman's method with

$$c_k = \frac{1 + \cos \beta_k - \sin \beta_k - 4c/T^2}{1 - \cos \beta_k - \sin \beta_k + 4c/T^2}$$

$$d_k = \frac{1 - \cos \beta_k + \sin \beta_k - 4c/T^2}{1 - \cos \beta_k - \sin \beta_k + 4c/T^2}$$

$$c_k = \frac{1 + \cos \beta_k + \sin \beta_k + 4c/T^2}{1 - \cos \beta_k - \sin \beta_k + 4c/T^2}$$

23.4.2 Realization of HP, BP, and BS Filters

Consider two zero-phase rotated subfilters that were obtained from a 1-D analog high-pass transfer function using rotating angles $-\beta_1$ and β_1, where $0° < \beta_1 < 90°$. The idealized contour plots of the two subfilters are shown in Figure 23.14a and b. If these two subfilters are cascaded, the amplitude response of the combination is obtained by multiplying the amplitude responses of the subfilters at corresponding points. The idealized contour plot of the composite filter is thus obtained as illustrated in Figure 23.14c. As can be seen, the contour plot does not represent the amplitude response of a 2-D circularly symmetric high-pass filter, and, therefore, the design of high-pass filters cannot readily be achieved by simply cascading rotated subfilters as in the case of low-pass filters. However, the design of these filters can be accomplished through the use of a combination of cascade and parallel subfilters [16].

If the above rotated subfilters are connected in parallel, we obtain a composite filter whose contour plot is shown in Figure 23.14d. By subtracting the output of the cascade filter from the output of the parallel filter, we achieve an overall filter whose contour plot is depicted in Figure 23.14e. Evidently, this plot

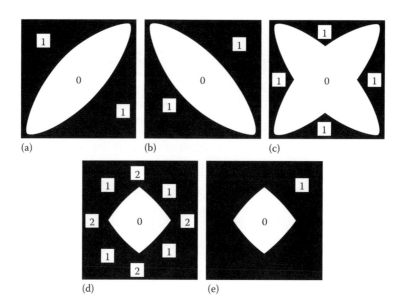

FIGURE 23.14 Derivation of 2D high-pass filter. (a) Contour plot of subfilter rotated by $-\beta_1$. (b) Contour plot of subfilter rotated by β_1. (c) Contour plot of subfilters in (a) and (b) connected in cascade. (d) Contour plot of subfilters in (a) and (b) connected in parallel. (e) Contour plot obtained by subtracting the amplitude response in (c) from that of (d).

resembles the idealized contour plot of a 2-D circularly symmetric high-pass filter, and, in effect, following this method a filter configuration is available for the design of high-pass filters.

The transfer function of the 2-D high-pass filter is then given by

$$\hat{H}_1 = \overline{H}_1 = H^{\beta_1} + H^{-\beta_1} - H^{\beta_1} H^{-\beta_1} \tag{23.68}$$

where

$$H^{\beta_1} = H_1(z_1, z_2) H_1(z_1^{-1}, z_2^{-1})$$

and

$$H^{-\beta_1} = H_1\left(z_1, z_2^{-1}\right) H_1\left(z_1^{-1}, z_2\right)$$

represent zero-phase subfilters rotated by angle β_1 and $-\beta_1$, respectively.

The above approach can be extended to two or more rotation angles in order to improve the degree of circularity. For N rotation angles, $\hat{H}_N$ is given by the recursive relation

$$\hat{H}_N = \hat{H}_{N-1} + \overline{H}_N - \hat{H}_{N-1}\overline{H}_N \tag{23.69}$$

where

$$\overline{H}_N = H^{\beta_N} + H^{-\beta_N} - H^{\beta_N} H^{-\beta_N}$$

and $\hat{H}_{N-1}$ can be obtained from $\overline{H}_{N-1}$ and $\hat{H}_{N-2}$. The configuration obtained is illustrated in Figure 23.15, where the realization of $\hat{H}_{N-1}$ is of the same as that of $\hat{H}_N$.

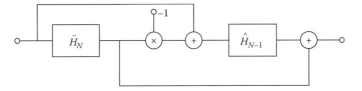

FIGURE 23.15   2D high-pass filter configuration.

As can be seen in Figure 23.15, the complexity of the high-pass configuration tends to increase rapidly with the number of rotations, and consequently, the number of rotations should be kept to a minimum. It should also be mentioned that the coefficients of the rotated filters must be properly adjusted, by using Equation 23.65, to ensure that the zero-phase is approximated. However, the use of this transformation leads to another problem: the 2-D digital transfer function obtained has spurious zeros at the Nyquist points. These zeros are due to the fact that the transformation in Equation 23.65 does not have type preservation in the neighborhoods of the Nyquist points but their presence does not appear to be of serious concern.

With the availability of circularly symmetric low-pass and high-pass filters, bandpass and bandstop filters with circularly symmetric amplitude responses can be readily obtained. A bandpass filter can be obtained by connecting a low-pass filter and a high-pass filter with overlapping passbands in cascade, whereas a bandstop filter can be realized by connecting a low-pass filter and a high-pass filter with overlapping passbands in parallel.

23.4.3 Design of Filters Satisfying Prescribed Specifications

A similar approach to that described in Section 23.3.3 can be used for the design of circularly symmetric filters satisfying prescribed specifications. Assume that the maximum/minimum passband and the maximum stopband gain of the 2-D filter are $(1 \pm \Delta p)$ and Δa, respectively, if K rotated filter sections are cascaded where half of the rotations are in the range of $180°–270°$ and the other half are in the range $270°–360°$. Then, we can assign the passband ripple δ_p and the stopband loss δ_a to be [2]

$$\delta_p = \frac{\Delta p}{K} \tag{23.70}$$

and

$$\delta_a = \Delta a^{2/K} \tag{23.71}$$

The lower (or upper) bound of the passband gain would be achieved if all the rotated sections were to have minimum (or maximum) passband gains at the same frequency point. Although it is possible for all the rotated sections to have minimum (or maximum) gains at the origin of the (ω_1, ω_2) plane, the gains are unlikely to be maximum (or minimum) together at some other frequency point and, in effect, the preceding estimate for δ_p is low. A more realistic value for δ_p is

$$\delta_p = \frac{2\Delta p}{K} \tag{23.72}$$

If Δp and Δa are prescribed, then the passband ripple and minimum stopband attenuation of the analog filter can be obtained from Equations 23.72 and 23.71 as

$$A_p = 20 \log \left[\frac{K}{K - 2\Delta p} \right] \tag{23.73}$$

and

$$A_a = \frac{40}{K} \log\left[\frac{1}{\Delta a}\right] \tag{23.74}$$

If the passband Ω_p and stopband Ω_a are also prescribed, the minimum order and the transfer function of the analog filter can be determined using the method in preceding sections.

Example 23.4

Using the method of Costa and Venetsanopoulos, design a circularly symmetric low-pass filter satisfying the following specifications:

$$\omega_{s1} = \omega_{s2} = 2\pi \text{ rad/s}$$
$$\omega_p = 0.4\pi \text{ rad/s}, \quad \omega_a = 0.6\pi \text{ rad/s}$$
$$\delta_p = \delta_a = 0.1$$

Solution

The filter satisfying prescribed specifications can be designed through the following steps:

1. Select a prototype of approximation and suitably select the number of rotations K.
2. Calculate rotation angles by Equation 23.51.
3. Determine A_p and A_a from δ_p and δ_a, respectively, using Equations 23.73 and 23.74. Calculate the prewarped Ω_p and Ω_a, from ω_p and ω_a, respectively.
4. Use above calculated specifications to obtain the prewarped 1-D analog transfer function.
5. Apply the transformations of Equations 23.52, 23.53, 23.57, and 23.58 to obtain K rotated subfilters.
6. Cascade all the rotated subfilters.

The 3-D plot of the amplitude response of the resulting filter is shown in Figure 23.16, where $K = 10$.

23.5 Design of 2-D IIR Filters by Optimization

In the preceding sections, several methods for the solution of approximation problems in 2-D IIR filters have been described. These methods lead to a complete description of the transfer function in closed form, either in terms of its zeros and poles or its coefficients. They are, as a consequence, very efficient and lead to very precise designs. Their main advantage is that they are applicable only for the design of filters with piecewise-constant amplitude responses. In the following sections, the optimization methods for the design of 2-D IIR filters are considered. In these methods, a discrete transfer function is assumed and an error function is formulated on the basis of some desired amplitude and/or phase response. These methods are iterative and, as a result, they usually involve a large amount of computation. However, unlike the closed-form methods, they are suitable for the design of filters having arbitrary amplitude or phase responses.

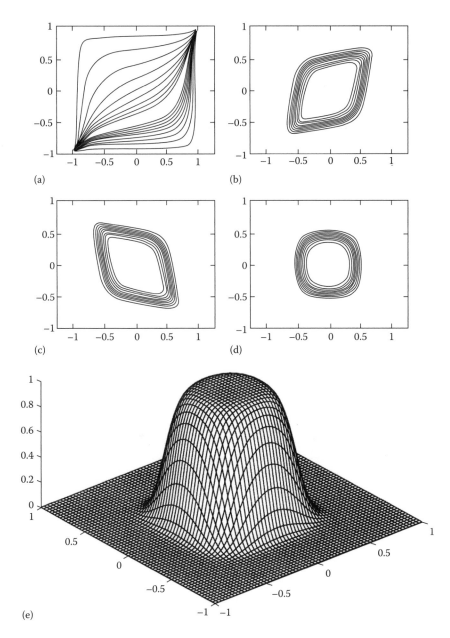

FIGURE 23.16 Amplitude response of a circularly symmetric filter in Example 23.4. (a) Subfilter for rotation angle of 243°. (b) Subfilters for rotation angles of 189°, 207°, 225°, 243°, and 261° in cascade. (c) Subfilters for rotation angles of 279°, 297°, 315°, 333°, and 351° in cascade. (d) All 10 subfilters in cascade. (e) 3D plot of the amplitude response of the resulting 2D low-pass filter.

23.5.1 Design by Least pth Optimization

The least pth optimization method has been used quite extensively in the past in a variety of applications. In this approach, an objective function in the form of a sum of elemental error functions, each raised to the pth power, is first formulated and is then minimized using any one of the available unconstrained optimization methods [17].

23.5.1.1 Problem Formulation

Consider the transfer function

$$H(z_1, z_2) = H_0 \prod_{k=1}^{K} \frac{N_k(z_1, z_2)}{D_k(z_1, z_2)}$$

$$= H_0 \prod_{k=1}^{K} \frac{\sum_{l=0}^{L_{1k}} \sum_{m=0}^{M_{1k}} a_{lm}^{(k)} z_1^{-1} z_2^{-m}}{\sum_{l=0}^{L_{2k}} \sum_{m=0}^{M_{2k}} b_{lm}^{(k)} z_1^{-1} z_2^{-m}} \quad (23.75)$$

where $N_k(z_1, z_2)$ and $D_k(z_1, z_2)$ are polynomials of order equal to or less than 2 and H_0 is a constant, and let

$$\mathbf{x} = [\mathbf{a}^{\mathrm{T}} \quad \mathbf{b}^{\mathrm{T}} \quad H_0]^{\mathrm{T}} \quad (23.76)$$

where

$$\mathbf{a} = \left[a_{10}^{(1)} \quad a_{20}^{(1)} \quad \cdots \quad a_{L_{11}M_{11}}^{(1)} \quad a_{10}^{(2)} \quad a_{20}^{(2)} \quad \cdots \quad a_{L_{12}M_{12}}^{(2)} \quad \cdots \quad a_{L_{1K}M_{1K}}^{(K)} \right]^{\mathrm{T}}$$

and

$$\mathbf{b} = \left[b_{10}^{(1)} \quad b_{20}^{(1)} \quad \cdots \quad b_{L_{11}M_{11}}^{(1)} \quad b_{10}^{(2)} \quad a_{20}^{(2)} \quad \cdots \quad b_{L_{12}M_{12}}^{(2)} \quad \cdots \quad b_{L_{2K}M_{2K}}^{(K)} \right]^{\mathrm{T}}$$

are row vectors whose elements are the coefficients of $N_k(z_1, z_2)$ and $D_k(z_1, z_2)$, respectively. An objective function can be defined in terms of the L_p norm of $\mathbf{E}(\mathbf{x})$ as

$$J(\mathbf{x}) = L_p = \| \mathbf{E}(\mathbf{x}) \|_p = \left[\sum_{i=1}^{K} |E_{mn}(\mathbf{x})|^p \right]^{1/p} \quad (23.77)$$

where p is an even positive integer

$$E_{mn}(\mathbf{x}) = \mathbf{M}(\mathbf{m}, \mathbf{n}) - \mathbf{M}_1(m, n) \quad (23.78)$$

and

$$M(m, n) = \left| H(e^{j\omega_{1m}T_1}, e^{j\omega_{2n}T_2}) \right|, \quad m = 1, \ldots, M, \, n = 1, \ldots, N$$

are samples of the amplitude response of the filter at a set of frequency pairs $(\omega_{1m}, \omega_{2n})$ ($m = 1, \ldots, M,$ $n = 1, \ldots, N$) with

$$\omega_{1m} = \frac{\omega_{s1}(m-1)}{2(M-1)} \quad \text{and} \quad \omega_{2n} = \frac{\omega_{s2}(n-1)}{2(N-1)}$$

$M_1(m, n)$ represents the desired amplitude response at frequencies $(\omega_{1m}, \omega_{2n})$.

Several special cases of the L_p norm are of particular interest. The L_1 norm, namely

$$L_1 = \sum_{i=1}^{K} |E_{mn}(x)|$$

is the sum of the magnitudes of the elements of $\mathbf{E}(\mathbf{x})$; the L_2 norm given by

$$L_2 = \left[\sum_{i=1}^{K} |E_{mn}(\mathbf{x})|^2 \right]^{1/2}$$

is the well-known Euclidean norm; and L_2^2 is the sum of the squares of the elements of $\mathbf{E}(\mathbf{x})$. In the case where $p \to \infty$ and

$$E_M(\mathbf{x}) = \max_{m,n} \{E_{mn}(\mathbf{x})\} \neq 0$$

we can write

$$L_\infty = \lim_{p \to \infty} \left\{ \sum_{k=1}^{K} |E_{mn}(\mathbf{x})|^p \right\}^{1/p}$$

$$= E_M(\mathbf{x}) = \lim_{p \to \infty} \left\{ \sum_{k=1}^{K} \left| \frac{E_{mn}(\mathbf{x})}{E_M(\mathbf{x})} \right|^p \right\}^{1/p}$$

$$= E_M(\mathbf{x}) \tag{23.79}$$

The design task at hand amounts to finding a parameter vector $\mathbf{x}$ that minimizes the least pth objective function $J(\mathbf{x})$ defined in Equation 23.77. If $J(\mathbf{x})$ is defined in terms of L_2^2, a least-squares solution is obtained; if the L_∞ norm is used, a so-called minimax solution is obtained, since in this case the largest element in $\mathbf{E}(\mathbf{x})$ is minimized.

23.5.1.2 Quasi-Newton Algorithms

The design problem described above can be solved by using any one of the standard unconstrained optimization algorithms. A class of such algorithms that has been found to be very versatile, efficient, and robust is the class of quasi-Newton algorithms [17–19]. These are based on the principle that the minimum point $\mathbf{x}^*$ of a quadratic convex function $J(\mathbf{x})$ of N variables can be obtained by applying the correction

$$\delta = -\mathbf{H}^{-1}\gamma$$

to an arbitrary point $\mathbf{x}$, that is

$$\mathbf{x}^* = \mathbf{x} + \mathbf{d}$$

where vector

$$\mathbf{g} = \nabla J(\mathbf{x}) = \left[\frac{\partial J}{\partial x_1}, \frac{\partial J}{\partial x_2}, \ldots, \frac{\partial J}{\partial x_N} \right]^{\mathrm{T}}$$

and $N \times N$ matrix

$$\mathbf{H} = \begin{vmatrix} \dfrac{\partial^2 J(\mathbf{x})}{\partial x_1^2} & \dfrac{\partial^2 J(\mathbf{x})}{\partial x_1 \partial x_2} & \cdots & \dfrac{\partial^2 J(\mathbf{x})}{\partial x_1 \partial x_N} \\ \dfrac{\partial^2 J(\mathbf{x})}{\partial x_2 \partial x_1} & \dfrac{\partial^2 J(\mathbf{x})}{\partial x_2^2} & \cdots & \dfrac{\partial^2 J(\mathbf{x})}{\partial x_2 \partial x_N} \\ \cdots & \cdots & \cdots & \\ \dfrac{\partial^2 J(\mathbf{x})}{\partial x_N \partial x_1} & \dfrac{\partial^2 J(\mathbf{x})}{\partial x_N \partial x_2} & \cdots & \dfrac{\partial^2 J(\mathbf{x})}{\partial x_N^2} \end{vmatrix}$$

are the gradient vector and Hessian matrix of $J(\mathbf{x})$ at point $\mathbf{x}$, respectively.

The basic quasi-Newton algorithm as applied to the 2-D IIR filter design problem is as follows [20].

ALGORITHM 1: Basic Quasi-Newton Algorithm

Step 1. Input $\mathbf{x}_0$ and ε. Set $\mathbf{S}_0 = \mathbf{I}_N$, where $\mathbf{I}_N$ is the $N \times N$ unity matrix and N is the dimension of $\mathbf{x}$, and set $k = 0$. Compute $\mathbf{g}_0 = -J(\mathbf{x}_0)$.

Step 2. Set $\mathbf{d}_k = -\mathbf{S}_k \mathbf{g}_k$ and find α_k, the value of α that minimizes $J(\mathbf{x}_k + \alpha \mathbf{d}_k)$, using a line search.

Step 3. Set $\mathbf{d}_k = \alpha_k \mathbf{d}_k$ and $\mathbf{x}_{k+1} = \mathbf{x}_k + \mathbf{d}_k$.

Step 4. If $\|\delta_k\| < \varepsilon$, then output $\mathbf{x}^* = \mathbf{x}_{k+1}$, $J(\mathbf{x}^*) = J(\mathbf{x}_{k+1})$ and stop, else go to step 5.

Step 5. Compute $\mathbf{g}_{k+1} = J(\mathbf{x}_{k+1})$ and set $\gamma_k = \mathbf{g}_{k+1} - \mathbf{g}_k$.

Step 6. Compute $\mathbf{S}_{k+1} = \mathbf{S}_k + \mathbf{C}_k$, where $\mathbf{C}_k$ is a suitable matrix correction.

Step 7. Check $\mathbf{S}_{k+1}$ for positive definiteness and if it is found to be nonpositive definite force it to become positive definite.

Step 8. Set $k = k + 1$ and go to step 2.

The correction matrix C_k required in step 6 can be computed by using either the Davidon–Fletcher–Powell (DFP) formula

$$C_k = \frac{\mathbf{d}_k \mathbf{d}_k^T}{\mathbf{g}_k^T \mathbf{g}_k} \frac{\mathbf{S}_k \mathbf{g}_k \mathbf{g}_k^T \mathbf{S}}{\mathbf{g}_k^T \mathbf{S}_k \mathbf{g}_k} \tag{23.80}$$

or the Broyden–Fletcher–Goldfarb–Shanno (BFGS) formula

$$C_k = \left(1 + \frac{\mathbf{g}_k^T \mathbf{S}_k \mathbf{g}_k}{\mathbf{d}_k^T \mathbf{d}_k}\right) \frac{\mathbf{d}_k \mathbf{d}_k^T}{\mathbf{g}_k^T \mathbf{d}_k} - \frac{\mathbf{d}_k \mathbf{g}_k^T \mathbf{S}_k + \mathbf{S}_k \mathbf{g}_k \mathbf{d}_k^T}{\mathbf{g}_k^T \mathbf{d}_k} \tag{23.81}$$

Algorithm 1 eliminates the need to calculate the second derivatives of the objective function; in addition, the matrix inversion is unnecessary. However, matrices $\mathbf{S}_1$, $\mathbf{S}_2, \ldots,$ $\mathbf{S}_k$ need to be checked for positive definiteness and may need to be manipulated. This can be easily done in practice by diagonalizing $\mathbf{S}_{k+1}$ and then replacing any nonpositive diagonal elements with corresponding positive ones. However, this would increase the computational burden quite significantly. The amount of computation required to complete a design is usually very large, due with the large numbers in 2-D digital filters and the large number of sample points needed to construct the objective function. Generally, the computational load can often be reduced by starting with an approximate design based on some closed-form solution. For example, the design of circularly or elliptical symmetric filters may start with filters that have square or rectangular passbands and stopbands.

Example 23.5 [20]

Design a circularly symmetric low-pass filter of order $(2, 2)$ with $\omega_{p1} = \omega_{p2} = 0.08\pi$ rad/s and $\omega_{a1} = \omega_{a2} = 0.12\pi$ rad/s, assuming that $\omega_{s1} = \omega_{s2} = 2\pi$ rad/s.

Solution

1. Construct the ideal discrete amplitude response of the filter

$$M_I(m, n) = \begin{cases} 1 & \text{for } \left(\omega_{1m}^2 + \omega_{2n}^2\right) \leq 0.008\pi \\ 0.5 & \text{for } 0.08\pi \leq \left(\omega_{1m}^2 + \omega_{2n}^2\right) \leq 0.12\pi \\ 0 & \text{otherwise} \end{cases}$$

where

$$\{\omega_{1m}\} = \{\omega_{2n}\} = 0, 0.02\pi, 0.04\pi, 0.2\pi, 0.4\pi, 0.6\pi, 0.8\pi, \pi$$

2. To reduce the amount of computation, a 1-D low-pass filter with passband edge $\omega_p = 0.08\pi$ and stopband edge $\omega_a = 0.1\pi$ is first obtained with the 1-D transfer function being

$$H_1(z) = 0.11024 \frac{1 - 1.64382z^{-1} + z^{-2}}{1 - 1.79353z^{-1} + 0.84098z^{-2}}$$

and then a 2-D transfer function with a square passband is obtained as

$$H(z_1, z_2) = H_1(z_1)H_1(z_2)$$

3. Construct the objective function of Equation 23.77, using algorithm 1 to minimize the objective function $J(\mathbf{x})$. After 20 more iterations the algorithm converges to

$$H(z_1, z_2) = 0.00895 \times \frac{\begin{bmatrix} 1 & z_1^{-1} & z_1^{-2} \end{bmatrix} \begin{vmatrix} 1.0 & -1.62151 & 0.99994 \\ -1.62151 & 2.6370 & -1.62129 \\ 0.99994 & -1.62129 & 1.00203 \end{vmatrix} \begin{vmatrix} 1 \\ z_2^{-1} \\ z_2^{-2} \end{vmatrix}}{\begin{bmatrix} 1 & z_1^{-1} & z_1^{-2} \end{bmatrix} \begin{vmatrix} 1.0 & -1.78813 & 0.82930 \\ -1.78813 & 3.20640 & -1.49271 \\ 0.82930 & -1.49271 & 0.69823 \end{vmatrix} \begin{vmatrix} 1 \\ z_2^{-1} \\ z_2^{-2} \end{vmatrix}}$$

The amplitude response of the final optimal filter is depicted in Figure 23.17.

23.5.1.3 Minimax Algorithms

Least pth Minimax Algorithm. When a objective function is formulated in terms of the L_p norm of the error function and then minimizing $J(\mathbf{x})$ for increasing values of p, such an objective function can be obtained as

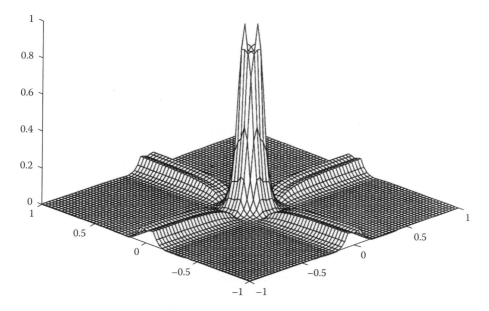

FIGURE 23.17 Amplitude response of the 2D optimal filter in Example 23.5.

$$J(x) = E_M(x) \left\{ \sum_{i=1}^{m} \left| \frac{E(x, \omega_{1i}, \omega_{2i})}{E_M(x)} \right|^p \right\}^{1/p} \tag{23.82}$$

where

$$E_M(x) = \max_{1 \le i \le m} \{E_i(x)\} = \max_{1 \le i \le m} \{E_i(x, \omega_{1i}, \omega_{2i})\} \tag{23.83}$$

A minimax algorithm based on $J(x)$ is as follows [21].

ALGORITHM 2: Least-pth Minimax Algorithm

Step 1. Input x_0 and ε. Set $k = 1$, $p = 1$, $\mu = 2$, $E_0 = 10^{99}$.
Step 2. Initialize frequencies ω_{1i}, ω_{2i} for $i = 1, 2, \ldots, m$.
Step 3. Using point x_{k-1} as initial point, minimize $J(x)$ with respect to x to obtain x_k. Set $E_k = E(x_k)$.
Step 4. If $|E_{k-1} - E_k| < \varepsilon$, then output $x^* = x_k$ and E_k and stop. Else, set $p = \mu p$, $k = k + 1$ and go to step 3.

The minimization in step 3 can be carried out using algorithm 1 or any other unconstrained optimization algorithms.

Charalambous Minimax Algorithm. The preceding algorithm gives excellent results except that it requires a considerable amount of computation. An alternative and much more efficient algorithm is the minimax algorithm proposed by Charalambous and Antoniou [22,23]. This algorithm is based on principles developed by Charalambous [24] and involves the minimization of the objective function $J(x, \zeta, \lambda)$, defined by

$$J(x, \zeta, \lambda) = \sum_{i \in I_1} \frac{1}{2} \lambda_i [J_i(x, \zeta)]^2 + \sum_{i \in I_2} \frac{1}{2} [J_i(x, \zeta)]^2 \tag{23.84}$$

where ζ and λ_i for $i = 1, 2, \ldots, m$ are constants,

$$J_i(x, \zeta) = E_i(x) - \zeta$$
$$I_1 = \{i: J_i(x, \zeta) > 0 \text{ and } \lambda_i > 0\}$$

and

$$I_2 = \{i: J_i(x, \zeta) > 0 \text{ and } \lambda_i > 0\}$$

The factor $1/2$ in Equation 23.84 is included for the purpose of simplifying the gradient that is given by

$$\nabla J(x, \zeta, \lambda) = \sum_{i \in I_1} \lambda_i J_i(x, \zeta) \nabla J_i(x, \zeta) + \sum_{i \in I_2} J_i(x, \zeta) \nabla J_i(x, \zeta) \tag{23.85}$$

It can be shown that, if

1. The second-order sufficient conditions for a minimum hold at x^*,
2. $\lambda_i = \lambda_i^*$, $i = 1, 2, \ldots, m$, where λ_i^* are the minimax multipliers corresponding to a minimum optimum solution x^*, and
3. $E(x^*) - \xi$ is sufficiently small then x^* is a strong local minimum point of $J(x, \xi, \lambda)$.

Condition 1 is usually satisfied in practice. Therefore, a local minimum point x* can be found by forcing λ_i to approach λ_i^* $(i=1,2,\ldots,m)$ and making $E(\mathbf{x}^*) - \xi$ sufficiently small. These two constraints can be simultaneously satisfied by applying the following algorithm.

ALGORITHM 3: Charalambous Minimax Algorithm

Step 1. Set $\xi = 0$ and $\lambda_i = 1$ for $i = 1, 2, \ldots, m$. Initialize **x**.
Step 2. Minimize function $J(\mathbf{x}, \xi, \lambda)$ to obtain **x**.
Step 3. Set

$$S = \sum_{i \in I_1} \lambda_i j_i(\mathbf{x}, \xi) + \sum_{i \in I_2} J_i(\mathbf{x}, \xi)$$

and update λ_i and ξ as

$$\lambda_i = \begin{cases} \lambda_i J_i(\mathbf{x}, \xi)/S & \text{if } J_i(\mathbf{x}, \xi) \geq 0, \, \lambda_i \geq 0 \\ J_i(\mathbf{x}, \xi)/S & \text{if } J_i(\mathbf{x}, \xi) \geq 0, \, \lambda = 0 \\ 0 & \text{if } J_i(\mathbf{x}, \xi) < 0 \end{cases}$$

$$\xi = \sum_{i=1}^{m} \lambda_i E_i(\mathbf{x})$$

Step 4. Stop if

$$\frac{E_M(\mathbf{x}) - \xi}{E_M(\mathbf{x})} \leq \varepsilon$$

otherwise go to step 2.

The parameter ε is a prescribed termination tolerance. When the algorithm converges, conditions 2 and 3 are satisfied and $\mathbf{x} = \mathbf{x}^*$. The unconstrained optimization in step 2 can be accomplished by applying a quasi-Newton algorithm.

Example 23.6 [23]

Design a 2-D circularly symmetric filter with the same specifications as in Example 23.5, using algorithm 3.

Solution

1. Construct the ideal discrete amplitude response of the filter. Since the passband and stopband contours are circles, the sample points can be placed on arcs of a set of circles centered at the origin. Five circles with radii

$$r_1 = 0.3\omega_p, \quad r_2 = 0.6\omega_p, \quad r_3 = 0.8\omega_p, \quad r_4 = 0.9\omega_p, \quad \text{and} \quad r_5 = \omega_p$$

are placed in the passband and five circles with radii

$$r_6 = \omega_a, \quad r_7 = \omega_a + 0.1(\pi - \omega_a), \quad r_8 = \omega_a + 0.2(\pi - \omega_a), \quad r_9 = \omega_a + 0.55(\pi - \omega_a), \quad \text{and} \quad r_{10} = \pi$$

are placed in the stopband.

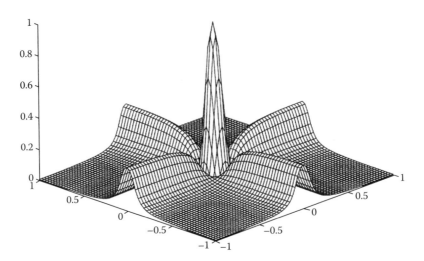

FIGURE 23.18 3D plot of the amplitude response of the filter in Example 23.6.

For circularly symmetric filters, the amplitude is uniquely specified by the amplitude response in the sector $[0°, 45°]$. Therefore, six equally spaced points on each circle described above between $0°$ and $45°$ are chosen. These points plus the origin $(\omega_1, \omega_2) = (0,0)$ form a set of 61 sample points.

2. Select the 2-D transfer function. Because a circularly symmetric filter has a transfer function with separable denominator [24], we can select the transfer function to be of the form

$$H(z_1, z_2) = H_0(z_1, z_2)^{-1} \times \prod_{k=1}^{k} \frac{z_1 z_2 + z_1^{-1} z_2^{-1} + a_k(z_1 + z_1^{-1} + z_2 + z_2^{-1}) + z_1^{-1} z_2 + z_1 z_2^{-1} + b_k}{(1 + c_k z_1^{-1} + d_k z_1^{-2})(1 + c_k z_2^{-1} + d_k z_2^{-2})} \quad (23.86)$$

with parameter H_0 fixed as $H_0 = (0.06582)^2$, $K = 1$, $\varepsilon = 0.01$.

3. Starting from

$$a_1^{(0)} = -1.514, \quad b_1^{(0)} = (a_1^{(0)})^2, \quad c_1^{(0)} = -1.784, \quad d_1^{(0)} = 0.8166$$

and algorithm 3 yields the solution

$$a_1^* = 1.96493, \quad b_1^* = -10.9934, \quad c_1^* = -1.61564, \quad d_1^* = 0.66781$$

$$E_M(\mathbf{x}) = 0.37995$$

The 3-D plot of the amplitude response of the resulting filter is illustrated in Figure 23.18.

23.5.2 Design by Singular-Value Decomposition

As will be seen, an important merit of the design methods of 2-D IIR filters based on singular-value decomposition (SVD) is that the required 2-D filter is decomposed into a set of 1-D digital subfilters, which are much easier to design by optimization than the original 2-D filters.

23.5.2.1 Problem Formulation

In a quadrantally symmetric filter, $H(z_1, z_2)$ has a separable denominator [25]. Therefore, it can be expressed as

$$H(z_1, z_2) = \sum_{i=1}^{K} f_i(z_1) g_i(z_2) \quad (23.87)$$

In effect, a quadrantally symmetric filter can always be realized using a set of K parallel sections where the ith section is a separable subfilter characterized by the transfer function $f_i(z_1) \, g_i(z_2)$.

Consider the desired amplitude response sample of 2-D filter $H(z_1, z_2)$, we form a 2-D amplitude specification matrix $\mathbf{A}$ as

$$
\mathbf{A} = \begin{vmatrix}
a_{11} & a_{12} & \cdots & a_{1L} \\
a_{21} & a_{22} & \cdots & a_{2L} \\
\vdots & \vdots & & \vdots \\
a_{M1} & a_{M1} & \cdots & a_{ML}
\end{vmatrix}
\tag{23.88}
$$

where $\{a_{ml}\}$ is a desired amplitude response sampled at frequencies $(\omega_{1l}, \omega_{2m}) = (\pi\mu_l/T_i, \pi\nu_m/T_2)$, with

$$
\mu_l = \frac{l-1}{L-1}, \quad \nu_m = \frac{m-1}{M-1} \quad \text{for } 1 \le l \le L, \quad 1 \le m \le M
$$

that is,

$$
a_{ml} = \left| H\left(e^{j\pi\mu_l}, e^{j\pi\nu_m}\right) \right|
$$

If the matrix $\mathbf{A}$ can be decomposed into the form of

$$
\mathbf{A} = \sum_{i=1}^{r} \mathbf{F}_i \mathbf{G}_i
\tag{23.89}
$$

then, by using the column vectors $\mathbf{F}_i$ and row vectors $\mathbf{G}_i$, we can construct matrices

$$
\mathbf{F} = [\mathbf{F}_1 \quad \mathbf{F}_2 \quad \cdots \quad \mathbf{F}_r]
\tag{23.90}
$$

$$
\mathbf{G} = [\mathbf{G}_1 \quad \mathbf{G}_2 \quad \cdots \quad \mathbf{G}_r]^{\mathrm{T}}
\tag{23.91}
$$

If all elements of $\mathbf{F}$ and $\mathbf{G}$ are nonnegative then they can be regarded as the amplitude specifications matrices of an r-input/1-output 1-D filter $F(z_1)$ and a 1-input/r-output 1-D filter $G(z_2)$,

$$
F(z_1) = [f_1(z_1), f_2(z_1), \ldots, f_r(z_1)]
\tag{23.92}
$$

$$
G(z_2) = [g_1(z_2), g_2(z_2), \ldots, g_r(z_2)]^{\mathrm{T}}
\tag{23.93}
$$

Therefore, the 2-D filter of Equation 23.87 can be approximated by

$$
H(z_1, z_2) = F(z_1)G(z_2)
\tag{23.94}
$$

In this section, two design procedures are described that can be applied to the design of 2-D IIR filters whose amplitude responses are quadrantally symmetric.

23.5.2.2 Method of Antoniou and Lu

Antoniou and Lu proposed a method of 2-D IIR filter based on the SVD of the amplitude response matrix $\mathbf{A}$ [26]. The SVD of matrix $\mathbf{A}$ gives [27]

$$
\mathbf{A} = \sum_{i=1}^{r} \sigma_i \mathbf{u}_i \mathbf{v}_i^{\mathrm{T}} = \sum_{i=1}^{r} \mathbf{f}_i \mathbf{g}_i^{\mathrm{T}}
\tag{23.95}
$$

where σ_i are the singular values of $\mathbf{A}$ such that $\sigma_1 \le \sigma_2 \le \cdots \le \sigma_r \le 0$ is the rank of $\mathbf{A}$, $\mathbf{u}_i$, and $\mathbf{v}_i$ are the ith eigenvector of $\mathbf{A}\mathbf{A}^T$ and $\mathbf{A}^T\mathbf{A}$, respectively, $\boldsymbol{\phi}_i = \sigma_i^{1/2}\mathbf{u}_i$, $\boldsymbol{\gamma}_i = \sigma_i^{1/2}\mathbf{v}_i$, and $\{\boldsymbol{\Phi}_i\colon 1 \le i \le r\}$ and $\{\boldsymbol{\gamma}_i\colon 1 \le i \le r\}$ are sets of orthogonal L-dimensional and M-dimensional vectors, respectively.

An important property of the SVD can be stated as

$$\left\| \mathbf{A} - \sum_{i=1}^{K} \mathbf{F}_i \mathbf{g}_i^T \right\| = \min_{\bar{\phi}, \bar{\gamma}} \left\| \mathbf{A} - \sum_{i=1}^{K} \overline{\mathbf{F}}_i \overline{\mathbf{g}}_i^T \right\| \quad \text{for } 1 \le K \le r \tag{23.96}$$

where $\overline{\phi}_i \in R^L$, $\overline{\gamma}_i \in R^M$

To design a 2-D IIR filter by SVD, two steps are involved, namely

Step 1. Design of the main section.
Step 2. Design of the error correction sections as will be detailed below.

Design of the Main Section. Note that Equation 23.95 can be written as

$$\mathbf{A} = \mathbf{f}_1 \mathbf{g}_1^T + \varepsilon_1 \tag{23.97}$$

where $\varepsilon_1 = \Sigma_i^{\gamma} \Phi_i \gamma_i^T$. And since all the elements of $\mathbf{A}$ are nonnegative, if follows that all elements of ϕ_1 and γ_1 are nonnegative.

On comparing Equation 23.97 with Equation 23.95 and assuming that $K = 1$ and that ϕ_1, γ_1 are sampled versions of the desired amplitude responses for the 1-D filters characterized by $f_1(z_1)$ and $g_1(z_2)$, respectively, a 2-D filter can be designed through the following procedures:

1. Design 1-D filters F_1 and G_1 characterized by $f_1(z_1)$ and $g_1(z_2)$.
2. Connect filters F_1 and G_1 in cascade, i.e.,

$$H_1(z_1, z_2) = f_1(z_1)g_1(z_2)$$

Step 1 above can be carried out by using an optimization algorithm such as the quasi-Newton algorithm or the minimax algorithm.

Since $f_1(z_1)g_1(z_2)$ corresponds to the largest singular value σ_1, the subfilter characterized by $f_1(z_1)\,g_1(z_2)$ is said to be the main section of the 2-D filter.

Design of the Error Correction Sections. The approximation error of $H_1(z_1, z_2)$ can be reduced by realizing more of the terms in Equation 23.95 by means of parallel filter sections. From Equation 23.97, we can write

$$\mathbf{A} = \mathbf{f}_1 \mathbf{g}_1^T + \mathbf{f}_2 \mathbf{g}_2^T + \varepsilon_{21} \tag{23.98}$$

Since ϕ_2 and λ_2 may have some negative components, a careful treatment in Equation 23.98 is necessary. Let ϕ_2^- and γ_2^- be the absolute values of the most negative components of ϕ_2 and γ_2, respectively. If

$$\mathbf{e}_\phi = [1\ 1 \cdots 1]^T \in R^L \quad \text{and} \quad \mathbf{e}_\gamma = [1\ 1 \cdots 1]^T \in R^M$$

then all components of

$$\phi_{2p} = \phi_2 + \phi_2^- \varepsilon_\phi \text{ ανδ } \gamma_{2p} = \gamma_2 + \gamma_2^- \varepsilon_\gamma$$

are nonnegative. If it is possible to design 1-D linear-phase or zero-phase filters characterized by $f_1(z_1)$, $g_1(z_2)$, $f_{2p}(z_1)$, and $g_{2p}(z_2)$, such that

$$f_1(e^{j\pi\mu_i}) = |f_1(e^{j\pi\mu_i})|e^{j\alpha_1\mu_i}$$
$$g_1(e^{j\pi v_m}) = |g_1(e^{j\pi v_m})|e^{j\alpha_2 v_m}$$

and

$$f_{2p}(e^{j\pi\mu_i}) = |f_{2p}(e^{j\pi\mu_i})|e^{j\alpha_1\mu_i}$$
$$g_{2p}(e^{j\pi v_m}) = |g_{2p}(e^{j\pi v_m})|e^{j\alpha_2 v_m}$$

for $1 \leq l \leq L, \leq m \leq M$, where

$$|f_1(e^{j\pi\mu_1})| \approx \phi_{1l}$$
$$|g_1(e^{j\pi v_m})| \approx \gamma_{1m}$$
$$|f_{2p}(e^{j\pi\mu_l})| \approx \phi_{2lp}$$
$$|g_{2p}(e^{j\pi v_m})| \approx \gamma_{2mp}$$

In above $\phi_{1l}, \phi_{2lp}, \gamma_{1m}, \gamma_{2mp}$ represent the lth component of ϕ_1, ϕ_{2p} and mth component of γ_1 and γ_{2p}, respectively. α_1 and α_2 are constants that are equal to zero if zero-phase filters are to be designed. Let

$$\alpha_1 = -\pi n_1, \quad \alpha_2 = -\pi n_2 \quad \text{with integers } n_1, n_2 \geq 0 \tag{23.99}$$

and define

$$f_2(z_1) = f_{2p}(z_1) - \phi_2^- z_1^{-n_1} \tag{23.100}$$

$$g_2(z_2) = g_{2p}(z_2) - \gamma_2^- z_2^{-n_2} \tag{23.101}$$

It follows that

$$f_2(e^{j\pi\mu_l}) = \left[f_{2p}(e^{j\pi\mu_l}) - \phi_2^-\right] e^{-j\pi n_1\mu_l} \approx \phi_{2l}e^{-j\pi\mu_l n_1}$$
$$g_2(e^{j\pi v_m}) = \left[g_{2p}(e^{j\pi v_m}) - \gamma_2^-\right] e^{-j\pi n_2 v_m} \approx \gamma_{2m}e^{-j\pi\gamma_m n_2}$$

Furthermore, if we form

$$H_2(z_1, z_2) = f_1(z_1)g_1(z_2) + f_2(z_1)g_2(z_2) \tag{23.102}$$

then

$$\left|H_2(e^{j\pi\mu_l}, e^{j\pi v_m})\right| = \left|f_1(e^{j\pi\mu_l})g_1(e^{j\pi v_m}) + f_2(e^{j\pi\mu_l})g_2(e^{j\pi v_m})\right|$$
$$\approx |\phi_{1l}\gamma_{1m} + \phi_{2l}\gamma_{2m}| \tag{23.103}$$

Follow this procedure, $K-1$ correction sections characterized by $k_2(z_1)g_2(x_2), \ldots, g_2(x_1)g_1(x_2)$ can be obtained, and $H_k(z_1, z_2)$ can be formed as

$$H_k(z_1, z_2) = \sum_{i=1}^{K} f_i(z_i)g_i(z_2) \tag{23.104}$$

and from Equation 23.96 we have

$$\left\| \mathbf{A} - \left| H_k(e^{j\pi\mu_i}, e^{j\pi v_m}) \right| \right\| \approx \left\| \mathbf{A} - \sum_{i=1}^{K} \mathbf{f}_i \mathbf{g}_i^T \right\|$$

$$\leq \left\| \varepsilon_K \right\| = \min_{\bar{\phi}_i, \bar{\gamma}_i} \left\| \sum_{i=1}^{K} \bar{\mathbf{f}}_i \bar{\mathbf{g}}_i^{\ T} \right\| \tag{23.105}$$

In effect, a 2-D filter consisting of K sections is obtained whose amplitude response is a minimal mean-square-error approximation to the desired amplitude response.

The method leads to an asymptotically stable 2-D filter, provided that all 1-D subfilters employed are stable. The general configuration of the 2-D filter obtained is illustrated in Figure 23.19, where the various 1-D subfilters may be either linear-phase or zero-phase filters.

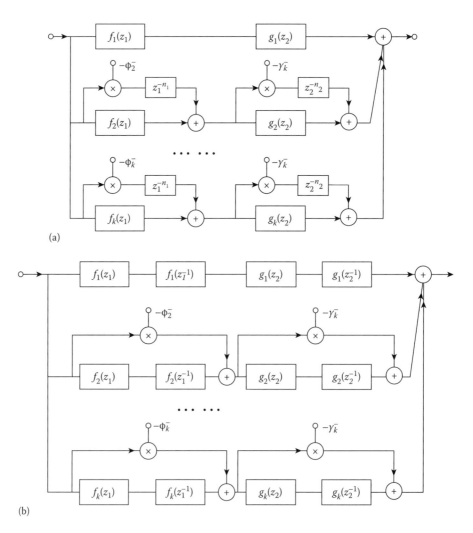

FIGURE 23.19 Configurations of 2D IIR filter by SVD. (a) General structure of 2D filter. (b) Structure using zero-phase IIR filters.

If linear-phase subfilters are to be used, the equalities in Equation 23.99 must be satisfied. This implies that the subfilters must have constant group delays. If zero-phase subfilters are employed, where $f_i(z_1)$ and $f_i(z_1^{-1})$, and $g_i(z_2)$ and $g_i(z_2^{-1})$ contribute equally to the amplitude response of the 2-D filter. The design can be accomplished by assuming that the desired amplitude responses for subfilters F_i, G_i are $\phi_i^{1/2}$, $\gamma_i^{1/2}$, for $i = 1, 2, k, \ldots, K$, respectively.

Error Compensation Procedure. When the main section and correction sections are designed by an optimization procedure as described above, approximation errors inevitably occur that will accumulate and manifest themselves as the overall error. The accumulation of error can be reduced by the following compensation procedure.

When filters F_1 and G_1 are designed, the approximation error matrix $\mathbf{E}_1$ can be calculated as

$$\mathbf{E}_1 = \mathbf{A} - f_1(e^{j\pi\mu_i})g_1(e^{j\pi\nu_m}) \tag{23.106}$$

and then perform SVD on $\mathbf{E}_1$ to obtain

$$\mathbf{E}_1 = S_{22}\mathbf{f}_{22}\mathbf{g}_{22}^T + \cdots + S_{r2}\mathbf{f}_{r2}\mathbf{g}_{r2}^T \tag{23.107}$$

Data ϕ_{22} and γ_{22} can be used to deduce filters $f_2(z_1)$ and $g_2(z_2)$. Thus, the first correction section can be designed. Next, form the error matrix $\mathbf{E}_2$ as

$$\mathbf{E}_2 = \mathbf{E}_1 - S_{22}f_2(e^{j\pi\mu_i})g_2(e^{j\pi\nu_m}) \tag{23.108}$$

and then perform SVD on $\mathbf{E}_2$ to obtain

$$\mathbf{E}_2 = S_{33}\,\mathbf{f}_{33}\,\mathbf{g}_{33}^T + \cdots + S_{r3}\,\mathbf{f}_{r3}\,\mathbf{g}_{r3}^T \tag{23.109}$$

and use data $\mathbf{f}_{33}$ and $\mathbf{g}_{33}$ to design the second correction section. The procedure is continued until the norm of the error matrix becomes sufficiently small that a satisfactory approximation to the desired amplitude response is reached.

Design of 1-D filters by using optimization can sometimes yield unstable filters. This problem can be eliminated by replacing poles outside the unit circle of the z plane by their reciprocals and simultaneously adjusting the multiplier constant to compensate for the change in gain [19].

Example 23.7 [26]

Design a circularly symmetric, zero-phase 2-D filter specified by

$$|H(\omega_1, \omega_2)| = \begin{cases} 1 & \text{for } (\omega_1^2 + \omega_2^2)^{1/2} < 0.35\pi \\ 0 & \text{for } (\omega_1^2 + \omega_2^2)^{1/2} \ge 0.65\,\pi \end{cases}$$

assuming that $\omega_{s1} = \omega_{s2} = 2\pi$.

Solution

1. Construct a sampled amplitude response matrix. By taking $L = M = 21$ and assuming that the amplitude response varies linearly with the radius in the transition band, the amplitude response matrix can be obtained as

$$A = \begin{vmatrix} A_1 & 0 \\ 0 & 0 \end{vmatrix}_{21 \times 21}$$

where

$$A = \begin{vmatrix}
1 & 1 & 1 & 1 & 1 & 1 & 1 & 1 & 1 & 0.75 & 0.5 & 0.25 \\
1 & 1 & 1 & 1 & 1 & 1 & 1 & 1 & 0.75 & 0.5 & 0.25 & 0 \\
1 & 1 & 1 & 1 & 1 & 1 & 1 & 1 & 0.75 & 0.5 & 0.25 & 0 \\
1 & 1 & 1 & 1 & 1 & 1 & 1 & 0.75 & 0.5 & 0.25 & 0 & 0 \\
1 & 1 & 1 & 1 & 1 & 1 & 1 & 0.75 & 0.5 & 0.25 & 0 & 0 \\
1 & 1 & 1 & 1 & 1 & 1 & 1 & 0.75 & 0.5 & 0.25 & 0 & 0 \\
1 & 1 & 1 & 1 & 1 & 0.75 & 0.5 & 0.25 & 0 & 0 & 0 & 0 \\
1 & 1 & 1 & 0.75 & 0.75 & 0.5 & 0.25 & 0 & 0 & 0 & 0 & 0 \\
1 & 0.75 & 0.75 & 0.5 & 0.5 & 0.25 & 0 & 0 & 0 & 0 & 0 & 0 \\
0.75 & 0.5 & 0.5 & 0.25 & 0.25 & 0 & 0 & 0 & 0 & 0 & 0 & 0 \\
0.5 & 0.25 & 0.25 & 0 & 0 & 0 & 0 & 0 & 0 & 0 & 0 & 0 \\
0.25 & 0 & 0 & 0 & 0 & 0 & 0 & 0 & 0 & 0 & 0 & 0
\end{vmatrix}$$

The ideal amplitude response of the filter is illustrated in Figure 23.20a.

2. Perform SVD to matrix **A** to obtain the amplitude response of the main section of the 2-D filter. It is worth noting that when a circularly symmetric 2-D filter is required, matrix **A** is symmetric and, therefore, Equation 23.95 becomes

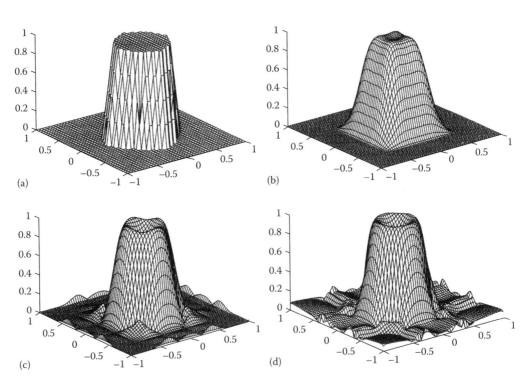

FIGURE 23.20 Amplitude responses of (a) the ideal circularly symmetric 2D filter; (b) the main section; (c) the main section plus the first correction; (d) the main section plus the first and second correction sections.

TABLE 23.3 Design Based on Fourth-Order Subfilters (Example 23.7)

The main section	$f_1(z) = 0.1255 \dfrac{(z^2 + 0.7239z + 1)\,(z^2 + 1.6343z + 1)}{(z^2 + 0.1367z + 1)\,(z^2 - 0.5328z + 0.2278)}$
The first correction section	$f_2(z) = 0.6098 \dfrac{(z^2 + 1.1618z + 0.1661)\,(z^2 + 0.8367z + 0.9958)}{(z^2 + 0.9953z)\,(z^2 + 0.5124z + 0.32)}$
	$S_2 = -1, \phi_2^- = 0.6266$
The second correction section	$f_3(z) = 0.4630 \dfrac{(z^2 + 1.5381z + 0.4456)\,(z^2 - 1.397z + 1.1191)}{(z^2 + 2.0408z - 1)\,(z^2 - 0.7092z + 0.6961)}$
	$S_3 = +1, \phi_3^- = 0.2764$

$$\mathbf{A} = \sum_{i=1}^{r} S_i \mathbf{f}_i \mathbf{f}_i^T \tag{23.110}$$

where $S_1 = 1$ and $S_i = \pm 1$ or -1, for $2 \le i \le r$. This implies that each parallel section requires only one 1-D subfilter to be designed. As a consequence the design work is reduced by 50%.

When vector f_1 is obtained, by selecting a fourth-order approximation and after optimization, the transfer function of the main section $f_1(z)$ is obtained.

3. Design the correction sections. Successively perform SVD to the error $\mathbf{E}_1$ and $\mathbf{E}_2$, and apply the preceding design technique; the transfer functions of the first and second correction sections can be obtained.

The transfer function of the main section and the first and second correction sections are listed in Table 23.3. And the amplitude responses of (1) the main section, (2) the main section plus the first correction, and (3) the main section plus the first and second correction sections are depicted in Figure 23.20b through d.

23.5.2.3 Method of Deng and Kawamata

In decomposing 2-D amplitude specifications into 1-D ones, the conventional SVD cannot avoid the problem that the 1-D amplitude specifications that result are often negative. Since negative values cannot be viewed as amplitude response, the problem of 1-D digital filter design becomes intricate. Deng and Kawamata [28] proposed a procedure that guarantees all the decomposition results to be always nonnegative and thus simplifies the design of correction sections.

The method decomposes the matrix $\mathbf{A}$ into the form

$$\mathbf{A} = \sum_{i=1}^{r} S_i \mathbf{F}_i \mathbf{G}_i \tag{23.111}$$

where all the elements of $\mathbf{F}_i$ and $\mathbf{G}_i$ are nonnegative and the decomposition error

$$\mathbf{E} = \left\| \mathbf{A} - \sum_{i=1}^{r} S_i \mathbf{F}_i \mathbf{G}_i \right\| \tag{23.112}$$

is sufficiently small, and $S_i = 1$ or -1 for $i = 1, 2, \ldots, r$.

The design procedure can be desired as follows.

Step 1. Let $A_1' = A, A_1^- = 0$, and perform the SVD on A_1^+ as

$$A_1^+ = \sum_{i=1}^{r_i} \sigma_{1i} \mathbf{u}_{1i} \mathbf{v}_{1i} \approx \mathbf{F}_1^+ \mathbf{G}_1^+ \tag{23.113}$$

where σ_{1i} is the ith singular value of $A_1^+ (\sigma_{11} \geq \sigma_{12} \geq \cdots \geq \sigma_{1r_1})$ and $F_1^+ = u_{11}\sigma_{11}^{1/2}, G_1^+ = u_{11}\sigma_{11}^{1/2}$.
Let

$$F_1 = F_1^+, \quad G_1 = G_1^+, \quad \text{and} \quad S_1 = 1 \tag{23.114}$$

all the elements of F_1 and G_1 are nonnegative.

Step 2. Calculate the approximation error matrix A_2 and decompose it into the sum of A_2^+ and A_2^- as

$$A_2 = A - S_1 F_1 G_1 = A_2^+ + A_2^- \tag{23.115}$$

where

$$A_2^+(m,n) = \begin{cases} A_2(m,n) & \text{if } A_2(m,n) \geq 0 \\ 0 & \text{otherwise} \end{cases} \tag{23.116}$$

and

$$A_2^-(m,n) = \begin{cases} A_2(m,n) & \text{if } A_2(m,n) \leq 0 \\ 0 & \text{otherwise} \end{cases} \tag{23.117}$$

To determine S_2 and F_2 and G_2 for approximating A_2 as accurately as possible, the following three steps are involved.

1. Perform the SVD on A_2^+ and approximate it as

$$A_2^+ = \sum_{i=1}^{r_2} s_{2i} u_{2i} v_{2i} \approx F_2^+ G_2^+ \tag{23.118}$$

where $F_2^+ = u_{21}\sigma_{21}^{1/2}, G_2^+ = \sigma_{21}^{1/2} v_{21}$. All the elements of F_2^1 and G_2^1 and G^* are nonnegative. If $F_2 = F_2^+, G_2 = G_2^+$, and $S_2 = 1$, the approximation error is

$$E_2^+ = \left\| A - \sum_{i=1}^{2} S_i F_i G_i \right\| \tag{23.119}$$

2. Perform the SVD on $-A_2^-$ and approximate it as

$$-A_2^- = \sum_{i=1}^{r_{2-}} \sigma_{2i-} u_{2i-} v_{2i-} \approx F_2^- G_2^- \tag{23.120}$$

where $F_2^- = u_{21-}\sigma_{21-}^{1/2}, G_2^- = \sigma_{21-}^{1/2} v_{21-}$, and r_{2-} is the rank of A_2^-. All the elements of F_2^- and G_2^- are nonnegative. If $F_2 = F_2^-, G_2 = G_2^-$, and $S_2 = 1$, the approximation error E_2^- is

$$E_2^- = \left\| A - \sum_{i=1}^{2} S_i F_i G_i \right\| \tag{23.121}$$

3. According to the results from steps 1 and 2, the optimal vectors F_2 and G_2 for approximating A are determined as

$$F_2 = F_2^+, \quad G_2 = G_2^+, \quad S_2 = 1 \quad \text{if } E_2^+ \leq E_2^-$$
$$F_2 = F_2^-, \quad G_2 = G_2^-, \quad S_2 = -1 \quad \text{if } E_2^+ \geq E_2^-$$

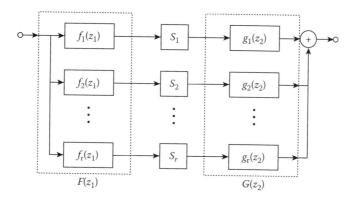

FIGURE 23.21 2D filter realization based on iterative SVD.

Successively decomposing the approximation error matrices $\mathbf{A}j$ $(j=3,4,\ldots,r)$ in the same way above described, a good approximation of matrix $\mathbf{A}$ can be obtained as in Equation 23.111.

Step 3. With the matrix $\mathbf{A}$ being decomposed into nonnegative vectors, the 1-D subfilters are designed through an optimization procedure, and a 2-D filter can then be readily realized, as shown in Figure 23.21.

It is noted that, in addition to SVD based methods as described in this section, design method based on other decomposition is also possible [29].

23.5.3 Design Based on Two-Variable Network Theory

Ramamoorthy and Bruton proposed a design method of 2-D IIR filters that always guarantees the stability of a filter and that involves the application of two-variable (2-V) strictly Hurwitz polynomials [30]. A 2 V polynomial $b(s_1, s_2)$ is said to be strictly Hurwitz if

$$b(s_1, s_2) \neq 0 \quad \text{for } Re\{s_1\} \geq 0 \text{ and } Re\{s_2\} \geq 0 \tag{23.122}$$

In their method, a family of 2-V strictly Hurwitz polynomials is obtained by applying network theory [31,32] to the frequency-independent, 2-V lossless network illustrated in Figure 23.22. The network has

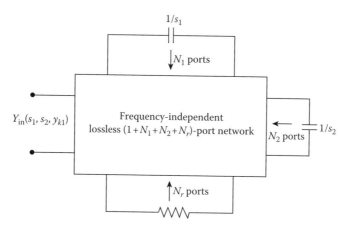

FIGURE 23.22 A $(1+N_1+N_2+N_r)$-port 2-V lossless network.

$1 + N_1 + N_2 + N_r$ ports and N_1 and N_2 are terminated in unit capacitors in complex variables s_1 and s_2, respectively, and N_r is terminated in unit resistors. Since the network is lossless and frequency independent, its admittance matrix $\mathbf{Y}$ is a real and skew-symmetric matrix given by

$$
\mathbf{Y} = \begin{vmatrix}
0 & y_{12} & y_{12} & \cdots & y_{1N} \\
-y_{12} & 0 & y_{23} & \cdots & y_{2N} \\
-y_{13} & -y_{23} & 0 & \cdots & y_{3N} \\
\vdots & \vdots & \vdots & & \vdots \\
-y_{1N} & -y_{2N} & -y_{3N} & \cdots & 0
\end{vmatrix}
$$

$$
= \begin{vmatrix} \mathbf{Y}_{11} & \mathbf{Y}_{12} \\ -\mathbf{Y}_{12}^T & \mathbf{Y}_{22} \end{vmatrix} \quad N = 1 + N_1 + N_2 + N_r \tag{23.123}
$$

If we define

$$
\phantom{Y_{22}(s_1,s_2,y_{kl}) = } \overset{N_r \quad N_1 \quad N_2}{} \tag{23.124}
$$
$$
\mathbf{Y}_{22}(s_1, s_2, y_{kl}) = \mathbf{Y}_{22} + \mathrm{diag}\ \{1 \cdots 1\ s_1 \cdots s_1\ s_2 \cdots s_2\}
$$

and

$$
\Delta(s_1, s_2, y_{kl}) = \det\left[\mathbf{\Psi}_{22}(s_1, s_2, y_{kl})\right] \tag{23.125}
$$

where $\mathrm{diag}(1 \cdots 1\ s_1 \cdots s_1\ s_2 \cdots s_2)$ represents a diagonal matrix in which each of the first N_r elements is unity, each of the next N_1 elements is s_1, and each of the last N_2, elements is s_2. Then, from the network theory, the input admittance at port 1 is given by

$$
\mathbf{Y}_{in}(s_1, s_2, y_{kl}) = \frac{\mathbf{Y}_{12}\mathrm{adj}\ [\mathbf{Y}_{22}(s_1, s_2, y_{kl})]\mathbf{Y}_{12}^T}{\det\ [\mathbf{Y}_{22}(s_1, s_2, y_{kl})]}
$$
$$
= \frac{p(s_1, s_2, y_{kl})}{\Delta(s_1, s_2, y_{kl})} \tag{23.126}
$$

where $\Delta(s_1, s_2, y_{kl})$ is defined by Equation 23.125 and is a strictly Hurwitz polynomial for any set of real values of the $(N-1)(N-2)/2$ independent parameters $\{y_{kl}: 1 < k < l \le N\}$. Table 23.4 lists polynomial $\Delta(s_1, s_2, y_{kl})$ for $N_r = 1$ $(N_1, N_2) = (2, 1)$ and $(N_1, N_2) = (2, 2)$ [30].

Having obtained the parameterized strictly Hurwitz polynomial $\Delta(s_1, s_2, y_{kl})$ the design procedure of a 2-D IIR filter can be summarized as follows.

TABLE 23.4 2-V Strictly Hurwitz Polynomials

N_1	N_2	N	$\Delta(s_1, s_2, y_{kl})$
1	1	4	$s_1 s_2 + y_{24}^2 s_1 + y_{23}^2 s_2 + y_{34}^2$
2	1	5	$s_1^2 s_2 + y_{25}^2 s_1^2 + (y_{23}^2)(y_{23}^2 + y_{24}^2)s_1 s_2 + (y_{35}^2 + y_{45}^2)s_1 + y_{34}^2 s_2$
			$+ (y_{23}y_{46} - y_{24}y_{35} + y_{25}y_{34})^2$
2	2	6	$s_1^2 s_2^2 + (y_{23}^2 + y_{24}^2)s_1 s_2^2 + (y_{25}^2 + y_{26}^2)s_1^2 s_2 + y_{56}^2 s_1^2 + y_{34}^2 s_2^2$
			$+ (y_{35}^2 + y_{36}^2 - y_{45}^2 + y_{46}^2)s_1 s_2 + y_{34}^2 s_2 + [(y_{23}y_{56} - y_{25}y_{36} + y_{26}y_{35})^2$
			$+ (y_{24}y_{56} - y_{25}y_{46} + y_{26}y_{45})^2]s_1 + [(y_{23}y_{45} - y_{24}y_{35} + y_{25}y_{34})^2$
			$+ (y_{23}y_{46} - y_{24}y_{36} + y_{26}y_{34})^2]s_2 + (y_{34}y_{56} - y_{35}y_{46} + y_{36}y_{45})^2$

Step 1. Construct a parameterized analog transfer function of the 2-D IIR filter, by using the Hurwitz polynomial $\Delta(s_1, s_2, y_{kl})$

$$H(s_1, s_2, y_{kl}, a_{ij}) = \frac{p(s_1, s_2)}{\Delta(s_1, s_2, y_{kl})} \tag{23.127}$$

where

$$p(s_1, s_2) = \sum_{i=1}^{N_1} \sum_{j=1}^{N_2} a_{ij} s_1^i s_2^j$$

is an arbitrary 2-V polynomial in s_1 and s_2 with degree in each variable not greater than the corresponding degree of the denominator.

Step 2. Perform the double bilinear transformation to the parameterized analog transfer function obtained in step 1.

$$H(z_1, z_2, y_{kl}, a_{ij}) = \frac{p(s_1, s_2)}{\Delta(s_1, s_2, y_{kl})} \Big|_{s_i = 2(z_i - 1)/T_i(z_i + 1), i=1, 2} \tag{23.128}$$

Step 3. Construct an objective function according to the given design specifications and the parameterized discrete transfer function obtained in step 2.

$$J(\mathbf{x}) = \sum_{n_1} \sum_{n_2} [M(n_1, n_2) - M_I(n_1, n_2)]^p \tag{23.129}$$

where p is an even positive, $M(n_1, n_2)$ and $M_I(n_1, n_2)$ are the actual and desired amplitude responses, respectively, of the required filter at frequencies $(\omega_{1n1}, \omega_{1n2})$, and $\mathbf{x}$ is the vector consisting of parameters $\{y_{kl}: 1 < k < l \le N\}$ and $\{a_{ij}, 0 \le i \le N_1, 0 \le j \le N_2\}$.

Step 4. Apply an optimization algorithm to find the optimal vector $\mathbf{x}$ that minimizes the objective function and substitute the resulting x into Equation 23.128 to obtain the required transfer function $H(z_1, z_2)$.

Example 23.8 [33]

By using the preceding approach, design a 2-D circularly symmetric low-pass filter of order (5, 5) with $\omega_p = 0.2\pi$, assuming that $\omega_{s1} = \omega_{s2} = 1.2\pi$.

Solution

1. Construct the desired amplitude response of the desired filter

$$M_I(\omega_{1n_1}, \omega_{2n_2}) = \begin{cases} 1 & \text{for } \left(\omega_{1n_1}^2 + \omega_{2n_2}^2\right) \le 0.2\pi \\ 0 & \text{otherwise} \end{cases}$$

where

$$\omega_{1n_1} = \begin{cases} 0.01\pi n_2 & \text{for } 0 \le n_1 \le 20 \\ 0.01\pi n_1 & \text{for } 21 \le n_1 \le 24 \end{cases}$$

and

$$\omega_{2n_2} = \omega_{1(24-n_2)} \quad \text{for } 0 \le n_2 \le 24$$

2. Construct the 2-D analog transfer function and perform double bilinear transformation to obtain the discrete transfer function. The analog transfer function at hand is assumed to be an all-pole transfer function of the form

$$\mathbf{H}(s_1, s_2, \mathbf{x}) = \frac{1}{\mathbf{D}(s_1, s_2, y_{kl})}$$

Therefore, the corresponding discrete transfer function can be written as

$$H(z_1, z_2, \mathbf{x}) = \frac{A(z_1 + 1)^5 (z_2 + 1)^5}{\displaystyle\sum_{i=0}^{5} \sum_{j=0}^{5} b_{ij} z_1^i z_2^j} \tag{23.130}$$

TABLE 23.5 Coefficients of Transfer Function in Equation 23.130
$[A = 0.28627, b_{ij}: 0 \le i \le 5, 0 \le j \le 5]$

0.0652	−0.6450	3.3632	−4.8317	0.3218	−0.1645
−0.7930	7.8851	−25.871	23.838	3.4048	3.4667
4.2941	−28.734	61.551	−29.302	−13.249	−25.519
−6.3054	28.707	−33.487	−7.2275	−22.705	83.011
0.7907	1.4820	−7.4214	−33.313	136.76	−128.43
−0.4134	6.0739	−36.029	101.47	140.20	78.428

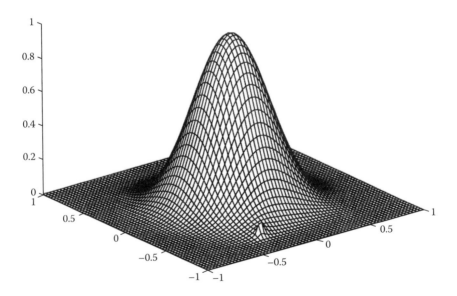

FIGURE 23.23 Amplitude response of circularly symmetric low-pass filter of Example 23.8.

where

$$\sum_{i=0}^{5}\sum_{j=0}^{5} b_{ij}z_1^i z_2^j = (z_1 + 1)^5 (z_2 + 1)^5 \Delta(s_1, s_2, y_{kl})\big|_{s_i=(z_i-1/z_i+1), i=1,2}$$

contains $(N-1)(N-2)/2 = 36$ parameters.

3. Optimization: A conventional quasi-Newton algorithm has been applied to minimize the objective function in Equation 23.129 with $p = 2$. The resulting coefficients are listed in Table 23.5. The amplitude response of the resulting filter is depicted in Figure 23.23.

References

1. J. L. Shanks, S. Treitel, and J. H. Justice, Stability and synthesis of two-dimensional recursive filters, *IEEE Trans. Audio Electroacoustic.*, AU-20, 115–128, June 1972.
2. W. S. Lu and A. Antoniou, *Two-Dimensional Digital Filters*, New York: Marcel Dekker, 1992.
3. M. Ahmadi, A. G. Constantinides, and R. A. King, Design technique for a class of stable two-dimensional recursive digital filters, in *Proc. 1976 IEEE Int. Conf. Acoust., Speech, Signal Process.*, Philadelphia, Pennsylvania, 1976, 145–147.
4. A. M. Ali, A. G. Constantinides, and R. A. King, On 2-variable reactance functions for 2-dimensional recursive filters, *Electron. Lett.*, 14, 12–13, January 1978.
5. R. King et al., *Digital Filtering in One and Two Dimensions: Design and Applications*, New York: Plenum, 1989.
6. R. A. King and A. H. Kayran, A new transformation technique for the design of 2-dimensional stable recursive digital filters, in *Proc. IEEE Int. Symp. Circuits Syst.*, Chicago, pp. 196–199, April 1981.
7. J. M. Costa and A. N. Venetsanopoulos, A group of linear spectral transformations for two-dimensional digital filters, *IEEE Trans. Acoust., Speech, Signal Process.*, ASSP-24, 424–425, October 1976.
8. K. P. Prasad, A. Antoniou, and B. B. Bhattacharyya, On the properties of linear spectral transformations for 2-dimensional digital filters, *Circuits Syst. Signal Process.*, 2, 203–211, 1983.
9. A. H. Kayran and R. A. King, Design of recursive and nonrecursive fan filters with complex transformations, *IEEE Trans. Circuits Syst.*, CAS-30, 849–857, 1983.
10. A. G. Constantinides, Spectral transformations for digital filters, *IEEE Proc.*, 117, 1585–1590, August 1970.
11. N. A. Pendergrass, S. K. Mitra, and E. I. Jury, Spectral transformations for two-dimensional digital filters, *IEEE Trans. Circuits Syst.*, CAS-23, 26–35, January 1976.
12. K. Hirano and J. K. Aggarwal, Design of two-dimensional recursive digital filters, *IEEE Trans. Circuits Syst.*, CAS-25, 1066–1076, December 1978.
13. S. K. Mitra and K. Hirano, Digital all-pass networks, *IEEE Trans. Circuits Syst.*, CAS-21, 688–700, September 1974.
14. J. M. Costa and A. N. Venetsanopoulos, Design of circularly symmetric two-dimensional recursive filters, *IEEE Trans. Acoust., Speech, Signal Process.*, ASSP-22, 432–443, December 1974.
15. D. M. Goodman, A design technique for circularly symmetric low-pass filters, *IEEE Trans. Acoust., Speech, Signal Process.*, ASSP-26, 290–304, August 1978.
16. G. V. Mendonca, A. Antoniou, and A. N. Venetsanopoulos, Design of two-dimensional pseudorotated digital filters satisfying prescribed specifications, *IEEE Trans. Circuits Syst.*, CAS-34, 1–10, January 1987.
17. R. Fletcher, *Practical Methods of Optimization*, 2nd edn., New York: Wiley, 1990.
18. S. Chakrabarti and S. K. Mitra, Design of two-dimensional digital filters via spectral transformations, *Proc. IEEE*, 6, 905–914, June 1977.

19. A. Antoniou, *Digital Filters: Analysis, Design and Applications*, 2nd edn., New York: McGraw-Hill, 1993.
20. G. A. Maria and M. M. Fahmy, An l_p design technique for two-dimensional digital recursive filters, *IEEE Trans. Acoust., Speech, Signal Process.*, ASSP-22, 15–21, February 1974.
21. C. Charalambous, A unified review of optimization, *IEEE Trans. Microwave Theory Tech.*, MTT-22, 289–300, March 1974.
22. C. Charalambous and A. Antoniou, Equalization of recursive digital filters, *IEE Proc., Pt. G*, 127, 219–225, October 1980.
23. C. Charalambous, Design of 2-dimensional circularly-symmetric digital filters, *IEE Proc., Pt. G*, 129, 47–54, April 1982.
24. C. Charalambous, Acceleration of the least pth algorithm for minimax optimization with engineering applications, *Math Program*, 17, 270–297, 1979.
25. P. K. Rajan and M. N. S. Swamy, Quadrantal symmetry associated with two-dimensional digital transfer functions, *IEEE Trans. Circuit Syst.*, CAS-29, 340–343, June 1983.
26. A. Antoniou and W. S. Lu, Design of two-dimensional digital filters by using the singular value decomposition, *IEEE Trans. Circuits Syst.*, CAS-34, 1191–1198, Oct. 1987.
27. G. W. Stewart, *Introduction to Matrix Computations*, New York: Academic, 1973.
28. T. B. Deng and M. Kawamata, Frequency-domain design of 2-D digital filters using the iterative singular value decomposition, *IEEE Trans. Circuits Syst.*, CAS-38, 1225–1228, 1991.
29. T. B. Deng and T. Soma, Successively linearized non-negative decomposition of 2-D filter magnitude design specifications, *Digital Signal Process.*, 3, 125–138, 1993.
30. P. A. Ramamoorthy and L. T. Bruton, Design of stable two-dimensional analog and digital filters with applications in image processing, *Circuit Theory Appl.*, 7, 229–245, 1979.
31. T. Koga, Synthesis of finite passive networks with prescribed two-variable reactance matrices, *IEEE Trans. Circuit Theory*, CT-13, 31–52, 1966.
32. H. G. Ansel, On certain two-variable generalizations of circuit theory, with applications networks of transmission lines and lumped reactance, *IEEE Trans. Circuit Theory*, CT-11, 214–233, 1964.
33. P. A. Ramamoorthy and L. T. Bruton, Design of stable two-dimensional recursive filters, in *Topics in Applied Physics*, 42, T. S. Huang, Ed., New York: Springer-Verlag, 1981, pp. 41–83.

24

1-D Multirate Filter Banks

Nick G. Kingsbury
University of Cambridge

David B. H. Tay
Latrobe University

24.1 Introduction: Why Use Filter Banks?

An important class of digital filter system is the multirate filter bank. In this chapter we shall only be considering the one-dimensional (1-D) type of filter bank, as might be applied to typical signals that evolve with time, such as audio waveforms and communications signals.

The reason that filter banks are important is that we can often achieve useful functionality by separating signals into various different frequency bands and applying different processing to each band. Typical examples of this are the widely used MP3 digital audio compression systems. In order to achieve a low coded data file size for digital music tracks, the MP3 coding standard specifies that the audio signal should be split into many frequency bands and that separate adaptive quantization should be applied to each band. The quantization is designed to take maximum advantage of the noise masking properties of the human auditory system, such that frequency bands containing substantial audio energy are quantized quite coarsely (because the quantizing noise gets masked by the signal here), whereas bands with low levels of audio are quantized more finely (since the masking is only effective at frequencies close to those containing most of the audio energy).

Filter banks can operate with filtered outputs being sampled at the same rate as the input signal. However, with many filters operating in parallel, this can lead to an unacceptably large amount of output data being generated. It is therefore sensible to subsample the outputs of the filters in a filter bank so that the total output data rate from all the filters is similar to that of the input. Such filters are called multirate filters, and the complete system is a multirate filter bank. When signals are subsampled, aliasing can occur and cause degradation of signal quality, but, with careful design, aliasing effects can be eliminated in multirate filter banks as long as the total output data rate is no less than the input rate. Multirate filters can be implemented with much less computational cost than the equivalent full rate filters, and, in the case of compression systems, they generate much less data to be coded by the adaptive quantizers.

An alternative way to view the advantages of filter banks is that, with careful design, they can encourage signal sparsity, i.e., most of the energy of an input signal can be concentrated in a small proportion of the output samples from the filter bank. Sparsity has been shown to be a key element in successfully performing many signal processing tasks, such as compression, denoising, signal separation, and other enhancement techniques. Sparsity can be achieved if, at any given time, the input signal can be well approximated by a weighted sum of the impulse responses from just a few of the filters in the filter bank. This occurs when the filters are matched to typical components of the signal.

Digital filter banks have been actively studied since the 1960s. However their use achieved a considerable boost with the development of wavelet theory in the 1980s. The theory of wavelet transforms was developed principally by French and Belgian mathematicians, notably A. Grossman, J. Morlet, Y. Meyer, I. Daubechies, and S. Mallat, and efficient implementation of the wavelet transform is usually achieved with multirate filter banks. The two topics are now firmly linked and are of great importance for signal analysis and compression.

The discrete wavelet transform (DWT) may be used to analyze a wide variety of signals, particularly those that combine long low-frequency events with much shorter high-frequency events (e.g., transients). It has perhaps achieved its greatest success with images. Although these are two-dimensional (2-D) signals, the 1-D filters considered in this chapter are still highly relevant since 2-D wavelet transforms are usually achieved using separable 1-D processing along the rows and then down the columns of the image (or vice versa).

In this chapter we shall be introducing some of the basic ideas behind 1-D filter banks, and then will concentrate much of our coverage on the 2-band multirate filter bank, which is the workhorse of the DWT. In the final sections we shall extend the discussion to M-band ($M > 2$) filter banks and also to Hilbert pairs of filter banks, which lead to the dual-tree complex wavelet transform.

Most of the following discussions in this chapter will assume that the filters are finite-impulse-response (FIR) filters and that all samples of the input signal are available in the memory of the signal processing hardware. Hence causality of the filters is not an issue and filter taps corresponding to negative delays (positive powers of z in the z-transform) pose no implementation problems. In this situation, the most natural way to design filters is for zero overall delay (the zero-phase condition).

If these assumptions are not valid (e.g., in the case of a continuously evolving audio signal), then appropriate delays can usually be inserted so that the filters are still implementable. Only in the case of recursive infinite-impulse-response (IIR) filters is the causality issue a real potential problem; and so more care is required on this issue in Section 24.7. Elsewhere we shall ignore problems of causality.

24.2 2-Band Filter Bank

A simple analysis filter bank with just two bands is shown in Figure 24.1a. It comprises two filters H_0 and H_1 which split the input signal X into its lower and higher frequency components, respectively. If X is sampled at a rate of f_s Hz, then its frequency spectrum can occupy the range 0 to $\frac{1}{2}f_s$, while still satisfying Nyquist's rule; and so H_0 will normally be designed to pass frequencies from 0 to $\frac{1}{4}f_s$, and H_1 from $\frac{1}{4}f_s$ to $\frac{1}{2}f_s$. Hence, H_0 will be a low-pass filter and H_1 will be high-pass.

Let us initially consider the simplest practical form of 2-band filter bank in which H_0 and H_1 are both 2-tap FIR filters with coefficient vectors $\mathbf{h}_0 = [\frac{1}{\sqrt{2}}, \frac{1}{\sqrt{2}}]$ and $\mathbf{h}_1 = [\frac{1}{\sqrt{2}}, \frac{-1}{\sqrt{2}}]$ (corresponding to the well-known Haar wavelet basis). Hence the filter outputs may be expressed as

$$y_0(n) = \frac{1}{\sqrt{2}}x(n-1) + \frac{1}{\sqrt{2}}x(n)$$

$$y_1(n) = \frac{1}{\sqrt{2}}x(n-1) - \frac{1}{\sqrt{2}}x(n)$$

(24.1)

As z-transforms, Equation 24.1 becomes

$$Y_0(z) = H_0(z)X(z) \quad \text{where} \quad H_0(z) = \frac{1}{\sqrt{2}}(z^{-1} + 1)$$

$$Y_1(z) = H_1(z)X(z) \quad \text{where} \quad H_1(z) = \frac{1}{\sqrt{2}}(z^{-1} - 1)$$

(24.2)

(By substituting $z = e^{j\omega}$, the reader may check that these filters are indeed low-pass and high-pass, respectively. We shall later extend the filters to be more complicated.)

In practice, $y_0(n)$ and $y_1(n)$ are only calculated at alternate (say even) values of n so that the total number of samples in vectors $\mathbf{y}_0$ and $\mathbf{y}_1$ is the same as in the input vector $\mathbf{x}$. This is shown by the 2:1 downsamplers on the right in Figure 24.1a.

It is straightforward to invert Equation 24.1 to obtain the two samples of x back from the filter output samples, as follows:

$$x(n-1) = \frac{1}{\sqrt{2}}y_0(n) + \frac{1}{\sqrt{2}}y_1(n)$$

$$\text{for } n \text{ even.}$$

$$x(n) = \frac{1}{\sqrt{2}}y_0(n) - \frac{1}{\sqrt{2}}y_1(n)$$

(24.3)

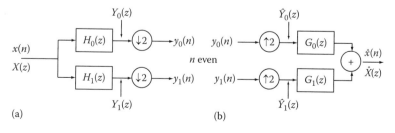

FIGURE 24.1   Two-band filter banks for analysis (a) and reconstruction (b).

Assuming that the missing samples of y_0 and y_1 are zero at odd values of n, we may combine the equations in Equation 24.3 into a single expression for $x(n)$, valid for all n:

$$x(n) = \frac{1}{\sqrt{2}}[y_0(n+1) + y_0(n)] + \frac{1}{\sqrt{2}}[y_1(n+1) - y_1(n)] \tag{24.4}$$

or as z-transforms

$$X(z) = G_0(z)Y_0(z) + G_1(z)Y_1(z) \tag{24.5}$$

where

$$G_0(z) = \frac{1}{\sqrt{2}}(z+1) \quad \text{and} \quad G_1(z) = \frac{1}{\sqrt{2}}(z-1) \tag{24.6}$$

Note that the factors of $\frac{1}{\sqrt{2}}$ in the coefficients of the H and G filters are chosen to ensure that the l_2-norm (energy) of each vector of filter coefficients is unity, so that total energy is preserved from the X-domain to the Y-domain and vice versa.

In Equation 24.5 the signals $Y_0(z)$ and $Y_1(z)$ are not really the same as $Y_0(z)$ and $Y_1(z)$ in Equation 24.2 because those in Equation 24.2 have not had alternate samples set to zero. Also, in Equation 24.5 $X(z)$ is the reconstructed output whereas in Equation 24.2 it is the input signal.

To avoid confusion we shall use $\hat{X}$, $\hat{Y}_0$, and $\hat{Y}_1$ for the signals in Equation 24.5, so it becomes

$$\hat{X}(z) = G_0(z)\hat{Y}_0(z) + G_1(z)\hat{Y}_1(z) \tag{24.7}$$

We may show this reconstruction operation as upsampling followed by two filters, as in Figure 24.1b, forming a 2-band reconstruction filter bank.

If $\hat{Y}_0$ and $\hat{Y}_1$ are not the same as Y_0 and Y_1, it is important to know how they do relate to each other. Now

$$\hat{y}_0(n) = y_0(n) \quad \text{for } n \text{ even}, \quad \hat{y}_0(n) = 0 \quad \text{for } n \text{ odd} \tag{24.8}$$

Therefore its z-transform, $\hat{Y}_0(z)$, is a polynomial in z, comprising *only* the terms in even powers of z from $Y_0(z)$. This may be written as

$$\hat{Y}_0(z) = \sum_{\text{even } n} y_0(n)z^{-n} = \sum_{\text{all } n} \frac{1}{2}[y_0(n)z^{-n} + y_0(n)(-z)^{-n}] = \frac{1}{2}[Y_0(z) + Y_0(-z)] \tag{24.9}$$

Similarly

$$\hat{Y}_1(z) = \frac{1}{2}[Y_1(z) + Y_1(-z)] \tag{24.10}$$

This is our general model for downsampling by two, followed by upsampling by two as defined in Equation 24.8.

Substituting Equations 24.9 and 24.10 into Equation 24.7 and then using Equation 24.2, we get

$$\hat{X}(z) = \frac{1}{2}G_0(z)[Y_0(z) + Y_0(-z)] + \frac{1}{2}G_1(z)[Y_1(z) + Y_1(-z)]$$

$$= \frac{1}{2}G_0(z)H_0(z)X(z) + \frac{1}{2}G_0(z)H_0(-z)X(-z)$$

$$+ \frac{1}{2}G_1(z)H_1(z)X(z) + \frac{1}{2}G_1(z)H_1(-z)X(-z)$$

$$= \frac{1}{2}X(z)[G_0(z)H_0(z) + G_1(z)H_1(z)]$$

$$+ \frac{1}{2}X(-z)[G_0(z)H_0(-z) + G_1(z)H_1(-z)] \tag{24.11}$$

This result will be used in Section 24.6.

24.3 Multirate Filtering

In order to be able to calculate the characteristics of more complicated configurations, for example, comprising cascaded filter banks such as used for wavelet transforms, it is necessary to derive some key results for multirate filter systems.

24.3.1 Multirate Filtering Theorem

To calculate the impulse and frequency responses for a multistage filter network with downsampling/upsampling between stages, we derive an important theorem for multirate filters.

THEOREM 24.1

The downsample–filter–upsample operation of Figure 24.2a is equivalent to either the filter–downsample–upsample operation of Figure 24.2b or the downsample–upsample–filter operation of Figure 24.2c, if the filter is changed from $H(z)$ to $H(z^2)$.

PROOF: Expressing the convolution and down + upsampling of Figure 24.2a in full:

$$\hat{y}(n) = \sum_i x(n - 2i)h(i) \quad \text{for } n \text{ even}$$

$$= 0 \qquad\qquad \text{for } n \text{ odd} \tag{24.12}$$

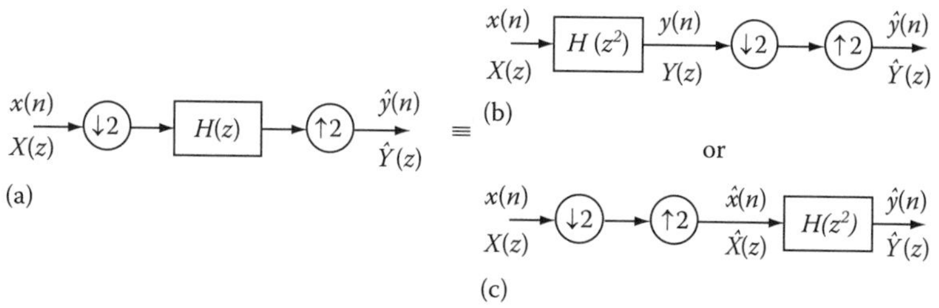

FIGURE 24.2   Multirate filtering (a). This figure shows the result of shifting a filter ahead of a downsampling operation (b) or after an upsampling operation (c).

Taking z-transforms

$$\hat{Y}(z) = \sum_n \hat{y}(n)z^{-n} = \sum_{\text{even } n}\sum_i x(n - 2i)h(i)z^{-n} \qquad (24.13)$$

Reversing the order of summation and letting $m = n - 2i$:

$$\hat{Y}(z) = \sum_i h(i) \sum_{\text{even } m} x(m)z^{-m}z^{-2i}$$

$$= \sum_i h(i)z^{-2i} \sum_{\text{even } m} x(m)z^{-m}$$

$$= H(z^2)\frac{1}{2}[X(z) + X(-z)] \qquad (24.14)$$

$$= \frac{1}{2}[H(z^2)X(z) + H((-z)^2)X(-z)] \quad \text{since } (-z)^2 = z^2$$

$$= \frac{1}{2}[Y(z) + Y(-z)] \quad \text{where } Y(z) = H(z^2)X(z) \qquad (24.15)$$

Equation 24.15 describes the operations of Figure 24.2b. Hence the first result is proved. The result from Equation 24.14 gives

$$\hat{Y}(z) = H(z^2)\frac{1}{2}[X(z) + X(-z)] = H(z^2)\hat{X}(z) \qquad (24.16)$$

This shows that the filter $H(z^2)$ may be placed after the down + upsampler as in Figure 24.2c, which proves the second result.

24.3.2 General Results for M:1 Subsampling

The results above may be extended to the case of M:1 down and upsampling as follows:

- $H(z)$ becomes $H(z^M)$ if shifted ahead of an M:1 downsampler or following an M:1 upsampler. (These two results are known as the Noble identities.)
- M:1 down + upsampling of a signal $X(z)$ produces

$$\hat{X}(z) = \frac{1}{M}\sum_{m=0}^{M-1} X(ze^{j2\pi m/M}) \qquad (24.17)$$

These results will now be used to analyze binary filter trees.

24.4 Binary Filter Trees

For applications such as signal compression, the purpose of the 2-band filter bank is to compress most of the signal energy into the samples representing the low-frequency half of the signal band. Hence 50% of the filter-bank output samples (the low-pass half) may well contain 90% or more of the signal energy (if it is a signal with dominant low-frequency components such as an image or audio signal).

We may achieve greater compression if the low band is further split into two. This may be repeated a number of times to give the binary filter tree. An example with four levels of decomposition is shown in Figure 24.3.

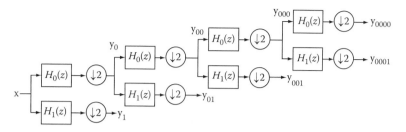

FIGURE 24.3   Extension of the 2-band filter bank into a binary filter tree. H_0 and H_1 are low-pass and high-pass filters.

For an N-sample input vector $\mathbf{x}$, sampled at a frequency f_s, the sizes and bandwidths of the signals of the 4-level filter tree are

Signal	No. of Samples	Approximate Pass Band
$\mathbf{x}$	N	$0 \rightarrow \frac{1}{2}f_s$
$\mathbf{y}_1$	$N/2$	$\frac{1}{4} \rightarrow \frac{1}{2}f_s$
$\mathbf{y}_{01}$	$N/4$	$\frac{1}{8} \rightarrow \frac{1}{4}f_s$
$\mathbf{y}_{001}$	$N/8$	$\frac{1}{16} \rightarrow \frac{1}{8}f_s$
$\mathbf{y}_{0001}$	$N/16$	$\frac{1}{32} \rightarrow \frac{1}{16}f_s$
$\mathbf{y}_{0000}$	$N/16$	$0 \rightarrow \frac{1}{32}f_s$

Because of the downsampling by 2 at each level, the total number of output samples equals N, regardless of the number of levels in the tree; so the process is nonredundant.

The H_0 filter is normally designed to be a low-pass filter with a passband from 0 to approximately $\frac{1}{4}$ of the input sampling frequency for that stage; and H_1 is a high-pass (bandpass) filter with a pass band approximately from $\frac{1}{4}$ to $\frac{1}{2}$ of the input sampling frequency.

When formed into a 4-level tree, the filter outputs have the approximate pass bands given in the above table. The final output $\mathbf{y}_{0000}$ is a low-pass signal, while the other outputs are all bandpass signals, each covering a band of approximately one octave.

An inverse tree, mirroring Figure 24.3, may be constructed using filters G_0 and G_1 instead of H_0 and H_1, as shown for just one level in Figure 24.1b. If the PR conditions of Equations 24.40 and 24.41 are satisfied, then the output of each level will be identical to the input of the equivalent level in Figure 24.3, and the final output will be a PR of the input signal.

24.4.1 Transformation of the Filter Tree

Using the result of Equation 24.15, Figure 24.3 can be redrawn as in Figure 24.4 with all down-samplers moved to the outputs. (Note Figure 24.4 requires much more computation than Figure 24.3,

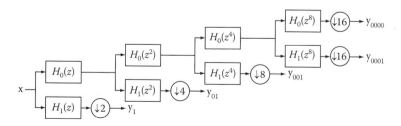

FIGURE 24.4   Binary filter tree, transformed so that all downsampling operations occur at the outputs.

but is a useful analysis aid.) We can now calculate the transfer function to each output (before the downsamplers) as

$$H_{01}(z) = H_0(z)H_1(z^2)$$
$$H_{001}(z) = H_0(z)H_0(z^2)H_1(z^4)$$
$$H_{0001}(z) = H_0(z)H_0(z^2)H_0(z^4)H_1(z^8)$$ (24.18)
$$H_{0000}(z) = H_0(z)H_0(z^2)H_0(z^4)H_0(z^8)$$

In general the transfer functions to the two outputs at level k of the tree are given by

$$H_{k,1} = \left(\prod_{i=0}^{k-2} H_0(z^{2^i})\right)H_1(z^{2^{k-1}}) \quad \text{and} \quad H_{k,0} = \prod_{i=0}^{k-1} H_0(z^{2^i}) \tag{24.19}$$

For the Haar filters of Equation 24.2, the transfer functions to the outputs of the 4-level tree become

$$H_{01}(z) = \frac{1}{2}[(z^{-3} + z^{-2}) - (z^{-1} + 1)]$$

$$H_{001}(z) = \frac{1}{2\sqrt{2}}[(z^{-7} + z^{-6} + z^{-5} + z^{-4}) - (z^{-3} + z^{-2} + z^{-1} + 1)]$$

$$H_{0001}(z) = \frac{1}{4}[(z^{-15} + z^{-14} + z^{-13} + z^{-12} + z^{-11} + z^{-10} + z^{-9} + z^{-8}) \tag{24.20}$$
$$- (z^{-7} + z^{-6} + z^{-5} + z^{-4} + z^{-3} + z^{-2} + z^{-1} + 1)]$$

$$H_{0000}(z) = \frac{1}{4}(z^{-15} + z^{-14} + z^{-13} + z^{-12} + z^{-11} + z^{-10} + z^{-9} + z^{-8}$$
$$+ z^{-7} + z^{-6} + z^{-5} + z^{-4} + z^{-3} + z^{-2} + z^{-1} + 1)$$

These transfer functions are illustrated in Figure 24.5, simply by interpreting the coefficients of the z-transform polynomials as samples of the corresponding impulse responses; and by substituting $z = e^{j2\pi f/f_s}$ into the transfer functions and evaluating their magnitudes as a function of frequency, f, to obtain the frequency responses.

24.5 Wavelets and Scaling Functions

The process of creating the outputs y_1 to y_{0000} from x in Figure 24.3 is known as the DWT; and the reconstruction process is the inverse DWT [1—3]. The term wavelet refers to the impulse response of the cascade of filters which leads to a given bandpass output.

Since the frequency responses of the bandpass bands are scaled down by 2:1 at each level (see Figure 24.5), their impulse responses become longer by the same factor at each level, but their shapes remain very similar. The basic impulse response wave shape is almost independent of scale and is known as the mother wavelet. The impulse response to a low-pass output $H_{k,0}$ is called the scaling function at level k.

Figure 24.5 shows these effects using the impulse responses and frequency responses for the five outputs of the 4-level tree of Haar filters, based on the z-transforms given in Equation 24.20. Notice the abrupt transitions in the midddle and at the ends of the Haar wavelets. These result in noticeable blocking artifacts in decompressed images and other signal types. For this reason we are interested in developing a more general theory of wavelets, with improved impulse and frequency responses.

There are several forms of the wavelet transform but the focus here will be on the form that is related to the 2-band filter bank. This form, which is the most popular in applications, is known as the DWT. Unless otherwise stated, it is assumed that the coefficients of all low-pass filters are scaled such that they

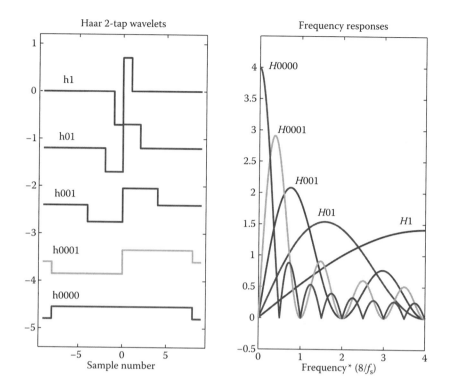

FIGURE 24.5 Impulse responses and frequency responses of the 4-level tree of Haar filters.

sum to the value $\sqrt{2}$, i.e., the DC frequency response is $H_0(e^{j0}) = H_0(1) = \sum_n h_0(n) = \sqrt{2}$. (This scaling has the effect of making the DWT tend to be an energy preserving transformation.)

Central to the principle of the DWT are the two-scale equations that link the discrete time filters of the filter bank to continuous time functions known as the scaling function and the mother wavelet. These equations exist both for the analysis side of the filter bank and for the synthesis side:

1. Analysis side equations:

$$\tilde{\phi}(t) = \sqrt{2} \sum_k h_0(k)\tilde{\phi}(2t - k) \tag{24.21}$$

$$\tilde{\psi}(t) = \sqrt{2} \sum_k h_1(k)\tilde{\phi}(2t - k) \tag{24.22}$$

where $\tilde{\phi}(t)$ and $\tilde{\psi}(t)$ are the analysis scaling function and mother wavelet, respectively. The coefficients of filters $H_0(z)$ and $H_1(z)$ are $h_0(k)$ and $h_1(k)$: i.e.,

$$H_0(z) = \sum_k h_0(k)z^{-k} \quad \text{and} \quad H_1(z) = \sum_k h_1(k)z^{-k} \tag{24.23}$$

2. Synthesis side equations:

$$\phi(t) = \sqrt{2} \sum_k g_0(k)\phi(2t - k) \tag{24.24}$$

$$\psi(t) = \sqrt{2} \sum_k g_1(k)\phi(2t - k) \tag{24.25}$$

where $\phi(t)$ and $\psi(t)$ are the synthesis scaling function and mother wavelet, respectively. The coefficients of filters $G_0(z)$ and $G_1(z)$ are $g_0(k)$ and $g_1(k)$:
i.e.,

$$G_0(z) = \sum_k g_0(k)z^{-k} \quad \text{and} \quad G_1(z) = \sum_k g_1(k)z^{-k} \tag{24.26}$$

Note that the scaling functions ($\tilde{\phi}(t)$ or $\phi(t)$) are effectively the result of a cascade of low-pass filters (H_0 or G_0) across very many progressively finer scales. The mother wavelets ($\tilde{\psi}(t)$ or $\psi(t)$) are the result of cascading H_1 or G_1 at the coarsest scale, with very many low-pass filters at finer scales.

In the Fourier domain on the analysis side, this gives the infinite product formulae, achieved by transforming Equations 24.21 and 24.22 as follows:

$$\tilde{\Phi}(\omega) = \sqrt{2} \sum_k h_0(k) \int_{-\infty}^{\infty} \tilde{\phi}(2t - k)e^{-j\omega t}\, dt$$

$$= \sqrt{2} \sum_k h_0(k) \int_{-\infty}^{\infty} \tilde{\phi}(\tau)e^{-j\omega\tau/2}e^{-j\omega k/2}\frac{d\tau}{2} \quad \text{if } \tau = 2t - k$$

$$= \frac{1}{\sqrt{2}} H_0(e^{j\omega/2})\tilde{\Phi}(\omega/2)$$

$$= \frac{1}{\sqrt{2}} H_0(e^{j\omega/2})\frac{1}{\sqrt{2}} H_0(e^{j\omega/4})\tilde{\Phi}(\omega/4)$$

$$\vdots$$

$$= \prod_{k=1}^{\infty} \left\{ \frac{1}{\sqrt{2}} H_0\left(e^{j\omega/2^k}\right) \right\} \tilde{\Phi}(0) \tag{24.27}$$

Similarly

$$\tilde{\Psi}(\omega) = \frac{1}{\sqrt{2}} H_1(e^{j\omega/2})\tilde{\Phi}(\omega/2)$$

$$= \frac{1}{\sqrt{2}} H_1(e^{j\omega/2}) \prod_{k=2}^{\infty} \left\{ \frac{1}{\sqrt{2}} H_0\left(e^{j\omega/2^k}\right) \right\} \tilde{\Phi}(0) \tag{24.28}$$

where $\tilde{\Phi}(\omega)$ and $\tilde{\Psi}(\omega)$ are the Fourier transforms of $\tilde{\phi}(t)$ and $\tilde{\psi}(t)$, respectively.

The above equations explicitly define the analysis scaling function and mother wavelet in terms of the analysis filters. Similar infinite product formulae exist linking the synthesis scaling function and wavelet to the synthesis filters.

Equations 24.27 and 24.28 are reminiscent of the equivalent transfer functions defined in Equation 24.19 for the binary filter tree. The relationships can be made more precise as follows. The transfer functions (repeated here for convenience) for the k level tree are

$$H_{k,0}(z) \equiv \prod_{i=0}^{k-1} H_0(z^{2^i}) = \sum_n h_{k,0}(n)z^{-n} \tag{24.29}$$

$$H_{k,1}(z) \equiv H_1(z^{2^{k-1}}) \prod_{i=0}^{k-2} H_0(z^{2^i}) = \sum_n h_{k,1}(n) z^{-n} \tag{24.30}$$

where $h_{k,0}(n)$ and $h_{k,1}(n)$ are the coefficients of $H_{k,0}(z)$ and $H_{k,1}(z)$.

From the coefficients of the equivalent filters, we construct the following piecewise constant functions:

$$\tilde{\phi}^{(k)}(t) \equiv 2^{k/2} h_{k,0}(n), \quad \frac{n}{2^k} \le t < \frac{n+1}{2^k} \tag{24.31}$$

$$\tilde{\psi}^{(k)}(t) \equiv 2^{k/2} h_{k,1}(n), \quad \frac{n}{2^k} \le t < \frac{n+1}{2^k} \tag{24.32}$$

Each coefficient defines a rectangular pulse of width 2^{-k}, and both $\tilde{\phi}^{(k)}(t)$ and $\tilde{\psi}^{(k)}(t)$ are made up of sequences of such pulses. The width of the pulse is halved with each increase in tree level (scale) k but the $2^{k/2}$ amplitude normalization ensures that the total energy of each function remains constant. When the number of levels k tends to infinity, the scaling function and mother wavelet are obtained:

$$\tilde{\phi}(t) = \lim_{k \to \infty} \tilde{\phi}^{(k)}(t) \quad \text{and} \quad \tilde{\psi}(t) = \lim_{k \to \infty} \tilde{\psi}^{(k)}(t) \tag{24.33}$$

i.e., the shape of the scaling function and wavelet are the shape of the impulse response of equivalent filters of the binary tree as $k \to \infty$. In practice, the shape of the scaling function and wavelet is obtained quite accurately after $k \simeq 6$ levels. Similar relationships exist on the synthesis side.

The discussion above involving infinite products assumes convergence but this is not always guaranteed. The filters must be properly designed to ensure convergence. One condition that is necessary for convergence is

$$H_0(e^{j\pi}) = H_0(-1) = 0 \quad \text{and} \quad G_0(e^{j\pi}) = G_0(-1) = 0 \tag{24.34}$$

i.e., the low-pass filter responses must vanish at the aliasing (half-sampling) frequency. This means that $(1 + z^{-1})$ must be a factor of both $H_0(z)$ and $G_0(z)$.

A simple degree-1 factor, i.e., $(1 + z^{-1})$, may not necessarily guarantee convergence; and even if convergence is achieved, the resultant scaling and wavelet functions may not be smooth. In general, higher order factors are usually imposed on the low-pass filters, such that

$$H_0(z) = 2^{-L_H}(1 + z^{-1})^{L_H} R_H(z) \tag{24.35}$$

$$G_0(z) = 2^{-L_G}(1 + z^{-1})^{L_G} R_G(z) \tag{24.36}$$

where $R_H(z)$ and $R_G(z)$ are the remainder factors of $H_0(z)$ and $G_0(z)$, respectively.

The orders L_H and L_G determine the numbers of vanishing moments (VM) of the corresponding wavelet function, such that

$$\int_{-\infty}^{+\infty} t^n \tilde{\psi}(t) dt = 0 \quad \text{for } n = 0, \dots, L_H - 1 \tag{24.37}$$

$$\int_{-\infty}^{+\infty} t^n \psi(t) dt = 0 \quad \text{for } n = 0, \dots, L_G - 1 \tag{24.38}$$

In general the higher the number of VM, the smoother will be the scaling and wavelet functions.

24.6 Good FIR Filters and Wavelets

24.6.1 Perfect Reconstruction Condition

One of the most important requirements for most filter banks is to be able to reconstruct perfectly the input signal $X(z)$ at the reconstruction filter bank output $\hat{X}(z)$ (see Figure 24.1).

We now wish to find the constraints on arbitrary filters, $\{H_0, H_1, G_0, G_1\}$, such that perfect reconstruction (PR) occurs.

Repeating the result from Equation 24.11, the input–output relationship for the pair of filter banks in Figure 24.1 is

$$\hat{X}(z) = \frac{1}{2}X(z)[G_0(z)H_0(z) + G_1(z)H_1(z)]$$
$$+ \frac{1}{2}X(-z)[G_0(z)H_0(-z) + G_1(z)H_1(-z)] \tag{24.39}$$

If we require $\hat{X}(z) \equiv X(z)$—the PR condition—then

$$G_0(z)H_0(z) + G_1(z)H_1(z) \equiv 2 \tag{24.40}$$

and

$$G_0(z)H_0(-z) + G_1(z)H_1(-z) \equiv 0 \tag{24.41}$$

Equation 24.41 is known as the antialiasing condition because the term in $X(-z)$ in Equation 24.39 is the unwanted aliasing term caused by the 2 : 1 downsampling of y_0 and y_1.

It is straightforward to show that the expressions for $\{H_0, H_1, G_0, G_1\}$, given in Equations 24.2 and 24.6 for the filters based on the Haar wavelet basis, satisfy Equations 24.40 and 24.41. They are the simplest set of filters which do.

24.6.2 Good Filters/Wavelets

Our main aim now is to search for better filters which result in wavelets and scaling functions that are smoother than the Haar functions (i.e., which avoid the discontinuities evident in the waveforms of Figure 24.5).

We start our search with the two PR identities, Equations 24.40 and 24.41.

The usual way of satisfying the antialiasing condition Equation 24.41, while permitting H_0 and G_0 to have low-pass responses (passband where $\mathrm{Re}[z] > 0$) and H_1 and G_1 to have high-pass responses (passband where $\mathrm{Re}[z] < 0$), is with the following relations in which k must be an odd integer:

$$H_1(z) = z^{-k}G_0(-z) \quad \text{and} \quad G_1(z) = z^k H_0(-z) \tag{24.42}$$

Hence:

$$\begin{aligned} G_0(z)H_0(-z) + G_1(z)H_1(-z) &= G_0(z)H_0(-z) + z^k H_0(-z)(-z)^{-k}G_0(z) \\ &= G_0(z)H_0(-z) + (-1)^{-k}H_0(-z)G_0(z) \\ &\equiv 0 \quad \text{if } k \text{ is odd} \end{aligned} \tag{24.43}$$

and so Equation 24.41 is satisfied, as required.

Now we define the low-pass product filter:

$$P(z) = H_0(z)G_0(z) \tag{24.44}$$

and substitute relations in Equation 24.42 into Equation 24.40 to get

$$G_0(z)H_0(z) + G_1(z)H_1(z) = G_0(z)H_0(z) + H_0(-z)G_0(-z)$$
$$= P(z) + P(-z) \equiv 2 \tag{24.45}$$

Hence satisfying Equation 24.40 requires that all $P(z)$ terms in even powers of z be zero, except the z^0 term which must be 1. However the $P(z)$ terms in odd powers of z may take any desired values since they cancel out in Equation 24.45. If P is low-pass, this type of filter is often known as a halfband filter since $P(e^{j\omega}) + P(e^{j(\omega+\pi)}) = $ a constant, and so the bandwidth of P must be half of the input bandwidth.

A further commonly applied constraint on $P(z)$ is that it should be zero phase,* in order to minimize the magnitude of any distortions due to samples from the high-pass filters being suppressed (perhaps as a result of quantization or denoising). Hence $P(z)$ should be of the form:

$$P(z) = \cdots + p_5 z^5 + p_3 z^3 + p_1 z + 1 + p_1 z^{-1} + p_3 z^{-3} + p_5 z^{-5} + \cdots \tag{24.46}$$

The design of a set of PR filters H_0, H_1 and G_0, G_1 can now be summarized as

1. Choose a set of coefficients $p_1, p_3, p_5 \cdots$ to give a zero-phase low-pass product filter $P(z)$ with desirable characteristics. (This is nontrivial and is discussed below.)
2. Factorize $P(z)$ into $H_0(z)$ and $G_0(z)$, preferably so that the two filters have similar low-pass frequency responses.
3. Calculate $H_1(z)$ and $G_1(z)$ from Equation 24.42.

It can help to simplify the tasks of choosing $P(z)$ and factorizing it if, based on the zero-phase requirement, we transform $P(z)$ into $P_t(Z)$ such that

$$P(z) = P_t(Z) = 1 + p_{t,1}Z + p_{t,3}Z^3 + p_{t,5}Z^5 + \cdots \quad \text{where} \quad Z = \frac{1}{2}(z + z^{-1}) \tag{24.47}$$

To calculate the frequency response of $P(z) = P_t(Z)$, let $z = e^{j\omega T_s}$. Therefore

$$\therefore Z = \frac{1}{2}(e^{j\omega T_s} + e^{-j\omega T_s}) = \cos(\omega T_s) \tag{24.48}$$

This is a purely real function of ω, varying from 1 at $\omega = 0$ to -1 at $\omega T_s = \pi$ (half the sampling frequency). Hence we may substitute $\cos(\omega T_s)$ for Z in $P_t(Z)$ to obtain its frequency response directly.

24.6.3 Some Simple Filters/Wavelets (Haar and LeGall)

As discussed in Section 24.5, in order to achieve smooth wavelets after many levels of the binary tree, the low-pass filters $H_0(z)$ and $G_0(z)$ must both have a number of zeros at half the sampling frequency (at $z = -1$). These will also be zeros of $P(z)$, and so $P_t(Z)$ will have zeros at $Z = -1$ (each of which will correspond to a pair of zeros at $z = -1$).

---

* See the end of Section 24.1 for a discussion of causality assumptions related to this.

The simplest case is a single zero at $Z = -1$, so that $P_t(Z) = 1 + Z$.

$$\therefore P(z) = \frac{1}{2}(z + 2 + z^{-1}) = \frac{1}{2}(z + 1)(1 + z^{-1}) = G_0(z)H_0(z)$$

which gives the familiar Haar filters.

As we have seen in Figure 24.5, the Haar wavelets have significant discontinuities so we need to add more zeros at $Z = -1$. However to maintain PR, we must also ensure that all terms in even powers of Z in $P_t(Z)$ are zero, so the next more complicated P_t must be third-order and of the form

$$P_t(Z) = (1 + Z)^2(1 + aZ) = 1 + (2 + a)Z + (1 + 2a)Z^2 + aZ^3$$

$$= 1 + \frac{3}{2}Z - \frac{1}{2}Z^3 \quad \text{if } a = -\frac{1}{2} \text{ to suppress the term in } Z^2 \qquad (24.49)$$

Allocating the factors of P_t such that $(1 + Z)$ gives H_0 and $(1 + Z)(1 + aZ)$ gives G_0:

$$H_0(z) = \frac{1}{2}(z + 2 + z^{-1})$$

$$G_0(z) = \frac{1}{8}(z + 2 + z^{-1})(-z + 4 - z^{-1})$$

$$= \frac{1}{8}(-z^2 + 2z + 6 + 2z^{-1} - z^{-2}) \qquad (24.50)$$

Using Equation 24.42 with $k = 1$, the corresponding high-pass filters then become

$$G_1(z) = zH_0(-z) = \frac{1}{2}z(-z + 2 - z^{-1})$$

$$H_1(z) = z^{-1}G_0(-z) = \frac{1}{8}z^{-1}(-z^2 - 2z + 6 - 2z^{-1} - z^{-2}) \qquad (24.51)$$

This is often known as the *LeGall 3,5-tap filter set*, since it was first published in the context of 2-band filter banks by Didier LeGall in 1988.

The wavelets of the LeGall 3,5-tap filters, H_0 and H_1 above, and their frequency responses are shown in Figure 24.6. The scaling function (bottom left) converges to a pure triangular pulse and the wavelets are the superposition of two triangular pulses of opposing polarity.

The triangular scaling function produces linear interpolation between consecutive lowband coefficients and also causes the wavelets to be linear interpolations of the coefficients of the H_1 filter, $-1, -2, +6, -2, -1$ (scaled appropriately).

These wavelets have quite desirable properties for signal compression (note the absence of waveform discontinuities and the much lower sidelobes of the frequency responses), and they are one of the simplest useful wavelet types. Unfortunately there is one drawback—the inverse wavelets are not very good. These are formed from the LeGall 5,3-tap filter pair, G_0 and G_1 above, whose wavelets and frequency responses are shown in Figure 24.7

The main problem here is that the wavelets do not converge after many levels to a smooth function and hence the frequency responses have large unwanted sidelobes. For example, the jaggedness of the scaling function and wavelets causes highly visible coding artifacts if these filters are used for reconstruction of a compressed image.

However the allocation of the factors of $P_t(Z)$ to H_0 and G_0 is a free design choice, so we may swap the factors (and hence swap G and H) from the choice made in Equation 24.50 in order that the

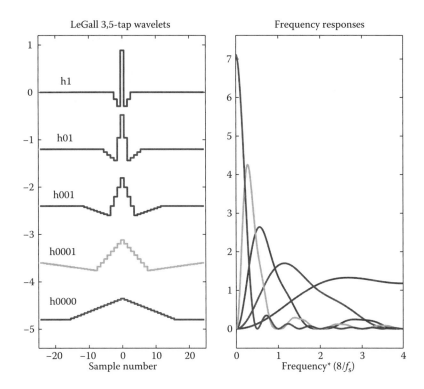

FIGURE 24.6 Impulse responses and frequency responses of the 4-level tree of LeGall 3,5-tap filters.

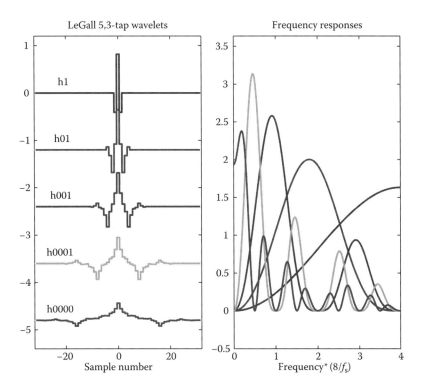

FIGURE 24.7 Impulse responses and frequency responses of the 4-level tree of LeGall 5,3-tap filters.

smoother 3,5-tap filters become G_0, G_1 and are used for reconstruction. It turns out that this leads to a good low-complexity solution for image compression and that the jaggedness of the analysis filters is not critical.

Unbalance between analysis and reconstruction filters/wavelets is nevertheless often regarded as being undesirable, particularly as it prevents the filtering process from being represented as an orthonormal transformation of the input signal and hence from preserving energy through the filter bank. An unbalanced PR filter system is often termed a biorthogonal filter bank.

We now consider ways to reduce this unbalance.

24.6.4 Filters with Balanced *H* and *G* Frequency Responses (but Nonlinear Phase Responses)—Daubechies Wavelets

In the above analysis, we used the factorization of $P_t(Z)$ to give us H_0 and G_0. This always gives unbalanced factors if terms of P_t in even powers of Z are zero, since $P_t(Z)$ must then be of odd-order.

However, each of these factors in Z may itself be factorized into a pair of factors in z, since

$$(\alpha z + 1)(1 + \alpha z^{-1}) = \alpha z + (1 + \alpha^2) + \alpha z^{-1}$$

$$= (1 + \alpha^2) + 2\alpha Z \quad \text{if } Z = \frac{1}{2}(z + z^{-1})$$

$$= (1 + \alpha^2)(1 + \beta Z) \quad \text{where } \beta = \frac{2\alpha}{1 + \alpha^2} \tag{24.52}$$

For each factor of $P_t(Z)$, we may allocate one of its z subfactors to $H_0(z)$ and the other to $G_0(z)$. Where roots of $P_t(Z)$ are complex, the subfactors must be allocated in conjugate pairs so that H_0 and G_0 remain purely real.

Since the subfactors occur in reciprocal pairs (roots at $z = \alpha$ and α^{-1} for any given β, as shown above), we find that

$$G_0(z) = H_0(z^{-1}) \tag{24.53}$$

This means that the impulse response of G_0 is the time-reverse of that of H_0.

Therefore, using the substitution $z = e^{j\omega T_s}$, their frequency responses are related by

$$G_0(e^{j\omega T_s}) = H_0(e^{-j\omega T_s}) \tag{24.54}$$

Hence, if H_0 and G_0 are real functions of z, the magnitudes of their frequency responses are the same, and their phases are opposite. It may be shown that this is sufficient to obtain orthogonal wavelet basis functions and scaling functions, but unfortunately the separate filters can no longer be zero (or linear) phase, apart from the unique case of the Haar wavelet. When the filters satisfy Equation 24.53, in addition to Equations 24.42 and 24.45, the filter bank is known as a conjugate quadrature filter bank (CQF).

The well-known Daubechies wavelets may be generated in this way, if we include the added constraint that the maximum number of zeros of $P_t(Z)$ are placed at $Z = -1$ (producing pairs of zeros of $P(z)$ at $z = -1$), consistent with terms in even powers of Z in $P_t(Z)$ being zero.

If $P_t(Z)$ is of order $2K - 1$, then it may have K zeros at $Z = -1$ such that

$$P_t(Z) = (1 + Z)^K R_t(Z) \tag{24.55}$$

where $R_t(Z)$ is of order $K - 1$ and its $K - 1$ roots may be chosen such that terms of $P_t(Z)$ in the $K - 1$ even powers of Z are zero.

Equation 24.49 is the $K = 2$ solution to Equation 24.55.

In this case, $R_t(Z) = 1 - \frac{1}{2}Z$, so $\beta = -\frac{1}{2}$ and, from Equation 24.52, the factors of $R(z)$ are

$$R(z) = \frac{(\alpha z + 1)(1 + \alpha z^{-1})}{1 + \alpha^2} \quad \text{where } \alpha = -2 \pm \sqrt{3}$$

since α must be a solution of $1 + \alpha^2 = 2\alpha/\beta$.

Also

$$(1 + Z)^2 = \frac{1}{2}(z + 1)^2 \frac{1}{2}(1 + z^{-1})^2$$

Hence

$$H_0(z) = \frac{1}{2\sqrt{1 + \alpha^2}} (1 + z^{-1})^2 \, (1 + \alpha z^{-1})$$
$$= 0.4830 + 0.8365 z^{-1} + 0.2241 z^{-2} - 0.1294 z^{-3} \tag{24.56}$$

and $G_0(z)$ is the time-reverse of this (replacing z^{-1} by z). Therefore

$$H_1(z) = z^{-3} G_0(-z) = z^{-3} H_0(-z^{-1})$$
$$= 0.1294 + 0.2241 z^{-1} - 0.8365 z^{-2} + 0.4830 z^{-3} \tag{24.57}$$

The wavelets and frequency responses for these 4-tap filters are shown in Figure 24.8. It is clear that the wavelets and scaling function are no longer linear phase (symmetric about a central point) and

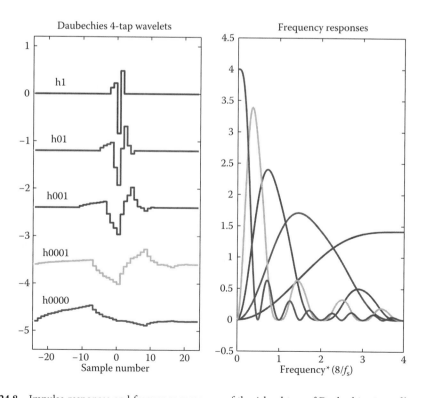

FIGURE 24.8 Impulse responses and frequency responses of the 4-level tree of Daubechies 4-tap filters.

are less smooth than those for the LeGall 3,5-tap filters. The frequency responses also show worse sidelobes. The G_0, G_1 filters give the time-reverse of these wavelets and identical frequency (magnitude) responses.

Higher order Daubechies filters achieve smoother wavelets but they still suffer from nonlinear phase. This tends to result in more visible image coding artefacts than linear phase filters, which distribute any artefacts equally on either side of sharp edges in the image.

When there is more than one root of the residual polynomial $R_t(Z)$, the designer may choose how the factors, $(1 + \alpha z)$ or $(1 + \alpha z^{-1})$, for each root are distributed to $H_0(z)$ and $G_0(z)$. While the distribution may be chosen to give maximum-phase or minimum-phase H_0 and G_0 filters, it is usual to attempt the distribution so as to approximate linear phase more closely. Depending on the choice of product filter order $2K - 1$, different levels of approximation to linear phase become possible.

A fairly good approximation to linear phase, often known as the Daubechies symlet, is obtained for the case when $K = 4$. This gives four zeros of $P_t(Z)$ at $Z = -1$ (and hence four pairs of zeros of $P(z)$ at $z = -1$, which can be shared equally between $H_0(z)$ and $G_0(z)$). The remainder polynomial $R_t(Z)$ is then third order. When converted back to being a function of z, the product filter is given by

$$P(z) = H_0(z)G_0(z) = \frac{z^{-4}}{2048}(z + 1)^8 R(z) \tag{24.58}$$
$$\text{where } R(z) = 5z^3 - 40z^2 + 131z - 208 + 131z^{-1} - 40z^{-2} + 5z^{-3}$$

$R(z)$ has real zeros at 0.3289 and 3.0407, and complex zeros at $0.2841 \pm 0.2432j$ and $2.0311 \pm 1.7390j$. To get the closest approximation to linear phase, we group the smaller conjugate pair with the larger real zero, and combine these with four of the zeros at $z = -1$ to obtain $H_0(z)$, while the remaining zeros of $P(z)$ give $G_0(z)$ (or vice versa). H_0 and G_0 are then both seventh-order polynomials (8-tap filters). The impulse and frequency responses of the wavelets and scaling functions from these filters are shown in Figure 24.9, from which we see the somewhat improved symmetry of the waveforms about their highest point (hence the name symlet), as compared with the 4-tap waveforms in Figure 24.8, as well as their greater smoothness and lower sidelobes in the frequency domain.

A useful way to obtain equations of the form of Equation 24.55, for arbitrary odd order $N \geq 2K - 1$ of the transformed product filter $P_t(z)$, is to use the Bernstein polynomial [4,5]:

$$B_N(x; \alpha) = \sum_{i=0}^{(N-1)/2} (1 - \alpha_i) \binom{N}{i} x^i (1 - x)^{N-i} + \sum_{i=(N+1)/2}^{N} \alpha_{N-i} \binom{N}{i} x^i (1 - x)^{N-i} \tag{24.59}$$

where $\alpha = [\alpha_0 \ \alpha_1 \ \ldots \ \alpha_{(N-1)/2}]^T$.

This polynomial has the property that, for any choice of x and α

$$B_N(x; \alpha) + B_N(1 - x; \alpha) = 1 \tag{24.60}$$

For PR, from Equations 24.45 and 24.47, it is required that

$$P_t(Z) + P_t(-Z) = 2 \tag{24.61}$$

Hence, if we let $P_t(Z) = 2B_N(x; \alpha)$, where $x = (1 - Z)/2$ so that x goes from 0 to 1 as Z goes from $+1$ to -1, then $P_t(-Z) = 2B_N(1 - x; \alpha)$ and, from Equation 24.60, $P_t(Z)$ will satisfy the above PR condition (Equation 24.61) for any choice of α.

To obtain K zeros of $P_t(Z)$ at $Z = -1$, we require $B_N(x; \alpha)$ to have K zeros at $x = 1$, which may be easily obtained by setting $\alpha_i = 0$ for $i = 0 \ldots K - 1$, as long as $N \geq 2K - 1$.

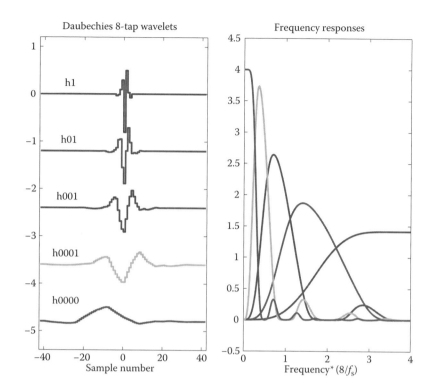

FIGURE 24.9 Impulse responses and frequency responses of the 4-level tree of Daubechies 8-tap symlet filters.

When $N = 2K - 1$, we have the maximum number of zeros at $Z = -1$ for a given N, as assumed above for the cases $K = 2$ and $K = 4$. Then are then no further degrees of freedom available to the designer. We shall see below that, in some situations, at least one additional degree of freedom can be very helpful, and this is achieved when $N = 2K + 1$ and the free parameter is α_K. More degrees of freedom may be achieved by increasing N in steps of two while keeping K constant.

24.6.5 Filters with Linear Phase and Nearly Balanced Frequency Responses

We have seen how to create filters with balanced frequency responses and an approximation to linear phase. Now we consider filters with exact linear phase and approximately balanced frequency responses.

Linear phase filters allow an elegant technique, known as symmetric extension, to be used at the outer edges of finite datasets such as images, where wavelet filters would otherwise require the size of the transformed data to be increased to allow for convolution with the filters. Symmetric extension assumes that the data is reflected by mirrors at each endpoint or edge. For images in 2-D, an infinitely tesselated plane of reflected images is generated. Reflections avoid unwanted edge discontinuities. If the filters are linear phase, then the resulting DWT coefficients also form reflections at the endpoints and no increase in size of the transformed data is necessary to accommodate convolution effects.

To ensure that the filters H_0, H_1 and G_0, G_1 are linear phase, the factors in Z must be allocated to H_0 or G_0 as a whole and not be split, as was done for the Daubechies filters. In this way, the symmetry between z and z^{-1} is preserved in all filters. Perfect balance of frequency responses between H_0 and G_0 is then not possible, if PR is preserved.

A popular linear-phase option, which works very well for image coding and is part of the JPEG 2000 standard, is a linear phase factorization of the same $P(z)$ product filter, used for the 8-tap Daubechies symlet and given in Equation 24.58. This factorization still distributes the eight zeros at $z = -1$ equally

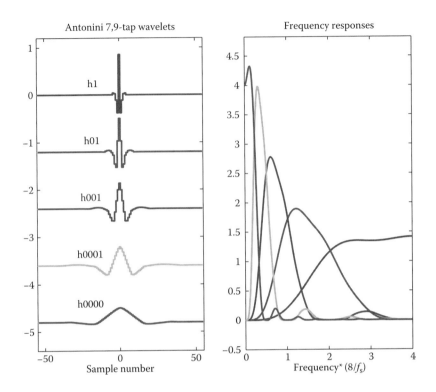

FIGURE 24.10   Impulse responses and frequency responses of the 4-level tree of Daubechies 7,9-tap filters.

between H_0 and G_0; but, to get linear phase and keep the filter coefficients purely real, it is necessary to allocate the two real zeros of $R(z)$ to one filter (say G_0) and the four complex zeros of $R(z)$ to the other filter (H_0). Linear phase is obtained because the zeros are in reciprocal pairs, and real coefficients are obtained because the complex zeros are also in conjugate pairs. The resulting 7-tap and 9-tap filters are known as the Daubechies biorthogonal 7,9-tap wavelet filters (and also known as Antonini filters) [6]. These 7,9-tap and 9,7-tap wavelets and scaling functions are shown in Figures 24.10 and 24.11.

The 7-tap filter from this pair tends to be somewhat smoother than the 9-tap one, and so is normally chosen for the reconstruction low-pass filter G_0 (and hence also generates the analysis high-pass filter H_1, Equation 24.42). In compression applications, where many higher frequency wavelet coefficients are set to zero, this tends to result in a smoother reconstructed signal. It may be seen that the smoother waveforms, with lower sidelobes in the frequency domain, occur in Figure 24.10 while the less smooth option is in Figure 24.11.

We have found a factorization of $P_t(Z)$ which achieves a much closer balance of the responses, by reducing K the number of zeros at $Z = -1$. This is

$$P_t(Z) = (1 + Z)(1 + aZ + bZ^2)(1 + Z)(1 + cZ) \tag{24.62}$$

This is a fifth order polynomial, and if the terms in Z^2 and Z^4 are to be zero, there are two constraints on the three unknowns $[a, b, c]$ so that one of them (say c) may be regarded as a free parameter. These constraints require that

$$a = -\frac{(1 + 2c)^2}{2(1 + c)^2} \quad \text{and} \quad b = \frac{c(1 + 2c)}{2(1 + c)^2} \tag{24.63}$$

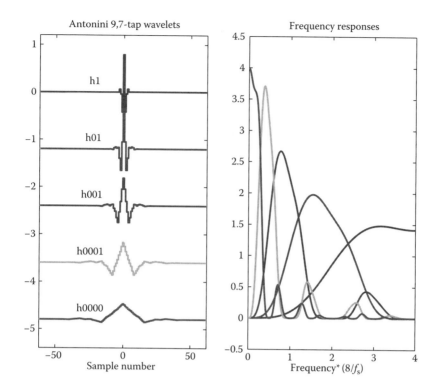

FIGURE 24.11   Impulse responses and frequency responses of the 4-level tree of Daubechies 9,7-tap filters.

c may then be adjusted to give maximum similarity between the left and right pairs of factors in Equation 24.62 as Z varies from 1 to -1 (i.e., as ωT_s varies from 0 to π).

It turns out that $c = -\frac{2}{7}$ gives good similarity and when substituted into Equations 24.62 and 24.63 gives

$$P_t(Z) = \frac{1}{50}(1 + Z)(50 - 9Z - 6Z^2)\frac{1}{7}(1 + Z)(7 - 2Z)$$

$$= \frac{1}{50}(50 + 41Z - 15Z^2 - 6Z^3)\frac{1}{7}(7 + 5Z - 2Z^2) \qquad (24.64)$$

We get $G_0(z)$ and $H_0(z)$ by substituting $Z = \frac{1}{2}(z + z^{-1})$ into these two polynomial factors. This results in 5,7-tap filters whose wavelets and frequency responses are shown in Figure 24.12.

The near balance of the analysis and reconstruction responses may be seen from Figure 24.13 which shows the alternative 7,5-tap versions (i.e., with H and G swapped). It is quite difficult to spot the minor differences between these figures and it makes little difference which way round the filters are allocated.

These filters may also be obtained using the Bernstein polynomial method of equation (Equation 24.59), using $N = 5$, $K = 2$, and α_K as the free parameter. The value which corresponds to $c = -\frac{2}{7}$ is $\alpha_2 = \frac{477}{1750} = 0.2726$.

24.6.6 Smoother Wavelets

In all of the above designs we have used the substitution $Z = \frac{1}{2}(z + z^{-1})$. However other substitutions may be used to create improved wavelets. To preserve PR, the substitution should contain only odd powers of z (so that odd powers of Z produce only odd powers of z); and to produce zero phase, the coefficients of the substitution should be symmetric about z^0.

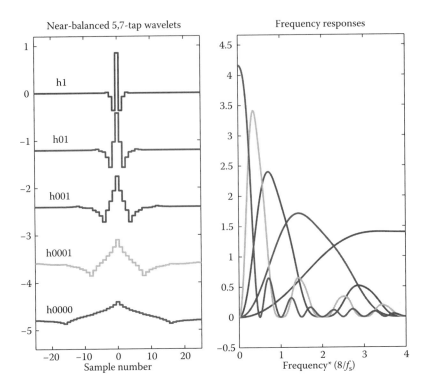

FIGURE 24.12   Impulse responses and frequency responses of the 4-level tree of near-balanced 5,7-tap filters.

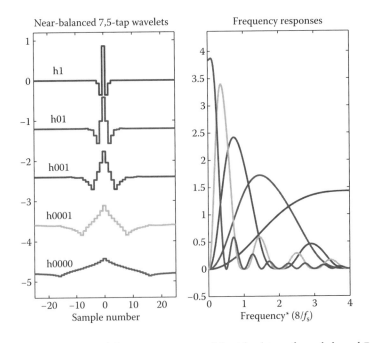

FIGURE 24.13   Impulse responses and frequency responses of the 4-level tree of near-balanced 7,5-tap filters.

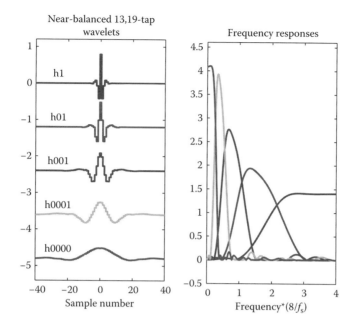

FIGURE 24.14 Impulse responses and frequency responses of the 4-level tree of near-balanced 13,19-tap filters.

A substitution, which can give much greater flatness near $z = \pm 1$ while still satisfying $Z = \pm 1$ when $z = \pm 1$, is

$$Z = pz^3 + (\tfrac{1}{2} - p)(z + z^{-1}) + pz^{-3} \tag{24.65}$$

Z then becomes the following function of frequency when $z = e^{j\omega T_s}$:

$$Z = (1 - 2p)\cos(\omega T_s) + 2p\cos(3\omega T_s) \tag{24.66}$$

Maximal flatness is achieved near $\omega T_s = 0$ and π, when $p = -\frac{1}{16}$. This is equivalent to more zeros at $z = -1$ for each $(Z + 1)$ factor than when $Z = \frac{1}{2}(z + z^{-1})$ is used. A high degree of flatness (with some ripple) and sharper transition bands is obtained if $p = -\frac{3}{32}$ instead.

The second order factor of $P_t(Z)$ in Equation 24.64 now produces terms from z^6 to z^{-6} and the third-order factor produces terms from z^9 to z^{-9}. Hence the filters become 13- and 19-tap filters, although two taps of each are zero and the outer two taps of the 19-tap filter are very small ($\sim 10^{-4}$).

Figure 24.14 shows the wavelets and frequency responses of the 13,19-tap filters, obtained by substituting Equation 24.65 into Equation 24.64 with $P = -3/32$. Note the smoother wavelets and scaling function and the much lower sidelobes in the frequency responses from these higher order filters.

Figure 24.15 demonstrates that the near-balanced properties of Equation 24.64 are preserved in the higher order filters.

There are many other types of wavelets with varying features and complexities, but we have found the examples given to be near optimum for applications such as image compression.

24.7 IIR Filter Banks

In general, for a given frequency response specification (e.g., transition bandwidth, stopband ripple magnitude), an IIR filter can achieve that specification with a lower computational complexity (e.g., number of

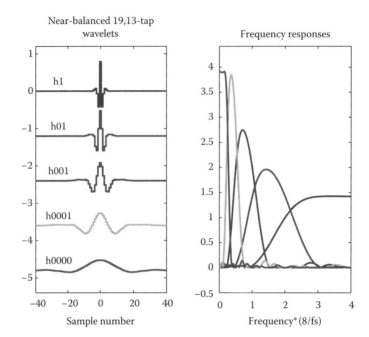

FIGURE 24.15 Impulse responses and frequency responses of the 4-level tree of near-balanced 19,13-tap filters.

multiplies per output sample) compared to an FIR filter. This is the potential advantage of IIR filter banks but the disadvantage is that their design and their implementation are usually more complicated. The main issue that needs to be considered when using IIR filters is stability. The well-known criterion for stability is that the poles of the IIR transfer function must lie inside the unit circle. This criterion, however, assumes that the IIR filter is implemented using a difference equation in the forward time direction, i.e., causal filtering. In some situations, in which the data is fully stored in memory and may therefore be processed backward as well as forward, anticausal filtering is feasible too. With anticausal filtering the poles need to be outside the unit circle for stability. There are therefore two types of IIR filter banks:

1. Causal stable IIR filter banks: used in situations where only causal filtering is allowed. Examples are real-time applications such as speech processing and echo-cancellation. All filters must then have poles inside the unit circle.
2. Noncausal IIR filter banks: used in situations which can allow off-line processing or when the signal is finite in duration (e.g., still images). Suppose an IIR filter $F(z)$ has poles both inside and outside the unit circle. The filter may be factored into a product of its minimum and maximum-phase components:

$$F(z) = F_{\min}(z)F_{\max}(z) \tag{24.67}$$

where the poles of $F_{\min}(z)$ are inside the unit circle, and those of $F_{\max}(z)$ are outside it. The filtering consists of two stages, in which causal filtering is first performed forward to implement $F_{\min}(z)$ and anticausal filtering is then performed backward to implement $F_{\max}(z)$.

Just as in FIR filter banks, the scaling functions and wavelets corresponding to IIR filter banks can be defined through the two-scale equation and infinite product formula. The main difference in the IIR case is that the corresponding continuous functions have support width that is infinite, i.e., the function duration is infinite. For the causal stable case the function extends to positive infinity and for the noncausal case the function extends to infinity in both directions.

A versatile building block for IIR filters is the all-pass filter $a(z)$ which has the following two properties:

$$\frac{1}{a(z)} = a(z^{-1}) \quad \text{and} \quad |a(e^{j\omega})| = 1 \tag{24.68}$$

The all-pass filter $a(z)$ can be considered a generalization of the unit delay z^{-1} since $|e^{-j\omega}| = |a(e^{j\omega})| = 1$, i.e., both have the same magnitude response. The unit delay has a linear phase response but the all-pass filter can have a variety of phase responses. A wide family of IIR filters can be obtained by various interconnections of all-pass filters [7–9]. The simplest interconnection (described below) can give a large class of useful IIR filter banks.

24.7.1 All-Pass Filter Design Methods

Consider analysis filters obtained by the weighted sum and difference of two all-pass filters, a_0 and a_1:

$$H_0(z) = \frac{1}{\sqrt{2}} [a_0(z) + a_1(z)]$$
$$\tag{24.69}$$
$$H_1(z) = \frac{1}{\sqrt{2}} [a_0(z) - a_1(z)]$$

This can be considered an extension of the FIR Haar filter pair in Equation 24.2 to IIR filters. The terms 1 and z^{-1} (simple all-pass) have been replaced with a_0 and a_1, general all-pass filters. The Haar filters are quadrature mirror versions of each other. If the two all-pass filters are chosen to be of the form $a_0(z) = A_0(z^2), a_1(z) = z^{-1}A_1(z^2)$, then the analysis filters, given by

$$H_0(z) = \frac{1}{\sqrt{2}} \left[A_0(z^2) + z^{-1}A_1(z^2) \right]$$
$$\tag{24.70}$$
$$H_1(z) = \frac{1}{\sqrt{2}} \left[A_0(z^2) - z^{-1}A_1(z^2) \right]$$

will have the quadrature mirror characteristics:

$$|H_1(e^{j\omega})| = |H_0(e^{j(\pi-\omega)})|, \quad \text{i.e., be mirrored about } \omega = \frac{\pi}{2} \tag{24.71}$$

This means that if $H_0(e^{j\omega})$ has a low-pass frequency response, $H_1(e^{j\omega})$ will automatically have a high-pass response.

How should the all-pass filters be designed so that $H_0(e^{j\omega})$ is a good low-pass filter with half of the full bandwidth?

By denoting the phase response of the all-pass filters as $\phi_0(\omega) = \angle A_0(e^{j\omega})$ and $\phi_1(\omega) = \angle A_1(e^{j\omega})$, the magnitude response of the low-pass filter can be written as

$$|H_0(e^{j\omega})| = \frac{1}{\sqrt{2}} \left| e^{j\phi_0(2\omega)} + e^{j(\phi_1(2\omega)-\omega)} \right|$$

$$= \frac{1}{\sqrt{2}} \left| e^{j\phi_0(2\omega)} \right| \left| 1 + e^{j(\phi_1(2\omega)-\phi_0(2\omega)-\omega)} \right|$$

$$= \frac{1}{\sqrt{2}} \left| 1 + e^{j(\theta(2\omega)-\omega)} \right| \quad \text{where} \quad \theta(\omega) \equiv \phi_1(\omega) - \phi_0(\omega) \tag{24.72}$$

The term $e^{j(\theta(2\omega)-\omega)}$ can be considered a frequency dependent phasor that is added to the constant phasor 1. If the two phasors add constructively (i.e., if $e^{j(\theta(2\omega)-\omega)} \approx 1$), then $|H_0(e^{j\omega})| \approx \sqrt{2}$; and if the two phasors add destructively (i.e., if $e^{j(\theta(2\omega)-\omega)} \approx -1$), then $|H_0(e^{j\omega})| \approx 0$. Therefore the ideal criterion for the phase is

$$\theta(2\omega) - \omega = \begin{cases} 0, & 0 < \omega < \pi/2 \\ \pm\pi, & \pi/2 < \omega < \pi \end{cases} \tag{24.73}$$

What about the synthesis filters?

For the Haar filters, the synthesis filters were obtained from the analysis filters by replacing z^{-1} with z as shown in Equation 24.6, i.e., a reciprocal term is used. The situation now with all-pass filters (instead of delays) is, however, more complicated and there are two cases to consider:

1. No reciprocals are used and the synthesis filters are chosen as

$$G_0(z) = \frac{1}{\sqrt{2}}\left[A_0(z^2) + z^{-1}A_1(z^2)\right]$$

$$G_1(z) = -\frac{1}{\sqrt{2}}\left[A_0(z^2) - z^{-1}A_1(z^2)\right] \tag{24.74}$$

and are essentially the same as the analysis filters (apart from a sign change for G_1). Aliasing cancellation is achieved (because $G_0(z)H_0(-z) + G_1(z)H_1(-z) = 0$), but not PR. The transfer function between the input and output of the entire filter bank, i.e., the reconstruction function, is given from Equation 24.11 by

$$T(z) \equiv \frac{\hat{X}(z)}{X(z)} = \frac{1}{2}[G_0(z)H_0(z) + G_1(z)H_1(z)] = z^{-1}A_0(z^2)A_1(z^2) \tag{24.75}$$

which is an all-pass function. If phase-distortion can be tolerated, e.g., for speech, then this may be acceptable as it stands. Alternatively some postfiltering to equalize the phase of $T(z)$ may be needed.

If the poles of $A_0(z)$ and $A_1(z)$ are inside the unit circle then the whole filter bank is causal stable.

2. If the synthesis filters are instead chosen as

$$G_0(z) = \frac{1}{\sqrt{2}}\left[\frac{1}{A_0(z^2)} + z\frac{1}{A_1(z^2)}\right] = \frac{1}{\sqrt{2}}\left[A_0(z^{-2}) + zA_1(z^{-2})\right]$$

$$G_1(z) = \frac{1}{\sqrt{2}}\left[\frac{1}{A_0(z^2)} - z\frac{1}{A_1(z^2)}\right] = \frac{1}{\sqrt{2}}\left[A_0(z^{-2}) - zA_1(z^{-2})\right] \tag{24.76}$$

then PR is achieved, i.e., $T(z) = 1$. However, the filter bank then becomes noncausal as the inversion of the all-pass filters causes poles that were originally inside the unit circle now to be outside the unit circle.

24.7.2 Transformation-Based Design Methods

The frequency transformation technique, described in Section 24.6 for FIR filter banks, can also be used to construct IIR filter banks [10].

Recall that in the transformation technique, given the prototype linear phase low-pass filters, $H_t(Z)$ and $F_t(Z)$, the following transformation is applied:

$$Z = M(z) \tag{24.77}$$

where $M(z) = pz^3 + (\frac{1}{2} - p)(z + z^{-1}) + pz^{-3}$ (from Equation 24.65) and $p = 0$ or $-\frac{3}{32}$ depending on the desired complexity/smoothness required. It is necessary that the transformation is of the form $M(z) \equiv zT(z^2)$ (i.e., it must have terms in only odd powers of z).

Hence the transformation function $M(z)$ for the FIR case is an FIR filter that is related to a halfband filter,* $H_{HB}(z) \equiv \frac{1}{2}[1 + M(z)]$. IIR filter banks can be obtained by using an IIR transformation function instead of an FIR function and this is equivalent to using an IIR halfband filter. The transformation function should ideally approximate $+1$ in the passband and -1 in the stopband, i.e.,

$$M(e^{j\omega}) = e^{j\omega} T(e^{j2\omega}) \approx \begin{cases} +1, & 0 < \omega < \pi/2 \\ -1, & \pi/2 < \omega < \pi \end{cases} \tag{24.78}$$

Therefore a straightforward choice is for $M(z)$ to be an all-pass filter approximating the following ideal phase condition:

$$\phi_{ideal}(\omega) = \begin{cases} 0, & 0 < \omega < \pi/2 \\ \pm\pi, & \pi/2 < \omega < \pi \end{cases} \tag{24.79}$$

If the poles of $M(z)$ are inside the unit circle then the IIR filter bank is causal stable. Furthermore, PR is achieved.

A simple example for $M(z)$ is given by

$$M(z) = z \frac{k_1 + z^{-2}}{1 + k_1 z^{-2}} \frac{k_2 + z^{-2}}{1 + k_2 z^{-2}} \tag{24.80}$$

with $k_1 = -0.0991$ and $k_2 = 0.5426$ and the phase response is shown in Figure 24.16.

In Equation 24.49, we saw that a simple $H_t(Z)$ and $F_t(Z)$ are given by

$$H_t(Z) = (1 + Z)$$
$$F_t(Z) = \frac{1}{2}(1 + Z)(2 - Z) \tag{24.81}$$

Using an all-pass filter for $Z = M(z)$ in Equation 24.81 will however result in a 4 dB overshoot (bump) in the frequency response of the filter $F_0(z) = F_t(M(z))$. With a more general filter for $M(z)$ (non all-pass) the overshoot can be reduced but the design becomes more complicated.

Alternatively, a different set of prototype filters may lead to a lower overshoot. The following set of prototype filters from the product filter in Equation 24.64:

$$H_t(Z) = \frac{1}{7}(1 + Z)(7 - 2Z)$$
$$F_t(Z) = \frac{1}{50}(1 + Z)(50 - 9Z - 6Z^2) \tag{24.82}$$

---

* Halfband filters are defined as filters that satisfy $H_{HB}(z) + H_{HB}(-z) = 1$. The even index coefficients are zero except for the center (zero index) coefficient, which is one half.

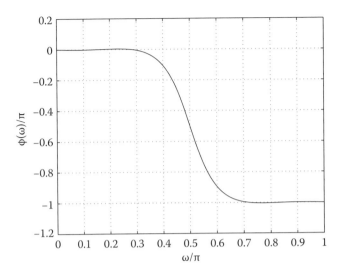

FIGURE 24.16   Phase response of IIR all-pass transformation in Equation 24.80.

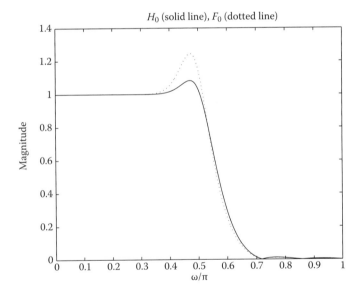

FIGURE 24.17   Magnitude responses of causal stable IIR filter bank given by Equation 24.82 with the transformation of Equation 24.80.

with an all-pass $M(z)$ will lead to IIR filters with overshoots of 0.68 dB and 1.90 dB, respectively. Using these prototypes with the transformation in Equation 24.80 gives low-pass filters with magnitude responses shown in Figure 24.17.

24.8 Polyphase Representations and the Lifting Scheme

The polyphase representation is an important tool in multirate systems, from both theoretical and application perspectives. Multirate systems are (periodically) linear time-varying systems due to the presence of downsamplers and upsamplers. The polyphase decomposition allows a multirate system to be represented as a sum of linear time-invariant systems, thus simplifying its analysis and design [9,11,12]. The form of the representation depends on the downsampling/upsampling factor M, but the discussion here will focus on the simplest case of $M = 2$ which is relevant for 2-band filter-banks. Unless otherwise stated, all filters are assumed to be FIR in this section.

24.8.1 Basic Polyphase Concepts

The basic idea of an $M = 2$ polyphase representation is to separate the coefficients of a filter or a sequence into even-indexed and odd-indexed coefficients:

$$H(z) = \sum_n h(2n)z^{-2n} + \sum_n h(2n + 1)z^{-2n-1} \tag{24.83}$$

By defining the polyphase components as

$$E_0(z) \equiv \sum_n h(2n)z^{-n} \quad \text{and} \quad E_1(z) \equiv \sum_n h(2n + 1)z^{-n} \tag{24.84}$$

the filter function $H(z)$ can be written as

$$H(z) = E_0(z^2) + z^{-1}E_1(z^2) \tag{24.85}$$

The separation of $H(z)$ into $E_0(z)$ and $E_1(z)$ is known as the polyphase decomposition of the filter. The decomposition in Equation 24.85 is known as a type 1 decomposition. An alternative is the type 2 decomposition defined as

$$H(z) = z^{-1}R_0(z^2) + R_1(z^2) = z^{-1}\left[R_0(z^2) + zR_1(z^2)\right] \tag{24.86}$$

The polyphase components of both types are related as follows:

$$R_0(z) = E_1(z) \quad R_1(z) = E_0(z)$$

i.e., type 2 is a permutation of type 1.

By applying the type 1 polyphase decomposition to the 2-band analysis filter bank of Figure 24.1a, the following is obtained:

$$\begin{bmatrix} H_0(z) \\ H_1(z) \end{bmatrix} = \begin{bmatrix} E_{0,0}(z^2) & E_{0,1}(z^2) \\ E_{1,0}(z^2) & E_{1,1}(z^2) \end{bmatrix} \begin{bmatrix} 1 \\ z^{-1} \end{bmatrix} \tag{24.87}$$

The matrix

$$\mathbf{E}(z) \equiv \begin{bmatrix} E_{0,0}(z) & E_{0,1}(z) \\ E_{1,0}(z) & E_{1,1}(z) \end{bmatrix}$$

is known as the analysis polyphase matrix of the filter bank. The first row contains the polyphase components of the low-pass analysis filter H_0, and the second row those of the high-pass filter H_1.

For the synthesis filters, the type 2 polyphase decomposition is applied:

$$[G_0(z)\ G_1(z)] = [z^{-1}\ 1]\begin{bmatrix} R_{0,0}(z^2) & R_{1,0}(z^2) \\ R_{0,1}(z^2) & R_{1,1}(z^2) \end{bmatrix} \tag{24.88}$$

and the synthesis polyphase matrix is defined as

$$\mathbf{R}(z) \equiv \begin{bmatrix} R_{0,0}(z) & R_{1,0}(z) \\ R_{0,1}(z) & R_{1,1}(z) \end{bmatrix}$$

The first column contains the polyphase components of the low-pass synthesis filter G_0, and the second column those of the high-pass filter G_1.

Now suppose the input signal is decomposed as

$$X(z) = X_0(z^2) + zX_1(z^2)$$

which is similar to the type 2 polyphase decomposition except for absence of the delay z^{-1} (compare with Equation 24.86). The output of the analysis filter bank can then be written compactly as

$$\begin{bmatrix} Y_0(z) \\ Y_1(z) \end{bmatrix} = \begin{bmatrix} E_{0,0}(z) & E_{0,1}(z) \\ E_{1,0}(z) & E_{1,1}(z) \end{bmatrix}\begin{bmatrix} X_0(z) \\ X_1(z) \end{bmatrix} = \mathbf{E}(z)\begin{bmatrix} X_0(z) \\ X_1(z) \end{bmatrix} \tag{24.89}$$

By using a type 2 polyphase decomposition of the output signal

$$\hat{X}(z) = z^{-1}\hat{X}_0(z^2) + \hat{X}_1(z^2)$$

the output of the synthesis filter bank can be written compactly as

$$\begin{bmatrix} \hat{X}_0(z) \\ \hat{X}_1(z) \end{bmatrix} = \begin{bmatrix} R_{0,0}(z) & R_{1,0}(z) \\ R_{0,1}(z) & R_{1,1}(z) \end{bmatrix}\begin{bmatrix} Y_0(z) \\ Y_1(z) \end{bmatrix} = \mathbf{R}(z)\begin{bmatrix} Y_0(z) \\ Y_1(z) \end{bmatrix} \tag{24.90}$$

Equations 24.89 and 24.90 are the equivalent time-invariant representation of the multirate filter bank. Figure 24.18 shows the equivalent representation of the filter bank in polyphase form. The combination of the delay chain and downsampler on the analysis side provides a polyphase decomposition of the input signal. The combination of the upsamplers, delay chain, and adder on the synthesis side provides a reconstruction of the output signal from the polyphase components. The polyphase filter bank is operating at the lower sampling rate of the system and is thus more efficient than the original filter bank.

Combining Equations 24.89 and 24.90 gives the polyphase form of the input/output relationship of the entire filter bank:

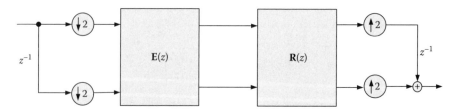

FIGURE 24.18   Polyphase form of filter bank.

$$\begin{bmatrix} \hat{X}_0(z) \\ \hat{X}_1(z) \end{bmatrix} = \mathbf{R}(z)\mathbf{E}(z)\begin{bmatrix} X_0(z) \\ X_1(z) \end{bmatrix}$$

The total system polyphase matrix is defined as

$$\mathbf{P}(z) \equiv \mathbf{R}(z)\mathbf{E}(z)$$

and if

$$\mathbf{P}(z) = \mathbf{R}(z)\mathbf{E}(z) = \mathbf{I} \tag{24.91}$$

then

$$\begin{bmatrix} \hat{X}_0(z) \\ \hat{X}_1(z) \end{bmatrix} = \begin{bmatrix} X_0(z) \\ X_1(z) \end{bmatrix}$$

Therefore

$$\hat{X}(z) = z^{-1}X_0(z^2) + X_1(z^2) = z^{-1}\left[X_0(z^2) + zX_1(z^2)\right] = z^{-1}X(z)$$

and PR (with one sample delay) is achieved. This is a convenient alternative formulation for PR to that given in Equations 24.40 and 24.41.

Equation 24.91 provides a concise condition for PR and the result is useful in the theoretical study and formulation of design techniques for filter banks. Conceptually, if synthesis filter bank is chosen such that

$$\mathbf{R}(z) = \mathbf{E}^{-1}(z) = \frac{\text{adj}(\mathbf{E}(z))}{\det(\mathbf{E}(z))} \tag{24.92}$$

where adj($\cdot$) and det($\cdot$) denote the adjoint and determinant of the matrix, respectively, then PR is assured. However, in general, this would lead to IIR synthesis filters which may be unstable (poles outside the unit circle) due to the presence of det ($\mathbf{E}(z)$) in the denominator. Therefore to ensure that all filters are FIR in the filter bank, it is required that

$$\det(\mathbf{E}(z)) = cz^K \tag{24.93}$$

where c is a nonzero constant and K an integer.

An important property of determinants is that

$$\text{if } \mathbf{A} = \mathbf{A}_1\mathbf{A}_2 \quad \text{then } \det(\mathbf{A}) = \det(\mathbf{A}_1)\det(\mathbf{A}_2)$$

Therefore, a straightforward way to build polyphase matrices satisfying Equation 24.93 is through a cascade (product) of simpler matrices that satisfy Equation 24.93, i.e.,

$$\mathbf{E}(z) = \mathbf{E}_1(z)\mathbf{E}_2(z)\cdots\mathbf{E}_L(z) \tag{24.94}$$

where

$$\det(\mathbf{E}_i(z)) = c_i z^{K_i} \quad i = 1,\dots,L$$

The cascade of polyphase matrices is the essence of virtually all design techniques of filter banks that are polyphase based.

24.8.2 Orthogonal Lattice Structures

This type of structure is also known as the lossless or paraunitary structure [13,14]. The polyphase matrix is given by

$$E(z) = R_L D(z) R_{L-1} \cdots D(z) R_0 = R_L \prod_{i=0}^{L-1} [D(z) R_i] \tag{24.95}$$

where

$$D(z) \equiv \begin{bmatrix} 1 & 0 \\ 0 & z^{-1} \end{bmatrix} \quad \text{and} \quad R_i \equiv \begin{bmatrix} \cos \theta_i & \sin \theta_i \\ -\sin \theta_i & \cos \theta_i \end{bmatrix}$$

and $\theta_0 \ldots \theta_{L-1}$ are the L free parameters that define the filters. In the design process, these parameters are optimized with respect to some criterion, e.g., minimizing the stopband energy of the low-pass filter, which is given by

$$J = \int_{\omega_s}^{\pi} |H_0(e^{j\omega})|^2 d\omega$$

where ω_s is the stopband edge. The matrices R_i are also known as rotation matrices and are orthogonal, i.e., $R_i R_i^T = R_i^T R_i = I$ (where T denotes transpose).
Since $D(z) D^T(z^{-1}) = D^T(z^{-1}) D(z) = I$, the polyphase matrix satisfies

$$E(z) E^T(z^{-1}) = E^T(z^{-1}) E(z) = I$$

Therefore the synthesis polyphase matrix can be obtained as

$$R(z) = E^T(z^{-1}) = R_0^T D(z^{-1}) \cdots R_{L-1}^T \cdots D(z^{-1}) R_L^T$$

The structure for such filter banks is shown in Figure 24.19

24.8.3 Linear Phase Structures

There are two types of structure that lead to linear phase filters [15,16]:

1. Type A: where all filters in the bank have even length. The low-pass analysis filter coefficients are symmetric, while the synthesis coefficients are antisymmetric.
   The polyphase matrix is given by

$$E(z) = \begin{bmatrix} 1 & 1 \\ 1 & -1 \end{bmatrix} D(z) S_{L-1} \cdots D(z) S_0 = \begin{bmatrix} 1 & 1 \\ 1 & -1 \end{bmatrix} \prod_{i=0}^{L-1} [D(z) S_i] \tag{24.96}$$

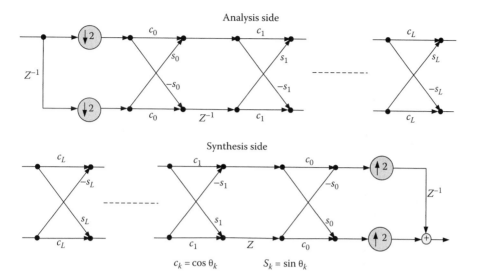

FIGURE 24.19 Orthogonal lattice structure.

where

$$\mathbf{D}(z) \equiv \begin{bmatrix} 1 & 0 \\ 0 & z^{-1} \end{bmatrix} \quad \text{and} \quad \mathbf{S}_i \equiv \begin{bmatrix} 1 & \alpha_i \\ \alpha_i & 1 \end{bmatrix}$$

and $\alpha_i \neq 0$ are the free parameters that define the filters. The structure here is similar to the orthogonal lattice structure except for a sign change in one of the element of $\mathbf{S}_i$.

2. Type B: where all filters in the bank have odd length. All filter coefficients are then symmetric. The polyphase matrix is given by

$$\mathbf{E}(z) = \prod_{i=1}^{L} \mathbf{B}_i(z) \tag{24.97}$$

where

$$\mathbf{B}_i(z) = \begin{bmatrix} 1 + z^{-1} & \alpha_i \\ 1 + \beta_i z^{-1} + z^{-2} & \alpha_i(1 + z^{-1}) \end{bmatrix}$$

and $\alpha_i \neq 0, \beta_i \neq 2$ are the free parameters that define the filters.

This structure can be used to implement all of the linear-phase odd-length wavelet filter banks discussed in Sections 24.6.3, 24.6.5, and 24.6.6, such as the (5,3), (7,5), (9,7), and (19,13)-tap filters and their inverses.

24.8.4 Lifting Scheme

The basic idea behind lifting [17–19] is simple:

> Whatever change was performed to the signal by some process of addition can be undone by the equivalent process of subtraction.

This idea ties in nicely with the concept of PR where the synthesis filter bank is supposed to undo whatever processing was done by the analysis filter bank. There are two attractive features of lifting: (1) the construction of filter banks, because PR can be guaranteed at every step of the design stage and (2) the implementation of filter banks, because computational efficiency gains of up to 2:1 are possible when compared with separate filters for each band of the filter bank. The former aspect will be considered first.

24.8.4.1 Construction of Filter Banks

The basic building block in lifting is called the lifting step and is related to concept of elementary matrices which is fundamental in matrix analysis. Elementary matrices can be used to perform elementary operations on an arbitrary matrix by either the pre- or post-multiplication of the former with the latter. In the context of filter banks, the arbitrary matrix is the analysis or synthesis polyphase matrix (which is a matrix of Laurent series*). There are three types of elementary matrices (operations) but the type that is relevant in lifting is the one that performs the addition to the elements of a certain row (or column) of a multiple (factor) of the elements of another row (or column). For the analysis polyphase matrix there two types of lifting step:

1. Primal lifting: with elementary matrix given by

$$\begin{bmatrix} 1 & p(z) \\ 0 & 1 \end{bmatrix}$$

   where $p(z)$ is an arbitrary FIR factor. Premultipication by this matrix will change the first row of an arbitrary matrix.
2. Dual lifting: with elementary matrix given by

$$\begin{bmatrix} 1 & 0 \\ d(z) & 1 \end{bmatrix}$$

   where $d(z)$ is an arbitrary FIR factor. Premultipication by this matrix will change the second row of an arbitrary matrix.

Note that both of these lifting matrices have determinants of unity.

Now suppose that there is an existing set of analysis filters with corresponding polyphase matrix $\mathbf{E}^0(z)$; i.e.,

$$\begin{bmatrix} H_0^0(z) \\ H_1^0(z) \end{bmatrix} = \mathbf{E}^0(z^2) \begin{bmatrix} 1 \\ z^{-1} \end{bmatrix} = \begin{bmatrix} E_{0,0}^0(z^2) & E_{0,1}^0(z^2) \\ E_{1,0}^0(z^2) & E_{1,1}^0(z^2) \end{bmatrix} \begin{bmatrix} 1 \\ z^{-1} \end{bmatrix} \tag{24.98}$$

1. If the primal lifting is applied to $\mathbf{E}^0(z)$

$$\mathbf{E}^P(z) \equiv \begin{bmatrix} 1 & p(z) \\ 0 & 1 \end{bmatrix} \mathbf{E}^0(z) \tag{24.99}$$

   then the new set of analysis filters becomes

$$H_0^P(z) = H_0^0(z) + H_1^0(z)p(z^2) \quad \text{and} \quad H_1^P(z) = H_1^0(z)$$

   i.e., the low-pass filter is changed but the high-pass is unchanged.

---

* Like polynomials but with both positive and negative powers of z.

2. If the dual lifting is applied to $\mathbf{E}^0(z)$

$$\mathbf{E}^d(z) \equiv \begin{bmatrix} 1 & 0 \\ d(z) & 1 \end{bmatrix} \mathbf{E}^0(z) \tag{24.100}$$

then the new set of analysis filters becomes

$$H_0^d(z) = H_0^0(z) \quad \text{and} \quad H_1^d(z) = H_1^0(z) + H_0^0(z)d(z^2)$$

i.e., the high-pass filter is changed but the low-pass is unchanged.

Since the determinant of the lifting matrix is equal to one, the new polyphase matrix, $\mathbf{E}^p(z)$ or $\mathbf{E}^d(z)$, will satisfy the PR condition (Equation 24.93) if the original matrix $\mathbf{E}^0(z)$ also satisfies the condition.

Starting with the trivial polyphase matrix $\mathbf{E}^0(z) = \mathbf{I}$ (identity), alternate primal and dual lifting can be used to construct a whole range of PR filter banks:

$$\begin{aligned}
\mathbf{E}(z) &= \begin{bmatrix} K_0 & 0 \\ 0 & K_1 \end{bmatrix} \begin{bmatrix} 1 & 0 \\ d_M(z) & 1 \end{bmatrix} \begin{bmatrix} 1 & p_M(z) \\ 0 & 1 \end{bmatrix} \cdots \begin{bmatrix} 1 & 0 \\ d_1(z) & 1 \end{bmatrix} \begin{bmatrix} 1 & p_1(z) \\ 0 & 1 \end{bmatrix} \\
&= \begin{bmatrix} K_0 & 0 \\ 0 & K_1 \end{bmatrix} \prod_{i=M}^{1} \begin{bmatrix} 1 & 0 \\ d_i(z) & 1 \end{bmatrix} \begin{bmatrix} 1 & p_i(z) \\ 0 & 1 \end{bmatrix}
\end{aligned} \tag{24.101}$$

The factors K_0 and K_1 are for normalizing the gains of the low-pass and high-pass filters, respectively.

Now consider the synthesis filter banks. First note that the inverse of the lifting step matrix can be easily obtained by negating the sign of the arbitrary factors, $p(z)$ or $d(z)$, since

$$\begin{bmatrix} 1 & p(z) \\ 0 & 1 \end{bmatrix}^{-1} = \begin{bmatrix} 1 & -p(z) \\ 0 & 1 \end{bmatrix}$$

and

$$\begin{bmatrix} 1 & 0 \\ d(z) & 1 \end{bmatrix}^{-1} = \begin{bmatrix} 1 & 0 \\ -d(z) & 1 \end{bmatrix}$$

Suppose the initial synthesis polyphase matrix of the analysis and synthesis banks, $\mathbf{E}^0(z)$ and $\mathbf{R}^0(z)$, satisfy the PR condition:

$$\mathbf{E}^0(z)\mathbf{R}^0(z) = \mathbf{I}$$

If the equation above is pre- and post-multiplied, respectively, by the primal and inverse primal lifting matrix, then

$$\left(\begin{bmatrix} 1 & p(z) \\ 0 & 1 \end{bmatrix} \mathbf{E}^0(z) \right) \left(\mathbf{R}^0(z) \begin{bmatrix} 1 & -p(z) \\ 0 & 1 \end{bmatrix} \right) = \begin{bmatrix} 1 & p(z) \\ 0 & 1 \end{bmatrix} \begin{bmatrix} 1 & -p(z) \\ 0 & 1 \end{bmatrix} = \mathbf{I}$$

$$\Rightarrow \mathbf{E}^p(z)\mathbf{R}^p(z) = \mathbf{I} \tag{24.102}$$

where $\mathbf{E}^p(z)$ is the new analysis polyphase matrix by primal lifting as in Equation 24.99 and $\mathbf{R}^p(z)$ is the new synthesis polyphase matrix by inverse primal lifting:

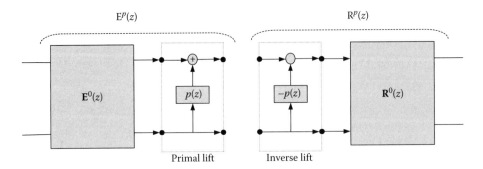

FIGURE 24.20   Primal lifting.

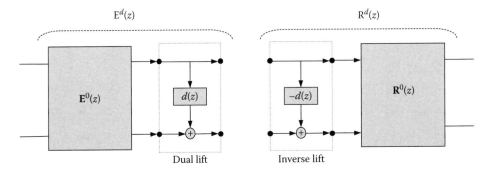

FIGURE 24.21   Dual lifting.

$$\mathbf{R}^p(z) \equiv \mathbf{R}^0(z) \begin{bmatrix} 1 & -p(z) \\ 0 & 1 \end{bmatrix} \tag{24.103}$$

With the combination of the primal/inverse-primal lifting, PR is preserved and this can also been easily seen in the structure in Figure 24.20: whatever was added in the analysis side has been exactly subtracted in the synthesis side.

   Wherever the dual lifting is applied in the analysis side, as in Equation 24.100, then the inverse dual lifting should be applied in synthesis side:

$$\mathbf{R}^d(z) \equiv \mathbf{R}^0(z) \begin{bmatrix} 1 & 0 \\ -d(z) & 1 \end{bmatrix} \tag{24.104}$$

with the corresponding structure shown in Figure 24.21. Therefore, if the analysis polyphase matrix is as shown in Equation 24.101, the corresponding synthesis polyphase matrix is given by

$$
\begin{aligned}
\mathbf{R}(z) &= \begin{bmatrix} 1 & -p_1(z) \\ 0 & 1 \end{bmatrix} \begin{bmatrix} 1 & 0 \\ -d_1(z) & 1 \end{bmatrix} \cdots \begin{bmatrix} 1 & -p_M(z) \\ 0 & 1 \end{bmatrix} \begin{bmatrix} 1 & 0 \\ -d_M(z) & 1 \end{bmatrix} \begin{bmatrix} 1/K_0 & 0 \\ 0 & 1/K_1 \end{bmatrix} \\
&= \prod_{i=1}^{M} \begin{bmatrix} 1 & -p_i(z) \\ 0 & 1 \end{bmatrix} \begin{bmatrix} 1 & 0 \\ -d_i(z) & 1 \end{bmatrix} \cdot \begin{bmatrix} 1/K_0 & 0 \\ 0 & 1/K_1 \end{bmatrix}
\end{aligned} \tag{24.105}
$$

24.8.4.2 Implementation of Filter Banks

The lifting scheme is also useful for the efficient implementation of filter banks. The analysis or synthesis polyphase matrix of a PR filter bank can be factorized into a product of lifting matrices. The factorization can be achieved through the use of the Euclidean algorithm for Laurent series which is described next.

Given two Laurent series:

$$a(z) = \sum_{n=-L_a}^{M_a} a(n)z^{-n} \quad \text{and} \quad b(z) = \sum_{n=-L_b}^{M_b} b(n)z^{-n}$$

with degrees $|a(z)|$ and $|b(z)|$ defined as

$$|a(z)| = M_a + L_a \quad \text{and} \quad |b(z)| = M_b + L_b$$

The degree of a Laurent series in z is the length of the equivalent FIR filter minus one.

Without loss of generality, we assume $|a(z)| \geq |b(z)|$. Dividing $a(z)$ by $b(z)$ will yield a quotient $q(z)$ and remainder $r(z)$, such that

$$a(z) = b(z)q(z) + r(z) \quad \text{where} \quad |q(z)| = |a(z)| - |b(z)| \text{ and } |r(z)| < |b(z)| \tag{24.106}$$

In Matlab, this operation may be performed using the function $[q, r] = \text{deconv}(a, b)$.

Unlike the case of regular polynomials (with only positive powers of z), the quotient and remainder in Equation 24.106 are not unique and there is some freedom in choosing $q(z)$, depending on how $r(z)$ is aligned within the range of the indices of $a(z)$. This means that in the following algorithm, there are multiple valid factorizations of a given pair of filter polynomials. The Matlab function, deconv($\cdot$), chooses to align $r(z)$ with the right-hand end of $a(z)$, assuming row vectors are used for the polynomial coefficients.

The Euclidean algorithm iteratively performs the division process on two Laurent series, $a(z)$ and $b(z)$, as follows:

$$[a_{i+1}(z) \quad b_{i+1}(z)] = [a_i(z) \quad b_i(z)] \begin{bmatrix} 0 & 1 \\ 1 & -q_i(z) \end{bmatrix} \tag{24.107}$$

starting with $a_0(z) = a(z)$ and $b_0(z) = b(z)$. At each iteration, $i + 1$, the higher degree series a_i is divided by the lower degree series b_i and the remainder is recorded as b_{i+1} for the next iteration, while b_i becomes a_{i+1}. The process continues until iteration n when the remainder b_n becomes zero. The whole process can be expressed as

$$[a_n(z) \quad 0] = [a(z) \quad b(z)] \prod_{i=1}^{n} \begin{bmatrix} 0 & 1 \\ 1 & -q_i(z) \end{bmatrix} \tag{24.108}$$

where $a_n(z)$ is the greatest common factor (gcf) of $a(z)$ and $b(z)$, i.e., $a_n(z) = \text{gcf}(a(z), b(z))$.

Inverting the terms in the product gives

$$[a(z) \quad b(z)] = [a_n(z) \quad 0] \prod_{i=n}^{1} \begin{bmatrix} q_i(z) & 1 \\ 1 & 0 \end{bmatrix}$$

$$= [a_n(z) \quad 0] \prod_{i=n/2}^{1} \begin{bmatrix} q_{2i}(z) & 1 \\ 1 & 0 \end{bmatrix} \begin{bmatrix} q_{2i-1}(z) & 1 \\ 1 & 0 \end{bmatrix} \tag{24.109}$$

where n is assumed to be even and the product is grouped into pairs of even and odd terms. If n is odd, it is increased by 1 and we set $q_n(z) = 0$.

Each term in the product can be expressed in two forms that involve a primal or a dual lifting matrix

$$\begin{bmatrix} q_i(z) & 1 \\ 1 & 0 \end{bmatrix} = \begin{bmatrix} 1 & q_i(z) \\ 0 & 1 \end{bmatrix} \begin{bmatrix} 0 & 1 \\ 1 & 0 \end{bmatrix} = \begin{bmatrix} 0 & 1 \\ 1 & 0 \end{bmatrix} \begin{bmatrix} 1 & 0 \\ q_i(z) & 1 \end{bmatrix}$$

Using the first form for even terms in Equation 24.109 and the second form for odd terms gives

$$[a(z) \quad b(z)] = [a_n(z) \quad 0] \prod_{i=n/2}^{1} \begin{bmatrix} 1 & q_{2i}(z) \\ 0 & 1 \end{bmatrix} \begin{bmatrix} 1 & 0 \\ q_{2i-1}(z) & 1 \end{bmatrix} \qquad (24.110)$$

For the special case when $\gcf(a(z), b(z))$ is a constant (i.e., there is no common factor), the following factorization in terms of lifting steps exists:

$$[a(z) \quad b(z)] = [K \quad 0] \prod_{i=n/2}^{1} \begin{bmatrix} 1 & q_{2i}(z) \\ 0 & 1 \end{bmatrix} \begin{bmatrix} 1 & 0 \\ q_{2i-1}(z) & 1 \end{bmatrix} \cdot \qquad (24.111)$$

Now suppose there is an analysis polyphase matrix, given by

$$\mathbf{E}(z) \equiv \begin{bmatrix} E_{0,0}(z) & E_{0,1}(z) \\ E_{1,0}(z) & E_{1,1}(z) \end{bmatrix}$$

and satisfying

$$\det(\mathbf{E}(z)) \equiv E_{0,0}(z)E_{1,1}(z) - E_{0,1}(z)E_{1,0}(z) = 1 \qquad (24.112)$$

i.e., Equation 24.93 with $c = 1$ and $K = 0$ (no delay).* Equation 24.112 implies that $\gcf(E_{0,0}(z), E_{0,1}(z))$ is a constant or else the determinant cannot be constant.

Using the Euclidean algorithm described above with $a(z) = E_{0,0}(z)$ and $b(z) = E_{0,1}(z)$ gives

$$[E_{0,0}(z) \quad E_{0,1}(z)] = [K \quad 0] \prod_{i=n/2}^{1} \begin{bmatrix} 1 & q_{2i}(z) \\ 0 & 1 \end{bmatrix} \begin{bmatrix} 1 & 0 \\ q_{2i-1}(z) & 1 \end{bmatrix}$$

Form the following matrix

$$\mathbf{E}^0(z) = \begin{bmatrix} E_{0,0}(z) & E_{0,1}(z) \\ E_{1,0}^0(z) & E_{1,1}^0(z) \end{bmatrix} = \begin{bmatrix} K & 0 \\ 0 & 1/K \end{bmatrix} \prod_{i=n/2}^{1} \begin{bmatrix} 1 & q_{2i}(z) \\ 0 & 1 \end{bmatrix} \begin{bmatrix} 1 & 0 \\ q_{2i-1}(z) & 1 \end{bmatrix}$$

The matrix $\mathbf{E}^0(z)$ has unit determinant ($\det(\mathbf{E}^0(z)) = 1$), i.e. PR, and the corresponding low-pass filter is the same as that in matrix $\mathbf{E}(z)$ under consideration. As explained earlier, the dual lifting (Equation 24.100) can be used to change the high-pass filter while leaving the low-pass filter unchanged. Therefore

---

* There is no loss of generality here as any general matrix satisfying Equation 24.93 can be made into a matrix with unit determinant by muliplying the first (second) row with z^{-K}/c. This is equivalent to the introducing a scaling factor of $1/c$ and the delay z^{-2K} in the low-pass (high-pass) filter which does not significantly change its filtering properties.

by applying an appropriate dual lifting step, $E^0(z)$ can be changed to any valid $E(z)$ which produces PR, using

$$E(z) = \begin{bmatrix} 1 & 0 \\ d(z) & 1 \end{bmatrix} E^0(z)$$

$$= \begin{bmatrix} 1 & 0 \\ d(z) & 1 \end{bmatrix} \begin{bmatrix} K & 0 \\ 0 & 1/K \end{bmatrix} \prod_{i=n/2}^{1} \begin{bmatrix} 1 & q_{2i}(z) \\ 0 & 1 \end{bmatrix} \begin{bmatrix} 1 & 0 \\ q_{2i-1}(z) & 1 \end{bmatrix}$$

$$= \begin{bmatrix} K & 0 \\ 0 & 1/K \end{bmatrix} \begin{bmatrix} 1 & 0 \\ K^2 d(z) & 1 \end{bmatrix} \prod_{i=n/2}^{1} \begin{bmatrix} 1 & q_{2i}(z) \\ 0 & 1 \end{bmatrix} \begin{bmatrix} 1 & 0 \\ q_{2i-1}(z) & 1 \end{bmatrix} \qquad (24.113)$$

Note that, if necessary, the scaling matrix can also be factorized into a sequence of lifting steps (but this is not usually helpful computationally):

$$\begin{bmatrix} K & 0 \\ 0 & 1/K \end{bmatrix} = \begin{bmatrix} 1 & K - K^2 \\ 0 & 1 \end{bmatrix} \begin{bmatrix} 1 & 0 \\ -1/K & 1 \end{bmatrix} \begin{bmatrix} 1 & K - 1 \\ 0 & 1 \end{bmatrix} \begin{bmatrix} 1 & 0 \\ 1 & 1 \end{bmatrix} \qquad (24.114)$$

Therefore it can be concluded that every perfect reconstructing polyphase matrix can be factorized into a sequence of lifting steps.

Example 24.1

The analysis polyphase matrix corresponding to the Haar wavelet filters and its factorization is given by

$$E(z) = \begin{bmatrix} \frac{1}{2} & \frac{1}{2} \\ -\frac{1}{2} & \frac{1}{2} \end{bmatrix} = \begin{bmatrix} \frac{1}{2} & 0 \\ 0 & 1 \end{bmatrix} \begin{bmatrix} 1 & 0 \\ -\frac{1}{2} & 1 \end{bmatrix} \begin{bmatrix} 1 & 1 \\ 0 & 1 \end{bmatrix}$$

which consist of one primal lift, one dual lift, and one scaling.

Example 24.2

The analysis polyphase matrix corresponding to the LeGall 3/5 filters and its factorization is given by

$$E(z) = \begin{bmatrix} \frac{1}{2} & \frac{1}{4}(1+z) \\ -\frac{1}{4}(1+z^{-1}) & \frac{3}{4} - \frac{1}{8}(z^{-1}+z) \end{bmatrix} = \begin{bmatrix} \frac{1}{2} & 0 \\ 0 & 1 \end{bmatrix} \begin{bmatrix} 1 & 0 \\ -\frac{1}{4}(1+z^{-1}) & 1 \end{bmatrix} \begin{bmatrix} 1 & \frac{1}{2}(1+z) \\ 0 & 1 \end{bmatrix}$$

which consist of one primal lift, one dual lift, and one scaling.

24.9 Nonlinear Filter Banks

In general it is difficult to achieve PR if the linear filters H_0, H_1, G_0, and G_1 are replaced with nonlinear filters. Furthermore, there is no generic framework which can be used to analyze the properties of filter banks with nonlinear filters. The lifting scheme described earlier, which works in the polyphase domain, however provides a convenient way to generalize linear filter banks to nonlinear ones. If the primal (dual) lifting filter $p(z)(d(z))$, which is linear, is replaced with a nonlinear filter, PR will still be

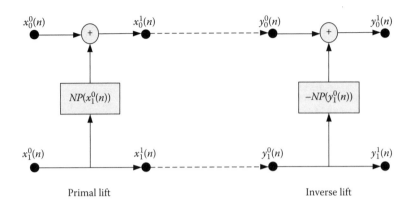

FIGURE 24.22   Nonlinear primal lifting.

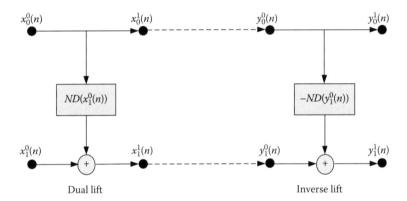

FIGURE 24.23   Nonlinear dual lifting.

maintained as can be easily seen in signal flow graph in Figures 24.22 and 24.23. Note that since the filter is nonlinear, the z-transform transfer function cannot (in general) be defined for the filter. The symbols $NP(\cdot)$ $(ND(\cdot))$ are used instead to denote the primal (dual) nonlinear operator on the signal in the time domain.

A particularly important type of nonlinear filter is the combination of a linear filter and a quantizer [20], such as

$$NP(x_1^0(n)) = \left\lfloor \sum_k p(k)x_1^0(n-k) + \frac{1}{2} \right\rfloor$$

$$ND(x_0^1(n)) = \left\lfloor \sum_k d(k)x_0^1(n-k) + \frac{1}{2} \right\rfloor$$

where $\lfloor x \rfloor$ denotes the largest integer that is no bigger than x, i.e., the floor operation, and $p(k)$ $(d(k))$ are the coefficients of the linear primal (dual) filter. If the inputs $(x_0^0(n), x_1^0(n))$ are integer valued, then the outputs with the primal lifting are given by

$$x_0^1(n) = x_0^0(n) + \left\lfloor \sum_k p(k) x_1^0(n-k) + \frac{1}{2} \right\rfloor$$

$$x_1^1(n) = x_1^0(n)$$

will also be integer valued as the final addition operation involves integers. The same applies with the dual lifting and any combinations of primal and dual liftings. Any linear lifting-based filter bank can therefore be converted into a filter bank that maps integer inputs to integer outputs by using the floor operation to the outputs of the lifting filters and the transform is known as an integer wavelet transform (IWT). The IWT shares most properties of the parent (nonquantized) transform, as long as the floor operations do not contribute a significant level of distortion to the processed signals. This transform is particularly useful for implementing lossless compression schemes, since it retains its PR properties despite the rounding distortions of each primal or dual filter.

Some other nonlinear filters that can be useful in lifting stages are

Median filters (or more general order-statistic filters), which can provide robustness to noise and outliers in the data

Motion-compensated prediction filters, which can be used to improve the performance of video compression systems that employ wavelets along the time axis as well as in the two spatial directions.

24.10 *M*-Band Filter Banks

The 2-band filter bank can be generalized to an arbitrary number of channel as shown in Figure 24.24. The input signal is ideally split into M nonoverlapping contiguous frequency subbands with equal bandwidth. Since there is a factor of M reduction in the bandwidth, downsampling by a factor of M can be applied to preserve the data rate. M-band filter banks are most conveniently analyzed by using the polyphase representation [3,9,12].

The type 1 polyphase decomposition of the analysis filters is given by

$$H_i(z) = E_{i,0}(z^M) + z^{-1} E_{i,1}(z^M) + \cdots + z^{-(M-1)} E_{i,M-1}(z^M) \tag{24.115}$$

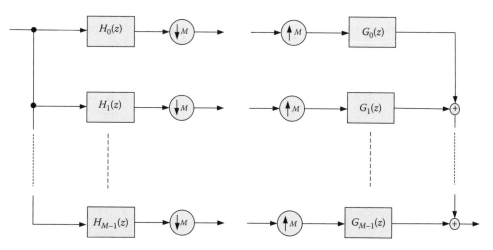

FIGURE 24.24   *M*-band filter bank.

for $i = 0, \ldots, M - 1$; and the corresponding analysis polyphase matrix of size $M \times M$ is

$$\mathbf{E}(z) = \begin{bmatrix} E_{0,0}(z) & E_{0,1}(z) & \cdots & E_{0,M-1}(z) \\ E_{1,0}(z) & E_{1,1}(z) & \cdots & E_{1,M-1}(z) \\ \vdots & \vdots & \ddots & \vdots \\ E_{M-1,0}(z) & E_{M-1,1}(z) & \cdots & E_{M-1,M-1}(z) \end{bmatrix} \tag{24.116}$$

The type 2 polyphase decomposition of the synthesis filters is given by

$$G_i(z) = z^{-(M-1)} R_{i,0}(z^M) + z^{-(M-2)} R_{i,1}(z^M) + \cdots + R_{i,M-1}(z^M) \tag{24.117}$$

for $i = 0, \ldots, M - 1$; and the corresponding synthesis polyphase matrix of size $M \times M$ is

$$\mathbf{R}(z) = \begin{bmatrix} R_{0,0}(z) & R_{1,0}(z) & \cdots & R_{M-1,0}(z) \\ R_{0,1}(z) & R_{1,1}(z) & \cdots & R_{1,M-1}(z) \\ \vdots & \vdots & \ddots & \vdots \\ R_{0,M-1}(z) & R_{1,M-1}(z) & \cdots & R_{M-1,M-1}(z) \end{bmatrix} \tag{24.118}$$

Figure 24.25 shows the equivalent representation of the M-band filter bank in polyphase form. Just as in the 2-band case, the total polyphase matrix is defined to be $\mathbf{P}(z) \equiv \mathbf{R}(z)\mathbf{E}(z)$ and the condition for PR is

$$\mathbf{P}(z) = \mathbf{R}(z)\mathbf{E}(z) = \mathbf{I} \tag{24.119}$$

For an FIR filter bank it is required that

$$\det(\mathbf{E}(z)) = cz^K \tag{24.120}$$

where c is a nonzero constant and K an integer.

As in the 2-band case, a product (cascade) of matrices satisfying Equation 24.120 will give a matrix that also satisfies Equation 24.120. Lifting can also be defined in the M-band setting but instead of just the

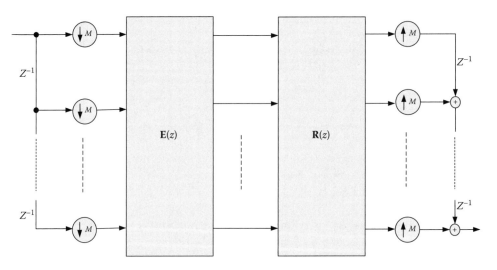

FIGURE 24.25   Polyphase form of the M-band filter bank.

primal and dual liftings, there are in general $M(M - 1)$ types of lifting as there is a choice M bands that can be used to modify the signal in the other $M - 1$ bands.

Special types of M-band filter banks are

1. Discrete fourier transform (DFT). Although the DFT is usually considered a transform that works on a block of data, it can also be viewed as an M-band filter bank. The analysis polyphase matrix in this case is the DFT matrix with elements given by

$$E_{i,k}(z) = \frac{1}{\sqrt{M}} W_M^{ik} \quad \text{where } W_M = \exp\left(-\frac{j2\pi}{M}\right) \quad \text{for } i, k = 0 \dots M - 1 \tag{24.121}$$

2. Discrete cosine transform (DCT). This can also be viewed as an M-band filter bank. The analysis polyphase matrix in this case is the DCT matrix with elements given by

$$E_{0,k}(z) = \frac{1}{\sqrt{M}} \qquad\qquad \text{for } k = 0 \dots M - 1$$

$$E_{i,k}(z) = \sqrt{\frac{2}{M}} \cos\left(\frac{(2k+1)i\pi}{2M}\right) \quad \text{for } \begin{cases} i = 1 \dots M - 1 \\ k = 0 \dots M - 1 \end{cases} \tag{24.122}$$

3. Cosine modulated filter bank. The idea here is to design a single prototype low-pass filter for the low-pass subband and to use modulation on the prototype filter to obtain the other filters in the filter bank. The analysis/synthesis filter coefficients are given by

$$h_i(n) = g_i(n) = p(n)\sqrt{\frac{2}{M}} \cos\left(\frac{(2i+1)(2n+M+1)\pi}{4M}\right) \quad \text{for } i = 0 \dots M - 1 \tag{24.123}$$

The prototype filter is $p(n)$ and if it is of length $2M$ and satisfies

$$p(n) = p(2M - n) \quad \text{and} \quad p^2(n) + p^2(n + M) = 1 \tag{24.124}$$

then PR is achieved.

24.11 Hilbert Pairs of Filter Banks (the Dual Tree)

In this final section on 1-D multirate filter banks, we return to the case of the 2-band system, formed into a wavelet tree of filters, as shown in Figure 24.3. While the wavelet transform, implemented in this form, has been found to be remarkably effective as a tool for signal compression (coding) algorithms, it suffers from a rather serious drawback when used as a signal analysis tool. This problem is known as *shift dependence*, or a lack of shift invariance. It is caused by the 2:1 downsampling operations, shown in Figure 24.3, and manifests itself as aliased signal components which are generated by the downsamplers at each stage of the filter tree.

Although these aliased components do indeed cancel each other out when all the subbands are recombined in the reconstruction filter tree of the inverse wavelet transform (provided that the anti-aliasing condition of Equation 24.41 is satisfied), the aliased components are still very significant in the wavelet coefficients (subband outputs) themselves and tend to result in poor performance if the wavelet transform is used as a front-end for more complicated tasks such as signal classification and recognition, or for denoising.

One way to eliminate the aliasing distortion in each subband and obtain *shift invariance* is to use the modified filter tree of Figure 24.4, but without any of the downsamplers. This is known as the *à trous* algorithm of Mallat [21]. However, for a K-level tree, it results in $(K + 1)$ times as many wavelet coefficients being generated and also requires a similar increase in computation load to implement the filters, when compared with the standard nonredundant tree of Figure 24.3 (shown for $K = 4$). These factors become significantly worse for higher-dimensional signals such as images or 3-D datasets.

We now consider a more efficient alternative to the *à trous* algorithm.

24.11.1 Dual-Tree Wavelet Transform

The dual-tree wavelet transform [22] achieves approximate shift invariance with only 2:1 redundancy of coefficients and twofold increase in computation load, relative to the nonredundant tree. It simply employs a second tree of wavelet filters in parallel with the first tree, as shown in Figure 24.26.

The key to obtaining shift invariance from the dual-tree structure lies in designing the relative filter delays at each stage, such that the low-pass filter outputs in tree b are effectively sampled at points midway between the sampling points of the equivalent filters in tree a. This requires a delay difference between the a and b low-pass filters of 1 sample period at tree level 1, and of $\frac{1}{2}$ sample period at subsequent levels (because at these later levels half of the required delay has already been provided by earlier stages). At level 1 we can use any standard orthogonal or biorthogonal wavelet filters and produce the required delay shift trivially by insertion or deletion of unit delays. However, at subsequent levels, the $\frac{1}{2}$ sample delay difference is more difficult to achieve. In Ref. [23], Selesnick showed that these low-pass delay constraints produce a Hilbert pair relationship between the wavelet bases for the two trees. This leads naturally to the interpretation of the outputs from trees a and b as the real and imaginary parts, respectively, of complex wavelet coefficients. The two filter banks are then known as a Hilbert pair.

The $\pm\frac{1}{2}$ sample delay difference between filters $H_{0a}(z)$ and $H_{0b}(z)$ is difficult to express in the time domain as the filters are discrete, but it can be easily expressed in the frequency domain as

$$H_{0b}(e^{j\omega}) \simeq H_{0a}(e^{j\omega})e^{\pm j\omega/2} \tag{24.125}$$

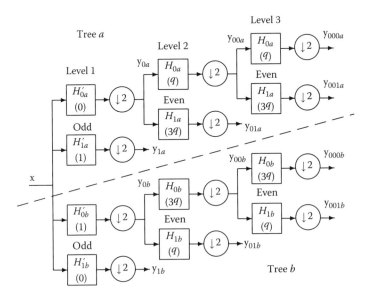

FIGURE 24.26 Dual-tree complex wavelet transform over three levels. The numbers in brackets indicate the delay of each filter (q = one quarter of a sample).

The approximation is necessary here assuming finite-order filters are being used. Solutions are possible if the phase responses are exact and the magnitude responses are approximate, or if the magnitude responses are exact and the phase responses are approximate. In the former case, odd-length and even-length linear-phase filters can be used for H_{0a} and H_{0b}, but better overall characteristics can usually be achieved with the latter case, which allows both filter banks to be CQFs, satisfying Equations 24.42, 24.45, and 24.53.

By convention, we select tree b to lag relative to tree a, so the polarity of the final exponent term in Equation 24.125 is selected to be negative.

24.11.2 Common-Factor Dual-Tree Filter Design

Selesnick [24] proposed a method of satisfying Equation 24.125 by letting

$$H_{0a}(z) = F(z)D(z) \quad \text{and} \quad H_{0b}(z) = F(z)z^{-L}D(z^{-1}) \tag{24.126}$$

where

$$A(z) = \frac{z^{-L}D(z^{-1})}{D(z)} \tag{24.127}$$

is an all pass filter, with group delay designed to approximate the required $\frac{1}{2}$ sample delay difference. Hence we require that

$$A(e^{j\omega}) \simeq e^{-j\omega/2} \tag{24.128}$$

Once $D(z)$ has been selected to satisfy Equations 24.127 and 24.128, we can design $F(z)$ such that both H_{0a} and H_{0b} result in CQFs. This is possible because the product filter $P(z)$ is the same in both cases, since

$$P_a(z) = H_{0a}(z)H_{0a}(z^{-1})$$
$$= F(z)F(z^{-1})D(z)D(z^{-1})$$
$$= H_{0b}(z)H_{0b}(z^{-1}) = P_b(z) \tag{24.129}$$

Hence $F(z)$ can be designed to contain the required number of zeros at $z = -1$ (to give the desired number of VM) and a factor $R(z)$ of sufficient order to eliminate all the nonzero even powers of z in $P(z)$. This is very similar to the method of choosing $R(z)$ used in Section 24.6.4.

There is a straightforward way to select $D(z)$ of a given order L using Thiran's filter:

$$D(z) = 1 + \sum_{n=1}^{L} \binom{L}{n} \left[\prod_{k=0}^{n-1} \frac{\tau - L + k}{\tau + 1 + k} \right] (-z)^{-n} \tag{24.130}$$

where $\tau = 0.5$ to get $\angle A(e^{j\omega}) \simeq \omega/2$. A range of filters designed with this method are given in Ref. [24].

The one disadvantage of this method of designing a Hilbert pair of filter banks is that the resulting complex basis functions will not be linear phase (i.e., conjugate-symmetric about their midpoint). To achieve this, $F(z)$ must be linear phase, which requires a biorthogonal factorization of $P(z)$ instead of the orthogonal CQF factorization, assumed above. The penalty then is that the reconstruction wavelets will differ from the analysis wavelets.

24.11.3 Q-Shift Dual-Tree Filter Design

A suitable way to employ CQFs and to obtain linear phase complex basis functions from the dual tree was proposed in Ref. [25]. The key here is to make the tree b filters equal to the time-reverse of the tree a filters. Hence

$$H_{0b}(z) = z^{-L}H_{0a}(z^{-1}) \tag{24.131}$$

The integer delay L is normally chosen so that H_{0a} and H_{0b} cover the same range of z indices.
The filters have real coefficients, so Equation 24.131 automatically ensures that

$$|H_{0b}(e^{j\omega})| = |H_{0a}(e^{j\omega})| \quad \text{and} \quad \angle H_{0b}(e^{j\omega}) = -L\omega - \angle H_{0a}(e^{j\omega})$$

It is also required, from Equation 24.125 (11.1), that $\angle H_{0b}(e^{j\omega}) - \angle H_{0a}(e^{j\omega}) \simeq -\frac{\omega}{2}$, thus giving

$$\angle H_{0a}(e^{j\omega}) \simeq -\frac{L\omega}{2} + \frac{\omega}{4} \quad \text{and} \quad \angle H_{0b}(e^{j\omega}) \simeq -\frac{L\omega}{2} - \frac{\omega}{4} \tag{24.132}$$

This shows that H_{0a} and H_{0b} must have delays that approximate $\pm\frac{1}{4}$ of a sample period relative to the midpoint of the filters at $L/2$ sample periods. Hence this type of 2-band filter system is called a Q-shift (quarter-shift) filter pair. A neat way to visualize the delay property of the Q-shift pair is shown in Figure 24.27, in which the filter taps of the two filters are interleaved to form a single smooth low-pass filter of even length and with symmetric filter coefficients (linear phase), given by

$$H_{L2}(z) = H_{0a}(z^2) + zH_{0b}(z^2) \tag{24.133}$$

The sample rate of the new filter H_{L2} is double that of H_{0a} and H_{0b} and its delay is therefore $(L - \frac{1}{2})$ sample periods. (In Figure 24.27, $L = 1$.)

As well as possessing the approximate Q-shift delay property, H_{0a} and H_{0b} must also satisfy the usual 2-band filter bank constraints of no aliasing and PR. Additionally it is advantageous in many emerging applications of complex wavelets that the Q-shift filters should be orthogonal (i.e., be CQFs) to make the transform a tight frame which conserves energy from the signal in the transform domain. This has the further advantage that the same filters (and their time-reverses) can be used in the forward and inverse transforms, as well as in the two trees a and b. Finally it is usually highly desirable that the filters have good smoothness properties when iterated over scale.

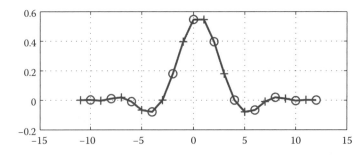

FIGURE 24.27 Impulse response of $H_{L2}(z)$ for $n = 6$. The H_{0a} and H_{0b} filters have $2n = 12$ taps each, shown as circles and crosses, respectively.

Two main design approaches have emerged to meet all of these constraints for Q-shift filter pairs:

Zero-forcing methods with search for good delay properties: The Daubechies orthonormal filters, described in Section 24.6.4, have the maximal number of zeros in the filters H_0 and G_0 at $z = -1$ and there are no further degrees of freedom. These are not able to produce filter pairs with good approximations to Q-shift delays. However the Bernstein method of equation (Equation 24.59) can produce much better approximations if $N = 2K + 1$, which allows one degree of freedom (the parameter α_K) that can be adjusted to optimize the Q-shift delays.

In Ref. [5], all even-length CQFs were evaluated from length 4 to 22. It was found that good approximations to the required Q-shift delay could be obtained for CQF lengths of 8, 12, 18, and 22. The optimal values of α_K in each case were found to be 0.0460, 0.1827, 0.2405, and 0.0245, respectively.

However, despite its good delay properties, the length-8 CQF, with 3 zeros at $z = -1$, was much less smooth than the longer filters, with 5, 8, and 10 zeros at $z = -1$, respectively. Figure 24.28 shows the impulse and frequency responses at level-4 of the DT-CWT, using length-12 Q-shift CQFs, designed with this method. Designs for other filter lengths are shown in Ref. [5]. In all the results of this section, the standard Daubechies 7,9-tap (Antonini) filters were used at level 1 of the dual tree (see Figure 24.10).

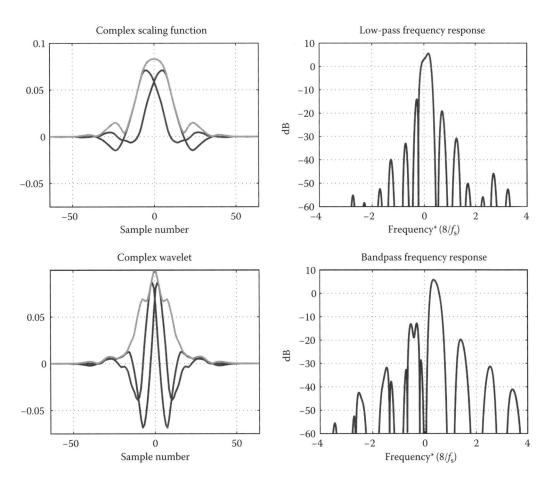

FIGURE 24.28 Impulse and frequency responses at level-4 of the DT-CWT, using length-12 Q-shift CQFs, designed with the zero-forcing method of Ref. [5]. In the impulse responses the two darker curves show the real and imaginary parts of the impulse responses, and the light gray curves show their magnitudes.

Iterative energy minimization methods: The other main design approach for Q-shift filter pairs has been presented in Ref. [26]. The aim is to design the filter $H_{0a}(z)$ (and hence also its time-reverse $H_{0b}(z)$) such that it satisfies the standard CQF conditions and also gives a smooth low-pass filter when formed into $H_{L2}(z)$ using Equation 24.133, which automatically produces the approximate $\frac{1}{4}$-sample (Q-shift) delay property if H_{L2} is smooth enough.

In Ref. [26] it is shown that the PR property of CQFs can be obtained if the product filter $H_{L2}(z)H_{L2}(z^{-1})$ has the property that the term in z^0 is unity and all terms in z^{4k} (k integer) are zero. Smoothness of the filter $H_{L2}(z)$ can be formulated as a need for all components of its spectrum to be approximately zero above about one-sixth of its sample rate ($\omega > \pi/3$). So the ideal design conditions for the length $4n$ symmetric low-pass filter H_{L2} have now been reduced to

1. Zero amplitude for all the terms of $H_{L2}(z)H_{L2}(z^{-1})$ in z^{4k} except the term in z^0, which must be unity
2. Zero (or near-zero) amplitude of $H_{L2}(e^{j\omega})$ for the stopband, $\frac{\pi}{3} \leq \omega \leq \pi$

Condition 1 is a set of quadratic constraints on the elements of the filter tap vector $\mathbf{h}_{L2}$, while condition 2 is a set of linear constraints on $\mathbf{h}_{L2}$, evaluated at a sufficiently fine set of frequencies covering the stopband. Together they form an overdetermined set of equations for the $2n$ unknowns that form one-half of the symmetric vector $\mathbf{h}_{L2}$ (where $2n$ is the length of filters H_{0a} and H_{0b}). If the constraints were all linear, the least mean square (LMS) error solution could be found in the standard way using the pseudo-inverse of the matrix which defines the equations.

To deal with the quadratic constraints, the problem is linearized using an iterative solution. If $\mathbf{h}_{L2}$ at iteration i is given by $\mathbf{h}_i = \mathbf{h}_{i-1} + \Delta\mathbf{h}_i$, then, since convolution (*) is commutative,

$$\begin{aligned}
\mathbf{h}_i{}^*\mathbf{h}_i &= (\mathbf{h}_{i-1} + \Delta\mathbf{h}_i)*(\mathbf{h}_{i-1} + \Delta\mathbf{h}_i) \\
&= \mathbf{h}_{i-1}*(\mathbf{h}_{i-1} + 2\Delta\mathbf{h}_i) + \Delta\mathbf{h}_i*\Delta\mathbf{h}_i
\end{aligned} \tag{24.134}$$

If the incremental update $\Delta\mathbf{h}_i$ is assumed to become small as i increases, the final term can be neglected and the convolution be expressed as a linear function of $\Delta\mathbf{h}_i$.

Hence the design problem can now be expressed as

$$\text{Solve for } \Delta\mathbf{h}_i \text{ such that:} \quad \mathbf{C}(\mathbf{h}_{i-1} + 2\Delta\mathbf{h}_i) = [\,0\ldots0\quad 1\,]^{\mathrm{T}} \tag{24.135}$$

$$\mathbf{F}(\mathbf{h}_{i-1} + \Delta\mathbf{h}_i) \simeq [\,0\ldots0\,]^{\mathrm{T}} \tag{24.136}$$

where

$\mathbf{C}$ is a matrix which calculates every fourth term in the convolution with $\mathbf{h}_{i-1}$

$\mathbf{F}$ is a matrix which evaluates the Fourier transform at M discrete frequencies ω from $\frac{\pi}{3}$ to π (typically $M \simeq 8n$ to ensure that all sidelobe maxima and minima are captured reasonably accurately)

Note that only one side of the convolution is needed in $\mathbf{C}$, since the result is symmetric about the central term. Also, the columns of $\mathbf{C}$ and $\mathbf{F}$ can be combined in pairs so that only the first half of the symmetric $\Delta\mathbf{h}_i$ need be solved for.

In typical applications of complex wavelets, it is often more important to ensure high accuracy in the PR condition than to produce highly smooth wavelets. This is why Equation 24.135 is shown as an equality while Equation 24.136 is only an approximation. Within an iterative LMS framework, high accuracy solutions to some equations can be produced by scaling these up by a factor, β_i, which is progressively increased with i. Hence the optimization may now be expressed as the iterative LMS solution of

$$\begin{bmatrix} 2\beta_i\mathbf{C} \\ \mathbf{F} \end{bmatrix} \Delta\mathbf{h}_i = \begin{bmatrix} \beta_i(\mathbf{c} - \mathbf{C}\,\mathbf{h}_{i-1}) \\ -\mathbf{F}\,\mathbf{h}_{i-1} \end{bmatrix} \tag{24.137}$$

$$\mathbf{h}_i = \mathbf{h}_{i-1} + \Delta\mathbf{h}_i \tag{24.138}$$

where $\mathbf{c} = [0\ldots01]^\mathrm{T}$. Typically one may choose $\beta_i = 2^i$, and iterate over about 20 iterations, so that reconstruction errors become of the order of $2^{-20} \simeq 10^{-6}$ of smoothness errors. There are two further refinements to this method, described in Ref. [26], which allow for transition band effects and can insert predefined zeros (e.g., VM) in the filters.

Figures 24.29 and 24.30 show some typical scaling functions and wavelets designed using the above method for filter lengths of 14 and 24, respectively. Note the high degree of smoothness that is achievable if the length of the filters is increased to 24. The frequency responses in Figures 24.28 through 24.30, with vertical scales in decibels, clearly show the much lower sidelobe levels and greater ability to reject negative frequencies, obtainable with longer filters. Many other intermediate results are possible with this energy minimization design method, and the method works satisfactorily for filter lengths up to 50 or more, which can be useful if sharp transition bands and/or very low sidelobe levels are required, such as for high-quality audio applications.

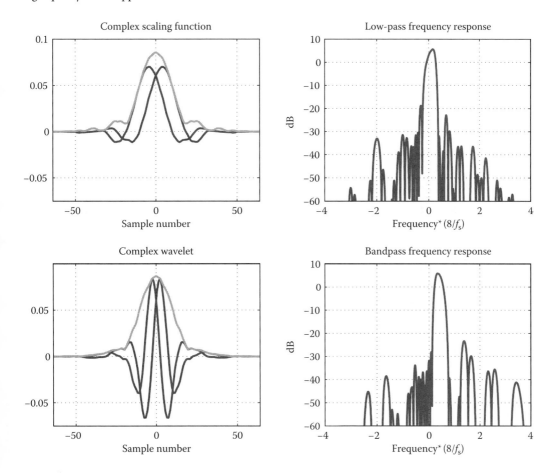

FIGURE 24.29 Impulse and frequency responses at level-4 of the DT-CWT, using length-14 Q-shift CQFs, designed with the energy minimization method of Kingsbury, N. G., Design of Q-shift complex wavelets for image processing using frequency domain energy minimization. In *Proc. IEEE Int. Conf. Image Processing*, Barcelona, September 2003, pp. 1013–1016.

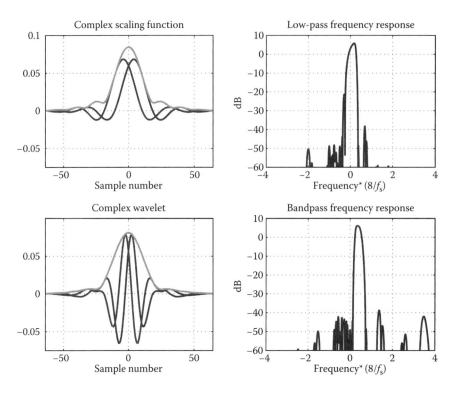

FIGURE 24.30 Impulse and frequency responses at level-4 of the DT-CWT, using length-24 Q-shift CQFs, designed with the energy minimization method of Kingsbury, N. G., Design of Q-shift complex wavelets for image processing using frequency domain energy minimization. In *Proc. IEEE Int. Conf. Image Processing*, Barcelona, September 2003, pp. 1013–1016.

Earlier papers on the dual-tree method [25,27] proposed some more heuristic methods for designing Q-shift filters. Most notable of these results are a very simple length-6 filter with fair performance and a length-14 filter with generally good performance (similar to Figure 24.29).

24.11.4 Metrics for Shift Dependence of a Filter Bank

Probably the most important feature of Hilbert pairs of wavelets, in addition to those of normal wavelet decompositions, is their approximate *shift invariance* or, expressed another way, their low level of *shift dependence*. This is discussed in detail in Ref. [27] and here we summarize the main technique from that paper for measuring shift dependence.

In order to examine the shift invariant properties of the dual tree, consider what happens when we choose to retain the coefficients of just one type (wavelet or scaling function) from just one level of the dual tree of Figure 24.26. For example, we might choose to retain only the level-3 wavelet coefficients y_{001a} and y_{001b}, and set all others to zero. If the signal $\hat{x}$, reconstructed from just these coefficients, is free of aliasing then we define the transform to be shift invariant at that level. This is because absence of aliasing implies that a given subband has a unique z-transfer function and so its impulse response is linear and time (shift) invariant. In this context we define a subband as comprising all coefficients from *both* trees at a given level and of a given type (either wavelet or scaling function).

Figure 24.31 shows the simplified analysis and reconstruction parts of the dual tree when coefficients of just one type and level are retained. All downsampling and upsampling operations are moved to the outputs of the analysis filter banks and the inputs of the reconstruction filter banks, respectively, using the transformation of Figure 24.4 on both trees, and the cascaded filter transfer functions are combined.

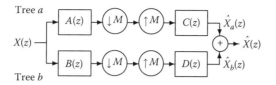

FIGURE 24.31 Basic configuration of the dual tree if either wavelet or scaling-function coefficients from just level *m* are retained ($M = 2^m$).

$M = 2^m$ is the total down/upsampling factor. For example if y_{001a} and y_{001b} from Figure 24.26 are the only sets of retained coefficients, then the subsampling factor $M = 8$, and $A(z) = H'_{0a}(z)H_{0a}(z^2)H_{1a}(z^4)$, the transfer function from x to y_{001a}. The transfer function $B(z)$ (from x to y_{001b}) is obtained similarly using $H_{...b}(z)$; as are the inverse functions $C(z)$ and $D(z)$ from $G_{...a}(z)$ and $G_{...b}(z)$, respectively.

It is a standard result of multirate analysis that a signal $U(z)$, which is downsampled by M and then upsampled by the same factor (by insertion of zeros), becomes $\frac{1}{M} \sum_{k=0}^{M-1} U(W^k z)$, where $W = e^{j2\pi/M}$. Applying this result to Figure 24.31 gives

$$\hat{X}(z) = \hat{X}_a(z) + \hat{X}_b(z) = \frac{1}{M} \sum_{k=0}^{M-1} X(W^k z)\left[A(W^k z)C(z) + B(W^k z)D(z)\right] \tag{24.139}$$

The aliasing terms in this summation correspond to those for which $k \neq 0$, because only the term in $X(z)$ (when $k = 0$ and $W^k = 1$) corresponds to a linear time (shift) invariant response. For shift invariance, the aliasing terms must be negligible, so we must design $A(W^k z)C(z)$ and $B(W^k z)D(z)$ either to be very small or to cancel each other when $k \neq 0$. Now W^k introduces a frequency shift equal to kf_s/M to the filters A and B (where f_s is the input sampling frequency), so for larger values of k the shifted and unshifted filters have negligible passband overlap and it is quite easy to design the functions $B(W^k z)D(z)$ and $A(W^k z)C(z)$ to be very small over all frequencies, $z = e^{j\theta}$. But at small values of k (especially $k = \pm 1$) this becomes virtually impossible due to the significant width of the transition bands of short-support filters. In these cases it is necessary to design for cancellation when the two trees are combined, and this is what is achieved by the half-sample delay difference strategy outlined above.

A useful way of quantifying the shift dependence of a transform is to examine Equation 24.139 and determine the ratio of the total energy of the unwanted aliasing transfer functions (the terms with $k \neq 0$) to the energy of the wanted transfer function (when $k = 0$), as given by the aliasing energy ratio:

$$R_a = \frac{\sum_{k=1}^{M-1} \mathcal{E}\{A(W^k z)C(z) + B(W^k z)D(z)\}}{\mathcal{E}\{A(z)C(z) + B(z)D(z)\}} \tag{24.140}$$

where $\mathcal{E}\{U(z)\}$ calculates the energy, $\sum_r |u_r|^2$, of the impulse response of a z-transfer function, $U(z) = \sum_r u_r z^{-r}$. Note that $\mathcal{E}\{U(z)\}$ may alternatively be calculated as the integral of the squared magnitude of the frequency response, $\frac{1}{2\pi} \int_{-\pi}^{\pi} |U(e^{j\theta})|^2 d\theta$ from Parseval's theorem. Since R_a is an energy ratio, it is convenient to measure it in decibels.

In Table 3 of Ref. [27], values of R_a are given for various single-tree and dual-tree filter combinations. It is shown that R_a is typically only $-3.5\,\text{dB}$ for a single-tree DWT, whereas it may be improved to approximately $-31\,\text{dB}$ with 18-tap Q-shift filters in the dual-tree complex wavelet transform (DT-CWT). Longer filters can reduce the aliasing energy much further if needed, while shorter filters tend to produce more aliasing energy and hence somewhat greater shift dependence.

Other metrics for shift dependence of filter-bank transforms can be based on the amount of shift-dependent variation in the impulse or step responses of the transform, as illustrated in Figure 8

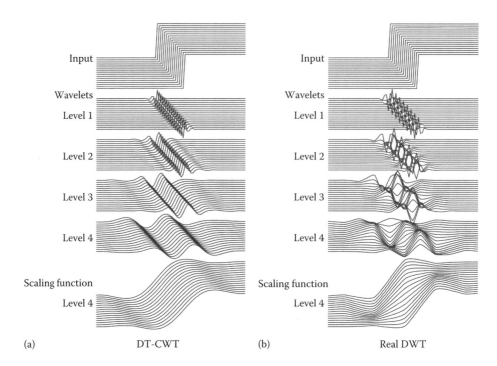

FIGURE 24.32 Wavelet and scaling-function components at levels 1–4 of 16 shifted step responses of the DT-CWT (a) with 18-tap Q-shift filters and the real DWT (b) with 13,19-tap filters.

of Ref. [27], part of which is reproduced here in Figure 24.32. Good transforms have responses that are visually almost identical (Figure 24.32a), whereas poor transforms (e.g., the DWT) have dramatically fluctuating responses (Figure 24.32b).

In this relatively short discussion of dual-tree/Hilbert pair ideas, we have not discussed the important extension of these ideas to two- and three-dimensional datasets, in which a second and very important advantage of the complex nature of the wavelet coefficients lies in their ability to produce strongly directionally selective wavelets while still retaining the computational advantages of separable filtering [22]. In addition, extension of the dual-tree ideas to M-band filter banks has been achieved by Chaux et al. [28], making even greater directional selectivity possible in two and three dimensions.

References

1. I. Daubechies. *Ten Lectures on Wavelets*. Society for Industrial and Applied Mathematics Philadelphia, PA, 1992.

2. M. Vetterli and J. Kovacevic. *Wavelets and Subband Coding*. Prentice-Hall, Englewood Cliffs, NJ, 1995.

3. G. Strang and T. Nguyen. *Wavelets and Filter Banks*. Wellesley-Cambridge Press, Wellesley, MA, 1996.

4. H. Caglar and A. N. Akansu. A generalized parametric PR-QMF design technique based on Bernstein polynomial approximation. *IEEE Trans. Signal Proc.*, 41(7):2314–2321, July 1993.

5. D. B. H. Tay, N. G. Kingsbury, and M. Palaniswami. Orthonormal Hilbert-pair of wavelets with (almost) maximum vanishing moments. *IEEE Signal Proc. Lett.* 13(9):533–536, September 2006.

6. M. Antonini, M. Barlaud, P. Mathieu, and I. Daubechies. Image coding using wavelet transform. *IEEE Trans. Image Proc.*, 1(2):205–220, April 1992.

7. P. A. Regalia, S. K. Mitra, and P. P. Vaidyanathan. The digital all-pass filter: A versatile signal processing building block. *Proc. IEEE*, 76(1):19–37, January 1988.

8. M. J. T. Smith and S. L. Eddins. Analysis/synthesis techniques for sub-band coding. *IEEE Trans. Acoust. Speech Signal Proc.*, 38(8):1446–1456, August 1990.

9. P. P. Vaidyanathan. Multirate digital filters, filter banks, polyphase networks and applications: A tutorial. *IEEE*, 78(1):56–93, January 1990.

10. D. B. H. Tay and N. G. Kingsbury. Design of 2-D perfect reconstruction filter banks using transformations of variables: IIR case. *IEEE Trans. Circuits Syst.—II: Analog and Digital Signal Processing*, 43(3):274–279, March 1996.

11. R. E. Crochiere and L. R. Rabiner. *Multirate Digital Signal Processing*. Prentice-Hall, Englewood Cliffs, NJ, 1983.

12. P. P. Vaidyanathan. *Multirate Systems and Filter Banks*. Prentice-Hall, Englewood Cliffs, NJ, 1993.

13. P. P. Vaidyanathan. Theory and design of M-channel maximally decimated quadrature mirror filters with arbitrary M, having the perfect-reconstruction property. *IEEE Trans. Acoust. Speech Signal Proc.*, 35(4):476–492, April 1987.

14. P. P. Vaidyanathan and P. Q. Hoang. Lattice structures for optimal design and robust implementation of two-channel perfect-reconstruction QMF banks. *IEEE Trans. Acoust. Speech Signal Proc.*, 36(1):81–94, January 1988.

15. T. Q. Nguyen and P. P. Vaidyanathan. Two-channel perfect-reconstruction FIR QMF structures which yield linear-phase analysis and synthesis filters. *IEEE Trans. Acoust. Speech Signal Proc.*, 37(5):676–690, May 1989.

16. M. Vetterli and D. LeGall. Perfect reconstruction FIR filter banks: Some properties and factorizations. *IEEE Trans. Acoust. Speech Signal Proc.*, 37(7):1057–1071, July 1989.

17. W. Sweldens. The lifting scheme: A custom-design of biorthogonal wavelets. *Appl. Comput. Harmonic Anal.*, 3(2):186–200, 1996.

18. W. Sweldens. The lifting scheme: A construction of second generation wavelets. *SIAM J. Math. Anal.*, 29(2):511–546, 1997.

19. I. Daubechies and W. Sweldens. Factoring wavelet transforms into lifting steps. *J. Fourier Anal. Appl.*, 4(3):247–269, 1998.

20. A. R. Calderbank, I. Daubechies, W. Sweldens, and B. L. Yeo. Wavelet transforms that map integers to integers. *App. Comput. Harmonic Anal.*, 5(3):332–369, 1998.

21. S. Mallat. *A Wavelet Tour of Signal Processing*. Academic Press, San Diego, CA, 1998.

22. I. W. Selesnick, R. G. Baraniuk, and N. G. Kingsbury. The dual-tree complex wavelet transform. *IEEE Signal Process. Mag.*, 22(6):123–151, November 2005.

23. I. W. Selesnick. Hilbert transform pairs of wavelet bases. *IEEE Signal Proc. Lett.*, 8(6):170–173, June 2001.

24. I. W. Selesnick. The design of approximate Hilbert transform pairs of wavelet bases. *IEEE Trans. Signal Proc.*, 50(5):1144–1152, May 2002.

25. N. G. Kingsbury. A dual-tree complex wavelet transform with improved orthogonality and symmetry properties. In *Proc. IEEE Int. Conf. Image Processing*, Vancouver, September 2000, paper 1429.

26. N. G. Kingsbury. Design of Q-shift complex wavelets for image processing using frequency domain energy minimization. In *Proc. IEEE Int. Conf. Image Processing*, Barcelona, September 2003.

27. N. G. Kingsbury. Complex wavelets for shift invariant analysis and filtering of signals. *Appl. Comput. Harmonic Anal.*, 10(3):234–253, May 2001.

28. C. Chaux, L. Duval, and J.-C. Pesquet. Image analysis using a dual-tree M-band wavelet transform. *IEEE Trans. Image Proc.*, 15(8):2397–2412, August 2006.

25

Directional Filter Banks

Jose Gerardo Rosiles
The University of Texas at El Paso

Mark J. T. Smith
Purdue University

25.1 Introduction

Directional filter banks (DFBs), which were introduced by Bamberger in 1989 [1–3], are digital analysis–synthesis filter banks that allow an image to be represented as a collection of subbands. In the analysis section, filters succeeded by downsampling matrix operators decompose an image into a set of subbands. In the synthesis section, these subbands are upsampled via a complementary set of matrix operators, then filtered and merged to reconstruct the original image. Such a description is strikingly reminiscent of a conventional 2-D analysis–synthesis filter bank. The distinguishing feature is that the DFB subbands embody angular information (a different angle for each subband), as opposed to the traditional low-, mid-, and high-frequency information. The directional subbands in the DFB are maximally decimated so that the total number of pixels in the full set of subbands is equal to the number of pixels in the original image. Thus, the subbands are nonredundant or, equivalently stated, they form a critically sampled representation.

The DFB has the property that it can achieve exact reconstruction. That is, if the constituent filters are properly designed, the output of the synthesis section can reconstruct the original image from the decimated subbands exactly. This property is important in applications where modifications are performed in the subband domain for the purpose of enhancing image quality in synthesis. In addition, DFBs are very efficient from a computational perspective and are typically designed for implementation in a separable two-hand filter bank tree structure. This is a noteworthy characteristic as it enables the DFB to have arithmetic complexity comparable to (though a little higher than) the popular 2-D transforms, like discrete Fourier transforms (DFTs) and discrete cosine transforms (DCTs).

DFBs are attractive for many image-processing applications that can benefit from directional analysis such as noise reduction, edge sharpening, feature enhancement, compression, object recognition, and texture synthesis, to mention a few. Furthermore, DFBs are also attractive from a visual information-processing perspective. Research has shown that the first layer of cells in the human visual cortex responds to different orientations and scales. Consequently, DFBs can model in an approximate way the human visual system. All of these reasons have motivated much of the study and interest in directional analysis–synthesis systems in recent years.

Since the introduction of the DFB in 1989, there have been a number of extensions and variations that have been considered, such as relaxation of the maximum decimation condition (leading to directional pyramids), multidimensional extensions of the DFB, nonseparable implementations, and mixed low-, mid-, high-frequency angular decompositions [4–7]. To avoid confusion among these variations, we will refer to the original DFB as the Bamberger DFB (BDFB) in our subsequent discussion and use the term DFB to refer more generically to the broad class of directional decompositions.

25.2 Basic Theory of 2-D Multirate Systems

To set the foundation for the later discussion on the DFB, we start by reviewing the basic concepts of 2-D multirate filter banks. The reader seeking a more comprehensive treatment on the subject is referred to Refs. [8,9].

First, we assume that the digital input signal $x[n_0, n_1]$ is a sampled version of a continuous time signal $x_a(t_0, t_1)$ such that

$$x[n_0, n_1] = x_a(n_0 T_0, n_1 T_1),$$

where T_0 and T_1 are the sampling periods. In matrix notation,* this equates to $x[\mathbf{n}] = x_a(\mathbf{Vn})$ where

$$\mathbf{n} = \begin{bmatrix} n_0 \\ n_1 \end{bmatrix}, \quad \mathbf{V} = \begin{bmatrix} T_0 & 0 \\ 0 & T_1 \end{bmatrix}.$$

More generally, $\mathbf{V}$ can be any nonsingular real matrix, which implies that there are an infinite number of ways to sample a 2-D signal. Each sampling matrix gives rise to a different sampling geometry or lattice in the 2-D plane. The "lattice" generated by the sampling matrix $\mathbf{V}$ is the set of all points $\mathbf{t}$, which implies that $\mathbf{t} = \mathbf{Vn}$, where $\mathbf{n} \in \mathcal{N}$ (the set of all integer vectors). The lattice associated with $\mathbf{V}$ is denoted as $LAT(\mathbf{V})$. Such a lattice corresponds to all the "integer" linear combinations of the columns of $\mathbf{V}$, which can be written as $\mathbf{V} = [\mathbf{v}_0 | \mathbf{v}_1]$. It should be noted that the relationship between $\mathbf{V}$ and $LAT(\mathbf{V})$ is not one-to-one. Hence different sampling matrices can give rise to the same lattice, a fact that will be useful in later discussion.

An important component of multirate 2-D systems is the unimodular matrix. An integer matrix $\mathbf{E}$ is unimodular if and only if its inverse is an integer matrix, or equivalently if $|\det \mathbf{E}| = 1$. A couple of useful unimodular matrix properties are summarized in the following theorem [8]:

---

* Boldface capital letters denote matrices and boldface lowercase letters denote vectors.

THEOREM 25.1

Let V be a 2 × 2 real nonsingular matrix generating a lattice LAT(V).

1. *Let $\hat{V} = VE$, where E is a 2 × 2 unimodular matrix. Then $LAT(V) = LAT(VE) = LAT(\hat{V})$.*
2. *Let $\hat{V}$ also be a basis for LAT(V). Then there exists a unimodular matrix E such that $\hat{V} = VE$.*

Note that $LAT(V) = LAT(\hat{V})$ does not imply $x_a(\mathbf{Vn}) = x_a(\hat{V}n)$. Although the lattice points are the same for both matrices, they do not occur in the same order, i.e., the samples of $x_a(\mathbf{Vn})$ are rearranged samples of $x_a(\hat{V}n)$. This property is useful and was employed in Ref. [3] as a "change-of-variables" procedure to simplify the DFB design.

To illustrate the idea of lattice resampling with unimodular matrices, consider the lattice generated by the matrix

$$V = \begin{bmatrix} 1 & 1 \\ -1 & 1 \end{bmatrix} = [\mathbf{v}_0|\mathbf{v}_1]. \tag{25.1}$$

It generates what is known as the quincunx lattice, shown in Figure 25.1a, where the heavy dark dots correspond to elements of $LAT(\mathbf{V})$. The geometry of the lattice is established by vectors $\mathbf{v}_0$ and $\mathbf{v}_1$. A set of trivial unimodular matrices can be obtained by choosing variations of the matrices

$$\begin{bmatrix} \pm 1 & 0 \\ 0 & \pm 1 \end{bmatrix} \quad \text{and} \quad \begin{bmatrix} 0 & \pm 1 \\ \pm 1 & 0 \end{bmatrix}. \tag{25.2}$$

Resampling with these matrices, we can produce rotations and reflections of any lattice. In particular, there are eight variations/permutations of the quincunx sampling matrix.

Another useful type of unimodular matrix is

$$E = \begin{bmatrix} 1 & 1 \\ 0 & 1 \end{bmatrix}. \tag{25.3}$$

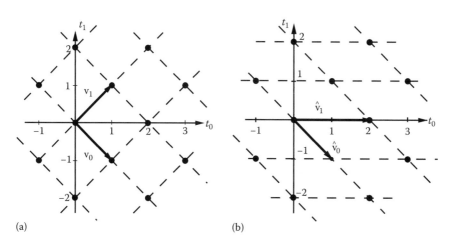

(a)   (b)

FIGURE 25.1 Examples of 2-D lattices. (a) Lattice generated by sampling with a quincunx matrix. (b) Lattice generated by resampling a quincunx matrix with a unimodular matrix. Note that the sample density of the lattice is same in both figures, but the geometry of the lattice is different.

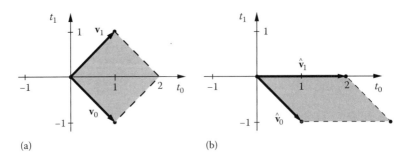

FIGURE 25.2 Fundamental parallepipeds for (a) the quincunx lattice and (b) the resampled quincunx lattice.

It generates the sampling matrix

$$\hat{V} = VE = \begin{bmatrix} 1 & 2 \\ -1 & 0 \end{bmatrix}, \tag{25.4}$$

whose lattice is illustrated in Figure 25.1. It is easy to see that $LAT(V) = LAT(\hat{V})$, but the geometry imposed by $\hat{v}_0$ and $\hat{v}_1$ is different than that of V, i.e., the samples in the lattice have been permuted or reindexed. In the frequency domain, the reindexing of the lattice induces skewings and rotations in the frequency domain of the 2-D signal but does not alter the sample density.

The fundamental parallepiped of V, denoted by $FPD(V)$, is defined by

$$FPD(V) = \{ \text{Set of all points } Vx \text{ with } x \in [0,1)^2 \}. \tag{25.5}$$

Graphically, this is represented by the parallelogram formed by the sampling vectors v_0 and v_1, as shown in Figure 25.2 for V and $\hat{V}$.

25.2.1 2-D Decimation and Interpolation

As in the 1-D case, decimation is performed by filtering (to reduce aliasing) followed by downsampling. Here, the downsampling is performed with a resampling matrix M. Keep in mind that a 2-D resampling matrix does not only alter the sampling rate, but it also introduces a geometric modification to the sampling lattice, as illustrated in the previous figures. Thus a 2-D downsampler is implemented by

$$y[n] = x[Mn], \tag{25.6}$$

where M is a 2×2 nonsingular integer matrix. The signal $y[n]$ retains only those samples of $x[n]$ that reside on $LAT(M)$. The dark dots in Figure 25.3a illustrate this point for the case of quincunx resampling. The dark samples of $x[n]$ are the ones retained by resampling with the quincunx matrix Q defined as

$$Q = \begin{bmatrix} 1 & -1 \\ 1 & 1 \end{bmatrix} = [v_0 | v_1]. \tag{25.7}$$

Figure 25.3b illustrates how the samples are rearranged over the (n_0, n_1) plane. For this quincunx downsampling, the samples experience a clockwise rotation of $45°$. For 2-D resampling matrices, the decimation ratio is the determinant of the matrix. Thus for the quincunx resampling case above, the decimation ratio is $|\det Q| = 2$, as is evident from Figure 25.3.

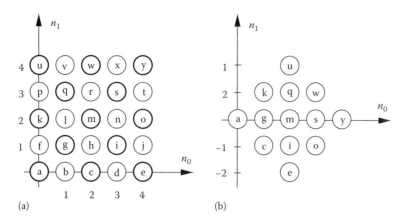

FIGURE 25.3 Downsampling with a quincunx matrix. (a) Original position of samples in the 2-D plane; samples in bold are retained by the downsampling operation. (b) Resulting data after quincunx downsampling with matrix $\mathbf{Q}$ in Equation 25.7.

To show that this property is true in general, consider the set

$$\mathcal{N}(\mathbf{M}) = \{\text{Set of integer vectors in } FPD(\mathbf{M})\} = \{\text{Set of integers of the form } \mathbf{Mx}, \mathbf{x} \in [0,1)^2\}.$$

It is easy to see that the cardinality of the set $J(\mathbf{M})$ is $|\mathcal{N}(\mathbf{M})| = |\det \mathbf{M}|$. For instance, if we consider quincunx downsampling, then

$$\mathcal{N}(\mathbf{Q}) = \left\{ \begin{bmatrix} 0 \\ 0 \end{bmatrix}, \begin{bmatrix} 1 \\ 0 \end{bmatrix} \right\} = \{\mathbf{k}_0, \mathbf{k}_1\}. \tag{25.8}$$

The vectors $\mathbf{k}_i$ are known as coset vectors.

Assuming the signal $x(\mathbf{t})$ has been sampled with $\mathbf{V} = \mathbf{I}$ (i.e., a rectangular uniform grid over the 2-D plane), then any other sublattice $LAT(\mathbf{M})$ obtained by downsampling $LAT(\mathbf{I})$ is known as a sublattice of $\mathbf{I}$. For matrix $\mathbf{M}$, we can obtain $J(\mathbf{M})$ distinct, nonoverlapping sublattices from $\mathbf{I}$, expressed as

$$x_i[\mathbf{n}] = x(\mathbf{Mn} + \mathbf{k}_i). \tag{25.9}$$

Each sublattice is also known as a coset of $LAT(\mathbf{M})$. Figure 25.3a shows graphically that there are two sublattices, $x_0[\mathbf{n}]$ and $x_1[\mathbf{n}]$ for the quincunx matrix $\mathbf{Q}$. From a multirate systems perspective, cosets are the equivalent of polyphase components.

25.2.2 Alias-Free 2-D Decimation

In the 1-D case, an antialiasing filter is generally applied to the signal prior to downsampling. For decimation by a factor of M, the filter has an ideal cutoff frequency of π/M, where M is the decimation ratio. In two dimensions there is an analogous cutoff frequency requirement but also a prescribed region of support determined by the geometry of $LAT(\mathbf{M})$, where $\mathbf{M}$ is the downsampling matrix.

Let $X(\omega)$ be the 2-D DTFT of $x[\mathbf{n}]$, which is 2π-periodic in ω over the 2-D frequency plane with $\omega = [\omega_0 \omega_1]^T$. For convenience, with abuse of notation, $X(\omega)$ will be used in place of $X(e^{j\omega_0}, e^{j\omega_1})$.

The region of support for an antialiasing filter needed to ideally downsample by $\mathbf{M}$ is given by [8]

$$\omega = \pi\mathbf{M}^{-T}\mathbf{x} + 2\pi\mathbf{m}, \quad \mathbf{x} \in [-1,1)^2, \quad \mathbf{m} \in \mathcal{N}. \tag{25.10}$$

The term "$2\pi\mathbf{m}$" represents the periodicity of the antialias region of support, while the term "$\pi\mathbf{M}^{-T}\mathbf{x}$" corresponds to the fundamental frequency cell (or region of support) that is periodically replicated over the frequency plane. If

$$\mathbf{M} = \begin{bmatrix} m_{00} & m_{01} \\ m_{10} & m_{11} \end{bmatrix}, \tag{25.11}$$

the frequency cell can be expressed as

$$-\pi \le m_{00}\omega_0 + m_{10}\omega_1 < \pi \ \cap \ -\pi \le m_{01}\omega_0 + m_{11}\omega_1 < \pi. \tag{25.12}$$

It is also illuminating to examine the 2-D decimator in the frequency domain. The key here is understanding the 2-D downsampling process. Suppose a signal $X(\omega)$ is downsampled with the matrix $\mathbf{M}$. The frequency-domain expression for the downsampled signal $Y(\omega)$ in terms of $X(\omega)$ is (analogous to the 1-D case)

$$Y(\omega) = \frac{1}{J(\mathbf{M})} \sum_{\mathbf{k} \in \mathcal{N}(\mathbf{M}^T)} X(\mathbf{M}^{-T}(\omega - 2\pi\mathbf{k})), \tag{25.13}$$

where $X(\mathbf{M}^{-T}\omega)$ is a "stretched" version of $X(\omega)$. For the case of the signal $X(\omega)$ shown in Figure 25.4a that undergoes quincunx downsampling, we have

$$Y(\omega) = \frac{1}{2}X(\mathbf{Q}^{-T}\omega) + \frac{1}{2}X(\mathbf{Q}^{-T}(\omega - 2\pi\mathbf{k}_1)), \tag{25.14}$$

where $\mathbf{k}_1 = [1 \ 0]^T$ is the second coset vector of $\mathbf{Q}$. Figure 25.4b shows the result after the downsampling operation. In this case, from Equation 25.10, the support for the term $X(\mathbf{Q}^{-T}\omega)$ is $\omega = \pi\mathbf{x} + 2\pi\mathbf{Q}^T\mathbf{m}$,

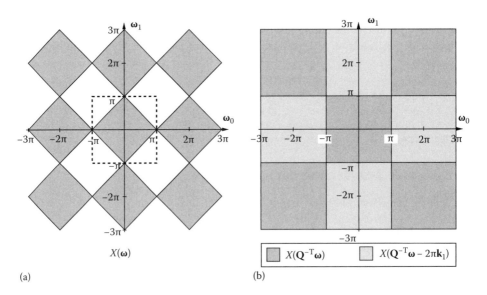

(a)   (b)

FIGURE 25.4 Effect of downsampling by $\mathbf{Q}$. (a) Diamond band limited signal. (b) The diamond region is stretched by a factor of two and rotated by 45° covering the full unit cell. The empty regions in (a) are covered by $X(\mathbf{Q}^{-T}(\omega - 2\pi\mathbf{k}_1))$.

which corresponds to the full $[-\pi, \pi)^2$ support periodically replicated with periodicity matrix $2\pi\mathbf{Q}^{\mathrm{T}}$. Hence the frequency support of $X(\omega)$ has been stretched by a factor of 2 and rotated by 45°. The complementary regions are filled by the term $X(\mathbf{Q}^{-\mathrm{T}}(\omega - 2\pi\mathbf{k}_1))$.

25.2.3 Effect of Unimodular Matrices in the Frequency Domain

Previously, we used the unimodular matrix $\mathbf{E}$ from Equation 25.3 to modify $LAT(\mathbf{V})$. Similar properties follow for the case of resampling a discrete signal using unimodular matrices. For instance, we can modify the quincunx lattice generated by $\mathbf{Q}$ using $\mathbf{E}$ such that $LAT(\mathbf{QE})$ contains the same points as $LAT(\mathbf{Q})$ but with a different sample ordering. From Equation 25.12 the frequency support for a signal lying over $LAT(\mathbf{QE})$ is given by

$$-\pi \le \omega_0 - \omega_1 < \pi \ \cap \ -\frac{\pi}{2} \le \omega_0 < \frac{\pi}{2}. \tag{25.15}$$

This represents a skewing of the frequency cell associated with $LAT(\mathbf{Q})$.

25.2.4 2-D Upsamplers and Interpolators

Complementary to the decimation operation is the interpolation operation, which is composed of an upsampler followed by an anti-imaging filter. For a 2-D signal $x[\mathbf{n}]$, the upsampling operation with matrix $\mathbf{M}$ generates the signal

$$y[\mathbf{n}] = \begin{cases} x(\mathbf{M}^{-1}\mathbf{n}) & \text{if } \mathbf{n} \in LAT(\mathbf{M}), \\ 0 & \text{otherwise.} \end{cases} \tag{25.16}$$

This means that a matrix $\mathbf{M}$ with $J(\mathbf{M})$ coset vectors maps all the samples of $x[\mathbf{n}]$ to the sublattice produced by $\mathbf{k}_0$, while the remaining sublattices are populated by zeros.

The frequency domain expression for the interpolator is

$$Y(\omega) = X(\mathbf{M}^{\mathrm{T}}\omega), \tag{25.17}$$

which implies $Y(\omega)$ has a periodicity matrix $2\pi\mathbf{M}^{-\mathrm{T}}$, and there are $|\det \mathbf{M}|$ compressed images of $X(\omega)$ in the $[-\pi, \pi]^2$ frequency cell. Decimators and interpolators are the basic elements of multirate systems and will be used extensively in the discussions that follow.

25.3 2-D Maximally Decimated Filter Banks

A more general condition than Equation 25.10 to derive the support of the antialias filter is

$$\omega = \mathbf{c} + \pi\mathbf{M}^{-\mathrm{T}}x + 2\pi\mathbf{m}, \quad x \in [-1,1)^2, \quad \mathbf{m} \in \mathcal{N}, \tag{25.18}$$

where $\mathbf{c}$ is some arbitrary constant vector. This implies there are an infinite number of possible antialiasing filter support regions for sampling matrix $\mathbf{M}$. There are, however, a few choices of $\mathbf{c}$ that are of more practical importance as our intuition can suggest.

Going back to the 1-D case, an M-channel maximally decimated filter bank requires having M nonoverlapping (ideal) filters with bandwidth $\frac{\pi}{M}$, so that the range $[-\pi, \pi)$ is completely covered by the M channels. This concept can be generalized to two dimensions by finding the nonoverlapping regions whose union covers the frequency cell $[-\pi, \pi)^2$. Such regions are obtained by choosing

$$\mathbf{c} = 2\pi\mathbf{M}^{-\mathrm{T}}\mathbf{k}, \quad \mathbf{k} \in \mathcal{N}(\mathbf{M}^{\mathrm{T}}) \tag{25.19}$$

in Equation 25.18. It is easy to show that for each of the $J(\mathbf{M})$ coset vectors $\mathbf{k}_i$, there is a different nonoverlapping frequency region. From the filter bank perspective, each of these regions can be used as the frequency-domain region for a subband whose information can be extracted with the properly designed subband filter. This concept is illustrated next with a couple of examples.

25.3.1 Diamond and Fan Filter Banks

Two popular 2-D filter banks are the diamond and fan filter banks (FFBs) shown in Figure 25.5. They form the building blocks for a broad class of directional decompositions, which we will discuss later. The diamond filter has a passband as shown in Figure 25.5a in the $[-\pi, \pi)^2$ region. These filters can provide alias-free decimation for the quincunx resampling matrix $\mathbf{Q}$. Using the corresponding coset vectors (Equation 25.8) in Equations 25.18 and 25.19, we obtain the support regions

$$\Omega_0: \ -\pi \leq \omega_0 - \omega_1 < \pi \ \cap \ -\pi \leq \omega_0 + \omega_1 < \pi,$$
$$\Omega_1: \ \pi \leq \omega_0 - \omega_1 < 3\pi \ \cap \ -\pi \leq \omega_0 + \omega_1 < \pi. \tag{25.20}$$

Related to the diamond filter is the fan filter which has the passband shown in Figure 25.5b. It is identical to the diamond filter except that it is frequency shifted (either in ω_0 or ω_1) by π. We can show that fan support is a valid antialias passband for the quincunx matrix by finding the vector $\mathbf{c}$ that satisfies Equation 25.18. If we let

$$\mathbf{c}_0 = \begin{bmatrix} \pi \\ 0 \end{bmatrix}, \quad \mathbf{c}_1 = \begin{bmatrix} 0 \\ \pi \end{bmatrix},$$

we get the two fundamental complementary fan-shaped regions for $F_0(\omega)$ and $F_1(\omega)$:

$$\Omega_0: \ 0 \leq \omega_0 - \omega_1 < 2\pi \ \cap \ 0 \leq \omega_0 + \omega_1 < 2\pi,$$
$$\Omega_1: \ -2\pi \leq \omega_0 - \omega_1 < 0 \ \cap \ 0 \leq \omega_0 + \omega_1 < 2\pi. \tag{25.21}$$

Again, the diamond filter bank and the FFB play a crucial role in the construction of DFBs.

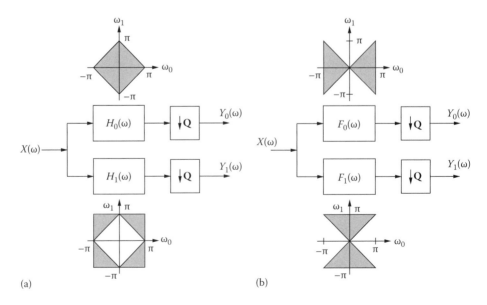

(a)   (b)

FIGURE 25.5   Block diagram for the diamond and FFBs showing the ideal support for the 2-D filters.

25.4 Bamberger Directional Filter Bank

In this chapter, we present the theory of the DFB introduced in Refs. [2,3]. The BDFB splits the frequency cell $[-\pi, \pi)^2$ into an even number of wedge-shaped regions as shown in Figure 25.6. The BDFB employs a tree-structured 2-D filter bank that diagrammatically is analogous to a 1-D tree-structured filter bank.

Using this approach, Bamberger introduced BDFBs with different numbers of subbands such as 6, 10, 18, and more [2]. However, the BDFBs that have received most attention over the years are the uniform N-stage tree structure filter banks that generate $M = 2^N$ subbands. Given the extensive number of applications in which $M = 2^N$ BDFBs are employed, we focus on this case in the reminder of this chapter.

The BDFB is implemented with a small set of well-defined building blocks, each with low computational complexity. Without loss of generality, we present the BDFB for the $M = 8(N = 3)$ case, which achieves the frequency plane partitioning shown in Figure 25.6d. The extension to 16 bands, 32 bands, and higher follows by a straightforward extension of the tree structure. The block diagram for an eight-band BDFB analysis stage is shown in Figure 25.7. The primary building block is the two-channel FFB.

The third stage of the BDFB includes additional resampling matrices $\mathbf{U}_i$ and $\mathbf{B}_i$, which are unimodular. The matrices $\mathbf{U}_i$ resample the four subbands from the second stage to remap their frequency support to a fan-shaped region. This allows the use of the FFB on all stages. The function of the $\mathbf{B}_i$ matrices is to adjust the sampling lattice of the resulting subbands so they have a rectangular geometry (i.e., the overall sampling matrix is diagonal with a downsampling rate of M). The use of $\mathbf{U}_i$ matrices was introduced by

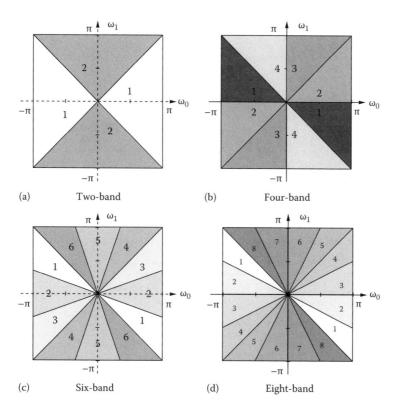

FIGURE 25.6 Frequency band partitions achieved by the BDFB. (a) Two-band. (b) Four-band. (c) Six-band. (d) Eight-band.

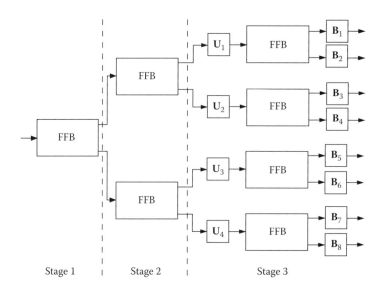

FIGURE 25.7   Implementation of an eight-band BDFB using a tree structure with FFBs and backsampling matrices.

Park et al. [10] as a way to achieve a more geometrically accurate subband representation. According to the selection rules developed in Ref. [10], we have chosen the following matrix values for $\mathbf{U}_i$:

$$\mathbf{U}_1 = \begin{bmatrix} 1 & -1 \\ 0 & 1 \end{bmatrix}, \quad \mathbf{U}_2 = \begin{bmatrix} -1 & 1 \\ 0 & 1 \end{bmatrix}, \quad \mathbf{U}_3 = \begin{bmatrix} 0 & 1 \\ -1 & 1 \end{bmatrix}, \quad \mathbf{U}_4 = \begin{bmatrix} 1 & 0 \\ -1 & 1 \end{bmatrix}. \tag{25.22}$$

Similarly, the values for the $\mathbf{B}_i$ matrices are given by

$$\mathbf{B}_1 = \mathbf{B}_2 = \begin{bmatrix} -1 & 1 \\ 0 & -1 \end{bmatrix}, \quad \mathbf{B}_3 = \mathbf{B}_4 = \begin{bmatrix} -1 & -1 \\ 0 & 1 \end{bmatrix},$$

$$\mathbf{B}_5 = \mathbf{B}_6 = \begin{bmatrix} 1 & 1 \\ -1 & 0 \end{bmatrix}, \quad \mathbf{B}_7 = \mathbf{B}_8 = \begin{bmatrix} -1 & 1 \\ -1 & 0 \end{bmatrix}. \tag{25.23}$$

A detailed stage-by-stage analysis of the BDFB structure is presented in Refs. [2,11]. Here, we analyze the BDFB by collapsing the tree structure into the M-channel parallel structure shown in Figure 25.8. Using the multirate identity that states "downsampling by $\mathbf{M}$ followed by $H(\omega)$ is equivalent to $H(\mathbf{M}^T\omega)$ followed by downsampling by $\mathbf{M}$," it is possible to migrate all the filters in the tree structure to the left and the downsampling operations to the right of the tree structure. For instance, for an eight-band BDFB, the overall analysis filters are given by

$$G_\ell(\omega) = F_j(\omega)F_k(\mathbf{Q}_1^T\omega)F_m(\mathbf{U}_i^T\mathbf{Q}_2^T\mathbf{Q}_1^T\omega), \tag{25.24}$$

where
   $j, k, m \in \{0, 1\}$
   $i \in \{1, 2, 3, 4\}$
   $\ell = 1, 2, \ldots, 8$

Looking at the tree structure as a binary tree, we have $\ell = j + 2k + 4m + 1$. The multiplication of the three frequency responses in Equation 25.24 results in the filters $G_\ell(\omega)$ with the wedge-shaped passbands

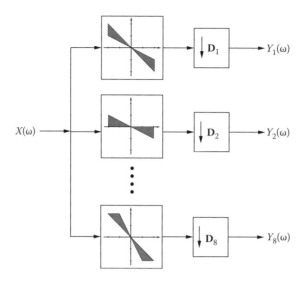

FIGURE 25.8 Parallel implementation of a maximally decimated eight-band DFB. Note that $|\det(\mathbf{D}_i)| = 8$.

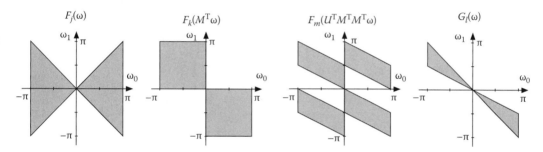

FIGURE 25.9 Illustration of the synthesis of wedge passband for an eight-band DFB. The wedge support is obtained by taking the product of the frequency responses as described by Equation 25.24.

shown in Figure 25.6d. As an illustration, Figure 25.9 shows graphically how the cascade of filters depicted in Equation 25.24 produces a wedge-shaped passband. The fan-shaped support for $F_0(\omega)$ and $F_1(\omega)$ is rotated, skewed, or scaled according to the resampling operations indicated by Equation 25.24. Analogous operations can be implemented to generate the remaining wedge filters.

Viewing the analysis tree structure as a parallel form structure, the eight-band BDFB has overall downsampling matrices given by

$$\mathbf{D}_\ell = \mathbf{Q}_1 \mathbf{Q}_2 \mathbf{U}_i \mathbf{Q}_3 \mathbf{B}_i, \tag{25.25}$$

for $\ell = 1, 2, \ldots, 8$ and $i = \lceil \frac{\ell}{2} \rceil$. This implies that $\mathbf{D}_{2i-1} = \mathbf{D}_{2i}$. It is easy to show that $\mathbf{D}_\ell$ is diagonal with one of the following values:

$$\mathbf{C}_1 = \begin{bmatrix} 2 & 0 \\ 0 & 4 \end{bmatrix} \quad \text{and} \quad \mathbf{C}_2 = \begin{bmatrix} 4 & 0 \\ 0 & 2 \end{bmatrix}. \tag{25.26}$$

We can conclude three things from the above result:

1. Decimation ratio for all channels is $|\det(\mathbf{C}_1)| = |\det(\mathbf{C}_2)| = 8$, giving us a maximally decimated system as expected.
2. Since the $\mathbf{D}_\ell$ matrices are diagonal, the corresponding sampling grids and fundamental parallelepipeds are rectangular. We should note that this is a consequence of the judicious selection of the unimodular matrices $\mathbf{U}_i$ and $\mathbf{B}_i$. Having the data lie on a rectangular lattice makes further processing easier.
3. Half of the bands ($\ell = 1, 2, 3, 4$) are subsampled by two in the horizontal direction and by four in the vertical direction. The remaining bands ($\ell = 5, 6, 7, 8$) have the opposite structure.

For brevity we focussed on the analysis stage of the BDFB in our discussion. However, it should be noted that the same multirate concepts can be applied to the synthesis stage in an analogous manner.

The generation of BDFBs with $16, 32, \ldots, 2^N$ subbands is achieved by replicating the third stage in Figure 25.7. For an N-stage BDFB, the subbands will have an overall downsampling matrix given by

$$\mathbf{C}_1 = \begin{bmatrix} 2 & 0 \\ 0 & 2^{N-1} \end{bmatrix} \quad \text{or} \quad \mathbf{C}_2 = \begin{bmatrix} 2^{N-1} & 0 \\ 0 & 2 \end{bmatrix}.$$

In summary, we have shown how to construct directional filters using multirate operations and FFBs as building blocks. In Section 25.5, we show how to design the FFB filters, $F_0(\omega)$ and $F_1(\omega)$.

25.5 Design of 2-D Two-Channel Fan Filter Banks

Filter banks are usually designed to either satisfy aliasing-cancelation (AC) or perfect-reconstruction (PR) constraints in addition to the usual frequency response specifications. For the 2-D case, a commonly used approach for filter bank design is to derive separable filters from a 1-D systems. The typical consequence of separable filter banks is rectangular tilings of the frequency plane. Implementations of multidimensional filter banks for nonrectangular filters have been reported in the literature [8,9,12–15,58,59]. However, the resulting filters are, in general, nonseparable and require a high order to achieve good frequency selectivity. A more efficient approach is to use low-complexity FFBs. FFBs can be designed using the change-of-variable method, which involves a 1-D to 2-D mapping. Similar to separable implementations, we can take advantage of well-known 1-D filter bank design techniques. The resulting filter banks have efficient polyphase implementations that only require separable filtering operations.

25.5.1 FFB Design Using 1-D Quadrature Mirror Filters

Thus far, the BDFB theory has been developed using filters with ideal frequency responses. In practice, one designs the FFB filters to meet either AC or PR constraints while approximating good passband characteristics. Designing 2-D filter banks with diamond, fan, and other 2-D geometries has been studied by several authors [9,15–17,59,60] where different alias-free and PR 2-D filter banks have been proposed. Bamberger's approach [2] was different in that he presented a design method for FIR PR nonseparable systems using McClellan transformations. In addition, he generalized the change-of-variables scheme introduced by Ansari [18] to produce Quadrature Mirror Filter (QMF) finite impulse response (FIR) and infinite impulse response (IIR) implementations, which provides an efficient polyphase implementation. A drawback of this scheme is that the FIR implementations are not PR. More recently, Rosiles and Smith [19,20] reported the use of the ladder-based filter banks [16,17] to implement the BDFB with FIR PR filters. Ladder networks also have an efficient implementation and consequently are attractive. In this section, we discuss the QMF and ladder structure implementation of the FFB.

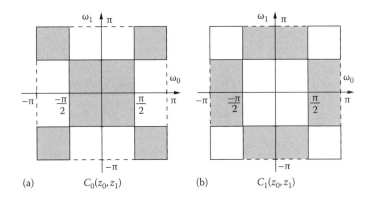

FIGURE 25.10 Ideal checkerboard magnitude responses needed for the generation of the FFB using 1-D QMFs.

The development that follows starts with the design of 2-D diamond filter banks. This system consists of two filters $H_0(\omega_0, \omega_1)$ and $H_1(\omega_0, \omega_1)$ with complementary diamond passband/stopband regions, as shown in Figure 25.5. Fan-shaped filters are obtained by a modulation operation, which in the frequency domain is expressed by $F_0(\omega_0, \omega_1) = H_0(\omega_0 - \pi, \omega_1)$ and $F_1(\omega_0, \omega_1) = H_1(\omega_0 - \pi, \omega_1)$.

The change-of-variables design method [2,18] requires the availability of a 2-D filter $C_0(\omega_0, \omega_1)$ with checker board geometry and the complementary filter $C_1(\omega_0, \omega_1)$. The required passbands and stopbands are shown in Figure 25.10. To achieve the desired filter geometries, the sampling matrix **M** with the corresponding FPD is used to apply the following change of variables in the frequency domain

$$\begin{bmatrix} \omega_0' \\ \omega_1' \end{bmatrix} = \frac{\mathbf{Q}^T}{|\det(\mathbf{Q})|} \begin{bmatrix} \omega_0 \\ \omega_1 \end{bmatrix}. \tag{25.27}$$

For the case of diamond (and fan) filters, the quincunx matrix **Q** provides the appropriate mapping, which simplifies to $\omega_0' = (\omega_0 + \omega_1)/2$ and $\omega_1' = (\omega_1 - \omega_0)/2$. In the z-domain, we can express the change of variables as*

$$z_0 \rightarrow z_0^{\frac{1}{2}} z_1^{\frac{1}{2}} \quad \text{and} \quad z_1 \rightarrow z_0^{-\frac{1}{2}} z_1^{\frac{1}{2}}. \tag{25.28}$$

We should note that this scheme is general and works for other sampling matrices. The effect of the change of variables is to scale, rotate, and skew the passband regions of $C_0(\omega_0, \omega_1)$ and $C_1(\omega_0, \omega_1)$ to obtain the desired geometries.

To design an FFB with an efficient implementation, select a pair of 1-D prototypes that satisfy the QMF relationship

$$H_0(z) = H_1(-z). \tag{25.29}$$

It is well known that the polyphase representation for these filters is

$$H_0(z) = E_0(z^2) + z^{-1} E_1(z^2),$$
$$H_1(z) = E_0(z^2) - z^{-1} E_1(z^2). \tag{25.30}$$

---

* In this chapter z_0 and z_1 are used to denote the 2-D z-transform complex variables.

It is easy to see that the checker board filters from Figure 25.10 can be designed from the 1-D prototypes using separable filtering operations. In the z-domain this is expressed as

$$
\begin{aligned}
C_0(z_0, z_1) &= H_0(z_0)H_0(z_1) + H_0(-z_0)H_0(-z_1) = 2[E_0(z_0^2, z_1^2) + z_0^{-1}z_1^{-1}E_1(z_0^2, z_1^2)], \\
C_1(z_0, z_1) &= H_0(z_0)H_1(z_1) + H_0(-z_0)H_1(-z_1) = 2[E_0(z_0^2, z_1^2) - z_0^{-1}z_1^{-1}E_1(z_0^2, z_1^2)].
\end{aligned}
\tag{25.31}
$$

The 2-D polyphase filters are defined as

$$
\begin{aligned}
E_0(z_0, z_1) &= E_0(z_0)E_0(z_1), \\
E_1(z_0, z_1) &= E_1(z_0)E_1(z_1).
\end{aligned}
\tag{25.32}
$$

Finally, the diamond filters are obtained by applying the change of variables in Equation 25.28 to $C_0(z_0, z_1)$ and $C_1(z_0, z_1)$. This step produces the geometric transformation on the checkerboard support to obtain the desired diamond support filters. In the frequency domain, we have

$$
\begin{aligned}
H_0(\boldsymbol{\omega}) &= E_0(\mathbf{Q}^\mathrm{T}\boldsymbol{\omega}) + e^{-j\boldsymbol{\omega}^\mathrm{T}\mathbf{k}_1}E_1(\mathbf{Q}^\mathrm{T}\boldsymbol{\omega}), \\
H_1(\boldsymbol{\omega}) &= E_0(\mathbf{Q}^\mathrm{T}\boldsymbol{\omega}) - e^{-j\boldsymbol{\omega}^\mathrm{T}\mathbf{k}_1}E_1(\mathbf{Q}^\mathrm{T}\boldsymbol{\omega}),
\end{aligned}
\tag{25.33}
$$

where $\mathbf{Q}$ is the quincunx matrix and $k_1 = [1\ 0]^\mathrm{T}$. These expressions are similar to those of Equation 25.30 for the 1-D case, implying that the filters are implementable in the polyphase domain. Moreover, the 2-D polyphase components are separable, providing a 2-D structure with low computational complexity. The resulting two-channel analysis structure is shown in Figure 25.11. The corresponding synthesis stage is analogous.

For the 1-D prototype $H_0(z)$, one can use the Johnston QMFs [21], which are aliasing-free but not PR. Another choice is to use the IIR linear phase filters proposed by Smith and Eddins in Ref. [22]. These filters have the form

$$
H_0(z) = \frac{b_0 + b_1 z^{-1} + b_2 z^{-2} + b_2 z^{-3} + b_1 z^{-4} + b_0 z^{-5}}{a_0 + a_1 z^{-2} + a_0 z^{-4}}
\tag{25.34}
$$

and can achieve exact reconstruction. The IIR filters are noncausal, but can be used for the case of finite length signals (like images), using forward and backward difference equations. Moreover, the implementation is very efficient, requiring only 5.1 multiplies and 5.6 adds per output sample, while achieving similar passband characteristics to that of the 32 tap Johnston QMF.

To construct fan filters, we apply the change of variables $z_0 \rightarrow -z_0$ in $H_0(z_0, z_1)$ and $H_1(z_0, z_1)$ to induce a shift by π along the ω_0-axis. Thus "off-the-shelf" QMFs can be used to generate high-quality

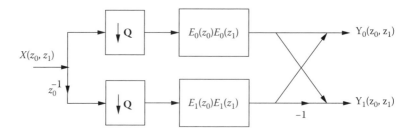

FIGURE 25.11 Efficient polyphase domain implementation of diamond filter bank. The filtering operations are separable.

fan filters. Similarly these QMFs can be applied in successive stages as discussed previously to produce directional filters. To illustrate this point, four directional filters are shown in Figure 25.12. These filters were designed using the Johnston 32D QMF, and are the constituent filters of the eight-band BDFB. Such filter banks can be used to decompose images, an example of which is shown in Figure 25.13.

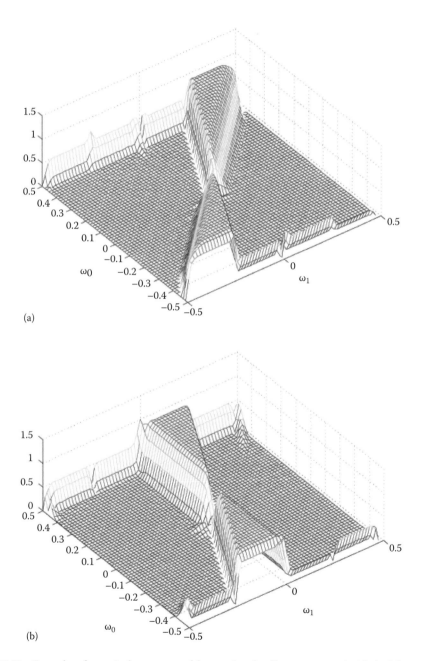

(a)

(b)

FIGURE 25.12 Examples of magnitude responses of the actual wedge filters constructed with the Johnston 32D FIR filters.

(continued)

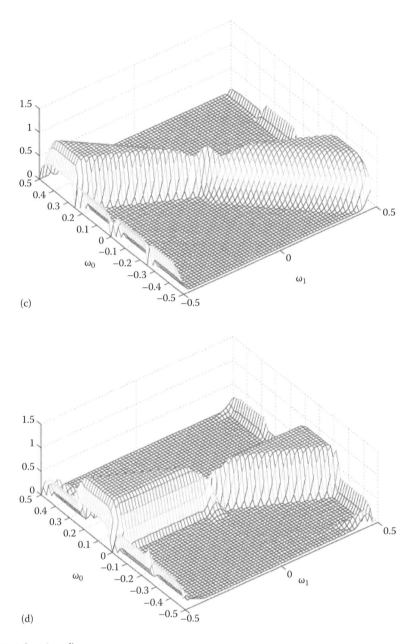

(c)

(d)

FIGURE 25.12 (continued)

25.5.2 BDFB Design Using Ladder Structures

Ladder structures (also known as lifting schemes [23]) are flexible structures for implementing filter banks and discrete wavelet transforms. The construction of wavelets is achieved by cascading simple filtering elements successively in ladder steps. The PR property is imposed structurally without the explicit need to meet the conditions used in the early PR filter bank design methodology. The construction of biorthogonal systems using ladder networks has been reported in various articles [23–25].

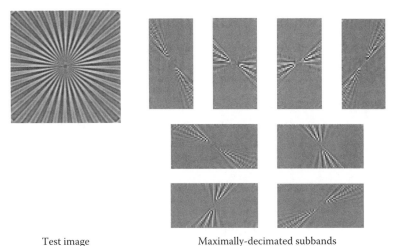

Test image   Maximally-decimated subbands

FIGURE 25.13 Example of an eight-band BDFB using a test image with localized directional structure.

Additionally, ladder structures allow straightforward implementation of nonlinear filter banks like integer-to-integer wavelet transforms [26], and adaptation to irregular sampling grids [27].

In this section, ladder structures are used as the framework for the design of digital fan-shaped filters. These structures have been explored for 2-D and multidimensional filter bank implementations [16,17,24,25,28], where the design method involves transforming a 1-D filter to a 2-D filter with a simple mapping. This 1-D to 2-D mapping was introduced by Kim and Ansari [61] and Phoong et al. [16] in the context of a two-stage ladder structure and was extended to a three-stage structure by Ansari et al. [17]. The three-stage design was shown to improve the control of filter frequency responses.

To begin, consider the 1-D case. The 1-D system is designed by constructing the analysis polyphase matrix as the product of ladder steps given by

$$E(z) = \begin{bmatrix} 1 & 0 \\ -p_2\beta_2(z) & 1 \end{bmatrix} \begin{bmatrix} 1 & zp_1\beta_1(z) \\ 0 & 1/(1+p) \end{bmatrix} \begin{bmatrix} p_0 & 0 \\ -p\beta_0(z) & 1 \end{bmatrix}, \tag{25.35}$$

which generates the analysis filters

$$H_0(z) = p_0 + p_1 z\beta_1(z^2)A(-z) \quad \text{and} \quad H_1(z) = A(-z) - p_2 H_0(z)\beta_2(z^2), \tag{25.36}$$

where

$$A(z) = \frac{(1 + z\beta_0(z^2))}{1+p} \tag{25.37}$$

and the constants p_0, p_1, p_2 are given by $p_0 = p_1 = (1+p)/2$ and $p_2 = (1-p)/(1+p)$.

This 1-D filter bank structure is illustrated graphically in Figure 25.14. The inherent structure associated with ladder filters implies that the filter bank is biorthogonal and consequently the relationship between the analysis and synthesis filters is given by

$$G_0(z) = -z^{-1}H_1(-z), \quad G_1(z) = z^{-1}H_0(-z), \tag{25.38}$$

where

$H_0(z)$ and $H_1(z)$ are the analysis filters
$G_0(z)$ and $G_1(z)$ are the synthesis filters

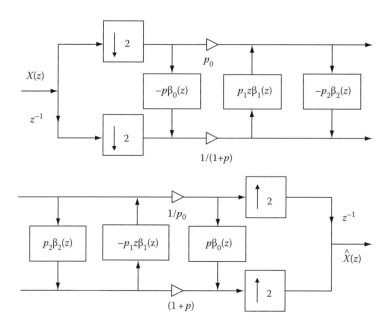

FIGURE 25.14 Analysis–synthesis ladder structure with three ladder steps, as proposed in Ref. [17].

It is easy to show that $H_0(z)$ and $G_0(z)$ are halfband filters. In order to obtain ideal lowpass filters, the following condition for the $\beta_i(z)$ functions must be met

$$\beta_i(e^{j2\omega}) = \begin{cases} e^{j(-2N+1)\omega} & \text{for} \quad 0 \leq \omega \leq \pi/2, \\ -e^{j(-2N+1)\omega} & \text{for} \quad \pi/2 < \omega \leq \pi, \end{cases} \tag{25.39}$$

which implies $\beta_i(e^{j\omega})$ has allpass behavior. An FIR solution that approximates Equation 25.39 is obtained by designing an even length, linear phase function with a magnitude response optimized to approximate unity. This is a very simple requirement that can be satisfied with widely available filter design algorithms. It is possible to design the $\beta_i(z)$ filters separately or to choose the same function by making $\beta(z) = \beta_1(z) = \beta_2(z) = \beta_3(z)$, which significantly simplifies the design procedure. We note that $H_0(z)$ and $H_1(z)$ inherit the linear phase property of $\beta(z)$. Analysis filters obtained by approximating $\beta_i(z)$ using the Parks–McClellan algorithm with $L = 8$ are shown in Figure 25.16a. Additionally, these design constraints can be met by IIR designs, however here we focus on the FIR case [16].

Maximally flat 1-D ladder filters can be obtained using the closed-form Lagrange formula

$$v_k = \frac{(-1)^{N-k-1} \prod_{i=0}^{2N} (N + 1/2 - i)}{2(N-k)!(N-1+k)!(2k-1)}, \tag{25.40}$$

where
  $N = L/2$ is the half length of $\beta(z)$
  $\beta_{N-k} = \beta_{N+k-1} = v_k$ for $k = 1, 2, \ldots, N$

The maximally flat design method is relevant for the generation of regular biorthogonal wavelets.

For our purposes, the most attractive feature of ladder structures is the implementation of 2-D FFBs using a simple 1-D to 2-D change of variables. This transformation reported in Ref. [16] is applied to the entries of $\mathbf{E}(z)$ expressed in terms of lifting steps. First, the 1-D transfer function $\beta(z)$ is replaced with the

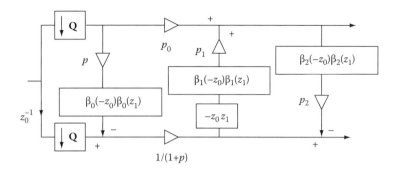

FIGURE 25.15 Ladder structure for the implementation of a 2-D two-channel biorthogonal analysis filter bank.

separable 2-D transfer function $\beta(z_0)\beta(z_1)$. Then, the 1-D delays z^{-1} are replaced with the 2-D delays $z_0^{-1}z_1^{-1}$. The resulting 2-D filters $H_0(z_0, z_1)$ and $H_1(z_0, z_1)$ have diamond-shaped support. In order to obtain the fan-shaped filters $F_0(z_0, z_1)$ and $F_1(z_0, z_1)$ the diamond filters are modulated by π letting $z_0 \to -z_0$. Following this procedure, the three-stage ladder structure from Figure 25.14 is transformed to the 2-D structure in Figure 25.15. The 2-D fan filter responses $|F_0(z_0, z_1)|$ and $|F_1(z_0, z_1)|$ obtained with this transformation are presented in Figure 25.16 using the same $\beta(z)$ function for all ladder stages.

An important property of this FFB implementation is its low computational complexity. The filtering operations can be implemented in the polyphase domain and only involve separable operations. Moreover since $\beta(z)$ is even length with linear phase, the number of multiplies per output is half the length of $\beta(z)$.

As in Section 25.5.1, the tree-structured BDFB from Figure 25.7 can be realized as a three-stage ladder FFB. The biorthogonal property is preserved by the 1-D to 2-D mapping and the resulting BDFBs have the attractive properties that (1) PR is achieved for FIR systems, (2) the filtering is separable in the polyphase domain and hence is efficient, (3) the 2-D filter quality can be controlled by a single function, (4) any 1-D filter design technique can be employed to generate $\beta(z)$, and (5) any set of

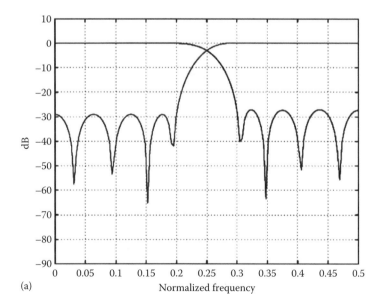

FIGURE 25.16 Magnitude response of the analysis fan filters obtained with a three-stage ladder structure. (a) Lowpass and highpass 1-D filter prototypes.

(continued)

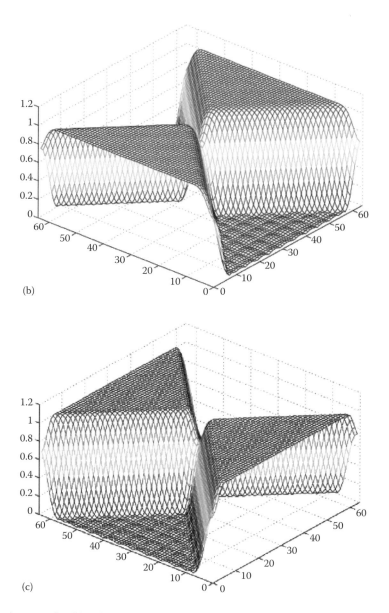

(b)

(c)

FIGURE 25.16 (continued)   (b) and (c) Corresponding 2-D fan shaped filters.

biorthogonal wavelet filters can be used to implement the BDFB by factoring them into ladder steps and applying the 1-D to 2-D mapping.

25.6 Undecimated Directional Filter Bank

In many applications like pattern recognition and image enhancement, shift invariance is considered an important property. To achieve shift invariance the subbands cannot be maximally decimated. In the case of the 2-D discrete wavelet transform (DWT), shift invariance is achieved by removing the downsampling operations from the filter bank structure. This is the so-called undecimated DWT. It can be shown that PR can still be preserved under conditions of partial decimation and in the complete absence of decimation [29]. While the undecimated representation is shift invariant, the coefficients are not decorrelated anymore. Thus, there is a trade-off.

For a number of applications it is desirable to have an undecimated version of the BDFB. A shift-invariant BDFB can be obtained from Figure 25.8 by removing the downsampling matrices D_ℓ and using the filters $G_\ell(\omega)$ as defined by Equation 25.24. This undecimated BDFB has a high computational cost since the filters are nonseparable and, in order to achieve good frequency selectivity, filters generally need to have order of 120×120 or higher. Although frequency domain implementations are possible, the boundary effects are more difficult to control.

Fortunately, there are ways to address efficiency. The BDFB tree structure from Figure 25.7 can be modified to obtain an UDFB. The derivation of the UDFB is explained in detail in Refs. [20,30]. The UDFB is obtained from the BDFB through the use of multirate identities and results in a modified tree structure. To illustrate this, the UDFB tree structure for the eight-band case is shown in Figure 25.19. In this case two building blocks are needed, the undecimated FFB (UFFB) and an undecimated checkerboard filter bank (UCFB). The UCFB is a two-channel structure with the complementary passbands and stopbands as shown in Figure 25.17.

For additional efficiency the UFFB can be implemented in a ladder structures as shown in Figure 25.18 where

$$\mathbf{Q} = \begin{bmatrix} -1 & 1 \\ 1 & 1 \end{bmatrix}.$$

The UCFB is obtained from the UFFB structure by simply removing the resampling operations in Figure 25.18. This relationship has been described in Ref. [18] where fan filters are derived from checkerboard

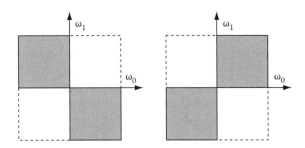

FIGURE 25.17 Ideal magnitude responses for the two-channel UCFB.

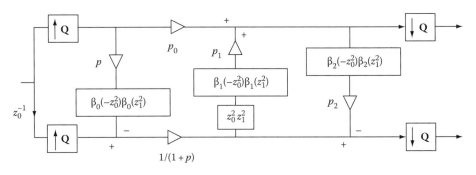

FIGURE 25.18 Ladder structure implementation for the UFFB.

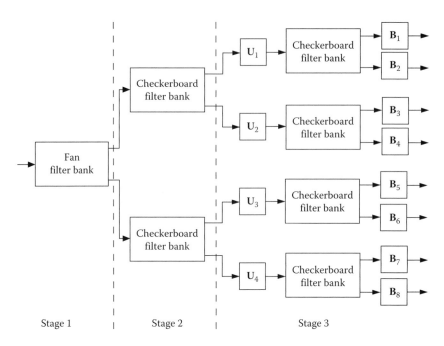

FIGURE 25.19 Implementation of an eight-band UDFB using a tree structure with UFFBs and UCFBs.

filters by a change-of-variables transformation. Finally, the filtering operations in Figure 25.18 remain separable and the $\beta_i(z)$ functions are changed to $\beta_i(z^2)$, which implies that each function has been upsampled by a factor of 2. As a result, low computational complexity of the BDFB is retained in the undecimated case.

The overall UDFB tree structure is shown in Figure 25.19. Similar to the maximally decimated case, it is possible to generate a 2^N-band UDFB by cascading UFFB and UCFB structures. The eight-band UDFB is implemented with a UFFB in the first stage and UCFBs in the remaining stages. UFFBs could be used in the third stage, but UCFBs are more efficient. As before, the third stage requires the use of unimodular resampling matrices $\mathbf{U}_i$. A possible set of matrices is given by

$$\mathbf{U}_1 = \begin{bmatrix} 1 & 1 \\ 0 & 1 \end{bmatrix}, \quad \mathbf{U}_2 = \begin{bmatrix} 1 & -1 \\ 0 & 1 \end{bmatrix}, \quad \mathbf{U}_3 = \begin{bmatrix} 1 & 0 \\ 1 & 1 \end{bmatrix}, \quad \mathbf{U}_4 = \begin{bmatrix} 1 & 0 \\ -1 & 1 \end{bmatrix}.$$

Finally, the matrices $\mathbf{B}_i$ are used in order to reestablish a rectangular sampling geometry. In this case $\mathbf{B}_i = \mathbf{U}_i^{-1}$. An example of an eight-band UDFB decomposition is shown in Figure 25.20. Note that all the subbands have the same size and that the decomposition is shift invariant.

25.7 Bamberger Pyramids

Other image decompositions like the 2-D DWT, the steerable pyramid [31], the complex-valued wavelet transform [32], and 2-D Gabor representations [33,34] separate information across different resolutions as well as directions. This multiresolution (MR) process could be an implicit part of the decomposition (e.g., the DWT), or could be implemented as a separate structure (e.g., the steerable pyramid [31]). In the latter case, we say that the decomposition is polar-separable, implying that a radial frequency decomposition (i.e., a pyramid) is performed independently of the angular decomposition.

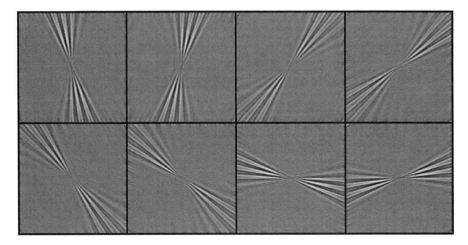

FIGURE 25.20 Example of an eight-band UDFB decomposition.

Given that many problems of interest in image processing and image analysis require MR directional analysis, extending the theory of the BDFB to polar-separable representations is desirable. As it turns out polar-separable versions of the BDFB and UDFB can be constructed and are straightforward to derive.

Different types of radial pyramids can be considered [30], but here we focus on a couple of examples associated with the BDFB. The simplest Bamberger pyramid consists of a "lowpass–highpass" decomposition. The image is first filtered with a lowpass filter $L_{\omega_c}(z_0, z_1)$ with cutoff frequency ω_c. A high frequency component is obtained by subtracting the filtered version from the original image. The resulting high frequency channel is decomposed into directional subbands using the BDFB or the UDFB, depending on the application.

Another straightforward Bamberger pyramid can be formed by combining a J-level Laplacian pyramid with the BDFB [4,30]. The analysis structure is presented in Figure 25.21. At the high- and mid-frequency levels the subbands can be processed with the BDFB. If required, the UDFB can be used in

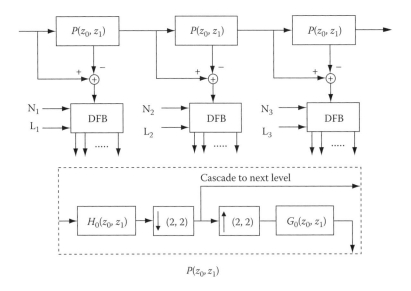

FIGURE 25.21 Bamberger pyramid using the Laplacian pyramid structure combined with the BDFB.

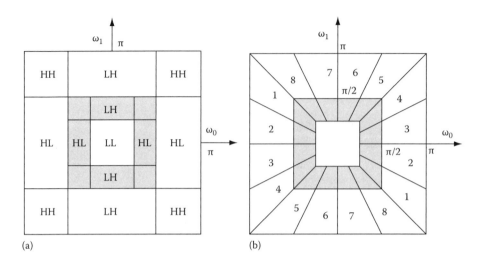

FIGURE 25.22 Comparison of the frequency plane partitioning of the 2-D separable DWT (a) and the proposed directional pyramids (b).

place of the BDFB. More generally the directional decomposition can be designed independently at each resolution. For instance, the number of subbands and the order of the $\beta_i(z)$ filters can be chosen independently. Since the polar components of Bamberger pyramids are invertible, it is easy to see that the overall system is PR.

For the maximally decimated case, the Laplacian-BDFB structure increases the data by approximately a factor of $\frac{4}{3}$. For the cases where the UDFB is used, the increase in data is significantly larger. For the case where all radial bands are decomposed into an N-band UDFB, the data increase is given by a factor of $\frac{4N}{3}$. Where shift invariance is needed at all resolutions and orientations, we can remove all downsampling operations from the Laplacian pyramid and modify the lowpass kernels resulting in $H_0(z_0^{2^j}, z_1^{2^j})$ and $G_0(z_0^{2^j}, z_1^{2^j})$ at each resolution level, where $j = 0, 1, \ldots, J - 1$. When combined with the BDFB, the resulting Bamberger pyramid is overcomplete by a factor of J. If the UDFB is used, the data increases by a factor of $N(J - 1) + 1$.

A comparison of the frequency plane partitioning obtained with the Bamberger pyramids described above and the traditional separable 2-D DWT (and its undecimated version) is shown in Figure 25.22. The 2-D DWT has limited angular sensitivity (mixed-diagonal, horizontal, and vertical directions), while Bamberger pyramids can have 2^N directional subbands.

25.8 Applications

The BDFB has proven to be a useful tool for different problems in image processing and analysis. In this section, we present a summary of different applications reported in the literature. Detailed development and discussion is out of the scope of this work and can be obtained from the references.

25.8.1 Texture Analysis and Segmentation

Texture is an important characteristic present in many images that often contains useful information, which can be exploited in analysis applications. Multichannel methods are among the best in exploiting texture information and include Gabor decompositions, DCTs, DSTs, wavelets, wavelet packets, and dual-tree complex wavelets [35–37]. Decompositions based on the BDFB provide some of the best results for texture classification and segmentation reported in the literature [5,19,30,38,39].

Texture information is distributed across different scales and orientations according to the structural and statistical properties of the image and thus useful texture components can often be separated with a multichannel transform. A simple way to obtain a texture characterization is to measure the subband energies and form a feature vector **f**. Such feature vectors can be used in combination with a classifier or clustering algorithm for recognition or segmentation.

In texture classification the objective is to assign an unknown texture sample to a particular class within a set of known texture classes. The Bayes minimum distance classifier has been used extensively in texture classification [29,35,38,40,41]. This classifier is a supervised scheme that requires the estimation of the mean feature vector and feature covariance matrix for each class. The BDFB was evaluated for texture classification using the Brodatz data set [42] and the methodology described in Ref. [35]. BDFB-based classifiers provide some of the best results for this data set, achieving 99.62% correct classification using 10 features [30,38]. The BDFB results are comparable to the tree-structured wavelet transform and discrete sine transform reported as the best performers in Ref. [35].

In practice, classification systems have to deal with geometrical distortions like rotation. For this problem, rotation-invariant features can be attractive. Multichannel decompositions that split the frequency plane into directionally selective channels can achieve rotation invariance using a DFT-encoding step [39,43,44]. In this approach, a feature vector **f** is formed from the set of directional subbands at each resolution. Then, a DFT is computed for the features at each resolution separately. The resulting DFT coefficients are grouped in a vector **F**. A rigid texture rotation is encoded in **f** as a circular shift which in **F** is encoded as a complex exponential factor. A rotation-invariant feature set is obtained by taking the magnitude of **F** and discarding half of its components.

For instance, in Ref. [38] a Bamberger pyramid is used as the front end of a rotation-invariant classification system. The pyramid consists of an undecimated J-level Laplacian pyramid and an N-band BDFB which achieve the frequency plane partitioning shown in Figure 25.22b. A rotation-invariant feature set is obtained after applying DFT encoding to each pyramid resolution separately. Using the Bayes distance classifier, the system was tested with the data set introduced by Haley and Manjunath [41], which consists of 13 texture classes, each scanned at rotations of $0°, 30°, 60°, 90°, 120°$, and $150°$. The BDFB-based system obtained correct classification rates of 96.96% with 15 features. These results compare favorably with those reported by Haley and Manjunath [41] using a class of 2-D analytic Gabor wavelets; in their system they achieve 96.8% correct classification using a feature vector with 204 components.

Multichannel schemes have also been applied to texture segmentation. In this case a feature vector is used for each pixel in order to capture the local rather than global texture cues. A generic segmentation system is presented in Figure 25.23. The nonlinear operations are used as limiters that control the presence of outliers. For each spatial location (n, m) the local energy is estimated across all subbands. A weighted sum of squared coefficients is typically used in this step, usually composed of a squaring operation followed by a Gaussian smoothing operator. The local energy estimates are then grouped to form feature vectors. In the final step, a classifier is used to produce a segmentation map that assigns a label to each (n, m) location. Bamberger pyramids are well suited for this purpose [19,30]. A segmentation example is presented in Figure 25.24 that corresponds to an undecimated Bamberger pyramid with a redundancy factor of JN where $J = 4$ and $N = 8$. The learning vector quantization (LVQ) algorithm from Kohonen [45] was selected as the classifier following the work in Ref. [36]. The segmentation error

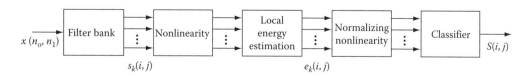

FIGURE 25.23 Classical segmentation system based on multichannel filtering.

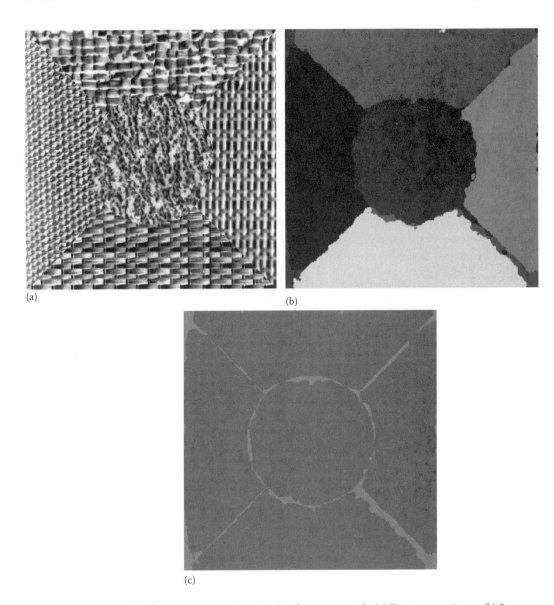

FIGURE 25.24 Example of texture segmentation using Bamberger pyramids. (a) Five texture mixture. (b) Segmentation map identifying the five texture classes. (c) Segmentation error, localized mainly along the texture boundaries.

for this example is 4.67%. A comparative study [30] showed that on average Bamberger pyramids provide the best performance compared to other multichannel decompositions like the DCT, DWT, undecimated uniform filter banks [36], and the DT-CWT [37].

25.8.2 Image Denoising Using Bamberger Pyramids

Denoising is a popular application in which wavelets have been employed. The goal in denoising is to enhance the quality of an image that has been contaminated by additive noise, often assumed to be Gaussian white noise. The basic wavelet denoising procedure consists of first decomposing the noisy image into subbands, after which each subband is processed by a subband-specific nonlinearity. In the last step, the processed subbands are recombined in a synthesis transform to reconstruct a noise suppressed version of the noisy original. Since the forward and inverse transforms amount to an identity system, it is evident that the noise suppression can be directly attributed to the nonlinear operations performed on the subbands.

Typically, the nonlinearity is a shrinking or coring operation which takes a subband coefficient and modifies its magnitude. Small coefficients tend to be suppressed and large coefficients maintain their values. A commonly used operator for images is soft thresholding, whereby a subband coefficient $x(n_0, n_1)$ is modified to $\hat{x}(n_0, n_1) = \text{sgn}(x(n_0, n_1))(|x(n_0, n_1)| - T)$ when the coefficient magnitude is greater than T, and the coefficient is set to zero otherwise. The value of the threshold is set explicitly within each subband or adapted individually to each coefficient based on some criterion, such as energy or statistical characteristics [46,47].

This approach was first explored with maximally decimated filter banks and later with undecimated transforms. It turns out that better results are generally achieved with undecimated (shift-invariant) decompositions.

Since the Bamberger pyramid, whose decomposition is shown in Figure 25.22, provides both radial and angular subband resolution, one might imagine that it would perform well in a denoising application [47,48]. In fact such is the case as shown by the comparisons in Figure 25.25. Bamberger pyramids can provide better directional selectivity across resolutions along with shift invariance when the UDFB is used. Shown in Figure 25.25c is the denoising result for an undecimated Bamberger pyramid with frequency plane partitioning similar to the Steerable pyramid [31]. The midband pyramid levels in this particular system are decomposed with an eight-band UDFB.

For threshold selection, the spatially adaptive wavelet thresholding (SAWT) algorithm [46] was used, where a threshold is computed for each subband coefficient using local statistics under a Bayesian framework.

(a)   (b)   (c)

FIGURE 25.25 Denoising results using Lenna (a) Image with additive white Gaussian noise with $\sigma = 22.5$. (b) Denoised image using the UDWT. (c) Denoised image using Bamberger pyramids.

Also considered is a similar system using the undecimated DWT (UDWT) instead of the UDFB. The denoised image from the DWT system is shown in Figure 25.25b. Both the DWT and DFB systems have good performance as can be seen in contrast to the original image with visible additive noise shown in Figure 25.25a. Both UDWTs and DFBs are good choices for denoising applications.

25.8.3 Fingerprint Enhancement and Recognition

The analysis of fingerprint images can be a challenging problem. The obvious goal is to identify a fingerprint image as belonging to a particular individual from among a huge set of candidates. DFBs can play a role in the recognition process. Park et al. [49] proposed a new image-based fingerprint matching method that is robust to diverse rotations and translations of an input fingerprint. This scheme does not require minutiae extraction, as is the typical approach. Rather, the scheme of Park represents the fingerprint in terms of directional energies. The area within a certain radius around a reference point is used as a region of interest (ROI) for feature extraction. Fingerprint features are then extracted from the ROI using the BDFB. More specifically, the ROI for each subband is divided into blocks and directional energy values are calculated for each block. Only the blocks with dominant energy are retained, while the rest of the directional energies are set to zero, which effectively treats them as noise. As part of the matching process, rotational and translational alignments between the input and template are performed through a normalized Euclidean distance. Experimental results reported by Park et al. [49] demonstrate that the proposed DFB method has comparable verification accuracy to the other leading techniques while having faster processing speed and greater robustness to positional variation.

25.8.4 Iris Recognition

Another biometric identification system that has received attention in recent years uses iris patterns. The physiological characteristics of the iris provides a biometric difficult to modify or reproduce by synthetic methods. The work pioneered by Daugman [50] generates an iris code based on localized energy measurements of the iris texture using 2-D multiscale directional Gabor filters. As is well known the implementation of Gabor filter banks is computationally expensive and the frequency selectivity is limited. The BDFB has been explored as an alternative to Gabor filter banks. Helen Sulochana and Selvan [51] reported on a system that produces a feature vector from the BDFB subbands by dividing each subband into 9×9 blocks and calculating the energy from each block. The energies are then thresholded to form a binary iris signature. The BDFB-based system has similar performance to the leading systems, but is less complex and faster.

25.8.5 Finite-Field DFBs

BDFBs can also be applied to binary images using binary arithmetic. An important property of the ladder structure is that it allows for a straightforward implementation of nonlinear filter banks with PR. Since ladder steps are added in the analysis and subtracted in lock step in the synthesis, exact reconstruction is preserved regardless of the kind of ladder step operation being performed. Consequently nonlinear operations can be accommodated, such as quantization, rounding, rank order filtering, and so on. For instance, true integer-to-integer wavelet transforms are possible by adding rounding operations after each filtering step.

Since FFBs can be realized with ladder structures, nonlinear BDFBs can be constructed. A particular case of interest is processing bilevel or binary images with binary arithmetic. Randolph and Smith [52] reported on a binary BDFB that produces directional subbands that are also binary. To achieve binary valued subbands, a threshold is applied after filtering by $\beta(Z_0)\beta(Z_1)$ in the two-stage ladder FFB. The threshold maps negative values to zero and positive values to one. All remaining operations in the ladder

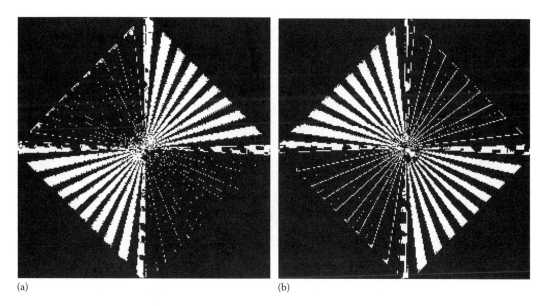

(a)   (b)

FIGURE 25.26   Output of a two-channel binary BDFB. The subbands are binary valued.

structure are implemented using modulo-two arithmetic. An example of the output for a two-channel binary BDFB is shown in Figure 25.26 for a binary version of the image in Figure 25.13.

The binary BDFB has been explored in the context of printed character enhancement and character recognition. Specifically, it has been applied to rotation and scale-invariant character recognition [53] and enhancement of low-resolution documents [54,55].

25.8.6 Velocity Selective Filter Banks

Motion analysis is a critical element in video compression, object tracking, computer vision, and situation analysis. Popular approaches for analyzing motion have included optical flow and block matching techniques. An alternative, explored more recently, has been the use of velocity tuning filters where motion is determined by looking at the spatiotemporal distribution of energy along planes in the 3-D frequency domain. As objects move, they trace a trajectory in time captured by their spatial displacement from frame to frame. In Ref. [56], object trajectories have been extracted using a 3-D continuous wavelet transform (CWT). This transform can be tuned to find motion using spatial translation, temporal translation, scale, velocity magnitude, and velocity orientation. Such flexibility comes with a high computational cost. However, velocity selective filter banks (VSFB) have been reported in Refs. [11,57] that provide an attractive alternative to the oversampled CWT. The implementation boils down to a 3-D generalization of the BDFB which requires the use of full-rate or undecimated subbands. The VSFB is spatiotemporally separable meaning that the BDFB is computed for each individual frame and then subbands along a specific orientation across different frames are grouped and processed temporally using the BDFB. The VSFB produces 3-D wedges as depicted in Figure 25.27a. This wedge captures motion for objects moving along the directions over the fixed velocities captured by the temporal aperture of the wedge. The CWT is capable of additionally separating information across scales as shown in Figure 25.27b. Compared to the CWT, the VSFB is constrained to the speed and position resolutions determined by the number of directional subbands. On the other hand, its computational complexity is an order of magnitude lower than the CWT.

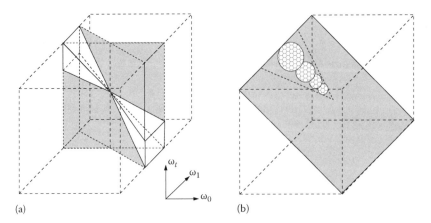

FIGURE 25.27   Comparison of the 3-D frequency supports generated by velocity tuning filters (a) VSFB (b) CWT.

25.9 Closing Remarks

DFBs have many properties that are both interesting and important. They can provide decompositions with a wide variety of angular resolutions, the filters can be designed to have good passband characteristics, and they can be designed to have exact reconstruction. Furthermore, DFBs can accommodate a range of decimation factors for the subbands, from maximally decimated to undecimated. This flexibility can be important in applications where either a compact representation is desirable, like image compression, or applications where shift invariance is deemed important, as is the case with image denoising.

This chapter is not intended as a comprehensive review of directional transforms, but rather an overview of the basic multidimensional theory, the DFB tree structure, the design and efficient implementation of the filter bank, and some applications that have been considered. The interested reader seeking a more comprehensive discussion of this material is directed to the references provided at the end of this chapter.

References

1. Roberto H. Bamberger and Mark J. T. Smith, Filter banks for directional filtering, in *Proc. 1989 IASTED Int. Conf. Adaptive Knowledge-Based Control Signal Process.*, Honolulu, Hawaii, August 1989, pp. 105–109.
2. Roberto H. Bamberger, The directional filterbank: A multirate filter bank for the directional decomposition of images, PhD thesis, Georgia Institute of Technology, November 1990.
3. Roberto H. Bamberger and Mark J. T. Smith, A filter bank for the directional decomposition of images, *IEEE Trans. Signal Process.*, 40(4), 882–893, April 1992.
4. M. N. Do and M. Vetterli, Pyramidal directional filter banks and curvelets, in *Proc. IEEE Int. Conf. Image Process.*, Thessaloniki, Greece, October 2001, pp. 158–161.
5. K. O. Cheng, N. F. Law, and W. C. Siu, Multiscale directional filter bank with applications to structured and random texture retrieval, *Pattern Recognit.*, 40(4), 1182–1194, April 2007.
6. Y. Lu and M. N. Do, Multidimensional directional filter banks and surfacelets, *IEEE Trans. Image. Process.*, 16, 918–931, April 2007.
7. R. Eslami and H. Radha, A new family of nonredundant transforms using hybrid wavelets and directional filter banks, *IEEE Trans. Image Process.*, 16, 1152–1167, April 2007.
8. P. P. Vaidyanathan, *Multirate Systems and Filter Banks*, Prentice Hall, Englewood Cliffs, NJ, 1933.

9. E. Viscito and J. P. Allebach, The analysis and design of multidimensional FIR perfect reconstruction filter banks for arbitrary sampling lattices, *IEEE Trans. Circuits Syst.*, 38(1), 29–41, January 1991.

10. Sang-Il Park, Mark J. T. Smith, and Russell M. Mersereau, A new directional filter bank for image analysis and classification, in *Proc. 1999 IEEE Int. Conf. Acoust., Speech Signal Process.*, October 1999, vol. 2, pp. 1286–1290, Chicago, IL.

11. Sang-Il Park, New directional filter banks and their application in image processing, PhD thesis, Georgia Institute of Technology, Atlanta, GA, November 1999.

12. I. Shah and A. Kalker, Theory and design of multidimensional QMF sub-band filters from 1-d filters using transforms, in *Proc. 4th Int. Conf. Image Process Appl.*, April 1992, pp. 474–477.

13. T. Chen and P. P. Vaidyanathan, Multidimensional multirate filters and filter banks derived from one dimensional filters, *IEEE Trans. Signal Process.*, 41(5), 1749–1765, May 1993.

14. D. Tay and N. Kingsbury, Flexible design of multidimensional perfect reconstruction FIR 2-band filters using transformations of variables, *IEEE Trans. Image Process.*, 2, 466–480, October 1993.

15. J. Kovacevic and M. Vetterli, Nonseparable two- and three-dimensional wavelets, *IEEE Trans. Signal Process.*, 43(5), 1269–1273, May 1995.

16. See-May Phoong, Chai W. Kim, P. P. Vidyanathan, and Rashid Ansari, A new class of two-channel biorthogonal filter banks and wavelet bases, *IEEE Trans. Signal Process.*, 43(3), 649–665, March 1995.

17. Rashid Ansari, Chai W. Kim, and M. Dedovic, Structure and design of two-channel filter banks derived from a triplet of halfband filters, *IEEE Trans. Circuits Syst.-II: Anal. Dig. Signal Process.*, 46(12), 1487–1496, December 1999.

18. R. Ansari, Efficient IIR and FIR fan filters, *IEEE Trans. Circuits Syst.*, CAS-34(8), 941–945, August 1987.

19. J. G. Rosiles and M. J. T. Smith, Texture segmentation using a biorthogonal directional decomposition, in *Systematics, Cybernetics and Informatics 2000*, July 2000, Orlando, FL.

20. J. G. Rosiles and M. J. T. Smith, A low complexity undecimated directional image decomposition, in *ICIP*, 2003, vol. 1, pp. 1049–1052.

21. J. D. Johnston, A filter family designed for use in quadrature mirror filter banks, in *Proc. IEEE Int. Conf. Acoust. Speech Signal Process.*, April 1980, 291–294.

22. Mark J. T. Smith and Steven L. Eddins, Analysis/synthesis techniques for subband image coding, *IEEE Trans. Acoust. Speech Signal Process.*, 38(8), 1446–1456, August 1990.

23. I. Daubechies and W. Sweldens, Factoring wavelet transforms into lifting steps, *J. Fourier Anal. Appl.*, 4(3), 245–267, 1998.

24. T. A. C. M. Kalker and I. A. Shah, Ladder structures for multidimensional linear phase perfect reconstruction filter banks and wavelets, in *SPIE Visual Communications and Image Processing'92*, 1992, vol. 1818, pp. 2–20.

25. L. Tolhuizen, H. Hollmann, and T. A. C. M. Kalker, On the realizability of biorthogonal, m-Dimensional two-band filter banks, *IEEE Trans. Signal Process.*, 43(3), 640–648, March 1995.

26. A. R. Calderbank, Ingrid Daubechies, Wim Sweldens, and Boon-Lock Yeo, Lossless image compression using integer to integer wavelet transforms, in *Proc. IEEE Int. Conf. Image Process.*, Washington, DC, October 1997, vol. 1, pp. 596–599.

27. V. Delouille, M. Jansen, and R. von Sachs, Second generation wavelet methods for denoising of irregularly spaced data in two dimensions, *Signal Process.*, 86(7), 1435–1450, 2006.

28. D. B. H. Tay and M. Palaniswami, A novel approach to the design of the class of triplet halfband filter banks, *IEEE Trans. Circ. Sys.-II*, 51(7), 378–383, July 2004.

29. M. Unser, Texture classification and segmentation using wavelet frames, *IEEE Trans. Image Process.*, 4(11), 1549–1560, November 1995.

30. Jose Gerardo Gonzalez Rosiles, Image and texture analysis using biorthogonal angular filter banks, PhD thesis, Georgia Institute of Technology, July 2004.

31. E. P. Simoncelli, W. T. Freeman, E. H. Adelson, and D. J. Heeger, Shiftable multi-scale transforms, *IEEE Trans. Inf. Theory Special Issue Wavelets*, 38(2), 587–607, March 1992.

32. Nick G. Kingsbury, The dual-tree complex wavelet transform: A new efficient tool for image restoration and enhancement, in *Proc. Euro. Signal Process. Conf., EUSIPCO 98*, September 1998, vol. 1, pp. 319–322, Rhodes.

33. Alan C. Bovik, M. Clark, and W. S. Geisler, Multichannel texture analysis using localized spatial filters, *IEEE Trans. PAMI*, 12, 55–73, 1990.

34. A. K. Jain and F. Farrokhnia, Unsupervised texture segmentation using gabor filters., *Pattern Recognit.*, 24(12), 1167–1186, 1991.

35. Tianhorng Chang and C.-C. Jay Kuo, Texture analysis and classification with tree-structured wavelet transform, *IEEE Trans. Image Process.* 2(4), 429–441, 1993.

36. T. Randen and J. H. Husøy, Filtering for texture classification: A comparative study, *IEEE Trans. PAMI*, 21(4), Apr. 1999.

37. Peter De Rivaz, Complex wavelet based image analysis and synthesis, PhD thesis, University of Cambridge, Cambridge, U.K., 2000.

38. J. G. Rosiles and M. J. T. Smith, Texture classification with a biorthogonal directional filter bank, in *ICASSP*, 2001, 3, 1549–1552.

39. J. G. Rosiles, M. J. T. Smith, and R. M. Mersereau, Rotation invariant texture classification using bamberger pyramids, in *IEEE Int. Conf. Multimedia Exp.*, July 2005.

40. M. Unser, Local linear transforms for texture measurements, *Signal Process.*, 11(1), 61–79, July 1986.

41. G. M. Haley and B. S. Manjunath, Rotation-invariant texture classification using a complete space-frequency model, *IEEE Trans. Image Process.*, 2(8), 255–269, February 1999.

42. P. Brodatz, *Textures, A Photographic Album for Artists and Designers*, Dover Publications, New York, 1966.

43. H. Greenspan, S. Belongie, R. Goodman, and P. Perona, Rotation invariant texture recognition using a steerable pyramid, in *IEEE Proc. Int. Conf. Image Process.*, Jerusalem, Israel, October 1994.

44. P. R. Hill, D. R. Bull, and C. N. Canagarajah, Rotationally invariant texture features using the dual-tree complex wavelet transform, in *IEEE Proc. Int. Conf. Image Process.*, September, Vancouver, Canada, 2000.

45. T. Kohonen, The self-organizing map, *Proc. IEEE*, vol. 78, pp. 1464–1480, September 1990.

46. S. Grace Chang, Bin Yu, and Martin Vetterli, Spatially adaptive wavelet thresholding with context modeling for image denoising, *IEEE Trans. Image Process.*, 9(9), 1522–1531, September 2000.

47. J. Portilla, V. Strela, M. Wainright, and E. Simoncelli, Image denoising using gaussian scale mixtures in the wavelet domain, *IEEE Trans. Image Process.*, 12(11), 1338–1351, November 2003.

48. A. Achim, P. Tsakalides, and A. Bezerianos, Sar image denoising via bayesian wavelet shrinkage based on heavy-tailed modeling, *IEEE Trans. Geo. Rem. Sen.*, 41(8), 1773–1784, August 2003.

49. C.-H. Park, J.-J. Lee, M. J. T. Smith, S.-I. Park, and K.-H. Park, Directional filter bank-based fingerprint feature extraction and matching, *IEEE Trans. Circuits Syst. Video Tech.*, 14(1), 74–85, January 2004.

50. J. Daugman, High confidence visual recognition of persons by a test of statistical independence., *IEEE Trans. Pattern Anal. Mach. Intell.*, 15(11), 1148–1161, November 1993.

51. C. Helen Sulochana and S. Selvan, Iris feature extraction based on directional image representation, *ICGST Int. J. Graph., Vis. Image Process.*, 06, pp. 55–62, 2007.

52. T. R. Randolph and M. J. T. Smith, A directional representation for binary images, in *Ninth DSP Workshop (DSP 2000)*, October 2000.

53. T. Randolph, Image compression and classification using nonlinear filter banks, PhD thesis, Georgia Institute of Technology, May 2001.

54. T. R. Randolph and M. J. T. Smith, Enhancement of fax documents using a binary angular representation, in *Proc. 2001 Int. Symp. Intell. Multimedia, Video Speech Process.*, May 2001, pp. 125–128.

55. T. R. Randolph and M. J. T. Smith, Fingerprint image enhancement using a binary angular representation, in *ICASSP*, May 2001.

56. S.-I. Park, M. J. T. Smith, and R. M. Mersereau, A new motion parameter estimation algorithm based on the continuous wavelet, *IEEE Trans. Image Process.*, 9(5), 873–888, May 2000.

57. F. A. Mujica, R. Murenzi, M. J. T. Smith, and S-I Park, Motion estimation using the spatio-temporal continuous wavelet transform: new results and alternative implementations, in *Proc. IEEE Int. Conf. Image Process.*, September 2000, vol. 2, pp. 550–553.

58. R. Ansari and C.-L. Lau, Two-dimensional IIR filters for exact reconstruction in tree-structured sub-band decomposition, *Elec. Letters*, 23(12), 633–634, June 1987.

59. R. Ansari and C. Guillemot, Exact reconstruction filter banks using diamond FIR filters, in *Proc. Bilcon 1990*, Elsevier Press, Amsterdam, The Netherlands, pp. 1412–1424, July 1990.

60. R. Ansari, A. E. Cetin, and S. H. Lee, Subband coding of images using nonrectangular filter banks, in *Proc. SPIE Conf. Appl. Digital Image Process*, August 1988, vol. 974, pp. 315–323.

61. C. W. Kim and R. Ansari, Subband decomposition procedure for Quincunx sampling grids, in *Proc. SPIE Conf. Visual Comm. Image Process.*, pp. 112–123, November 1991.

26

Nonlinear Filtering Using Statistical Signal Models

Kenneth E. Barner
University of Delaware

Tuncer C. Aysal
Cornell University

Gonzalo R. Arce
University of Delaware

26.1 Introduction

Linear methods that satisfied the principle of superposition dominate current signal processing theory and practice. Linear signal processing is founded in the rich theory of linear systems, and in many applications linear signal processing methods prove to be optimal. Moreover, linear methods are inherently simple to implement, with their low computational cost, perhaps the dominant reason for their widespread use in practice. While linear methods will continue to play a leading role in signal processing applications, nonlinear methods are emerging as viable alternative solutions.

The rapid emergence of nonlinear signal processing algorithms is motivated by the growth of increasingly challenging applications, for instance in the areas of multimedia processing and communications. Such applications require the use of increasingly sophisticated signal processing algorithms. The growth of challenging applications is coupled with continual gains in digital signal processing hardware, in terms of speed, size, and cost. These gains enable the practical deployment of more sophisticated and computationally intensive algorithms. Thus, nonlinear algorithms and filtering methods are being developed and employed to address an increasing share of theoretical problems and practical applications.

A disadvantage of nonlinear approaches is that, unlike their linear counterparts, nonlinear methods lack a unified and universal set of tools for analysis and design. The lack of unifying theory has led to the development of hundreds of nonlinear signal processing algorithms. These algorithms range from theoretically derived broad filter classes, such as polynomial and rank–order based methods [1–9], to boutique methods tailored to specific applications. Thus the dynamic growth of nonlinear methods and lack of unifying theory makes covering the entirety of such operators in a single chapter impossible. Still, large classes of nonlinear filtering algorithms can be derived and studied through fundamentals that are well founded.

The fundamental approach adopted in this chapter is that realized through the coupling of statistical signal modeling with optimal estimation-based filter development. This general approach leads to a number of well-established filtering families, with the specific filtering scheme realized depending on the estimation methodology adopted and the particular signal model deployed. Particularly amenable to filter development is the maximum likelihood estimation (*M*-estimation) approach. Originally developed in the theory of robust statistics [10], *M*-estimation provides a framework for the development of statistical process location estimators, which, when employed with sliding observation windows, naturally extend to statistical filtering algorithms.

The characteristics of a derived family of filtering operators depend not only on the estimation methodology upon which the family is founded, but also on the statistical model employed to characterize a sequence of observations. The most commonly employed statistical models are those based on the Gaussian distribution. Utilization of the Gaussian distribution is well founded in many cases, for instance due to the central limit theorem, and leads to computationally simple linear operations that are optimal for the assumed environment. There are many applications, however, in which the underlying processes are decidedly non-Gaussian. Included in this broad array of applications are important problems in wireless communications, teletrafic, networking, hydrology, geology, economics, and imaging [11–15]. The element common to these applications, and numerous others, is that the underlying processes tend to produce more large magnitude observations, often referred to as outliers or impulses, than is predicted by Gaussian models. The outlier magnitude and frequency of occurrence predicted by a model is governed by the decay rate of the distribution tail. Thus, many natural sequences of interest are governed by distributions that have heavier tails (e.g., lower tail decay rates) than that exhibited by the Gaussian distribution. Modeling such sequences as Gaussian processes leads not only to a poor statistical fit, but also to the utilization of linear operators that suffer serious degradation in the presence of outliers.

Couplings, an estimation (filtering) methodology with a statistical model appropriate for the observed sequence, significantly improves performance. This is particularly true in heavy tailed environments. As an illustrative example, consider the restoration of an image corrupted by (heavy tailed) salt and pepper noise. Typical sources of salt and pepper include flecks of dust on the lens or inside the camera, or, in digital cameras, faulty CCD elements. Figure 26.1 shows a sample corrupted image, the results of two-dimensional linear and nonlinear filtering, and the true underlying (desired) image. It is clear that the linear filter, unable to exploit the characteristics of the corrupting noise, provides an unacceptable result. On the other hand, the nonlinear filter, utilizing the statistics of the image, provides a very good result. The nonlinear filtering utilized in this example is the median, which is derived to be optimal for certain heavy tailed processes. This appropriate statistical modeling results in performance that is far superior to linear processing, which is inherently based on the processing of light tailed samples.

To formally address the processing of heavy tailed sequences, this chapter first considers sequences of samples drawn from the generalizes Gaussian distribution (GGD). This family generalizes the Gaussian distribution by incorporating a parameter that controls the rate of exponential tail decay. Setting this parameter to 2 yields the standard Gaussian distribution, while for values less than two the GGD tails decay slower than in the standard Gaussian case, resulting in heavier tailed distributions. Of particular interest is the first order exponential decay case, which yields the double exponential, or Laplacian, distribution. The Gaussian and Laplacian GGD special cases warrant particular attention due to their theoretical underpinnings, widespread use, and resulting classes of operators when deployed in an *M*-estimation framework. Specifically, it is shown here that *M*-estimation of Gaussian distributed

(a)   (b)

(c)   (d)

FIGURE 26.1   The original figure depicted in (a) is corrupted with transmission noise, the result of which is given in (b). The received image is then processed with (c) linear (mean) and (d) nonlinear (median) filters.

observations samples leads to traditional linear filtering, while the same framework leads to median filtering for Laplacian distributed samples. Thus as linear filters are optimal for Gaussian processes, the median filter and its weighted generalizations are optimal for Laplacian processes. Median type filters are more robust than linear filters and operate more efficiently in impulsive environments, characteristics that result directly from the heavy tailed characteristic of the Laplacian distribution.

Although the Laplacian distribution has a lower rate of tail decay than the Gaussian distribution, extremely impulsive processes are not well modeled as Laplacian. The GGD family, in fact, is limited in its ability to appropriately model extremely impulsive sequences due to the constraint that, while incorporating freedom in the detail decay rate, the tail decay rate is, nonetheless, restricted to be exponential. Appropriate modeling of such sequences is of critical importance, as a wide variety of extremely impulsive processes are observed in practice. Many such sequences arise as the superposition of numerous independent effects. Examples of which include radar clutter, formed as the sum of many signal reflections from irregular surfaces, the received sum of multiuser transmitted signals observed at a detector in a communications problem, the many impulses caused by the contact of rotating machinery parts in electromechanical systems, and atmospheric noise resulting from the superposition of lightning-based electrical discharges around the globe.

The superposition nature of many observed heavy tailed processes has led to the utilization of α-Stable distributions as signal models for such processes [16–19]. Indeed, the family of α-Stable distributions can be justified by the Generalized Central Limit Theorem [19–25]. Moreover, the distribution family (other than the Gaussian limiting case) possesses algebraic tales, making α-Stable modeling of heavy tailed processes more accurate than exponential tailed GGD family modeling. Although an accurate model for many heavy tailed processes, the utility of the α-Stable family is limited by the fact that only a single heavy tailed distribution in the family possesses a closed form, namely the Cauchy distribution. Thus heavy-tail focused theories and methods derived from the α-Stable family are limited, and based on a single distribution.

To overcome the drawbacks associated with GGD and α-Stable based approaches, we present methods derived from a robust extension to *M*-estimation referred to as *LM*-estimation. The *LM*-estimation formulation yields operators that are significantly more robust than traditional (GGD) *M*-estimation based methods. Moreover, *LM*-estimation is statistically related to, and derives its optimality from, the generalized Cauchy density (GCD). Utilization of GCD derived methods is particularly advantageous in that (1) the GCD is a family of distributions possessing algebraic tail decay rates and that (2) they have closed form expressions. Thus like the α-Stable family, the GCD is an appropriate model for extremely impulsive phenomena. But in contrast to the α-Stable family (and like the GGD), the GCD is a broad family of distributions that can be represented in closed form, thereby presenting a framework from which estimation and filtering procedures can be derived. Thus the GCD combines the advantages of the α-Stable and GGD families (accurate heavy tailed modeling and closed form expressions), while eliminating their respective disadvantages.

Much like the Gaussian and Laplacian distributions represent special cases of importance within the GGD, the GCD possesses several special cases that are worthy of thorough investigation. Specifically, we cover in depth the Cauchy and Meridian distribution special cases of the GCD. These distributions are coupled with the *LM*-estimation framework and shown to yield the Myriad and Meridian filtering operations [26–29]. These filtering classes are proven and shown to be significantly more robust than traditional linear, and even median, filtering. Additionally, they contain a free parameter that controls the level of robustness. This degree of freedom allows for a wide array of filtering characteristics including limiting cases that converge to traditional filtering algorithms. Specifically, the (least robust) limiting case of the Myriad filter is the traditional linear filter, while the median filter is the (least robust) limiting case of the Meridian filter. This illustrates the broad range of filtering characteristics exhibited by GCD-based methods and justifies their in-depth coverage within this chapter.

The remainder of the chapter is organized as follows. Section 26.2 introduces *M*-estimation and couples this approach with the GGD signal modeling. Particular attention is given to the Gaussian and Laplacian distribution special cases and the resulting linear and median filtering operations. The *LM*-estimation robust extension to *M*-estimation is covered in Section 26.3. The link between the GCD family and *LM*-estimation is covered along with in-depth coverage of the Myriad and Meridian special cases. Coverage includes an analysis of the filtering objective functions, limiting special cases, evaluations of properties and characteristics, as well as the presentation of optimization procedures. Applications and numerical examples illustrating and contrasting the capabilities of covered filtering methods are presented in Section 26.4. Specifically considered applications are basedband communications, recently emerging powerline communications, and highpass filtering. Finally, conclusions are drawn and future research directions noted in Section 26.5.

26.2 *M*-Estimation

To address the filtering problem, we begin by formally developing the *M*-estimation framework and the commonly employed GGD statistical model. This development allows the problem to be rigorously addressed, but also presents an intuitive approach from which filtering algorithms can be derived and understood. It is shown that combining the GGD statistical model and a special case of *M*-estimation referred to as maximum likelihood estimation (ML), results in simple norms that, among other applications, can be used to define filtering structures. The norms are distribution specific, and we consider in depth the Gaussian and Laplacian special cases of the GGD showing that the resulting norms are the commonly utilized L_2 and L_1 metrics, respectively. Moreover, the Gaussian distribution and L_2 metric directly lead to the class of linear filters, while the Laplacian and L_1 metric correspond to the family of median filters. The difference in norm and filtering characteristics is directly dependent on the distribution tail decay rates, with those derived from the heavier detailed Laplacian distribution being more robust than those derived from light tailed Gaussian distribution.

26.2.1 Generalized Gaussian Density and Maximum Likelihood Estimation

Perhaps the most fundamental form of estimation is the problem of location estimation. While fundamentally simple, location estimation is easily extended to the filtering problem through utilization of a sliding observation window. To develop location estimation based filtering operations, we first consider the slightly more general problem of ML estimation, which was developed as a special case of M-estimation within the theory of robust statistics [10,30].

To establish the estimation and filtering operators, consider a set of observations (input samples), $\{x(i) : i = 1, 2, \ldots, N\}$, formed from a signal $s(i; \theta)$, with and underlying parameter of interest θ, corrupted by additive noise, i.e.,

$$x(i) = s(i; \theta) + n(i), \tag{26.1}$$

where $n(i)$ represents the additive noise samples that are distributed as $n(i) \sim f_n(\cdot)$. The assumed model is quite general, as is the M-estimation formulation for the underlying parameter of interest, which is stated in the following definition.

Definition 26.1: Given the set of observations $\{x(i; \theta) : i = 1, 2, \ldots, N\}$, the M-estimate of θ is given by

$$\hat{\theta}_M = \arg\min_{\theta} \sum_{i=1}^{N} \rho(x(i) - s(i; \theta)), \tag{26.2}$$

where $\rho(\cdot)$ is defined as the cost function of the M-estimate.

Note that the cost function can take on many forms, and can be tailored to a particular signal model or application. One particularly simple single model is to assume that the observation samples are statistically independent. Coupling this assumption with a cost function tied directly to the statistics of the observed samples, namely $\rho(u) = -\log\{f_n(u)\}$, yields the class of ML estimators. Maximum likelihood estimators have received broad attention, and have been applied across a large array of sample distributions and to a vast number of applications.

The effectiveness of ML estimation depends on the statistical model employed in the cost function, and how well this model represents the observation sequence. The most commonly employed statistical model is the Gaussian distribution, which in its generalized density form is expressed as

$$f_{\text{GGD}}(u) = \frac{\beta}{2\alpha\Gamma(1/\beta)} \exp\left\{-\left(\frac{|u|}{\alpha}\right)^{\beta}\right\}, \tag{26.3}$$

where $\Gamma(\cdot)$ is the gamma function. The parameter α determines the width of the density peak (standard deviation), while β controls the tail decay rate. Generally, α is referred to as the scale parameter while β is called the shape parameter. The GGD can be used to model a broad range of noise processes. Moreover, taking the $n(i)$ terms in Equation 26.1 to be independent, identically distributed (i.i.d.) GGD samples yields a particularly simple form for the ML estimate of θ. That is, Equation 26.2 reduces to a compact, intuitive expression, which is given in the following theorem.

THEOREM 26.1

Consider the set of observations $\{x(i) : i = 1, 2, \ldots, N\}$ corrupted by i.i.d. GGD distributed noise. The ML estimate of θ is the solution to the following minimization problem:

$$\hat{\theta} = \arg\min_{\theta} \sum_{i=1}^{N} |x(i) - s(i; \theta)|^{\beta}. \tag{26.4}$$

This result is obtained by letting $\rho(u) = -\log\{f_{\text{GGD}}(u)\}$ and substituting the result into the *M*-estimation expression, Equation 26.2. To appreciate this result, it is instructive to consider the range of distributions within the GGD. An intuitive appreciation of the distribution family can be obtained through the examination of two special cases, namely the Gaussian and Laplacian distributions. The Gaussian distribution is realized for $\beta = 2$. For all $\beta < 2$ cases, the resulting distributions are heavier detailed than the Gaussian distribution. This is illustrated in Figure 26.2, which shows that density function for several values of β including the Laplacian distribution special case ($\beta = 1$). The figure clearly illustrates the relationship between β and the distribution tail decay rate—decreasing β increases the tail heaviness and vice versa.

A review of the ML estimate in the GGD case, Equation 26.4, makes clear that the distribution tail decay rate directly affects the estimate. Namely, the tail decay rate defines a norm under which the estimate is formulated. Thus, the GGD corresponds directly to norms of the form $\rho(u) = |u|^{\beta}$, where

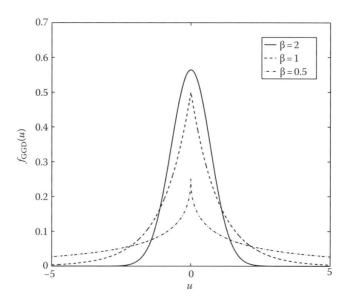

FIGURE 26.2   Generalized Gaussian density function for $\alpha = 1$ and $\beta \in \{2, 1, 0.5\}$. Note that the $\beta = 2$ and $\beta = 1$ cases correspond to the Gaussian and Laplacian distributions, respectively, and that the rate of the tail decay is proportional to β.

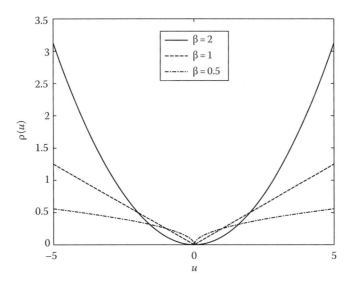

FIGURE 26.3 GCD derived $\rho(u) = |u|^{\beta}$ norms for $\beta \in \{2,1,0.5\}$. Note that the $\beta = 2,1$, and 0.5 cases correspond to the squared, absolute deviation, and fractional lower order moment formulations, respectively, and that the norm robustness is inversely proportional to β.

$\beta \in [0,2]$. The special cases $\beta = 2$, $\beta = 1$, and $\beta < 1$ thus reduce to least-squares, least-absolute deviations, and fractional lower order moment formulations. These norms are widely applied in a range of applications including curve fitting, segmentation, filtering, and the vast majority of optimization problems. The behavior of the L_{β} norms, as they are sometimes referred to, is illustrated in Figure 26.3. Of note is the influence outliers have on the norms. In the $\beta = 2$ (Gaussian/L_2) case, the effect of outliers is squared, while outliers have a linear effect in the $\beta = 1$ (Laplacian/L_1) case and even less influence in the fractional lower order moment case. That is to say that as β decreases, the error norms become more robust, behavior directly linked to the prevalence of outliers in GCD distributions with small β values.

Having established the GGD family of distributions, ML estimation for the additive noise model, and the norms that arise from the GGD–ML coupling, these tools can now be applied to the development of filtering algorithms for specific distribution cases.

26.2.2 Gaussian Statistics: Linear Filtering

To directly address the filtering problem, the observed signal model in Equation 26.1 is simplified to be a direct location function of θ. This yields observation samples of the form

$$x(i) = \theta + n(i) \tag{26.5}$$

and reduces the *M*-estimation expression to simply

$$\hat{\theta}_M = \arg\min_{\theta} \sum_{i=1}^{N} \rho(x(i) - \theta). \tag{26.6}$$

This simplified observation model and estimation structure yields standard operations when combined with commonly utilized distribution models. Consider first the case in which the $n(i)$ observation noise samples are Gaussian distributed,

$$f(u) = \frac{1}{2\sigma^2} \exp\left\{-\left(\frac{|u|^2}{2\sigma^2}\right)\right\}. \tag{26.7}$$

The following theorem addresses this case and shows that the resulting filter is a simple linear operator.

THEOREM 26.2

Consider a set of N independent samples $\{x(i) : i = 1, 2, \ldots, N\}$ each obeying the Gaussian distribution with location θ and variance σ^2. The ML estimate of location is given by

$$\hat{\theta} = \arg\min_{\theta} \left[\sum_{i=1}^{N} (x(i) - \theta)^2\right] = \frac{1}{N} \sum_{i=1}^{N} x(i) = \text{mean}\{x(i) : i = 1, 2, \ldots, N\}. \tag{26.8}$$

This result follows from steps similar to those utilized in Section 26.2.1. The expression in Equation 26.8 shows clearly that the optimization criteria in this case reduces to the L_2 norm. Moreover, the resulting expression can be interpreted as a mean filtering structure, $y = 1/N \sum_{i=1}^{N} x(i)$, where y denotes the filter output. Windowing of the observation sequence can be used to form sliding observation sets, yielding an indexed output $y(i)$ and making the filtering operation explicit.

Although the results of the previous theorem can be interpreted as yielding a filtering operation, the realized operation is somewhat limited in that it does not utilize weights to control the characteristics of the filter. This is a direct result of the somewhat restrictive i.i.d. assumption imposed in the theorem. Fortunately, the identically distributed constraint can be relaxed. This results in a more general signal model and yields a more traditional linear filtering structure that incorporates sample weighting.

THEOREM 26.3

Consider a set of N independent samples $\{x(i) : i = 1, 2, \ldots, N\}$ each obeying the Gaussian distribution with common location θ and (possibly) different variances $\sigma^2(i)$. The ML estimate of location is given by

$$\hat{\theta} = \arg\min_{\theta} \left[\sum_{i=1}^{N} \frac{1}{\sigma^2(i)} (x(i) - \theta)^2\right] = \frac{\sum_{i=1}^{N} h(i)x(i)}{\sum_{i=1}^{N} h(i)} \triangleq \text{mean}\{h(i) \cdot x(i) : i = 1, 2, \ldots, N\}, \tag{26.9}$$

where $h_i = 1/\sigma^2(i) > 0$.

This is simply a linear filtering structure, $y = \sum_{i=1}^{N} \tilde{h}(i)x(i)$, where the $\tilde{h}(i) = h(i)/\sum_{i=1}^{N} h(i)$ terms are the normalized filter weights. As derived in this development, the weights are inversely proportional to individual sample variances. This is an intuitive result, as samples with high variability will be given small weight and have minimal influence on the result. The positivity constraint, however, restricts the resulting operators to the class of smoothers. In practice, this constraint is relaxed enabling the resulting class of linear finite impulse response (FIR) filters to employ both positive and negative weights that provide a wide array of spectral characteristics.

26.2.3 Laplacian Statistics: Median Filtering

The ML-based filter development can be extended to any distribution within the GGD, or, in fact, any valid distribution. Although not all distributions yield compact filtering expressions, specific special cases do correspond to simple, effective filtering structures. To derive such a structure that is more robust to sample outliers than the Gaussian distribution optimal linear filter, consider the heavier tailed Laplacian distribution ($\beta = 1$) special case of the GGD,

$$f(u) = \frac{1}{2\sigma} \exp\left\{-\left(\frac{|u|}{\sigma}\right)\right\}. \tag{26.10}$$

The following theorem shows that the median filter is the optimal operator for Laplacian distributed samples.

THEOREM 26.4

Consider a set of N independent samples $\{x(i) : i = 1, 2, \ldots, N\}$ each obeying the Laplacian distribution with common location θ and variance σ^2. The ML estimate of location is given by

$$\hat{\theta} = \arg\min_{\theta} \left[\sum_{i=1}^{N} |x(i) - \theta|\right] = \text{median}\{x(i) : i = 1, 2, \ldots, N\}. \tag{26.11}$$

The arguments utilized previously, with the appropriate distribution substitution, prove the result. The expression in Equation 26.11 shows that, in this case, the optimization criteria reduces to the more robust L_1 norm. Moreover, the resulting expression is simply a median filter structure, $y = \text{median}\{x(i) : i = 1, 2, \ldots, N\}$, where y denotes the filter output. This operation is clearly nonlinear as the output is formed by sorting the observation samples and taking the middle, or median, value as the output.*

Similarly to the mean filtering case, the median filtering operation can be generalized to admit weights. The theoretical motivation for this generalization is, like in the previous case, the relaxation of the identically distributed constraint placed on the observation samples in the above theorem.

THEOREM 26.5

Consider a set of N independent samples $\{x(i) : i = 1, 2, \ldots, N\}$ each obeying the Laplacian distribution with common location θ and (possibly) different variances $\sigma^2(i)$. The ML estimate of location is given by

$$\hat{\theta} = \arg\min_{\theta} \left[\sum_{i=1}^{N} \frac{1}{\sigma^2(i)} |x(i) - \theta|\right] = \text{median}\{h(i) \diamond x(i) : i = 1, 2, \ldots, N\}. \tag{26.12}$$

where $h_i = 1/\sigma^2(i) > 0$ and $\diamond$ is the replication operator defined as $h(i) \diamond x(i) = \overbrace{x(i), x(i), \ldots, x(i)}^{h(i)\text{times}}$.

The weighting operation in this case is achieved through repetition, rather than the scaling employed in the linear filter. But like the linear case, sample weights are inversely proportional to the sample variances, indicating again that samples with large variability contribute less to the determination of the output than well behaved (smaller variance) samples. This magnitude relationship between a sample's weight and its influence holds even for the relaxed case of positive and negative weights. This relaxation on the weights employs sign coupling and enables a broader range of filtering characteristics to be realized by weighted median (WM) filters [31]:

$$y = \text{median}\{|h(i)| \diamond \text{sgn}(h(i))x(i) : i = 1, 2, \ldots, N\}, \tag{26.13}$$

---

* For cases in which the number of observation samples is an even number, the median value is set as the average of the two central samples in the ordered set.

where

$$\text{sgn}(x) = \begin{cases} 1 & \text{if } x > 0, \\ 0 & \text{if } x = 0, \\ -1 & \text{if } x < 0. \end{cases} \tag{26.14}$$

Considerable analysis is available in the literature on the detail preservation and outliers rejection characteristics of WM filters [8,31–36].

To contrast the performance of linear and WM filters, consider the simple problem of running (constant) location estimation from noisy measurements. Figure 26.4 shows such an example for two noise processes, Laplacian and α-Stable distributed samples. The Laplacian distribution is within the

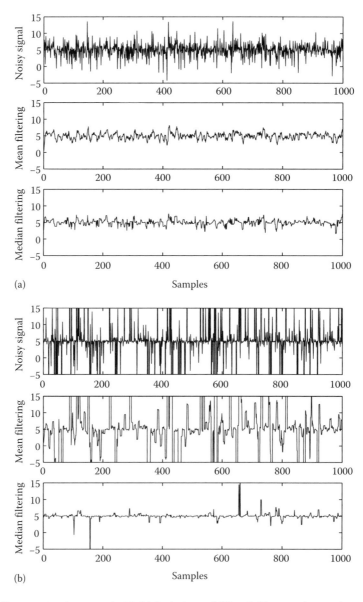

FIGURE 26.4 Constant signal corrupted with (a) Laplacian and (b) α-Stable noise (α = 0.5) processed with mean and median filters.

GGD family, and as such the linear and WM filters perform well in this case, with the Laplacian noise optimal WM returning the best performance. The α-Stable distribution, however, has significantly heavier tails and both linear and WM filters breakdown in this environment. This indicates that GGD-based methods are not well suited to extremely impulsive environments, and that more sophisticated methods for addressing samples characterized by very heavy (algebraic) tailed distributions must be developed and employed.

26.3 *LM*-Estimation

Many contemporary applications contain samples with very heavy tailed statistics including the aforementioned powerline communications, economic forecasting, network traffic processing, and biological signal processing problems [15,37–44]. The GGD family of distributions, while representing a broad class of statistics with varying tail parameters, is, nevertheless, restricted to distributions with an exponential rate of tail decay. Distributions with exponential rates of tail decay are generally considered light tailed, and do not accurately model the prevalence or magnitude of outliers in true heavy-tailed processes. Such processes are often modeled utilizing the α-Stable family of distributions [16–19]. While α-Stable distributions do possess tails with algebraic decay rates, and are thus appropriate models for impulsive sequences, the distribution lacks a full-family closed form expression and it is therefore not easily coupled with estimation techniques such as ML.

To overcome the drawbacks of GGD and α-Stable-based techniques, we derive a generalization of the *M*-estimation framework that exhibits a spectrum of optimality characteristics including greater robustness. This generalization is referred to as *LM*-estimation, the general form of which is given in the following definition.

Definition 26.2: Given the set of independent observations $\{x(i) : i = 1, 2, \ldots, N\}$ formed as $x(i) = s(i; \theta) + n(i)$, the LM-estimate of θ is defined as

$$\hat{\theta}_{LM} = \arg \min_{\theta} \sum_{i=1}^{N} \log\{\delta + \rho(x(i) - s(i; \theta))\}, \qquad (26.15)$$

where $\delta > 0$ and $\rho(\cdot)$ are the robustness parameter and cost function, respectively.

In the following, we show that *LM*-estimation is statistically related to the GCD. The GCD family consists of algebraic detailed distributions with closed form expressions, and is therefore an appropriate model for heavy tailed sequences and a family from which estimation and filtering techniques can be derived. We consider GCD and *LM*-estimation based filters, focusing on the Cauchy and Meridian distribution special cases and their resulting filter structures. Properties of the filters are detailed along with optimization procedures. While the GGD-based results are well established and reported in numerous works, the *LM*-estimation and GCD material presented represents the newest developments in this area, and as such proofs for many results are included.

26.3.1 Generalized Cauchy Density and Maximum Likelihood Estimation

As the previous section covering GGD-based methods shows, the robustness of error norms, estimation techniques, and filtering algorithms derived from a density is directly related to the density tail decay rate. The robustness of *LM*-estimation derives from its statistical relation to the GCD. The GCD function is defined by

$$f_{GCD}(u) = \frac{\lambda \Gamma(2/\lambda)}{2[\Gamma(1/\lambda)]^2} \frac{v}{(v^\lambda + |u|^\lambda)^{2/\lambda}}. \qquad (26.16)$$

As in the GGD case, v is referred to as the scale parameter while λ is called the shape parameter.

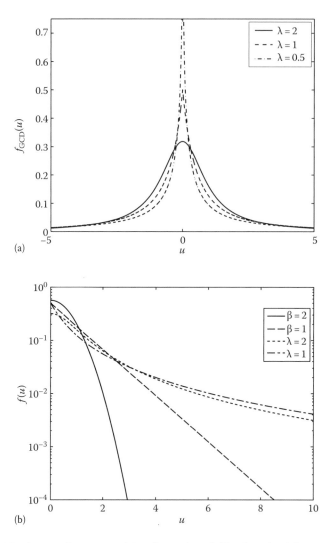

FIGURE 26.5   (a) GCD function for $\nu = 1$ and $\lambda \in \{2,1,0,5\}$; and (b) enlarged tail function for select GGD and GCD distributions.

The Generalized Cauchy distribution was first proposed by Rider in 1957 [45], and "rediscovered" under a different parametrization by Miller and Thomas in 1972 [14]. Distributions within the GCD family have algebraically decaying tails and are thus appropriate models of impulsive sequences. Indeed, the GCD is used in several studies of impulsive radio noise [14,46,47]. Of note within the GCD family are two special cases, namely the Cauchy and Meridian distributions that are realized for $\lambda = 2$ and $\lambda = 1$, respectively [8,28,29]. The GCD is depicted in Figure 26.5a for the $\lambda \in \{2,1,0.5\}$ cases, all with $\nu = 1$. The slow rate of GCD tail decay, which is inversely proportional to λ, is clearly seen in the figure. To make plain the difference in GGD and GCD tail decay rates, Figure 26.5b plots enlargements of tail sections from distributions in each family.

The utility of a heavy-tailed distribution family defined by closed forum expressions is that it can readily be applied in ML estimation, as detailed for the GCD in the following theorem.

THEOREM 26.6

Consider the set of observations $\{x(i) : i = 1, 2, \ldots, N\}$ corrupted by i.i.d. GCD distributed noise. The ML estimate of θ is the solution to the following minimization problem:

$$\hat{\theta} = \arg\min_{\theta} \sum_{i=1}^{N} \log\{v^{\lambda} + |x(i) - s(i; \theta)|^{\lambda}\}. \tag{26.17}$$

Proof: The ML estimate is defined by

$$\hat{\theta} = \arg\max_{\theta} \prod_{i=1}^{N} f_x(x(i)). \tag{26.18}$$

Substituting the GCD expression and denoting $C(\lambda) = \lambda\Gamma(2/\lambda)/(2[\Gamma(1/\lambda)]^2)$ yields

$$\hat{\theta} = \arg\max_{\theta} \prod_{i=1}^{N} C(\lambda) \frac{v}{(v^{\lambda} + |x(i) - s(i; \theta)|^{\lambda})^{2/\lambda}}. \tag{26.19}$$

Taking the natural $\log\{\cdot\}$ and noting that $C(\lambda)$ and v are constant with respect to the maximization of θ gives

$$\hat{\theta} = \arg\max_{\theta} \sum_{i=1}^{N} -\frac{2}{\lambda} \log\{v^{\lambda} + |x(i) - s(i; \theta)|^{\lambda}\}. \tag{26.20}$$

Finally, noting that maximizing $g(u)$ is equivalent of minimizing $-g(u)$, and that $2/\lambda$ is constant with respect to θ, gives the desired result.

  This theorem shows that the ML estimate for GCD distributed samples is simply a special case of LM estimation. Comparing Equations 26.15 and 26.17 shows the equivalence holds for $\rho(\cdot) = |\cdot|^{\lambda}$ and $\delta = v^{\lambda}$. Thus LM estimation is a more general framework, but one that derives its optimality from GCD–ML estimation. Moreover, *M*-estimation is a (least robust) limiting case of *LM*-estimation, as is shown in the following proposition. This theorem utilizes the somewhat simplified case of location estimation (i.e., $s(i; \theta) = \theta$), which we again consider from this point forward as it is most directly related to filter development.

PROPOSITION 26.1

The LM-estimator reduces to an M-estimator as δ tends to infinity, i.e.,

$$\lim_{\delta \to \infty} \hat{\theta}_{LM} = \arg\min_{\theta} \sum_{i=1}^{N} \rho(x(i) - \theta) \tag{26.21}$$

with cost function $\rho(\cdot)$.

Proof: Utilizing the properties of arg min and log functions, we have the following equalities

$$\lim_{\delta \to \infty} \hat{\theta}_{LM} = \lim_{\delta \to \infty} \arg\min_{\theta} \sum_{i=1}^{N} \log\{\delta + \rho(x(i) - \theta)\} \tag{26.22}$$

$$= \lim_{\delta \to \infty} \arg\min_{\theta} \sum_{i=1}^{N} \log\left\{1 + \frac{\rho(x(i) - \theta)}{\delta}\right\} + \log\{\delta\} \qquad (26.23)$$

$$= \lim_{\delta \to \infty} \arg\min_{\theta} \sum_{i=1}^{N} \log\left\{1 + \frac{\rho(x(i) - \theta)}{\delta}\right\} \qquad (26.24)$$

$$= \lim_{\delta \to \infty} \arg\min_{\theta} \sum_{i=1}^{N} \delta \log\left\{1 + \frac{\rho(x(i) - \theta)}{\delta}\right\} \qquad (26.25)$$

$$= \lim_{\delta \to \infty} \arg\min_{\theta} \sum_{i=1}^{N} \log\left\{1 + \frac{\rho(x(i) - \theta)}{\delta}\right\}^{\delta}. \qquad (26.26)$$

Applying the fact that

$$\lim_{\delta \to \infty} \log\left\{1 + \frac{u}{\delta}\right\}^{\delta} = u \qquad (26.27)$$

yields the desired result.

The fact that *M* estimation is a limiting case of *LM* estimation indicates that the latter is a more general family of estimators and that methods derived under this framework are, consequently, more general and subsume those emanating from *M* estimation. Thus, *LM* estimation based methods are inherently more efficient than (or at least equally efficient to) *M* estimation based methods. It should also be recalled that in the *LM* estimation definition, Equation 26.15, δ is referred to as the robustness parameter. Thus the above shows equality between *LM* and *M* estimation at the robustness limit. It is easy to see that this is the least robust limit of *LM* estimation. To make this plain, consider the error norm defined by the GCD in case of Equation 26.15, which can be expressed in simplified form as

$$\rho(u) = \log\left\{1 + \frac{|u|^{\lambda}}{\delta}\right\}^{\delta}. \qquad (26.28)$$

Figure 26.6 plots this norm, showing the effects of varying δ and λ. The impact of δ on robustness is clear, justifying the naming of this parameter and indicating that the $\delta \to \infty$ case is indeed the least robust case. Also evident is the greater robustness of GCD norms over their GGD counterparts, with equality occurring at the limit point (up to a scaling factor). Having established *LM* estimation and the related GCD norms, these tools are now employed to develop filtering algorithms that arise from the consideration of specific distribution cases, namely the Cauchy and Meridian distributions.

26.3.2 Running Myriad Smoothers: Myriad Filtering

Consider first the Cauchy distribution special case ($\lambda = 2$). The following theorem sets the sample myriad as the optimal estimate of location for samples obeying this distribution.

THEOREM 26.7

Consider a set of N independent samples $\{x(i) : i = 1, 2, \ldots, N\}$ each obeying the Cauchy distribution with common location θ and scale γ. The ML estimate of location, or sample myriad, is given by

$$\hat{\theta} = \arg\min_{\theta} \left[\sum_{i=1}^{N} \log\{\gamma^2 + (x(i) - \theta)^2\}\right] = \text{myriad}\{x(i) : i = 1, 2, \ldots, N; \gamma\}, \qquad (26.29)$$

where γ is the linearity parameter.

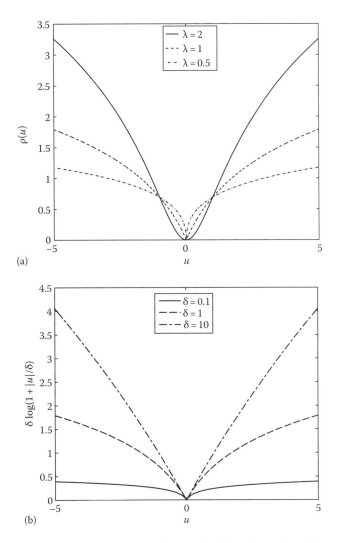

FIGURE 26.6 The error norms defined by the GCD function for (a) $\lambda \in \{0.5,1,2\}$ with $\delta = 1$ and (b) $\lambda = 1$ and $\delta \in \{0.1,1,10\}$.

Note the slight change in notation, γ replacing ν, which we employ to be consistent with the literature [26–28]. Also, the reference to γ as the linearity parameter is made clear in subsequent properties. An appreciation of the myriad operator is obtained through an investigation of the cost function defining the operator. Thus, let $Q(\theta)$ denote the objective function minimized in Equation 26.29, i.e.,

$$Q(\theta) \overset{\Delta}{=} \sum_{i=1}^{N} \log\{\gamma^2 + (x(i) - \theta)^2\}. \tag{26.30}$$

The following proposition brings together a few key properties of the myriad cost function. The properties are illustrated by Figure 26.7, which shows the form of a typical objective function.

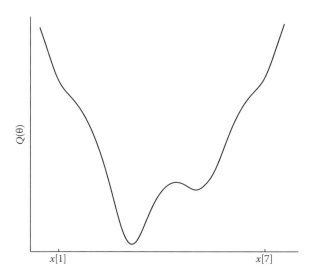

FIGURE 26.7 Typical sketch of the objective function minimized by the myriad operator (for $\gamma = 0.5$ and randomly generated $N = 7$ samples).

PROPOSITION 26.2

Let x[i] signify the order statistics (samples sorted in increasing order of amplitude) of the input samples $\{x(i) : i = 1, 2, \ldots, N\}$, with the smallest x[1] and the largest x[N]. The following statements hold:

(1) *Objective function $Q(\theta)$ has a finite number (at most N) of local extrema.*
(2) *Myriad is one of the local minima of $Q(\theta)$: $Q'(\theta) = 0$.*
(3) *$Q'(\theta) > 0$ [$Q(\theta)$ strictly increasing] for $\theta > x[N]$, $Q'(\theta) < 0$ and [$Q(\theta)$ strictly decreasing] for $\theta < x[1]$.*
(4) *All the local extrema of $Q(\theta)$ lie within the range [x[1], x[N]] of the input samples.*
(5) *Myriad is in the range of input samples: $x[1] \leq \hat{\theta} \leq x[N]$.*

Note that, unlike the mean or median, the definition of the myriad involves the free-tunable parameter γ. Importantly, Proposition 26.1 shows that in the limit of this parameter (or equivalently δ) *LM*-estimation converges to *M*-estimation. A particularly interesting realization of this general result holds for the myriad case. Namely, the myriad converges to the mean in the limiting case, as formally defined in the following.

COROLLARY 26.1

Given a set of samples $\{x(i): i = 1, 2, \ldots, N\}$, the sample myriad $\hat{\theta}$ converges to the sample mean as γ tends to infinity. That is,

$$\lim_{\gamma \to \infty} \hat{\theta} = \frac{1}{N} \sum_{i=1}^{N} x(i). \tag{26.31}$$

The fact that an infinite value of γ converts the nonlinear myriad operation to the linear sample average illustrates why γ is aptly named the linearity parameter: the larger the value of γ, the closer the behavior of the myriad is to the (linear) mean estimator. As γ is decreased, the myriad becomes more robust. In the limit, when γ tends to zero, the estimator treats every observation as a possible outlier, assigning more credibility to the most repeated observation values. This "mode-type" characteristic is reflected in the name mode–myriad given this limiting case.

COROLLARY 26.2

Given a set of samples $\{x(i) : i = 1, 2, \ldots, N\}$, the sample myriad $\hat{\theta}$ converges to a mode–type estimator as $\gamma \to 0$. That is,

$$\lim_{\gamma \to 0} \hat{\theta} = \arg \min_{x(j) \in \mathcal{M}} \left[\prod_{i=1, x(i) \neq x(j)}^{N} |x(i) - x(j)| \right], \qquad (26.32)$$

where $\mathcal{M}$ is the set of most repeated values.

The linearity parameter also allows the meridian filter to address three special cases of the α-Stable distribution family. Those three cases are (1) $\alpha = 1$, which yields the Cauchy distribution for which the myriad is optimal, (2) $\alpha = 2$, which yields the Gaussian distribution under which optimal filtering is realized by letting $\gamma \to \infty$, and (3) $\alpha \to 0$, in which case the distribution is extremely impulsive and $\gamma = 0$ yields the optimal results. These three optimality points have been complemented with a simple empirical formula relating γ to the characteristic exponent (α) and dispersion parameter (κ) of an α-Stable asymmetric distribution,

$$\gamma(\alpha) = \sqrt{\frac{\alpha}{2 - \alpha}} \kappa^{1/\alpha}, \qquad (26.33)$$

which is plotted in Figure 26.8.

Having established the myriad filter, the role of the linearity parameter, and two limiting cases (sample mean and mode–myriad), we present three myriad filter properties that are of importance in signal processing, namely, no under-shoot/overshoot, shift and sign invariance, and unbiasedness. To simplify the notation in the properties, the myriad output is written compactly as $\hat{\theta}(\mathbf{x}) = \text{myriad}\{x(i) : i = 1, 2, \ldots, N; \gamma\}$, where $\mathbf{x} \overset{\Delta}{=} [x(1), x(2), \ldots, x(N)]^T$.

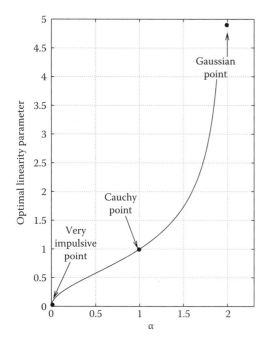

FIGURE 26.8 Empirical α–γ curve for α-Stable distributions. The curve values at $\alpha = 0, 1$, and 2 constitute the optimality points of the α-Stable triplet. (From Gonzalez, J. G. and Arce, G. R., *IEEE Trans. Signal Process.*, 49, 438, 2001. With permission.)

PROPERTY 26.1 (No Undershoot/ Overshoot)

The output of the myriad estimator operating on samples $\{x(i) : i = 1, 2, \ldots, N\}$ is always bounded by

$$x[1] \leq \hat{\theta}(\mathbf{x}) \leq x[N]. \qquad (26.34)$$

PROPERTY 26.2 (Shift and Sign Invariance)

Consider the observation set $\{x(i) : i = 1, 2, \ldots, N\}$ and let $z(i) = x(i) + b$. Then,

(1) $\hat{\theta}(\mathbf{z}) = \hat{\theta}(\mathbf{x}) + b$;
(2) $\hat{\theta}(\mathbf{z}) = -\hat{\theta}(-\mathbf{z})$.

PROPERTY 26.3 (Unbiasedness)

Given a set of samples $\{x(i) : i = 1, 2, \ldots, N\}$ that are independent and symmetrically distributed around a symmetry center c, $\hat{\theta}(\mathbf{x})$ is also symmetrically distributed around c. In particular, if $\mathbb{E}\{\hat{\theta}(\mathbf{x})\}$ exists, then $\mathbb{E}\{\hat{\theta}(\mathbf{x})\} = c$.

The shift invariance and unbiasedness of the myriad filter show that the operator can be applied without concerns to an overall change in location (e.g., a shift in the mean luminance of an image) and that the operator preserves the mean. The undershoot/overshoot property shows that the myriad can be applied without concerns of introducing amplifying artifacts, such as ringing effects.

While the myriad filter offers increased robustness over GGD-based methods and possesses several properties of importance, the equal treatment of samples within the observation set limits the filter, as defined above, to a single operation. Weighting of samples provides a much broader range of filtering operations. This is theoretically justified in a fashion analogous to that utilized in the generalizations of the mean and median operators. Specifically, the following theorem shows that considering samples with a common location but varying scale factors leads to the weighted myriad operator.

THEOREM 26.8

Consider a set of N independent samples $\{x(i) : i = 1, 2, \ldots, N\}$ each obeying the Cauchy distribution with common location θ and (possibly) varying scale factors $s(i) = \gamma/\sqrt{w(i)}$ [8,28]. The ML estimate of location, or weighted myriad, is given by,

$$\hat{\theta} = \arg\min_{\beta} \left[\sum_{i=1}^{N} \log\{\gamma^2 + w(i)(x(i) - \beta)^2\} \right] = \text{myriad}\{w(i) \circ x(i) : i = 1, 2, \ldots, N; \gamma\}, \qquad (26.35)$$

where $\circ$ denotes the weighting operation in the minimization problem.

As in the weighted linear and median filter cases, the weight applied to a sample in the myriad filter is inversely proportional to the sample's variability. This not only minimizes the effect of unreliable samples (those with large scale), but can also be used to exploit correlations between samples. This is the case, for instance, when a sliding window is employed and higher weights are assigned to samples at spatial locations that are most highly correlated with the desired output, e.g., samples further away in time and less correlated are given smaller weight. Weighting thus allows myriad filters to take on a wide array of characteristics and to be tuned to the needs of specific applications. Moreover, the properties detailed above for the unweighted myriad filter also hold in the weighted case. The properties are not restated for the weighted case since they are identical. We do, however, formally state corollaries governing the behavior of the weighted myriad operator at the limiting cases of the linearity parameter.

COROLLARY 26.3

Given a set of samples $\{x(i): i = 1, 2, \ldots, N\}$ and corresponding (positive) weights $\{w(i): i = 1, 2, \ldots, N\}$, the weighted myriad $\hat{\theta}$ converges to a normalized linear estimate as γ tends to infinity. That is,

$$\lim_{\gamma \to \infty} \hat{\theta} = \frac{\sum_{i=1}^{N} w(i)x(i)}{\sum_{i=1}^{N} w(i)}. \tag{26.36}$$

COROLLARY 26.4

Given a set of samples $\{x(i): i = 1, 2, \ldots, N\}$ and corresponding (positive) weights $\{w(i): i = 1, 2, \ldots, N\}$, the weighted myriad $\hat{\theta}$ converges to a weighted mode-type estimate as γ tends to 0. That is,

$$\lim_{\gamma \to 0} \hat{\theta} = \arg \min_{x(j) \in \mathcal{M}} \left(\frac{1}{w(j)} \right)^{r/2} \left[\prod_{i=1, x(i) \neq x(j)}^{N} |x(i) - x(j)| \right], \tag{26.37}$$

where $\mathcal{M}$ is the set of most repeated values and r is the number of times a member of $\mathcal{M}$ is repeated in the sample.

The linearity parameter again controls the behavior of the weighted myriad, ranging between a normalized linear operator, in the least robust limit, and a weighted mode operator, in the most robust limit. Thus weighted myriad filters are more general than linear filters, including the latter as a special case. But unlike linear filters, or even median-based filters, the output of the weighted myriad is not available in explicit form. Computation of the output is therefore nontrivial, requiring minimization of the weighted myriad objective function, $Q(\theta)$. Fortunately, $Q(\theta)$ has a number of characteristics that can be exploited to construct fast iterative methods for computing the objective function minimum.

Recall that the weighted myriad is given by

$$\hat{\theta} = \arg \min_{\theta} Q(\theta)$$

$$= \arg \min_{\theta} \sum_{i=1}^{N} \log \left[1 + \left(\frac{x(i) - \theta}{s(i)} \right)^2 \right], \tag{26.38}$$

where the change in notation, $s(i) = \gamma/\sqrt{w(i)}$, introduced in Theorem 26.8 is employed. As the output is a local minima of $Q(\theta)$, these points can be identified by determining $Q'(\hat{\theta})$, which, after some manipulations, can be written as

$$Q'(\theta) = 2 \sum_{i=1}^{N} \frac{\left(\frac{\theta - x(i)}{s(i)^2} \right)}{1 + \left(\frac{x(i) - \theta}{s(i)} \right)^2}. \tag{26.39}$$

Defining $\psi(v) \triangleq \frac{2v}{1+v^2}$, the following equation is obtained for the local extrema of $Q(\theta)$:

$$Q'(\theta) = -\sum_{i=1}^{N} \frac{1}{s(i)} \cdot \psi \left(\frac{x(i) - \theta}{s(i)} \right) = 0. \tag{26.40}$$

By introducing the positive functions

$$h(i;\theta) \overset{\Delta}{=} \frac{1}{s(i)^2} \cdot \varphi\left(\frac{x(i) - \theta}{s(i)}\right) > 0, \tag{26.41}$$

for $i = 1, 2, \ldots, N$, where $\varphi(v) \overset{\Delta}{=} \frac{\psi(v)}{v} = \frac{2}{1+v^2}$, the local extrema of $Q(\theta)$ in Equation 26.40 can be formulated as

$$Q'(\theta) = -\sum_{i=1}^{N} h(i;\theta) \cdot (x(i) - \theta) = 0. \tag{26.42}$$

This formulation implies that the *sum of weighted deviations* of the samples is zero, with the (positive) weights themselves being functions of θ. This property, in turn, leads to a simple fixed point iterative procedure for computing the myriad weights.

To develop the fixed point algorithm, note that Equation 26.42 can be written as

$$\theta = \frac{\sum_{i=1}^{N} h(i;\theta) \cdot x(i)}{\sum_{i=1}^{N} h(i;\theta)}, \tag{26.43}$$

which is a weighted mean of the input samples, $x(i)$. Since the weights $h(i;\theta)$ are always positive, the right-hand side of Equation 26.43 is in $[x[1], x[N]]$, confirming that all the local extrema lie within the range of the input samples. By defining the mapping

$$T(\theta) \overset{\Delta}{=} \frac{\sum_{i=1}^{N} h(i;\theta) \cdot x(i)}{\sum_{i=1}^{N} h(i;\theta)}, \tag{26.44}$$

the local extrema of $Q(\theta)$, or the roots of $Q'(\theta)$, are seen to be the *fixed points* of $T(\cdot)$:

$$\theta^\star = T(\theta^\star). \tag{26.45}$$

The following fixed point iteration results in an efficient algorithm to compute these fixed points:

$$\theta_{m+1} \overset{\Delta}{=} T(\theta_m) = \frac{\sum_{i=1}^{N} h(i;\theta_m) \cdot x(i)}{\sum_{i=1}^{N} h(i;\theta_m)}. \tag{26.46}$$

In the classical literature, this is also called the *method of successive approximation* for the solution of the equation $\theta = T(\theta)$. This method is proven to converge to a fixed point of $T(\cdot)$, indicating that

$$\lim_{m \to \infty} \theta_m = \theta^\star = T(\theta^\star). \tag{26.47}$$

The speed of convergence depends on the initial value θ_0. A simple approach to initializing the algorithm is to set $\hat{\theta}_0$ equal to the input sample $x(i)$ that yields the smallest cost $P(x(i))$, where $\log(P(\theta)) \overset{\Delta}{=} Q(\theta)$, i.e., $P(\theta) = \prod_{i=1}^{N} \left(\gamma^2 + w(i)(x(i) - \theta)^2\right)$. The fixed point weighted myriad optimization can now be summarized as follows.

Fixed point weighted myriad search

Step 1: Select the initial point $\hat{\theta}_0$ among the values of the input samples:

$$\hat{\theta}_0 = \arg\min_{x(i)} P(x(i)).$$

Step 2: Using $\hat{\theta}_0$ as the initial value, perform L iterations of the fixed point recursion $\theta_{m+1} = T(\theta_m)$, the full expression of which is given in Equation 26.46. The final value of these iterations is then chosen as the weighted myriad output, $\hat{\theta} = T^{(L)}(\hat{\theta}_0)$.

This algorithm is compactly written as

$$\hat{\theta} = T^{(L)}\left(\arg\min_{x(i)} P(x(i)) \right). \tag{26.48}$$

Note that for the special case $L = 0$, no fixed point iterations are performed and the above algorithm computes the selection weighted myriad.

26.3.3 Running Meridian Smoothers: Meridian Filtering

Having established the myriad as the optimal filtering operation that arises from the location estimation problem in Cauchy distributed samples, we consider a second special case of the GCD, namely the Meridian distribution. This special case ($\lambda = 1$) has even heavier tails than the Cauchy distribution, and is an appropriate model for the most impulsive sequences seen in practice. As the Cauchy and Meridian distributions fall within the GCD, the development for the Meridian is similar to that for the Cauchy and so many proofs with similar steps are omitted.

THEOREM 26.9

Consider a set of N independent samples $\{x(i) : i = 1, 2, \ldots, N\}$ each obeying the Myriad distribution with common location θ and scale δ. The ML estimate of location, or sample meridian, is given by

$$\hat{\theta} = \arg\min_{\theta}\left[\sum_{i=1}^{N} \log\{\delta + |x(i) - \theta|\} \right] = \text{meridian}\{x(i) : i = 1, 2, \ldots, N; \delta\}, \tag{26.49}$$

where δ is referred to as the medianity parameter.

As in all previous cases, the performance of the meridian filter is directly related to the defining objective function. Properties describing the objective function are given in the following proposition. The properties described therein are illustrated in Figure 26.9, which plots an example of the objective function, $Q(\theta)$, that results from a set of typical observation samples in the $N = 7$ case.

PROPOSITION 26.3

Let

$$Q(\theta) \stackrel{\Delta}{=} \sum_{i=1}^{N} \log\{\delta + |x(i) - \theta|\}. \tag{26.50}$$

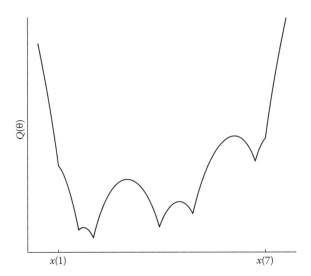

FIGURE 26.9 Typical plot of the objective function minimized by the meridian operator (for $\delta = 0.5$ and randomly generated $N = 7$ samples). (From Aysal, T. C. and Barner, K. E., *IEEE Trans. Signal Process.*, 55, 3949, 2007. With permission.)

The following statements hold:

(1) $Q'(\theta) > 0$ ($Q(\theta)$ is strictly increasing) for $\theta > x[N]$, and $Q'(\theta) < 0$ ($Q(\theta)$ is strictly decreasing) for $\theta < x[1]$.

(2) All local extrema of $Q(\theta)$ lie within the range of input samples, $[x[1], x[N]]$.

(3) Objective function $Q(\theta)$ has a finite number of local minima (at most equal to the number of input samples, N).

(4) Meridian $\hat{\theta}$ is one of the local minima of $Q(\theta)$, i.e., one of the input samples.

The meridian estimator output is hence the input sample that yields the smallest $Q(\theta)$ function value. The selective nature of the meridian estimator, shared with the median estimator, facilitates the filter output computation which is formulated as

$$\hat{\theta} = \arg\min_{\theta \in \mathbf{x}} \sum_{i=1}^{N} \log\{\delta + |x(i) - \theta|\}. \tag{26.51}$$

The behavior of the meridian operator is markedly dependent on the value of δ, which is referred to as the *medianity* parameter. As the name suggests, we show in the following that the sample meridian is equivalent to the sample median for large values of δ, whereas the estimator acquires the form of the sample mode for small δ.

COROLLARY 26.5

Given a set of samples $\{x(i): i = 1, 2, \ldots, N\}$, the sample meridian $\hat{\theta}$ converges to the sample median as $\delta \to \infty$. That is,

$$\lim_{\delta \to \infty} \hat{\theta} = \lim_{\delta \to \infty} \text{meridian}\{x(i) : i = 1, 2, \ldots, N; \delta\} = \text{median}\{x(i): i = 1, 2, \ldots, N\}. \tag{26.52}$$

Thus the family of meridian estimators subsumes the sample median as a limiting case. This simple fact makes the meridian filter class inherently more efficient than (or at least equally efficient to) median filters over all noise distribution including the Laplacian. The opposite limiting case, which results in the mode–meridian operator, is addressed in the following.

COROLLARY 26.6

Given a set of samples $\{x(i) : i = 1, 2, \ldots, N\}$, the sample meridian $\hat{\theta}$ converges to a mode–type estimator as $\delta \to 0$. That is,

$$\lim_{\delta \to 0} \hat{\theta} = \arg\min_{x(j) \in \mathcal{M}} \left[\prod_{i=1, x(i) \neq x(j)}^{N} |x(i) - x(j)| \right], \tag{26.53}$$

where $\mathcal{M}$ is the set of most repeated values.

Thus the myriad and meridian operators converge to a common mode operator as their linearity/medianity parameters go to 0.

Since all the operators covered in this chapter belong to the class of *M-estimators* [10], many robust statistics tools are available for evaluating their robustness [10,48]. *M-estimators* are formulated by a set of implicit functions, where $\rho(\cdot)$ is an arbitrary objective function. Assuming that $\psi(x) = \partial(\rho(x))/\partial x$ exists, the *M-estimator* is obtained by solving

$$\sum_{i=1}^{N} \psi(x(i) - \theta) = 0, \tag{26.54}$$

where $\psi(\cdot)$ is proportional to the so-called influence function. The influence function of an estimator is important in that it determines the effect of contamination on the estimator.

To further characterize *M-estimates*, it is useful to list the desirable features of a robust influence function [10,48]:

- B-robustness. An estimator is *B-robust* if the supremum of the absolute value of the influence function is finite.
- Rejection Point. The *rejection point* defined as the distance from the center of the influence function to the point where the influence function becomes negligible, should be finite. The rejection point measures whether the estimator rejects outliers and, if so, at what distance.

The influence functions for the sample mean, median, myriad, and meridian can be shown to be

$$\psi(x) = 2x, \tag{26.55}$$

$$\psi(x) = \text{sgn}(x), \tag{26.56}$$

$$\psi(x) = \frac{2x}{\gamma^2 + x^2}, \tag{26.57}$$

$$\psi(x) = \frac{\text{sgn}(x)}{\delta + |x|}, \tag{26.58}$$

respectively. Figure 26.10 plots each of these for comparison. Since the influence function of the mean is unbounded, a gross error in the observations leads to severe distortion in the estimate. The mean is

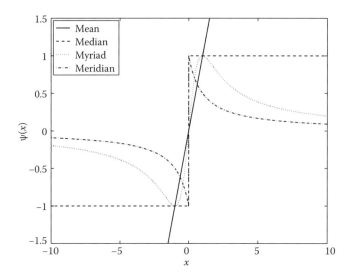

FIGURE 26.10   Influence functions for (solid:) the sample mean, (dashed:) the sample median, (dotted:) the sample myriad, and (dash–dotted:) the sample meridian. (From Aysal, T. C. and Barner, K. E., *IEEE Trans. Signal Process.*, 55, 3949, 2007. With permission.)

clearly not B-robust and its rejection point is infinite. On the other hand, a similar gross error has a limited effect on the median estimate.

While the median is B-robust, its rejection point, like the mean, is not finite. Thus the median estimate is always affected by outliers. The myriad estimate is clearly B-robust and the effect of the errors decreases as the error increases. The meridian estimate is also B-robust, and, in addition, its rejection point is smaller than that of myriad. This indicates that the operators can be ordered from least to most robust as: linear, median, myriad, and meridian.

In addition to desirable influence function characteristics, the meridian possesses the following properties important in signal processing applications.

PROPERTY 26.4

Given a set of samples $\{x(i) : i = 1, 2, \ldots, N\}$, let $\delta < \infty$. Then the meridian output is such that

$$\lim_{x(N) \to \pm\infty} \hat{\theta}(x(1), x(2), \ldots, x(N)) = \hat{\theta}(x(1), x(2), \ldots, x(N-1)). \tag{26.59}$$

According to Property 26.4, large errors are efficiently eliminated by a meridian estimator with a finite medianity parameter. Note that this is not the case for the median estimate, as large positive or negative values can always shift the output. In robust statistics, M-estimators satisfying the outlier rejection property are called redescending. It can be proven that a necessary and sufficient condition for an M-estimator to be redescending is that $\lim_{x \to \pm\infty} \partial/\partial x(\rho(x)) = 0$. Note that this condition holds for meridian and myriad operators, whereas it does not for the mean and median. The meridian also possesses the same no undershoot/overshoot, shift and sign invariance, and unbiasedness properties as the myriad.

PROPERTY 26.5   (No Undershoot/Overshoot)

The output of the meridian estimator operating on samples $\{x(i) : i = 1, 2, \ldots, N\}$ is always bounded by

$$x[1] \leq \hat{\theta}(\mathbf{x}) \leq x[N]. \tag{26.60}$$

PROPERTY 26.6   (Shift and Sign Invariance)

Consider the observation set $\{x(i) : i = 1, 2, \ldots, N\}$ and let $z(i) = x(i) + b$. Then,

(1) $\hat{\theta}(\mathbf{z}) = \hat{\theta}(\mathbf{x}) + b$;
(2) $\hat{\theta}(\mathbf{z}) = -\hat{\theta}(-\mathbf{z})$.

PROPERTY 26.7   (Unbiasedness)

Given a set of samples $\{x(i) : i = 1, 2, \ldots, N\}$ that are independent and symmetrically distributed around a symmetry center c, $\hat{\theta}(\mathbf{x})$ is also symmetrically distributed around c. In particular, if $\mathbb{E}\{\hat{\theta}(\mathbf{x})\}$ exists, then $\mathbb{E}\{\hat{\theta}(\mathbf{x})\} = c$.

The meridian characteristics can be broadened through the introduction of weights. The weighted meridian possesses the same properties as the unweighted version and converges to the expected special cases in the limit of the medianity parameter. The weighted case is formally defined and the limiting cases stated, but the properties are omitted due to their direct similarity to the previous formulations.

THEOREM 26.10

Consider a set of N independent samples $\{x(i) : i = 1, 2, \ldots, N\}$ each obeying the Meridian distribution with common location θ and (possibly) varying scale parameters $v(i) = \delta/w(i)$. The ML estimate of location, or weighted meridian, is given by

$$\hat{\theta} = \arg\min_{\beta} \left[\sum_{i=1}^{N} \log\{\delta + w(i)|x(i) - \theta|\} \right] = \text{meridian}\{w(i) \star x(i) : i = 1, 2, \ldots, N\}, \tag{26.61}$$

where $\star$ denotes the weighting operation in the minimization problem.

COROLLARY 26.7

Given a set of samples $\{x(i) : i = 1, 2, \ldots, N\}$ and corresponding (positive) weights $\{w(i) : i = 1, 2, \ldots, N\}$, the weighted meridian $\hat{\theta}$ converges to the weighted median as $\delta \to \infty$. That is,

$$\lim_{\delta \to \infty} \hat{\theta} = \lim_{\delta \to \infty} \text{meridian}\{w(i) \star x(i) : i = 1, 2, \ldots, N; \delta\} = \text{median}\{w(i) \diamond x(i) : i = 1, 2, \ldots, N\}. \tag{26.62}$$

COROLLARY 26.8

Given a set of samples $\{x(i): i = 1, 2, \ldots, N\}$ and corresponding (positive) weights $\{w(i): i = 1, 2, \ldots, N\}$, the weighted meridian $\hat{\theta}$ converges to one of the most repeated values in the sample set as $\delta \to 0$. That is,

$$\lim_{\delta \to 0} \hat{\theta} = \arg\min_{x_j \in \mathcal{M}} \left[\left(\frac{1}{w(j)} \right)^r \prod_{i=1, x(i) \neq x(j)}^{N} |x(i) - x(j)| \right], \tag{26.63}$$

where $\mathcal{M}$ is the set of most repeated values and r is the number of times a member of $\mathcal{M}$ is repeated in the sample set.

Table 26.1 summarizes the *M*-estimators and filters covered in this chapter. The GGD derived operators, for $\beta = 2$ and $\beta = 1$ are reported on table rows 1, 2 and 5, 6, while the GCD optimal operators, for $\lambda = 2$ and $\lambda = 1$, are reported on rows 3, 4 and 7, 8. This table brings together and contrast the objective functions for the GGD-based linear and median filters and the GCD-based myriad and meridian filters.

26.3.4 Introduction of Real-Valued Weights and Optimization

The coverage to this point has enforced the positivity constraint that arises naturally in each of the operators' estimation-based development. Only in the median case is the sign coupling approach utilized to allow the introduction of real-valued (positive and negative) weights. See Equations 26.13 and 26.14. For convenience, Equation 26.13 is repeated here:

$$\hat{\theta} = \arg\min_{\theta} \left[\sum_{i=1}^{N} |h(i)| \cdot |\text{sgn}(h(i))x(i) - \theta| \right] = \text{median}\{|h(i)| \diamond \text{sgn}(h(i))x(i)|_{i=1}^{N}\}. \tag{26.64}$$

TABLE 26.1   *M*-Estimator and *M*-Smoother (Weighted *M*-Estimator) Objective Functions and Outputs for Various Filter Families

Smoother	Cost Function	Filter Output
Mean	$\sum_{i=1}^{N} (x(i) - \theta)^2$	Mean$\{x(i): i = 1, 2, \ldots, N\}$
Median	$\sum_{i=1}^{N} |x(i) - \theta|$	Median$\{x(i): i = 1, 2, \ldots, N\}$
Myriad	$\sum_{i=1}^{N} \log\{\gamma^2 + (x(i) - \theta)^2\}$	Myriad$\{x(i): i = 1, 2, \ldots, N; \gamma\}$
Meridian	$\sum_{i=1}^{N} \log\{\delta + |x(i) - \theta|\}$	Meridian$\{x(i): i = 1, 2, \ldots, N; \delta\}$
Weighted mean	$\sum_{i=1}^{N} w(i)(x(i) - \theta)^2$	Mean$\{w(i) \cdot x(i): i = 1, 2, \ldots, N\}$
Weighted median	$\sum_{i=1}^{N} w(i)|x(i) - \theta|$	Median$\{w(i) \diamond x(i): i = 1, 2, \ldots, N\}$
Weighted myriad	$\sum_{i=1}^{N} \log\{\gamma^2 + w(i)(x(i) - \theta)^2\}$	Myriad$\{w(i) \circ x(i): i = 1, 2, \ldots, N; \gamma\}$
Weighted meridian	$\sum_{i=1}^{N} \log\{\delta + w(i)|x(i) - \theta|\}$	Meridian$\{w(i) \star x(i): i = 1, 2, \ldots, N; \delta\}$

Source: Aysal, T. C. and Barner, K. E., *IEEE Trans. Signal Process.*, 55, 3949, 2007. With permission.

The most general cases of the myriad and meridian are, by direct analogy, given by

$$\hat{\theta} = \arg\min_{\theta} \left[\sum_{i=1}^{N} \log\{\gamma^2 + |h(i)| \cdot (\text{sgn}(h(i))x(i) - \theta)^2\} \right] = \text{myriad}\{|h(i)| \circ \text{sgn}(h(i))x(i)|_{i=1}^{N}\} \quad (26.65)$$

and

$$\hat{\theta} = \arg\min_{\theta} \left[\sum_{i=1}^{N} \log\{\delta + |h(i)| \cdot |\text{sgn}(h(i))x(i) - \theta|\} \right] = \text{meridian}\{|h(i)| \star \text{sgn}(h(i))x(i)|_{i=1}^{N}\}, \quad (26.66)$$

respectively. Under this formulation, the operators are no longer restricted to be smoothers and can, therefore, be set to exhibit a wide range of frequency-selective filtering operations, including bandpass and highpass filtering.

Also of importance is that the filter weights can be optimized. This is most easily done through a training process, typically under the mean absolute error (MAE). For instance, using a steepest descent approach yields a relatively simple weight update for the myriad case,

$$h(i; n + 1) = h(i; n) - \mu e(n) \left[\frac{\gamma^2 \text{sgn}(h(i))\left(\hat{\theta} - \text{sgn}(h(i))x(i)\right)}{\left(\gamma^2 + |h(i)| \cdot \left(\hat{\theta} - \text{sgn}(h(i))x(i)\right)^2\right)^2} \right], \quad (26.67)$$

where $e(n)$ is simply the error between the filter output and the desired signal and μ is the iteration step size.

It is important to note that the optimal filtering action is independent of the choice of γ; the filter only depends on the value of $w/(\gamma^2)$. Note that similar updates can be derived for the median and meridian filters—all are analogous to the LMS update for the linear filter.

For cases in which no training sequence is available, there are simple, suboptimal approaches to setting the operator parameters. Recall that the myriad has the linearization free parameter (γ), while the meridian has medianization free parameter (δ). These parameters control the robustness of the operators. Numerous optimization techniques exist for linear and median filters. Thus one approach to setting the filter parameters is to design weights for the simpler linear or median filter that address the problem at hand, then decrease the free parameter value to achieve the desired level of additional robustness. This is referred to as myriadization in the myriad case and meridianization in the meridian case. This approach allows the rich set of design tools in the more traditional filtering domains to be applied to the newer, more robust operators.

26.4 Applications of Robust Nonlinear Estimators/Filters

The covered filtering methods are evaluated in two communications problems, standard baseband communications and power line communications, as well as an illustrative frequency selective filtering problem. In each of these applications, an underlying desired signal is to be extracted. Complicating the extraction is additive noise. While many different noise distributions can be investigated and justified by various problem physics, we avoid direct use of either GGD or GCD densities. Employing, for instance, Gaussian noise will yield the obvious result of linear filtering providing the best performance, while the myriad filter is clearly optimal for applications in which the noise is Cauchy distributed. The presentation of such obvious results provides little information or insight into the characteristics of the individual algorithms. Rather, results are presented for the commonly utilized α-Stable density family. This provides a fairer comparison and illustrates the performance of the filters operating on a widely used heavy-tailed distribution, but one that shares only tail order with the GCD.

Utilization of α-Stable distributed samples, like Gaussian distributed samples, is very appealing for the modeling of real-life phenomena because they constitute the only variables that can be described as the superposition of many small independent and identically distributed effects [15]. The class of symmetric α-Stable distributions is usually described by its characteristic function:

$$\phi(\omega) = \exp\left(-\kappa|\omega|^{\alpha}\right). \tag{26.68}$$

The parameter κ, which is commonly called the *dispersion*, is a positive constant related to the scale of the distribution and $\kappa^{1/\alpha}$ is the scale parameter of the distribution. In order for Equation 26.68 to define a characteristic function, the value of α must be restricted to the interval $[0, 2]$. Importantly, α determines the impulsiveness, or tail heaviness, of the distribution, where smaller values of α indicate increased levels of impulsiveness. The limit $\alpha = 2$ case corresponds to the zero-mean Gaussian distribution with variance 2κ. All other values of α correspond to impulsive distributions with infinite variance and algebraic tail behavior [15]. Indeed, α-Stable distributions are successfully utilized to model impulsive environments accross a wide array of applications [8,12,15,23,28,31,34,46,49].

26.4.1 Baseband Communications

Consider first the baseband communication model given in Figure 26.11 [50]. Suppose that A (real) is to be communicated over the channel. Denote $s(t)$ as the combined impulse response of the transmitter and channel, and take the pulse $As(t)$ to be corrupted by additive white noise. The received pulse is then given by

$$r(t) = As(t) + n(t), \tag{26.69}$$

which, after sampling at rate $1/T$, corresponds to the sequence

$$r(kT) = As(kT) + n(kT). \tag{26.70}$$

Taking the common case assumption that $s(kT) \neq 0$ only for k an integer, the communications goal is to estimate A using the samples $As(kT) + n(kT), k = 1, 2, \ldots, K$.

The formulated problem is well known to be optimally addressed through matched filtering. Moreover, each of the covered filters can be formulated in matched form. Thus to compare the performances of the matched linear, median, myriad, and meridian filters, 10,000 Gaussian distributed $\{A(i) : i = 1, 2, \ldots, 10,000\}$ parameters are generated, sent through the baseband communication channel, sampled with $K = 21$, and filtered with each of the matched filters to obtain the estimates $\{\hat{A}(i) : i = 1, 2, \ldots, 10,000\}$. The corrupting channel noise is α-Stable distributed with $\alpha = 0.4$. The

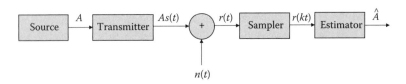

FIGURE 26.11   Baseband communication model: $s(t)$ denotes the combined impulse response of the transmitter and the channel, and $As(t)$ is corrupted by additive white noise $n(t)$.

TABLE 26.2 Matched Linear, Median, Myriad and Meridian Filters Output MAEs and MSEs

Performance Criteria	Matched Linear	Matched Median	Matched Myriad	Matched Meridian
MAE	5.9215×10^3	0.2080	0.1605	0.1121
MSE	3.2949×10^7	0.1687	0.0520	0.0380

Source: Aysal, T. C. and Barner, K. E., *IEEE Trans. Signal Process.*, 55, 3949, 2007. With permission.

pulse carrying the symbol is taken to be rectangular. The mean absolute and squared errors of the matched filter outputs are tabulated in Table 26.2. The tabulated results show that the linear filter completely breaks down in the presence of the heavy tailed corruption. The median is significantly more robust, but still suffers from its inability to completely discount outliers. The myriad and meridian operators, both derived for such algebraic tailed environments, provide significantly better results than the GGD-based methods. The noise parameter utilized in this example, $\alpha = 0.4$, results in extremely impulsive corruption, which is why the best performance is provided by the most robust operator covered, the matched meridian filter.

26.4.2 Power Line Communications

Consider next the problem of power line communications (PLCs). The use of existing power lines for transmitting data and voice has received considerable interest in recent years [51–54]. The advantages of PLCs are obvious due to the ubiquity of power lines and power outlets. The potential of power lines to deliver broadband services, such as fast Internet access, telephone, fax services, and home networking is emerging in new communications industry technology. However, there remain considerable challenges for PLCs, such as communications channels that are hampered by the presence of large amplitude noise superimposed on top of traditional white Gaussian noise. The overall interference is appropriately modeled as an algebraic tailed process, with α-Stable ($\alpha \approx 1$) often chosen as the parent distribution [51].

To compare the various filtering algorithms, consider a PLC problem in which there are a set of voltage levels, $\mathcal{V} = \{-2, -1, 0, 1, 2\}$ unknown at the receiver.* A signal randomly composed of these voltage levels, an example of which is shown in Figure 26.12a, is transmitted through a powerline. The observed signal is given in Figure 26.12b, where the noise is modeled as α-Stable distributed with $\alpha = 1.25$. Note that this results in corruption that is somewhat less impulsive than in the previously considered example. The observed powerline signal is processed with mean, median, myriad, and meridian filters, each utilizing window length $N = 9$, the results of which are given in Figures 26.12c through, respectively.

Inspection of the figures shows that the mean is vulnerable to impulsive noise even when α is relatively large. The median, myriad, and meridian, in contrast, all perform relatively well. The fact that the median performs well in this example indicates that the noise impulsivity is reasonably close to the GGD Laplacian distribution special case. The meridian is, perhaps, overly robust for the level of corruption observed in this example, but still performs nearly as well as the myriad and median. This example shows that the myriad and meridian give up little in terms of performance in relatively light tailed environments while, as the previous example shows, offering considerably better performance in the presence of truly heavy tailed observations.

26.4.3 Highpass Filtering of a Multitone Signal

As a final example, consider the problem of preserving a high-frequency tone while removing all low frequency terms. This requires highpass filtering, which employs both positive and negative weights within the filter window. Such a problem demonstrates not only the utilization of positive and negative

---

* In the case where the signal alphabet is known at the receiver, the linear, median, myriad, and meridian estimators are readily extended to detectors.

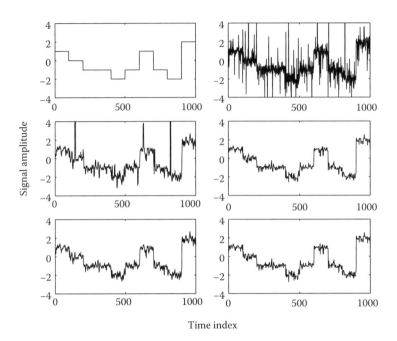

Time index

FIGURE 26.12 Power line communications signal enhancement: (a) transmitted signal, (b) observed signal corrupted by $\alpha = 1.25$ noise, output of the (c) mean [2.5388]{8.6216}, (d) median [2.4267]{7.4393}, (e) myriad [2.4385]{7.4993}, and (f) meridian filters [2.4256]{7.4412}, where [·] and {·} denotes the mean absolute and squared error for the corresponding filtering structure.

weights, but also the ability of robust nonlinear filters to perform frequency selectivity that is traditionally accomplished through linear filtering. Figure 26.13a depicts a two-tone signal with normalized frequencies 0.02 and 0.4 Hz. The signal is 1000 samples long, although a cropped version of the original signal {0,200} is shown for presentation purposes. Figure 26.13b shows the multitone signal filtered by a 40-tap linear FIR filter designed by the MATLAB fir1 command with a normalized cutoff frequency of 0.3 Hz. The myriad and median filter weights are optimized in this case utilizing the adaptive methods detailed in the previous section.* Noting that the median filter is a limiting case of the meridian filter, the meridian filter weights are set equal to those of the median.

The clean multitone input signal is corrupted by additive α-Stable noise with $\alpha = 0.4$ to form a corrupted observation from which the single high frequency tone must be extracted, Figure 26.13c. The noisy multitone signal is processed by the linear, median, myriad, and meridian filters with the results given in Figures 26.13d through g, respectively. Similarly to the first example, the observation signal in this case contains very heavy tailed outliers. These outliers cause the linear filter to, once again, break down. Each of the nonlinear filters, in contrast, offers more robust processing, with the outputs more closely reflecting the range and frequency content of the desired signal. The increasing robustness of the operators is again apparent, with the myriad being more robust to outliers than the median and the meridian being the most robust. It should be noted that the nonlinear filters not only minimize the influence of outliers, but their flexible weighted structures are able to effectively pass desired frequency content while rejecting content outside the desired frequency band. The ability to simultaneously reject

---

* The clean multitone signal and the desired high-frequency signal are utilized as the input and desired signals, respectively. For more detail see Refs. [28,29,31].

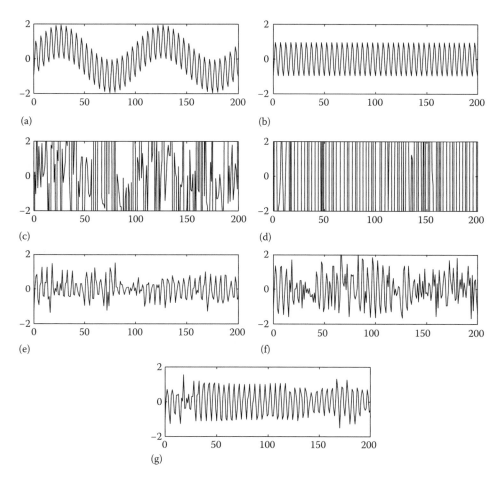

FIGURE 26.13 High frequency–selection filtering in impulsive noise: (a) clean two-tone input signal, (b) (desired) highpass component of the input signal (processed with lowpass FIR filter), (c) two-tone signal corrupted with stable noise ($\alpha = 0.4$), noisy two-tone input signal processed with (d) linear, (e) median, (f) myriad, and (g) meridian filters. (From Aysal, T. C. and Barner, K. E., *IEEE Trans. Signal Process.*, 55, 3949, 2007. With permission.)

outliers and perform frequency selectivity enables the nonlinear median, myriad, and meridian filters to be deployed in a wide range of applications, including those previously dominated by, but not necessarily well addressed by, linear filtering.

26.5 Concluding Remarks and Future Directions

The development and application of nonlinear filters is necessitated by the complexity and challenging environments of many contemporary problems. In this chapter we have taken a fundamental approach to filter development, considering operators from the first principles of location estimation. This development focused on two important broad classes of distributions, generalized Gaussian and generalized Cauchy distribution families. Within the GGD family, particular focus was placed on the Gaussian and Laplacian distribution special cases and their resulting linear and median filtering operations. It was shown that while the GGD family constitutes a broad array of distributions with varying tail parameters, distributions within the family are not capable of accurately modeling the most impulsive and environment seen in practice due to the constraint that GGD density tail decay rates be exponential. The consequence of this tail decay rate constraint is that the linear and median filters deriving their optimality from the

GGD are not effective in the processing of very impulsive sequences. The GCD family, in contrast, consists of distributions with algebraic tail decay rates that do accurately modeled these most demanding impulsive environments. Within this family we again focus on two special cases, the Cauchy and Meridian distributions and their resulting myriad and meridian filtering operations. As these filters are derived from heavy tailed distributions, they are well suited for applications dominated by very impulsive statistics.

Properties and optimization procedures are presented for each of the filters covered. In particular, we show that the operators can be ordered in terms of their robustness, from least to most robust, as linear, median, myriad, and meridian. Moreover, myriad filters contain linear filters as special cases, while meridian filters contain median operators as special cases. Thus the myriad and meridian operators are inherently more efficient than their traditional (subset) counterparts. Simulations presented in communications and frequency selective filtering applications show and contrast the performances of the filters in applications with varying levels of heavy tailed statistics. As expected, linear operators breakdown in these environments while the robust, nonlinear operators yield desirable results.

Although the presentation in this chapter ranges from theoretical development through properties, optimization, and applications, the coverage is, in fact, simply an overview of one segment within the broad array of nonlinear filtering algorithms. To probe further, the interested reader is referred to the cited articles, as well as numerous other works in this area. Additionally, there are many other areas of research in nonlinear methods that are under active investigation. Research areas of importance include (1) order-statistic based signal processing, (2) mathematical morphology, (3) higher order statistics and polynomial methods, (4) radial basis function and kernel methods, and (5) emerging nonlinear methods. Researchers and practitioners interested in the broader field of nonlinear signal processing are encouraged to see the many good books, monographs, and research papers covering these, and other, areas in nonlinear signal processing.

References

1. B. I. Justusson, Median filtering: statistical properties, *Two Dimensional Digital Signal Processing II.* New York: Springer Verlag, 1981.
2. I. Pitas and A. Venetsanopoulos, *Nonlinear Digital Filters: Principles and Application.* New York: Kluwer Academic, 1990.
3. K. E. Barner and G. R. Arce, Eds., *Nonlinear Signal and Image Processing: Theory, Methods, and Applications.* Boca Raton, FL: CRC Press, 2004.
4. M. Schetzen, *The Volterra and Wiener Theories of Nonlinear Systems.* New York: Wiley, 1980.
5. C. L. Nikias and A. P. Petropulu, *Higher Order Spectra Analysis: A Nonlinear Signal Processing Framework.* Englewood Cliffs, NJ: Prentice-Hall, 1993.
6. V. J. Mathews and G. L. Sicuranza, *Polynomial Signal Processing.* New York: John Wiley & Sons, Inc., 2000.
7. M. B. Priestrley, *Non-linear and Non-stationary Time Series Analysis.* New York: Academic, 1988.
8. G. R. Arce, *Nonlinear Signal Processing: A Statistical Approach.* New York: John Wiley & Sons, Inc., 2005.
9. H. A. David and H. N. Nagaraja, *Order Statistics.* New York: John Wiley & Sons, Inc., 2003.
10. P. Huber, *Robust Statistics.* New York: John Wiley & Sons, Inc., 1981.
11. H. Hall, A new model for impulsive phenomena: application to atmospheric noise communications channels, Stanford University Electronics Laboratories Technical Report, no. 3412–8 and 7050–7, 1966, sU-SEL-66-052.
12. J. Ilow, Signal proceeing in alpha–stable noise environments: Noise modeling, detection and estimation, PhD dissertation, University of Toronto, Toronto, ON, 1995.
13. B. Mandelbrot, Long-run linearity, locally Gaussian processes, H-spectra, and infinite variances, *International Economic Review*, 10, 82–111, 1969.

14. J. H. Miller and J. B. Thomas, Detectors for discrete-time signals in non-Gaussian noise, *IEEE Transactions on Information Theory*, 18, 241–250, Mar. 1972.

15. C. L. Nikias and M. Shao, *Signal Processing with Alpha–Stable Distributions and Applications*. New York: Wiley, 1995.

16. G. A. Tsihrintzis and C. L. Nikias, Fast estimation of the parameters of alpha-stable impulsive interference, *IEEE Transactions on Signal Processing*, 44(6), 1492–1503, Jun. 1996.

17. A. G. Dimakis and P. Maragos, Phase-modulated resonances modeled as self-similar processes with application to turbulent sounds, *IEEE Transactions on Signal Processing*, 53(11), 4261–4272, Nov. 2005.

18. G. A. Tsihrintzis and C. L. Nikias, Data-adaptive algorithms for signal detection in sub-Gaussian impulsive interference, *IEEE Transactions on Signal Processing*, 45(7), 1873–1878, Jul. 1997.

19. J. P. Nolan, *Stable Distributions: Models for Heavy Tailed Data*. Boston, MA: Birkhuser, 2005.

20. G. A. Tsihrintzis and C. L. Nikias, Incoherent receivers in alpha-stable impulsive noise, *IEEE Transactions on Signal Processing*, 43(9), 2225–2229, Sep. 1995.

21. J. Ilow and D. Hatzinakos, Analytic alpha-stable noise modeling in a Poisson field of interferers or scatterers, *IEEE Transactions on Signal Processing*, 46(6), 1601–1611, Jun. 1998.

22. R. F. Brcich, D. Iskander, and A. Zoubir, The stability test for symmetric alpha-stable distributions, *IEEE Transactions on Signal Processing*, 53(3), 997–986, Mar. 2005.

23. E. Masry, Alpha-stable signals and adaptive filtering, *IEEE Transactions on Signal Processing*, 48(11), 3011–3016, 2000.

24. R. Durrett, *Probability: Theory and Examples*, 2nd edn., Belmont, CA: Duxbury Press, 1996.

25. V. M. Zolotarev, *One Dimensional Stable Distributions*. Translations of Mathematical Monographs 65, American Mathematical, Providence, Rhode Island, 1986.

26. S. Kalluri and G. R. Arce, Adaptive weighted myriad filter algorithms for robust signal processing in α–Stable environments, *IEEE Transactions on Signal Processing*, 46(2), 322–334, Feb. 1998.

27. J. G. Gonzalez and G. R. Arce, Optimality of the myriad filter in practical impulsive-noise environments, *IEEE Transactions on Signal Processing*, 49(2), 438–441, Feb. 2001.

28. S. Kalluri and G. R. Arce, Robust frequency–selective filtering using weighted myriad filters admitting real-valued weights, *IEEE Transactions on Signal Processing*, 49(11), 2721–2733, Nov. 2001.

29. T. C. Aysal and K. E. Barner, Meridian filtering for robust signal processing, *IEEE Transactions on Signal Processing*, 55(8), 3949–3962, Aug. 2007.

30. S. A. Kassam and H. V. Poor, Robust techniques for signal processing, *Proceedings of IEEE*, 73, 433–481, Mar. 1985.

31. G. R. Arce, A general weighted median filter structure admitting negative weights, *IEEE Transactions on Signal Processing*, 46, 3195–3205, Dec. 1998.

32. R. Yang, L. Yin, M. Gabbouj, J. Astola, and Y. Neuvo, Optimal weighted median filters under structural constraints, *IEEE Transactions on Signal Processing*, 43(3), 591–604, 1995.

33. L. Yin, R. Yang, M. Gabbouj, and Y. Neuvo, Weighted median filters: A tutorial, *IEEE Transactions on Circuits and Systems*, 41, 157–192, May 1996.

34. K. E. Barner and T. C. Aysal, Polynomial weighted median filtering, *IEEE Transactions on Signal Processing*, 54(2), 636–650, Feb. 2006.

35. T. C. Aysal, Filtering and estimation theory: First-order, polynomial and decentralized signal processing, PhD dissertation, University of Delaware, DE, 2007.

36. T. C. Aysal and K. E. Barner, Hybrid polynomial filters for gaussian and non-gaussian noise environments, *IEEE Transactions on Signal Processing*, 54(12), 4644–4661, Dec. 2006.

37. R. Adler, R. Feldman, and E. M. S. Taqqu, *A Practical Guide to Heavy Tails: Statistical Techniques for Analyzing Heavy-Tailed Distributions*. Boston, MA: Birkhauser, 1997.

38. E. J. Wegman, S. G. Schwartz, and J. Thomas, *Topics in Non–Gaussian Signal Processing*. New York: Academic Press, 1989.

39. G. R. Wilson and D. R. Powell, Experimental and modelled density estimates of underwater acoustic returns, In *Statistical Signal Processing*, E. J. Wegman and G. Smith, Eds., Marcel Decker, New York, pp. 223–239, 1984.

40. A. Briassouli and M. G. Strintzis, Locally optimum nonlinearities for DCT watermark detection, *IEEE Transactions on Image Processing*, 13(12), 1604–1617, Dec. 2004.

41. M. N. Do and M. Vetterli, Wavelet-based texture retrieval using generalized Gaussian density and kullback-leibler distance, *IEEE Transactions on Image Processing*, 11(2), 146–158, Feb. 2002.

42. J. M. Leski, Robust weighted averaging [of biomedical signals], *IEEE Transactions on Biomedical Engineering*, 49(8), 796–804, Aug. 2002.

43. P. Sun, Q. H. Wu, A. M. Weindling, A. Finkelstein, and K. Ibrahim, An improved morphological approach to background normalization of ECG signals, *IEEE Transactions on Biomedical Engineering*, 50(1), 117–121, Jan. 2003.

44. J. M. Leski and A. Gacek, Computationally effective algorithm for robust weighted averaging, *IEEE Transactions on Biomedical Engineering*, 51(7), 1280–1284, Jul. 2004.

45. P. R. Rider, Generalized cauchy distributions, *Annals of the Institute of Statistical Mathematics*, 9, 215–223, 1957.

46. S. A. Kassam, *Signal Detection in Non-Gaussian Noise*. New York: Springer-Verlag, 1985.

47. J. G. Gonzalez, Robust techniques for wireless communications in non-Gaussian environments, PhD dissertation, University of Delaware, New York, DE, 1997.

48. F. Hampel, E. Ronchetti, P. Rousseeuw, and W. Stahel, *Robust Statistics: The Approach Based on Influence Functions*. New York: Wiley, 1986.

49. T. C. Aysal and K. E. Barner, Second-order heavy-tailed distributions and tail analysis, *IEEE Transactions on Signal Processing*, 54(7), 2827–2835, Jul. 2006.

50. J. Astola and Y. Neuvo, Matched median fitlering, *IEEE Transactions on Communications*, 40(4), 722–729, Apr. 1992.

51. Y. H. Ma, P. L. So, and E. Gunawan, Performance analysis of OFDM systems for broadband power line communications under impulsive noise and multipath effects, *IEEE Transactions on Power Delivery*, 20(2), 674–682, Apr. 2005.

52. O. G. Hooijen, A channel model for the residential power circuit used as a digital communications medium, *IEEE Transactions on Electromagnetic Compatibility*, 40(4), 331–336, Nov. 1998.

53. O. G. Hooijen, On the channal capacity of the residential power circuit used as a digital communications medium, *IEEE Communcations Letter*, 2(10), 267–268, Oct. 1998.

54. M. Zimmerman and K. Dostert, Analysis and modeling of impulsive noise in broadband power line communications, *IEEE Transactions on Electromagnetic Compatibility*, 44(1), 249–258, Feb. 2002.

27

Nonlinear Filtering for Image Denoising*

Nasir M. Rajpoot
University of Warwick

Zhen Yao
University of Warwick

Roland G. Wilson
University of Warwick

Denoising is one of the fundamental problems in digital image processing and can be defined as the removal of noise from observed image data, for example in broadcast and surveillance as well as in medical imaging applications. Often performed as a preprocessing step prior to any analysis or modeling of the images, denoising improves the image quality thus reducing the effects of noise on the output of any subsequent operations. Let $f(\mathbf{r})$ denote some original image data corrupted by additive white Gaussian noise $\nu(\mathbf{r})$, resulting in observed image data $g(\mathbf{r})$, where $\mathbf{r}$ denotes the n-dimensional coordinates of the image data, and $n \in \{2, 3, 4\}$. For $n = 2, \mathbf{r}$ usually represents the horizontal and vertical coordinates (often denoted by x and y), whereas for $n = 4$ for instance in case of real-time capture of three-dimensional (3-D) images (such as fMRI image data), $\mathbf{r}$ may be composed of the three spatial coordinates (x, y, and z) and a time coordinate. Corruption of the original image data can, therefore, be expressed in the following form,

---

* Parts of this chapter have appeared in the Proceedings of Baiona SPC'2003, ICIP'2004, and WIAMIS'2004, and ICIP'2005. Much of the material in Section 27.2 appears in Yao's PhD thesis, submitted to the University of Warwick in October 2007, and is in preparation for submission to a journal.

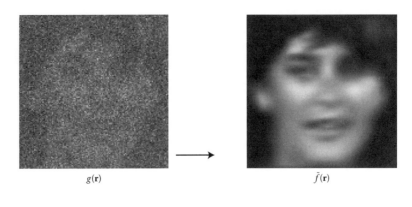

$$g(\mathbf{r})$$
$$\tilde{f}(\mathbf{r})$$

$$g(\mathbf{r}) = f(\mathbf{r}) + \nu(\mathbf{r}) \tag{27.1}$$

where $\nu = \mathcal{N}(0, \sigma^2)$ denotes uncorrelated, zero-mean, Gaussian noise with a standard deviation of σ. The addition of white Gaussian noise often takes place at the imaging side, for instance due to the finite exposure time of the imaging devices, thermal noise in the CCD arrays, quantization noise, or a combination of some or all of these. The problem of image denoising can be stated as follows.

PROBLEM STATEMENT:

Given the observed image data $g(\mathbf{r})$ and our knowledge about the noise process $\nu(\mathbf{r})$ being the white Gaussian noise with the level of noise being unknown, find an approximation $\tilde{f}(\mathbf{r})$ of the original image data such that the following criterion of total distortion

$$\sum_{\mathbf{r} \in \Omega} \| f(\mathbf{r}) - \tilde{f}(\mathbf{r}) \|^2 \tag{27.2}$$

is minimized, where $\Omega \subset \mathcal{R}^n$ is the set of all coordinates for the image data.

It is evident from the above problem statement that denoising is an *inverse problem* as well as an ill-posed one, since the objective here is to find a mapping $\Phi : g(\mathbf{r}) \mapsto \tilde{f}(\mathbf{r})$ which minimizes the sum of squared errors between $f(\mathbf{r})$ and $\tilde{f}(\mathbf{r})$, without having any prior knowledge of $f(\mathbf{r})$.

27.1 Filtering for Image Denoising

Denoising methods proposed in the literature can be categorized into two main classes: spatial (or spatiotemporal) and transform domain. Spatial or spatiotemporal denoising methods modify the intensity of the observed image data elements by applying a *linear* filter (e.g., averaging or Gaussian smoothing) or *nonlinear* (median filtering, edge-preserving filter such as Kuwahara filter or anisotropic diffusion) filtering operation on intensity values of the observed image data. Denoising methods that operate in the transform domain aim to amend the transform coefficients in such a way that prominent features of the original image are retained or highlighted and noise-related coefficients or noisy components of the transform coefficients are suppressed. The modification of transform coefficients can also be done in a linear, for example, multiplication or convolution, or nonlinear, thresholding or manipulation of coefficients using a nonlinear function such as $\tanh(\cdot)$, fashion.

Linear filters in both spatial and transform domains usually face the dilemma that although they can be computationally inexpensive, they often result in smoothing (also known as *blurring*) of important features, such as edges in images.

Nonlinear filters, on the other hand, aim at preserving image features that are crucial to the fidelity of reconstructed image data. Furthermore, nonlinear filtering in the transform domain offers the added advantage that it can faithfully reconstruct as well as highlight prominent image features, provided an appropriate transform for representing the image data is employed. Assuming the representation is sparse, filtering can be performed by simply thresholding the transform coefficients. The overall mapping function $\Phi(g)$ can then be defined as follows:

$$\Phi(g) = \tilde{f} = \Gamma^{-1}(\mathcal{T}\{\Gamma(g)\}), \tag{27.3}$$

where
  Γ and Γ^{-1}, respectively, denote the forward and inverse transforms
  $\mathcal{T}\{\cdot\}$ denotes the thresholding operation

The denoising function Φ is a nonlinear filter, even when a unitary transform Γ is employed, due to the presence of the thresholding operator $\mathcal{T}$, which is a highly nonlinear operator.

27.1.1 Nonlinear Filtering in the Transform Domain

The use of multiresolution image representations in the restoration of noisy images dates back to the late 1980s [1], in which a statistical estimator based on a simple quadtree image model was used in smoothing images corrupted by additive Gaussian noise. For image denoising, the transform Γ in Equation 27.3 ought to be chosen such that it is an appropriate representation of the most prominent spatial or spatiotemporal features present in $g(\mathbf{r})$ with the help of basis functions with suitable shape and localization. An additional advantage of the transform-based approach to filtering is that it allows one to extract important features for analysis purposes; for example, curvilinear features in still images, moving edges in video sequences, tissue interfaces in medical images, or fault surfaces in seismic images.

But what criteria should one use to choose an appropriate transform? Sparsity of a representation is widely believed to be one of the criteria used to judge its effectiveness for a particular type of feature. A representation can be regarded as being *efficient* if its basis functions resemble closely the type of feature being sought in the image data, thus resulting in its sparse representation. It is believed that during the millenia of evolution, the external visual stimuli of complex natural scenarios have influenced the human visual system (HVS), so that cells in the primary visual cortex can respond to certain important spatiotemporal features. Hubel and Wiesel [2] first showed that biological visual systems in mammals actually analyze images along dimensions such as orientation, scale, and frequency. More recently, it was shown by Olshausen and Field [3,4] that the sparse components generated by image-dependent linear transformation known as independent component analysis (ICA) resemble simple-cell receptive fields. This matches the hypothesis that the HVS captures essential information with a minimum number of excited neurons, which can be understood as a biological form of sparse representation.

27.1.2 Transform: Separable or Nonseparable?

The ordinary separable wavelet transform offers most of the desiderata that a suitable representation mimicking the HVS is required to satisfy multiresolution, sparse, localized. However, the fact that it is computed separably, i.e., on rows and columns in 2-D, for instance, implies that its basis functions are simply tensor products of 1-D wavelet basis functions along the rows and columns, resulting in capturing edges along only three orientations in 2-D: horizontal, vertical, and diagonal. Since the separable 2-D wavelet basis functions can only capture features of three possible orientations, the ordinary wavelet transform results in a nonsparse representation of edges in other directions. This limited *orientation selectivity* can be a severe limitation in applications where detection, coding, or denoising where reconstruction of strong directional features is important. In higher dimensions also, separability seriously limits the ability of wavelets to efficiently represent higher dimensional features (such as lines in images or planes in 3-D image volumes). To overcome this limitation, several nonseparable geometric wavelets (mostly in 2-D) have been proposed in the last decade or so. These are commonly referred to as the *X-lets* in the literature, such as ridgelets [5], curvelets [6,7], wedgelets [8], beamlets [9], contourlets [10,11], brushlets [12], and arclets [13].

27.1.3 Transform: Fixed or Adaptive?

Most of the nonseparable *X-lets* mentioned above employ fixed basis functions. Some researchers believe that since the type of features that would excite the retinal neurons is fixed, it also suggests that the HVS is more likely a representation with a fixed basis rather than an adaptive one like the ICA. However, in some cases, image-dependent adaptive bases have been shown to be superior in terms of their ability to represent prominent image features, particularly when the representation is not *overcomplete*.* They offer the advantage of compactly representing the image features using basis functions that are adapted to the contents of the image data. In summary, while the fixed transform idea is appealing in terms of its computational complexity and in its comparison to the HVS, the adaptive transform may be better suited in some cases due to its adaptability but may be prohibitively expensive computation-wise.

27.1.4 Chapter Organization

In this chapter, we describe three wavelet-like representations employed to filter noisy 2-D and 3-D image data:

- Polar cosine transform (PCT) is a nonseparable 2-D transform designed to capture two of the most important features in 2-D images: curvilinear edges and texture. The development of PCT is motivated by the fact that natural images contain not only piecewise smooth regions separated by edges, but also regions with strong textural characteristics. Two constructions of PCT, one using Radon transform and another using DCT and DST in a butterfly fashion, are described in Section 27.2. The former PCT construction can also be tailored to be image-dependent such that it can be adapted to both directional linear features as well as directional textures in a unified manner according to the "*image = edge + texture*" model.
- Planelets are nonseparable 3-D basis functions, resembling planar wave functions and having compact support in space–time as well as spatiotemporal frequency. Locally planar structures can be found in many 3-D applications, including video sequences, where they represent sweeping luminance edges, and medical volume data, where they represent the interfaces between different tissue types, for example. Planelets can be regarded as an extension of the complex wavelet bases [14] which have been increasing in popularity in recent years. Because transformation to the new basis is efficient computationally, the overall denoising algorithm is highly efficient. The representation is translation invariant, offers good directional selectivity, and can be computed efficiently. The success of the approach is demonstrated using noisy video sequence and a noisy human knee MR image volume.

---

* An overcomplete representation is one that employs more basis functions than are required for an exact reconstruction of the original image data.

- Adaptive 3-D wavelet packet (A3-DWP) representation, which is separable but uses basis functions that are adapted to the contents of a given 3-D image data, a video sequence, or an image volume. The A3-DWP representation is optimal in terms of compactly representing local, in both space and time, spatiotemporal frequencies in the given image data. In order to reduce the effects of Gibbs phenomenon in the restored image data, translation dependence is removed by averaging the restored instances of the shifted data in all three directions.

27.2 Polar Cosine Transform

It is widely recognized that natural images consist of piecewise smooth regions separated by edges as well as textured regions. While various different models for curved edges exist, including some of the *X-lets* mentioned in Section 27.1, texture representation has faced difficulties due to lack of a single widely accepted mathematical definition of texture patterns that can be found in images. There are, however, certain assumptions that can be made for most natural textures:

- Oriented. Many spatial textures have a prominent direction, for example wood patterns and seismograms, or a combination of several such components.
- Periodic. The texture tends to repeat certain basic elements at a given frequency and tends to repeat itself with a degree of affine invariance.
- Localized. In natural images, the textures are usually confined within a particular region. In other words, they are spatially localized.

Over the years, several wave packet bases such as local cosines [15,16], wavelet packets [17,18], and brushlets [19] have been proposed as suitable representations for oscillatory textures. However, they either lack orientation selectivity, or their spatial-frequency localization is not satisfactory. Moreover, due to the Uncertainty Principle, a sparse representation for both singularities and periodic oscillations in a fixed basis is a fundamental dilemma [20]. The resort is to find an image-dependent *adaptive* basis which can accommodate both directional linear features as well as directional textures in a unified manner according to the "*image = edge + texture*" model.

27.2.1 Continuous Transform

A prototypical oriented texture pattern can be seen as a higher dimensional function which has a waveform constant along a fixed direction. Such functions are usually referred to as "planar waves" or "ridge functions" in approximation related literature [21]. Formally, a ridge function is a multivariate function of the form $\psi(\vec{\xi} \cdot \vec{\theta})$ where $\psi(\cdot)$ is a univariate function usually referred to as the *ridge profile*, with spatial coordinate vector $\vec{\xi} \in \mathbb{R}^d$ and $\vec{\theta} \in \mathbf{S}^{d-1}$. This means $\vec{\theta}$ is on the unit sphere $\mathbf{S}^{d-1}$ in dimension $d > 1$ and indicates the orientation.

In order to model oriented texture, a real-to-complex ridge profile $\psi: \mathbb{R} \to \mathbb{C}$ can be defined as

$$\psi_\omega(\xi) = e^{i\omega\xi}. \tag{27.4}$$

which is basically the 1-D Fourier basis function, and the corresponding multivariate ridge function by definition is

$$\psi_\omega(\vec{\xi} \cdot \vec{\theta}) = e^{i\omega(\vec{\xi} \cdot \vec{\theta})}. \tag{27.5}$$

The ridge function gives a complex trigonometric oscillatory constant along the direction $\vec{\theta}$. Taking the inner product of it with a function $f(\vec{\xi})$ gives

$$\hat{f}(\omega, \vec{\theta}) = \langle f(\vec{\xi}), \psi_\omega(\vec{\xi} \cdot \vec{\theta}) \rangle = \int f(\vec{\xi}) e^{-i\omega(\vec{\xi} \cdot \vec{\theta})} d\vec{\xi}. \tag{27.6}$$

This is exactly the d-dimensional Fourier transform expressed in polar coordinate form, or just simply *polar Fourier transform*. The conversion to Cartesian coordinates can be done by separating the radial frequency into a frequency vector:

$$\vec{\omega} = \omega\vec{\theta}. \tag{27.7}$$

With a slight rearrangement of Equation 27.6, the equivalence becomes obvious:

$$\hat{f}(\omega, \vec{\theta}) = \int f(\vec{\xi}) e^{-i\vec{\xi} \cdot (\omega\vec{\theta})} \, d\vec{\xi} = \int f(\vec{\xi}) e^{-i\vec{\xi} \cdot \vec{\omega}} \, d\vec{\xi} = \hat{f}(\vec{\omega}). \tag{27.8}$$

Meanwhile, the definition for the continuous Radon transform in d-dimensions is

$$\mathcal{R}f(t, \vec{\theta}) = \int f(\vec{\xi}) \delta(\vec{\xi} \cdot \vec{\theta} - t) d\vec{\xi}. \tag{27.9}$$

The relationship between the Radon transform and the polar Fourier transform is stated as the Fourier Slice Theorem [22]:

THEOREM 27.1   (Fourier Slice Theorem)

The 1-D Fourier transform with respect to t of the projection $\mathcal{R}f(t, \vec{\theta})$ is equal to a central slice, at a given orientation $\vec{\theta}$, of the higher dimensional Fourier transform of the function $f(\vec{\xi})$, that is,

$$\widehat{\mathcal{R}f(t, \vec{\theta})} = \hat{f}(\omega, \vec{\theta}). \tag{27.10}$$

If some boundary condition is to be posed at each polar orientation, instead of being defined on $\mathbb{R}$, the Radon slice $\mathcal{R}_{\vec{\theta}} f(t)$ is assumed to be an even function $\mathcal{R}_{\vec{\theta}} f(t) = \mathcal{R}_{\vec{\theta}} f(-t)$, then its Fourier transform can be written as follows:

$$\widehat{\mathcal{R}f(t, \vec{\theta})} = \sqrt{\frac{2}{\pi}} \int_0^{+\infty} \mathcal{R}f(t, \vec{\theta}) \cos(\omega t) dt. \tag{27.11}$$

It is noted that only the real cosine part remains in the above equation due to the even boundary assumption on the slices. This is equivalent to a cosine tranform with ridge-type basis functions defined as

$$\psi_\omega^C(\vec{\xi} \cdot \vec{\theta}) = \sqrt{\frac{2}{\pi}} \cos(\omega\vec{\xi} \cdot \vec{\theta}). \tag{27.12}$$

In the same way by assuming $\mathcal{R}_\theta f(t) = -\mathcal{R}_\theta f(-t)$, the sine counterpart is

$$\psi^S_\omega(\vec{\xi} \cdot \vec{\theta}) = \sqrt{\frac{2}{\pi}} \sin(\omega \vec{\xi} \cdot \vec{\theta}). \tag{27.13}$$

Correspondingly, the continuous polar cosine and sine transform are denoted, respectively, as

$$\mathcal{C}f(\omega, \vec{\theta}) = \langle f(\vec{\xi}), \psi^C_\omega(\vec{\xi} \cdot \vec{\theta}) \rangle = \sqrt{\frac{2}{\pi}} \int_0^{+\infty} f(\vec{\xi}) \cos(\omega \vec{\xi} \cdot \vec{\theta}) d\vec{\xi} \tag{27.14}$$

and

$$\mathcal{S}f(\omega, \vec{\theta}) = \langle f(\vec{\xi}), \psi^S_\omega(\vec{\xi} \cdot \vec{\theta}) \rangle = \sqrt{\frac{2}{\pi}} \int_0^{+\infty} f(\vec{\xi}) \sin(\omega \vec{\xi} \cdot \vec{\theta}) d\vec{\xi}. \tag{27.15}$$

These two transforms, along with the polar Fourier transform, can be regarded as a family of *Polar Trigonometric Transforms*. In the context of image processing, the PCT is the transform of interest in this work, due to its better approximation convergence properties attributed to the even boundary extension [23]. Since it is essentially the Fourier transform, the completeness of the transform makes the transform operator $\mathcal{C}$ unitary which means $\mathcal{C}^{-1} = \mathcal{C}^*$. Since the transform is real, the operator is self-conjugate, making the inverse transform exactly the same as the forward transform:

$$\mathcal{C}^{-1} = \mathcal{C}. \tag{27.16}$$

The same also holds for the continuous polar sine transform and its inverse.

27.2.2 Discrete Transform

The continuous PCT essentially replaces the polar Fourier transform's complex basis with a real cosine sinusoid as the ridge function, by assuming the projected Radon slice is even. In a discrete case, the Fourier-related transforms that operate on a function over a finite domain can be thought of as implicitly defining an extension of that function outside the domain. The discrete Fourier transform (DFT) implies a periodic extension of the original function. A discrete cosine transform (DCT), like the continuous cosine transform, implies an even extension of the original function.

However, when the transform is to operate on finite, discrete sequences, two issues arise that are not relevant in case of the continuous cosine transform. First, one has to specify whether the function is even or odd at both the left and right boundaries of the domain. Second, one has to specify around what point the function is even or odd. These issues result in several different versions of DCTs, a full list of which along with corresponding DSTs is given in Table 27.1. For a discrete polar cosine transform, any of the DCTs can be adopted to form the basis ridge function. In particular, the so-called DCT-II is widely used in many applications such as the famed JPEG compression [24] due to its even boundary extensions on both ends, which gives better approximation convergence. The 1-D DCT-II function is defined by the ridge profile:

$$\psi_k[n] = \cos\left[\frac{\pi}{N}\left(n + \frac{1}{2}\right)k\right]. \tag{27.17}$$

where
   N is the possible number of frequencies as well as the length of the Radon projection slice
   $k = 0, 1, \ldots, N - 1$ is the frequency index

TABLE 27.1   List of DCTs and DSTs

DCT-I	$\frac{1}{2}(x_0 + (-1)^k x_{N-1}) + \sum_{n=1}^{N-2} x_n \cos\left[\frac{\pi}{N-1} nk\right]$
DCT-II	$\sum_{n=0}^{N-1} x_n \cos\left[\frac{\pi}{N}(n + \frac{1}{2})k\right]$
DCT-III	$\frac{1}{2}x_0 + \sum_{n=1}^{N-1} x_n \cos\left[\frac{\pi}{N}n(k + \frac{1}{2})\right]$
DCT-IV	$\sum_{n=0}^{N-1} x_n \cos\left[\frac{\pi}{N}(n + \frac{1}{2})(k + \frac{1}{2})\right]$
DST-I	$\sum_{n=0}^{N-1} x_n \sin\left[\frac{\pi}{N+1}(n + 1)(k + 1)\right]$
DST-II	$\sum_{n=0}^{N-1} x_n \sin\left[\frac{\pi}{N}(n + \frac{1}{2})(k + 1)\right]$
DST-III	$\sum_{n=0}^{N-2} x_n \sin\left[\frac{\pi}{N}(n + 1)(k + \frac{1}{2})\right]$
DST-IV	$\sum_{n=0}^{N-1} x_n \sin\left[\frac{\pi}{N}(n + \frac{1}{2})(k + \frac{1}{2})\right]$

In the context of image processing, the discussion of discretization of the PCT will be further restricted to the 2-D case, which means the unit orientation vector becomes

$$\vec{\theta} = \begin{bmatrix} \cos\theta \\ \sin\theta \end{bmatrix}. \tag{27.18}$$

With $\vec{\xi} = [\xi_1, \xi_2]$, the discrete cosine ridge function is

$$\psi_k[\xi_1 \cos\theta + \xi_2 \sin\theta] = \cos\left[\frac{\pi}{N}\left(\xi_1 \cos\theta + \xi_2 \sin\theta + \frac{1}{2}\right)k\right]. \tag{27.19}$$

Then the discrete polar cosine transform on an $M \times M$ 2-D image $f[\vec{\xi}]$ can be defined as

$$Cf[k, \theta] = \left\langle f[\vec{\xi}], \psi_k[\xi_1 \cos\theta + \xi_2 \sin\theta] \right\rangle$$
$$= \sum_{\xi_1=0}^{M-1} \sum_{\xi_2=0}^{M-1} f[\xi_1, \xi_2] \cos\left[\frac{\pi}{N}\left(\xi_1 \cos\theta + \xi_2 \sin\theta + \frac{1}{2}\right)k\right]. \tag{27.20}$$

The inverse transform for DCT-II is the DCT-III transform. The corresponding cosine ridge function for the inverse transform is

$$\psi_k^{-1}[\xi_1 \cos\theta + \xi_2 \sin\theta] = \lambda_k \cos\left[\frac{\pi}{N}\left(\xi_1 \cos\theta + \xi_2 \sin\theta + \frac{1}{2}\right)\left(k + \frac{1}{2}\right)\right], \tag{27.21}$$

where

$$\lambda_k = \begin{cases} 1/2 & \text{if } k = 0, \\ 1 & \text{if } k \neq 0, \end{cases}$$

and the discrete inverse polar cosine transform operator C^{-1} is given by

$$C^{-1}f[k, \theta] = \langle f[\vec{\xi}], \psi_k[\xi_1 \cos\theta + \xi_2 \sin\theta] \rangle$$
$$= \sum_{\xi_1=0}^{M-1} \sum_{\xi_2=0}^{M-1} f[\xi_1, \xi_2] \lambda_k \cos\left[\frac{\pi}{N}\left(\xi_1 \cos\theta + \xi_2 \sin\theta + \frac{1}{2}\right)\left(k + \frac{1}{2}\right)\right]. \tag{27.22}$$

27.2.3 Radon-Based Digital Implementation

Equations 27.10 and 27.11 suggest that the discrete PCT can be implemented by a Radon transform, that is, taking the 1-D DCT on Radon slices:

$$\mathcal{C}f[k,\ \theta] = \sum_{t=0}^{N-1} \mathcal{R}f[t,\theta]\cos\left[\frac{\pi}{N}\left(t+\frac{1}{2}\right)k\right]. \tag{27.23}$$

As the Fourier Slice Theorem shows, the Radon transform can be implemented by taking the central slice in the Fourier spectrum and then performing a 1-D inverse Fourier transform on it. Applying DCT on the Radon slices, we can obtain the discrete PCT.

This gives a means for a digital implementation of the discrete polar cosine transform. For the Radon transform, various ways to implement it in a discrete fashion have been attempted, among which the Fast Slant Stack [25] was chosen, which is based on a pseudo-polar Fast Fourier Transform (FFT) [26]. According to Ref. [25], the transform is computationally efficient, algebraically exact, geometrically faithful, and its inversion is numerically stable.

With the implementation described above, the PCT basis vectors for an 8×8 image block can be seen in Figure 27.1. The basis vectors are indexed by frequency k and orientation θ. As a result of the Cartesian-to-polar conversion in the frequency domain, the low-frequency basis vectors are clearly being over-sampled, resulting in a redundant frame where the number of coefficients is four times the original data size.

27.2.4 Butterfly-Based Digital Implementation

The above digital implementation of the PCT based on the Radon transform is akin to the digital ridgelet transform [27]. Such a Radon-based approach is flexible in constructing various directional ridge-type transforms, but it suffers from several drawbacks of the underlying Radon transform. First is the high computational requirement both for the forward and inverse transforms. The forward transform includes a pseudo-polar Fourier transform, 1-D inverse Fourier transforms on polar slices, and forward cosine

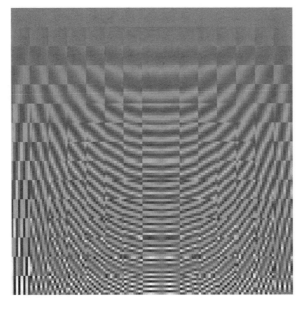

FIGURE 27.1 8×8 discrete polar cosine basis vectors.

transforms. The inverse Fast Slant Stack uses an iterative algorithm in the reconstruction to minimize numerical instability. Secondly, the redundancy factor could be an issue in some applications. While it seems implausible to construct a digital ridgelet transform without invoking the Radon transform, it does not necessarily hold true for the PCT. A possible construction of such directional real trigonometric transform in the context of modulated lapped transforms was discussed in Ref. [28]. The same construction is adopted here in implementing a digital PCT.

It can be observed from Table 27.1 that the DCT-IV and DST-IV share the same function parameters. In 2-D, Cartesian separable basis functions can be formed by the tensor product of 1-D bases:

$$\psi^{\mathcal{C}}_{k_1,k_2}[\xi_1, \xi_2] = \cos\left[\frac{\pi}{N}\left(\xi_1 + \frac{1}{2}\right)\left(k_1 + \frac{1}{2}\right)\right] \cos\left[\frac{\pi}{N}\left(\xi_2 + \frac{1}{2}\right)\left(k_2 + \frac{1}{2}\right)\right]. \tag{27.24}$$

$$\psi^{\mathcal{S}}_{k_1,k_2}[\xi_1, \xi_2] = \sin\left[\frac{\pi}{N}\left(\xi_1 + \frac{1}{2}\right)\left(k_1 + \frac{1}{2}\right)\right] \sin\left[\frac{\pi}{N}\left(\xi_2 + \frac{1}{2}\right)\left(k_2 + \frac{1}{2}\right)\right]. \tag{27.25}$$

By setting $A = \frac{\pi}{N}(\xi_1 + \frac{1}{2})(k_1 + \frac{1}{2})$ and $B = \frac{\pi}{N}(\xi_2 + \frac{1}{2})(k_2 + \frac{1}{2})$, the expression is simplified as follows:

$$\psi^{\mathcal{C}}_{k_1,k_2}[\xi_1, \xi_2] - \psi^{\mathcal{S}}_{k_1,k_2}[\xi_1, \xi_2] = \cos A \cos B - \sin A \sin B$$

$$= \cos[A + B]$$

$$= \cos\left[\frac{\pi}{N}\left(\left(\xi_1 + \frac{1}{2}\right)\left(k_1 + \frac{1}{2}\right) + \left(\xi_2 + \frac{1}{2}\right)\left(k_2 + \frac{1}{2}\right)\right)\right]$$

$$= \cos\left[\frac{\pi}{N}\left(\xi_1 k_1 + \xi_2 k_2 + \frac{1}{2}(\xi_1 + \xi_2 + k_1 + k_2 + 1)\right)\right]. \tag{27.26}$$

Therefore the difference between the 2-D basis functions of DCT-IV and DST-IV at given frequencies can be reduced into a cosine component. Furthermore, since $k_1 = k\cos\theta$, $k_2 = k\sin\theta$, where $k = \sqrt{k_1^2 + k_2^2}$ is the polar frequency index and $\theta = \arctan(k_2/k_1)$, Equation 27.26 can be rewritten in the polar cosine form:

$$\psi^{\mathcal{X}}_{k,\theta}[\xi_1, \xi_2] = \psi^{\mathcal{C}}_{k\cos\theta, k\sin\theta}[\xi_1, \xi_2] - \psi^{\mathcal{S}}_{k\cos\theta, k\sin\theta}[\xi_1, \xi_2]$$

$$= \cos\left[\frac{\pi}{N}\left((\xi_1\cos\theta + \xi_2\sin\theta)k + \frac{1}{2}(\xi_1 + \xi_2 + k\cos\theta + k\sin\theta + 1)\right)\right]. \tag{27.27}$$

However, it should be noted that compared with the PCT-II defined in Equation 27.20, although the basis vectors do not represent a strict form of ridge functions, they are directional and can be arranged in polar form. The above is set for the case when $k_1 = 0, \ldots, N-1$ and $k_2 = 0, \ldots, N-1$. For negative frequencies, it is clear that

$$\psi^{\mathcal{X}}_{k_1,-k_2-1}[\xi_1, \xi_2] = \psi^{\mathcal{C}}_{k_1,-k_2-1}[\xi_1, \xi_2] - \psi^{\mathcal{S}}_{k_1,-k_2-1}[\xi_1, \xi_2]$$

$$= \cos A \cos[-B] - \sin A \sin[-B]$$

$$= \cos A \cos B + \sin A \sin B$$

$$= \psi^{\mathcal{C}}_{k_1,k_2}[\xi_1, \xi_2] + \psi^{\mathcal{S}}_{k_1,k_2}[\xi_1, \xi_2]. \tag{27.28}$$

This suggests that for $-k_2 - 1 = -1, \ldots, -N$, the basis function is just the sum of DCT-IV and DST-IV. In the same way, it is not difficult to see that when $k_1 = -1, \ldots, -N$ and $k_2 = -N, \ldots, N-1$, which is the basis functions for the lower half-plane of the frequency spectrum:

$$\psi^{\mathcal{X}}_{k_1,k_2}[\xi_1, \xi_2] = \psi^{\mathcal{X}}_{-k_1-1,-k_2-1}[\xi_1, \xi_2]. \tag{27.29}$$

Thus the corresponding transform domain exhibits the same Hermite symmetry as the Fourier transform, with basis functions very close to the FFT's basis functions. As a result, the transform

$$\chi f[k_1, k_2] = \langle f[\xi_1, \xi_2], \psi^\chi_{k_1, k_2}[\xi_1, \xi_2] \rangle \tag{27.30}$$

is only twice redundant, by removing a half-plane of the spectrum and can be efficiently implemented by a butterfly computation. A schematic illustration of the transform can be seen in Figure 27.2. For the sake of convenience, this particular digital implementation will be referred to as PCT-X. A complete basis for size 8×8 is shown in Figure 27.3.

It is observed that the PCT-X is closely related to the Fourier transform itself, despite the new transform being real. The expansion of PCT-X, therefore, can be expected to have similarities with the Fourier magnitude spectrum.

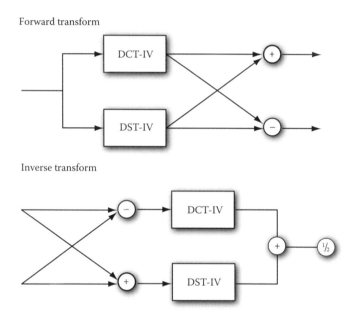

FIGURE 27.2 Forward PCT-X transform and the inverse transform.

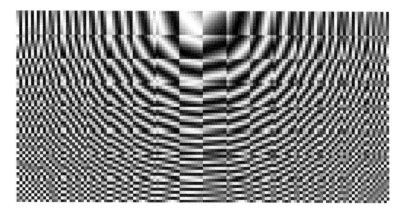

FIGURE 27.3 8×8 discrete polar cosine basis vectors.

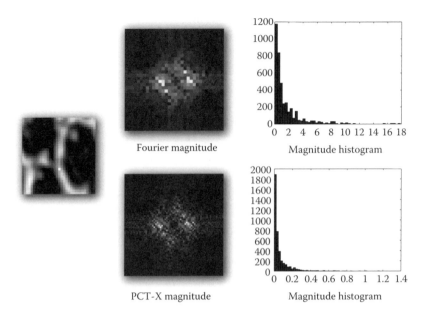

FIGURE 27.4   Spectrum comparison between the PCT-X and the Fourier transform.

In Figure 27.4, a comparison between the PCT-X spectrum and the Fourier spectrum is made. A reptile texutre patch is chosen as the test data and the two transforms yield very similar magnitude responses. However, the histograms suggest that the coefficients of the PCT-X expansion are sparser than those of the Fourier transform, attributed to the symmetric boundary extension of the underlying DCT-IV and DST-IV.

27.2.5 Nonlinear Approximation

As previously mentioned in Section 27.2, with the difficulty in mathematically formulating texture, it is hard to approach the nonlinear approximation problem quantitatively. The theoretical study of the transform in the context of computational harmonic analysis is beyond the scope of this thesis, where the main concern is on examining its effectiveness in providing sparsity in the transformed expansion. The approach taken here is more empirical, where the nonlinear approximation results are presented in the form of output from numerical experiments.

Two digital PCT implementations were selected for nonlinear approximation experiments: the Radon-based PCT with DCT-II as ridge profile (PCT-II) and the butterfly PCT-X implementation as shown in Figure 27.2. They are compared with three other transforms: the FFT; the DCT-II, which is a Cartesian separable real-to-real implementation of FFT; and the *fast curvelet transform* (FCT) which is implemented by wrapping of specially selected subbands of Fourier samples [29].

Two different image patches are tested for nonlinear approximation. The first one contains a portion of a fingerprint image, which consists of repeated curved ridges. The second one is a simple straight line. The purpose of including such a linear singularity is that although it is not ideal to be represented by an oscillatory basis, it would be still worth comparing with other trigonometric transforms. All the patches are sized 256×256, cropped from the original 512×512 images. They are prewhitened by taking the highpass subband Laplacian pyramid.

The nonlinear approximation results are represented in Figures 27.5 and 27.6 as PSNR curves. The PSNR values are plotted against the top percentage of coefficients, from 2% to 50%. The percentage

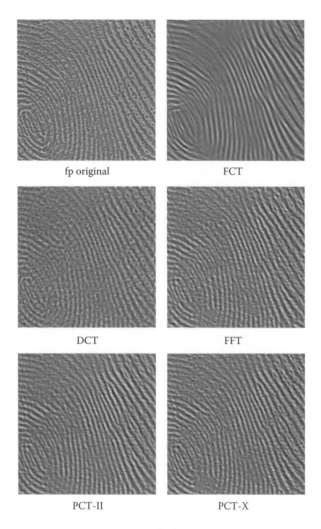

fp original   FCT

DCT   FFT

PCT-II   PCT-X

FIGURE 27.5 Nonlinear approximation PSNR curves for fp.

instead of the number of coefficients is used due to the fact that different transforms yield different number of coefficients. For the sake of convenience, a table of percentages and the corresponding retained numbers of coefficients for different transforms are listed in Table 27.2. It is observed that for the textured image **fp**, the PCTs outperform the other candidates consistently. With fewer coefficients, the PCT-X's PSNR is usually close to the PCT-II, occasionally outperforms it (on **fp**). This is due to the fact that PCT-II's reconstruction is less stable with fewer coefficients while PCT-X's inverse transform is exact. However, the goodness of DCT-II's even symmetric boundaries eventually yields better PSNR with relatively more coefficients involved. While low PSNR performances are to be expected from the trigonometric transforms, it is also reassuring that the two PCTs give better PSNR than the DCT and FFT.

Figure 27.7 shows the reconstructions from the top 2% coefficients of these four transforms on **fp**. The reconstruction from FCT looks artificial, due to its inability to capture directional harmonics of the image, an oscillatory pattern has to be described as several directional singularities. The DCT reconstruction, while being closer to the original, does not highlight the oriented patterns as well as

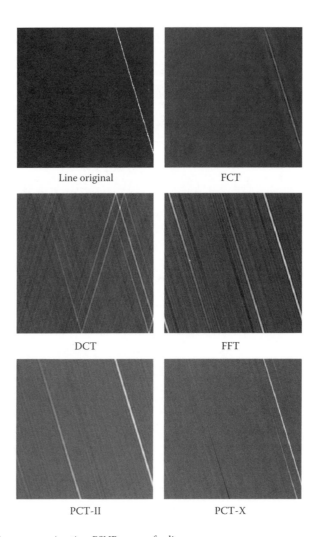

Line original   FCT

DCT   FFT

PCT-II   PCT-X

FIGURE 27.6   Nonlinear approximation PSNR curves for line.

TABLE 27.2   Percentage versus Number of Retained Coefficients
for Different Transforms

Percentage (%)	FCT	DCT	FFT	PCT-II	PCT-X
0.1	185	66	66	262	132
2	3700	1311	1311	5243	2622
15	27748	9830	9830	39322	19660
100	184985	65536	65536	262144	131072

the PCTs. The results from FFT and PCTs are very similar, while the PCTs give slightly better reconstructions with the directional regularities on the fingerprint ridges, particularly by PCT-X.

Another example is the line in Figure 27.8. The percentage of coefficients is kept as little as 0.1%. It is not surprising that the FCT restores the line faithfully even at such a low ratio. The reconstructions from the four Fourier-type bases (DCT, FFT, and the two PCTs) all exhibit the "ghosting" artifacts sometimes referred to as the Gibbs phenomenon, due to suppressing too many high frequency

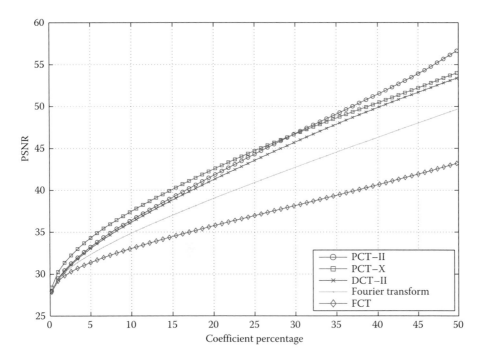

FIGURE 27.7   Illustrative results for nonlinear approximation on fp.

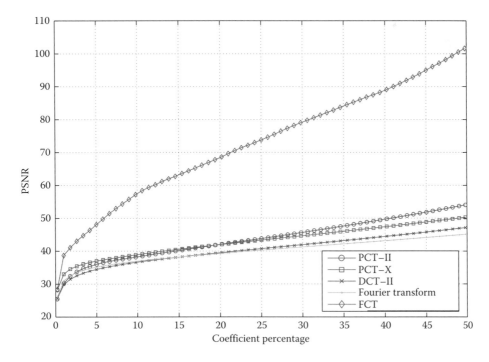

FIGURE 27.8   Illustrative results for nonlinear approximation on line.

coefficients. The DCT shows the ghosting artifact on two different orientations, the FFT has two ghost lines appearing on the reconstruction. The PCT-II, on the other hand, due to the boundary extension being assumed on the Radon slices, shows only one significant ghost line. The ghosting artifact is further reduced in PCT-X's reconstruction.

27.2.6 Polar Cosine Packets

As previously discussed, the Radon transform can reduce the PCT into a matter of performing the cosine transform on Radon slices. This also makes it possible to analyze directional features other than sinusoidal ridges. This is achieved by taking the transform in separate spatial windows. Local trigonometric bases proposed by Coifman and Meyer [30] and by Malvar [31] use smooth window functions to split the signal and to fold overlapping parts back into the original pieces so that the orthogonality is preserved. This treatment can avoid the discontinuity artifacts caused by a rectangular window, while introducing no redundancy.

The local cosine basis is composed of basis functions of the form

$$\varphi_{p,k}(t) = w_p(t)\sqrt{\frac{2}{|I_p|}}\cos\left[\pi\left(k + \frac{1}{2}\right)\frac{t - c_p}{|I_p|}\right]. \tag{27.31}$$

The Cosine-VI basis is used and modulated by a window function $w_p(t)$ which lies on an interval $[a_{p-1}, a_p]$ with $c_p = (a_p + a_{p-1})/2$, and $|I_p| = c_{p+1} - c_p$ being the length of the window, with the overlapping part included. With a careful choice of the window (see Ref. [15]), the set $\{a_p\} \subset \mathbb{R}$ forms a partition of unity and the local cosine basis associated to such partition is said to form a library of orthonormal bases usually referred to as the cosine packets.

The polar cosine packets can be implemented by placing the 1-D cosine packets on the Radon slices; the operator is denoted as

$$\mathcal{P}f[p, k, \theta] = \langle \mathcal{R}f[t, \theta], \varphi_{p,k}[t] \rangle. \tag{27.32}$$

Unfortunately, due to its implementation, it is not possible to construct similar polar cosine packets with the PCT-X.

27.2.7 Best Basis Selection

The arbitrary choice over the library of local trigonometric bases over a compact interval U is of an extremely large range. To seek a feasible "best basis," the library of cosine packets is reduced to only taking dyadically partitioned decomposition of U only. This organization is depicted schematically in Figure 27.9. Then I_{00} is a cosine basis on the entire U and $I_{p,s}$ will correspond to the local cosine basis over interval p of the 2^s intervals at level s of the tree. The best basis can be found by induction on s.

Let $B_{p,s}$ denote the cosine basis vectors corresponding to interval $I_{p,s}$, and $A_{p,s}$ be the best basis. For $s = 0$, the best basis $A_{p,0} = B_{p,0}$, otherwise

$$A_{p,s+1} = \begin{cases} B_{p,s+1} & \text{if } \mathcal{M}(B_{p,s+1}x) < \mathcal{M}(A_{2p,s}x) + \mathcal{M}(A_{2p+1,s}x), \\ A_{2p,s} \oplus A_{2p+1,s} & \text{otherwise.} \end{cases} \tag{27.33}$$

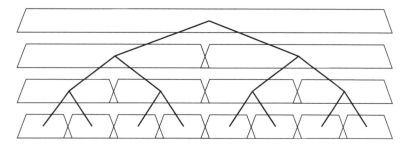

FIGURE 27.9   Binary tree of the localization of partitioned intervals.

where

⊕ denotes a concatenation operation

$\mathcal{M}(\cdot)$ is a certain cost function

The resulting best basis is optimal relative to the cost function $\mathcal{M}$, which means different choices of the cost function reflecting certain criterion can result in different selections of the best basis. Such an approach is called "entropy-based best basis selection." For a complete treatment on the subject, see Ref. [32].

The best basis can be sought with the local cosine basis on the Radon slices. The resulting transform is called the polar cosine packet transform (PCPT), while the choice of the cost function $\mathcal{M}$ depends on the application.

27.2.8 Multiscale Polar Cosine Transform

It was recognized, as early as in 1978 [33], that, for an image representation to be useful, the transform should be well spatially localized. More importantly, the transform should be multiscale, in order to capture patterns of interest at different scales. While the PCT is able to provide a sparse expansion for directional patterns with good frequency resolution, putting it into a multiresolution, spatially localized manner is required. The result is a new wavelet-type transform called the multiscale polar cosine transform (MPCT), which is discussed in the following sections.

27.2.8.1 Construction of MPCT

A prototypical multiscale polar cosine function has the form

$$\Psi_{k,\vec{\theta},s,\vec{\eta}}(\vec{\xi}) = w\left(\frac{\vec{\xi} - \vec{\eta}}{s}\right)\psi_k\left(\frac{(\vec{\xi} - \vec{\eta}) \cdot \vec{\theta}}{s}\right). \tag{27.34}$$

where

$k, \vec{\theta}, \vec{\eta}$, and s denote the frequency, orientation, location, and scale parameters of the function, respectively

$w(\cdot)$ is the smooth window function chosen along with the sampling interval to ensure invertibility of the discrete form of the transform

Effectively the transform can be viewed as a stack of windowed polar cosine transforms at a range of scales, with different block sizes for windowing. The coarser the scale s is, the larger the window becomes. The digital implementation of such a multiscale lapped transform requires two operators: a level operator which decomposes the signal into different scales, and a local operator which handles the decomposition locally within a block—in this case could be either the PCT operator $\mathcal{C}$ or the Polar cosine packet operator $\mathcal{P}$, which gives us the multiscale polar cosine packet transform (MPCPT).

A desirable level operator should have the following two properties:

1. Operator should be able to separate the signal into different frequency subbands.
2. Decomposed subbands should be isotropic, which then can be exploited well by the high-frequency resolution and the directional selectivity of the PCT.

A reasonable choice of the level operator is the Laplacian pyramid [34]. As in Chapter 3, for a particular level x_s of subband, it is computed as

$$x_s(\vec{\eta}) = (\mathbf{I} - \mathbf{G}_{s,s+1}\mathbf{G}_{s+1,s})x_s'. \tag{27.35}$$

where
   $\mathbf{I}$ is the identity operator
   x_s' is the Gaussian lowpass pyramid representation of $x(\vec{\eta})$

$$x_s'(\vec{\eta}) = \sum_{l=0}^{s-1} \mathbf{G}_{l+1,l}x(\vec{\eta}). \tag{27.36}$$

   $\mathbf{G}_{s,s+1}, \mathbf{G}_{s+1,s}$ are the raising and lowering operators associated with transitions between levels in the Gaussian pyramid, as defined previously in Chapter 3

Due to the dyadic decimation, the pyramid is known to have some 33% extra redundancy in 2-D, which is acceptable for general use in many image processing tasks. This also allows us to use a constant window size over different scales, which is equivalent to a dyadic increment of the window size.

27.2.8.2 Basis Functions and Frequency Tiling

The basis functions of the MPCT are 2-D cosine ridges with certain frequency, orientation confined to certain spatial location and scale. Several MPCT basis functions are shown in both spatial and frequency domains in Figure 27.10. These are three members of the MPCT basis at increasing scales on a 256×256 grid. Every function oscillates coherently in a preferred direction and frequency and confined by a

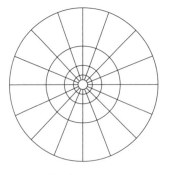
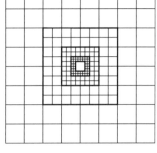

Continuous case   Discrete case

FIGURE 27.10   Basis functions at different scales and their frequency responses.

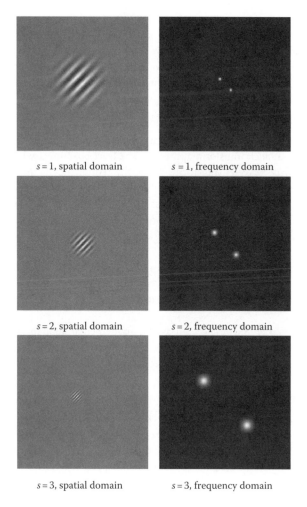

s = 1, spatial domain   s = 1, frequency domain

s = 2, spatial domain   s = 2, frequency domain

s = 3, spatial domain   s = 3, frequency domain

FIGURE 27.11 Frequency tilings of the MPCT.

cosine-squared window. In the frequency domain, the Fourier transform of these functions is well-localized as well, exhibiting two symmetric blobs at an orientation perpendicular to that of the spatial functions. The size of the frequency blob increases with the scale, while the spatial window size decreases dyadically.

Figure 27.11 schematically demonstrates how the frequency plane is partitioned by the MPCT decomposition, both in continuous and discrete cases. It is clear that the polar orientation resolutions at different scales are the same and each of the subbands from the level operator is partitioned into equal number of boxes.

27.2.9 Relationship with Other Transforms

The construction of the MPCT presented here follows a long legacy in both harmonic analysis and image processing, and is directly connected with many other transforms proposed in the literature.

27.2.9.1 Multiresolution Fourier Transform

The multiresolution Fourier transform (MFT) is a windowed Fourier transform at different resolutions and the digital implementation is usually done by a similar lapped block transform like the MPCT with a local FFT operator. Due to the close relationship between PCT and FFT, the MPCT can also be seen as a real version of the MFT, and the assumption of polar symmetric boundary extension increases the sparsity of the transformed coefficients. Unlike the conventional 2-D DCT, which is constructed by tensor product of the 1-D transform, the local operator PCT of MPCT is a true polar transform, combining the merits of both FFT and DCT into one setting.

27.2.9.2 Ridgelets and Curvelets

The first digital curvelet transform, the so-called curvelet-99 implementation proposed in Ref. [7] also adopts a multiscale lapped block transform approach. The level operator used in that specific transform was an undecimated *á trous* wavelet transform. While the redundancy is useful in the denoising task, substituting it with a less redundant operator like the Laplacian pyramid would still retain the curvelet notion. The local operator is the digital ridgelet transform which is a 1-D wavelet transform on Fast Slant Stack Radon slices. Therefore, the essential difference between the curvelet-99 transform and the MPCT transform is just the 1-D transform performed on Radon slices: one being the wavelet and another being the cosine transform.

In Ref. [35], implementations of possible digital ridgelet packets were discussed. One of these uses a basis on the Radon domain from a wavelet packet or cosine packet dictionary. This coincides with the polar cosine packets discussed before, which can be used to deal with features like both ridgelet and polar cosine basis functions.

It is clear that the Fourier Slice Theorem is the fundamental tool in implementing the Radon transform, and it links the curvelets, polar cosine, and Fourier transforms. The relationship between these multiscale lapped transforms can be illustrated schematically as in Figure 27.12.

27.2.9.3 Brushlets and Wave Atoms

The brushlets, proposed by Meyer and Coifman [19], partition the Fourier frequency plane by local cosine windowing, achieving basis functions of directional oscillating patterns localized in orientation, frequency, location, and scale. However, due to the nature of frequency implementation and the choice of the window, the basis functions are not well localized in space, with significant spreads. The recently introduced 2-D wave atoms frame [36] as implemented by using the Villemoes wavelet packets [37] in the frequency domain. The resulting basis functions are much like brushlets, but with only two bumps in the spatial domain.

The MPCT's basis functions are in essence same as the brushlets and wave atoms, but the transform is implemented by spatial windowing instead of frequency windowing. The MPCT and these two transforms are spatial-frequency duals of the same idea, like wavelet packets and the local trigonometric bases.

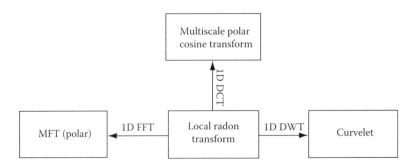

FIGURE 27.12   Relationships between MFT, local radon, curvelet, and MPCT.

27.2.10 Image Denoising in the PCT Domain

In order to demonstrate the effectiveness of the MPCT transform, the transform is applied to the task of noise removal in 2-D by thresholding the transformed coefficients.

The MPCT implementation used in the denoising has the following configuration:

1. Level of decomposition, or the total number of scales is set to $J = 5$.
2. Local window is 16×16, modulated with a squared cosine.
3. Windows are overlapped by 50%.
4. Threshold for each scale is computed as

$$T_s = a \frac{\sqrt{2 \log M^2} \sigma}{1.23^{J-s}} \tag{27.37}$$

with a set to $a = 0.08$ for the MPCT and $a = 0.062$ for the MPCPT. In the above equation, σ is the standard deviation of the noise. The best basis is computed according to the cost function [38]:

$$\mathcal{M}(f, T_s) = \sum_{i=1}^{N} \Phi\left(|\langle f, \psi_k[i] \rangle|^2\right), \tag{27.38}$$

where

$$\Phi(u) = \begin{cases} u - \sigma^2 & \text{if } u \leq T_s^2, \\ \sigma^2 & \text{if } u > T_s^2. \end{cases} \tag{27.39}$$

27.2.11 Denoising Results

The potential of nonlinear filtering in the PCT domain is demonstrated here for two standard images: barbara which contains some directional and nondirectional periodic textures and lena which can be regarded as one of the "curvelet-friendly" images, since it mainly consists of linear singularities at different scales. These images are given in Figure 27.13.

It can be seen from the denoising results in Figures 27.14 and 27.15 that the wave atoms approach produces unpleasant artifacts on lena, due to the fact that its basis functions are not well localized

Barbara   Lena

FIGURE 27.13   Two standard images for denoising experiments.

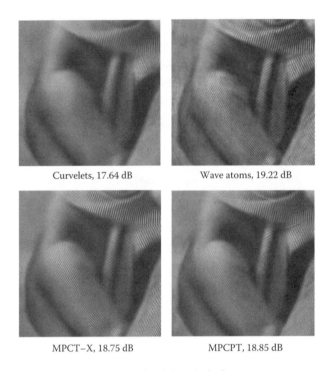

FIGURE 27.14   Comparative denoising results on detailed 10 dB barbara.

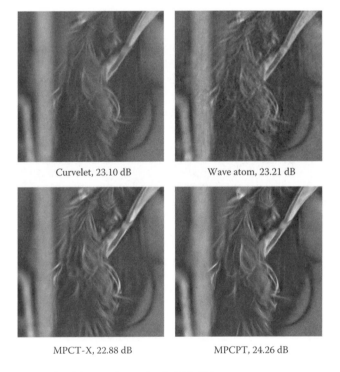

FIGURE 27.15   Comparative denoising results on detailed 15 dB lena.

in space and due to its inability to model edges well. The curvelet transform, MPCT, and MPCPT are all constructed by block-based local transforms on top of the Laplacian pyramid. The MPCT gives consistently better numerical results than curvelets, as well as well-matched visual results. Although the underlying local operator has an oscillatory pattern instead of anisotropic ridge forms, the overlapping blocks effectively compensate for the drawback in edge areas by confining the oscillations locally. In textural regions, the MPCT shows significant gains over the curvelets. The MPCT packet denoising, as expected, combines the merits of both curvelet transform and the MPCTs into one setting.

The difference in the numerical results between the MPCT-II and MPCT-X is compatible with the nonlinear approximation results in Section 27.2.5. The MPCT-X generally performs better with heavier noise, and MPCT-II can outperform MPCT-X by a significant margin when there is less noise. However, visually, the two implementation do not differ significantly. Overall, the MPCPT can be regarded as the winner for its better visual quality as well as numerical error measures.

27.3 Planelets

We now turn our attention to planelets, a special-purpose nonseparable representation for 3-D image data.

27.3.1 Introduction

As mentioned earlier in this chapter, locally planar structures, such as moving edges or interfaces between volumes, convey most of the information in 3-D image data. Preservation of such features requires a basis in which they are sparsely represented. An obvious choice might, therefore, seem to be the Fourier basis, since a planar surface in a volume corresponds to a line in the Fourier domain. Of course, this misses the key epithet: *local*; all image data show only local planarity, with curvature a significant feature at larger scales.

Planelets are designed specifically to efficiently represent planar singularities in 3-D image data. Extraction of such planar features may be useful in various applications, such as video denoising, video coding, geometry estimation [39], and tracking of objects in video sequences. Planelet representation offers translation invariance, good orientation-selectivity, localization in both space and time, invertibility, and efficient computation. The planelet basis has a combination of scaling, locality, and directional characteristics which are well matched to the locally planar surfaces of interest in applications.

Planelets can be regarded as a modification of the complex wavelet bases proposed in Refs. [40,41]. The computational complexity of a planelet transform is $O(n)$, where n is the number of points in analysis window. The computational complexity of planelet denoising is essentially equivalent to a windowed Fourier transform (WFT) and it does not require any motion estimation. In its current form, the representation provides a nonorthogonal basis and is redundant by less than 14%.

27.3.2 Continuous Planelet Transform

A prototypical planelet basis function in 1-D is of the following form:

$$f_{\xi,\omega,a}(x) = win\left(\frac{x-\xi}{a}\right) \exp\left[-j\frac{\omega(x-\xi)}{a}\right], \tag{27.40}$$

where ξ, ω, and a are, respectively, the location, frequency, and scale parameters of the function. The function $win(\cdot)$ is a window function, chosen alongwith the sampling interval to ensure invertibility of the discrete form of the transform. In 2-D, the planelet basis can be regarded as a modification of the complex wavelet bases proposed in Refs. [40,41], which show both translation invariance and directional selectivity, and may be used as an alternative to the ridgelet representation. In 3-D, the basis comprises of the set of Cartesian products over ξ, ω at each scale a. That the continuous transform defined by Wilson et al. [40] is invertible follows directly from the observation that it is simply the MFT [40].

27.3.3 Discrete Planelet Transform

The discrete form, however, is significantly different from that described in Ref. [40]. The discrete planelet transform (DPT) is a combination of two well-known image transforms: the Laplacian pyramid [42] and the WFT. In some ways, it is similar to the octave band Gabor representation proposed in Ref. [43], but avoids some of the more unpleasant numerical properties of the Gabor functions. The DPT of a video sequence x, in vector form, at scale m is given by

$$\hat{X}_m = \mathcal{F}_n(I - G_{m,m+1}G_{m+1,m})x_m, \tag{27.41}$$

where
   $\hat{X}_m$ denotes the DPT at scale m
   $\mathcal{F}_n$ is the WFT operator with window size $n \times n \times n$
   I is the identity operator
   x_m is the Gaussian pyramid representation of x at level m

$$x_m = \prod_0^{m-1} G_{l+1,l}x \tag{27.42}$$

$G_{m,m+1}, G_{m+1,m}$ are the raising and lowering operators associated with transitions between levels in the Gaussian pyramid

Invertibility follows directly from Equations 27.41 and 27.42.

THEOREM 27.2

The representation defined by Equation 27.41 is invertible.

Proof: First we note that the WFT operator $\mathcal{F}_n$ has an inverse, which can be denoted by $\mathcal{F}_n^{-1}$. Second, we know from Burt and Adelson that the Laplacian pyramid is invertible, since, trivially,

$$x_m = x_m - G_{m,m+1}x_{m+1} + G_{m,m+1}x_{m+1} \tag{27.43}$$

and the proof is completed by induction on m.

Importantly, although both the pyramid and WFT operators are Cartesian separable, the closeness of the Burt and Adelson filter to a Gaussian function gives the pyramid virtually isotropic behavior, which can be exploited well by the high-frequency resolution of a Fourier basis. The planelet basis functions resemble planar structures and have compact support in both space–time and spatiotemporal frequency.

27.3.4 Denoising Algorithm

The denoising algorithm works in three steps: forward DPT, adaptive thresholding, and inverse DPT. The coefficients of D forward DPT of the image volume are computed using the algorithm outlined in Section 27.3.3. Since the presence of additive Gaussian white noise means that almost all the DPT coefficients are affected by it, soft thresholding would reduce the contribution of noise to the restored image volume. Based on the assumption that the coefficients relatively small in magnitude at each resolution (i.e., below a certain threshold) are most probably due to the noise variation, coefficients with magnitude above a certain threshold are kept while the remaining ones are discarded. The inverse DPT, therefore, provides an estimation of the original uncorrupted image volume.

The choice of threshold is crucial to the performance of this type of transform domain denoising. Donoho and Johnstone [44] have shown that an adaptive threshold θ given by

$$\theta = \sigma\sqrt{2\log n} \tag{27.44}$$

is an asymptotically optimal choice for threshold value when denoising a 1-D noisy signal, where n denotes the number of samples in the signal and σ is standard deviation of the additive Gaussian white noise. Our experiments showed that using an adaptive threshold for coefficients at different resolutions gives better denoising results as compared to using same threshold value for transform coefficients at all resolutions. We use threshold value θ_i for coefficients at level i of the pyramid as given by

$$\theta_i = \mathcal{L}(\sigma)\sqrt{2\log n_i}, \tag{27.45}$$

where
$\quad n_i$ denotes the number of pixels at level i of the pyramid
$\quad \mathcal{L}(\sigma)$ is a suitably chosen function of σ

The following expression for $\mathcal{L}(\sigma)$ was empirically chosen for our experiments:

$$\mathcal{L}(\sigma) = a\log_{10}\sigma + b, \tag{27.46}$$

where $a, b \in \Re$ and $b = 2a$.

27.3.5 Planar Feature Extraction

Planelets provide an ideal tool for representing local planes in a video sequence (or an image volume, in general) due to their ability to localize planar surfaces which correspond to lines in the Fourier domain. The presence of planar surface in a local analysis window can be inferred by computing the eigenvalues of the local inertia tensor in the window and analysing them. The parameters for orientation of the local planar surface and translation from center of the window can also be estimated by analyzing the most significant coefficients in the locality. Consider a video sequence synthesized by moving the center of a circle on a sinusoidal wave in the time direction. Nonlinear approximations of this sequence using only 0.07% of the wavelet and planelet coefficients are shown in Figures 27.16a and b, respectively. It is clear from this example that the planelet approximation of a video sequence containing locally planar surfaces can result in a smaller approximation error as compared to that using wavelets. Planelets, therefore, can also be used for a piecewise planar

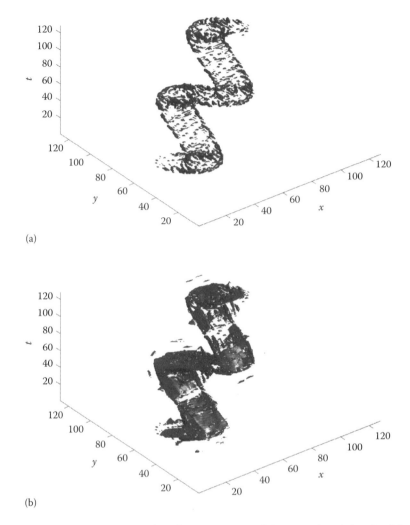

FIGURE 27.16 Nonlinear approximation of a video sequence containing a moving circle using (a) wavelets and (b) planelets.

approximation of a video sequence. Moreover, the approximation can be made to be adaptive to the local scale of the planar surfaces.

27.4 3-D Wavelet Packets

The case for thresholding in spatiotemporal wavelet domain is supported by the fact that certain errors in motion estimation can be overcome by including the temporal direction in the realm of wavelet domain. Recent attempts to solve the video restoration problem have included combined spatial and temporal wavelet denoising [45], and the use of thresholding in nonseparable transform domains, such as oriented 3-D ridgelets [46] and 3-D complex wavelets [47].

Although wavelet shrinkage performs significantly better than most other commonly used denoising methods, visual quality of the restored video can sometimes suffer from *ringing* type of artifacts, valleys around the edges, due to the Gibbs phenomenon. The shift-variant nature of the wavelet transform worsens the effect of the Gibbs phenomenon, resulting in unpleasant artifacts. Translation invariant (TI) wavelet denoising of Coifman and Donoho [48] was developed to counter such artifacts by averaging out the translation dependence. Another feature of wavelet denoising is that it imposes a fixed dyadic wavelet basis on all types of input signals. Not only can the use of dyadic wavelets result in a blurred reconstruction, it can also limit the analysis of a locally occurring phenomenon in the spatiotemporal frequency domain. The solution to this problem lies in the use of basis functions which are well localized in spatiotemporal frequency as well as in space and time.

27.4.1 Adaptive 3-D WP Transform

The ability of wavelet packets to capture locally occurring frequency phenomena in a signal has led to their successful application to many problems including image coding [49,50]. The fundamental idea is to relax the restricted decomposition of only the lowpass subband and allow the exploration of all frequency bands up to the maximum depth. The discrete wavelet packet transform (DWPT) of a 1-D signal x of length N can be computed as follows:

$$w_{2n,d,l} = \sum_k g_{k-2l} w_{n,d-1,k} \quad l = 0, 1, \ldots, N2^{-d} - 1,$$

$$w_{2n+1,d,l} = \sum_k h_{k-2l} w_{n,d-1,k} \quad l = 0, 1, \ldots, N2^{-d} - 1,$$

$$w_{0,0,l} = x_l \qquad\qquad l = 0, 1, \ldots, N - 1,$$

where $d = 1, 2, \ldots, J - 1$ is the scale index, with $J = \log_2 N$, n and l, respectively, denote the frequency and position indices, $\{h_n\}$ and $\{g_n\}$ correspond to the lowpass and highpass filters, respectively, for a two-channel filter bank and the transform is invertible if appropriate dual filters $\{\tilde{h}_n\}, \{\tilde{g}_n\}$ are used on the synthesis side. These equations can be used to compute full wavelet packet (FWP) tree of the signal decomposition. However, this implies that a large number of combinations of basis functions is now available to completely represent the signal. A tree-pruning approach such as Ref. [51] can be used to efficiently select the *best basis* with respect to a cost function.

The 3-D DWPT can be computed by applying above equations separably in all three directions to get the FWP decomposition up to the coarsest resolution of subbands. The best basis can be selected in $O(N \log N)$ time, where N denotes the number of samples (frame resolution times the number of frames) in the video sequence. Given the goal here is to capture the significant spatiotemporal frequency phenomena in a video sequence, we used the Coifman–Wickerhauser entropy [51] as a cost function to select the best basis.

27.4.2 Restoration Algorithm

The effect of the Gibbs phenomenon can be weakened by averaging the restored signal over a range of circular shifts [48]. For this reason, we apply soft threhsolding to the 3-D wavelet packet coefficients of the shifted (in all three directions) noisy video sequence. A modified BayesShrink [52] method is used to compute the optimal value of threshold adaptively for each subband. Threshold θ_b for a subband of length N in an L-level WP decomposition is given by

$$\theta_b = \sqrt{\log N/L} \left(\frac{\sigma^2}{\sqrt{\max\left(\sigma_b^2 - \sigma^2, 0\right)}} \right),$$

where

σ_b^2 is the subband variance

σ^2 is the noise variance

If σ^2 is not known, a robust median estimate for noise standard deviation $\hat{\sigma}$ is obtained as follows:

$$\hat{\sigma} = \mathcal{E}\{\hat{\Sigma}\}, \quad \hat{\sigma}_i = \frac{\text{Median}(|Y_i|)}{0.6745},$$

where $\hat{\sigma}_i \in \hat{\Sigma}$, $Y_i \in \{\mathcal{Y}\}$, set of all HHH bands in the decomposition tree, and the mean $\mathcal{E}$ is taken only on the smaller half of the sorted $\hat{\Sigma}$ excluding the smallest value.

27.5 Discussion and Conclusion

The denoising algorithms presented in Sections 27.3.4 and 27.4.2 were tested against a number of other algorithms for restoration of several standard video sequences, three of which are included here: *Miss America, Hall,* and *Football,* all at a resolution of 128^3. The video sequences were corrupted with additive white Gaussian noise, with the SNR of the noisy sequences being 0, 5, 10, and 15 dB. Table 27.3 gives denoising results in terms of SNR for these noisy sequences using the following algorithms: TI hard thresholding in 3-D wavelet domain (TIW3-D) 3-D wavelet packet (WP3-D) with BayesShrink [52], both non-TI and TI 3-D wavelet packet (TIWP3-D) with the modified form of

TABLE 27.3   SNR Results for Three Standard Video Sequences

Video Sequence	Noise (dB)	TIW3D Hard	WP3D Bayes	WP3D Proposed	TIWP3D Proposed	Planelet SURE
		\multicolumn{5}{c}{Denoising Algorithm (Transform + Thresholding)}				
Miss America	0	17.9	17.2	17.4	**18.9**	17.3
	5	19.5	19.0	19.3	**20.7**	19.6
	10	21.5	21.1	21.5	**23.0**	21.5
	15	23.9	23.1	23.8	**25.2**	23.5
Hall	0	14.7	14.8	15.0	**16.7**	14.8
	5	16.6	17.2	17.3	**18.9**	17.2
	10	19.0	19.5	19.8	**21.3**	19.5
	15	21.7	22.1	22.5	**24.1**	21.8
Football	0	11.9	11.9	12.0	**12.8**	12.1
	5	13.1	12.9	13.3	**14.3**	13.2
	10	15.0	13.9	15.2	**16.9**	14.7
	15	18.0	15.3	17.9	**20.0**	16.6

BayesShrink described in Section 27.4.2, and nonseparable planelet [53] domain thresholding using SUREShrink [54] method. Comparative SNR curves for individual frames for two of the test sequences are provided in Figure 27.17. For all our experiments, the proposed algorithm produces by far the best results in terms of both overall and individual SNR. Some of the frames of the test sequences restored by our algorithm and TIW3-D-Hard, a 3-D realization of the algorithm in Ref. [48], are shown in Figure 27.18. While TIW3-D restores clean and smooth version of the original frames, some of the details are restored by TIWP3-D.

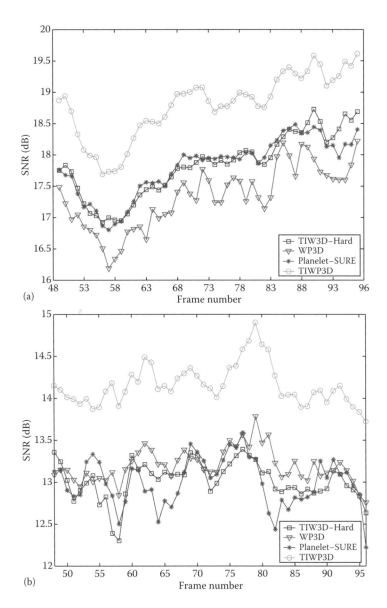

(a)

(b)

FIGURE 27.17 Frame-by-frame comparative results (a) *Miss America* and (b) *Football*.

FIGURE 27.18   Denoising results for three standard video sequences. Frame# 90 of *Miss America*: (a) Original, (b) Noisy (SNR = 0 dB), (c) TIW3D-Hard (SNR = 17.9 dB), and (d) TIWP3D (SNR = 18.9 dB); Frame# 106 of *Hall*: (e) Original, (f) Noisy (SNR = 10 dB), (g) TIW3D-Hard (SNR = 19.0 dB), and (h) TIWP3D (SNR = 21.3 dB); Frame# 60 of *Football*: (i) Original, (j) Noisy (SNR = 5 dB), (k) TIW3D-Hard (SNR = 13.1 dB), and (l) TIWP3D (SNR = 14.3 dB).

For comparison purposes, computational complexity for each of the algorithms considered is also provided in Table 27.4. It is clear from this table that the planelet algorithm of Ref. [53] is the least computationally expensive, whereas the TI implementations of 3-D wavelet and 3-D WP are towards the more expensive side with TIWP3-D being the most expensive due to the additional one-off cost of best basis selection.

TABLE 27.4   Computational Complexity of the Tested Algorithms

Algorithm	Complexity
TIW3D Hard	$O(N + l^3 N)$
WP3D Bayes	$O(N \log N)$
WP3D Proposed	$O(N \log N)$
TIWP3D Proposed	$O(N \log N + l^3 N)$
Planelet SURE	$O(n)$

Note: N and n, respectively, denote sequence size and the planelet window size, and l denotes length of the wavelet filter.

In conclusion, planelets are an efficient representation tool for 3-D functions with planar singularities. Such singularities are commonly found in video sequences in the form of moving luminance edges. A piecewise planar approximation of a video sequence can be obtained by using a very small fraction of transform coefficients in the planelet domain. The ability of planelets to extract planar features from a video sequence makes them an attractive tool for analysis in various applications. It is worth noting that while being the least expensive, the planelet method [53] produces SNR results which are still comparable to those of TIW3-D-Hard. These results also suggest that the localization of spatiotemporal frequency, achieved by TIWP3-D, is a desirable feature of the domain in which video sequences are represented.

References

1. S. Clippingdale and R. Wilson. Least squares image restoration based on a multiresolution model. In *Proceedings ICASSP-89,* Glasgow, U.K., 1989.
2. D. H. Hubel and T. N. Wiesel. Receptive fields, binocular interaction and functional architecture in the cat's visual cortex. *Journal of Physiology,* 160:106–154, 1962.
3. B. A. Olshausen and D. J. Field. Emergence of simple-cell receptive field properies by learning a sparse code for natural images. *Nature,* 381:607–609, 1996.
4. B. A. Olshausen and D. J. Field. Sparse coding with an overcomplete basis set: A strategy employed by V1? *Vision Research,* 37:3311–3325, 1997.
5. E. J. Candés. Ridgelets: Theory and applications. PhD thesis, Department of Statistics, Stanford, University, Stanford, CA, 1998.
6. E. J. Candés and D. L. Donoho. Curvelets—A suprisingly effective non-adaptive representation for objects with edges. In C. Rabut, A. Cohen, and L. L. Schumaker, editors, *Curves and Surfaces,* pp. 105–120. Vanderbilt University Press, Nashville, TN, 2000.
7. J. Starck, E. J. Candés, and D. L. Donoho. The curvelet transform for image denoising. *IEEE Transactions on Image Processing,* 11(6):670–684, June 2002.
8. D. Donoho. Wedgelets: Nearly-minimax estimation of edges. *Annals of Statistics,* 27:353–382, 1999.
9. D. L. Donoho and X. Huo. Beamlets and multiscale image analysis, *Multiscale and Multiresolution Methods: Theory and Applications,* 20:149–196, 2002.
10. M. N. Do and M. Vetterli. Contourlet. In G. V. Welland, editor, *Beyond Wavelets.* Academic Press, New York, 2003.
11. M. N. Do and M. Vetterli. The contourlet transform: An efficient directional multiresolution image representation. *IEEE Transactions Image Processing,* 14(12):2091–2106, December 2005.
12. F. G. Meyer and R. R. Coifman. Brushlets: A tool for directional image analysis and image compression. *Applied and Computational Harmonic Analysis,* 4(2):147–187, 1997.
13. P. Pongpiyapaiboon. Development of efficient algorithms for geometrical representation based on arclet decomposition. Master's thesis, Technische Universität München, Germany, 2005.
14. N. G. Kingsbury. Image processing with complex wavelets. *Philosophical Transactions of the Royal Society London,* A(357):2543–2560, September 1999.
15. P. Auscher, G. Weiss, and M. V. Wickerhauser. Local sine and cosine bases of Coifman and Meyer and the construction of smooth wavelets. In C. K. Chui, editor, *Wavelets: A Tutorial in Theory and Applications,* pp. 237–256. Academic Press, San Diego, 1992.
16. G. Aharoni, R. Coifman A. Averbuch, and M. Israeli. Local cosine transform—A method for the reduction of the blocking effect in JPEG. *Journal of Mathematical Imaging and Vision,* 3:7–38, 1993.
17. R. R. Coifman and Y. Meyer. Orthonormal wave packet bases. Technical report, Department of Mathematics, Yale University, New Haven, 1990.
18. K. Ramchandran and M. Vetterli. Best wavelet packet bases in a rate distortion sense. *IEEE Transactions on Image Processing,* 2(2):160–175, April 1993.

19. F. G. Meyer and R. R. Coifman. Brushlets: Steerable wavelet packets. In J. Stoeckler and G. V. Welland, editors, *Beyond Wavelets*, pp. 1–25. Academic Press Inc., New York, 2001.

20. R. Wilson and G. H. Granlund. The uncertainty principle in image processing. *IEEE Transactions on Pattern Analysis and Machine Intelligence*, 6(11):758–767, 1984.

21. A. Pinkus. Approximating by ridge functions. In A. Le Mehaute, C. Rabut, and L. L. Schumaker, editors, *Surface Filling and Multiresolution Methods*, pp. 279–292, Vanderbilt University Press, Nashville, TN, 1997.

22. R. N. Bracewell. Numerical transforms. *Science*, 248:697–704, 1990.

23. N. Ahmed, T. Natarajan, and K. R. Rao. Discrete cosine transform. *IEEE Transactions on Computers*, 100(23):90–93, January 1974.

24. G. K. Wallace. The JPEG still picture compression standard. *Consumer, Electronics, IEEE Transactions*, 38(1):30–44, April 1992.

25. A. Averbuch, R. Coifman, D. L. Donoho, and M. Israeli. Fast slant stack: A notion of Radon transform for data in a Cartesian grid which is rapidly computible, algebraically exact, geometrically faithful and invertible. To appear in *SIAM Scientific Computing*.

26. A. Averbuch, R. Coifman, D. Donoho, M. Israeli, and J. Walden. The pseudopolar FFT and its applications. Technical report, University of Yale, New Haven, CI 1999. YaleU/DCS/RR-1178.

27. E. J. Candés and D. L. Donoho. Ridgelets: A key to higher-dimensional intermittency? *Philosophical Transactions of the Royal Society of London A*, 357(1760): 2495–2509, 1999.

28. T. Aach and D. Kunz. A lapped directional transform for spectral image analysis and its application to restoration and enhancement. *Signal Processing*, 80:2347–2364, 2000.

29. E. J. Candés, L. Demanet, D. Donoho, and L. Ying. Fast discrete curvelet transforms. Technical report, California Institute of Technology Pasadena, CA, July 2006.

30. R.R. Coifman and Y. Meyer. Remarques sur l'analyse de Fourier à fenêtre. *Comptes Rendus de l'Academie des Sciences, Série 1, Mathematique*, 312(3):259–261, 1991.

31. H. S. Malvar. Lapped transforms for efficient transform/subband coding. *IEEE Transactions on Acoustics, Speech, and Signal Processing*, 38:969–978, 1990.

32. R. R. Coifman and M. V. Wickerhauser. Entropy-based algorithms for best basis selection. *IEEE Transactions on Information Theory*, 38(2):713–718, 1992.

33. G. H. Granlund. In search of a general picture processing operator. *Computer Graphics and Image Processing*, 8:155–173, 1978.

34. P. J. Burt and E. H. Adelson. The Laplacian pyramid as a compact image code. *IEEE Transactions on Communications*, 31:532–540, 1983.

35. A. G. Flesia, H. Hel-Or, A. Averbuch, E. J. Candés, R. R. Coifman, and D. L. Donoho. Digital implementation of ridgelet packets. In G. Welland, editor, *Beyond Wavelets*, pp. 31–60. Academic Press, New York, September 2003.

36. L. Demanet and L. Ying. Wave atoms and sparsity of oscillatory patterns. *Applied Computational Harmonic Analysis*, 23(3):368–387, 2007.

37. L. Villemoes. Wavelet packets with uniform time–frequency localization. *Comptes-Rendus Mathematique*, 335(10):793–796, 2002.

38. H. Krim, D. Tucker, S. Mallat, and D. Donoho. On denoising and best signal representation. *IEEE Transactions on Information Theory*, 45(7):2225–2238, November 1999.

39. A. Bhalerao and R. Wilson. A Fourier approach to 3-D local feature estimation from volume data. In *Proceedings British Machine Vision Conference*, Manchester, U.K., 2001.

40. R. G. Wilson, A. D. Calway, and E. R. S. Pearson. A generalized wavelet transform for Fourier analysis: The multiresolution Fourier transform and its applications to image and audio signal analysis. *IEEE Transancation on Information Theory*, 38(2):674–690, March 1992.

41. N. Kingsbury. Image processing with complex wavelets. *Philosophical Transactions of the Royal Society A: Mathematical, Physical and Engineering Sciences*, 357(1760): 2543–2560, 1999.

42. P. J. Burt and E. H. Adelson. The Laplacian pyramid as a compact image code. *IEEE Transactions on Communications*, 31:532–540, 1983.

43. M. Porat and Y. Y. Zeevi. The generalized Gabor scheme of image representation in biological and machine vision. *IEEE Transactions on PAMI*, 10:452–468, 1988.

44. D. L. Donoho and I. M. Johnstone. Ideal spatial adaptation via wavelet shrinkage. *Biometrika*, 31:425–455, 1994.

45. A. Pizurica, V. Zlokolika, and W. Philips. Combined wavelet domain and temporal video denoising. In *Proceedings of the IEEE International Conference on Advanced Video and Signal Based Surveillance (AVSS)*, July 2003.

46. P. Carre, D. Helbert, and E. Andres. 3-D fast ridgelet transform. In *Proceedings of the IEEE International Conference on Image Processing (ICIP)*, September 2003.

47. I. W. Selesnick and K. Y. Li. Video denoising using 2-D and 3-D dual-tree complex wavelet transforms. In *Proceedings SPIE Wavelets X*, August 2003.

48. R. R. Coifman and D. L. Donoho. Translation-invariant denoising. In *Wavelets and Statistics. Lecture Notes in Statistics*, 1995.

49. F. G. Meyer, A. Z. Averbuch, and J.-O. Strömberg. Fast adaptive wavelet packet image compression. *IEEE Transactions on Image Processing*, 9:792–800, May 2000.

50. N. M. Rajpoot, R. G. Wilson, F. G. Meyer, and R. R. Coifman. Adaptive wavelet packet basis selection for zerotree image coding. *IEEE Transactions on Image Processing*, 12(12):1460–1472, December 2003.

51. R. R. Coifman and M. V. Wickerhauser. Entropy-based algorithms for best basis selection. *IEEE Transactions on Information Theory*, 38(2):713–718, March 1992.

52. G. Chang, B. Yu, and M. Vetterli. Adaptive wavelet thresholding for image denoising and compression. *IEEE Transactions on Image Processing*, 9(9):1532–1546, September 2000.

53. N. M. Rajpoot, R. G. Wilson, and Z. Yao. Planelets: A new analysis tool for planar feature extraction. In *Proceedings of the 5th International Workshop on Image Analysis for Multimedia Interactive Services (WIAMIS)*, April 2004.

54. M. Jansen. *Noise Reduction by Wavelet Thresholding*. Springer-Verlag, New York, 2001.

28

Video Demosaicking Filters

Bahadir K. Gunturk
Louisiana State University

Yucel Altunbasak
Georgia Institute of Technology

28.1 Introduction

Consumer-level digital cameras were introduced in mid-1990s; in about a decade, the digital camera market has grown rapidly to exceed film camera sales. Today, there are point-and-shoot cameras with more than 8 million pixels; professional digital single lens reflex (SLR) cameras with more than 12 million pixels are available; resolution, light sensitivity, and dynamic range of the sensors have been improved significantly. Image quality of digital cameras has become comparable to that of film cameras.

During an image capture process, a digital camera performs a significant amount of processing to produce a viewable image. This processing includes auto focus, white balance adjustment, color interpolation, color correction, compression, and more. A very important part of the imaging pipeline is color filter array interpolation or demosaicking.

To produce a high-quality color image, there should be at least three color samples at each pixel location. One approach is to use beam-splitters along the optical path to project the image onto three separate sensors as illustrated in Figure 28.1. Using a color filter in front of each sensor, three full-channel color images are obtained. This is a costly approach as it requires three sensors and these sensors should be aligned precisely. A more convenient approach is to put a color filter array (CFA) in front of the sensor to capture one color component at a pixel and then interpolate the missing two color components. Because of the mosaic pattern of the CFA, this interpolation process is known as demosaicking.

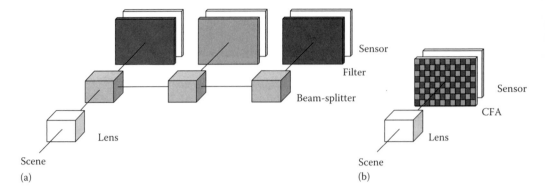

FIGURE 28.1   Illustration of optical paths for multichip and single-chip digital cameras.

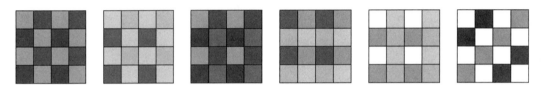

FIGURE 28.2   Several CFA designs are illustrated. From left to right: (a) This is the most commonly used CFA pattern: the Bayer CFA pattern. It consists of red, green, and blue samples. It leads to very good color reproduction performance. (b) This is the Bayer pattern with subtractive primaries: yellow, magenta, and cyan. The color filters have high transmittance values; therefore, good performance in low-light conditions is expected. (c) This pattern uses red, green, blue, and emerald. It is recently used in some Sony cameras. (d) This is a pattern commonly used in video cameras; it consists of yellow, magenta, cyan, and green. (e) This pattern consists of yellow, cyan, and green filters, and unfiltered pixels. The unfiltered pixels improve light sensitivity. (f) This is a pattern that is introduced very recently by Kodak. It has red, green, blue, and unfiltered pixels.

A variety of patterns exist for the color filter array. Some of these patterns are illustrated in Figure 28.2. Among these, the most common array is the Bayer color filter array. The Bayer array measures the green image on a quincunx grid and the red and blue images on rectangular grids. The green image is measured at a higher sampling rate because the peak sensitivity of the human visual system lies in the medium wavelengths, corresponding to the green portion of the spectrum (see Figure 28.3). Although this chapter discusses the demosaicking problem with reference to the Bayer CFA, the discussions and algorithms can in general be extended to other patterns.

The simplest solution to the demosaicking problem is to apply a standard image interpolation technique to each channel separately. However, this neglects the correlation among color channels and results in visible artifacts. For example, in Figure 28.4, Bayer sampling is applied on a full-color image and later bicubic interpolation is applied on each channel. The resulting image suffers from color artifacts. This result motivates the need to find a specialized algorithm for the demosaicking problem. There have been many algorithms published on this topic; this chapter surveys the main approaches.

28.2 Imaging Model

Most demosaicking algorithms model the imaging process as subsampling from a full-color image to a mosaicked data. This is a sufficient model when the goal is only to estimate the missing color samples. (When the goal is to obtain a higher resolution image, then the modulation transfer function of the camera should also be taken into account.)

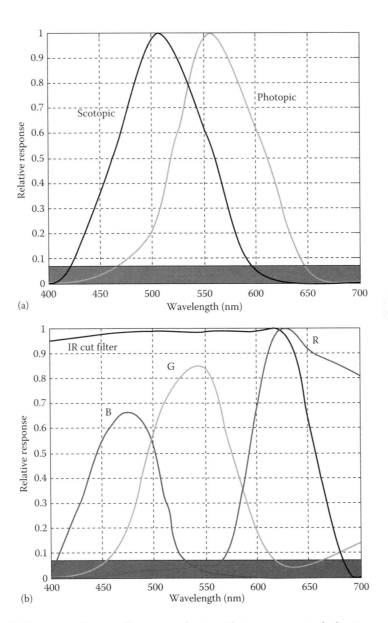

FIGURE 28.3 (a) Luminous efficiency of human visual system. Photopic response is the luminance response of the cone receptors. Scotopic response is the luminance response of the rod receptors, that is, response in low-light conditions. (b) Typical color filter responses in a digital camera.

According to this model, a full-color channel S, where $S = R$ for red, $S = G$ for green, and $S = B$ for blue, is converted to a mosaicked observation z according to a CFA sampling pattern:

$$z = \sum_{S=R,G,B} z_S = \sum_{S=R,G,B} M_S S, \tag{28.1}$$

where
   z_R, z_G, z_B are the subsampled color channels
   mask M_S takes a color sample at a pixel according to the CFA pattern

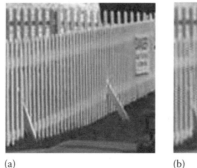

(a)   (b)

FIGURE 28.4 Bicubic interpolation used for color filter array interpolation results in numerous artifacts. (a) Original image and (b) bicubic interpolation.

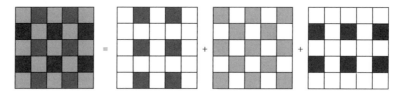

FIGURE 28.5 The mosaicked image z and the sampled components z_R, z_G, and z_B for the Bayer pattern.

For example, at a red pixel location, $[M_R, M_G, M_B]$ is $[1, 0, 0]$.

In Ref. [1], the masks are explicitly written for a Bayer CFA in terms of cosine functions:

$$z_R(i,j) = M_R(i,j)R(i,j) = \frac{1}{4}(1 - \cos \pi i)(1 + \cos \pi j)R(i,j),$$

$$z_G(i,j) = M_G(i,j)G(i,j) = \frac{1}{2}(1 + \cos \pi i \cos \pi j)G(i,j), \qquad (28.2)$$

$$z_B(i,j) = M_B(i,j)B(i,j) = \frac{1}{4}(1 + \cos \pi i)(1 - \cos \pi j)B(i,j),$$

where (i, j) indicate the pixel coordinates, starting with $(0, 0)$. Figure 28.5 illustrates the CFA image z and the sampled components z_R, z_G, and z_B.

28.3 Demosaicking Methods

28.3.1 Single-Channel Interpolation

Single-channel interpolation methods treat each channel separately without utilizing any interchannel correlation. Standard image interpolation techniques, such as bilinear interpolation, bicubic interpolation, spline interpolation, and adaptive methods (e.g., edge-directed interpolation) are applied to each color channel individually. These methods, in general, do not perform as well as the methods that use interchannel correlation.

Among these methods, bilinear interpolation is commonly used as a part of other demosaicking methods. Figure 28.6 provides the linear filters used to perform bilinear interpolation.

$$
\text{(a)} \quad
\begin{bmatrix}
0 & 1/4 & 0 \\
-1/4 & 1 & 1/4 \\
0 & 1/4 & 0
\end{bmatrix}
\qquad
\text{(b)} \quad
\begin{bmatrix}
1/4 & 1/2 & 1/4 \\
1/2 & 1 & 1/2 \\
1/4 & 1/2 & 1/4
\end{bmatrix}
$$

FIGURE 28.6 Filters for bilinear interpolation. (a) Filter applied on z_G to obtain the green channel and (b) filter applied on z_R/z_B to obtain the red/blue channels.

28.3.2 Constant-Hue-Based Interpolation

One commonly used assumption in demosaicking is that the hue (color ratios or differences) within an object in an image is constant. Although this is a simplification of image formation, it is a reasonable assumption within small neighborhoods of an image. This perfect interchannel correlation assumption is formulated such that the color ratios or differences within small neighborhoods are constant. This prevents abrupt changes in color intensities, and has been extensively used for the interpolation of the chrominance (red and blue) channels [2–6]. This approach is called the constant-hue-based interpolation approach.

As a first step, these algorithms interpolate the luminance (green) channel, which is typically done using bilinear or edge-directed interpolation. The chrominance (red and blue) channels are then estimated from the interpolated "red hue" (red-to-green ratio) and "blue hue" (blue-to-green ratio). To be more explicit, the interpolated "red hue" and "blue hue" values are multiplied by the green value to determine the missing red and blue values at a particular pixel location. The hues can be interpolated with any method (bilinear, bicubic, edge-directed, etc.).

As mentioned, instead of interpolating the color ratios, it is also possible to interpolate the color differences or the logarithm of the color ratios. This is illustrated in Figure 28.7.

It is also possible to update all color channels iteratively. That is, the green channel is interpolated first. The red/blue channels are interpolated using constant-hue-based interpolation. The green channel is then updated using the interpolated red/blue channels; and so on [7].

The constant difference idea is sometimes combined with median filtering and used as a postprocessing step to reduce color artifacts [5,8,9]. For example, in Ref. [9], the interpolated color channels are updated as follows:

$$
\begin{aligned}
G'(i,j) &= \frac{\left(R(i,j) - \underset{(i,j)}{\mathrm{median}}(R - G) \right) + \left(B(i,j) - \underset{(i,j)}{\mathrm{median}}(B - G) \right)}{2}, \\
R'(i,j) &= G'(i,j) + \underset{(i,j)}{\mathrm{median}}(R - G), \\
B'(i,j) &= G'(i,j) + \underset{(i,j)}{\mathrm{median}}(B - G),
\end{aligned}
\tag{28.3}
$$

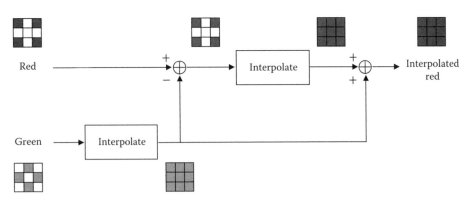

FIGURE 28.7 Constant-hue-based interpolation is illustrated for the interpolation of red channel. (From Gunturk, B. K., et al., *IEEE Signal Process. Mag.*, 22, 44, 2005. With permission.)

where median$_{(i,j)}(\cdot)$ returns the median within a small neighborhood of (i,j). In Ref. [8], red and blue channels are updated first as given in the above equation, followed by the green channel update. This procedure is repeated several times.

28.3.3 Edge-Directed Interpolation

Although nonadaptive algorithms can provide satisfactory results in smooth regions of an image, they usually fail in textured regions and edges. Edge-directed interpolation is an adaptive approach, where edge detection is performed for each pixel in question, and interpolation is done along the edges rather than across them.

In the demosaicking problem, edge-directed interpolation is first applied to the green channel, which is sampled more densely and therefore is less likely to be aliased. Red and blue channel interpolations follow, based on the edge directions found for the green channel. A simple way of performing edge detection is to compare the absolute difference among the neighboring pixels [10]. Referring to Figure 28.8, horizontal and vertical gradients at a missing green location can be calculated from the horizontally and vertically adjacent green pixels. If the horizontal gradient is larger than the vertical gradient, suggesting a possible edge in the horizontal direction, interpolation is performed along the vertical direction. If the vertical gradient is larger than the horizontal gradient, interpolation is performed only in the horizontal direction. When the horizontal and vertical gradients are equal, the green value is obtained by averaging its four neighbors. It is also possible to compare the gradients against a predetermined threshold value [10].

The edge-directed interpolation approach in Ref. [10] can be modified by using larger regions (around the pixel in question) with more complex predictors and by exploiting the texture similarity in different color channels. In Ref. [4], the red and blue channels (in the 5×5 neighborhood of the missing pixel) are used instead of the green channel to determine the gradients. In order to determine the horizontal and vertical gradients at a blue (red) sample, second-order derivatives of blue (red) values are computed in the corresponding direction. This algorithm is illustrated in Figure 28.9.

Once the missing samples of the green channel are computed, the red and blue channels are interpolated. A typical approach for the red/blue interpolation is constant-hue-based interpolation, which was explained in Section 28.3.2.

In the algorithms explained so far, the edge direction is determined first, and then the missing sample is estimated by interpolating along the edge. This is a "hard" decision process. Instead, the likelihood of an edge in a certain direction can be found, and the interpolation can be done based on the edge likelihoods. Such

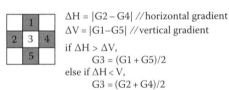

$\Delta H = |G2 - G4|$ //horizontal gradient
$\Delta V = |G1 - G5|$ //vertical gradient
if $\Delta H > \Delta V$,
    $G3 = (G1 + G5)/2$
else if $\Delta H < V$,
    $G3 = (G2 + G4)/2$
else
    $G3 = (G1 + G5 + G2 + G4)/4$

FIGURE 28.8 Edge-directed interpolation in Ref. [10] is illustrated. $G1$, $G2$, $G4$, and $G5$ are measured green values; $G3$ is the estimated green value at pixel 3. (From Gunturk, B. K., et al., *IEEE Signal Process. Mag.*, 22, 44, 2005. With permission.)

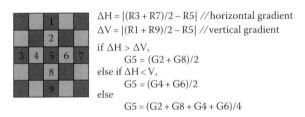

$\Delta H = |(R3 + R7)/2 - R5|$ //horizontal gradient
$\Delta V = |(R1 + R9)/2 - R5|$ //vertical gradient
if $\Delta H > \Delta V$,
    $G5 = (G2 + G8)/2$
else if $\Delta H < V$,
    $G5 = (G4 + G6)/2$
else
    $G5 = (G2 + G8 + G4 + G6)/4$

FIGURE 28.9 Edge-directed interpolation in Ref. [4] is illustrated for estimating the green (G) value at pixel 5. The red (R) values are used to determine the edge direction. When the missing green pixel is at a blue pixel, the blue values are used to determine the edge direction. (From Gunturk, B. K., et al., *IEEE Signal Process. Mag.*, 22, 44, 2005. With permission.)

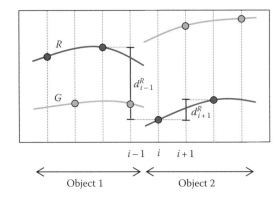

Define:

$$d_i^S = S(i+1) - S(i-1)$$

$$e_i^S = \frac{1}{\sqrt{1 + \left(d_i^S\right)^2}}$$

- Interpolate the green at the missing locations:

$$G(i) = \frac{e_{i-1}^R G(i-1) + e_{i+1}^R G(i+1)}{e_{i-1}^R + e_{i+1}^R}$$

- Repeat for three times:
  - Interpolate the red using the ratio rule:

$$R(i) = G(i) \; \frac{e_{i-1}^G \dfrac{R(i-1)}{G(i-1)} + e_{i+1}^G \dfrac{R(i+1)}{G(i+1)}}{e_{i-1}^G + e_{i+1}^G}$$

- Correct the green to fit the ratio rule:

$$G(i) = R(i) \; \frac{e_{i-1}^R \dfrac{G(i-1)}{R(i-1)} + e_{i+1}^R \dfrac{G(i+1)}{R(i+1)}}{e_{i-1}^R + e_{i+1}^R}$$

FIGURE 28.10 Reference [6] is illustrated for a one-dimensional signal. S is a generic symbol for red (R) and green (G). d_i^S is the gradient for channel S at location i; and e_i^S is the corresponding edge indicator.

an algorithm is presented in Ref. [6]. The algorithm defines edge indicators in several directions as measures of edge likelihood in those directions, and determines a missing pixel intensity as a weighted sum of its neighbors. If the likelihood of an edge crossing in a particular direction is high, the edge indicator returns a small value, which results in less contribution from the neighboring pixel of that direction. The algorithm for one-dimensional signals is illustrated in Figure 28.10. The green channel is interpolated first; the red and blue channels are interpolated from the red/green and blue/green ratios. The color channels are then updated iteratively to obey the color-ratio rule. The extension to two-dimensional images is given in Ref. [6].

A similar algorithm is proposed in Ref. [9], where edge indicators are determined in a 7×7 window for the green and a 5×5 window for the red/blue channels. In this case, the edge indicator function is based on the L_1 norm (absolute difference) as opposed to the L_2 norm of Ref. [6]. In Ref. [11], gradients are filtered adaptively using local means and variances before deciding edge directions. Based on the edge directions, interpolation is done horizontally, vertically, or bidirectionally. This edge direction selection procedure is illustrated in Figure 28.11.

28.3.4 Using Gradients as Correction Terms

Linear interpolation methods have much less computational complexity compared to nonlinear methods, but do not perform as well. When interchannel correlation is included in the linear interpolation, much better performance can be achieved. Recently, the authors of Ref. [12] demonstrated that gradients in one

Calculate horizontal gradients

$$\Delta H(i) = 2\big(G(i+1) - G(i-1)\big) + \big(R(i+2) - R(i-2)\big)$$

Calculate mean and variance of horizontal gradients

$$\mu_i = \frac{1}{3} \sum_{j=i-1}^{i+1} \Delta H(j)$$

$$\sigma_i^2 = \frac{1}{3} \sum_{j=i-1}^{i+1} \big(\Delta H(j) - \mu_j\big)^2$$

Filter the horizontal gradients

$$\Delta H^*(i) = \mu_{i-1} + \frac{\sigma_{i-1}^2}{\sigma_{i-1}^2 + \sigma_{i+1}^2} \big(\mu_{i+1} - \mu_{i-1}\big)$$

Repeat the same procedure to obtain vertical gradients $\Delta V^*(i)$

If $\big|\Delta H^*(i)\big| < \alpha \big|\Delta V^*(i)\big|$, then interpolate horizontally

Else if $\big|\Delta V^*(i)\big| < \alpha \big|\Delta H^*(i)\big|$, then interpolate vertically

Else interpolate bidirectionally

FIGURE 28.11 Reference [11] is illustrated. Gradients are calculated, and then adaptively filtered. The interpolation direction at a pixel is selected based on the relative magnitude of filtered gradients. α is a number in the range [0–1].

channel can improve the interpolation performance in another one; and it can be put in a linear interpolation framework: Suppose, the green value at a red location will be estimated by adding the gradient of the red channel to an initial estimate:

$$\hat{G}(i,j) = \hat{G}_{\text{bilinear}}(i,j) + \alpha\Delta_R(i,j), \tag{28.4}$$

where

$\hat{G}_{\text{bilinear}}(i,j)$ is the initial estimate obtained through bilinear interpolation
α is a scale factor that controls the amount of correction

$\Delta_R(i,j)$ is the gradient of the red channel defined as follows:

$$\Delta_R(i,j) = R(i,j) - \frac{1}{4} \sum_{(m,n)=\left\{\begin{matrix}(0,-2),\,(0,2),\\(-2,0),\,(2,0)\end{matrix}\right\}} R(i+m, j+n). \tag{28.5}$$

Similarly, the red pixel values at green and blue locations can be estimated using green and blue gradients:

$$\hat{R}(i,j) = \hat{R}_{\text{bilinear}}(i,j) + \beta\Delta_G(i,j) \tag{28.6}$$

and

$$\hat{R}(i,j) = \hat{R}_{\text{bilinear}}(i,j) + \gamma\Delta_B(i,j). \tag{28.7}$$

Specific green and blue gradient definitions are given in Ref. [12]. (Similar equations can be written for the green pixels at blue locations and for the blue pixels at green and red locations.) To determine appropriate values for the gain parameters α, β, γ, Ref. [12] uses training images to find the least squares estimates for the gain parameters. These gain parameters are then approximated by integer multiples of

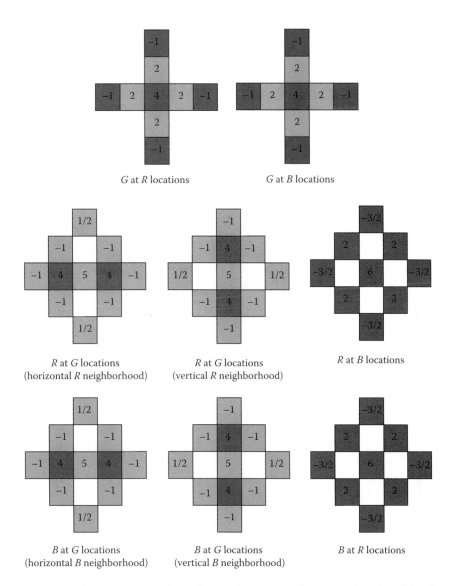

FIGURE 28.12 Filters (From Malvar, H. S., et al., *Proc. IEEE Int. Conf. Acoustics, Speech and Signal Processing*, Montreal, Canada, 3, 485–488, 2004. With permission.)

small powers of $1/2$. The final results are $\alpha = 1/2$, $\beta = 5/8$, and $\gamma = 3/4$. The equivalent linear FIR filter coefficients for each interpolation case are shown in Figure 28.12.

It is possible to incorporate edge-directed interpolation idea to this approach. Reference [13] presents such an algorithm. The gradient terms are added in either horizontal or vertical directions. The direction is chosen adaptively based on edge direction estimates. (Historically earlier than Ref. [12], Ref. [13] does not use training to obtain the optimal coefficients.) Figure 28.13 illustrates this algorithm.

28.3.5 Frequency-Domain Approach

There are two observations that are important for the demosaicking problem. The first is that for natural images, there is a high correlation among the red, green, and blue channels. All three channels are very

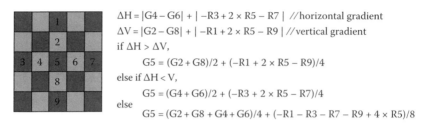

$\Delta H = |G4 - G6| + |-R3 + 2 \times R5 - R7|$ // horizontal gradient
$\Delta V = |G2 - G8| + |-R1 + 2 \times R5 - R9|$ // vertical gradient
if $\Delta H > \Delta V$,
$\qquad G5 = (G2 + G8)/2 + (-R1 + 2 \times R5 - R9)/4$
else if $\Delta H < V$,
$\qquad G5 = (G4 + G6)/2 + (-R3 + 2 \times R5 - R7)/4$
else
$\qquad G5 = (G2 + G8 + G4 + G6)/4 + (-R1 - R3 - R7 - R9 + 4 \times R5)/8$

FIGURE 28.13   Graphical illustration. (From Gunturk, B. K., et al., *IEEE Signal Process. Mag.*, 22, 44, 2005. With permission.)

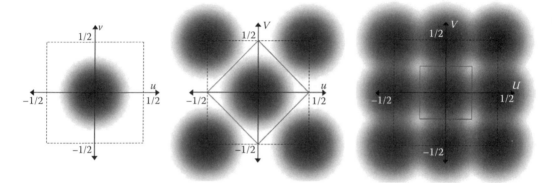

FIGURE 28.14   Frequency domain analysis of CFA sampling. (a) Suppose this is the frequency spectrum of the red, green, and blue channels; (b) frequency spectrum of the sampled green channel; and (c) frequency spectrum of the red/blue channels. Note that while there is no aliasing for the green channel, red and blue channels are aliased. The green channel can be fully recovered with a low-pass filter whose pass-band is outlined in the middle figure. For the red/blue channels, such a low-pass filtering operation cannot eliminate the aliasing.

likely to have the same texture and edge locations. (Because of the similar edge content, we expect this interchannel correlation to be even higher when it is measured between the high-frequency components of the channels [14].) The second observation is that digital cameras often use a CFA in which the luminance (green) channel is sampled at a higher rate than the chrominance (red and blue) channels. Therefore, the green channel is less likely to be aliased, and details are preserved better in the green channel than in the red and blue channels. This is illustrated in Figure 28.14.

28.3.5.1 Alias Canceling Interpolation

In Ref. [15], it is assumed that the high-frequency contents of green and red/blue channels are identical; therefore, high-frequency content of the green image is used to remove aliasing in the red and blue images. First, the red and blue images are interpolated with a rectangular low-pass filter according to the rectangular sampling grid. This fills in the missing values in the grid, but allows aliasing distortions into the red and blue output images. These output images are also the missing high-frequency components needed to produce a sharp image. However, because the green image is sampled at a higher rate, the high-frequency information can be taken from the green image to improve an initial interpolation of the red and blue images. A horizontal high-pass filter and a vertical high-pass filter are applied to the green image. This provides the high-frequency information that the low sampling rate of the red and blue images cannot preserve. Aliasing occurs when high-frequency components are shifted into the low-frequency portion of the spectrum, so if the outputs of the high-pass filters are modulated into the low-frequency regions, an estimate of the aliasing in the red and blue images can be found. This estimate is used to reduce the aliasing in the red and blue images, as illustrated in Figure 28.15. This method relies

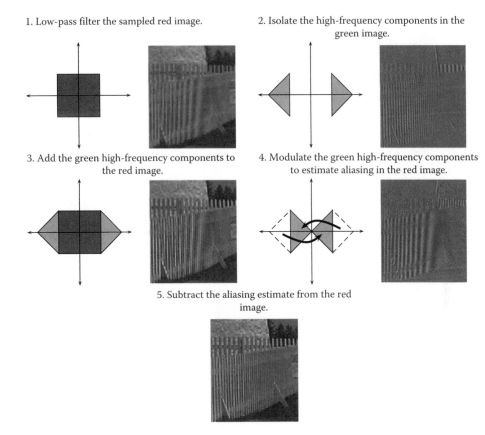

1. Low-pass filter the sampled red image.

2. Isolate the high-frequency components in the green image.

3. Add the green high-frequency components to the red image.

4. Modulate the green high-frequency components to estimate aliasing in the red image.

5. Subtract the aliasing estimate from the red image.

FIGURE 28.15 An illustration of Ref. [15] by John Glotzbach. High-frequency information from the green image is modulated and used to cancel aliasing in the red image. (From Gunturk, B. K., et al., *IEEE Signal Process. Mag.*, 22, 44, 2005. With permission.)

on the assumption that the high-frequency information in the red, green, and blue images is identical. If this assumption does not hold, the addition of the green information into the red and blue images can add unwanted distortions. This method also makes the assumption that the input image is band-limited within the diamond-shaped Nyquist region of the green quincunx sampling grid (the region outlined in Figure 28.14). When this assumption fails, the aliasing artifacts are enhanced instead of reduced because the green image also contains aliasing. This system is composed entirely of linear filters, making it efficient to implement.

28.3.5.2 Frequency-Domain Filtering

In Ref. [1], the CFA-sampled image is reorganized into newly defined luminance and chrominance components. Analyzing the CFA sampling in Fourier domain, filters that recover these components are designed. The derivation is as follows. Using Equation 28.2, the mosaicked image can be written as

$$
\begin{aligned}
z(i,j) &= z_R(i,j) + z_G(i,j) + z_B(i,j) \\
&= \tfrac{1}{4}(R(i,j) + 2G(i,j) + B(i,j)) \\
&\quad + \tfrac{1}{4}(B(i,j) - R(i,j))(\cos \pi i - \cos \pi j) \\
&\quad + \tfrac{1}{4}(-R(i,j) + 2G(i,j) - B(i,j))\cos \pi i \cos \pi j.
\end{aligned} \tag{28.8}
$$

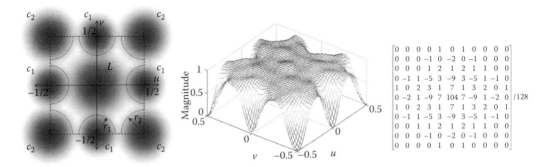

FIGURE 28.16 Filter design in Ref. [1]. Left: Luminance and chrominance components when the Fourier transform of the mosaicked image is taken. The red line outlines the passband of the filter. The radii r_1 and r_2 are design parameters, whose values are determined empirically. Middle and Right: Fourier domain and spatial domain representations of the filter used in Ref. [1].

In this equation, $L(i,j) = (1/4)[R(i,j) + 2G(i,j) + B(i,j)]$ is defined as the luminance component, and $C_1(i,j) = (1/4)[B(i,j) - R(i,j)]$ and $C_2(i,j) = (1/4)[-R(i,j) + 2G(i,j) - B(i,j)]$ as the chrominance components. With these definitions, Equation 28.8 becomes

$$z(i,j) = L(i,j) + C_1(i,j) \, (\cos \pi i - \cos \pi j) + C_2(i,j) \cos \pi : \cos \pi j, \tag{28.9}$$

where

$$\begin{bmatrix} L \\ C_1 \\ C_2 \end{bmatrix} = \begin{bmatrix} 1/4 & 1/2 & 1/4 \\ -1/4 & 0 & 1/4 \\ -1/4 & 1/2 & -1/4 \end{bmatrix} \begin{bmatrix} R \\ G \\ B \end{bmatrix}. \tag{28.10}$$

When the Fourier transform of Equation 28.9 is taken, the luminance component stays in the baseband, the C_1 component will be modulated at frequencies (0.5, 0) and (0, 0.5), and the C_2 component will be modulated at frequency (0.5, 0.5). This is illustrated in Figure 28.16. By designing appropriate filters, the luminance and chrominance components can be recovered. Using the inverse of the matrix given in Equation 28.10, the red/green/blue values are obtained. Figure 28.16 also shows the filter to recover the luminance component.

An extension of this approach is given in Ref. [16], where it is noticed that the C_1 component suffers from spectral overlap, and the overlap often occurs in only one of the (horizontal/vertical) directions. Therefore, a weighted sum of the horizontally and vertically filtered C_1 components is taken, where the weight is less for the one with least crosstalk.

Reference [17] gives another extension of Ref. [1]. This time adaptive filtering is applied on the luminance component. The luminance values at green locations are estimated using a filter similar to the one in Ref. [1]; while the values at red/blue locations are estimated as a weighted sum of neighboring luminance values, where the weights are selected according to the horizontal and vertical gradients (edge indicators).

28.3.6 Homogeneity-Directed Interpolation

Instead of choosing the interpolation direction based on edge indicators, Ref. [8] uses local homogeneity as an indicator. The homogeneity-directed interpolation imposes the similarity of the luminance and chrominance values within small neighborhoods, and it leads to very good perceptual results. The underlying idea is to interpolate color channel horizontally and vertically, and to pick up either the horizontally interpolated pixel values or the vertically interpolated pixel values at every pixel location based on the homogeneity.

Reference [8] defines the homogeneity as follows. Suppose that $R(i,j)$, $G(i,j)$, $B(i,j)$ are the values in the RGB space, and $L(i,j)$, $a(i,j)$, $b(i,j)$ are the corresponding luminance and chrominance values in the CIELab space. Three neighbor sets of (i,j) are defined. The first one is the set of pixel locations that are close in space:

$$N_D(i,j) = \left\{ (m,n) \mid \left[(m-i)^2 + (n-j)^2 \right]^{1/2} \leq \varepsilon_D \right\}. \qquad (28.11)$$

The other two neighbor sets are the sets of pixel locations with similar luminance and chrominance values:

$$N_L(i,j) = \left\{ (m,n) \mid |L(m,n) - L(i,j)| \leq \varepsilon_L \right\} \qquad (28.12)$$

and

$$N_C(i,j) = \left\{ (m,n) \mid \left([a(m,n) - a(i,j)]^2 + [b(m,n) - b(i,j)]^2 \right)^{1/2} \leq \varepsilon_C \right\}. \qquad (28.13)$$

Then the homogeneity is defined as

$$H(i,j) = \frac{\text{size}[N_D(i,j) \cap N_L(i,j) \cap N_C(i,j)]}{\text{size}[N_D(i,j)]}. \qquad (28.14)$$

Referring to Figure 28.17, the algorithm works as follows. The RGB data is first interpolated horizontally and vertically. (The green channel is interpolated using red and blue data as correction terms, as in Ref. [13]. The red and blue channels are interpolated from the interpolated red–green difference and blue–green difference as shown in Figure 28.7.) The interpolated images are then converted to the CIELab space. The homogeneity maps for horizontally and vertically interpolated images are found. The homogeneity maps are smoothed with a 3×3 averaging filter. At each pixel, either the horizontally or the vertically interpolated color values are taken depending on which has the largest homogeneity.

In Ref. [8], the neighborhood parameter ε_D is kept constant, while ε_L and ε_C are determined adaptively at each pixel such that they reflect typical variations among pixels of the same object. This

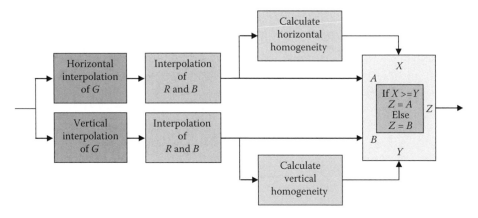

FIGURE 28.17 Block diagram of the homogeneity-directed interpolation (From Hirakawa, K. and Parks, T. W., *IEEE Trans. Image Process.*, 14, 360, 2005. With permission.)

is achieved by analyzing the four nearest neighbors of the pixel in question. ε_L at location (i, j) is calculated as follows:

$$
\begin{aligned}
\varepsilon_{LH}(i,j) &= \max\{|L(i-1,j) - L(i,j)|, \ |L(i+1,j) - L(i,j)|\}, \\
\varepsilon_{LV}(i,j) &= \max\{|L(i,j-1) - L(i,j)|, \ |L(i,j+1) - L(i,j)|\}, \\
\varepsilon_{L}(i,j) &= \min\{\varepsilon_{LH}(i,j), \varepsilon_{LV}(i,j)\},
\end{aligned}
\tag{28.15}
$$

where the first two equations give the maximum variations in horizontal and vertical directions, and the last equation picks up the minimum of these, by which the maximum variation within the object is determined. ε_C is determined similarly.

28.3.7 Projections onto Convex Sets Approach

The algorithm presented in Ref. [15] proposes to decompose the green channel into its frequency components and then add the high-frequency components of the green channel to the low-pass filtered red and blue channels. This is based on the observation that the high-frequency components of the red, blue, and green channels are similar and the fact that the green channel is less likely to be aliased. One problem with this approach is that the high-frequency components of the red, green, and blue channels may not be identical. Therefore, replacement of the high-frequency components of the red and blue channels with those of the green channel may not work well. Reference [14] proposes an algorithm that ensures data consistency at the cost of higher computational complexity. The algorithm defines two constraint sets, one ensuring that the restored images are consistent with the measured data and the other imposing similar high-frequency components in the color channels, and reconstructs the color channels using the projections onto convex sets (POCS) technique.

28.3.7.1 Constraint Sets

The first constraint set guarantees that the restored color channels are consistent with (are identical to) the color samples captured by the digital camera.

The second constraint set is a result of the high interchannel correlation. Reference [14] shows that color channels have very similar detail (high-frequency) subbands. This information would not be enough to define constraint sets if all channels lost the same amount of information in sampling. However, the red and blue channels lose more information (details) than the green channel when captured with a color filter array. Therefore, it is reasonable to define constraint sets on the red and blue channels that force their high-frequency components to be similar to the high-frequency components of the green channel. The similarity is imposed such that the detail coefficients are within a fixed proximity.

28.3.7.2 Alternating Projections Algorithm

The block diagram of the POCS algorithm is given in Figure 28.18. The projection onto the observation constraint set inserts the observed data into their corresponding locations in the current image. This is illustrated in Figure 28.19. The projection operation onto the detail constraint set is illustrated in

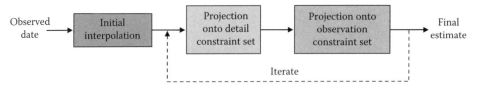

FIGURE 28.18   Block diagram of the algorithm given in Ref. [14]. Red/green/blue channels are interpolated first. The red/blue channels are then updated by iteratively projecting on the detail and observation constraint sets.

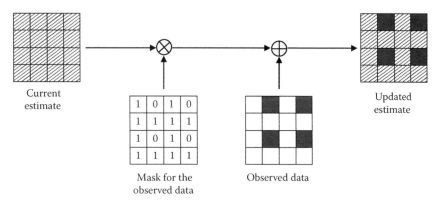

Current estimate

Mask for the observed data

Observed data

Updated estimate

FIGURE 28.19 Projection onto observation constraint set (From Gunturk, B. K., Altunbasak, Y. and Mersereau, R. M., *IEEE Trans. Image Process.*, 11, 997, 2002. With permission.)

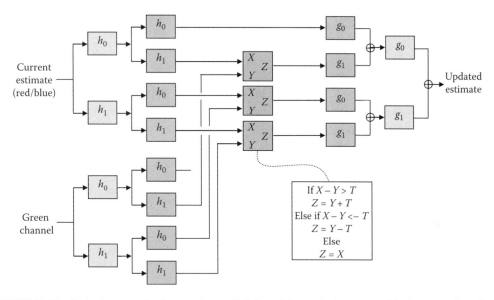

FIGURE 28.20 Projection onto detail constraint set [14]. h_0 and h_1 are the low-pass and high-pass analysis filters. g_0 and g_1 are the corresponding synthesis filters.

Figure 28.20. By alternately applying these two projections onto the initial red and blue channel estimates, these channels are enhanced.

28.3.8 Spectral Response Modeling

The last group of methods models the image formation process and formulates the demosaicking problem based on this model. To understand this approach, we first need to understand the imaging process. The image acquisition process is usually modeled as a linear process between the light radiance arriving at the camera and the pixel intensities produced by the sensors. An image sensor has a specific spectral response $L_S(\lambda)$, which is a function of the spectral wavelength λ, and a spatial response $h_S(x, y)$, which results from optical blur and the spatial integration at each sensor site. (Typical spectral sensor sensitivities are illustrated in Figure 28.3. The λ space is typically modeled as a 7–11 dimensional space.) The imaging process can be formulated as

$$S(x, y) = \iiint L_S(\lambda) h_S(x - u, y - v) r(u, v, \lambda) du \ dv \ d\lambda + N_S(x, y), \tag{28.16}$$

where
  $S(x, y)$ is the pixel value at spatial location (x, y)
  $r(x, y, \lambda)$ is the incident radiance
  $N_S(x, y)$ is the additive noise that is a result of thermal/quantum effects and quantization

There are couple of assumptions in this formulation: (1) the input–output relation is assumed to be linear; (2) the spatial blur $h_S(x, y)$ is assumed to be space-invariant and independent of wavelength; and (3) only the additive noise is considered. These assumptions are reasonable for practical purposes.
  Since we are dealing with digital data, we need to have the discrete version of Equation 28.16

$$S(i, j) = \sum_l \sum_{m, n} L_S(l) h_S(i - m, j - n) r(m, n, l) + N_S(i, j). \tag{28.17}$$

The color filters sample the signal $S(i, j)$ to produce a mosaicked data $z(i, j)$ as given in Equation 28.21. Therefore, the observation model is a linear system, which can be written in the compact form

$$\mathbf{z} = \mathbf{Hr} + \mathbf{N}, \tag{28.18}$$

where
  $\mathbf{r}$, $\mathbf{z}$, and $\mathbf{N}$ are the stacked forms of $r(m, n, l)$, $z(i, j)$, and CFA-sampled $N_S(i, j)$, respectively
  $\mathbf{H}$ is the matrix that includes the combined effects of optical blur, sensor blur, spectral response, and CFA sampling

In Refs. [18–20] the minimum mean square error (MMSE) solution of Equation 28.18 is given:

$$\mathbf{r}^{\text{MMSE}} = E[\mathbf{rz}^{\text{T}}] \left(E[\mathbf{zz}^{\text{T}}] \right)^{-1} \mathbf{z}, \tag{28.19}$$

where $E[\cdot]$ is the expectation operation. In Ref. [20], the point spread function (PSF) is taken as an impulse function; and $\mathbf{r}$ is represented as a weighted sum of spectral basis functions to reduce the dimensionality of the problem. (Later, Ref. [21] extended Ref. [20] to include the PSF in the reconstruction.) In Ref. [19], adaptive reconstruction and ways to reduce computational complexity are discussed. Reference [18] constructs a FIR filter based on a wide sense stationary assumption.

28.4 Demosaicking of Video and Super-Resolution Reconstruction

When there are multiple images, it is possible to estimate the missing samples than in the case of a single image. Even if a color sample does not exist in an image as a result of Bayer sampling, that particular sample could have been captured in another frame (due to motion). By warping all captured samples onto the common frame to be demosaicked, a better estimate of a missing sample can be obtained.
  Figure 28.21 illustrates this multiframe interpolation idea. In the figure, red channels of three Bayer-sampled images are shown. The grid locations with "triangles," "circles," and "squares" show the red samples in these images. We would like to estimate the missing samples in the middle image. The other two images are warped onto the middle image. The estimation problem now becomes an interpolation problem from a set of nonuniformly sampled data. Taking the weighted of these samples, where the weights are inversely proportional with the distance to the pixel in question, the missing sample can be estimated.

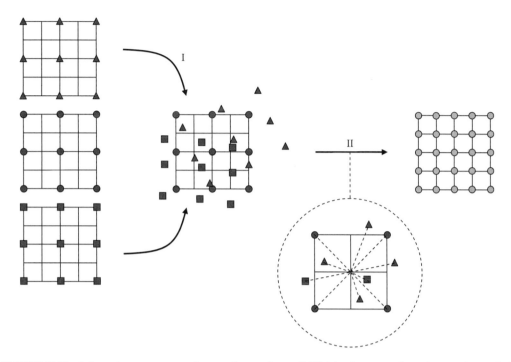

FIGURE 28.21 I: Input images are warped onto reference frame. II: Weighted average of samples are taken to find values on the reference sampling grid.

This idea can be combined with the other demosaicking ideas, such as constant-hue-based interpolation, to formulate a solution. When these multiple frame are available, we can now talk about a spatiotemporal neighborhood instead of a spatial neighborhood of a pixel.

It is also possible to obtain subpixel resolution by combining multiple images. This is known as superresolution reconstruction, and it was recently applied to CFA sampled color images. The imaging model for super-resolution reconstruction starts with a high-resolution image: Let x_S be a color channel of a high-resolution image, where a channel can be red (x_R), green (x_G), or blue (x_B). The ith observation, $S^{(i)}$, is obtained from this high-resolution image through spatial warping, blurring, and downsampling operations:

$$S^{(i)} = DCW^{(i)}x_S, \quad \text{for } S = R, G, B, \quad \text{and} \quad i = 1, 2, \ldots, K, \tag{28.20}$$

where

$\quad$ K is the number of input images
$\quad$ $W^{(i)}$ is the warping operation (to account for the relative motion between observations)
$\quad$ C is the convolution operation (to account for the point spread function of the camera)
$\quad$ D is the downsampling operation (to account for the spatial sampling of the sensor)

The full-color image ($R^{(i)}, G^{(i)}, B^{(i)}$) is then converted to a mosaicked observation $z^{(i)}$ according to a CFA sampling pattern:

$$z^{(i)} = \sum_{S=R,G,B} M_S y_S^{(i)} \tag{28.21}$$

as we explained earlier. Then, the super-resolution problem becomes estimation of the high-resolution image $x_S, S = R, G, B$, from low-resolution mosaicked data $z^{(i)}, i = 1, 2, \ldots, K$. A typical flowchart of super-resolution reconstruction is illustrated in Figure 28.22.

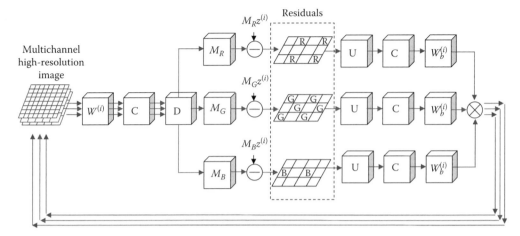

FIGURE 28.22 Typical super-resolution reconstruction algorithm is illustrated. The algorithm starts with a high-resolution image estimate. Simulated observations are obtained by forward imaging operations, including the CFA sampling. The residuals are computed on Bayer pattern samples for each channel, and then back-projected. The notation in the figure is as follows. $W^{(i)}$: Spatial warping onto ith observation, C: Convolution with the PSF, D: Downsampling by the resolution enhancement factor, U: Upsampling by zero insertion, $W_b^{(i)}$: Back-warping to the reference grid.

In Ref. [22], data fidelity and regularization terms are combined to produce high-resolution images. The data fidelity term is based on a cost function that consists of the the sum of residual differences between actual observations and high-resolution image projected onto observations (simulated observations). Regularization functions are added to this cost function to eliminate color artifacts and preserve edge structures. These additional constraints are defined as luminance, chrominance, and orientation regularization [22]. A similar algorithm is presented in Ref. [23]. Reference [24] extends the POCS algorithm of Ref. [14] to multiple frames, where observation constraint set is obtained through bilateral filtering of samples.

28.5 Related Research Problems

In this chapter, we have covered some of the basic demosaicking approaches. There are various others, such as the Bayesian estimation based [25,26] and the neural network based [27]. There are also research problems that are related to the CFA sampling, some of which are listed below:

Denoising: Denoising becomes critical in a digital camera pipeline, especially for images captured in low-light conditions and with high ISO speed. Although standard denoising algorithms can be combined with standard demosaicking algorithms, denoising and demosaicking can be done jointly. Reference [28] proposes such a joint technique based on the total least squares denoising method. The technique reports superior image quality compared to sequential applications of standard denoising and demosaicking algorithms.

Compression: Compression is the last process in a digital camera pipeline. Typically, demosaicking is done on raw CFA data to obtain full-color images, which are later compressed with a compression algorithm, such as the JPEG for still images or the MPEG for videos. The image/video compression algorithms generally decompose images into luminance and chrominance channels and then down-samples chrominance channels to achieve higher compression rates. Because of this downsampling, it seems redundant to do demosaicking before compression. Compression could be done on CFA data; and demosaicking could be added to the end of the decoding processing. Recent studies [29,30] show that, with this alternative processing chain, higher image quality can be achieved at low compression rates. Also, the processing cost is reduced at the camera side.

Camera identification and forgery detection: Due to the demosaicking process, pixels in a digital image are correlated; since different cameras use different demosaicking algorithms, the correlation among the pixels can be used to identify the camera or to detect image forgeries, such as cropping and pasting a region from one image to another [31,32].

Optimal spectral sensitivity functions: Selection of the spectral sensitivity functions is an important part of the digital camera design. The Bayer RGB CFA is known for its superior color reproduction; while the CMYG CFA has better signal-to-noise ratio performance in low-light conditions. Methods for designing optimal spectral sensitivities for color reproduction have been studied earlier [33]. Recently, importance of spectral sensitivities on demosaicking [34] and design of optimal sensitivities for both color and spatial reproduction [35] are discussed.

28.6 Evaluation of Demosaicking Algorithms

Since raw CFA data is not available for most digital cameras, demosaicking methods have been compared based on simulations: Full color images are sampled according to a CFA pattern; the original and the restored images are later compared quantitatively. (This neglects many of the processes in the camera pipeline, most importantly, optical low-pass filtering. Therefore, simulation results may be misleading;

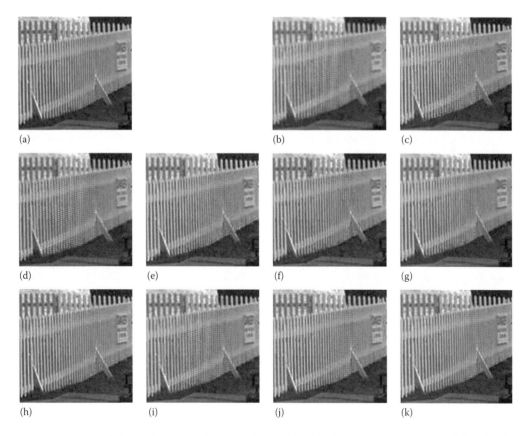

FIGURE 28.23 Result images for example lighthouse image. (a) Original image, (b) bilinear interpolation, (c) edge-directed interpolation in Ref. [10], (d) constant hue-based interpolation in Ref. [3], (e) weighted sum in Ref. [6], (f) second-order gradients as correction terms in Ref. [13], (g) Bayesian approach in Ref. [25], (h) homogeneity-directed in Ref. [8], (i) pattern matching (Chang) in Ref. [39], (j) Alias cancellation in Ref. [15], and (k) POCS in Ref. [14].

the demosaicking algorithms should be tested on raw data; different cameras may have different demosaicking algorithms working best for them.) Two commonly used quantitative measures for evaluating demosaicking algorithms are mean square error and peak signal-to-noise ratio. Euclidean distances in the perceptually uniform CIELab and CIELuv spaces and the s-CIELab [36] are better measures considering the human visual perception.

For the demosaicking algorithms, the Kodak color image database [37] has become a standard test image set. Many algorithms evaluate their performances using these images, enabling comparison among different algorithms. Figure 28.23 provides a visual comparison of several demosaicking algorithms available in 2005 [38]. Recent papers present improvement over these methods.

In addition to the restoration performance, computational complexity is also an important factor for demosaicking algorithms due to the limited resources of a digital camera. The trade-off between the image quality and computational time should be considered in designing the camera pipeline.

28.7 Conclusions and Future Directions

As the sensor technology and processing power of digital cameras advance, the image quality of digital still and video cameras will continue to improve. Eventually, the image quality of consumer level digital cameras will catch and exceed the quality of film cameras. Digital video cameras will be able to produce images with the quality of digital still cameras.

Although the manufacturers are able to fit more and more pixels in a fixed size chip, higher pixel count does always not correspond to higher image quality. As the dimensions of the photosensitive region of a pixel decrease, the dynamic range and noise performance of the sensor decreases. The solution is therefore to produce larger size sensor chips and to increase the fill ratio (photosensitive region area/pixel area).

There is also research to develop alternative sensor technologies. For example, a recent technology exploits the absorption characteristics of silicon to eliminate CFA and increase resolution. The blue portion of the light is absorbed at the surface of a silicon, while the red portion penetrates deeper. By putting detectors at various depths, color information can be extracted without the need of a CFA. However, for such a sensor, the color components are convoluted, the color reproduction and noise performance could be problematic. Another approach is to place microgratings above pixels. These microgratings diffract light according their spectral content, and detectors capture different spectral information.

Although these technologies look promising, single-chip image capture, and therefore, demosaicking seem to remain essential for some time. Soon HDTVs will become widespread, and higher image quality from digital video cameras will be expected. Some of the demosaicking algorithms that we explained in this article cannot be included in a camera due to the limited resources. However, the computational power of cameras will increase, and advanced image processing algorithms will be utilized in digital cameras. The research on the inter-channel correlation can also help in modeling and restoration of multi-frame and hyper-spectral data.

Acknowledgments

This chapter is an extension of Ref. [38]. We would like to acknowledge the vital contributions of John Glotzbach, Ronald W. Schafer, and Russell M. Mersereau. This work is supported in part by the National Science Foundation under Grant 0528785.

References

1. D. Alleysson, S. Susstrunk, and J. Herault, Linear demosaicing inspired by the human visual system, *IEEE Trans. Image Process.*, 14(4), 439–449, April 2005.
2. D. R. Cok, Signal processing method and apparatus for producing interpolated chrominance values in a sampled color image signal, U.S. Patent 4,642,678, February 1986.

3. J. E. Adams, Interactions between color plane interpolation and other image processing functions in electronic photography, *Proc. SPIE Technology for Electronic Imaging Systems*, 2416, 144–151, February 1995.

4. C. A. Laroche and M. A. Prescott, Apparatus and method for adaptively interpolating a full color image utilizing chrominance gradients, U.S. Patent 5,373,322, December 1994.

5. W. T. Freeman, Method and apparatus for reconstructing missing color samples, U.S. Patent 4,774,565, 1988.

6. R. Kimmel, Demosaicing: Image reconstruction from ccd samples, *IEEE Trans. Image Process.*, (9), 1221–1228, September 1999.

7. X. Li, Demosaicing by successive approximation, *IEEE Trans. Image Process.*, 14(3), 370–379, March 2005.

8. K. Hirakawa and T. W. Parks, Adaptive homogeneity-directed demosaicing algorithm, *IEEE Trans. Image Process.*, 14(3), 360–369, March 2005.

9. W.-M. Lu and Y.-P. Tan, Color filter array demosaicking: New method and performance measures, *IEEE Trans. Image Process.*, 12(10), 1194–1210, October 2003.

10. R. H. Hibbard, Apparatus and method for adaptively interpolating a full color image utilizing luminance gradients, U.S. Patent 5,382,976, January 1995.

11. C.-Y. Tsai and K.-Tai Song, Heterogeneity-projection hard-decision color interpolation using spectral–spatial correlation, *IEEE Trans. Image Process.*, 16(1), 78–91, January 2007.

12. H. S. Malvar, L.-W. He, and R. Cutler, High-quality linear interpolation for demosaicing of color images, in *Proc. IEEE Int. Conf. Acoustics, Speech and Signal Processing*, Montreal, Canada, 2004, vol. 3, pp. 485–488.

13. J. E. Adams and J. F. Hamilton, Design of practical color filter array interpolation algorithms for digital cameras, *Proc. SPIE Real Time Imaging II*, 3028, 117–125, February 1997.

14. B. K. Gunturk, Y. Altunbasak, and R. M. Mersereau, Color plane interpolation using alternating projections, *IEEE Trans. Image Process.*, 11(9), 997–1013, September 2002.

15. J. W. Glotzbach, R. W. Schafer, and K. Illgner, A method of color filter array interpolation with alias cancellation properties, in *Proc. IEEE Int. Conf. Image Processing*, Thessaloniki, Greece, 2001, vol. 1, pp. 141–144.

16. E. Dubois, Frequency-domain methods for demosaicking of bayer-sampled color images, *IEEE Trans. Signal Process. Lett.*, 12(12), 847–850, December 2005.

17. N.-X. Lian, L. Chang, Y.-P. Tan, and V. Zagorodnov, Adaptive filtering for color filter array demosaicking, *IEEE Trans. Image Process.*, 16(10), 2515–2525, October 2007.

18. D. Taubman, Generalized Wiener reconstruction of images from colour sensor data using a scale invariant prior, in *Proc. IEEE Int. Conf. Image Processing*, Vancouver, Canada, 2000, vol. 3, pp. 801–804.

19. H. J. Trussell and R. E. Hartwig, Mathematics for demosaicking, *IEEE Trans. Image Process.*, 3(11), 485–492, April 2002.

20. D. H. Brainard, Bayesian method for reconstructing color images from trichromatic samples, *Proc. IS & T 47th Annual Meeting*, Rochester, New York, pp. 375–380, 1994.

21. P. Longere, X. Zhang, P. B. Delahunt, and D. H. Brainard, Perceptual assessment of demosaicing algorithm performance, *Proc. IEEE*, 90(1), 123–132, January 2002.

22. S. Farsiu, M. Elad, and P. Milanfar, Multiframe demosaicing and super-resolution of color images, *IEEE Trans. Image Process.*, 15(1), 141–159, January 2006.

23. T. Gotoh and M. Okutomi, Direct super-resolution and registration using raw CFA images, in *Proc. IEEE Computer Vision and Pattern Recognition*, Washington, D.C., July 2004, vol. 2, pp. 600–607.

24. M. Gevrekci, B. K. Gunturk, and Y. Altubasak, POCS-based restoration of bayer-sampled image sequences, in *Proc. IEEE Int. Conf. Acoustics, Speech, and Signal Processing*, Honolulu, HI, April 2007, vol. 1, pp. 753–756.

25. J. Mukherjee, R. Parthasarathi, and S. Goyal, Markov random field processing for color demosaicing, *Pattern Recognit. Lett.*, 22(3–4), 339–351, March 2001.

26. Y. Hel-Or and D. Keren, Image demosaicing method utilizing directional smoothing, U.S. Patent 6,404,918, July 2002.
27. J. Go, K. Sohn, and C. Lee, Interpolation using neural networks for digital still cameras, *IEEE Trans. Consumer Elec.*, 4(3), 610–616, August 2000.
28. K. Hirakawa and T. W. Parks, Joint demosaicing and denoising, *IEEE Trans. Image Process.*, 15(8), 2146–2157, August 2006.
29. N.-X. Lian, L. Chang, V. Zagorodnov, and Y.-P. Tan, Reversing demosaicking and compression in color filter array image processing: Performance analysis and modeling, *IEEE Trans. Image Process.*, 15(11), 3261–3278, November 2006.
30. C. C. Koh, New efficient methods of image compression in digital cameras with color filter array, *IEEE Trans. Consumer Elec.*, 49(4), 1448–1456, November 2003.
31. S. Bayram, H. Sencar, N. Memon, and I. Avcibas, Source camera identification based on CFA interpolation, in *Proc. IEEE Int. Conf. Image Processing*, Genoa, Italy, September 2005, vol. 3, pp. 69–72.
32. Y. Long and Y. Huang, Image based source camera identification using demosaicking, in *Proc. IEEE 8th Workshop on Multimedia Signal Processing*, Victoria, Canada, October 2006, pp. 419–424.
33. Mathematical methods for the design of color scanning filters, *IEEE Trans. Image Process.*, 6(2), 312–320, February 1997.
34. D. Alleysson, S. Susstrunk, and J. Marguier, Influence of spectral sensitivity functions on demosaicing, in *Proc. 11th IS&T Color Imaging Conference*, Scottsdale, AZ, November 2003, pp. 351–357.
35. M. Parmar and S. J. Reeves, Selection of optimal spectral sensitivity functions for color filter arrays, in *Proc. IEEE Int. Conf. Image Processing*, Atlanta, GA, October 2006, pp. 1005–1008.
36. X. Zhang and B. A. Wandell, A spatial extension of cielab for digital color image reproduction, *J. Soc. Inf. Display*, 5(1), 61–63, March 1997.
37. Eastman Kodak Company, Kodak Photo CD PCD0992. Available online at http://r0k.us/graphics/kodak/
38. B. K. Gunturk, J. Glotzbach, Y. Altunbasak, R. W. Schafer, and R. M. Mersereau, Demosaicking: Color filter array interpolation, *IEEE Signal Process. Mag.*, 22, 44–54, 2005.
39. E. Chang, S. Cheung, and D. Y. Pan, Color filter array recovery using a threshold-based variable number of gradients, *Proc. SPIE Image Sensor Design and Characterization*, 3650, 36–43, January 1999.

Index